T0189549

Communications
in Computer and Information Science   1160

Commenced Publication in 2007
Founding and Former Series Editors:
Phoebe Chen, Alfredo Cuzzocrea, Xiaoyong Du, Orhun Kara, Ting Liu,
Krishna M. Sivalingam, Dominik Ślęzak, Takashi Washio, Xiaokang Yang,
and Junsong Yuan

Editorial Board Members

Simone Diniz Junqueira Barbosa ⓘ
Pontifical Catholic University of Rio de Janeiro (PUC-Rio),
Rio de Janeiro, Brazil

Joaquim Filipe ⓘ
Polytechnic Institute of Setúbal, Setúbal, Portugal

Ashish Ghosh
Indian Statistical Institute, Kolkata, India

Igor Kotenko ⓘ
St. Petersburg Institute for Informatics and Automation of the Russian
Academy of Sciences, St. Petersburg, Russia

Lizhu Zhou
Tsinghua University, Beijing, China

More information about this series at http://www.springer.com/series/7899

Linqiang Pan · Jing Liang · Boyang Qu (Eds.)

Bio-inspired Computing: Theories and Applications

14th International Conference, BIC-TA 2019
Zhengzhou, China, November 22–25, 2019
Revised Selected Papers, Part II

 Springer

Editors
Linqiang Pan (iD)
Huazhong University of Science
and Technology
Wuhan, China

Jing Liang
Zhengzhou University
Zhengzhou, China

Boyang Qu
Zhongyuan University of Technology
Zhengzhou, China

ISSN 1865-0929   ISSN 1865-0937 (electronic)
Communications in Computer and Information Science
ISBN 978-981-15-3414-0   ISBN 978-981-15-3415-7 (eBook)
https://doi.org/10.1007/978-981-15-3415-7

© Springer Nature Singapore Pte Ltd. 2020
This work is subject to copyright. All rights are reserved by the Publisher, whether the whole or part of the material is concerned, specifically the rights of translation, reprinting, reuse of illustrations, recitation, broadcasting, reproduction on microfilms or in any other physical way, and transmission or information storage and retrieval, electronic adaptation, computer software, or by similar or dissimilar methodology now known or hereafter developed.
The use of general descriptive names, registered names, trademarks, service marks, etc. in this publication does not imply, even in the absence of a specific statement, that such names are exempt from the relevant protective laws and regulations and therefore free for general use.
The publisher, the authors and the editors are safe to assume that the advice and information in this book are believed to be true and accurate at the date of publication. Neither the publisher nor the authors or the editors give a warranty, expressed or implied, with respect to the material contained herein or for any errors or omissions that may have been made. The publisher remains neutral with regard to jurisdictional claims in published maps and institutional affiliations.

This Springer imprint is published by the registered company Springer Nature Singapore Pte Ltd.
The registered company address is: 152 Beach Road, #21-01/04 Gateway East, Singapore 189721, Singapore

Preface

Bio-inspired computing is a field of study that abstracts computing ideas (data structures, operations with data, ways to control operations, computing models, artificial intelligence, etc.) from biological systems or living phenomena such as cells, tissues, neural networks, the immune system, an ant colony, or evolution. The areas of bio-inspired computing include neural networks, brain-inspired computing, neuromorphic computing and architectures, cellular automata and cellular neural networks, evolutionary computing, swarm intelligence, fuzzy logic and systems, DNA and molecular computing, membrane computing, and artificial intelligence, as well as their application in other disciplines such as machine learning, image processing, computer science, and cybernetics. Bio-Inspired Computing: Theories and Applications (BIC-TA) is a series of conferences that aims to bring together researchers working in the main areas of bio-inspired computing to present their recent results, exchange ideas, and cooperate in a friendly framework.

Since 2006, the conference has taken place in Wuhan (2006), Zhengzhou (2007), Adelaide (2008), Beijing (2009), Liverpool and Changsha (2010), Penang (2011), Gwalior (2012), Anhui (2013), Wuhan (2014), Anhui (2015), Xi'an (2016), Harbin (2017), and Beijing (2018). Following the success of previous editions, the 14th International Conference on Bio-Inspired Computing: Theories and Applications (BIC-TA 2019) was held in Zhengzhou, China, during November 22–25, 2019, and was organized by Zhongyuan University of Technology with the support of Zhengzhou University, Zhengzhou University of Light Industry, Henan Normal University, Henan University of Technology, North China University of Water Resources and Electric Power, Pingdingshan University, Nanyang Institute of Technology, Peking University, Huazhong University of Science and Technology, Henan Electrotechnical Society, and Operations Research Society of Hubei.

We would like to thank the President of Zhongyuan University of Technology, Prof. Zongmin Wang, and Academician of the Chinese Academy of Engineering, Prof. Xiangke Liao, for commencing the opening ceremony.

Thanks are also given to the keynote speakers for their excellent presentations: Mitsuo Gen (Tokyo University of Science, Japan), Yaochu Jin (University of Surrey, UK), Derong Liu (Guangdong University of Technology, China), Ponnuthurai Nagaratnam Suganthan (Nanyang Technological University, Singapore), Kay Chen Tan (City University of Hong Kong, China), Mengjie Zhang (Victoria University of Wellington, New Zealand), and Ling Wang (Tsinghua University, China).

We gratefully acknowledge Zongmin Wang, Qingfu Zhang, Jin Xu, Haibin Duan, Zhoufeng Liu, Xiaowei Song, Jinfeng Gao, Yanfeng Wang, Yufeng Peng, Dexian Zhang, Hongtao Zhang, Xichang Xue, Qinghui Zhu, and Xiaoyu An for their contribution in organizing the conference.

A special thanks goes to Prof. Guangzhao Cui for his extensive guidance and assistance in the local affairs and financial support of the conference.

BIC-TA 2019 attracted a wide spectrum of interesting research papers on various aspects of bio-inspired computing with a diverse range of theories and applications. 121 papers were selected for inclusion in the BIC-TA 2019 proceedings, publish by Springer Nature in the series *Communications in Computer and Information Science* (CCIS).

We are grateful to the external referees for their careful and efficient work in the reviewing process, and in particular the Program Committee chairs Maoguo Gong, Rammohan Mallipeddi, Ponnuthurai Nagaratnam Suganthan, Zhihui Zhan, and the Program Committee members. The warmest thanks are given to all the authors for submitting their interesting research work.

We thank Lianghao Li, Wenting Xu, Taosheng Zhang, et al. for their help in collecting the final files of the papers and editing the volume. We thank Zheng Zhang and Lianlang Duan for their contribution in maintaining the website of BIC-TA 2019 (http://2019.bicta.org/). We also thank all the other volunteers, whose efforts ensured the smooth running of the conference.

Special thanks are due to Springer Nature for their skilled cooperation in the timely production of these volumes.

December 2019

Linqiang Pan
Jing Liang
Boyang Qu

Organization

Steering Committee

Atulya K. Nagar	Liverpool Hope University, UK
Gheorghe Paun	Romanian Academy, Romania
Giancarlo Mauri	Università di Milano-Bicocca, Italy
Guangzhao Cui	Zhengzhou University of Light Industry, China
Hao Yan	Arizona State University, USA
Jin Xu	Peking University, China
Jiuyong Li	University of South Australia, Australia
Joshua Knowles	The University of Manchester, UK
K. G. Subramanian	Liverpool Hope University, UK
Kalyanmoy Deb	Michigan State University, USA
Kenli Li	University of Hunan, China
Linqiang Pan (Chair)	Huazhong University of Science and Technology, China
Mario J. Perez-Jimenez	University of Sevilla, Spain
Miki Hirabayashi	National Institute of Information and Communications Technology, Japan
Robinson Thamburaj	Madras Christian College, India
Thom LaBean	North Carolina State University, USA
Yongli Mi	Hong Kong University of Science and Technology, Hong Kong

Honorable Chair

Zongmin Wang	Zhongyuan University of Technology, China

General Chairs

Qingfu Zhang	City University of Hong Kong, China
Jin Xu	Peking University, China
Haibin Duan	Beihang University, China
Zhoufeng Liu	Zhongyuan University of Technology, China
Jing Liang	Zhengzhou University, China

Program Committee Chairs

Boyang Qu	Zhongyuan University of Technology, China
Linqiang Pan	Huazhong University of Science and Technology, China
Dunwei Gong	China University of Mining and Technology, China

Maoguo Gong	Xidian University, China
Zhihui Zhan	South China University of Technology, China
Rammohan Mallipeddi	Kyungpook National University, South Korea
P. N. Suganthan	Nanyang Technological University, Singapore

Organizing Chairs

Xiaowei Song	Zhongyuan University of Technology, China
Jinfeng Gao	Zhengzhou University, China
Yanfeng Wang	Zhengzhou University of Light Industry, China
Yufeng Peng	Henan Normal University, China
Dexian Zhang	Henan University of Technology, China
Hongtao Zhang	North China University of Water Resources and Electric Power, China
Xichang Xue	Pingdingshan University, China
Qinghui Zhu	Nanyang Institute of Technology, China
Xiaoyu An	Henan Electrotechnical Society, China

Special Session Chairs

| Yinan Guo | China University of Mining and Technology, China |
| Shi Cheng | Shaanxi Normal University, China |

Tutorial Chairs

| He Jiang | Dalian University of Technology, China |
| Wenyin Gong | China University of Geosciences, China |

Publicity Chairs

Ling Wang	Tsinghua University, China
Aimin Zhou	East China Normal University, China
Hongwei Mo	Harbin Engineering University, China
Ke Tang	Southern University of Science and Technology, China
Weineng Chen	South China University of Technology, China
Han Huang	South China University of Technology, China
Zhihua Cui	Taiyuan University of Science and Technology, China
Chaoli Sun	Taiyuan University of Science and Technology, China
Handing Wang	Xidian University, China
Xingyi Zhang	Anhui University, China

Local Chairs

Kunjie Yu	Zhengzhou University, China
Chunlei Li	Zhongyuan University of Technology, China
Xiaodong Zhu	Zhengzhou University, China

Publication Chairs

Yuhui Shi   Southern University of Science and Technology, China
Zhihua Cui   Taiyuan University of Science and Technology, China
Boyang Qu   Zhongyuan University of Technology, China

Registration Chairs

Xuzhao Chai   Zhongyuan University of Technology, China
Li Yan   Zhongyuan University of Technology, China
Yuechao Jiao   Zhongyuan University of Technology, China

Program Committee

Muhammad Abulaish   South Asian University, India
Chang Wook Ahn   Gwangju Institute of Science and Technology,
                                 South Korea
Adel Al-Jumaily   University of Technology Sydney, Australia
Bin Cao   Hebei University of Technology, China
Junfeng Chen   Hoahi University, China
Wei-Neng Chen   Sun Yat-sen University, China
Shi Cheng   Shaanxi Normal University, China
Tsung-Che Chiang   National Taiwan Normal University, China
Kejie Dai   Pingdingshan University, China
Bei Dong   Shanxi Normal University, China
Xin Du   Fujian Normal University, China
Carlos Fernandez-Llatas   Universitat Politecnica de Valencia, Spain
Shangce Gao   University of Toyama, Japan
Wenyin Gong   China University of Geosciences, China
Shivaprasad Gundibail   MIT, Manipal Academy of Higher Education (MAHE),
                                 India
Ping Guo   Beijing Normal University, China
Yinan Guo   China University of Mining and Technology, China
Guosheng Hao   Jiangsu Normal University, China
Shan He   University of Birmingham, UK
Tzung-Pei Hong   National University of Kaohsiung, China
Florentin Ipate   University of Bucharest, Romania
Sunil Jha   Banaras Hindu University, India
He Jiang   Dalian University of Technology, China
Qiaoyong Jiang   Xi'an University of Technology, China
Liangjun Ke   Xian Jiaotong University, China
Ashwani Kush   Kurukshetra University, India
Hui L.   Xi'an Jiaotong University, China
Kenli Li   Hunan University, China
Yangyang Li   Xidian University, China
Zhihui Li   Zhengzhou University, China

Qunfeng Liu	Dongguan University of Technology, China
Xiaobo Liu	China University of Geosciences, China
Wenjian Luo	University of Science and Technology of China, China
Lianbo Ma	Northeastern University, China
Wanli Ma	University of Canberra, Australia
Xiaoliang Ma	Shenzhen University, China
Holger Morgenstern	Albstadt-Sigmaringen University, Germany
G. R. S. Murthy	Lendi Institute of Engineering and Technology, India
Akila Muthuramalingam	KPR Institute of Engineering and Technology, India
Yusuke Nojima	Osaka Prefecture University, Japan
Linqiang Pan (Chair)	Huazhong University of Science and Technology, China
Andrei Paun	University of Bucharest, Romania
Xingguang Peng	Northwestern Polytechnical University, China
Chao Qian	University of Science and Technology of China, China
Rawya Rizk	Port Said University, Egypt
Rajesh Sanghvi	G. H. Patel College of Engineering & Technology, India
Ronghua Shang	Xidian University, China
Zhigang Shang	Zhengzhou University, China
Ravi Shankar	Florida Atlantic University, USA
V. Ravi Sankar	GITAM University, India
Bosheng Song	Huazhong University of Science and Technology, China
Tao Song	China University of Petroleum, China
Jianyong Sun	University of Nottingham, UK
Yifei Sun	Shaanxi Normal University, China
Handing Wang	Xidian University, China
Yong Wang	Central South University, China
Hui Wang	Nanchang Institute of Technology, China
Hui Wang	South China Agricultural University, China
Gaige Wang	Ocean University of China, China
Sudhir Warier	IIT Bombay, China
Slawomir T. Wierzchon	Polish Academy of Sciences, Poland
Zhou Wu	Chongqing University, China
Xiuli Wu	University of Science and Technology Beijing, China
Bin Xin	Beijing Institute of Technology, China
Gang Xu	Nanchang University, China
Yingjie Yang	De Montfort University, UK
Zhile Yang	Shenzhen Institute of Advanced Technology, Chinese Academy of Sciences, China
Kunjie Yu	Zhengzhou University, China
Xiaowei Zhang	University of Science and Technology of China, China
Jie Zhang	Newcastle University, UK
Gexiang Zhang	Southwest Jiaotong University, China
Defu Zhang	Xiamen University, China

Peng Zhang	Beijing University of Posts and Telecommunications, China
Weiwei Zhang	Zhengzhou University of Light Industry, China
Yong Zhang	China University of Mining and Technology, China
Xinchao Zhao	Beijing University of Posts and Telecommunications, China
Yujun Zheng	Zhejiang University of Technology, China
Aimin Zhou	East China Normal University, China
Fengqun Zhou	Pingdingshan University, China
Xinjian Zhuo	Beijing University of Posts and Telecommunications, China
Shang-Ming Zhou	Swansea University, UK
Dexuan Zou	Jiangsu Normal University, China
Xingquan Zuo	Beijing University of Posts and Telecommunications, China

Contents – Part II

Neural Networks and Artificial Intelligence

Contents – Part I

Bioinformatics and Systems Biology

Correlated Protein Function Prediction with Robust Feature Selection

Dengdi Sun[1], Haifeng Sun[1], Hang Wu[1], Huadong Liang[2],
and Zhuanlian Ding[3]([⊠])

[1] Key Lab of Intelligent Computing and Signal Processing of Ministry of Education,
School of Computer Science and Technology, Anhui University, Hefei 230601, China
[2] iFlytek Co., Ltd., Hefei 230088, China
[3] School of Internet, Anhui University, Hefei 230039, China
dingzhuanlian@163.com

Abstract. Determining the functional roles of proteins is a vital task
to understand life at molecular level and has great biomedical and phar-
maceutical implications. With the development of novel high-throughput
techniques, enormous amounts of protein-protein interaction (PPI) data
are collected and provide an important and feasible way for studying
protein function predictions. According to this, many approaches assign
biological functions to all proteins using PPI networks directly. However,
due to the extreme complexity of the topology structure of real PPI networks, it is very difficult and time consuming to seek the global optimization or clustering on the networks. In addition, biological functions are
often highly correlated, which makes functions assigned to proteins are
not independent. To address these challenges, in this paper we propose
a two-stage function annotation method with robust feature selection.
First, we transform the network into the low-dimensional representations
of nodes via manifold learning. Then, we integrate the functional correlation into the framework of multi-label linear regression, and introduce
robust sparse penalty to achieve the function assignment and representative feature selection simultaneously. For the optimization, we design an
efficient algorithm to iteratively solve several subproblems with closed-
form solutions. Extensive experiments against other baseline methods
on Saccharomyces cerevisiae data demonstrate the effectiveness of the
proposed approach.

Keywords: Protein-protein interactions · Function annotation ·
Feature selection · Functional correlations

1 Introduction

Protein function prediction is a fundamental and active problem in bioinformatics and machine learning. It aims at identifying the possible function of unannotated proteins automatically via the vast amount of accumulated genomic and
proteomic data, and has received increasing attentions due to its wide range

© Springer Nature Singapore Pte Ltd. 2020
L. Pan et al. (Eds.): BIC-TA 2019, CCIS 1160, pp. 3–17, 2020.
https://doi.org/10.1007/978-981-15-3415-7_1

of applications in biology and medicine, such as disease treatment, new drug development, and crop improvement. Benefiting from major efforts in high-throughput mapping, today more and more potentially relevant interactions are predicted accurately by computational tools, and form gradually a large scale protein-protein interaction (PPI) networks. Meanwhile, in organisms, functions are often performed by proteins physically interacting with each other, or located closely in the same complex, which makes it a feasible and effective strategy to predict protein functions in silico by leveraging PPI networks.

Among numerous existing PPI networks based algorithms, the most natural and straightforward method is neighbor counting, that is, to label a protein with the functions occurring most frequently in its interacting partners [14,15]. Then Hishigaki et al. used χ^2 statistics to identify the functions that are over-represented in then interacting partners of a protein [5]. These methods only leverage the local information by general fact that closer proteins in the network are more likely to have the similar functions [1,8,12,13]. In contrast, several methods have been proposed toward global optimization by taking into account the full topology of the network. Some methods were proposed which predict protein function by random walks on a hybrid graph [9,17]. Gligorijevic et al. [4] proposed a network fusion method based on multimodal deep autoencoders to extract high-level features of proteins from multiple heterogeneous interaction networks. However, these methods rely on the whole topological structure of PPI network, which is extremely time consuming and difficult to obtain the unique optimal solution.

In fact, the complex network usually could be treated as a manifold structure contained in a high-dimensional space. Using the manifold learning technology, we can capture the low-dimensional representations of PPI networks, which are very useful to the follow-up structure analysis and function annotation. According to this theory, a plenty of recent literatures on function annotation follow this way. For example, You et al. utilizes isometric feature mapping (ISOMAP) to embed the PPI networks and access pairwise similarity [23]. Zhao et al. [22] applies locally linear embedding (LLE) to reconstruct the PPI networks into a low-dimensional subspace. Wang et al. [16] establishes a new knowledge representations by maximizing the consistency between the knowledge similarity upon annotations and the PPIs. It is worth noting that the embedded representation features are abstract and lack of specific physical meaning, and we cannot know the importance of different features for protein function prediction. Additionally, too many features will increase the burden of subsequent calculations. A common way to resolve this problem is feature selection, that is to select a subset of the most representative or discriminative features from the input feature set (feature dimensions). In recent years, inspired by brain cognitive science, sparse coding techniques have achieved great success in data compression and feature selection. This method is derived from the simulation of the simple cell receptive field in the V1 region of the main visual cortex in the mammalian visual system. Furthermore, the newly studied ℓ_{21} norm-based structural sparse coding technique has the rotation invariant feature and can obtain more robust feature representations.

In addition, most existing computational approaches usually assume that different kinds of labels are independent of each other, fundamentally, i.e., the annotation for each functional category is conducted independently. However, in reality most biological functions are highly correlated, and protein functions can be inferred from one another through their interrelatedness. As well known, one protein can perform more than one biological function, which means the function annotation can be seamlessly mapped to a typical multi-label learning problem in machine learning. Apparently, a valid multi-label model should be able to take into account the dependencies between labels, whereas the function category correlations, albeit useful, are not utilized adequately during the prediction process. Moreover, effective feature representations should also be highly correlated with the class labels, so incorporating the association between functions will help to further select functionally relevant features and improve prediction performance significantly.

To address the above two problems, we propose a method in this paper, named as **C**orrelated **A**nnotation with robust **Fe**ature **S**election (CAFES), to identify protein functions using PPI networks. In this method, we first develop the whole weighted PPI networks, and embed it into a low dimensional Euclidean space. Aiming at the co-occurrence and relevance among the functions, then we extend the multi-label linear regression model by incorporating the functional correlations into the optimization objective to infer the functions of unannotated protein. To make the novel CAFES method to take effect, we introduce a robust sparsity penalty in order to select the most related features for the multi-label learning task and devise a multiplicative updating rule with a convergence guarantee. Extensive experiments on real networks, in comparison with several state-of-the-art methods, are performed to assess the performance of CAFES.

2   Problem Formalization and Notations

To begin with, we first define the terms and notations which will be frequently used in this paper. Throughout this paper, lowercase letters $(u, v, \cdots)$ mean scalars and boldface lowercase letters $(\boldsymbol{u}, \boldsymbol{v}, \cdots)$ stand for vectors. u_i represents the i-th entry in a vector $\boldsymbol{u}$, and the ℓ_2-norm of $\boldsymbol{u}$ is defined as $\|\boldsymbol{u}\|_2 = \sqrt{\sum_i u_i^2}$. Besides, matrices are written as boldface uppercase letters. Given a matrix $\boldsymbol{U} = \{u_{ij}\}$, we denote its j-th column and i-th row as $\boldsymbol{u}_j$ and $\boldsymbol{u}^{(i)}$ respectively. The ℓ_2 norm of vector $\boldsymbol{u}$ is defined as $\|\boldsymbol{u}\|_2 = \sqrt{\sum_i u_i^2}$, and the Frobenius norm of matrix $\boldsymbol{U}$ is $\|\boldsymbol{U}\|_F = \sqrt{\sum_{ij} u_{ij}^2}$. Finally, we use decorated letter $\mathcal{U}$ to represent the set, and $\mathbf{1}$ means certain size column vector filled with 1.

Given m biological functions and n proteins, a PPI network can be simplified as a graph $\mathcal{G}(\mathcal{V}, \mathcal{E})$, where the set of n proteins (nodes) can be represented like $\mathcal{V} = \{v_1, \ldots, v_n\}$, and the edges $\mathcal{E}$ are weighted by an $n \times n$ similarity $\boldsymbol{W}$ with w_{ij} indicating the similarity between v_i and v_j. In simplest case, $\boldsymbol{W}$ is the binary adjacency matrix where $w_{ij} = 1$ if proteins v_i and v_j interact, and 0 otherwise. In this work, $\boldsymbol{W}$ is computed in Eq. (1) (See Sect. 3) to incorporate more useful information.

In protein function prediction, let $\mathcal{X}$ is the input space, $\mathcal{Y} = \{1, \cdots, m\}$ is the label space with m possible labels, and $\mathcal{D} = \{(\boldsymbol{x}_1, \boldsymbol{y}_1), \cdots, (\boldsymbol{x}_l, \boldsymbol{y}_l)\}$ denotes the training data that consists of l annotated proteins, where $1 \leq l < n$. Each protein v_i is represented a p dimensional feature vector $\boldsymbol{x}_i = [x_{i1}, \cdots, x_{ip}]^{\mathrm{T}}$, and associated with a set of labels represented by a function assignment indication vector $\boldsymbol{y}_i = [y_{i1}, \cdots, y_{im}]^{\mathrm{T}}$, such that y_{ik} is 1 if v_i has the k-th function and 0 otherwise. For convenience, we rewrite the input feature vectors and labels in matrix format $\boldsymbol{X} = [\boldsymbol{x}_1, \cdots, \boldsymbol{x}_l] \in \Re^{p \times l}$ and $\boldsymbol{Y} = [\boldsymbol{y}_1, \cdots, \boldsymbol{y}_l] = [\boldsymbol{y}^{(1)}, \cdots, \boldsymbol{y}^{(m)}]^{\mathrm{T}} \in \Re^{m \times l}$ respectively.

3 Methodology

Under the problem formalization in last section, we propose a two-stage function annotation method with robust feature selection. The PPI networks are firstly embedded into the low-dimensional space via manifold learning, and then the multi-label linear regression model is extended by incorporating the functional correlations and robust sparsity constrain into the optimization objective to assign the protein functions.

3.1 Network Embedding

Manifold learning is an important nonlinear dimensionality reduction technology proposed in recent years, which built on the following assumption: complex real data is actually distributed on potential low-dimensional manifolds. Leveraging this approach, we can embed the network topology into a low-dimensional feature space while maintaining the geometric metrics of the data to obtain the valid representation vectors of the nodes. In our work, we use ISOMAP to perform network embedding. Since the original PPI networks are binary without interaction strength, here we propose a structure-based network weighting strategy to capture better representations of nodes, that is, the greater the similarity, the closer the nodes are in embedded space. It's worth nothing that although the real-world PPI networks are generally non-fully connected, there is one largest connected sub-network, which contains most of the nodes and edges. Therefore, only the largest connected component is embedded by ISOMAP. The process of manifold embedding algorithm based on ISOMAP can be summarized as follows.

(1) First, we define the new weight adjacent matrix $\widetilde{\boldsymbol{W}} = \{\widetilde{w}_{ij}\}$ by edge betweenness as follows:

$$\widetilde{\boldsymbol{W}}_{ij} = \sum_{u \neq v \in V} \delta_{uv}(e_{ij}) = \sum_{u \neq v} \frac{\kappa_{uv}(e_{ij})}{\kappa_{uv}}. \tag{1}$$

where κ_{uv} indicates the number of all shortest path between arbitrary protein pair u and v, and $\kappa_{uv}(e_{ij})$ computes the number of these paths through the edge e_{ij}. As shown in Fig. 1, it is reasonable to take advantage of the edge betweenness to weight network, since the edges passed frequently tend to have a high credibility and interactivity in the real PPI network.

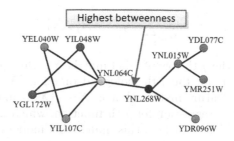

Fig. 1. Take advantage of the edge betweenness to weight original PPI network. The edges passed frequently which tend to have a high credibility and interactivity in the real PPI network.

(2) Moreover, although many of proteins are not neighbors each other, they also could be connected by a series of neighbor links. According to it, the geodesic distance d_{ij} is utilized to describe the hidden true relationship between two proteins v_i and v_j in the manifold approximately. In this paper, the geodesic distance is defined by the shortest path distances d_{ij}^G between all pairwise vertices in the weight adjacent matrix $\widetilde{W}$, which can be computed through the Dijkstra algorithm. If the protein data points are sampled from a probability distribution which is supported by the entire manifold, then, as the density of protein data points tends to infinity, it turns out that the estimated d_{ij}^G will converge to d_{ij} under the manifold assumption.

(3) Finally, the geodesic distance matrix $D^G = \{d_{ij}^G\}$ is embedded into a p-dimensional subspace as representation vectors $x_1, \cdots, x_n$ of n nodes, by minimizing the cost function: $\|\tau(D^G) - \tau(D^E)\|$, where $D^E = \{d_{ij}^E\}$ denotes the matrix of Euclidean distances in p-dimensional subspace, which $d_{ij}^E = \|x_i - x_j\|$; $\tau(D^G) = -\frac{H(D^G)^2 H}{2}$ and $\tau(D^E) = -\frac{H(D^E)^2 H}{2}$ are the centered matrix of squared graph distances D^G and squared embedding Euclidean distances D^Y respectively, and $H = I_n - 11^T/n$ is the centering matrix. The optimal solution is given by the $v_1, \cdots, v_p$ with corresponding to the largest eigenvalues, $\lambda_1 > \cdots > \lambda_p$ of $\tau(D^G)$, that is the p-dimensional vector y_i equal to $\sqrt{\lambda_p} v_p^i$, so that the estimated intrinsic geometry of the manifold is preserved accurately.

3.2   Multi-label Learning with Feature Selection

After network embedding, we can capture the p-dimensional representation vectors of all proteins (nodes), which contains l annotated proteins and $n-l$ unannotated proteins. As mentioned in the second section, we extract the representation vectors $x_1, \cdots, x_l$ of l annotated proteins and their corresponding labels Y as the training set $\mathcal{X}$, and use it to establish the following multi-label regression model.

$$\min_{\beta_1 \cdots \beta_m} \sum_{j=1}^{m} \|\boldsymbol{y}^{(j)} - \boldsymbol{\beta}_j^{\mathrm{T}} \boldsymbol{X}\|_2^2 \tag{2}$$

where $\boldsymbol{y}^{(j)} \in \Re^{1 \times l}$ is the j-th row of label matrix $\boldsymbol{Y}$, which indicates the annotations of all l proteins for the j-th biological function. $\boldsymbol{X} \in \Re^{(p+1) \times 1}$ denote the augmented input feature matrix, that is $\boldsymbol{X} \leftarrow [\boldsymbol{X}, 1]$ and $\boldsymbol{\beta}_j \in \Re^{(p+1) \times 1}$ is the coefficient vector of regression for j-th function, which also absorb the corresponding bias term. Thus, in fact, this multi-label model can be decomposed into m single-label linear regressions.

Since the features are abstracted from PPI networks, and have no clearly specific meaning, it is necessary to select a subset of the most representative or discriminative features from the input feature set for function assignment task. To choose $r < l$ features, we enforce that there are only r nonzero elements in $\boldsymbol{\beta}_j$. Thus, if $\beta_{ji} = 0$, the corresponding i-th row of $\boldsymbol{X}$ will not participate in the linear combination of $\boldsymbol{y}^{(j)}$, In other words, the i-th feature is eliminated and the remaining features are selected for the j-th functional assignment task. Therefore, the feature selection problem is converted into the framework of sparse linear regression. Furthermore, to select a common subset of features for all m functions, we introduce the structure sparsity constrain via ℓ_{21} norm as the regularizer in Eq. (2) as:

$$\min_{B} \sum_{j=1}^{m} \|\boldsymbol{y}^{(j)} - \boldsymbol{\beta}_j^{\mathrm{T}} \boldsymbol{X}\|_2^2 + \lambda \|\boldsymbol{B}\|_{21} \tag{3}$$

where $\boldsymbol{B} = [\boldsymbol{\beta}_1 \cdots \boldsymbol{\beta}_m] \in \Re^{(p+1) \times m}$ is the overall regression coefficient matrix. $\|\boldsymbol{B}\|_{21} = \sum_i \sqrt{\sum_j \beta_{ij}^2}$ denote the matrix ℓ_{21} norm, and when $\boldsymbol{B}$ is row-sparse enough, the objective function achieve the global minimum. In other words, the indices of the nonzero rows of $\boldsymbol{B}$ correspond to the indices of the rows of $\boldsymbol{X}$ which are chosen as the efficient features. As a result, we can rewrite the Eq. (3) in matrix format:

$$\min_{B} \|\boldsymbol{Y} - \boldsymbol{B}^{\mathrm{T}} \boldsymbol{X}\|_F^2 + \lambda \|\boldsymbol{B}\|_{21} \tag{4}$$

3.3 Function Correlation

As shown in biological experiments, proteins assigned to two different functions may overlap. Statistically, the bigger the overlap is, the more closely the two functions are related. Therefore, functions assigned to a protein are no longer independent, but can be inferred from one another, which makes functional correlations significant for improving prediction performance. Using cosine similarity, we define a function category correlation matrix, $\boldsymbol{C} \in \Re^{m \times m}$, where c_{jk} captures the correlation between the j-th and k-th functions as following [6,7]:

$$c_{jk} = \cos(\boldsymbol{y}^{(j)}, \boldsymbol{y}^{(k)}) = \frac{\langle \boldsymbol{y}^{(j)}, \boldsymbol{y}^{(k)} \rangle}{\|\boldsymbol{y}^{(j)}\| \|\boldsymbol{y}^{(k)}\|}. \tag{5}$$

Obviously, the selected features should also be consistent with the functional correlations. To this end, following [6], we expect to maximize $\mathrm{tr}(\boldsymbol{BCB}^\mathrm{T})$, where $\mathrm{tr}(\cdot)$ denotes the trace of matrix. However, the linear regression seeks to minimize the objective function, which contradicts the maximization of functional correlations. To address this problem, in this paper, we devise a regularization term with the assumption that if function j and k are related to each other, their corresponding regression coefficients (i.e., $\boldsymbol{\beta}_j$ and $\boldsymbol{\beta}_k$) should be more similar. To do this, we penalize a loss function with the correlation c_{jk} on the distance $\|\boldsymbol{\beta}_j - \boldsymbol{\beta}_k\|_2^2$, which can be transferred the following graph regularization:

$$\frac{1}{2} \sum_{j,k=1}^m c_{jk} \|\boldsymbol{\beta}_j - \boldsymbol{\beta}_k\|_2^2 = \mathrm{tr}(\boldsymbol{BLB}^\mathrm{T}) \tag{6}$$

where, $\boldsymbol{L}$ is the laplacian matrix of $\boldsymbol{C}$. In fact, due to $\boldsymbol{L} = \boldsymbol{D} - \boldsymbol{C}$ ($\boldsymbol{D}$ is the diagonal degree matrix of $\boldsymbol{C}$), Eq. (6) essentially turns the maximization of $\boldsymbol{BCB}^\mathrm{T}$ into the opposite side.

3.4   Correlated Annotation with robust Feature Selection (CAFES)

By integrating the above three goals into a unified framework, the final objective function of CAFES is defined as follows:

$$\min_{\boldsymbol{B}} \sum_{j=1}^m \|\boldsymbol{Y} - \boldsymbol{B}^\mathrm{T}\boldsymbol{X}\|_F^2 + \gamma\mathrm{tr}(\boldsymbol{BLB}^\mathrm{T}) + \lambda\|\boldsymbol{B}\|_{21} \tag{7}$$

where $\gamma > 0$ and $\lambda > 0$ are the tuning parameters.

In Eq. (7), the first two terms are designed to simultaneously achieve minimal regression error (via the first term) and preservation of function correlations (via the second term). The last term is designed to generate the row sparsity (via the ℓ_{21}-norm regularizer) to select adaptively the most representative or discriminative features. The optimization of Eq. (7) is shown in Sect. 4, and its pseudocode is listed in Algorithm 1.

3.5   Predicting Functions of Unannotated Proteins

After conducting the proposed CAFES model, we can get the regression coefficient matrix $\boldsymbol{B}$, and predict the functions of remaining $n-l$ unannotated protein. For the i-th unannotated protein, its prediction value $\widehat{\boldsymbol{y}}_i$ can be represented as following:

$$\widehat{\boldsymbol{y}}_i = \boldsymbol{B}^\mathrm{T}\boldsymbol{x}_i \tag{8}$$

where $\boldsymbol{x}_i$ is the representation vector of i-th unannotated protein and obtained in network embedding (Sect. 3.1). Since the prediction value $\widehat{\boldsymbol{y}}_i$ is continuous solution in certain range rather than binary 0 or 1. Thus, threshold h is adapted to discretize the prediction value in order to get the eventual sets of labels:

$$\boldsymbol{y}_i^{(k)} = \begin{cases} 1, & \text{if } \boldsymbol{y}_i^{(k)} \geq h, \\ 0, & \text{if } \boldsymbol{y}_i^{(k)} < h. \end{cases} \tag{9}$$

4   Optimization

Since the objective function in Eq. (7) is convex and nonsmooth, so B has a unique global optimum solution. We propose an iterative reweighed algorithm in this paper to optimize Eq. (7). The alternative iteration procedure is repeated until the algorithm converges.

First, let $J(B) = \|Y - B^T X\|_F^2 + \gamma tr(BLB^T) + \lambda\|B\|_{21}$. By setting the derivative of the objective function $J(B)$ with respect to B as zero, we have

$$\frac{\partial J(B)}{\partial B} = 2XX^T B - 2XY^T + 2\gamma BL + 2\lambda AB = 0 \tag{10}$$

$$\Rightarrow (XX^T + \lambda A)B + \gamma BL = XY^T$$

where A is a diagonal matrix with the i-th diagonal element as

$$a_{ii} = \frac{1}{2\|B^{(i)}\|_2} \tag{11}$$

By observing Eq. 10, we know that A and B are related to each other. Hence, we can optimize Eq. (7) by alternatively computing A and B. First, given a fixed B, it is easy to solve A using Eq. (11). Then, given a fixed A, the value of B is updated as follows.

Since both $(XX^T + \lambda A)$ and γL are positive semidefinite, we perform singular value decomposition on them to obtain

$$XX^T + \lambda A = U\Sigma_1 U^T$$

$$\gamma L = V\Sigma_2 V^T \tag{12}$$

where U and V are unitary matrices. Then, Eq. (10) can be expressed as

$$U\Sigma_1 U^T B + BV\Sigma_2 V^T = XY^T \tag{13}$$

Left multiplying the two sides of Eq. (13) by U^T, and right multiplying V on both sides of Eq. (13), respectively, we have

$$\Sigma_1 U^T BV + U^T BV\Sigma_2 = U^T XY^T V \tag{14}$$

By denoting

$$\widetilde{B} = U^T BV \tag{15}$$

$$\Omega = U^T XY^T V \tag{16}$$

Substituting Eqs. (15) and (16) into Eq. (14), we arrive at:

$$\Sigma_1\widetilde{B} + \widetilde{B}\Sigma_2 = \Omega \tag{17}$$

Note that $\Sigma_1 = diag(\sigma_1^{(1)}, \cdots, \sigma_1^{(p)})$ and $\Sigma_2 = diag(\sigma_2^{(1)}, \cdots, \sigma_2^{(m)})$ are positive definite, which means $\sigma_1^{(i)} > 0, i = 1, \cdots, d$ and $\sigma_2^{(j)} > 0, j = 1, \cdots, m$. Thus, each element in $\widetilde{B}$ can be obtained by

$$\widetilde{\beta}_{ij} = \frac{\omega_{ij}}{\sigma_1^{(i)} + \sigma_2^{(j)}} \tag{18}$$

Algorithm 1. An efficient iterative algorithm to solve the proposed CAFES model

---

Input: feature matrix $X \in \Re^{p \times l}$, label matrix $Y \in \Re^{m \times l}$, function correlation laplacian matrix $L \in \Re^{m \times m}$, parameter γ and λ.
 1: **Initialize:** $t = 0$, $A_t \in \Re^{p \times p} = I$ (identity matrix)
 2: **while** not converge **do**
    2.1: Conduct SVD on $XX^{\mathrm{T}} + \lambda A_t$ and γL as Eq. (12);
    2.2: update Ω_{t+1} according to Eq. (16);
    2.3: update $\widetilde{B}_{t+1}$ according to Eq. (18);
    2.4: update B_{t+1} according to Eq. (19);
    2.5: update A_{t+1} according to Eq. (11);
    2.6: $t = t + 1$;
    2.7: Check if the objective function Eq. (7) converges.
 3: **end while**
Output: coefficient matrix $B \in \Re^{p \times m}$.

After obtaining the $\widetilde{B}$ we can calculate the optimum B as

$$B = U\widetilde{B}V^{\mathrm{T}} \tag{19}$$

We have done the convergence analysis of the proposed algorithm with the following results: **Theorem 1.** The objective function in Eq. (7) decreases monotonically, $J(B_{t+1}) \leq J(B_t)$ with each update in Algorithm 1.

Due to the length limit of the paper, the proof of Theorem 1 is omitted here. In Fig. 2, we show the behavior of objective function value of Eq. (7) during iterations of our proposed algorithm on the real-world datasets. In all data sets, the objective decreases monotonically and rapidly, confirming the result of Theorem 1.

5   Experiments and Results

For the performance evaluation, we conduct the proposed approach on Saccharomyces cerevisiae PPI database, compared with several benchmark computational methods for protein function prediction. The results demonstrate that our CAFES model can achieve better performance under multiple assessment criteria.

5.1   Materials and Data Sets

There are two different types of data involved in the experimental evaluations for protein function prediction: (1) function annotation data sets; (2) protein interaction data sets.

Function Annotation Data Sets
The functional catalogue (FunCat) [10] is a project under the Munich Information Center for Protein Sequences (MIPS), which is an annotation scheme for

the functional description of proteins from prokaryotes, unicellular eukaryotes, plants and animals. Taking into account the board and highly diverse spectrum of the known protein functions, FunCat of version 2.1 consists of 27 main functional categories, in this sudy, 24 of which are used in annotation *S. cerevisiae*. Furthermore, there are still other protein annotation systems such as the Gene Ontology [2], we use the Funcat annotation system due to its clear tree-like hierarchical structure and supplement the protein functional annotation with the GO annotation system.

Protein Interaction Data Sets

To demonstrate the effectiveness of our methodology, we use the protein-protein interaction data that can be generated from the Krogan Lab Interactome Database and we focus on the *S. cerevisiae*. The database contains many protein interactions curated from Krogan et al., Gavin et al., Ho et al. combined. In this study, by using the the Krogan Lab Interactome Database and removing the proteins with self-interactions and repeated interactions, we end up with 4594 proteins annotation by Funcat annotation and GO annotation scheme with 74707 PPIs, as well as 883 unannotated proteins.

5.2   Results of Iteration.

Before going any futher, the convergence property of the optimization process is illustrated firstly. Figure 2 shows the change of objective function values in the convergence. Moreover, the average precession against iteration times is illustrated in Fig. 3.

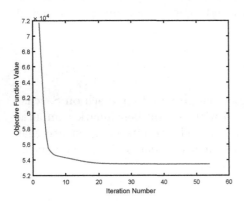

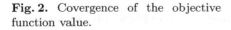

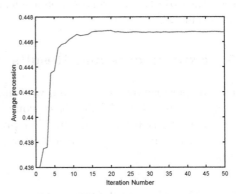

Fig. 2. Covergence of the objective function value.

Fig. 3. Average precession vs. iteration times.

5.3   Embedding and Low Representation for PPI Network

Here, we compare the embedding data reconstruction result for weighted PPI network with for original PPI data. Here we embed the original binary adjacency matrix W of PPI networks and weighted similarity matrix $\widetilde{W}$ into the same dimensional space, as mentioned in Sect. 3.1. The experiment result is shown in Fig. 4. Obviously, the residual error for embedding binary adjacent matrix is consistently larger within the whole dimensional interval as shown in Fig. 4, in contrast to the weight matrix. The result proves that the edge betweenness is more efficient in preserving the structure of PPI network, which firmly confirms the advantage of our embedding strategy. According to Fig. 4, in our experiments, the intrinsic dimension is set as 174.

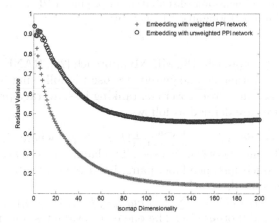

Fig. 4. Residual error of embedding information about two different weighted PPI network. The comparison residual error for embedding original PPI network and weighted PPI network.

By manifold learning, the topological structure of PPI networks can be faithfully preserved. Figure 5 vividly exhibits an example for local low-dimensional distribution and connection of the embedding PPI network. Contrast with Fig. 5(a) and (b), the low-dimensional distribution of proteins shared common functional class in weighted PPI (see Fig. 5(b)) is significantly closer than in the binary PPI network. The result demonstrates further the effectiveness of edge betweenness weighting.

5.4   Comparison and Evaluation for Different Multi-label Approaches

In this paper, we compare the performance of the proposed CAFES model with several state-of-the-art multi-label classification methods for protein function prediction, including BP-MLL approach [11], ML-RBF approach [19], ML-LOC

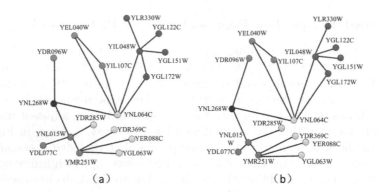

Fig. 5. Low-dimensional distribution information about two different weighted PPI network. (a) The low-dimensional distribution of original PPI network. (b) The low-dimensional distribution of weighted PPI network.

approach [21], LIFT approach [18], ML-NB approach [20], RANK-SVM approach [3]. To asses the performance adequately, we use the following 5 evaluation criteria: hamming loss, one-error, coverage, ranking loss, average precession. These evaluation measures are defined as follows:

- **hamming loss** $hloss(f) = \frac{1}{n}\sum_{i=1}^{n}\frac{1}{m}|f(x_i) \bigtriangleup y_i|$, where $\bigtriangleup$ stands for the symmetric difference between two sets. The hamming loss evaluates the fraction of misclassified instance-label pairs, i.e. a relevant label is missed or an irrelevant is predicted;
- **one-error** $one\text{-}error(f) = \frac{1}{n}\sum_{i=1}^{n}|[max_{y\in\mathcal{Y}}f(x_i,y)] \notin y_i|$, where $|\pi|$ equals to 1 if π holds and 0 otherwise; The one-error evaluates how many times the top-ranked label is not in the set of proper labels of the protein;
- **coverage** $coverage(f) = \frac{1}{n}\sum_{i=1}^{n}max_{y\in y_i}rank_f(x_i,y) - 1$, which evaluates how many steps are average need to move down the label list in order to cover all the proper labels of the protein;
- **ranking loss** $rloss(f) = \frac{1}{n}\sum_{i=1}^{n}\frac{1}{|y_i||\bar{y}_i|}|\mathcal{R}_i|$, $\mathcal{R}_i = \{(y^{(1)}, y^{(2)})|f(x_i,y^{(1)}) < f(x_i,y^{(2)})\}$, $\forall y^{(1)} \in y_i$, $\forall y^{(2)} \notin Y_i$, $(y^{(1)}, y^{(2)}) \in y_i \times \bar{y}_i$, and $\bar{y}_i$ is the complementary set of Y_i. The ranking loss evaluates the fraction of reversely ordered label pairs;
- **average precession** $avgprec(f) = \frac{1}{n}\sum_{i=1}^{n}\frac{1}{|Y_i|} \cdot \frac{|\mathcal{P}_i|}{rank_f(x_i,y)}$, where $\mathcal{P}_i = \{y'|rank_f(x_i,y') \leq rank_f(x_i,y), y' \in Y_i\}$. The average precision evaluates the average fraction of labels ranked above a particular label $y \in Y_i$ which actually are in Y.

In our experiment, for hamming-loss, one-error, coverage and ranking loss, the smaller the metric value, the better the performance with optimal value of $\frac{1}{n}\sum_{i=1}^{n}|y_i| - 1$ for coverage and 0 for one-error and ranking loss. For the average precision, on the other hand, the larger the metric value, the better the performance. Using these criteria, we measured the performance of each method by the standard 10-fold cross-validation. For our approach, adaptive decision boundary is used to compute the threshold to make prediction from the ranking list of the decision values for each function. Furthermore, for the six other approaches, we use their respective optimal parameters suggested in respective literatures. For neural network based methods, in BP-MLL approach, the number of hidden number of hidden neurons is set as 20% of the dimensionality and the number of training epochs is set as 100, and in ML-RBF approach, we set the cost=0.1 which gives the best performance. For svm based methods, in ML-LOC approach and RankSvm approach, we assign the penalty parameter according to the set suggested by [3,21]. For data distribution based methods, in ML-NB approach, smooth=1 gives the best performance, and in LIFT approach, ratio = 0.1 gives the best performance. The prediction performances of eight compared methods by five popular measures metrics in multi-label learning are recorded in Table 1. In Table 1, for each evaluation criterion, "↓" indicates "the smaller the better", while "↑" indicates "the larger the better". Furthermore, the best performance among the seven comparing algorithms is highlighted in boldface.

Table 1. Macro average of five evaluation metrics for the eight compared approaches over all main functional categories

Category	Algorithm	Evaluation metrics				
		Avg. precision ↑	Coverage ↓	Hamming loss ↓	One-error ↓	Ranking loss ↓
SVM model	ML-LOC	0.5485	10.2217	0.1674	0.4761	0.1979
	RANK-SVM	0.5076	11.7174	0.1754	0.5304	0.2433
Neural network	BP-MLL	0.5545	9.9217	0.1801	0.4609	0.1871
	ML-RBF	0.5607	10.9109	0.1563	0.4739	0.2061
Data distribution	ML-NB	0.5402	10.1000	0.1635	0.4826	0.1994
	LIFT	0.5406	9.9109	**0.1561**	0.4717	0.1887
Our methods	AFES	0.5621	9.9826	0.1573	**0.4391**	0.1882
	CAFES	**0.5650**	**9.7152**	0.1563	**0.4391**	**0.1820**

Due to placing the protein function prediction under the framework of multi-label learning, we report the over all prediction performance over all functions in order to address multi-label scenario. In Table 1, AFES denotes the multi-label learning with robust feature selection without function correlations, that is Eq. (3). The experiment results show that the prediction performance by AFES approach outperforms the other six approaches in most of evaluation criteria, which demonstrates the effectiveness of the robust feature selection in protein function prediction. Furthermore, the prediction performance by CAFES is relatively better than that of AFES, and clearly outperforms the six competing approaches in almost all evaluation criteria, only slightly weaker than LIFT in terms of hamming loss as well. The results reported in Table 1 also demonstrate that incorporating the inherent correlations among biological function can improve the prediction performance, which gives an evidence to support the advantages of the proposed CAFES approach.

6   Conclusions

In this paper, we propose a novel correlated protein function prediction method base on PPI networks, which can achieves the function assignment and representative feature selection simultaneously by taking the function correlation graph regularizer and the ℓ_{21}-norm sparsity into the multi-label regression framework. The experimental results show the effectiveness of the proposed method by comparing with the state-of-the-art methods. In the future, we will further discover graph embedding representation, including the most advanced graph deep learning model, to deal with more complex biological network systems.

Acknowledgements. This work was supported by the Key Natural Science Project of Anhui Provincial Education Department (KJ2018A0023), the Guangdong Province Science and Technology Plan Projects (2017B010110011), the Anhui Key Research and Development Plan (1804a09020101), the National Basic Research Program (973 Program) of China (2015CB351705) and the National Natural Science Foundation of China (61906002, 61402002, 61876002 and 61860206004).

References

1. Vazquez, A., Flammini, A., Maritan, A., Vespignani, A.: Global protein function prediction from protein-protein interaction networks. Nat. Biotechnol. **21**(6), 697–700 (2003)
2. Ashburner, M., et al.: Gene ontology: tool for the unification of biology. The gene ontology consortium. Nat. Genet. **25**, 25–29 (2000)
3. Elisseeff, A., Weston, J.: A kernel method for multi-labelled classification. In: Proceedings of the 14th International Conference on Neural Information Processing Systems: Natural and Synthetic, NIPS 2001, pp. 681–687. MIT Press, Cambridge (2001). http://dl.acm.org/citation.cfm?id=2980539.2980628
4. Gligorijevic, V., Barot, M., Bonneau, R.: deepNF: deep network fusion for protein function prediction. Bioinformatics (Oxford, England) **34**, 3873–3881 (2018). https://doi.org/10.1093/bioinformatics/bty440
5. Hishigaki, H., Nakai, K., Ono, T., Tanigami, A., Takagi, T.: Assessment of prediction accuracy of protein function from protein-protein interaction data. Yeast **18**(6), 523–531 (2001). https://doi.org/10.1002/yea.706
6. Wang, H., Huang, H., Ding, C.: Image annotation using multi-label correlated green's function. In: 2009 IEEE 12th International Conference on Computer Vision. pp. 2029–2034, September 2009. https://doi.org/10.1109/ICCV.2009.5459447
7. Wang, H., Huang, H., Ding, C.: Image annotation using bi-relational graph of images and semantic labels. In: CVPR 2011, pp. 793–800, June 2011. https://doi.org/10.1109/CVPR.2011.5995379
8. Karaoz, U., et al.: Whole-genome annotation by using evidence integration in functional-linkage networks. Proc. Natl. Acad. Sci. **101**, 2888–2893 (2004). https://doi.org/10.1073/pnas.0307326101
9. Liu, J., Wang, J., Yu, G.: Protein function prediction by random walks on a hybrid graph. Curr. Proteom. **13**, 130–142 (2016). https://doi.org/10.2174/1570164613021605140004307

10. Mewes, H., et al.: MIPS: a database for genomes and protein sequences. Nucleic Acids Res. **28**(1), 37–40 (2000). https://doi.org/10.1093/nar/28.1.37. http://europepmc.org/articles/PMC102494

11. Zhang, M.-L., Zhou, Z.-H.: Multilabel neural networks with applications to functional genomics and text categorization. IEEE Trans. Knowl. Data Eng. **18**(10), 1338–1351 (2006). https://doi.org/10.1109/TKDE.2006.162

12. Nabieva, E., Jim, K., Agarwal, A., Chazelle, B., Singh, M.: Whole-proteome prediction of protein function via graph-theoretic analysis of interaction maps. Bioinformatics **21**(1), 302–310 (2005)

13. Pizzuti, C.: GA-net: a genetic algorithm for community detection in social networks. In: Rudolph, G., Jansen, T., Beume, N., Lucas, S., Poloni, C. (eds.) PPSN 2008. LNCS, vol. 5199, pp. 1081–1090. Springer, Heidelberg (2008). https://doi.org/10.1007/978-3-540-87700-4_107

14. Schwikowski, B., Uetz, P., Fields, S.: A network of protein-protein interactions in yeast. Nat. Biotechnol. **18**(12), 1257–1261 (2000). https://doi.org/10.1038/82360

15. Sharan, R., Ulitsky, I., Shamir, R.: Network-based prediction of protein function. Mol. Syst. Biol. **3**, 88 (2007). https://doi.org/10.1038/msb4100129

16. Wang, H., Huang, H., Ding, C.: Correlated protein function prediction via maximization of data-knowledge consistency. In: International Conference on Research in Computational Molecular Biology (2014)

17. Yu, Z., Fu, G., Wang, J., Zhao, Y.: NewGOA: predicting new go annotations of proteins by bi-random walks on a hybrid graph. IEEE/ACM Trans. Comput. Biol. Bioinform. 1 (2017). https://doi.org/10.1109/TCBB.2017.2715842

18. Zhang, M., Wu, L.: LIFT: multi-label learning with label-specific features. IEEE Trans. Pattern Anal. Mach. Intell. **37**(1), 107–120 (2015). https://doi.org/10.1109/TPAMI.2014.2339815

19. Zhang, M.L.: ML-RBF: RBF neural networks for multi-label learning. Neural Process. Lett. **29**(2), 61–74 (2009). https://doi.org/10.1007/s11063-009-9095-3

20. Zhang, M.L., Peña, J.M., Robles, V.: Feature selection for multi-label naive bayes classification. Inform. Sci. **179**(19), 3218–3229 (2009). https://doi.org/10.1016/j.ins.2009.06.010. http://www.sciencedirect.com/science/article/pii/S0020025509002552

21. Zhang, M.L., Zhang, K.: Multi-label learning by exploiting label dependency. In: Proceedings of the 16th ACM SIGKDD International Conference on Knowledge Discovery and Data Mining, KDD 2010, pp. 999–1008. ACM, New York (2010). https://doi.org/10.1145/1835804.1835930

22. Zhao, H., Sun, D., Wang, R., Luo, B.: A network-based approach for protein functions prediction using locally linear embedding. In: International Conference on Bioinformatics and Biomedical Engineering (2010)

23. You, Z.H., Lei, Y.K., Gui, J., Huang, D.S., Zhou, X.: Using manifold embedding for assessing and predicting protein interactions from high-throughput experimental data. Bioinformatics **26**(21), 2744–2751 (2010)

Using of Processed Data to Design Genetic Circuits in GenoCAD

Mingzhu Li[1] and Yafei Dong[2(✉)]

[1] School of Computer Science, Shaanxi Normal University,
Xi'an 710119, Shaanxi, China
[2] College of Life Science, Shaanxi Normal University,
Xi'an 710119, Shaanxi, China
dongyf@snnu.edu.cn

Abstract. In recent years, synthetic biology develops rapidly, the purpose of which is to create beneficial products or organisms with special functions. Genetic circuits are dynamic regulating systems of organisms controlling their own life processes. According to specific rules and grammar, synthetic biologists try to design artificial genetic circuits to achieve their goals. In our research, we designed grammar rules on basis of the context-free grammar, imported a part library which we acquired from IGEM database and then we processed, and constructed a genetic circuit for expressing TetR in GenoCAD, which is a free web-based application for synthetic biology. Finally, we generated a plasmid map on the PlasMapper, and we can see the concrete information of the plasmid. We creatively designed the genetic circuit to express the TetR. The work provides a new dynamic regulating system for synthetic biology. We think the genetic circuit will help us predict the biological regulator process.

Keywords: Synthetic biology · Genetic circuit · GenoCAD

1 Introduction

Synthetic biology is a new field in biology, the purpose of which is to introduce the concepts and ideas of engineering into biology. It is a cross-discipline which includes many disciplines such as biology, chemistry, computer science, engineering, physics and so on. Its aim is to integrate the concepts and knowledge of these disciplines, to more efficiently, more economically and more accurately achieve the will of mankind, which is to create beneficial products or organisms with special functions [1, 2]. Part is an important concept in synthetic biology, and it's the component of the basic cellular activity. Various parts which are nucleotide or protein sequence with special functions, can be combined to synthesize various biological devices, moreover various devices can be combined to synthesize biological system [3]. In recent years, synthetic biology develops rapidly, which has brought new thoughts and new methods for the research of life science [4].

Genetic circuits are dynamic regulating systems of organisms controlling their own life processes [5]. There are a lot of natural genetic circuits in nature which compose

© Springer Nature Singapore Pte Ltd. 2020
L. Pan et al. (Eds.): BIC-TA 2019, CCIS 1160, pp. 18–25, 2020.
https://doi.org/10.1007/978-981-15-3415-7_2

three basic elements of life processes with substance metabolism and energy supply [6]. All organisms consist of substance, drive by energy, and control the metabolism of substance and the flow of energy by genetic circuit, to achieve multiple physiological activities such as cell division, individual morphological development and so on [7, 8]. The artificial genetic circuits consist of genetic switch, biological oscillator and logic gate to execute a good deal functions of regulation. The purpose of artificial genetic circuits is to carry out targeted control, to achieve given control logic and to play a role like that computer control chip [9, 10]. Hence, artificial genetic circuit is the iconic technology of synthetic biology, which is a concert embodiment of programmable control of life [11].

GenoCAD is a free web-based application for synthetic biology. It's can be used to design protein expression vectors, artificial gene networks and other genetic constructs with the parts in the GenoCAD part library on the basis of context-free grammar. GenoCAD allows users to create their own part library to design and synthesize DNA sequence [12].

TetR (tetracycline repressor gene) family is a transcription regulator family [13]. There are more than 2000 members in the family, but there are only 100 members containing all features. The TetR family is named on the basis of genetic and bio-chemical full sequence features of the TetR protein. The TetR protein control the expression of the tet gene, the products can repress the expression of tetracycline. In the absence of tetracycline, TetR binds to the tetracycline operon and repress the tran-scription of the tetracycline. When tetracycline enters the cell, tetracycline binds to TetR and changes its conformation, dissociating TetR from the tet operate sequence, thus releasing repression [14, 15]. TetR can be used extensively in bacterial as well as mammalian synthetic biology as a novel component.

In this research, we imported the most representative data into GenoCAD, created design grammar rules and designed a genetic circuit for expressing TetR.

2 Materials and Methods

2.1 Category

First of all, we defined several diverse categories which we need to use to design the artificial genetic circuit. There are some common categories, such as promoter, oper-ator, ribosome binding site(RBS), gene coding region and terminator [16]. The pro-moter is a specific DNA sequence which RNA polymerase recognizes, binds to and initiates transcription. It specifically binds to RNA polymerase, but its own sequence doesn't be transcribed. The operator is the functional unit of transcription the sequence of which is the binding site of prokaryotic repressor protein. When the operator sequence binds to repressor protein, it represses the bind of RNA polymerase and promoter sequence. The RBS is a untranslated region of the initiation codon, the region is the part of translating mRNA, and it can synthesize corresponding protein. The terminator is a specific DNA sequence with terminate function. We used one or two or more letters to express these categories, and described them with short sentences (see Table 1). GenoCAD allows user to choose an icon set to represent categories.

Table 1. Category set.

Symbol	Category	Description
S	Start	S is the symbol of start, all categories start from it
P	Promoter	Promoter
TC	Transcription	Transcription region
O	Operator	Operator
RBS	Ribosome binding site	Ribosome binding site
G	Gene codon region	Gene coding region
T	Terminator	Terminator

2.2 Rules

Next, we constructed context-free grammars rules. Before we constructed the rules, we had classified the categories into two types, rewritable and terminal. S is the start of the rules, every process of construction begins from S. The rewritable categories can be rewritten as one or more biological parts [17]. In addition, the terminal categories are the basic biological parts which can't be decomposed into smaller parts.

We used the letters to represent the categories, then wrote the rules. The design rules can be used to design our own protein express vectors, artificial genetic circuits and other genetic constructs. Here we show the rules which our research needed (see Table 2). For example, the mean of S → S S is the start symbol can be rewritten as two start symbol. Therefore, S → P TC T is we can rewrite the start symbol as a promoter, a transcript unit and a terminator. According to these rules, we can design what we want.

Table 2. Rules.

Rule code	Rule	Description
1	S → S S	Start (S)
2	S → P TC T	Promoter (P), transcript (TC), terminator (T)
3	TC → RBS G	Ribosome binding site (RBS), gene coding region (G)
4	P → P O	Promoter (P), operator (O)

2.3 Part Library

We already acquired more than 7000 parts from the Registry of Standard Biological Parts. At first, we divided these parts into more than 70000 features. Then we used SQL language to process the data and to delete the redundancy information, and obtained a part library includes 5148 features [18]. The data were used to construct the genetic circuit.

2.4 Biological Logic Gate

As we all know, the logic gate consists of electronic components, but in biology we can use biological parts to replace electronic components. In general, the output of logic gate only have two states, "0" and "1". We used "0" and "1" to represent the low and high state of the biological activity. Biological logic gate enables us to understand the state of life activity.

3 Result

We employed the parts shown in Table 1 to combine the construct. It's based on our grammar shown in Table 2. Here we will explain our design in more detail to help comprehend our design philosophy. The genetic circuit was divided into three parts (see Fig. 1). In the first part, LuxR protein is a kind of vital transcript regulator protein which was in the bacteria quorum-sensing machinery mediated by acylhomoserine lactone (AHL). When there exists AHL, LuxR protein binds to AHL to synthesize AHL-LuxR complex. Next, the AHL-LuxR complex actives the promoter PLuxR of the second part. The TetR protein doesn't be transcribed with no IPTG or lactose, because lacI codes repressor protein which binds to the operator sequence site on the downstream of the promoter, so that the promoter can't be activated. When we add IPTG or lactose into the system, IPTG or lactose binds to repressor protein, and the promoter is activated to express TetR. The expression of YFP is repressed by TetR, on the account of the binding of TetR and tetracycline operator (TO) within the tetracycline-inducible promoter PLtet. When the aTc (anhydro-tetracycline) is added, the TetR is replaced from TO, and the YFP was expressed. The third part was designed to verify whether the TetR was expressed. This is the full process of the genetic circuit.

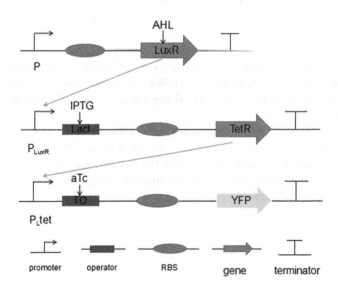

Fig. 1. Figure illustrating the regulator process of the genetic circuit, the icon of every part is shown below the figure.

Here, we can see the circuit as a three-input logic gate. Obviously, the output is whether the YFP expresses. If the YFP doesn't expresses, the output is "0"; if the YFP expresses, the output is "1". The AHL, IPTG and aTc can be regarded as three inputs. Therefore, we drew the figure of logic gate according to the above content (see Fig. 2).

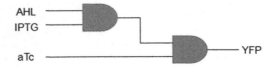

Fig. 2. The three-input logic gate represents the genetic circuit.

From the figure above, we created a truth table to better understand the regulator process (see Table 3). In the truth table, we know clearly only when the AHL, IPTG and aTc exist, the YFP can be expressed.

Table 3. Truth table.

Input 1: AHL	Input 2: IPTG	Input 3: aTc	Output: YFP
0	0	0	0
0	0	1	0
0	1	0	0
0	1	1	0
1	0	0	0
1	0	1	0
1	1	0	0
1	1	1	1

In the next, we constructed the sequence on the GenoCAD platform, and downloaded the nucleotide sequence (see Fig. 3). We can see the whole genetic circuit. The under symbol of every part is the selected feature the name of which was generated automatically by GenoCAD.

Finally, we inputted the nucleotide sequence into the input box of the software PlasMapper, after setting the parameters, we clicked the "Graphic Map" button. Whereupon, we acquired a plasmid map (see Fig. 4). From the figure, we can see the open reading frame, reporter gene, selectable marker, terminator and unique restriction site were identified and marked.

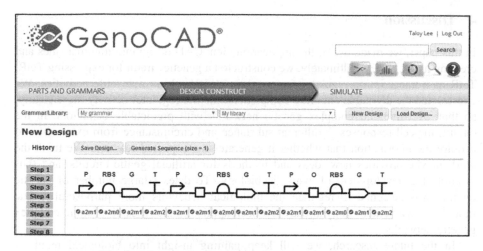

Fig. 3. The genetic circuit was shown on the GenoCAD.

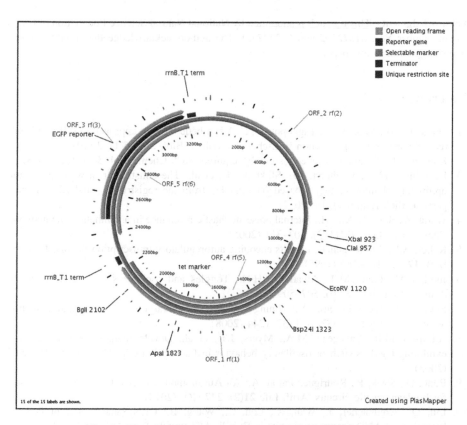

Fig. 4. The plasmid map was generated on the PlasMapper.

4 Discussion

In this paper, we made use of the application GenoCAD designing grammar rules and creating a part library. Ultimately, we constructed a genetic circuit for expressing TetR. TetR as a transcription regulator factor, plays a vital role in many life activities. We constructed the divinable circuit to comprehend the life regulator activity more clearly, so that we can build complex genetic networks with predictable behaviors. In the circuit, the cell responses to different substance and circumstance from externality, to acquire the information that whether it generates the expected reaction. We think the work maybe provides new ideas and methods for artificial gentic circuits, and thus promote the development of synthetic biology. To date, a large number of genetic circuits was designed to regulate the biological behaviors more purposefully. The thought of synthetic biology already has been applied to the research of more complex genetic networks.

In the future research, we will keep gaining insight into biological regulator activities, and do the experiments to verify the validity of the genetic circuit.

Acknowledgment. This research is supported by National Natural Science Foundation of China (Grant number Nos. 61572302 and 61272246). The authors acknowledge the anonymous referee's suggestion to this paper.

References

1. Dragosits, M., Nicklas, D., Tagkopoulos, I.: A synthetic biology approach to self-regulatory recombinant protein production in Escherichia coli. J. Biol. Eng. **6**(1), 2 (2012)
2. Kiessling, L.: Synthetic science: assembly required. ACS Chem. Biol. **3**(1), 1–2 (2008)
3. Hesselman, M.C., Koehorst, J.J., Slijkhuis, T., et al.: The constructor: a web application optimizing cloning strategies based on modules from the registry of standard biological parts. J. Biol. Eng. **6**(1), 14 (2012)
4. Wang, Z., Hou, Z., Xin, H.: Internal noise stochastic resonance in a synthetic gene network. Chem. Phys. Lett. **401**(1–3), 307–311 (2005)
5. Keller, A.D.: Model genetic circuits encoding autoregulatory transcription factors. J. Theor. Biol. **172**(2), 169–185 (1995)
6. Judd, E.M., Laub, M.T., Mcadams, H.H.: Toggles and oscillators: new genetic circuit designs. BioEssays **22**(6), 507–509 (2000)
7. Silva-Rocha, R., de Lorenzo, V.: Mining logic gates in prokaryotic transcriptional regulation networks. FEBS Lett. **582**(8), 1237–1244 (2008)
8. Atkinson, M.R., Savageau, M.A., Myers, J.T., et al.: Development of genetic circuitry exhibiting toggle switch or oscillatory behavior in Escherichia coli. Cell **113**(5), 597–607 (2003)
9. Bene, D., Sosík, P., Rodríguez-Patón, A.: An Autonomous in vivo dual selection protocol for boolean genetic circuits. Artif. Life **21**(2), 247–260 (2015)
10. Liu, Q., Schumacher, J., Wan, X., Lou, C., Wang, B.: Orthogonality and burdens of heterologous AND gate gene circuits in E. coli. ACS Synth. Biol. **7**(2), 553–564 (2018)
11. Marchisio, M.A., Colaiacovo, M., Whitehead, E., et al.: Modular, rule-based modeling for the design of eukaryotic synthetic gene circuits. BMC Syst. Biol. **7**, 42 (2013)

12. Czar, M.J., Cai, Y., Peccoud, J.: Writing DNA with GenoCADTM. Nucleic Acids Res. **37** (Web Server), W40–W47 (2009)
13. Zeng, Y., Jones, A.M., Thomas, E.E., et al.: A split transcriptional repressor that links protein solubility to an orthogonal genetic circuit. ACS Synth. Biol. **7**, 2126–2138 (2018). https://doi.org/10.1021/acssynbio.8b00129
14. Shi, W., Taylor, K., Kull, J.: Characterization and crystallization of BreR, the TetR family member bile response repressor of vibrio cholerae. Biophys. J. **102**(3), 72a–72a (2012)
15. Yang, M., Gao, C., Cui, T., et al.: A TetR-like regulator broadly affects the expressions of diverse genes in Mycobacterium smegmatis. Nucleic Acids Res. **40**(3), 1009–1020 (2012)
16. Cai, Y., Hartnett, B., Gustafsson, C., et al.: A syntactic model to design and verify synthetic genetic constructs derived from standard biological parts. Bioinformatics **23**(20), 2760–2767 (2007)
17. Coll, A., Wilson, L., Gruden, K., et al.: Rule-based design of plant expression vectors using GenoCAD. PLoS One **10**(7), e0132502 (2015)
18. Dong, Y., Shi, P., Lv, Q., et al.: Use of structured query language to simplify and analyze non-redundant data. J. Comput. Theor. Nanosci. **14**, 3741–3746 (2017)

Identifying Disease Modules Based on Connectivity and Semantic Similarities

Yansen Su, Huole Zhu, Lei Zhang, and Xingyi Zhang[✉]

Key Laboratory of Intelligent Computing and Signal Processing of Ministry
of Education, School of Computer Science and Technology,
Anhui University, Hefei, China
suyansen1985@163.com, xyzhanghust@gmail.com

Abstract. The identification of disease modules has attracted increasing attention due to the importance in comprehending pathogenesis of complex diseases. Most of the existing methods were based on the protein-protein interaction (PPI) networks with the incompleteness and incorrectness of the interactome, which results in many disease-related proteins and pathways not within their disease modules. In this paper, we propose a method named IDMCSS to effectively identify disease modules by adding some potential interactions and removing some incorrect interactions in the existing human PPI network. Firstly, the connective similarity and semantic similarity are calculated between the known disease proteins and each of their neighbors. Then, the network is adjusted by adding the interactions with large connective and semantic similarity, and removing those with small connective and semantic similarity. Further, the proteins are sorted in the descending order according to the similarity to the known disease proteins, and the sorted proteins are added into a candidate node set one by one until a certain biological information is not enriched in the candidate node set. Finally, the connected subnetwork with the largest number of nodes in the candidate node set is selected as the disease module. The disease modules identified by IDMCSS involve crucial biological processes of related diseases and can predict 12 targets for drug intervention. The experimental results on asthma demonstrate the effectiveness of the proposed method in comparison to existing algorithms for disease module identification.

Keywords: Disease module · Connectivity similarity · Semantic similarity · Protein-protein interaction network

Supported by National Natural Science Foundation of China (61822301, 61672033, U1804262, 61702200), Anhui Provincial Natural Science Foundation for Distinguished Young Scholars (1808085J06), Key Program of Natural Science Project of Educational Commission of Anhui Province (KJ2019A0029), and Recruitment program for Leading Talent Team of Anhui Province (2019-16).

© Springer Nature Singapore Pte Ltd. 2020
L. Pan et al. (Eds.): BIC-TA 2019, CCIS 1160, pp. 26–40, 2020.
https://doi.org/10.1007/978-981-15-3415-7_3

1   Introduction

It is widely accepted that complex diseases are usually caused by the interactions of genes. Moreover, gene products (e.g. proteins) linked to the same phenotype (e.g. a specific disease) are not randomly scattered within the PPI networks, but have significant interactions in the Human Interactome [2]. In addition, these gene products usually agglomerate in a specific regions to form a disease module associated with a specific disease phenotype [25]. The disorder of these disease modules will cause the occurrence of a certain disease. The identification of disease modules can help to uncover the molecular mechanisms of disease causation, identify new disease-related genes and pathways, and aid the rational drug target identification [7].

Recently, a large number of methods have been developed for predicting disease-related genes based on various topological features of PPI networks or dynamic properties on the network [30]. Although studying the interactions between the proteins (that are encoded by genes) in the human PPI network has become one of the most powerful approaches for elucidating the molecular mechanisms that underlie complex diseases [11], the above algorithms are based on the PPI networks whose interactions are assumed correct and fixed for further mining. However, the interactions in PPI network are incomplete. It is estimated that human protein interaction data is less than 20% of the actual interaction data [18]. In addition, high throughput experiments often produce large amounts of false-positive and false-negative data, producing large amounts of data with noise [6]. That is, some interactions in PPI network are incorrect. These will affect the accuracy of the disease module identification. Although some researchers have tried to use other biological or topological evidence to correct the PPI network for disease gene prediction [17], the performance is still far from satisfactory since the information they used is singular and they only consider the addition of the missing interactions while ignore the deletion of the wrong interactions in PPI network, or consider the deletion of the wrong interactions while ignore the addition of the missing interactions.

To solve the above shortcomings, in this paper by using both connective similarity and semantic similarity, we propose a method termed as IDMCSS (**I**dentifying **D**isease **M**odules based on **C**onnectivity and **S**emantic **S**imilarities) to effectively identify disease modules in PPI network by dynamically adjusting the PPI network in the mining process. To be specific, we first use the connective similarity to select candidate proteins which are related with the known disease proteins, and then several links (i.e. interactions) are added or removed based on the connective similarity and semantic similarity between the known disease proteins and the candidate proteins. Further, we measure the similarity between a protein and the known disease proteins based on the sum of the connective similarity and the semantic similarity, and the proteins in the network are prioritized according to the similarity between them and the known disease proteins. The protein with the largest similarity to the known disease proteins is added into a candidate disease protein set once at a time. Finally, the stopping criterion is set to define the boundary of the disease module. Compared with the state-of-the-art algorithms for disease module identification, the experimental results on the asthma demonstrate the effectiveness of the proposed method.

2   The Proposed IDMCSS Method

In this section, we present the proposed algorithm IDMCSS for identifying disease modules in detail. The main components of the proposed IDMCSS method are the strategies of adding and removing links based on the connective similarity and semantic similarity.

2.1   The Connective Similarity and Semantic Similarity

We combine both connective similarity and semantic similarity to measure the correlation between a protein and a set of disease proteins. Considering a protein p and a set of disease proteins $S = \{p_1, \cdots, p_t\}$, the similarity between the protein p and the set of disease proteins S, denoted as $sv(p, S)$, is the sum of the connective similarity and the semantic similarity, which is defined as Eq. 1.

$$sv(p, S) = cs(p, S) + ss(p, S), \tag{1}$$

where $cs(p, S)$ represents the connective similarity between p and S, and $ss(p, S)$ represents the semantic similarity between p and S. In what follows, we present the details of the connective similarity and the semantic similarity.

Connective Similarity. The connective significance proposed by Ghiassian represents the topological interaction patterns between disease proteins, which captures the correlation of proteins not only in locally dense communities but also in the whole PPI network, since the proteins associated with a particular disease do not reside in locally dense communities [7]. We measure the connective similarity between proteins in the protein-protein interaction network based on the connective significance.

Specifically, let us consider that a network has N nodes (i.e., proteins) and there are n_0 seed nodes associated with a particular disease (i.e., given known disease proteins). Suppose that there is a node p in the neighborhood of seed nodes, and the degree of p is k, where k_s out of k links are connected to seed nodes. The probability of hypergeometric distribution is defined as Eq. 2.

$$\rho(k, k_s) = \frac{C_{n_0}^{k_s} C_{N-n_0}^{k-k_s}}{C_N^k}. \tag{2}$$

Generally, if $\rho(k, k_s) < 0.01$, then the link between p and the seed nodes is statistically significant.

We define the connective similarity between a node p and a set of seed nodes S as Eq. 3, which means how closely the protein p connects to a set of disease proteins S.

$$cs(p, S) = 1 - \sum_{t=k_s}^{k} \rho(k, t). \tag{3}$$

Semantic Similarity. The calculation of semantic similarity is adopted from [31]. For N proteins, all of the terms which annotate the protein are in the term set **T**. For each term t, its information is calculated as Eq. 4.

$$I(t) = -log\ p(t), \tag{4}$$

where $p(t)$ denotes the probability of the occurrence of the term t in the term set **T**. For the protein p which is annotated by the term set $A_p = \{t_i | i = 1, \cdots, r\}$, the information of the protein p is calculated as Eq. 5.

$$I(p) = \sum_{i=1}^{r} I(t_i). \tag{5}$$

Suppose that two proteins p_1 and p_2 are respectively annotated by the term sets $A_{p_1} = \{t_{x_i} | i = 1, \cdots, m\}$ and $A_{p_2} = \{t_{y_j} | j = 1, \cdots, n\}$, where t_{x_i} and t_{y_j} represent terms. The semantic similarity of the proteins p_1 and p_2 is calculated as Eq. 6.

$$ss(p_1, p_2) = \frac{\sum_{t_i \in (A_{p_1} \cap A_{p_2})} I(t_i)}{I_{max}}, \tag{6}$$

where I_{max} denotes the largest value of the informations of N proteins.

Further, the semantic similarity between the protein p and diseases proteins $S = \{p_1, \cdots, p_t\}$ is calculated as Eq. 7.

$$ss(p, S) = \sum_{i=1}^{t} ss(p, p_i), \tag{7}$$

It is observed that the larger value of semantic similarity, the more functional similarities between a protein and disease proteins.

2.2   Adding and Removing Strategies

The existing protein-protein interaction network may have several missing links or incorrect links. For this reason, the proposed strategies need to locally adjust the protein-protein interaction network by adding the missing links which are likely to be related to the disease proteins and removing the links which have less correlation to the disease proteins. Specifically, the steps of adding and removing strategies in the protein-protein interaction network are performed as follows.

Given a network PPI with N nodes, where each node represents a protein, and a set of seed nodes which represent the disease proteins is $S = \{p_1, \cdots, p_t\}$, we select the neighbors of the seed nodes $NS = \{u_1, \cdots, u_\alpha\}$ based on the protein-protein interaction network PPI, where u_i $(i = 1, \cdots, \alpha)$ is a neighbor of one certain seed node in S.

Let u be a node in NS. The connective similarity between u and S is calculated according to Eq. 3. If the connective similarity between u and S is larger than 0.99, then the node u is considered as strong-linked nodes, which is closely correlated with the seed nodes from the aspect of topology. If the connective similarity between u and S is smaller than the average connective similarity of the nodes in NS, then the node u is considered as weak-linked nodes, which is weakly correlated with the seed nodes. In the following, the proposed strategy adds or removes several links related with these strong-linked or weak-linked nodes from the aspect of semantic relativity.

Adding Strategy: For a strong-linked node u', there may be several seed nodes which are not connected with it. Let $S'_1 = \{p_{i_1}, \cdots, p_{i_r}\}$ be the set of seed nodes which are connected with the node u'. $S'_2 = S/S'_1$ represents the set of seed nodes which are not connected with u'. To check whether the strong-linked node u' should be linked with a node in S'_2, we adopt a threshold $\varphi_1 = mean(ss(u', p_{i_1}), \cdots, ss(u', p_{i_r}))$. For each node p_{i_e} in S'_2, if the semantic similarity between u' and a node p_{i_e} is larger than φ_1, then a link between the node u' and the node p_{i_e} is added into the network.

Deleting Strategy: For a weak-linked node u'', the proposed deleting strategy checks whether we should remove some links to ensure that the weak-linked node is not connected to any seed node. Let $S''_1 = \{p_{j_1}, \cdots, p_{j_s}\}$ be the set of seed nodes which are not connected with the node u'', and $S''_2 = S/S''_1$ is the set of seed nodes which are connected with the node u''. The threshold $\varphi_2 = mean(ss(u'', p_{j_1}), \cdots, ss(u'', p_{j_s}))$ is used to check whether the link between the node u'' and the node p_{j_e} in S''_2 should be removed. Specifically, for each node p_{j_e} in S''_2, if the semantic similarity between u'' and p_{j_e} is smaller than φ_2, then we remove the link between u'' and p_{j_e}.

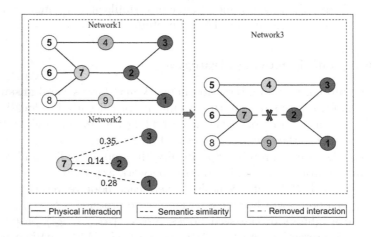

Fig. 1. An example of removing links.

Figure 1 shows an example for illustrating the strategy of deleting links, where the set of seed nodes is the set $S = \{1, 2, 3\}$, and the node **7** is a neighbor of the node **2**. For node **7**, $S_1'' = \{1, 3\}$ is the set of seed nodes which are not connected with node **7**, and $S_2'' = \{2\}$ is the set of seed nodes which are connected with node **7**. The node **7** is a weak-linked node, since the connective similarity between node **7** and set S is 0.9964, which is smaller than the average connective similarity of the nodes in NS, i.e., 0.9984. Because the semantic similarity between the nodes **7** and **2** in S_2'' is 0.14 which is smaller than the threshold $\frac{0.35+0.28}{2}$, the link between the nodes **7** and **2** is removed.

2.3   The Proposed IDMCSS

Based on the strategies of adding and removing links in a network, we develop an algorithm, termed as IDMCSS, for disease module detection in protein-protein interaction network. The general framework of the proposed IDMCSS is presented in Algorithm 1, which mainly consists of the following six steps.

Algorithm 1. General Framework of IDMCSS

Require: Protein-protein interaction network PPI, the set of seed nodes
   $S = \{s_1, \cdots, s_{n_0}\}$;
Ensure: A connected subnetwork G_{cs}
 1: Initialize the candidate node set $CS \leftarrow \emptyset$;
 2: Select the neighborhood set of the seed nodes $NS = \{b_1, \cdots, b_\alpha\}$ based on the protein-protein interaction network PPI, where $b_i(i = 1, \cdots, \alpha)$ is a neighbor of a certain seed node in S.
 3: The network is adjusted by adding or removing several links in PPI according to the proposed strategies in Section 2.2.
 4: The neighborhood set of the seed nodes NS is updated based on the adjusted PPI.
 5: Select the node in NS which has the largest similarity with the nodes in S, i.e., $b_k(k \in \{1, \cdots, \alpha\})$.
    $S \leftarrow S \cup \{b_k\}$, $CS \leftarrow CS \cup \{b_k\}$.
 6: The above steps 2-5 are repeated until a certain disease-related information (gene ontology, differential expression genes, pathways) is not significantly enriched in the set CS.
 7: A subnetwork G_s is made up of the nodes in S and the links between the nodes in S. The connected subnetwork with the largest number of nodes, i.e., G_{cs}, is extracted from the subnetwork G_s.

Firstly, we initialize a candidate node set CS to be empty. In the following, the node which is likely to be a disease protein will be preferentially added into the set CS. Secondly, the neighbors of seed nodes are selected based on the protein-protein interaction network. Thirdly, the protein-protein interaction network is locally adjusted by adding or removing several links which are associated with the seed nodes in S, according to the proposed strategies in Sect. 2.2. At the fourth step, the neighbors of the seed nodes are updated according to the adjusted network. At the fifth step, we select a neighbor of the seed nodes which has the largest similarity with them, and add the node into the set of the seed nodes S, and the candidate node set CS. The above 2–5 steps are repeated until a certain disease-related information (gene ontology, differential expression genes, pathways) is not significantly enriched in the set of seed nodes CS. Finally, we extract the largest connected subnetwork from the subnetwork consisting of the nodes in S and the links between the nodes in S.

3   Experimental Results

In this section, we analyze the results of the proposed IDMCSS for identifying the module of asthma, and verify the performance of the proposed IDMCSS by comparing it with four existing disease module detection algorithms.

3.1   Datasets

The proposed algorithm runs on the PPI network to detect the asthma-related modules, and the stopping criterion is set according to the asthma-related information, i.e., gene ontology, differential expression genes, and pathways. In the following, the PPI network, as well as the microarray expression data, asthma-related genes and pathways, are presented.

(1) **The PPI network.** The human interactome integrates seven kinds of physical interactions, i.e., regulatory interactions, biophysical interactions, literature curated interactions, metabolic enzyme-coupled interactions, protein complexes, kinase network and signaling interactions. To be specific, 1,335 regulatory interactions among 774 transcription factors and genes are extracted from the transcription factors database. We also obtain 28,653 biophysical interactions between 8,120 proteins by combining several yeast-two-hybrid high-throughput datasets [21] with binary interactions from the IntAct molecular interaction database (IntAct) [3] and the molecular interaction database (MINT) [4]. A total of 11,798 proteins are connected via 88,349 literature curated interactions from the IntAct, MINT, a general repository for interaction datasets [5] and human protein reference database [19]. We extract 5,325 metabolic enzyme-coupled interactions among 921 enzymes from the literature [15], where two enzymes are assumed to be coupled if they share adjacent reactions in the kyoto encyclopedia of genes and genomes [12] and biochemical genetic and genomic databases [23]. The comprehensive resource of mammalian protein complexes database supplies us 2,069 proteins connected by 31,276 links [22]. A total of 6,066 interactions among 1,843 kinases and substrates are extracted from the PhosphositePlus database [8]. In addition, we use the 32,706 signaling interactions among 6,339 proteins integrated by high-throughput analysis and literature curation [29].

The union of above seven kinds of interactions yields a network of 13,460 proteins that are interconnected by 141,296 physical interactions. Further, we extract 19,707 genes which are annotated with GO terms from the gene ontology annotation database (GOA) [10]. Finally, the resulting PPI network consists of 12,562 proteins with GO function annotation information via 130,390 physical interactions.

(2) **Microarray expression data sets.** In this paper, nine asthma-related microarray expression data sets are downloaded from the NCBI Gene Expression Omnibus database (GEO)[1], where the accession number of these microarray expression data sets are respectively GSE470, GSE2125, GSE3004, GSE4302, GSE16032, GSE31773, GSE35571, GSE41649, and GSE43696.

(3) **Known asthma-related genes.** The known genes associated with asthma in our work are compiled from previous literatures, asthma-related pathways and several datasets. Specifically, 59 genes are identified in the pervious studies [28]. The known asthma-related genes also include 68 genes which belong to 13 asthma pathways and 60 genes which are part of several reported pathways containing known drug targets for asthma. In addition, 45 asthma related genes are selected from the Online Mendelian Inheritance in Man (OMIM) database and the Medical Subject Headings (MeSH) database[2], where 19 asthma related genes catalogued in OMIM are considered and 26 genes associated with the MeSH term 'asthma' are selected from the Gene2MeSH database. Then, we remove duplicate genes from the above 232 selected genes and obtain 124 known asthma-related genes, since the genes selected from different sources may be duplicated. In this paper, we further select 107 known asthma-related genes which are included in the protein-protein interaction network for experimental analysis.

(4) **Known asthma-related pathways.** We collect 23 asthma-related pathways, where 20 are reported in the literature [26] and three pathways related with asthma are from the literature [24]. Specifically, the 20 asthma-related pathways in [26] are identified based on the genome-wide association study on asthma. These 20 asthma-related pathways include hsa05322, hsa05320, hsa5330, hsa5332, hsa04940, hsa04514, hsa04162, bbcell pathway, CSK Pathway, asbcell Pathway, mhc Pathway, blymphocyte Pathway, ctla4 Pathway, eosinophils Pathway, Th1/Th2 Pathway, IL5 Pathway, TCRA Pathway, hsa05310, hsa04640, inflamPathway. Three asthma-related pathways in [24] are respectively hsa04620, h_cdcPathway, and h_cytokinePathway.

3.2 Compared Algorithms

Because the proposed algorithm is a connective and semantic-based algorithm, we compare it with four algorithms, including two connective-based algorithms, i.e., disease module detection algorithm (DIAMOnD) [7] and random walk with restart algorithm (RWR) [14], a semantic-based algorithm, i.e., hybrid relative

---

[1] http://www.ncbi.nlm.nih.gov/geo/.
[2] http://gene2mesh.ncibi.org.

specificity similarity algorithm (HRSS), and an algorithm combining connective similarity and semantic similarity, i.e., combining topological similarity and semantic similarity algorithm (CTSS) [16].

The DIAMOnD algorithm is a seed-expanding method which can identify the disease module around a set of known disease proteins, where the connective significance is used for checking whether a node should be added into the module. The RWR, HRRS and CTSS algorithms are three sorting algorithms, which result in a ranking list of proteins. To be specific, RWR uses random walk analysis, which is a global network distance measure, to measure similarities among proteins in protein-protein interaction networks. HRRS ranks all nodes by calculating the relative specificity similarity of each node in the network to known disease nodes, where the relative specificity similarity is calculated by taking the global position of relevant gene ontology terms into account. CTSS improves the performance of RWR by using the relative specificity similarity between the candidate gene and the known disease gene to set the initial probability vector of the random walk. The similarity achieved by CTSS reflects both connectivity and semantic correlation of proteins. For the above comparison algorithms, the best parameters recommended in their original references are adopted.

3.3   Experiment Results

In this subsection, we first give the details of the disease module discovered by the proposed algorithm, and then present the disease-related pathways and genes involved in the disease module. Finally, the performance of the proposed algorithm is analyzed.

Disease Modules. In the experiments conducted in this paper, the final disease module of asthma is achieved by the proposed IDMCSS algorithm after 217 iterations. The reason for the stopping criterion is that, when one of the three biological attributes (the enrichment of gene ontology, the differential expression

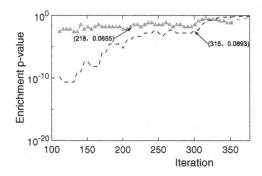

Fig. 2. The relationship of the biological significance and the iteration of the proposed IDMCSS.

genes, and the pathway related with asthma) is not significantly enriched in the candidate proteins associated with asthma, after IDMCSS iterates 217 times. Specifically, the enrichment of the differential expression genes and the pathway related with asthma are respectively not significantly enriched in the resulted candidate proteins associated with asthma when the proposed algorithm iterates 218 and 315 times (Fig. 2).

The disease module of asthma achieved by the proposed IDMCSS contains 279 nodes, where 62 nodes are seed nodes which represent the disease proteins associate with asthma reported in previous literatures, and the other 217 nodes are expanded nodes which are newly discovered to be asthma-related proteins. Note that not all the reported 107 disease proteins associate with asthma are included in the disease module of asthma. The reasons for above results are presented as follows. On one hand, the existing protein-protein interaction network may miss several links associated with disease proteins. On the other hand, several reported disease proteins may be not really related with asthma, resulting in a amount of false positives. In the disease module of asthma, there are 2819 links, where 489 links are added into the protein-protein interaction by the proposed IDMCSS. It is reasonable to add possible links into the protein-protein interaction network, since the protein-protein interaction network has been estimated to have missing links [18]. It is found that the proteins, which are involved in 116 newly added links, are statistically significantly in the same biological pathway (Fishers exact test, p-value = 0).

We measure the closeness of the disease module by the ratio of the number of inner-links to that of external-links. Specifically, the closeness of the disease module is 0.0592, where the number of inner-links of the disease module is 2819 and that of external-links is 47657. We find that the disease module is not a locally dense community, which is also in accordance with that proposed by Ghiassian [7]. Further, we compare the closeness of the disease module with those of random modules in the adjusted protein-protein interaction network for 30 times. It is found that the closeness of the disease module achieved by the proposed IDMCSS is statistically larger than those of the connected subnetworks which are randomly selected in the adjusted protein-protein interaction network (Student's t test, 6.46×10^{-32}).

We extract 72 pathways which have at least half of their genes in the disease module of asthma. In what follows, we will further predict asthma-related pathways, and the targets for therapy from the achieved 72 pathways and 279 proteins in the achieved disease module.

Asthma-Related Pathways in the Disease Module. From 304 human pathways from the Biocarta database, we extract 72 candidate pathways, of which at least half of the genes are in the disease module of asthma. It is found that these 72 pathways are statistically significantly enriched in the achieved disease module, which are considered as the asthma-related pathways. We further find that two of the asthma-related pathways obtained by the proposed algorithm are included in the 23 known asthma-related pathways, which have been reported in previous literatures. The other 70 pathways are the newly predicted asthma-related pathways. For example, five of the 70 pathways are associated with asthma in previous literatures, i.e., 'h-il7Pathway', 'h-pkcPathway', 'h-melanocytepathway', 'h-ngfPathway', and 'h-trkaPathway'. Specifically, the 'h-il7Pathway' pathway has contribution to allergen-induced eosinophilic airway inflammation in asthma [13]. Pkc signaling pathway causes psychological stress, which may lead to Acute exacerbation of asthma [9]. The alpha-melanocyte-stimulating hormone in the 'h-melanocytepathway' pathway inhibits allergic airway inflammation [20]. It is reported that NGF and TrkA increase the cell viability of isolated plasma cells of allergic asthma from inflamed airways via 'h-ngfPathway' and 'h-trkaPathway' pathways [1].

Asthma-Related Genes in the Disease Module. According to the disease module of asthma, we predict several targets of glucocorticoid which is an effective anti-inflammatory drug for asthma. In order to identify the drug targets of asthma, we select the effects of glucocorticoid, which are significantly present in the module. At first, we respectively select the genes which are differential expressed before and after the treatment of glucocorticoid in the asthma-specific cell lines and normal cell lines based on the experimental data supplied in [24]. It is found that the genes responding to the glucocorticoid treatment are different in normal and asthmatic subjects. Specifically, for the normal and asthmatic fibroblast cells, we respectively find 33 and 25 differential expression genes within the asthma module following the glucocorticoid treatment, compared with random expectations of 4.34×10^{-6} and 3.54×10^{-5} (Fishers exact test). Then, the genes are considered to be the targets of glucocorticoid in asthma, which are differentially expressed between asthmatic fibroblasts untreated and asthmatic fibroblast cells treated with glucocorticoid, but not between normal untreated fibroblast cells and normal fibroblasts treated with glucocorticoid. In our work, 12 genes are predicted to be targets of glucocorticoid, i.e., *acvrl1, ar, cdk1, ctgf, ddit3, icam1, jak1, rora, smad1, snca, tgfb2,* and *tlr4*.

We compare the enrichment of the differential expression genes in 62 seed nodes with that in 217 expanded nodes. To be specific, there are 23 (17) out of 217 expanded nodes are differentially expressed in normal (asthmatic) samples, while there are 10 (8) out of 62 seed nodes are differentially expressed in normal (asthmatic) samples. Thus, the enrichment of the differential expression genes in expanded nodes is respectively 3.8193×10^{-4} and 1.41×10^{-2} in normal and asthmatic by Fishers exact test. Further, the enrichment of the differential expression genes in expanded nodes is significantly higher than the enrichment of the seed genes, which is respectively 4.32×10^{-2} and 4.30×10^{-2}. It is found that the response to the glucocorticoid treatment is heavily localized in the genes which are added by the proposed IDMCSS algorithm (expanded nodes), rather than the consensus asthma genes (seeds nodes). Thus, it is indicated that the proposed algorithm is able to provide targets for therapeutic intervention.

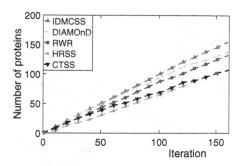

Fig. 3. Analysis of GO function annotations.

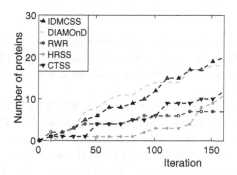

Fig. 4. Analysis of expression differences.

Comparison Results with Other Algorithms. In this subsection, we verify the detection quality of the proposed algorithm based on the protein-protein interaction network by comparing the enrichment of GO terms, differential expression genes and pathways in the disease module achieved by the proposed algorithm with those obtained by the compared algorithms. Let q be the number of nodes in the module that achieved by the proposed algorithm. For the three compared algorithms which sorts the proteins but fail to get the modules, the modules for comparison are the subnetworks in the protein-protein interaction network, where the nodes in the module are the first q nodes in the achieved list of proteins.

Figure 3 presents the number of proteins which significantly annotated by 940 asthma-related GO terms under different values of iteration, where the 940 asthma-related GO terms are those enriched in the 107 known asthma proteins. From the figure, it can be found that the proposed IDMCSS achieves the largest number of proteins which are significant enriched in asthma-related GO terms.

Figure 4 plots the number of differential expression genes of the proposed algorithm and four compared algorithms when the iteration ranges from 1 to 217. As can be seen from the figure, the proposed algorithm IDMCSS gains the largest number of differential expression genes when the iteration is larger

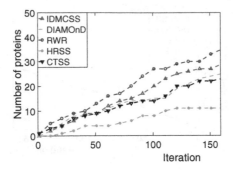

Fig. 5. Asthma pathways analysis.

than 111. The main reason may be attributed to the fact that by enhancing the structure of PPI, it becomes relatively easy to detect the differential expression genes, thus the proposed IDMCSS can achieve a competitive performance in detecting disease modules.

Figure 5 presents the number of proteins which belong to the 23 known asthma-related pathways. It is found that the proposed IDMCSS is slightly worse than RWR, but it is better than other algorithms. The main reason for the phenomenon is that the proteins linked by physical interactions tend to collaborate with each other in the same pathway [27]. The proteins obtained by RWR are always the known disease proteins' neighbors which are connected to the known disease proteins by physical interactions in the PPI network, while those obtained by IDMCSS may be the nodes which are not linked with the known disease proteins. Therefore, we can conclude that the proposed IDMCSS is a competitive disease module detection algorithm in terms of detection quality.

The Robustness of IDMCSS. We randomly delete 10%, 20%, and 30% of the known asthma disease genes, and repeat the experiment 30 times for different

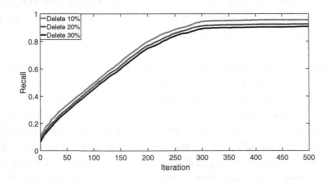

Fig. 6. The recall rate of disease module.

degrees of deletion to verify the recall rate of the disease module. As shown in Fig. 6, the removal of the initial known disease genes has little effects on the detection of disease modules in the 217 iterations. It is indicated that the proposed IDMCSS has little relatedness with the number of known disease genes, which also demonstrates the robustness of IDMCSS.

4   Conclusion and Future Work

In this paper, we have proposed a method named IDMCSS for effectively identifying disease modules. In IDMCSS, the connective similarity and semantic similarity between the given disease proteins and their neighboring proteins have been adopted to adjust the PPI network by adding several potential interactions and removing some incorrect interactions in the mining process. Then, the protein in the neighborhood of the known disease proteins with the best connective and semantic similarities can be extended into the disease module on the adjusted PPI network until the module spans across the entire network. Finally, the best extended disease module can be finally determined through the analysis of GO term, gene expression and pathways. Compared with the existing algorithms for disease modules identification, the experimental results on the validation of asthma disease have demonstrated the effectiveness of the proposed method IDMCSS. In the future, we will integrate phenotypic similarity information into the protein-protein interaction network to make more comprehensive and accurate predictions of disease modules.

References

1. Abram, M.: NGF increases cell viability of isolated plasma cells from inflamed airways via TrkA signalling in a mouse model of allergic asthma. J. Allergy Clin. Immunol. **121**(2), S200 (2008)
2. Albert-Lszl, B., Natali, G., Joseph, L.: Network medicine: a network-based approach to human disease. Nat. Rev. Genet. **12**(1), 56–68 (2011)
3. Aranda, B., Achuthan, P., Alam-Faruque, Y., Armean, I.M.: The IntAct molecular interaction dabase in 2010. Nucleic Acids Res. **38**, 525–531 (2009)
4. Ceol, A., Aryamontri, A.C., Licata, L., Peluso, D., et al.: MINT, the Molecular INTeraction database: 2009 update. Nucleic Acids Res. **35**, 572–574 (2007)
5. Chatraryamontri, A., Breitkreutz, B., Oughtred, R., et al.: The BioGRID interaction database: 2015 update. Nucleic Acids Res. **43**, 470–478 (2015)
6. Cho, Y., Montanez, G.: Predicting false positives of protein-protein interaction data by semantic similarity measures. Curr. Bioinform. **8**(3), 339–346 (2013)
7. Ghiassian, S.D., et al.: A DIseAse MOdule Detection (DIAMOnD) algorithm derived from a systematic analysis of connectivity patterns of disease proteins in the human interactome. PLoS Comput. Biol. **11**(4), e1004120 (2015)
8. Hornbeck, P., et al.: PhosphoSitePlus: a comprehensive resource for investigating the structure and function of experimentally determined post-translational modifications in man and mouse. Nucleic Acids Res. **40**, 261–270 (2012)

9. Hou, L., et al.: Participation of antidiuretic hormone (ADH) in asthma exacerbations induced by psychological stress via PKA/PKC signal pathway in airway-related vagal preganglionic neurons (AVPNs). Int. J. Exp. Cell. Physiol. Biochem. Pharmacol. **41**(6), 2230–2241 (2017)
10. Huntley, R.P., et al.: The GOA database: gene ontology annotation updates for 2015. Nucleic Acids Res. **43**, 1057–1063 (2015)
11. Igor, F., Andrey, R., Dennis, V.: Network properties of genes harboring inherited disease mutations. Proc. Nat. Acad. Sci. U.S.A. **105**(11), 4323–4328 (2008)
12. Kanehisa, M., Goto, S.: KEGG: kyoto encyclopedia of genes and genomes. Nucleic Acids Res. **28**(1), 27–30 (1999)
13. Kelly, E.A.B., et al.: Potential contribution of IL-7 to allergen-induced eosinophilic airway inflammation in asthma. J. Immunol. **182**(3), 1404–1410 (2009)
14. Kohler, S., et al.: Walking the interactome for prioritization of candidate disease genes. Am. J. Hum. Genet. **82**(4), 949–958 (2008)
15. Lee, D., Park, J.Y., Kay, K.A., Christakis, N.A., Oltvai, Z.N., Barabasi, A.: The implications of human metabolic network topology for disease comorbidity. Proc. Nat. Acad. Sci. U.S.A. **105**(29), 9880–9885 (2008)
16. Liu, B., Jin, M., Zeng, P.: Prioritization of candidate disease genes by combining topological similarity and semantic similarity. J. Biomed. Inform. **57**(C), 1–5 (2015)
17. Luo, J., Liang, S.: Prioritization of potential candidate disease genes by topological similarity of protein-protein interaction network and phenotype data. J. Biomed. Inform. **53**(7), 229–236 (2015)
18. Menche, J., et al.: Uncovering disease-disease relationships through the incomplete interactome. Science **347**(6224), 1257601 (2015)
19. Prasad, T.S.K., Goel, R., Kandasamy, K., Keerthikumar, S., et al.: Human protein reference database-2009 update. Nucleic Acids Res. **37**, 767–772 (2009)
20. Raap, U., Brzoska, T., Sohl, S., Path, G., et al.: Alpha-melanocyte-stimulating hormone inhibits allergic airway inflammation. J. Immunol. **171**(1), 353–359 (2003)
21. Rolland, T., Tasan, M., Charloteaux, B., Pevzner, S.J., et al.: A proteome-scale map of the human interactome network. Cell **159**(5), 1212–1226 (2014)
22. Ruepp, A., et al.: CORUM: the comprehensive resource of mammalian protein complexes. Nucleic Acids Res. **36**, 646–650 (2010)
23. Schellenberger, J., Park, J., Conrad, T.M., Palsson, B.O.: BiGG: a biochemical genetic and genomic knowledgebase of large scale metabolic reconstructions. BMC Bioinform. **11**(1), 213 (2010). https://doi.org/10.1186/1471-2105-11-213
24. Sharma, A., et al.: A disease module in the interactome explains disease heterogeneity, drug response and captures novel pathways and genes in asthma. Hum. Mol. Genet. **46**(7), 957–961 (2012)
25. Sol, A.D., Balling, R., Hood, L., Galas, D.: Diseases as network perturbations. Curr. Opin. Biotechnol. **21**(4), 566–571 (2010)
26. Song, G.G., Lee, Y.H.: Pathway analysis of genome-wide association study on asthma. Hum. Immunol. **74**(2), 256–260 (2013)
27. Venkatesan, K., Rual, J.F., Vazquez, A., Stelzl, U., et al.: An empirical framework for binary interactome mapping. Nat. Methods **6**(1), 83–90 (2008)
28. Vercelli, D.: Discovering susceptibility genes for asthma and allergy. Nat. Rev. Immunol. **8**(3), 169–182 (2008)
29. Vinayagam, A., et al.: A directed protein interaction network for investigating intracellular signal transduction. Sci. Signal. **4**(189), rs8 (2011)
30. Wang, X., Gulbahce, N., Yu, H.: Network-based methods for human disease gene prediction. Brief. Funct. Genomics **10**(5), 280–293 (2011)
31. Zhen, T., Maozu, G.: An improved calculation method of gene functional similarity. Intell. Comput. Appl. **7**(5), 123–126 (2017)

Complex Networks

Adaptive Synchronization of Nonlinear Complex Dynamical Networks with Time-Delays and Sampled-Data

Jiahui Bai[1](✉), Ningsheng Xu[1](✉), Yaoyao Ping[1](✉), and Xue Lu[2]

[1] College of Science, North China University of Technology, Beijing 100144, China
baijiahui1310@163.com, XuNingSheng1@163.com, 1126414491@qq.com
[2] Beijing Vocational College of Agriculture, Beijing 100144, China

Abstract. This paper investigates the adaptive synchronization of non-linear complex dynamical networks with time-delays and sampled-data by proposing a new adaptive strategy to coupling strengths and feedback gains. According to Lyapunov theorem, it is testified that the agents of sub-groups can converge those synchronous states respectively under some special conditions. In addition, some simulations are proposed to illustrate the theoretical results.

Keywords: Complex dynamical networks · Adaptive synchronization · Time-delays · Sampled-data

1 Introduction

Synchronization is a ubiquitous phenomenon in nature, such as the consistency of fireflies twinkling, Synchronized chirping of crickets and Synchronization of beating rhythm of cardiac myocytes. In recent years, the synchronization problems of nonlinear complex dynamical networks have attracted great attention and emerged a good deal of excellent works [1–4].

Synchronous methods of complex networks have emerged as the times require, and one of the most significant methods is to design advisable adaptive strategies for the relevant parameters, such as the coupling strengths and the feedback gains [3–8]. In [5], adaptive synchronization of complex dynamical networks was studied. Liu et al. [7] studied the adaptive synchronization of complex dynamical networks governed by local Lipschitz nonlinearlity on switching topology. For nonlinear complex dynamics networks, information interaction between agents can be considered as sampled information [9–12]. In [9], the author gave the

This work was supported in part by the National Natural Science Foundation of China Grant (No. 61773023), 2018 Beijing Education Commission Basic Science Research Expenses Project, "The-Great-Wall-Scholar" Candidate Training Plan of NCUT (XN070006), Construction Plan for Innovative Research Team of NCUT (XN018010), "Yu Xiu" Talents of NCUT.

© Springer Nature Singapore Pte Ltd. 2020
L. Pan et al. (Eds.): BIC-TA 2019, CCIS 1160, pp. 43–57, 2020.
https://doi.org/10.1007/978-981-15-3415-7_4

necessary and sufficient conditions for solving consensus problems of double-integrator dynamics via sampled control and group synchronization of nonlinear complex dynamics networks with sampled data was investigated in [11]. However, time-delay is widespread in communication information among agents in real life [1–3,5,10]. Sampled-data based consensus of continuous-time multi-agent systems with time-varying topology was studied in [10]. The authors divided the whole group into some sub-groups to research those synchronization, that is group synchronization [11–14]. In [13], the authors considered the group synchronization of complex network with nonlinear dynamics via pinning control.

Inspired by these literatures, we will consider the adaptive synchronization of nonlinear complex dynamical network with time-delays and sampled-data in this paper. The contribution of this paper are twofold. We first design effective adaptive strategies for the coupling strengths and the feedback gains, and present a stability analysis of adaptive synchronization of networks with time-delays and sampled-data. The second contribution is that the influence of adaptive strategies, time-delay and coupling on synchronization of nonlinear complex dynamical networks are considered.

An outline of this paper is organized as follows. Section 2 declares the model of nonlinear complex dynamics network and gives some preliminaries. In Sect. 3, we study the adaptive synchronization of nonlinear complex dynamics network with time-delays and sampled-data. The simulation results are presented in Sect. 4. Finally, Sect. 5 concludes this paper.

2   Model and Preliminaries

Consider a complex dynamical network of $N + M$ nodes with time-delays and sampled-data described by:

$$\dot{x}_i(t) = \begin{cases} f(x_i(t_k), x_i(t_k - \tau(t_k))) + \sum\limits_{j \in \mathcal{N}_{1i}} c_{ij}(t_k)a_{ij}(x_j(t_k - \tau(t_k)) - x_i(t_k - \tau(t_k))) \\ \quad + \sum\limits_{j \in \mathcal{N}_{2i}} d_{ij}(t_k)b_{ij}x_j(t_k - \tau(t_k)) + \mu_i, \quad \forall i \in \ell_1, \quad \forall t \in [t_k, t_{k+1}] \\ f(x_i(t_k), x_i(t_k - \tau(t_k))) + \sum\limits_{j \in \mathcal{N}_{2i}} c_{ij}(t_k)a_{ij}(x_j(t_k - \tau(t_k)) - x_i(t_k - \tau(t_k))) \\ \quad + \sum\limits_{j \in \mathcal{N}_{1i}} d_{ij}(t_k)b_{ij}x_j(t_k - \tau(t_k)) + \mu_i, \quad \forall i \in \ell_2, \quad \forall t \in [t_k, t_{k+1}] \end{cases} \tag{1}$$

where $x_i(t) = (x_{i1}(t), x_{i2}(t), \cdots, x_{in}(t))^T \in R^n$ denotes position vectors of the node i at time t, $f(\cdot) \in R^n$ describes the intrinsic dynamics of network and it is continuously differentiable, $c_{ij}(t_k), d_{ij}(t_k)$ represent the coupling strengths, and $\tau(t_k)$ denotes time-varying delays in transmission process. In this network, $\ell_1 = 1, 2, \cdots, N$, $\ell_2 = N + 1, N + 2, \cdots, N + M$, and $X_1 = \{x_i | i \in \ell_1\}$, $X_2 = \{x_i | i \in \ell_2\}$. $\mathcal{N}_i$ is the neighbor of node i, $\mathcal{N}_i \in \mathcal{N}_{1i} \cup \mathcal{N}_{2i}$, where $\mathcal{N}_{1i} \bigcap \mathcal{N}_{2i} = \emptyset$, $\mathcal{N}_{1i} = \{x_j \in X_1 | a_{ij} > 0, i, j \in \ell_1\}$, $\mathcal{N}_{2i} = \{x_j \in X_2 | a_{ij} > 0, i, j \in \ell_2\}$. If node i can get information from node j in the same group, then $a_{ij} > 0$; otherwise $a_{ij} = 0$; If node i can get information from node j between different groups, then

$b_{ij} \neq 0$; otherwise $b_{ij} = 0$. Thus the coupled matrix $A \in R^{(N+M) \times (N+M)}$ can be written as

$$A = \begin{bmatrix} A_{11}^{N \times N} & B_{12}^{N \times M} \\ B_{21}^{M \times N} & A_{22}^{M \times M} \end{bmatrix},$$

where let $S_i \triangleq a_{i1} + a_{i2} + \ldots + a_{iN}$, then $A_{11} = \begin{bmatrix} a_{11} - S_1 \cdots & a_{1N} \\ \vdots & \ddots & \vdots \\ a_{N1} & \cdots a_{NN} - S_N \end{bmatrix}_{N \times N}$,

and

$$A_{22} = \begin{bmatrix} a_{(N+1)(N+1)} - S_{N+1} \cdots & a_{(N+1)(N+M)} \\ \vdots & \ddots & \vdots \\ a_{(N+M)(N+1)} & \cdots a_{(N+M)(N+M)} - S_{N+M} \end{bmatrix}_{M \times M}$$

represent the coupling configuration of the subgroups, respectively.

In system (1), the controller is

$$\mu_i = \begin{cases} -c_i(t_k) h_i(x_i(t_k - \tau(t_k)) - \bar{x}_1(t_k - \tau(t_k))), & i \in \ell_1 \\ -c_i(t_k) h_i(x_i(t_k - \tau(t_k)) - \bar{x}_2(t_k - \tau(t_k))), & i \in \ell_2 \end{cases}, \tag{2}$$

where $\bar{x}_1(t)$, $\bar{x}_2(t) \in R^n$ are the synchronous states, h_i is an on-off control. If the system is sampled date, then $h_i = 1$; otherwise $h_i = 0$.

The adaptive strategies on coupling strengths and feedback gains designed as:

$$\dot{c}_{ij}(t_k) = \begin{cases} a_{ij} k_{ij} [(x_i(t_k - \tau(t_k)) - x_j(t_k - \tau(t_k)))^T (x_i(t_k - \tau(t_k)) - x_j(t_k - \tau(t_k))) \\ + (\dot{x}_i(t) - \dot{x}_j(t))^T (\dot{x}_i(t) - \dot{x}_j(t))] \ i, j \in \ell_1 \\ a_{ij} k_{ij} [(x_i(t_k - \tau(t_k)) - x_j(t_k - \tau(t_k)))^T (x_i(t_k - \tau(t_k)) - x_j(t_k - \tau(t_k))) \\ + (\dot{x}_i(t) - \dot{x}_j(t))^T (\dot{x}_i(t) - \dot{x}_j(t))] \ i, j \in \ell_2 \end{cases},$$

$$\dot{d}_{ij}(t_k) = \begin{cases} b_{ij} k_{ij} (x_j(t_k - \tau(t_k)) - \bar{x}_2(t_k - \tau(t_k)))^T (x_j(t_k - \tau(t_k)) - \bar{x}_2(t_k - \tau(t_k))) \\ i \in \ell_1, j \in \ell_2 \\ b_{ij} k_{ij} (x_j(t_k - \tau(t_k)) - \bar{x}_1(t_k - \tau(t_k)))^T (x_j(t_k - \tau(t_k)) - \bar{x}_1(t_k - \tau(t_k))) \\ i \in \ell_2, j \in \ell_1 \end{cases}, \tag{3}$$

$$\dot{c}_i(t_k) = \begin{cases} h_i k_i [(x_i(t_k - \tau(t_k)) - \bar{x}_1(t_k - \tau(t_k)))^T (x_i(t_k - \tau(t_k)) - \bar{x}_1(t_k - \tau(t_k))) \\ + (\dot{x}_i(t_k) - \dot{\bar{x}}_1(t_k))^T (\dot{x}_i(t_k) - \dot{\bar{x}}_1(t_k))] \ i \in \ell_1 \\ h_i k_i [(x_i(t_k - \tau(t_k)) - \bar{x}_2(t_k - \tau(t_k)))^T (x_i(t_k - \tau(t_k)) - \bar{x}_2(t_k - \tau(t_k))) \\ + (\dot{x}_i(t_k) - \dot{\bar{x}}_2(t_k))^T (\dot{x}_i(t_k) - \dot{\bar{x}}_2(t_k))] \ i \in \ell_2 \end{cases},$$

where $c_{ij} \geq 0$, $c_i \geq 0$, the constants $k_{ij} > 0$ and $k_i > 0$ are the weights of the $c_{ij}(t)$ and $c_i(t)$, respectively.

In the following, we will analyze the sampling period, which is an important factor in the sampling information. Given a positive real number α and a sample periodic T, we suppose that (see [12] in more detail)

$$t_{i+1} - t_i = \alpha T_i, \quad \forall i = 0, 1, 2, \cdots,$$

where $t_0 < t_1 < \cdots$ are the discrete times; the node j can obtain information from its neighbors and positive integer T_i is a sampled time about the ith time

$(\forall i = 0, 1, 2, \cdots)$ satisfying $T_i \leq T$. Under this condition, a linear consensus protocol based on a linear estimation-based sampling period is designed as follows:

$$
\begin{cases}
\dot{x}_i(t_k + \alpha) = \dot{x}_i(t_k) - \dfrac{1}{T}\dot{x}_i(t_k) = \left(1 - \dfrac{1}{T}\right)\dot{x}_i(t_k) \\[2mm]
\dot{x}_i(t_k + 2\alpha) = \dot{x}_i(t_k + \alpha) + (\dot{x}_i(t_k + \alpha) - \dot{x}_i(t_k)) = \left(1 - \dfrac{2}{T}\right)\dot{x}_i(t_k) \\[2mm]
\;\;\vdots \\[2mm]
\dot{x}_i(t_{k+1} - \alpha) = \dot{x}_i(t_k + T_k\alpha - \alpha) = \left(1 - \dfrac{T_k - 1}{T}\right)\dot{x}_i(t_k)
\end{cases}
\tag{4}
$$

Substituting system (4) into system (1), and let $h = 0, 1, \cdots, T_k - 1$, we can have

$$
\dot{x}_i(t) =
\begin{cases}
\begin{aligned}
&\left(1 - \dfrac{h}{T}\right) \times \Big[f(x_i(t_k), x_i(t_k - \tau(t_k))) + \sum_{j \in \mathcal{N}_{1i}} c_{ij}(t_k)a_{ij}(x_j(t_k - \tau(t_k)) - x_i(t_k - \tau(t_k))) \\
&+ \sum_{j \in \mathcal{N}_{2i}} d_{ij}(t_k)b_{ij}x_j(t_k - \tau(t_k)) + \mu_i \Big] \quad \forall i \in \ell_1, \quad \forall t \in [t_k, t_{k+1}];
\end{aligned} \\[6mm]
\begin{aligned}
&\left(1 - \dfrac{h}{T}\right) \times \Big[f(x_i(t_k), x_i(t_k - \tau(t_k))) + \sum_{j \in \mathcal{N}_{2i}} c_{ij}(t_k)a_{ij}(x_j(t_k - \tau(t_k)) - x_i(t_k - \tau(t_k))) \\
&+ \sum_{j \in \mathcal{N}_{1i}} d_{ij}(t_k)b_{ij}x_j(t_k - \tau(t_k)) + \mu_i \Big] \quad \forall i \in \ell_2, \quad \forall t \in [t_k, t_{k+1}],
\end{aligned}
\end{cases}
\tag{5}
$$

In order to solve the synchronization problem, we give the following assumptions and lemmas.

Assumption 1 [11]. *There exist nonnegative constants ρ_1 and ρ_2 such that*

$$
\|f(a,b) - f(c,d)\| \leq \rho_1\|a - c\| + \rho_2\|b - d\|, \quad \forall a, b, c, d \in R^n.
$$

Assumption 2. *The coupling strengths and feedback gains are bounded, which means that*

$$
\|c_{ij}(t_k)\| \leq c_{ij}, \qquad \|d_{ij}(t_k)\| \leq d_{ij}, \qquad \|c_i(t_k)\| \leq c_i.
$$

Definition 1. *Network is said to group synchronization if*

$$
\lim_{t \to \infty} \|(x_i(t) - x_j(t))\| = 0, \quad \forall i, j \in \ell_1, \qquad \lim_{t \to \infty} \|(x_i(t) - x_j(t))\| = 0, \quad \forall i, j \in \ell_2.
$$

Lemma 1 [3]. *Suppose that $x, y \in R^n$ are vectors, and in matrix M, the following inequality holds:*

$$
2x^T y \leq x^T M x + y^T M^{-1} y.
$$

Lemma 2 [5]. *For any real differentiable vector function $x(t) \in R^n$ and any $n \times n$ constant matrix $W = W^T > 0$, we have the following inequality:*

$$
\left[\int_{t-\tau(t)}^{t} x(s)ds \right]^T W \left[\int_{t-\tau(t)}^{t} x(s)ds \right] \leq \tau \int_{t-\tau(t)}^{t} x^T(s)Wx(s)ds, t \geq 0,
$$

where $0 \leq \tau(t) \leq \tau$.

Lemma 3 [9]. *If matrix* $A = (a_{ij}) \in R^{N \times N}$ *is the symmetric irreducible matrix, where* $a_{ii} = - \sum\limits_{j=1, j \neq i}^{N} a_{ij}$, *then all eigenvalues of matrix* $A - E$ *are negative numbers, where matrix* $E = diag(e, 0, \cdots, 0)$ *with* $e > 0$.

3   Main Results

In this section, we consider adaptive synchronization of nonlinear complex dynamics network with time-delays and sampled-data. We have the following theorems.

Theorem 1. *Under Assumptions 1–2 and Lemmas 1–3, if coupled matrix* A_{11}, A_{22} *are the symmetric irreducible matrices, suppose that topology graph of system (1) is connected and the time-delays are bound, then system (1) can be steered to the synchronous state* $\bar{x}_1(t)$ *and* $\bar{x}_2(t)$ *by the adaptive strategies (3).*

Proof. Define

$$
\begin{cases}
\tilde{x}_i(t_k) \triangleq x_i(t_k) - \bar{x}_1(t_k), & \tilde{x}_i(t_k - \tau(t_k)) \triangleq x_i(t_k - \tau(t_k)) - \bar{x}_1(t_k - \tau(t_k)), \\
\dot{\tilde{x}}_i(t_k) \triangleq \dot{x}_i(t_k) - \dot{\bar{x}}_1(t_k), & \dot{\tilde{x}}_i(t_k - \tau(t_k)) \triangleq \dot{x}_i(t_k - \tau(t_k)) - \dot{\bar{x}}_1(t_k - \tau(t_k)), \\
i = 1, 2, \ldots, N, \\
\\
\tilde{x}_i(t_k) \triangleq x_i(t_k) - \bar{x}_2(t_k), & \tilde{x}_i(t_k - \tau(t_k)) \triangleq x_i(t_k - \tau(t_k)) - \bar{x}_2(t_k - \tau(t_k)), \\
\dot{\tilde{x}}_i(t_k) \triangleq \dot{x}_i(t_k) - \dot{\bar{x}}_2(t_k), & \dot{\tilde{x}}_i(t_k - \tau(t_k)) \triangleq \dot{x}_i(t_k - \tau(t_k)) - \dot{\bar{x}}_2(t_k - \tau(t_k)), \\
i = N+1, N+2, \ldots, N+M,
\end{cases}
$$

then

$$
\dot{\tilde{x}}_i(t) = \begin{cases}
(1 - \dfrac{h}{T}) \times \Big[f(x_i(t_k), x_i(t_k - \tau(t_k))) - f(\bar{x}_1(t_k), \bar{x}_1(t_k - \tau(t_k))) \\
\quad + \sum\limits_{j \in \mathcal{N}_{1i}} c_{ij}(t_k) a_{ij}(\tilde{x}_j(t_k - \tau(t_k)) - \tilde{x}_i(t_k - \tau(t_k))) + \sum\limits_{j \in \mathcal{N}_{2i}} d_{ij}(t_k) b_{ij} \tilde{x}_j(t_k - \tau(t_k)) + \mu_i \Big] \\
\forall i \in \ell_1, \quad \forall t \in [t_k, t_{k+1}]; \\
\\
(1 - \dfrac{h}{T}) \times \Big[f(x_i(t_k), x_i(t_k - \tau(t_k))) - f(\bar{x}_2(t_k), \bar{x}_2(t_k - \tau(t_k))) \\
\quad + \sum\limits_{j \in \mathcal{N}_{2i}} c_{ij}(t_k) a_{ij}(\tilde{x}_j(t_k - \tau(t_k)) - \tilde{x}_i(t_k - \tau(t_k))) + \sum\limits_{j \in \mathcal{N}_{1i}} d_{ij}(t_k) b_{ij} \tilde{x}_j(t_k - \tau(t_k)) + \mu_i \Big] \\
\forall i \in \ell_2, \quad \forall t \in [t_k, t_{k+1}],
\end{cases}
\tag{6}
$$

Construct a Lyapunov function as follow:

$$
V(t_k) = V_1(t_k) + V_2(t_k) + V_3(t_k),
$$

where

$$V_1(t_k) = \frac{1}{2} \sum_{i \in \mathcal{N}_{1i}} \tilde{x}_i^T(t_k)\tilde{x}_i(t_k) + \sum_{i \in \mathcal{N}_{1i}} \sum_{j \in \mathcal{N}_{1i}} \frac{(c_{ij}(t_k) - 2c_{ij} - p)^2}{4k_{ij}}$$

$$+ \sum_{i \in \mathcal{N}_{1i}} \sum_{j \in \mathcal{N}_{2i}} \frac{(d_{ij}(t_k) - 2d_{ij} - 1)^2}{4k_{ij}} + \sum_{i \in \mathcal{N}_{1i}} \frac{(c_i(t_k) - \frac{3}{2}c_i - p)^2}{2k_i},$$

$$V_2(t_k) = \frac{1}{2} \sum_{i \in \mathcal{N}_{2i}} \tilde{x}_i^T(t_k)\tilde{x}_i(t_k) + \sum_{i \in \mathcal{N}_{2i}} \sum_{j \in \mathcal{N}_{2i}} \frac{(c_{ij}(t_k) - 2c_{ij} - p)^2}{4k_{ij}}$$

$$+ \sum_{i \in \mathcal{N}_{2i}} \sum_{j \in \mathcal{N}_{1i}} \frac{(d_{ij}(t_k) - 2d_{ij} - 1)^2}{4k_{ij}} + \sum_{i \in \mathcal{N}_{2i}} \frac{(c_i(t_k) - \frac{3}{2}c_i - p)^2}{2k_i},$$

$$V_3(t_k) = \tau \sum_{i \in \mathcal{N}_{1i}} \left(2\rho_1 + \rho_2 + \sum_{i \in \mathcal{N}_{1i}} c_{ij}a_{ij} \right.$$

$$+ \sum_{i \in \mathcal{N}_{2i}} d_{ij}b_{ij} + h_i c_i \right) \int_{t_k - \tau(t_k)}^{t_k} (s - t_k + \tau)\dot{\tilde{x}}_i^T(s)\dot{\tilde{x}}_i(s)ds$$

$$+ \tau \sum_{i \in \mathcal{N}_{2i}} \left(2\rho_3 + \rho_4 + \sum_{i \in \mathcal{N}_{2i}} c_{ij}a_{ij} \right.$$

$$+ \sum_{i \in \mathcal{N}_{1i}} d_{ij}b_{ij} + h_i c_i \right) \int_{t_k - \tau(t_k)}^{t_k} (s - t_k + \tau)\dot{\tilde{x}}_i^T(s)\dot{\tilde{x}}_i(s)ds,$$

Differentiating $V_1(t_k)$, we can know

$$\dot{V}_1(t_k) = \left(1 - \frac{h}{T}\right) \sum_{i \in \mathcal{N}_{1i}} \tilde{x}_i^T(t_k)\left[f(x_i(t_k), x_i(t_k - \tau(t_k))) - f(\bar{x}_1(t_k), \bar{x}_1(t_k - \tau(t_k)))\right]$$

$$+ \left(1 - \frac{h}{T}\right) \sum_{i \in \mathcal{N}_{1i}} \tilde{x}_i^T(t_k)\left[\sum_{j \in \mathcal{N}_{1i}} c_{ij}(t_k)a_{ij}(\tilde{x}_j(t_k - \tau(t_k)) - \tilde{x}_i(t_k - \tau(t_k))) \right.$$

$$+ \sum_{j \in \mathcal{N}_{2i}} d_{ij}(t_k)b_{ij}\tilde{x}_j(t_k - \tau(t_k)) - c_i(t_k)h_i(\tilde{x}_i(t_k - \tau(t_k))) \right]$$

$$+ \frac{1}{2} \sum_{i \in \mathcal{N}_{1i}} \sum_{j \in \mathcal{N}_{1i}} (c_{ij}(t_k) - 2c_{ij} - p)a_{ij}\left[(\dot{\tilde{x}}_i(t_k) - \dot{\tilde{x}}_j(t_k))^T(\dot{\tilde{x}}_i(t_k) - \dot{\tilde{x}}_j(t_k)) \right.$$

$$+ (\tilde{x}_i(t_k - \tau(t_k)) - \tilde{x}_j(t_k - \tau(t_k)))^T(\tilde{x}_i(t_k - \tau(t_k)) - \tilde{x}_j(t_k - \tau(t_k))) \right]$$

$$+ \frac{1}{2} \sum_{i \in \mathcal{N}_{1i}} \sum_{j \in \mathcal{N}_{2i}} (d_{ij}(t_k) - 2d_{ij} - 1)b_{ij}\tilde{x}_j^T(t_k - \tau(t_k))\tilde{x}_j(t_k - \tau(t_k))$$

$$+ \sum_{i \in \mathcal{N}_{1i}} (c_i(t_k) - \frac{3}{2}c_i - p)h_i\left[\tilde{x}_i^T(t_k - \tau(t_k))\tilde{x}_i(t_k - \tau(t_k)) + \dot{\tilde{x}}_i^T(t_k)\dot{\tilde{x}}_i(t_k) \right].$$

$$(7)$$

Under Assumption 1 and using Lemma 1, we can have

$$\dot{V}_1(t_k) \le (1 - \frac{h}{T})(\rho_1 + \frac{1}{2}\rho_2) \sum_{i \in \mathscr{N}_{1i}} \tilde{x}_i^T(t_k)\tilde{x}_i(t_k) + (1 - \frac{h}{T})\frac{1}{2}\rho_2 \sum_{i \in \mathscr{N}_{1i}} \tilde{x}_i^T(t_k - \tau(t_k)))\tilde{x}_i(t_k - \tau(t_k)))$$

$$+ (1 - \frac{h}{T})\frac{1}{2} \sum_{i \in \mathscr{N}_{1i}} \sum_{j \in \mathscr{N}_{1i}} c_{ij}(t_k)a_{ij}(\tilde{x}_j(t_k - \tau(t_k)))$$

$$- \tilde{x}_i(t_k - \tau(t_k)))^T (\tilde{x}_j(t_k - \tau(t_k)) - \tilde{x}_i(t_k - \tau(t_k)))$$

$$+ (1 - \frac{h}{T})\frac{1}{2} \sum_{i \in \mathscr{N}_{1i}} \sum_{j \in \mathscr{N}_{1i}} c_{ij}(t_k)a_{ij}\tilde{x}_i^T(t_k)\tilde{x}_i(t_k) + (1 - \frac{h}{T})\frac{1}{2} \sum_{i \in \mathscr{N}_{1i}} \sum_{j \in \mathscr{N}_{2i}} d_{ij}(t_k)b_{ij}\tilde{x}_i^T(t_k)\tilde{x}_i(t_k)$$

$$+ (1 - \frac{h}{T})\frac{1}{2} \sum_{i \in \mathscr{N}_{1i}} \sum_{j \in \mathscr{N}_{2i}} d_{ij}(t_k)b_{ij}\tilde{x}_j^T(t_k - \tau(t_k))\tilde{x}_j(t_k - \tau(t_k))$$

$$+ (1 - \frac{h}{T})\frac{1}{2} \sum_{i \in \mathscr{N}_{1i}} c_i(t_k)h_i\tilde{x}_i^T(t_k)\tilde{x}_i(t_k) + (1 - \frac{h}{T})\frac{1}{2} \sum_{i \in \mathscr{N}_{1i}} c_i(t_k)h_i\tilde{x}_i^T(t_k - \tau(t_k))\tilde{x}_i(t_k - \tau(t_k))$$

$$+ \frac{1}{2} \sum_{i \in \mathscr{N}_{1i}} \sum_{j \in \mathscr{N}_{1i}} (c_{ij}(t_k) - 2c_{ij} - p)a_{ij} [(\dot{\tilde{x}}_i(t_k) - \dot{\tilde{x}}_j(t_k))^T (\dot{\tilde{x}}_i(t_k) - \dot{\tilde{x}}_j(t_k))$$

$$+ (\tilde{x}_i(t_k - \tau(t_k)) - \tilde{x}_j(t_k - \tau(t_k)))^T (\tilde{x}_i(t_k - \tau(t_k)) - \tilde{x}_j(t_k - \tau(t_k)))]$$

$$+ \frac{1}{2} \sum_{i \in \mathscr{N}_{1i}} \sum_{j \in \mathscr{N}_{2i}} (d_{ij}(t_k) - 2d_{ij} - 1)b_{ij}\tilde{x}_j^T(t_k - \tau(t_k))\tilde{x}_j(t_k - \tau(t_k))$$

$$+ \sum_{i \in \mathscr{N}_{1,i}} (c_i(t_k) - \frac{3}{2}c_i - p)h_i [\tilde{x}_i^T(t_k - \tau(t_k))\tilde{x}_i(t_k - \tau(t_k)) + \dot{\tilde{x}}_i^T(t_k)\dot{\tilde{x}}_i(t_k)].$$

As a result of $h = 0, 1, \cdots, T_k - 1 < T$, and Assumption 2, we can get

$$\dot{V}_1(t_k) \le (\rho_1 + \frac{1}{2}\rho_2) \sum_{i \in \mathscr{N}_{1i}} \tilde{x}_i^T(t_k)\tilde{x}_i^T(t_k) + \frac{1}{2}\rho_2 \sum_{i \in \mathscr{N}_{1i}} \tilde{x}_i^T(t_k - \tau(t_k))\tilde{x}_i^T(t_k - \tau(t_k))$$

$$+ \frac{1}{2} \sum_{i \in \mathscr{N}_{1i}} \sum_{j \in \mathscr{N}_{1i}} c_{ij}a_{ij}\tilde{x}_i^T(t_k)\tilde{x}_i(t_k) + \frac{1}{2} \sum_{i \in \mathscr{N}_{1i}} \sum_{j \in \mathscr{N}_{2i}} d_{ij}b_{ij}\tilde{x}_i^T(t_k)\tilde{x}_i(t_k)$$

$$+ \frac{1}{2} \sum_{i \in \mathscr{N}_{1i}} c_ih_i\tilde{x}_i^T(t_k)\tilde{x}_i(t_k)$$

$$- \frac{p}{2} \sum_{i \in \mathscr{N}_{1i}} \sum_{j \in \mathscr{N}_{1i}} a_{ij}(\tilde{x}_i(t_k - \tau(t_k)) - \tilde{x}_j(t_k - \tau(t_k)))^T (\tilde{x}_i(t_k - \tau(t_k)) - \tilde{x}_j(t_k - \tau(t_k)))$$

$$- \frac{p}{2} \sum_{i \in \mathscr{N}_{1i}} \sum_{j \in \mathscr{N}_{1i}} a_{ij}(\dot{\tilde{x}}_i(t_k) - \dot{\tilde{x}}_j(t_k))^T (\dot{\tilde{x}}_i(t_k) - \dot{\tilde{x}}_j(t_k))$$

$$- \frac{1}{2} \sum_{i \in \mathscr{N}_{1i}} \sum_{j \in \mathscr{N}_{2i}} b_{ij}\tilde{x}_j^T(t_k - \tau(t_k))\tilde{x}_j(t_k - \tau(t_k))$$

$$- p \sum_{i \in \mathscr{N}_{1i}} h_i\tilde{x}_i^T(t_k - \tau(t_k))\tilde{x}_i(t_k - \tau(t_k)) - p \sum_{i \in \mathscr{N}_{1i}} h_i\dot{\tilde{x}}_i^T(t_k)\dot{\tilde{x}}_i(t_k).$$

Using the Leibniz-Newton formula:

$$x(t_k) - x(t_k - \tau(t_k)) = \int_{t_k - \tau(t_k)}^{t_k} \dot{x}(s)ds,$$

we can get that

$$x^T(t_k) = x^T(t_k - \tau(t_k)) + \left(\int_{t_k - \tau(t_k)}^{t_k} \dot{x}(s)ds \right)^T.$$

Using Lemma 1, we can have

$$
\begin{aligned}
\tilde{x}^T(t_k)\tilde{x}(t_k) &= \left[\tilde{x}^T(t_k - \tau(t_k)) + \left(\int_{t_k-\tau(t_k)}^{t_k} \dot{\tilde{x}}(s)ds\right)^T\right]\left[\tilde{x}(t_k - \tau(t_k)) + \left(\int_{t_k-\tau(t_k)}^{t_k} \dot{\tilde{x}}(s)ds\right)\right] \\
&\leq 2\tilde{x}^T(t_k - \tau(t_k))\tilde{x}(t_k - \tau(t_k)) + 2\left(\int_{t_k-\tau(t_k)}^{t_k} \dot{\tilde{x}}(s)ds\right)^T\left(\int_{t_k-\tau(t_k)}^{t_k} \dot{\tilde{x}}(s)ds\right) \\
&\leq 2\tilde{x}^T(t_k - \tau(t_k))\tilde{x}(t_k - \tau(t_k)) + 2\tau\int_{t_k-\tau(t_k)}^{t_k} \dot{\tilde{x}}_i^T(s)\dot{\tilde{x}}_i(s)ds.
\end{aligned}
\tag{8}
$$

Thus,

$$
\begin{aligned}
\dot{V}_1(t_k) &\leq \sum_{i\in\mathcal{N}_{1i}}\left[(2\rho_1 + \frac{3}{2}\rho_2) + \sum_{j\in\mathcal{N}_{1i}} c_{ij}a_{ij} + \sum_{j\in\mathcal{N}_{2i}} d_{ij}b_{ij} + c_i h_i\right]\tilde{x}^T(t_k - \tau(t_k))\tilde{x}(t_k - \tau(t_k)) \\
&+ \sum_{i\in\mathcal{N}_{1i}}\left[(2\rho_1 + \rho_2) + \sum_{j\in\mathcal{N}_{1i}} c_{ij}a_{ij} + \sum_{j\in\mathcal{N}_{2i}} d_{ij}b_{ij} + c_i h_i\right]\tau\int_{t_k-\tau(t_k)}^{t_k} \dot{\tilde{x}}_i^T(s)\dot{\tilde{x}}_i(s)ds \\
&+ p(\|\tilde{x}_1(t_k - \tau(t_k))\|, \|\tilde{x}_2(t_k - \tau(t_k))\|, \cdots, \|\tilde{x}_N(t_k - \tau(t_k))\|) \\
&\qquad (A_{11} - H_1)\begin{pmatrix} \|\tilde{x}_1(t_k - \tau(t_k))\| \\ \|\tilde{x}_2(t_k - \tau(t_k))\| \\ \cdots \\ \|\tilde{x}_N(t_k - \tau(t_k))\| \end{pmatrix} \\
&+ P(\|\dot{\tilde{x}}_1(t_k)\|, \|\dot{\tilde{x}}_2(t_k)\|, \cdots, \|\dot{\tilde{x}}_N(t_k)\|)(A_{11} - H_1)\begin{pmatrix} \|\dot{\tilde{x}}_1(t_k)\| \\ \|\dot{\tilde{x}}_2(t_k)\| \\ \cdots \\ \|\dot{\tilde{x}}_N(t_k)\| \end{pmatrix} \\
&- \frac{1}{2}\sum_{i\in\mathcal{N}_{1i}}\sum_{j\in\mathcal{N}_{2i}} b_{ij}(\tilde{x}_j(t_k - \tau(t_k)))^T(\tilde{x}_j(t_k - \tau(t_k))).
\end{aligned}
$$

Similarly, differentiating $V_2(t_k)$, we can have

$$
\begin{aligned}
\dot{V}_2(t_k) &\leq \sum_{i\in\mathcal{N}_{2i}}\left(2\rho_3 + \frac{3}{2}\rho_4 + \sum_{j\in\mathcal{N}_{2i}} c_{ij}a_{ij} + \sum_{j\in\mathcal{N}_{1i}} d_{ij}b_{ij} + c_i h_i\right)(\tilde{x}(t_k - \tau(t_k)))^T(\tilde{x}(t_k - \tau(t_k))) \\
&+ \sum_{i\in\mathcal{N}_{2i}}\left(2\rho_3 + \rho_4 + \sum_{j\in\mathcal{N}_{2i}} c_{ij}a_{ij} + \sum_{j\in\mathcal{N}_{1i}} d_{ij}b_{ij} + c_i h_i\right)\tau\int_{t_k-\tau(t_k)}^{t_k} \dot{\tilde{x}}_i^T(s)\dot{\tilde{x}}_i(s)ds \\
&+ p(\|\tilde{x}_{N+1}(t_k - \tau(t_k))\|, \|\tilde{x}_{N+2}(t_k - \tau(t_k))\|, \cdots, \|\tilde{x}_{N+M}(t_k - \tau(t_k))\|) \\
&\qquad (A_{22} - H_2)\begin{pmatrix} \|\tilde{x}_{N+1}(t_k - \tau(t_k))\| \\ \|\tilde{x}_{N+2}(t_k - \tau(t_k))\| \\ \cdots \\ \|\tilde{x}_{N+M}(t_k - \tau(t_k))\| \end{pmatrix} \\
&+ P(\|\dot{\tilde{x}}_{N+1}(t_k)\|, \|\dot{\tilde{x}}_{N+2}(t_k)\|, \cdots, \|\dot{\tilde{x}}_{N+M}(t_k)\|)(A_{22} - H_2)\begin{pmatrix} \|\dot{\tilde{x}}_{N+1}(t_k)\| \\ \|\dot{\tilde{x}}_{N+2}(t_k)\| \\ \cdots \\ \|\dot{\tilde{x}}_{N+M}(t_k)\| \end{pmatrix} \\
&- \frac{1}{2}\sum_{i\in\mathcal{N}_{2i}}\sum_{j\in\mathcal{N}_{1i}} b_{ij}(\tilde{x}_j(t_k - \tau(t_k)))^T(\tilde{x}_j(t_k - \tau(t_k))).
\end{aligned}
$$

Differentiating $V_3(t_k)$, we get

$$
\begin{aligned}
\dot{V}_3(t_k) = \ &\tau^2 \sum_{i \in \mathcal{N}_{1i}} \left(2\rho_1 + \rho_2 + \sum_{i \in \mathcal{N}_{1i}} c_{ij} a_{ij} + \sum_{i \in \mathcal{N}_{2i}} d_{ij} b_{ij} + h_i c_i\right) \dot{\tilde{x}}_i^T(t_k) \dot{\tilde{x}}_i(t_k) \\
&- \sum_{i \in \mathcal{N}_{1i}} \left(2\rho_1 + \rho_2 + \sum_{i \in \mathcal{N}_{1i}} c_{ij} a_{ij} + \sum_{i \in \mathcal{N}_{2i}} d_{ij} b_{ij} + h_i c_i\right) \int_{t_k - \tau(t_k)}^{t_k} \dot{\tilde{x}}_i^T(s) \dot{\tilde{x}}_i(s) ds \\
&+ \tau^2 \sum_{i \in \mathcal{N}_{2i}} \left(2\rho_3 + \rho_4 + \sum_{i \in \mathcal{N}_{2i}} c_{ij} a_{ij} + \sum_{i \in \mathcal{N}_{1i}} d_{ij} b_{ij} + h_i c_i\right) \dot{\tilde{x}}_i^T(t_k) \dot{\tilde{x}}_i(t_k) \\
&- \sum_{i \in \mathcal{N}_{2i}} \left(2\rho_3 + \rho_4 + \sum_{i \in \mathcal{N}_{2i}} c_{ij} a_{ij} + \sum_{i \in \mathcal{N}_{1i}} d_{ij} b_{ij} + h_i c_i\right) \int_{t_k - \tau(t_k)}^{t_k} \dot{\tilde{x}}_i^T(s) \dot{\tilde{x}}_i(s) ds,
\end{aligned}
$$

Count up $\dot{V}_1(t_k)$, $\dot{V}_2(t_k)$, $\dot{V}_3(t_k)$, we know that

$$
\begin{aligned}
\dot{V}(t_k) \leq \ &\sum_{i \in \mathcal{N}_{1i}} \left(2\rho_1 + \frac{3}{2}\rho_2 + \sum_{j \in \mathcal{N}_{1i}} c_{ij} a_{ij} + \sum_{j \in \mathcal{N}_{2i}} d_{ij} b_{ij} + c_i h_i\right) \\
&- \left(\frac{1}{2} + p\lambda_1\right)(\tilde{x}(t_k - \tau(t_k)))^T (\tilde{x}(t_k - \tau(t_k))) \\
&+ \tau^2 \sum_{i \in \mathcal{N}_{1i}} \left(2\rho_1 + \rho_2 + \sum_{i \in \mathcal{N}_{1i}} c_{ij} a_{ij} + \sum_{i \in \mathcal{N}_{2i}} d_{ij} b_{ij} + h_i c_i + p\lambda_1\right) \dot{\tilde{x}}_i^T(t_k) \dot{\tilde{x}}_i(t_k) \\
&+ \sum_{i \in \mathcal{N}_{2i}} \left(2\rho_3 + \frac{3}{2}\rho_4 + \sum_{j \in \mathcal{N}_{2i}} c_{ij} a_{ij} + \sum_{j \in \mathcal{N}_{1i}} d_{ij} b_{ij} + c_i h_i\right) \\
&- \left(\frac{1}{2} + p\lambda_2\right)(\tilde{x}(t_k - \tau(t_k)))^T (\tilde{x}(t_k - \tau(t_k))) \\
&+ \tau^2 \sum_{i \in \mathcal{N}_{2i}} \left(2\rho_3 + \rho_4 + \sum_{i \in \mathcal{N}_{2i}} c_{ij} a_{ij} + \sum_{i \in \mathcal{N}_{1i}} d_{ij} b_{ij} + h_i c_i + p\lambda_2\right) \dot{\tilde{x}}_i^T(t_k) \dot{\tilde{x}}_i(t_k),
\end{aligned}
$$

where λ_1, λ_2 are the minimum eigenvalue of $A_{11} - H_1$, $A_{22} - H_2$, respectively, with

$$
H_1 = diag\{h_i\} \ \forall i \in \ell_1, \qquad H_2 = diag\{h_i\} \ \forall i \in \ell_2.
$$

From the conditions of Theorem 1, we known that the matrix H_1, H_2 are diagonal matrices with at least one element equaling to 1. Since A_{11}, A_{22} are symmetric, all eigenvalues of $A_{11} - H_1$, $A_{22} - H_2$ are negative from Lemma 3. So $\lambda_1 < 0$, $\lambda_2 < 0$ and $p > 0$ is sufficiently large, therefore, $\dot{V}(t_k) < 0$, all the nodes of system (1) with time-delays and sampled-data can converge to their own synchronous states asymptotically. The proof is completed.

Theorem 1 shows that when coupled matrix A_{11}, A_{22} are the symmetric irreducible matrices, system (1) can synchronize. If coupled matrix A_{11}, A_{22} are the asymmetric irreducible matrices, the synchronization of the system (1) is given by the following result.

Theorem 2. *Under Assumptions 1–2 and Lemmas 1–3, if coupled matrix A_{11}, A_{22} are the asymmetric irreducible matrices, suppose that system (1) is connected and the time-delays are bound, then system (1) is steered to the synchronous state $\bar{x}_1(t)$ and $\bar{x}_2(t)$ by the adaptive strategies (3).*

Proof.

$$-\frac{1}{2}p\sum_{i\in\mathscr{N}_{1i}}\sum_{j\in\mathscr{N}_{1i}}a_{ij}[(\tilde{x}_i(t_k-\tau(t_k))-\tilde{x}_j(t_k-\tau(t_k)))^T(\tilde{x}_i(t_k-\tau(t_k))-\tilde{x}_j(t_k-\tau(t_k)))]$$

$$=-\frac{1}{2}p\sum_{i\in\mathscr{N}_{1i}}\sum_{j\in\mathscr{N}_{1i}}a_{ij}(\tilde{x}_i(t_k-\tau(t_k))^T(\tilde{x}_i(t_k-\tau(t_k)))$$

$$+\frac{1}{2}p\sum_{i\in\mathscr{N}_{1i}}\sum_{j\in\mathscr{N}_{1i}}a_{ij}(\tilde{x}_i(t_k-\tau(t_k)))^T(\tilde{x}_j(t_k-\tau(t_k)))$$

$$-\frac{1}{2}p\sum_{i\in\mathscr{N}_{1i}}\sum_{j\in\mathscr{N}_{1i}}a_{ij}(\tilde{x}_j(t_k-\tau(t_k)))^T(\tilde{x}_j(t_k-\tau(t_k)))$$

$$+\frac{1}{2}p\sum_{i\in\mathscr{N}_{1i}}\sum_{j\in\mathscr{N}_{1i}}a_{ij}(\tilde{x}_j(t_k-\tau(t_k)))^T(\tilde{x}_i(t_k-\tau(t_k)))$$

$$=p(\|\tilde{x}_1(t_k-\tau(t_k))\|,\|\tilde{x}_2(t_k-\tau(t_k))\|,\cdots,\|\tilde{x}_N(t_k-\tau(t_k))\|)(\frac{A_{11}+A_{11}^T}{2})\begin{pmatrix}\|\tilde{x}_1(t_k-\tau(t_k))\|\\\|\tilde{x}_2(t_k-\tau(t_k))\|\\\cdots\\\|\tilde{x}_N(t_k-\tau(t_k))\|\end{pmatrix}.$$

Similarly,

$$-\frac{1}{2}p\sum_{i\in\mathscr{N}_{2i}}\sum_{j\in\mathscr{N}_{2i}}a_{ij}[(\tilde{x}_i(t_k-\tau(t_k))-\tilde{x}_j(t_k-\tau(t_k)))^T(\tilde{x}_i(t_k-\tau(t_k))-\tilde{x}_j(t_k-\tau(t_k)))]$$

$$=p(\|\tilde{x}_{N+1}(t_k-\tau(t_k))\|,\|\tilde{x}_{N+2}(t_k-\tau(t_k))\|,\cdots,\|\tilde{x}_{N+M}(t_k-\tau(t_k))\|)$$

$$(\frac{A_{22}+A_{22}^T}{2})\begin{pmatrix}\|\tilde{x}_{N+1}(t_k-\tau(t_k))\|\\\|\tilde{x}_{N+2}(t_k-\tau(t_k))\|\\\cdots\\\|\tilde{x}_{N+M}(t_k-\tau(t_k))\|\end{pmatrix},$$

$$-\frac{1}{2}p\sum_{i\in\mathscr{N}_{1i}}\sum_{j\in\mathscr{N}_{1i}}a_{ij}[(\dot{\tilde{x}}_i(t_k)-\dot{\tilde{x}}_j(t_k))^T(\dot{\tilde{x}}_i(t_k)-\dot{\tilde{x}}_j(t_k))]$$

$$=P(\|\dot{\tilde{x}}_1(t_k)\|,\|\dot{\tilde{x}}_2(t_k)\|,\cdots,\|\dot{\tilde{x}}_N(t_k)\|)(\frac{A_{11}+A_{11}^T}{2})\begin{pmatrix}\|\dot{\tilde{x}}_1(t_k)\|\\\|\dot{\tilde{x}}_2(t_k)\|\\\cdots\\\|\dot{\tilde{x}}_N(t_k)\|\end{pmatrix},$$

$$-\frac{1}{2}p\sum_{i\in\mathscr{N}_{2i}}\sum_{j\in\mathscr{N}_{2i}}a_{ij}[(\dot{\tilde{x}}_i(t_k)-\dot{\tilde{x}}_j(t_k))^T(\dot{\tilde{x}}_i(t_k)-\dot{\tilde{x}}_j(t_k))]$$

$$=P(\|\dot{\tilde{x}}_{N+1}(t_k)\|,\|\dot{\tilde{x}}_{N+2}(t_k)\|,\cdots,\|\dot{\tilde{x}}_{N+M}(t_k)\|)(\frac{A_{22}+A_{22}^T}{2})\begin{pmatrix}\|\dot{\tilde{x}}_{N+1}(t_k)\|\\\|\dot{\tilde{x}}_{N+2}(t_k)\|\\\cdots\\\|\dot{\tilde{x}}_{N+M}(t_k)\|\end{pmatrix}.$$

Define the same Lyapunov function as Theorem 1, we can obtain

$$\dot{V}(t_k) \leq \sum_{i \in \mathcal{N}_{1i}} (2\rho_1 + \frac{3}{2}\rho_2 + \sum_{j \in \mathcal{N}_{1i}} c_{ij}a_{ij} + \sum_{j \in \mathcal{N}_{2i}} d_{ij}b_{ij} + c_i h_i - \frac{1}{2})(\tilde{x}(t_k - \tau(t_k)))^T(\tilde{x}(t_k - \tau(t_k)))$$

$$+ p(\|\tilde{x}_1(t_k - \tau(t_k))\|, \|\tilde{x}_2(t_k - \tau(t_k))\|, \cdots, \|\tilde{x}_N(t_k - \tau(t_k))\|)$$

$$(\frac{A_{11} + A_{11}^T}{2} - H_1) \begin{pmatrix} \|\tilde{x}_1(t_k - \tau(t_k))\| \\ \|\tilde{x}_2(t_k - \tau(t_k))\| \\ \cdots \\ \|\tilde{x}_N(t_k - \tau(t_k))\| \end{pmatrix}$$

$$+ \tau^2 \sum_{i \in \mathcal{N}_{1i}} (2\rho_1 + \rho_2 + \sum_{i \in \mathcal{N}_{1i}} c_{ij}a_{ij} + \sum_{i \in \mathcal{N}_{2i}} d_{ij}b_{ij} + h_i c_i)\dot{\tilde{x}}_i^T(t_k)\dot{\tilde{x}}_i(t_k)$$

$$+ P(\|\dot{\tilde{x}}_1(t_k)\|, \|\dot{\tilde{x}}_2(t_k)\|, \cdots, \|\dot{\tilde{x}}_N(t_k)\|)(\frac{A_{11} + A_{11}^T}{2} - H_1) \begin{pmatrix} \|\dot{\tilde{x}}_1(t_k)\| \\ \|\dot{\tilde{x}}_2(t_k)\| \\ \cdots \\ \|\dot{\tilde{x}}_N(t_k)\| \end{pmatrix}$$

$$+ \sum_{i \in \mathcal{N}_{2i}} (2\rho_3 + \frac{3}{2}\rho_4 + \sum_{j \in \mathcal{N}_{2i}} c_{ij}a_{ij} + \sum_{j \in \mathcal{N}_{1i}} d_{ij}b_{ij} + c_i h_i - \frac{1}{2})(\tilde{x}(t_k - \tau(t_k)))^T(\tilde{x}(t_k - \tau(t_k)))$$

$$+ p(\|\tilde{x}_{N+1}(t_k - \tau(t_k))\|, \|\tilde{x}_{N+2}(t_k - \tau(t_k))\|, \cdots, \|\tilde{x}_{N+M}(t_k - \tau(t_k))\|)$$

$$(\frac{A_{22} + A_{22}^T}{2} - H_2) \begin{pmatrix} \|\tilde{x}_{N+1}(t_k - \tau(t_k))\| \\ \|\tilde{x}_{N+2}(t_k - \tau(t_k))\| \\ \cdots \\ \|\tilde{x}_{N+M}(t_k - \tau(t_k))\| \end{pmatrix}$$

$$+ \tau^2 \sum_{i \in \mathcal{N}_{2i}} (2\rho_3 + \rho_4 + (\sum_{i \in \mathcal{N}_{2i}} c_{ij}a_{ij}) + (\sum_{i \in \mathcal{N}_{1i}} d_{ij}b_{ij}) + h_i c_i)\dot{\tilde{x}}_i^T(t_k)\dot{\tilde{x}}_i(t_k)$$

$$+ P(\|\dot{\tilde{x}}_{N+1}(t_k)\|, \|\dot{\tilde{x}}_{N+2}(t_k)\|, \cdots, \|\dot{\tilde{x}}_{N+M}(t_k)\|)(\frac{A_{22} + A_{22}^T}{2} - H_2) \begin{pmatrix} \|\dot{\tilde{x}}_{N+1}(t_k)\| \\ \|\dot{\tilde{x}}_{N+2}(t_k)\| \\ \cdots \\ \|\dot{\tilde{x}}_{N+M}(t_k)\| \end{pmatrix}$$

$$\leq \sum_{i \in \mathcal{N}_{1i}} (2\rho_1 + \frac{3}{2}\rho_2 + \sum_{j \in \mathcal{N}_{1i}} c_{ij}a_{ij} + \sum_{j \in \mathcal{N}_{2i}} d_{ij}b_{ij} + c_i h_i - \frac{1}{2} + p\lambda_1)\tilde{x}^T(t_k - \tau(t_k))\tilde{x}(t_k - \tau(t_k))$$

$$+ \tau^2 \sum_{i \in \mathcal{N}_{1i}} (2\rho_1 + \rho_2 + \sum_{i \in \mathcal{N}_{1i}} c_{ij}a_{ij} + \sum_{i \in \mathcal{N}_{2i}} d_{ij}b_{ij} + h_i c_i + p\lambda_1)\dot{\tilde{x}}_i^T(t_k)\dot{\tilde{x}}_i(t_k)$$

$$+ \sum_{i \in \mathcal{N}_{2i}} (2\rho_3 + \frac{3}{2}\rho_4 + \sum_{j \in \mathcal{N}_{2i}} c_{ij}a_{ij} + \sum_{j \in \mathcal{N}_{1i}} d_{ij}b_{ij} + c_i h_i - \frac{1}{2} + p\lambda_2)\tilde{x}^T(t_k - \tau(t_k))\tilde{x}(t_k - \tau(t_k))$$

$$+ \tau^2 \sum_{i \in \mathcal{N}_{2i}} (2\rho_3 + \rho_4 + \sum_{i \in \mathcal{N}_{2i}} c_{ij}a_{ij} + \sum_{i \in \mathcal{N}_{1i}} d_{ij}b_{ij} + h_i c_i + p\lambda_2)\dot{\tilde{x}}_i^T(t_k)\dot{\tilde{x}}_i(t_k),$$

where λ_1, λ_2 are the minimum eigenvalue of $\frac{A_{11} + A_{11}^T}{2} - H_1$, $\frac{A_{22} + A_{22}^T}{2} - H_2$, respectively, with $H_1 = diag\{h_i\}$ $\forall i \in \ell_1$, $H_2 = diag\{h_i\}$ $\forall i \in \ell_2$.

Even though matrix A_{11}, A_{22} are asymmetric, matrix $\frac{A_{11}+A_{11}^T}{2}, \frac{A_{22}+A_{22}^T}{2}$ are symmetric, thus, all eigenvalues of $\frac{A_{11}+A_{11}^T}{2} - H_1$, $\frac{A_{22}+A_{22}^T}{2} - H_2$ are negative from Lemma 3. So $\lambda_1 < 0$, $\lambda_2 < 0$ and $p > 0$ is sufficiently large, therefore, $\dot{V}(t_k) < 0$. Similar to Theorem 1, all the nodes of system (1) with time-delays and sampled-data can converge to their own synchronous states asymptotically. The proof is completed.

4   Simulations

Consider a complex dynamical network with $N+M$ nodes, where $N = 3$, $M = 3$. Let the initial value of the 6 nodes is $X(0) = [29\ 12\ 20\ 17\ 25\ -7\ 22\ 9]$, the initial values of the coupling strengths and the feedback gains are $c_{ij}(0) = d_{ij}(0) = c_i(0) = 0.01$.

Take A_{11}, A_{22} be symmetric as

$$A_{11} = \begin{bmatrix} -2 & 2 & 0 \\ 2 & -3 & 1 \\ 0 & 1 & -1 \end{bmatrix} * 0.1, \qquad A_{22} = \begin{bmatrix} -3 & 2 & 1 \\ 2 & -4 & 2 \\ 1 & 2 & -3 \end{bmatrix} * 0.05;$$

and A_{11}, A_{22} be asymmetric as,

$$A_{11} = \begin{bmatrix} -3 & 0 & 3 \\ 2 & -3 & 1 \\ 0 & 1 & -1 \end{bmatrix} * 0.1, \qquad A_{22} = \begin{bmatrix} -2 & 2 & 0 \\ 3 & -6 & 3 \\ 1 & 2 & -3 \end{bmatrix} * 0.05;$$

respectively.

$$\text{Take } B_{12} = \begin{bmatrix} 0 & 1 & 0 \\ 2 & 0 & 0 \\ 0 & 0 & 0.5 \end{bmatrix} * 0.1 \text{ and } B_{21} = \begin{bmatrix} 1 & 2 & 0 \\ 0 & 0 & 0 \\ 0 & 0 & 0 \end{bmatrix} * 0.1.$$

Figures 1 and 2 present that the effects of adaptive strategies for the synchronization of nonlinear complex dynamical networks. Figure 1 shows the simulation results on the synchronization of system (1) without adaptive strategies, where coupling matrices of sub-groups are symmetric. Figure 2 shows the simulation results on the adaptive synchronization of system (1) with $\tau = 0.2$, where coupling matrices of sub-groups are symmetric. Figures 2 and 3 show the simulation results on the group synchronization of system (1) with $\tau = 0.2$ and $\tau = 0.6$, where coupling matrices of sub-groups are symmetric as Figs. 2 and 3, respectively. From Figs. 2 and 3, we can see that all nodes of system achieve synchronization and the coupling strengths and the feedback gains also tend to be consistent. Figures 3 and 4 show the simulation results on the group synchronization of system (1) with $\tau = 0.6$, where coupling matrices of sub-groups are symmetric or symmetric presented as Figs. 3 and 4, respectively. Similarly, all nodes of system achieve synchronization and the coupling strengths and the feedback gains also tend to be consistent.

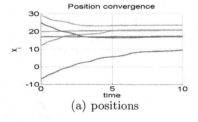
(a) positions

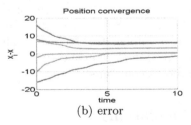
(b) error

Fig. 1. Without adaptive strategies and time delay $\tau = 0.2$ when intra-group coupling matrix is symmetric.

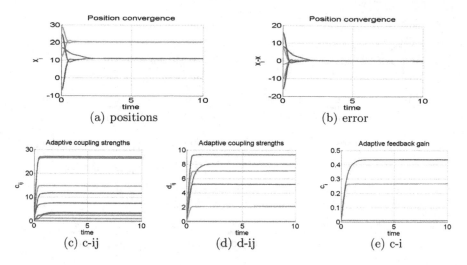

Fig. 2. Intra-group coupling matrix is symmetric and time delay $\tau = 0.2$.

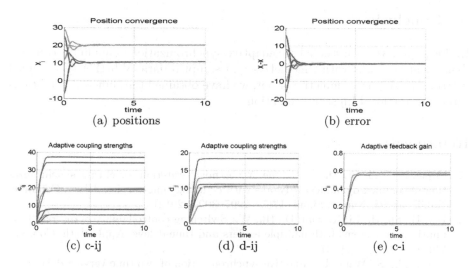

Fig. 3. Intra-group coupling matrix is symmetric and time delay $\tau = 0.6$.

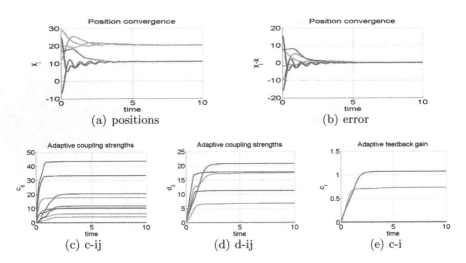

Fig. 4. Intra-group coupling matrix is asymmetric and time delay $\tau = 0.6$.

5 Conclusion

In this paper, we have studied the adaptive synchronization of nonlinear complex dynamical networks with time-delays and sampled-data. Whether the coupled matrix A_{11}, A_{22} are symmetric or not, we have obtained the sufficient conditions satisfying the local Lipschitz condition.

References

1. Wang, S., Yao, H., Zheng, S., Xie, Y.: A novel criterion for cluster synchronization of complex dynamical networks with coupling time-varying delays. Commun. Nonlinear Sci. Numer. Simul. **17**(7), 2997–3004 (2012)
2. Liu, B., Su, H., Li, R., Sun, D., Hu, W.: Switching controllability of discrete-time multi-agent systems with multiple leaders and time-delays. Appl. Math. Comput. **228**(9), 571–588 (2014)
3. Liu, B., Li, S., Wang, L.: Adaptive synchronization of two time-varying delay nonlinear coupled networks. In: Proceedings of the 33rd Chinese Control Conference, pp. 3800–3804. IEEE (2014)
4. Xu, Q., Zhuang, S., Zeng, Y., Xiao, J.: Decentralized adaptive strategies for synchronization of fractional-order complex networks. IEEE/CAA J. Autom. Sin. **4**(3), 543–550 (2017)
5. Liu, B., Wang, X., Su, H., Zhou, H., Shi, Y., Li, R.: Adaptive synchronization of complex dynamical networks with time varying delays. Circuits Syst. Signal Process. **33**(4), 1173–1188 (2014)
6. Du, L., Yang, Y., Lei, Y.: Synchronization in a fractional-order dynamic network with uncertain parameters using an adaptive control strategy. Appl. Math. Mech. (Engl. Ed.) **39**(3), 353–364 (2018)

7. Liu, B., Wang, X.L., Gao, Y.P., Xie, G.M., Su, H.S.: Adaptive synchronization of complex dynamical networks governed by local Lipschitz nonlinearlity on switching topology. J. Appl. Math. **2013**(1), 1–7 (2013)
8. Ding, D.W., Yan, J., Wang, N., Liang, D.: Adaptive synchronization of fractional order complex-variable dynamical networks via pinning control. Commun. Theor. Phys. **68**(9), 366–374 (2017)
9. Gao, Y., Wang, L.: Consensus of multiple dynamic agents with sampled information. IET Control Theory Appl. **4**(6), 945–956 (2010)
10. Gao, Y., Wang, L.: Sampled-data based consensus of continuous-time multi-agent systems with time-varying topology. IEEE Trans. Autom. Control **56**(5), 1226–1231 (2011)
11. Li, M., Liu, B., Zhu, Y.Q.: Group synchronization of nonlinear complex dynamics networks with sampled date. Hindawi Publ. Corp. Math. Probl. Eng. **2014**(6), 1–8 (2014)
12. Yu, Y.J., Yu, M., Hu, J.P., Liu, B.: Group consensus of multi agent systems with sampled data. In: Proceedings of the 32nd Chinese Control Conference, pp. 7168–7172. IEEE (2013)
13. Liu, B., Wei, P.E., Wang, X.F.: Group synchronization of complex network with nonlinear dynamics via pinning control. In: Proceedings of the 32nd Chinese Control Conference, pp. 235–240. IEEE (2013)
14. Yu, J., Wang, L.: Group consensus of multi-agent systems with directed information exchange. Int. J. Syst. Sci. **43**(2), 334–348 (2012)
15. Tang, Z., Huang, T., Shao, J., Hu, J.: Consensus of second-order multi-agent systems with nonuniform time-varying delays. Neurocomputing **97**(1), 410–414 (2012)

Controllability of Some Special Topologies with Time-Delays

Yaoyao Ping[1(✉)], Ningsheng Xu[1(✉)], Jiahui Bai[1(✉)], and Xue Lu[2]

[1] College of Science, North China University of Technology, Beijing 100144, China
pingyaoyaolb@163.com, xnsheng@139.com, baijiahui1310@163.com
[2] Beijing Vocational College of Agriculture, Beijing 100144, China

Abstract. This paper studies the controllability of some special topologies with time-delays and multiple leaders based on relative protocol. Some necessary and sufficient conditions are established for special topologies, respectively. An algorithm for controllability of multi-agent systems is introduced. Then the controllability of some special topologies are tested by algorithm. And some main conclusions can be obtained about different topologies. The results clearly indicate the controllability of different topologies based on relative protocol.

Keywords: Multi-agent systems · Controllability · Relative protocol · Time-delays

1 Introduction

The controllability problem of multi-agent systems with some different communication topologies has been concerned since controllability problem was put forward by Tanner in [1]. Controllability of multi-agent systems can be reflected by different communication topologies [2–9]. For example, the paper [5] presented an algorithm to design an optimal backbone network topology, which incorporates some real life constraints of reliability. The topologies construction and leaders location for multi-agent controllability based on consensus algorithm was studied in [6]. The paper [7] focused on the controllability of multi-agent systems with tree topology. A class of uncontrollable diffusively coupled multi-agent systems with a single leader and multi-chain topologies were constructed in [8]. A design method was proposed to uncover topology structures of graphs with any size in [10]. The paper [11] came up with a result that protocols design and uncontrollable topologies construction for multi-agent networks. [12] studied

This work was supported in part by the National Natural Science Foundation of China Grant (No. 61773023), 2018 Beijing Education Commission Basic Science Research Expenses Project, "The-Great-Wall-Scholar" Candidate Training Plan of NCUT (XN070006), Construction Plan for Innovative Research Team of NCUT (XN018010), "Yu Xiu" Talents of NCUT.

© Springer Nature Singapore Pte Ltd. 2020
L. Pan et al. (Eds.): BIC-TA 2019, CCIS 1160, pp. 58–67, 2020.
https://doi.org/10.1007/978-981-15-3415-7_5

the controllability of the dynamic systems and shown that there was an intricate relationship between the uncontrollability of the corresponding multi-agent systems and various graph-theoretic properties of the network.

In recent years, many results have also been obtained. The paper [13] investigated the characterization of the controllability of weighted and directed signed networks by graph partitions. A tree graph designing scheme for multi-group access control was studied in [14]. The controllability and the observability of multi-agent networks with strongly regular graphs or distance regular graphs were concerned in [15], respectively. A new control protocol was introduced in [16], that is affine formation control, and necessary and sufficient graphical conditions for affine formation control were obtained. [17] studied the consensus problems for a group of agents with switching topology and time varying communication delays. A graph-theoretic characterization of controllability for multi-agent systems was studied in [18]. The paper [19] investigated controllability of path graphs in undirected and connected networks. The problem of obtaining graph-theoretic characterizations of controllability for the Laplacian-based leader-follower dynamics is considered in [20, 21]. However, the controllability of multi-agent systems for some special topologies with time-delays is rarely studied. This paper aims to study the controllability of some special topologies with time-delays.

The remainder of this paper is organized as follows. Section 2 presents problem formulation and preliminaries. Section 3 gives the main results. Section 4 shows the controllability of some special topologies. Finally, the conclusions are written in Sect. 5.

2   Problem Formulation and Preliminaries

A weighted graph is composed of the note set $\mathscr{V} = \{1, 2, \cdots, n\}$ and the edge set $\mathscr{E} = \{\varepsilon_{ij} = (i, j) : i, j \in \mathscr{V}\}$ with the adjacency matrix $\mathscr{A} = [a_{ij}] \in \mathbb{R}^{n \times n}$, which is represented as $\mathscr{G} = (\mathscr{V}, \mathscr{E}, \mathscr{A})$. The neighbors set of node i is denoted by $\mathscr{N}_i = \{j \in \mathscr{V} | (j, i) \in \mathscr{E}\}$ and the corresponding Laplacian matrix is given by $L = D - \mathscr{A} \in \mathbb{R}^{n \times n}$, where $D = diag\{d_1, d_2, \cdots, d_n\} \in \mathbb{R}^{n \times n}$ is a diagonal matrix with $d_i = \sum_{j \in \mathscr{N}_i} a_{i,j}, i = 1, 2, \cdots, n$.

Consider a multi-agent system with $m + q$ agents, where m and q represent the numbers of followers and leaders, respectively. The followers are governed by the following protocol:

$$x_i(k+1) = x_i(k) - \sum_{i \in \mathscr{N}_{ij}} a_{ij}(x_i(k-h) - x_j(k-h)) - \sum_{p \in \mathscr{N}_{ip}} b_{ip}(x_i(k-h) - x_p(k-h)),$$

$$(1)$$

where $h > 0$ represents time-delay among agents, $x_i \in \mathbb{R}$ and $x_p \in \mathbb{R}$ are the state vectors of followers and leaders, respectively, for $x_p \in \mathbb{R}^1$, $i \in \{1, 2, \cdots, m\}$, and $p \in \{m+1, m+2, \cdots, m+q\}$, a_{ij} represents the weights of agent i and j. And $\mathscr{N}_i$ is the neighbor set of agent i, $\mathscr{N}_{ij}$ contains all followers, $\mathscr{N}_{ip}$ contains all leaders, and $\mathcal{N}_{ij} \bigcup \mathcal{N}_{ip} = \mathcal{N}_i$, $\mathcal{N}_{ij} \bigcap \mathcal{N}_{ip} = \emptyset$. k is step length of discrete time and $\mathcal{J}_k$ is a discrete time index set.

Let $x(k) = \{x_1(k), x_2(k), \cdots, x_m(k)\}$ be state set of all followers, and $y(k - h) = \{x_{m+1}(k - h), x_{m+2}(k - h), \cdots, x_{m+q}(k - h)\}$ be state set of all leaders. Then system (1) can be rewritten as

$$x(k + 1) = x(k) - (L + R)x(k - h) + By(k - h), \quad k \in \mathcal{J}_k, \tag{2}$$

where

$$R = diag(\sum_{p=m+1}^{m+q} b_{1p}, \sum_{p=m+1}^{m+q} b_{2p}, \cdots, \sum_{p=m+1}^{m+q} b_{mp}),$$

and $L = [l_{ij}]$ is Laplacian matrix, $B = [b_{ip}] \in \mathbb{R}^{m \times q}$ is a matrix with $b_{ip} \geq 0$.

Through special method, system (2) can be transformed to the classical formation of the discrete-time model as

$$
\begin{cases}
x(k + 1) = x(k) - (L + R)x(k - h) + By(k - h) \\
\quad x(k) = x(k) \\
\qquad \cdots \\
x(k - h + 1) = x(k - h + 1) \\
\quad y(k + 1) = y(k + 1) \\
\qquad y(k) = y(k) \\
\qquad \cdots \\
y(k - h + 1) = y(k - h + 1)
\end{cases}
$$

Denote $w(k) = (x(k)^T, x(k - 1)^T, \cdots, x(k - h)^T)^T$, $z(k) = (y(k)^T, y(k - 1)^T, \cdots, y(k - h)^T)^T$, then system (2) is rewritten as

$$w(k + 1) = \mathscr{A}w(k) + \mathscr{B}z(k), \quad k \in \mathcal{J}_k, \tag{3}$$

where

$$
\mathscr{A} = \begin{bmatrix}
I & 0 & \dots & 0 & -(L + R) \\
I & 0 & \dots & 0 & 0 \\
0 & I & \dots & 0 & 0 \\
\vdots & \vdots & \ddots & \vdots & \vdots \\
0 & 0 & \dots & I & 0
\end{bmatrix}_{(h+1)m \times (h+1)m}, \quad
\mathscr{B} = \begin{bmatrix}
0 & 0 & \cdots & B \\
0 & 0 & \cdots & 0 \\
\vdots & \vdots & \ddots & \vdots \\
0 & 0 & \cdots & 0
\end{bmatrix}_{(h+1)m \times (h+1)q},
$$

where I is an identity matrix with compatible dimensions, $\mathscr{A}$ and $\mathscr{B}$ are system matrix and control input matrix, respectively.

Remark 1. The controllability of system (1) is equivalent to that of system (3).

3   Main Results

Proposition 1. *System (3) is controllable if and only if matrix $\mathscr{Q}$ has full row rank, where*

$$\mathscr{Q} = [\mathscr{B} \ \ \mathscr{A}\mathscr{B} \ \ \cdots \ \ \mathscr{A}^{(h+1)m-1}\mathscr{B}].$$

Proposition 2. *System (3) is controllable if and only if system (3) satisfies one of the following conditions:*

(i) $Rank(sI - \mathscr{A}, \mathscr{B}) = (h+1)m$, $\forall s \in \mathbb{C}$;
(ii) $Rank(\lambda_i I - \mathscr{A}, \mathscr{B}) = (h+1)m$, where λ_i is the eigenvalue of matrix $\mathscr{A}$, $\forall i = 1, 2, \cdots, (h+1)m$.

Because of high dimension of matrix $\mathscr{A}$ and $\mathscr{B}$, Propositions 1–2 are very difficult to test the controllability of system (3). Thus an easier method is given for the controllability of system (3).

Theorem 1. *System (3) is controllable if and only if matrix $Y = [-(L+R) + \lambda^h I - \lambda^{h+1} I, B]$ has full row rank at every root of $det[-(L+R) + \lambda^h I - \lambda^{h+1} I] = 0$.*

Proof. From Proposition 2, system (3) is controllable if and only if matrix

$$(\mathscr{A} - \lambda I, \mathscr{B}) = \begin{bmatrix} I & 0 & \cdots\cdots & -(L+R) & 0 & \cdots & B \\ I & -\lambda I & \cdots\cdots & 0 & 0 & \cdots & 0 \\ 0 & I & \ddots & \ddots & \vdots & \vdots & \vdots & \vdots \\ \vdots & \ddots & \ddots & \ddots & \vdots & \vdots & \vdots & \vdots \\ 0 & \cdots & 0 & I & -\lambda I & 0 & \cdots & 0 \end{bmatrix} \rightarrow \begin{bmatrix} I & 0 & \cdots & 0 & 0 & \cdots & 0 \\ 0 & I & \cdots & 0 & 0 & \cdots & 0 \\ 0 & 0 & \cdots & 0 & 0 & \cdots & 0 \\ \vdots & \vdots & \ddots & \vdots & \vdots & \vdots \\ 0 & 0 & \cdots & T & B & \cdots & 0 \end{bmatrix}$$

has full row rank, where $T = -(L+R) + \lambda^h I - \lambda^{h+1} I$. Through elementary transformation of matrix, matrix $(\mathscr{A} - \lambda I, \mathscr{B})$ has full row rank if matrix $Y = [-(L+R) + \lambda^h I - \lambda^{h+1} I, B]$ has full row rank. Thus the proof is completed.

For controllability of multi-agent system, an algorithm is introduced to test controllability of some special topologies, it is written as follows.

Algorithm for testing controllability of multi-agent system

Input: h, m, C, Y_i
Output: r_i
1. Initialization: $h = 0, m = 0, i = 0, r_i = 0, \lambda_i = 0, C = [], Y_i = []$;
2: Input weights and parameters h, m;
3: Compute all roots λ_i of $detC = 0$, $i = 1, 2, \cdots, (h+1)m$;
while $(i \leq (h+1)m)$;
    $i = i + 1$;
    $r_i = rank(Y_i)$
end while
return r_i

---

4 Some Special Topologies

In the following, the controllability for some special topologies are shown by Theorem 1.

Theorem 2. *Path graph is controllable.*

Proof. For Fig. 1, let agent ν_{m+1} be leader, other agents be followers, and all weights be a, time-delay be h. Corresponding matrixes can be written as follows:

Fig. 1. Path graph.

$$L = \begin{bmatrix} a & -a & 0 & \cdots & 0 & 0 \\ -a & 2a & -a & \cdots & 0 & 0 \\ \vdots & \vdots & \vdots & \cdots & \vdots & \vdots \\ 0 & 0 & 0 & \cdots & 2a & -a \\ 0 & 0 & 0 & \cdots & -a & a \end{bmatrix}, R = \begin{bmatrix} 0 & \cdots & 0 \\ 0 & \cdots & 0 \\ \vdots & \cdots & \vdots \\ 0 & \cdots & a \end{bmatrix}, B = \begin{bmatrix} 0 \\ 0 \\ \vdots \\ a \end{bmatrix},$$

thus

$$C = -(L+R) + \lambda^h I - \lambda^{h+1} I = \begin{bmatrix} -a+t & a & 0 \cdots & 0 & 0 \\ a & -2a+t & a \cdots & 0 & 0 \\ \vdots & \vdots & \vdots \cdots & \vdots & \vdots \\ 0 & 0 & 0 \cdots -2a+t & a \\ 0 & 0 & 0 \cdots & a & -2a+t \end{bmatrix},$$

where $t = \lambda^h - \lambda^{h+1}$. Through using algorithm of testing controllability of multi-agent system. We choose the parameters a, h, m, and we put the every eigenvalue λ_i of matrix C into the $Y_i = [-(L+R) + \lambda_i^h I - \lambda_i^{h+1} I, B]$, then we can find every r_i is full row rank, where $i = 1, 2, \cdots, (h+1)m$. Therefore path graph is controllable.

Theorem 3. *Star graph is uncontrollable.*

Proof. For Fig. 2, corresponding matrixes can be written as follows:

$$L = \begin{bmatrix} (m-1)a & -a & -a & \cdots & -a & -a \\ -a & a & 0 & \cdots & 0 & 0 \\ \vdots & \vdots & \vdots & \cdots & \vdots & \vdots \\ -a & 0 & 0 & \cdots & a & 0 \\ -a & 0 & 0 & \cdots & 0 & a \end{bmatrix}, R = \begin{bmatrix} a & 0 & \cdots & 0 \\ 0 & a & \cdots & 0 \\ \vdots & \vdots & \cdots & \vdots \\ 0 & 0 & \cdots & a \end{bmatrix}, B = \begin{bmatrix} a \\ a \\ \vdots \\ a \end{bmatrix},$$

thus

$$C = -(L+R) + \lambda^h I - \lambda^{h+1} I = \begin{bmatrix} -ma+t & a & a \cdots & a & a \\ a & -2a+t & 0 \cdots & 0 & 0 \\ \vdots & \vdots & \vdots \cdots & \vdots & \vdots \\ a & 0 & 0 \cdots -2a+t & 0 \\ a & 0 & 0 \cdots & 0 & -2a+t \end{bmatrix},$$

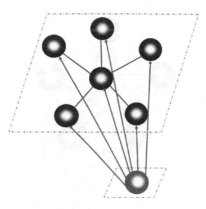

Fig. 2. Star graph.

where $t = \lambda^h - \lambda^{h+1}$. Through using algorithm of testing controllability of multi-agent system, and we can find r_i is not full row rank, where $i = 1, 2, \cdots, (h+1)m$. Therefore star graph is uncontrollable.

Theorem 4. *Cycle graph is uncontrollable.*

Proof. For Fig. 3, corresponding matrixes can be written as follows:

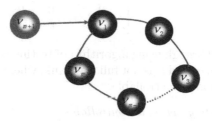

Fig. 3. Cycle graph.

$$
L = \begin{bmatrix} 2a & -a & 0 & 0 & \cdots & 0 & -a \\ -a & 2a & -a & 0 & \cdots & 0 & 0 \\ 0 & -a & 2a & -a & \cdots & 0 & 0 \\ \vdots & \vdots & \vdots & \vdots & \cdots & \vdots & \vdots \\ -a & 0 & 0 & 0 & \cdots & -a & 2a \end{bmatrix}, R = \begin{bmatrix} a & \cdots & 0 \\ 0 & \cdots & 0 \\ \vdots & \cdots & \vdots \\ 0 & \cdots & 0 \end{bmatrix}, B = \begin{bmatrix} a \\ 0 \\ \vdots \\ 0 \end{bmatrix},
$$

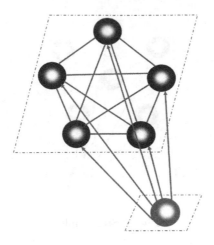

Fig. 4. Complete graph.

thus

$$C = -(L+R) + \lambda^h I - \lambda^{h+1} I = \begin{bmatrix} -3a+t & a & 0\cdots & 0 & a \\ a & -2a+t & a\cdots & 0 & 0 \\ \vdots & \vdots & \vdots\cdots & \vdots & \vdots \\ 0 & 0 & 0\cdots -2a+t & a \\ a & 0 & 0\cdots & a & -2a+t \end{bmatrix},$$

where $t = \lambda^h - \lambda^{h+1}$. Through using algorithm of testing controllability of multi-agent system, and we can find r_i is not full row rank, where $i = 1, 2, \cdots, (h+1)m$. Therefore cycle graph is uncontrollable.

Theorem 5. *Complete graph is uncontrollable.*

Proof. For Fig. 4, corresponding matrixes can be written as follows:

$$L = \begin{bmatrix} (m-1)a & -a & \cdots & -a & -a \\ -a & (m-1)a & \cdots & -a & -a \\ \vdots & \vdots & \cdots & \vdots & \vdots \\ -a & -a & \cdots (m-1)a & -a \\ -a & -a & \cdots & -a & (m-1)a \end{bmatrix}, R = \begin{bmatrix} a & 0 & \cdots & 0 \\ 0 & a & \cdots & 0 \\ \vdots & \vdots & \cdots & \vdots \\ 0 & 0 & \cdots & a \end{bmatrix}, B = \begin{bmatrix} a \\ a \\ \vdots \\ a \end{bmatrix},$$

thus

$$C = -(L+R) + \lambda^h I - \lambda^{h+1} I = \begin{bmatrix} -ma+t & a & \cdots & a & a \\ a & -ma+t & \cdots & a & a \\ \vdots & \vdots & \cdots & \vdots & \vdots \\ a & a & \cdots -ma+t & a \\ a & a & \cdots & a & -ma+t \end{bmatrix},$$

where $t = \lambda^h - \lambda^{h+1}$. Through using algorithm of testing controllability of multi-agent system, and we can find r_i is not full row rank, where $i = 1, 2, \cdots, (h+1)m$. Therefore complete graph is uncontrollable.

Theorem 6. *Binary tree graph is uncontrollable.*

Proof. For Fig. 5, corresponding matrixes can be written as follows:

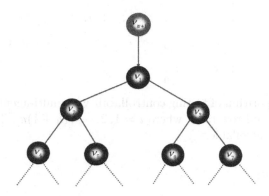

Fig. 5. Binary tree graph.

$$
L = \begin{pmatrix}
 & & & & \frac{m+1}{4} & \frac{m-1}{2} & \frac{m+1}{2} & & m \\
2a & -a & -a & \cdots & \cdots & \cdots & \cdots & \cdots & 0 \\
-a & 3a & 0 & \cdots & \cdots & \cdots & \cdots & \cdots & 0 \\
-a & 0 & 3a & \cdots & \cdots & \cdots & \cdots & \cdots & 0 \\
\vdots & \vdots & \vdots & \ddots & \cdots & \cdots & \cdots & \cdots & \vdots \\
\vdots & \vdots & \vdots & \vdots & \ddots & \cdots & \cdots & \cdots & \vdots \\
\vdots & \vdots & \vdots & \vdots & \vdots & \ddots & \cdots & \cdots & \vdots \\
0 & 0 & 0 & \cdots & -a & \cdots & a & \cdots & \vdots \\
\vdots & \vdots & \vdots & \cdots & \vdots & \cdots & \vdots & \ddots & \vdots \\
0 & 0 & 0 & \cdots & \cdots & \cdots & -a & \cdots & a
\end{pmatrix}_{m \times m} = L
$$

$$
R = \begin{bmatrix} a & 0 & \cdots & 0 \\ 0 & 0 & \cdots & 0 \\ \vdots & \vdots & \cdots & \vdots \\ 0 & 0 & \cdots & 0 \end{bmatrix}, B = \begin{bmatrix} a \\ 0 \\ \vdots \\ 0 \end{bmatrix},
$$

thus

$$C = -(L + R) + \lambda^h I - \lambda^{h+1} I$$

$$= \begin{bmatrix} -3a+\lambda^h-\lambda^{h+1} & a & a & \cdots & \cdots & \cdots & \cdots & \cdots & 0 \\ a & 3a+\lambda^h-\lambda^{h+1} & 0 & \cdots & \cdots & \cdots & \cdots & \cdots & 0 \\ a & 0 & 3a+\lambda^h-\lambda^{h+1} & \cdots & \cdots & \cdots & \cdots & \cdots & 0 \\ \vdots & \vdots & \vdots & \ddots & \cdots & \cdots & \cdots & \cdots & \vdots \\ \vdots & \vdots & \vdots & \vdots & \ddots & \cdots & \cdots & \cdots & \vdots \\ \vdots & \vdots & \vdots & \vdots & \vdots & \ddots & \cdots & \cdots & \vdots \\ 0 & 0 & 0 & \cdots & a & \cdots & -a+\lambda^h-\lambda^{h+1} & \cdots & \vdots \\ \vdots & \vdots & \vdots & \cdots & \vdots & \cdots & \vdots & \ddots & \vdots \\ 0 & 0 & 0 & \cdots & \cdots & a & \cdots & \cdots & -a+\lambda^h-\lambda^{h+1} \end{bmatrix}.$$

Through using algorithm of testing controllability of multi-agent system, and we can find r_i is not full row rank, where $i = 1, 2, \cdots, (h+1)m$. Therefore Binary tree graph is uncontrollable.

5   Conclusion

In this paper, we have studied the controllability of some special topologies with time-delays. The discrete-time system can be transformed to an equivalent system without time-delays. Through the theoretical analysis, we have derived an algorithm for controllability of multi-agent systems. The controllability of some special topologies have been tested by algorithm. The time-delays in the system may not always all equal. Some further research on the effect of time-delays in different topologies. And when communication time-delays are not entirely equal, the controllability of some special topologies with multiple time-delays can be studied in the further.

References

1. Tanner, H.G.: On the controllability of nearest neighbor interconnections. In: IEEE Conference on Decision Control, vol. 3, no. 3, pp. 2467–2472 (2004)
2. Tassiulas, L.: Scheduling and performance limits of networks with constantly changing topology. IEEE Trans. Inf. Theory **43**(3), 1067–1073 (1997)
3. Gasteier, M., Munch, M., Glesner, M.: Generation of interconnect topologies for communication synthesis. In: Proceedings Design, Automation and Test in Europe (1998)
4. Sayoud, H., Takahashi, K., Vaillant, B.: Designing communication network topologies using steady-state genetic algorithms. IEEE Commun. Lett. **5**(3), 113–115 (2001)
5. Mandal, S., Saha, D., Mukherjee, R., Roy, A.: An efficient algorithm for designing optimal backbone topology for a communication networks. In: International Conference on Communication Technology Proceedings (2003)
6. Ji, Z., Lin, H.: Topologies construction and leaders location in multi-agent controllability. In: Proceedings of the 31st Chinese Control Conference (2012)

7. Ji, Z., Lin, H.: Multi-agent controllability with tree topology. In: Proceedings of the 2010 American Control Conference (2010)

8. Cao, M., Zhang, S., Camlibel, M.K.: A class of uncontrollable diffusively coupled multiagent systems with multichain topologies. IEEE Trans. Autom. Control **58**(2), 465–469 (2013)

9. Liu, B., Chu, T., Wang, L., Xie, G.: Controllability of a leader-follower dynanic network with switching topology. IEEE Trans. Autom. Control **53**(4), 1009–1013 (2008)

10. Ji, Z., Yu, H.: A new perspective to graphical characterization of multi-agent controllability. IEEE Trans. Cybern. **47**(6), 1471–1483 (2017)

11. Ji, Z., Lin, H., Yu, H.: Protocols design and uncontrollable topologies construction for multi-agent networks. IEEE Trans. Autom. Control. **60**(3), 781–786 (2015)

12. Rahmani, A., Ji, M., Mesbahi, M., Egerstedt, M.: Controllability of multi-agent systems from a graph theoretic perspective. SIAM J. Control Optim. **48**(1), 162–186 (2009)

13. Liu, X., Ji, Z., Hou, T.: Graph partitions and the controllability of directed signed networks. Sci. China Inf. Sci. **62**(4), 1–11 (2019). https://doi.org/10.1007/s11432-018-9450-8

14. Koo, H.-S., Kwon, O.-H., Ra, S.-W.: A tree key graph design scheme for hierarchical multi-group access control. IEEE Commun. Lett. **13**(11), 874–876 (2009)

15. Alain, Y., Kibangou, C.C.: Observability in connected strongly regular graphs and distance regular graphs. IEEE Trans. Control Netw. Syst. **1**(4), 360–369 (2014)

16. Lin, Z., Wang, L., Chen, Z.: Necessary and sufficient graphical conditions for affine formation control. IEEE Trans. Autom. Control **61**(10), 2877–2891 (2016)

17. Jiang, F., Wang, L., Xie, G.: Consensus of high-order dynamic multi-agent systems with switching topology and time-varying delays. J. Control Theory Appl. **8**(1), 52–60 (2010)

18. Ji, M., Egersted, M.: A graph-theoretic characterization of controllability for multi-agent systems. In: American Control Conference (2007)

19. Yongcui, C., Zhijian, J., Yaowei, W.: Controllability of path graphs by using information of second-order neighbors. In: IEEE 2015 34th Chinese Control Conference, pp. 6619–6623 (2015)

20. Aguilar, C.O., Gharesifard, B.: Graph controllability classes for the Laplacian leader-follower dynamics. Autom. Control. IEEE Trans. **60**(6), 1611–1623 (2015)

21. Liu, B., Su, H., Li, R.: Switching controllability of discrete-time multi-agent systems with multiple leaders and time-delays. Appl. Math. Comput. **228**, 571–588 (2014)

Controllability of Second-Order Discrete-Time Multi-agent Systems with Switching Topology

Ningsheng Xu[1(✉)], Yaoyao Ping[1(✉)], Jiahui Bai[1(✉)], and Xue Lu[2]

[1] College of Science, North China University of Technology, Beijing 100144, China
xnsheng@139.com, pingyaoyaolb@163.com, baijiahui1310@163.com
[2] Beijing Vocational College of Agriculture, Beijing 100144, China

Abstract. This paper focuses on the controllability of second-order discrete-time multi-agent systems with switching topology. First, the controllability of the systems are investigated under the motion analysis of discrete-time linear systems and two necessary and sufficient conditions are established for controllability in terms of the system matrices. Second, it is shown that, if both the position and the velocity information interaction topologies are fixed, we convert the switching topology into fixed topology. Finally, simulation example is worked out to illustrate the validity of the theoretical result.

Keywords: Multi-agent systems · Controllability · Switching topology

1 Introduction

In recent years, distributed cooperative control of agent systems has attracted extensive interest of researchers because of the widespread in nature, engineer fields, migration of birds, learning of robotic fish, swarm action of animals and intelligent transportation systems, etc. Cooperation and control among multiple robots has become one of the main directions in the field of robotics. Hot topics in this area include consensus problem, controllability, structural controllability, reachability, observability, and so on [1–10].

Controllability is one of the basic characteristics that characterize the structure of systems from the perspective of control. The concept of controllability of multi-agent systems was first put forward by Kalman in the year of 1960. The goal of controllability is to steer the agents to form any configuration from any initial state by controlling a few of the agents (leaders). In terms of multi-agent systems, leader-follower structure was first put forward by Tanner [11],

This work was supported in part by the National Natural Science Foundation of China Grant (No. 61773023), 2018 Beijing Education Commission Basic Science Research Expenses Project, "The-Great-Wall-Scholar" Candidate Training Plan of NCUT (XN070006), Construction Plan for Innovative Research Team of NCUT (XN018010), "Yu Xiu" Talents of NCUT.

© Springer Nature Singapore Pte Ltd. 2020
L. Pan et al. (Eds.): BIC-TA 2019, CCIS 1160, pp. 68–76, 2020.
https://doi.org/10.1007/978-981-15-3415-7_6

who derive necessary and sufficient conditions according to the eigenvalues and eigenvectors of the partition of the Laplacian matrix. Inspired by this, a lot of research are based on this foundation. Authors studied the controllability of multi-agent systems from graphic [12] and directed topology perspective [13]. In [14], the authors investigated the controllability of multi-agent systems based on agreement protocols and proved that under the same topology, a network of high-order dynamic agents is controllable if and only if so is a network of single-integrator agents. Other results are also researched, such as structural controllability [15], target controllability [16], structural target controllability [17], and heterogenous dynamics [18]. The aforementioned research are all in continuous-time systems. For discrete-time multi-agent systems, the authors investigated controllability of discrete-time multi-agent systems with multiple leaders in [19]–[20]. Furthermore, Liu et al. [21] analyzed the controllability of a leader-follower dynamic network with switching topology. For multi-agent systems under switching topology, structural controllability of multi-agent systems was investigated in [22]. Lu et al. [23] shown that the multi-agent system with switching topology can be observable even if each of its subsystems is not observable. In [24], Tian et al. consider the controllability and observability of the continuous-time and discrete-time subsystems. On the controllability and observability of second-order continuous-time multi-agent systems also be studied in [25]. Inspired by the above work, this paper mainly studies the controllability of second-order discrete-time multi-agent systems with switching topology.

The rest of this paper is organized as follows. In Sect. 2, we introduced the problem formulation and preliminaries. Section 3 is the main results of this paper. In Sect. 4, one example is included to illustrate the theoretical result. Finally, the conclusion are written in Sect. 5.

2 Problem Formulation and Preliminaries

Here, we mainly introduce the knowledge of graph theory and notations. An weighted undirected graph denoted as $\mathscr{G} = (\mathscr{V}, \mathscr{E}, \mathscr{A})$, where $\mathscr{V} = \{v_1, v_2, \cdots, v_N\}$ is a *vertex set* and $\mathscr{E} = \{(v_i, v_j) : v_i, v_j \in \mathscr{V}\}$ is an *edge set*. The weighted adjacency matrix of $\mathscr{G}$ is expressed as $\mathscr{A} = [a_{ij}]$. a_{ij} donated that there exist an edge from v_j to v_i. Furthermore, the neighbors' set of agent v_i showed as $\mathscr{N}_i = \{v_j \in \mathscr{V} : (v_i, v_j) \in \mathscr{E}\}$. The Laplacian matrix is $L = \triangle - \mathscr{A}$, where $\triangle$ is the diagonal degree matrix and $\mathscr{A}$ is the weighted adjacency matrix.

Notations: The linear space $\mathbb{R}^n$ is the $n - dimensional$ vector space, such as $\mathbb{R}^{n \times p}$ is the matrix with n rows and p columns. I_n is the identity matrix with n rows and n columns. In this paper, e_i is denoted as the ith column of the identity matrix.

Consider a second-order multi-agent system included n agents. The dynamic of each agent can be described by

$$\begin{aligned} x_i(k+1) &= x_i(k) + v_i(k), \\ v_i(k+1) &= v_i(k) + u_i(k), \qquad i = 1, 2, \ldots, n, \end{aligned} \tag{1}$$

where $x_i \in \mathbb{R}^n$ and $v_i \in \mathbb{R}^n$ represent the position information and the velocity information of agent i, respectively. Leaders are actuated by the external inputs, and the rest agents are followers. The sets of leaders and followers are denoted by $\mathbb{V}_l = \{i_1, i_2, ..., i_p\}$ and $\mathbb{V}_f = \mathbb{V}/\mathbb{V}_l$, respectively. Then $\mathbb{V}_f \bigcup \mathbb{V}_l = \mathbb{V}$ and $\mathbb{V}_f \bigcap \mathbb{V}_l = \emptyset$. The control input $u_i(k)$ contents the following protocol:

$$
u_i(k) = \begin{cases} \displaystyle\sum_{j\in\mathcal{N}_i(k)} a_{ij}(x_j(k) - x_i(k)) + \sum_{j\in\mathcal{N}_i(k)} b_{ij}(v_j(k) - v_i(k)) + u_{o,i}, & i \in \mathbb{V}_l, \\ \displaystyle\sum_{j\in\mathcal{N}_i(k)} a_{ij}(x_j(k) - x_i(k)) + \sum_{j\in\mathcal{N}_i(k)} b_{ij}(v_j(k) - v_i(k)), & i \in \mathbb{V}_f, \end{cases}
$$

(2)

where $\mathcal{N}_i(k)$ is the neighbor set of agent i at time k and $u_{o,i}$ is the external input of leader i. We consider the case that the position and the velocity information interaction topologies. They are modeled by different graphs $\mathbb{G}_\chi$ and $\mathbb{G}_\nu$. Here we let $x(k) \triangleq [x_1, ..., x_n]^T$, $v(k) \triangleq [v_1, ..., v_n]^T$. Under protocol (7), the compact form of system (6) can be written as

$$
\begin{bmatrix} x(k+1) \\ v(k+1) \end{bmatrix} = \begin{bmatrix} I_n & I_n \\ -L^x_{\sigma(k)} & I_n - L^v_{\delta(k)} \end{bmatrix} \begin{bmatrix} x(k) \\ v(k) \end{bmatrix} + \begin{bmatrix} 0 \\ \tilde{B} \end{bmatrix} u(k),
$$

(3)

where $\tilde{B} \triangleq [e_{i_1}, ..., e_{i_p}] \in \mathbb{R}^{n\times p}$ and $u(k) \triangleq [u_1, u_2, ..., u_p]^T \in \mathbb{R}^p$ is the external input vector. $L^x_{\sigma(k)}$ and $L^v_{\delta(k)}$ are the Laplacian matrices of $\mathbb{G}_\chi$ and $\mathbb{G}_\nu$ at time instant k, respectively. We denoted $\tilde{\mathbb{G}} = \{\mathbb{G}_1, \mathbb{G}_2, ..., \mathbb{G}_N\}$ as the graphs corresponding to all the position and velocity information interaction topologies, where $\mathbb{N} = \{1, 2, ..., N\}$ is the index set. $\sigma(k)(resp.\delta(k)): \mathbb{R}^+ \longrightarrow \mathbb{N}$ ($\mathbb{R}^+$ stands for positive real number) is the switching signal of $\mathbb{G}_\chi$ and $\mathbb{G}_\nu$ to realize the position and the velocity information interaction topologies at time instant k. The matrix triple $(L^x_{i_m}, L^v_{j_m}, \tilde{B})$ represent the mth subsystem of system (3).

Definition 1. *The switching sequence of system (3) is defined as*

$$
\Pi = \{(i_1, j_1, h_1), (i_2, j_2, h_2), ..., (i_M, j_M, h_M)\},
$$

(4)

where M is the length of Π and $i_m, j_m \in \mathbb{N}$ stand for the indexes of the mth subsystem's position and velocity information topologies, respectively. $h_m > 0$ is the time of the mth subsystem $(L^x_{i_m}, L^v_{j_m}, \tilde{B})$ activated in $[k_{m-1}, k_m]$, where $h_m = k_m - k_{m-1}$ and $k_i = k_0 + \sum_{j=1}^{i} h_j$. k_0 is the initial time.

Definition 2. *For system (3), let the initial time be $k_0 = 0$. If for any nonzero initial state $z(0)$, there exists a switching sequence $\Pi = \{(i_M, j_M, h_M)\}_{m=1}^{M}$ and an external input $u(k)$, when the time is shifted from the initial time to $k_f = \sum_{l=1}^{M} h_l$, such that the final state $z(k_f) = 0$. We said the system (3) is controllable.*

3   Mainly Result

In this section, we discuss the controllability of the second-order discrete-time multi-agent system (3). Some conclusions are obtained.

For $m = 1, 2, \ldots, M$, we denote that

$$A_m = \begin{bmatrix} I_n & I_n \\ -L_{i_m}^x & I_n - L_{j_m}^v \end{bmatrix}, B = \begin{bmatrix} 0 \\ \tilde{B} \end{bmatrix}. \tag{5}$$

Matrix A_m is determined by the switching sequence $\Pi = \{(i_M, j_M, h_M)\}_{m=1}^{M}$. For switching sequence $\Pi = \{(i_M, j_M, h_M)\}_{m=1}^{M}$, the state of the system (3) at time instant k_M is

$$z(k_M) = A_M^{k_M - k_{M-1}} A_{M-1}^{k_{M-1} - k_{M-2}} \cdots A_2^{k_2 - k_1} A_1^{k_1} z(0) + A_M^{k_M - k_{M-1}} A_{M-1}^{k_{M-1} - k_{M-2}}$$

$$\cdots A_3^{k_3 - k_2} A_2^{k_2 - k_1} \sum_{l=k_0}^{k_1 - 1} A_1^{k_1 - 1 - l} Bu(l) + \ldots + A_M^{k_M - k_{M-1}} \sum_{l=k_{M-2}}^{k_{M-1} - 1} A_{M-1}^{k_{M-1} - 1 - l} Bu(l) +$$

$$\sum_{l=k_{M-1}}^{k_M - 1} A_M^{k_M - 1 - l} Bu(l).$$

Let $z(k_M) = 0$, we can get

$$T(\pi) = \{z | 0 = A_M^{k_M - k_{M-1}} A_{M-1}^{k_{M-1} - k_{M-2}} \cdots A_2^{k_2 - k_1} A_1^{k_1} z(0) + A_M^{k_M - k_{M-1}}$$

$$A_{M-1}^{k_{M-1} - k_{M-2}} \cdots A_3^{k_3 - k_2} A_2^{k_2 - k_1} \sum_{l=k_0}^{k_1 - 1} A_1^{k_1 - 1 - l} Bu(l) + \ldots + A_M^{k_M - k_{M-1}} \sum_{l=k_{M-2}}^{k_{M-1} - 1}$$

$$A_{M-1}^{k_{M-1} - 1 - l} Bu(l) + \sum_{l=k_{M-1}}^{k_M - 1} A_M^{k_M - 1 - l} Bu(l).$$

Simplify the above formula

$$z(0) = -\sum_{m=1}^{M} \prod_{j=1}^{m} A_j^{-h_j} \sum_{l=k_m-1}^{k_m - 1} A_m^{k_m - 1 - l} Bu(l). \tag{6}$$

Theorem 1. *If $A_m(m = 1, 2, \ldots, M)$ are all nonsingular matrix, the second-order discrete-time multi-agent system (3) is controllable if and only if there exists a switching sequence $\Pi = \{(i_M, j_M, h_M)\}_{m=1}^{M}$, such that $rank(Q_c) = [Q_{c,1}, A_2^{k_1 - k_2} Q_{c,2}, \ldots, A_2^{k_1 - k_2} \cdots A_M^{k_{M-1} - k_M} Q_{c,M}] = 2n$, where*

$$Q_{c,m} = \begin{cases} [A_m^{h_m - 1} B \ A_m^{h_m - 2} B \ldots A_m B \ B], & h_m \leq 2n, \\ [A_m^{2n - 1} B \ A_m^{2n - 2} B \ldots A_m B \ B], & h_m > 2n. \end{cases} \quad (m = 1, 2, 3, \ldots, M) \text{ is the}$$

controllability matrix of the mth subsystem $(L_{i_m}^x, L_{j_m}^v, \tilde{B})$.

Proof. According to the state of system (3), we can get

$$0 = A_M^{k_M - k_{M-1}} A_{M-1}^{k_{M-1} - k_{M-2}} \cdots A_2^{k_2 - k_1} A_1^{k_1} z(0) + A_M^{k_M - k_{M-1}} A_{M-1}^{k_{M-1} - k_{M-2}}$$

$$\cdots A_3^{k_3 - k_2} A_2^{k_2 - k_1} (A_1^{k_1 - 1} B, A_1^{k_1 - 2} B, \ldots, A_1 B, B) \begin{bmatrix} u(0) \\ u(1) \\ \vdots \\ u(k_1 - 1) \end{bmatrix} + \ldots +$$

$$A_M^{k_M - k_{M-1}} (A_{M-1}^{k_{M-1} - 1 - k_{M-2}} B, A_{M-1}^{k_{M-1} - 2 - k_{M-2}} B, \ldots, A_{M-1} B, B) \begin{bmatrix} u(k_{M-2}) \\ u(k_{M-2} + 1) \\ \vdots \\ u(k_{M-1} - 1) \end{bmatrix}$$

$$+ (A_M^{k_M - 1 - k_{M-1}} B, A_M^{k_M - 2 - k_{M-1}} B, \ldots, A_M B, B) \begin{bmatrix} u(k_{M-1}) \\ u(k_{M-1} + 1) \\ \vdots \\ u(k_M - 1) \end{bmatrix}.$$

In here, we let $u = [u(0), u(1), \cdots, u(k_1 - 1), u(k_1), \cdots, u(k_M - 1)]$. Then

$$z(0) = -A_1^{-k_1}[Q_{c,1}, A_2^{k_1 - k_2} Q_{c,2}, \ldots, A_2^{k_1 - k_2} \cdots A_M^{k_{M-1} - k_M} Q_{c,M}]u, \qquad (7)$$

where $\qquad Q_{c,m} \qquad = \qquad \begin{cases} [A_m^{h_m - 1} B \; A_m^{h_m - 2} B \ldots A_m B \; B], & h_m \le 2n, \\ [A_m^{2n - 1} B \; A_m^{2n - 2} B \ldots A_m B \; B], & h_m > 2n. \end{cases}$

$(m = 1, 2, 3, \ldots, M)$. Because of A_m are all nonsingular matrix, according to Definition 2, system (3) is controllable if $z(0) \ne 0$. This indicates, system (3) is controllable if and only if $rank(Q_c) = 2n$.

Corollary 1. *The second-order discrete-time multi-agent system (3) is controllable if system (3) satisfies one of the following conditions:*
(i) $rank(Q_{c,1}) = 2n$;
(ii) There exists no non-zero left eigenvalue α such that $\alpha^T A_i = \lambda \alpha^T$, $\alpha^T B = 0$, $i \in \{1, 2, \ldots, M\}$, where λ are all eigenvalues of the matrix A_i.

Proof. (i) If $rank(Q_{c,1}) = 2n$, than $rank(Q_c) = 2n$. According to Theorem 1, system (3) is controllable.

(ii) If there exists no non-zero left eigenvalue α such that $\alpha^T A_i = \lambda \alpha^T$, $\alpha^T B = 0$ $(i \in \{1, 2, \ldots, M\})$, we have $rank(Q_{c,1}) = 2n$. than $rank(Q_c) = 2n$, which proved that system (3) is controllable.

Theorem 2. *If the position and the velocity information interaction topologies are all fixed, system (3) is controllable if and only if $rank((\lambda - 1)^2 I_n + (\lambda - 1)L^v + L^x, B) = n$, where λ are the eigenvalues of A, L^x and L^v are the Laplacian matrices of the position and the velocity information interaction topologies, respectively.*

Proof. Suppose that the position and the velocity information interaction topologies are all fixed, i.e. $M = 1$ or $N = 1$. If $M = 1$, system (3) can written as

$$\begin{bmatrix} x(k+1) \\ v(k+1) \end{bmatrix} = \begin{bmatrix} I_n & I_n \\ -L^x & I_n - L^v \end{bmatrix} \begin{bmatrix} x(k) \\ v(k) \end{bmatrix} + \begin{bmatrix} 0 \\ \tilde{B} \end{bmatrix} u(k).$$

System (3) is controllable if and only if

$$rank \begin{bmatrix} \lambda I_n - I_n & -I_n & 0 \\ L^x & \lambda I_n - I_n + L^v & \tilde{B} \end{bmatrix} = 2n.$$

where λ are the eigenvalues of A. Then

$$rank \begin{bmatrix} \lambda I_n - I_n & -I_n & 0 \\ L^x & \lambda I_n - I_n + L^v & \tilde{B} \end{bmatrix}$$

$$= rank \begin{bmatrix} 0 & -I_n & 0 \\ (\lambda - 1)^2 I_n + (\lambda - 1)L^v + L^x & \lambda I_n - I_n + L^v & \tilde{B} \end{bmatrix}$$

$$= rank \begin{bmatrix} 0 & -I_n & 0 \\ (\lambda - 1)^2 I_n + (\lambda - 1)L^v + L^x & 0 & \tilde{B} \end{bmatrix}$$

$$= n + rank((\lambda - 1)^2 I_n + (\lambda - 1)L^v + L^x, \tilde{B}).$$

System (3) is controllable if and only if $rank((\lambda-1)^2 I_n + (\lambda-1)L^v + L^x, \tilde{B}) = n$. If $N = 1$, then $L^v = L^x$. The conclusion is similar to the above. Here we omitted.

4 Example and Simulation

This example is used for validate Theorem 1. For a topological graph with 3 agents, assumed that agent 1 is the single leader of system (3). The topologies are shown in Fig. 1.

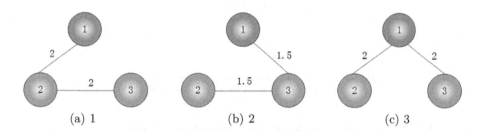

(a) 1   (b) 2   (c) 3

Fig. 1. The interaction topologies of system (3).

We can obtain

$$L_1 = \begin{bmatrix} 2 & -2 & 0 \\ -2 & 4 & -2 \\ 0 & -2 & 2 \end{bmatrix}, \quad L_2 = \begin{bmatrix} 1.5 & 0 & -1.5 \\ 0 & 1.5 & -1.5 \\ -1.5 & -1.5 & 3 \end{bmatrix},$$

$$L_3 = \begin{bmatrix} 4 & -2 & -2 \\ -2 & 2 & 0 \\ -2 & 0 & 2 \end{bmatrix}, \quad \tilde{B} = \begin{bmatrix} 1 \\ 0 \\ 0 \end{bmatrix}.$$

Design a switching sequence as $\Pi = \{(2,3,3),(2,1,4)\}$, then the position information interaction topology is fixed from $k_0 = 0$ to $k_f = 7$ and the velocity information interaction topology is switch L_3 to L_1. We get $Q_{c,1} = [B, A_1 B, A_1^2 B] =$

$$\begin{bmatrix} 0 & 1 & -2 \\ 0 & 0 & 3 \\ 0 & 0 & 3 \\ 1 & -3 & 28.5 \\ 0 & 3 & 3 \\ 0 & 3 & -10.5 \end{bmatrix}, Q_{c,2} = [B, A_2 B, A_2^2 B, A_2^3 B] = \begin{bmatrix} 0 & 1 & 0 & 12 \\ 0 & 0 & 3 & -6 \\ 0 & 0 & 1 & 6.5 \\ 1 & -1 & 12 & -35.5 \\ 0 & 3 & -9 & 84 \\ 0 & 1 & 5.5 & -21.5 \end{bmatrix} . \text{By comput-}$$

ing using Matlab, $rank(Q_c) = rank([Q_{c,1}, A_2^{-1} Q_{c,2}]) = 6$. Than the second-order discrete-time multi-agent system (3) is controllable. Figure 2 depicts the state trajectories of the three controllable agents by the switching sequence $\Pi = \{(2,3,3),(2,1,4)\}$. The initial states we randomly selected, denoted by circles and the final states denoted by asterisks are designed to form a line configuration.

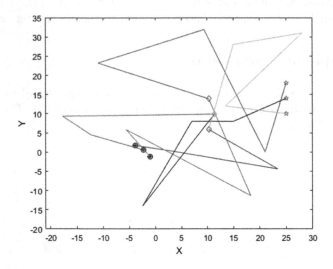

Fig. 2. A line configuration, where the circles and asterisks denote the initial state and the final desired configuration, respectively.

5   Conclusion

In this paper, we discussed the controllability of second-order discrete-time multi-agent systems with switching topology. We proved that the second-order discrete-time multi-agent systems is controllable if and only if there exists a switching sequence such that the controllability matrix full row rank. Also if both the position and the velocity information interaction topologies are fixed, we converted the switching topology into fixed topology. Compared with the switching topology, the criteria for multi-agent systems under fixed topology are simpler.

References

1. Reynolds, C.W.: Flocks, herds, and schools: a distributed behavioral model. Comput. Graph. **21**(4), 25–34 (1987)
2. Parrish, J.K., Viscido, S.V., Grunbaum, D.: Self-organized fish schools: an examination of emergent properties. Biol. Bull. **202**(3), 296–305 (2002)
3. Couzin, L.D., Krause, J., Franks, N.R., Levin, S.A.: Effective leadership and decision making in animal groups on the move. Nature **433**(7025), 513–516 (2005)
4. Olfati-Saber, R., Murray, R.M.: Consensus problems in networks of agents with switching topology and time-delays. IEEE Trans. Autom. Control **49**, 1520–1533 (2004)
5. Ren, W., Bresad, R., Atkins, E.: Information consensus in multivehicle cooperative control: collective group behavior through local interaction. IEEE Control Syst. Magzine **27**(2), 71–82 (2007)
6. Su, H., Wang, X., Lin, Z.: Flocking of multi-agents with a virtual leader. IEEE Trans. Autom. Control **54**(2), 293–307 (2009)
7. Liu, B., Chu, T., Wang, L., Zuo, Z., Chen, G., Su, H.: Controllability of switching networks of multi-agent systems. Int. J. Robust Nonlinear Control **22**, 630–644 (2012)
8. Ji, Z., Lin, H., Yu, H.: Protocols design and uncontrollable topologies construction for multi-agent networks. IEEE Trans. Autom. Control **60**(3), 781–786 (2015)
9. Liu, B., Han, Y., Jiang, F., Su, H., Zou, J.: Group controllability of discrete-time multi-agent systems. J. Franklin Inst. Eng. Appl. Math. **353**(14), 3524–3559 (2016)
10. Qu, J., Ji, Z., Lin, C., Yu, H.: Fast consensus seeking on networks with antagonistic interactions. Complexity **2018**, 1–15 (2018)
11. Tanner, H.G.: On the controllability of nearest neighbor interconnections. In: Proceedings of the 43rd IEEE Conference on Decision and Control, vol. 3, pp. 2467–2472 (2004)
12. Rahmani, A., Ji, M., Mesbahi, M.: Controllability of multi-agent systems from a graph-theoretic perspective. SIAM J. Control Optim. **48**(1), 162–186 (2009)
13. Guan, Y., Ji, Z., Zhang, L.: Controllability of multi-agent systems under directed topology. Sci. China Ser. F: Inf. Sci. **60**, 092203:1–092203:15 (2007)
14. Wang, L., Jiang, F., Xie, G., Ji, Z.: Controllability of multi-agent systems based on agreement protocols. Science China Ser. F: Inf. Sci. **52**(11), 2074–2088 (2009)
15. Guan, Y., Wang, L.: Structural controllability of multi-agent systems with absolute protocol under fixed and switching topologies. Sci. China Ser. F: Inf. Sci. **60**, 092203:1–092203:15 (2017). https://doi.org/10.1007/s11432-016-0498-8
16. Guan, Y., Wang, L.: Target controllability of multiagent systems under fixed and switching topologies. Int. J. Robust Nonlinear Control **29**(9), 1–17 (2019)
17. Czeizler, E., Wu, K.C., Gratie, C.: Structural target controllability of linear networks. IEEE/ACM Trans. Comput. Biol. Bioinf. **15**(4), 1 (2018)
18. Guan, Y., Ji, Z., Zhang, L.: Controllability of heterogeneous multi-agent systems under directed and weighted topology. Int. J. Control **89**(5), 1–25 (2015)
19. Liu, B., Su, H., Li, R., Sun, D., Hu, W.: Switching controllability of discrete-time multi-agent systems with multiple leaders and time-delays. Appl. Math. Comput. **228**, 571–588 (2014)

20. Liu, B., Feng, H., Wang, L.: Controllability of second-order multiagent systems with multiple leaders and general dynamics. Math. Prob. Eng. **2013**, 1–6 (2013)
21. Liu, B., Chu, T., Wang, L.: Controllability of a leader-follower dynamic network with switching topology. IEEE Trans. Autom. Control **53**(4), 1009–1013 (2008)
22. Zamani, M., Lin, H.: Structural controllability of multi-agent systems. In: 2009 American Control Conference, pp. 5743–5748 (2009)
23. Lu, Z., Zhang, L., Wang, L.: Observability of multi-agent systems with switching topology. IEEE Trans. Circ. Syst. II Express Briefs **64**(11), 1317–1321 (2017)
24. Tian, L., Guan, Y., Wang, L.: Controllability and observability of switched multi-agent systems. Int. J. Control **92**(8), 1–10 (2017)
25. Tian, L., Guan, Y., Wang, L.: Controllability and observability of multi-agent systems with heterogeneous and switching topologies. Int. J. Control **92**(80), 1–12 (2018)

DNA and Molecular Computing

Probe Machine Based Computing Model for Solving Satisfiability Problem

Jianzhong Cui[1,4], Zhixiang Yin[2(✉)], Jing Yang[3], Xianya Geng[3],
and Qiang Zhang[5]

[1] School of Electronic and Information Engineering,
Anhui University of Science and Technology, Huainan 232001, Anhui, China
[2] School of Mathematics, Physics and Statistics,
Shanghai University of Engineering Science,
Shanghai 201620, People's Republic of China
zxyin66@163.com
http://www.springer.com/lncs
[3] School of Mathematics and Big Data, Anhui University of Science and Technology,
Huainan 232001, Anhui, China
[4] Department of Computer, Huainan Union University,
Huainan 232038, Anhui, China
[5] School of Computer Science, Dalian University of Technology,
Dalian 116024, Liaoning, China

Abstract. Probe Machine (PM) is a recently reported theoretical model with massive parallelism. Particularly, Turing Machine (TM) had been proven to be the special case of PM. The paper proposed a PM based computing model for satisfiability problem. PM is capable of searching pairs of data fiber that are complementary to pre-devised probes, leading to the connection of pairs of data in parallel. We encoded the assignment to variables in the given 3-SAT formula into data fibers and devised probes between pairs of data fiber, the unique satisfying assignment of a hard 3-SAT formula could be generated within just one step of probe operation. More generally, for an arbitrary 3-SAT formula with variables and clauses, we presented a method for deciding the satisfiability using the concept of potential probe. Complexity analysis shows the encoding complexity and time complexity of the proposed model are $o(n)$ and $o(1)$, respectively. The distinguishing characteristics of the proposed model lie in two aspects. On one hand, solution to NP-complete problem was generated in just one step of probe operation rather than found in vast solution space. On the other hand, the proposed model is highly parallel. Most important of all, the parallelism increases with problem size. This marks a giant step in computational theory. With the parallel search capability inherited in PM, the size of NP-complete search problems is expected to be increased further.

Keywords: Probe Machine · NP-complete · Satisfiability · Search tree

© Springer Nature Singapore Pte Ltd. 2020
L. Pan et al. (Eds.): BIC-TA 2019, CCIS 1160, pp. 79–92, 2020.
https://doi.org/10.1007/978-981-15-3415-7_7

1   Introduction

In 1965, a formal definition of efficient computation (runs in time a fixed polynomial of the input size) was introduced [1]. The class of problems with efficient
solutions belongs to P (polynomial time). Apart of these, there is also a sizable
class of very applicable and significant problems that do not seem to have efficient algorithm for solving them. Such problems, for example, stable matching
problem [2,3], vertex coloring problem [4,5], analysis of rationality of consumption behavior [6–8], 0–1 programming problem [9–11], and so on, belong to the
class of NP-complete (nondeterministic polynomial time). The typical characteristic of these problems is: given a candidate solution it is quite easy to verify
efficiently for correctness, yet finding such a solution appears extremely difficult. The difficulty probably lies in two aspects. On one hand, solution space of
these problems is not well structured. With well structured solution space, for
example, sorting problem with elements can be efficiently solved. On the other
hand, the solution (solutions) of these problems is (are) rare, in contrast to non-
solutions. Almost all candidates are nonsolutions in the entire search space, and
consequently it is difficult to be found. That is to say, the distribution of the solution in the search space is extremely sparse. The extent of sparseness increases
exponentially with problem size. The searching occurs frequently in human brain
as well. The target information to be searched may be a tiny fraction of vast
information memorized, say, somebody's name, yet brain is capable of searching
them efficiently. Brain is the most complex organ in human body, which serves
as the center of nervous system. Cerebral cortex, the outside of the brain, contains approximately 15–33 billion neurons. A typical neuron consists of a cell
body, dendrites, and an axon. Each neuron can connect by means of synapses to
other neurons to form neural network or neural pathway. It is widely accepted
that the synapse plays an important role in the formation of memory. As neu-
rotransmitters activate receptor neuron across the synaptic cleft, the connection
between the two neurons is strengthened. The strength of two connected neural
pathways is thought to result in the storage of information, resulting in memory,
and resulting in searching the information memorized efficiently. This process
of synaptic strengthening is known as long-term potentiation. Inspired by this
similarity, probe machine (PM) [12] was introduced in 2016. PM is defined as
the following nine-tuple:

$$PM = (X, Y, \sigma_1, \sigma_2, \tau, \lambda, \eta, Q, C),$$

where each element denotes data library (X), probe library (Y), data controller
(σ_1), probe controller (σ_2), probe operation (τ), computing platform (λ), detector (η), true solution storage (Q), residue collector (C). Data library (X) is
the union of data sublibraries $X_i(i = 1, 2, \cdots, n)$. Each data sublibrary contains a unique type of data x_i. Data x_i consists of distinct types data fiber
$x_i^j(j = 1, 2, \cdots, p_i)$ and a data body, to which data fibers are attached. The
structure of data in data sublibrary originated and abstracted from the structure of neuron. Data fibers are an abstraction of dendrites, the data body is an
abstraction of the cell body, whereas an axon is omitted.

Probe library (Y) is the union of probe sublibraries $Y_{it}^{ab}, (a = 1, 2, \cdots, p_i, b = 1, 2, \cdots, p_t)$. Each probe sublibrary contains a unique type of probe $\overrightarrow{x_i^a x_t^b}$ (transitive) or $\overline{x_i^a x_t^b}$(connective), according to whether $Y_{it}^{ab} \neq Y_{ti}^{ba}$ or not. Probe $\overrightarrow{x_i^a x_t^b}$ $(\overline{x_i^a x_t^b})$ consists of two parts: the complement $\overrightarrow{x_i^a}(\overline{x_i^a})$ of data fiber x_i^a and the complement $\overrightarrow{x_t^b}(\overline{x_t^b})$ of data fiber x_t^b (the arrow or bar on the top represents complement.) The complements like this can be found elsewhere, for example, Waston-Crick complement in single-stranded DNA oligonucleotides. Probe serves as a media by means of which the target data fiber pair (x_i^a, x_t^b) can be connected together through complements. Since each data fiber of the pair is attached to an individual data body, the connection of data fiber pair (x_i^a, x_t^b) results in the connection of data pair (x_i, x_t). In such case, we call that data x_i and data x_t are potential probe. The probe is an abstraction of synapse which connects neuron pair to form neural network or neural pathway.

Data controller (σ_1) and probe controller (σ_2) are used to take data and probes from data library and probe library.

Probe operation (τ) is the process in the computing platform (λ), which distinct types of probe search target data fiber pairs, found, and connect them together.

Let $X' \subset X$ and $Y' \subset Y$, probe operation can be described as $\tau(X', Y')$.

Computing platform (λ) is an abstract of the environment, in which probes can smoothly find target data fiber pairs.

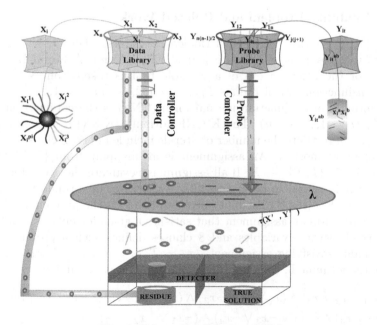

Fig. 1. Schematic diagram of PM.

Detector (η) determines whether generated products resulting from probe operation are solutions to the problem or not. The products with n data bodies are called n-aggregations. Particularly, a data X_i can be seen as 1-aggregation since it has a data body. Solutions are stored in true solution storage (Q), non-solutions are decomposed and stored in residue collector (C). Figure 1 below presented the schematic diagram of PM. For terminologies and notations not included in this paper, as well as the proof that TM is the special case of PM, readers may refer to [12] for detailed description.

In this paper, a computing model based on PM for satisfiability problem was proposed. Though well-researched and widely investigated, the problem remains the focus of continuing interest, wherever in theoretical or practical context. We encoded the assignment to variables in the given 3-SAT formula into data fibers, devised probes between pairs of data fiber, and the unique satisfying assignment of a hard 3-SAT formula was generated within just one step of probe operation. More generally, for an arbitrary 3-SAT formula with n variables, we presented a theorem for deciding the satisfiability of it using the concept of potential probe.

The rest of this paper was organized as follows. Section 2 presented the definition of satisfiability problem, related work, and the proposed model. Followed by, the encoding and time complexity analysis was presented. Conclusion and future work was presented in Sect. 4.

2   Materials and Methods

2.1   Satisfiability Problem and Related Work

Let $V = \{x_1, x_2, \cdots, x_n\}$ be a set of Boolean variables, where $x_i = F(x_i = T)$ represents variable x_i is assigned to logical False (True). A literal L is either a variable or the negation ($\neg$) of a variable that is taken from V. A clause C_i is the disjunction ($\vee$) of literals $L_{i,j}, (i = 1, 2, \cdots, m, j = 1, 2, \cdots, k)$. The Boolean formula in conjunction normal form (CNF) is the conjunction ($\wedge$) of clauses $C_i, (i = 1, 2, \cdots, m)$. The K-CNF formula (K-SAT), $F = \wedge_{i=1}^m C_i = \wedge_{i=1}^m (\vee_{j=1}^k L_{i,j})$, restricts the number of literals included in each clause of a CNF formula to be at most K. An assignment is an mapping from $\{F, T\}$ to each variable in V, $f : \{F, T\} \longrightarrow V$. If all assignments evaluate the given formula F, then the formula is said to be unsatisfiable, otherwise the formula is said to be satisfiable. The Boolean Satisfiability problem (SAT) is the problem of deciding whether there exists an assignment that satisfies a given formula. The following 3-SAT formula with 4 variables and 8 clauses defines such a problem. There exists a unique satisfying assignment, $(x_1 = T, x_2 = F, x_3 = T, x_4 = F)$ that evaluates the formula T, therefore, the given formula is satisfiable.

$$F = (\neg x_1 \vee x_2 \vee x_3) \wedge (x_1 \vee x_2 \vee \neg x_3) \wedge (x_1 \vee \neg x_2 \vee x_3) \wedge (x_1 \vee \neg x_2 \vee \neg x_3) \wedge$$
$$(x_1 \vee x_2 \vee x_3) \wedge (\neg x_1 \vee \neg x_3 \vee \neg x_4) \wedge (\neg x_1 \vee \neg x_2 \vee x_3) \wedge (\neg x_1 \vee \neg x_2 \vee \neg x_3)$$

SAT plays an important role in complexity theory and has extensive applications in industrial and other real-world problems, such as, model checking in formal verification and reasoning in artificial intelligence. In complexity theory, SAT problem belongs to NP-complete. The property of NP-completeness ensures that once an efficient algorithm for SAT is found, one would immediately have an efficient algorithm for all NP-complete problems, therefore $P = NP$. Whether $P = NP$ or not is the fundamental question that remains unsolved in complexity theory. The question was named as one of the seven Millennium Prize Problems by the Clay Math Institute in 2000. In electronic design automation context, SAT solver is employed as an indispensable model checking tool to guarantee the correctness of hardware design. Efficient SAT algorithm is the basis of automated verification techniques, such as bounded model checking [13], interpolation-based model checking [14], and IC3 [15]. In artificial intelligence context, SAT has received special concerns probably due to its direct relationship to deductive reasoning, a kind of top-down reasoning which reasons from one or more statements to reach a conclusion. For example, given a collection of facts $\sum$, a sentence α can be deduced iff $\sum \bigcup \{\neg \alpha\}$ is unsatisfiable.

The past half century has witnessed unceasing advances in the performance of algorithms for SAT, thanks to contributions made from different domains. By introducing novel branching heuristics, nonchronologically backtracking, restart heuristics, and so on, recent SAT algorithms are capable of handling instances of more and more variables and clauses with acceptable computational time. Here, in this paper, we focus on complete algorithms for SAT. The reason is that, on one hand complete algorithm can certainly find a solution iff the given formula is satisfiable. More importantly, it can provide the proof that the given formula is unsatisfiable on the other hand. These algorithms can be classified into two categories: bio-molecular based and Turing machine (TM) based, depending on the material or the tool they employed.

Bio-molecular based complete algorithms originated from [16]. In the paper, an instance of the directed Hamiltonian path problem was solved experimentally by manipulating on DNA molecules. Followed by the seminal paper, various algorithms employed biological molecules or techniques for SAT problem were successively proposed. In 2002, a 20 variables 3-SAT problem was reported to be solved [17]. The unique answer was found after an exhaustive search of 2^{20} possibilities. In 2003, three dimensional graph structure and DNA self-assembly [18] was proposed to solve 3-SAT problem in a constant number of laboratory steps. Generally speaking, early bio-molecular based complete algorithms required preconstructing an initial data pool which consisted of molecules encoding each possibilities of the assignment to variables. The data pool corresponded to the entire search space or complete search tree. The tree was subsequently pruned by either employing biological technologies to digest those molecules encoding unsatisfying assignments or separating molecules encoding unsatisfying assignments from satisfying ones. Therefore, the space complexity (the number of distinct bio-molecules) increased exponentially with problem size. It is commonly recognized that, adopting brute-force strategy, the upper bound for the number of variables that bio-molecule based algorithms can handle was 60 to 70. In 2000, a breadth-first search algorithm [19] was presented that could theoretically extend this bound to 120 variables. Followed by the work, a modified sticker model [20] and a DNA algorithm based on ligase chain reaction (LCR) [21] were presented to solve SAT problem. Their methods were similar in spirit, yet differed in biological technologies employed. Both of their methods started with an empty initial data pool, DNA molecules encoding satisfying assignments to variables were gradually generated, whereas unsatisfying ones were pruned by separation or LCR, clause by clause. That is to say, the search tree is not pre-constructed but partially constructed. By pruning unsatisfying assignments clause by clause, the size of the search tree is decreased. As a result, the bound was extended further. In 2016, P systems with proteins on membranes [22] were employed to solve SAT problem.

TM based complete algorithms may trace their history to 1960, the well-known DPLL algorithm [23], which has profound impact on modern SAT algorithms. The algorithm constructed a search tree by choosing a variable to make a decision. Each chosen variable, the decision variable, was a node of the tree. The order of decision variables corresponded to the depth of the tree. At each node, decision was made by assigning False (True) to current decision variable. Repeatedly applied unit propagation rule, monotone literal rule, and clause subsumption (Boolean Constraint Propagation), to make the tree grow top-down as deep as possible until a conflict resulting from assigning contrary values to decision variables took place. Then the algorithm backtracked to an upper level of the tree and re-assigned True (False) to the decision variable, attempting to correct the former decision leading to the conflict. If backtracking to the top of the tree, the conflict would inevitably occur, and the algorithm returned unsatisfiable. Since the algorithm had explored the entire tree, no satisfying assignment could be found. It is obvious that the efficiency of the algorithm is premised on both the size of the search tree and the number of backtracking. The later variations of DPLL algorithms [24,25] introduced conflict driven clause learning, nonchronologically backtracking, restart heuristics, and so on, in order to prune the tree as large portions of the tree as possible. In 2001, a new branching heuristics, backbone variables [26] was introduced. Utilizing this branching heuristics, they proved unsatisfiability of a hard random 3-SAT instance of 700 variables in around 25 days. This paper covered only parts of TM based complete algorithms, for including so many brilliant works may need a separate paper. For applications of SAT problem in model checking, reader may refer to [27] for a comprehensive overview. Recently, there are important contributions made from hardware accelerated SAT solvers [28,29], reads may refer to [30] for the trends, the challenges, and open questions. From the tool employed perspective, these algorithms are based on TM, hence are equivalent to one another, but differ in efficiency.

Recent years, progress in the improvement of TM based complete algorithms is gradually slow down. As described in [31], the gap between our ability to handle satisfiable and unsatisfiable instances has actually grown. This is probably due to that we are approaching to the minimal search tree in terms of the number of backtracking. The question is: Whether can we go beyond this bound?

2.2 PM-Based Computing Model

In this subsection, we presented PM-based computing model for SAT problem, taking an instance of hard 3-SAT problem presented in the preceding subsection 2.1 as an example. The satisfiability of random generated K-SAT problem around some critical point exhibits easy-hard-easy pattern, the so-called phase transition [31]. The critical point is formulated as the ratio of clauses to variables in the given formula, $c = m/n$. For random 3-SAT problem, the ratio is 4.25. The hardness means, around this ratio, the given formula has often a unique satisfying assignment out of all possibilities, which satisfying assignment is hard to be found. We started with the construction of data library X.

For the formula, each assignment is represented by a 4-digit binary number. A bit equals to 1 represents the corresponding Boolean variables in the formula is assigned to T. A bit equals to 0 represents the variables is assigned to F. From left to right, the ith bit of the binary number corresponds to Boolean variable x_i or its negation $\neg x_i$. For example, the satisfying assignment $(x_1 = T, x_2 = F, x_3 = T, x_4 = F)$ is represented by the binary number 1010.

Since the binary number has 4 bits and each bit can be either 0 or 1, we need 8 distinct types of data, $\{x_1, \neg x_2\}, (i = 1, 2, 3, 4)$ to represent the bits and their assigned values. Additionally, we need 2 distinct types of data x_0 and x_5 to represent the root and the leaf of the search tree, respectively. The 10 distinct types of data have the identical types of data body, but differ in the types of data fiber x_j^k. For data representing the root and the leaf, we need only 2 distinct types of data fibers, x_0^0 and x_5^5, to encode the bits, since the root and the leaf need not to be assigned a value. For data $x_i (i = 1, 2, 3, 4)$, we need 2 distinct types of data fibers, one type of which encodes the bit, the other encodes the assigned value. We called the two types of data fiber the bit encoding and the value encoding, for convenience.

For data $\neg x_i (i = 1, 2, 3, 4)$, the bit encoding is identical to the bit encoding of data x_i, due to they share the same bit. Thus, for data pair $(x_i, \neg x_i)$, we need 3 types of data fiber: one type of bit encoding, and two types of value encoding. The total number of types of distinct data fibers for the data pair $(x_i, \neg x_i)$ is 12. Therefore, the total number of types of distinct data fibers including bit encodings for the root and the leaf is 14. Next, we attached the data fibers to data bodies to fabricate 10 types of data. Figure 2(a),(b) below presented schematic diagram of the data pair $(x_i, \neg x_i)$, the data representing the root x_0(Fig. 2(c)), and the data representing the leaf x_5 (Fig. 2(d)). The solid circle positioned centrally is the identical data body. Curves connected to data bodies are data fibers encoding the bit or the assigned value.

For each type of data, we established a data sublibrary $x_0, x_i, \neg x_i, x_5 (i = 1, 2, 3, 4)$, which contained vast copies of data $x_0, x_i, \neg x_i, x_5 (i = 1, 2, 3, 4)$. Each sublibrary was associated with a controllable data controller $\sigma_{1,i}, (i = 1, \cdots, 10)$, in order to take data out from the corresponding data sublibrary. Finally, the data library X was constructed as follows:

$$X = x_0 \bigcup x_1 \bigcup \neg x_1 \bigcup x_2 \bigcup \neg x_2 \bigcup x_3 \bigcup \neg x_3 \bigcup x_4 \bigcup \neg x_4 \bigcup x_5$$

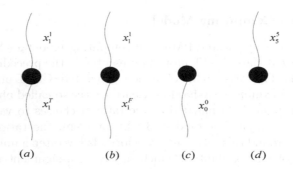

Fig. 2. Schematic diagram of data $x_1, \neg x_1, x_0, x_5$. (a) data x_1. (b) data $\neg x_1$. (c) data x_0. (d) data x_5.

We presented next, the construction of probe library Y. In order to figure out the type of probes, so that the satisfying assignment can be found, we presented the search tree illustrated in the following Fig. 3.

The leaf at the bottom 1010 italicized is the unique satisfying assignment.

For the first clause in the formula, $C_1 = (\neg x_1 \lor x_2 \lor x_3)$, the unsatisfying partial assignments to x_1, x_2, x_3 is $\{x_1 = 1, x_2 = 0, x_3 = 0\}$. Since the partial assignment evaluates what is clause C_1 F , hence evaluates the formula F. Therefore, the directed path passing through these partial assigned variables from the root to leafs, 1000 and 1001, must be unsatisfying assignments to the formula. That is to say, leafs 1000 and 1001 should be pruned from the search tree according to clause C_1. We marked these two paths with C_1 under leafs 1000 and 1001.

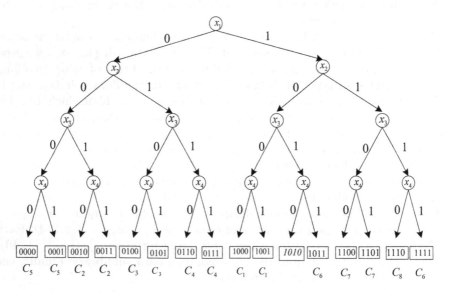

Fig. 3. The search tree of the given formula

For the second clause in the formula, $C_2 = (x_1 \vee x_2 \vee \neg x_3)$, the unsatisfying partial assignments to x_1, x_2, x_3 is $\{x_1 = 0, x_2 = 0, x_3 = 1\}$. The directed path passing through these partial assigned variables from the root to leafs, 0010 and 0011, must be unsatisfying assignments to the formula. Leafs 0010 and 0011 should be pruned from the search tree according to clause C_2. We marked these two paths with C_2 under leafs 0010 and 0011.

Similarly, leafs 0000, 0001, 0100, 0101, 0110, 0111,1011, 1100, 1101, 1110 and 1111 were pruned from the search tree clause by clause. It should be noted that the above pruning process can be executed concurrently. Ultimately, the unique satisfying assignment 1010 was thus obtained. The satisfying assignment is $(x_1 = T, x_2 = F, x_3 = T, x_4 = F)$. Therefore, based on the constructed data library, the types of probes were determined. There exists probe $\overrightarrow{x_0^0 x_1^1}$ between the bit encoding fiber x_0^0 of the data x_0 and the bit encoding fiber x_1^1 of the data x_1; probe $\overrightarrow{x_1^T x_2^2}$ between the value encoding fiber x_1^T of the data x_1 and the bit encoding fiber x_2^2 of the data x_2; probe $\overrightarrow{x_2^F x_3^3}$ between the value encoding fiber x_2^F of the data x_2 and the bit encoding fiber x_3^3 of the data x_3; probe $\overrightarrow{x_3^T x_4^4}$ between the value encoding fiber x_3^T of the data x_3 and the bit encoding fiber x_4^4 of the data x_4; probe $\overrightarrow{x_4^F x_5^5}$ between the value encoding fiber x_4^F of the data x_4 and the bit encoding fiber x_5^5 of the data x_5.

For each type of probes we established a probe sublibrary $Y_{i(i+1)}^{ab}$, which contained vast copies of probes $\overrightarrow{x_i^a x_{i+1}^b}$ for i=0,1,2,3,4, $ab \in \{01, T_2, F_3, T_4, F_5\}$. The established probe sublibraries were presented as follows:

$$Y_{01}^{01} = \{\overrightarrow{x_0^0 x_1^1}\}, Y_{12}^{T2} = \{\overrightarrow{x_1^T x_2^2}\}, Y_{23}^{F3} = \{\overrightarrow{x_2^F x_3^3}\}, Y_{34}^{T4} = \{\overrightarrow{x_3^T x_4^4}\}, Y_{45}^{F5} = \{\overrightarrow{x_4^F x_5^5}\}$$

The total number of distinct types of probe is 5. Each probe sublibrary was associated with a controllable probe controller $\sigma_{2,i}, (i = 1, \cdots, 5)$, in order to take probes out from corresponding probe sublibrary. Finally, the data library Y was constructed as follows:

$$Y = Y_{01}^{01} \bigcup Y_{12}^{T2} \bigcup Y_{23}^{F3} \bigcup Y_{34}^{T4} \bigcup Y_{45}^{F5}$$

We now present the following principles of devising probes for the construction of probe library.

Principe1: For each data x_i, there does not exist probe between data fiber pair belonging to the data.

Principe2: Probe possibly exists between data fiber belonging to data x_i and data fiber belonging to $x_{i+1}, (i = 0, 1, 2, 3, 4)$.

Data controllers $\sigma_{1,i}, (i = 1, \cdots, 10)$, took an appropriate amount of data $x_0, x_1, \neg x_1, x_2, \neg x_2, x_3, \neg x_3, x_4, \neg x_4, x_5$ from data sublibraries, $x_0, x_1, \neg x_1, x_2, \neg x_2, x_3, \neg x_3, x_4, \neg x_4, x_5$ placed them into the computing platform λ. Meanwhile, probe controllers $\sigma_{2,i}, (i = 1, \cdots, 5)$, took an appropriate amount of probes $\overrightarrow{x_0^0 x_1^1}, \overrightarrow{x_1^T x_2^2}, \overrightarrow{x_2^F x_3^3}, \overrightarrow{x_3^T x_4^4}, \overrightarrow{x_4^F x_5^5}$, from probe sublibraries $Y_{01}^{01}, Y_{12}^{T2}, Y_{23}^{F3}, Y_{34}^{T4}, Y_{45}^{F5}$ placed them into the computing platform λ, to perform probe operation τ. With the support of the computing platform, probe $\overrightarrow{x_0^0 x_1^1}$ searched target data fiber x_0^0 and data fiber x_1^1, found, and absorbed them, leading to the connection of data x_0 and x_1. Similarly, all probes found and absorbed their target data fiber pairs when probe operation terminated. All probe operations were performed parallel in the platform. We refered the above process as one step of probe operation. Obviously, the more types of probe and data involved in the probe operation, the higher parallelism the model has. That is to say, the parallelism of the model increases with the size of problem to be tackled. Eventually, the satisfying assignment to the formula is thus generated, as illustrated in Fig. 4

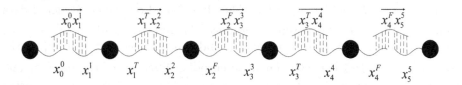

Fig. 4. Schematic diagram of result of probe operation

The 6-aggregations were separated and placed into true solution storage Q by the detector η. Others were placed into the residue collector C.

Let us now push the problem a step further. Instead of asking whether the given 3-SAT formula is satisfiable or not, we ask: What is the satisfying assignment?

In order to answer this question, we need additional 2 types of data fiber. One type of data fiber encodes the value T, the other encodes the value F. Attach the data fiber encoding T to data x_i, the data fiber encoding F to data $\neg x_i$. Do not devise probe between the 2 types of data fiber and other data fibers. Thus, when probe operation terminates, the 2 types of data fibers are left intact. The detector η detects the type of data fiber, which reports the assigned value to each variable in the formula.

More generally, for an arbitrary 3-SAT formula with n variables and m clauses, we proposed the following theorem deciding the satisfiability of it using the concept of probe potential. If there exists probe $\overrightarrow{x_i^a x_t^b}$ between data fiber x_i^a and data fiber x_t^b, then the data fiber pair (x_i^a, x_t^b) is called potential probe. Similarly, the data pair (x_1, x_t) is called potential probe.

Theorem
Begin Theorem
The given formula is satisfiable iff data pairs (x_i, x_{i+1}) are pairwise potential probe, for $i = 0, 1, \cdots, n$; or else, the given formula is unsatisfiable.
End Theorem
Proof. If the given formula is satisfiable, there must exists at least an satisfying assignment that evaluates the formula T. This means, there exists at least one directed path passing through each assigned variable from the root to the leaf in the search tree that must survive the pruning process. According to the structure of data and principles of devising probe, there must exist at least n+1

distinct types of probe between data fiber pairs (x_i, x_{i+1}) for $i = 0, 1, \cdots, n$, therefore, data pairs (x_i, x_{i+1}) are pairwise potential probe. If data pair (x_i, x_{i+1}) are pairwise potential probe, for $i = 0, 1, \cdots, n$, n distinct types value encoding data fiber form at least an assignment that must satisfy the formula, therefore, the given formula is satisfiable.

If the given formula is unsatisfiable, all directed paths passing through each assigned variable from the root to the leaf must be pruned in the search tree. There must not exist probes between any data fiber pairs. Therefore, data pairs are not pairwise potential probe, and the vice versa.

Based on the theorem, we can concluded that the satisfiability problem could be decided and solved by the proposed model.

3 Results and Discussion

For a given n variables and m clauses 3-SAT formula, the proposed computing model for satisfiability problem need encoding 3n+2 distinct types of data fiber, 1 type of data body, and fabricating 2n+2 distinct types of data. Hence, the encoding complexity for data is $o(n)$. As for the types of probe, in case of the given formula is satisfiable, since each variable has only 2 types of value encoding data fibers, there are at most 2 types of probe between data pair $(x_i, x_{i+1}), i = 0, 1, \cdots, n$. In addition, there is 1 type of probe between data pair (x_0, x_1). The total number of distinct types of probe is at most 2n+1. In case of the given formula is unsatisfiable, the number of distinct types of probe is zero, because there does not exist a satisfying assignment. Hence, the encoding complexity for probes is $o(n)$. In short, encoding complexity is $o(n)$. In the computing platform, the probe operation is performed in massive parallel manner. The satisfying assignment to variables in the given formula is generated within just one step of probe operation. Hence, time complexity is $o(1)$.

Compared with bio-molecular based complete algorithms, for an arbitrary unsatisfiable 3-SAT formula, the proposed PM based model accomplished to decide the unsatisfiability of the given formula in the step of devising probe (all leaves are pruned from the search tree, data pairs $(x_i, x_i + 1)$ are not pairwise potential probe.) For an arbitrary satisfiable 3-SAT formula, each specific type of probe devised serves to parallel generate the satisfying assignment (assignments) in the computing platform when performing probe operation, hence the cost of the searching that should be paid is the minimum. Besides, all satisfying assignments can be generated within just one step of probe operation. As the size of problem increases, the type of probe increases accordingly. The searching of satisfying assignments is more efficient. Therefore, the searching capacity of PM increases with problem size.

TM based complete algorithms are in essence serial, whereas the proposed computing model is highly parallel and the parallelism increases with problem size. With the parallel search capability inherited in PM, can the size of NP-complete search problems be increased further?

We believe the answer to this question is premised on the realization technology of PM. As we all know, Neumann employed semiconductor as component to realize Turing machine, giving birth to the modern general-purpose computer. In [12], employing nano-particle and single-stranded DNA oligonucleotide, nano-DNA, was proposed to realize PM. In this context, we discuss the upper bound for the number of variables the proposed model can handle for SAT problem.

Let I be the length of all probes. How many distinct types of probe are at most allowed to ensure the reliable generation of the satisfying assignments within one step of probe operation? In case of $I = 23$, if template DNA sequence design method and error-correcting Golay code is utilized [33], the distinct types of probe is $2^{12} = 4096$. These probes satisfy the following constraints: the nearly uniform G/C content; hamming distance of self-complement and overlap portion is at least I/3. Therefore, for $I = 23$, the bound is around 4097, although the actual number may be less than it. To pinpoint the exact number, further DNA sequence design methods and experiments are required. We are optimistic that, at certain length I, the number is not supposed to be small.

4   Conclusions

In this paper, a PM based computing model for satisfiability problem is proposed. We encoded the assignment to variables in the given 3-SAT formula into data fibers and devised probes between pairs of data fiber, and the unique satisfying assignment of a hard 3-SAT formula was generated within just one step of probe operation. More generally, for an arbitrary 3-SAT formula with n variables and m clauses, we presented a method for deciding the satisfiability of it using the concept of potential probe. Complexity analysis showed the encoding complexity and time complexity of the proposed model was $o(n)$ and $o(1)$, respectively.

The proposed PM based computing model is dramatically different from bio-molecular based or TM based algorithm. The distinguishing characteristics of the proposed model lie in two aspects. On one hand, solution to NP-complete problem was generated in just one step of probe operation rather than found in vast solution space. On the other hand, the proposed model is highly parallel. Most important of all, the parallelism increases with problem size. Theoretically, one step of probe operation can search 2^n possibilities. This search capability of PM is apparently superior to TM, which we think it marks a giant step in computational theory. With the parallel search capability inherited in PM, the size of NP-complete search problems is expected to be increased further.

Although PM has currently some obstacles to be overcome in realization technology, for example, the detection of solution, error-prone nature of biological experiments, etc. We believe that the advent of PM is an important contribution to computational theory. The research of PM will provide brand-new insight into the concept of NP-completeness from a quite different perspective, and lead to numerous research products. May PM become into reality soon.

Acknowledgments. The authors would like to thank every author appeared in the references. This work was supported by National Natural Science Foundation of China [61672001, 61702008], Natural Science Foundation of Anhui University [KJ2019A0538], and [18-163-ZT-005-009-01].

References

1. Edmonds, J.: Paths, trees, and flowers. Can. J. Math. **17**(3), 449–467 (1965)
2. Sobeyko, O., Monch, L.: Heuristic approaches for scheduling jobs in large-scale flexible job shops. Comput. Oper. Res. **68**, 97–109 (2016)
3. Cechlrov, K., Fleiner, T., Manlove, D.F., McBride, I.: Stable matchings of teachers to schools. Theoret. Comput. Sci. **653**, 15–25 (2016)
4. Lozin, V.V., Malyshev, D.S.: Vertex coloring of graphs with few obstructions. Discrete Appl. Math. **216**, 273–280 (2015)
5. Malyshev, D.S., Lobanova, O.O.: Two complexity results for the vertex coloring problem. Discrete Appl. Math. **219**, 158–166 (2016)
6. Li, H., Bai, Y., He, W., Sun, Q.: Vertex-distinguishing proper arc colorings of digraphs. Discrete Appl. Math. **209**, 276–286 (2016)
7. Karpiski, M.: Vertex 2-coloring without monochromatic cycles of fixed size is NP-complete. Theor. Comput. Sci. **659**, 88–94 (2017)
8. Shitov, Y.: A tractable NP-completeness proof for the two-coloring without monochromatic cycles of fixed length. Theor. Comput. Sci. **674**, 116–118 (2017)
9. Borrero, J.S., Gillen, C., Prokopyev, O.A.: A simple technique to improve linearized reformulations of fractional (hyperbolic) 0–1 programming problems. Oper. Res. Lett. **44**(4), 479–486 (2016)
10. Kodama, A., Nishi, T.: Petri net representation and reachability analysis of 0–1 integer linear programming problems. Inf. Sci. **400**, 157–172 (2017)
11. Yan, K., Ryoo, H.S.: 0–1 multilinear programming as a unifying theory for LAD pattern generation. Discrete Appl. Math. **218**, 21–39 (2017)
12. Clarke, E., Kroening, D., Ouaknine, J., Strichman, O.: Completeness and complexity of bounded model checking. In: Steffen, B., Levi, G. (eds.) VMCAI 2004. LNCS, vol. 2937, pp. 85–96. Springer, Heidelberg (2004). https://doi.org/10.1007/978-3-540-24622-0_9
13. Jhala, R., McMillan, K.L.: Interpolant-based transition relation approximation. In: Etessami, K., Rajamani, S.K. (eds.) CAV 2005. LNCS, vol. 3576, pp. 39–51. Springer, Heidelberg (2005). https://doi.org/10.1007/11513988_6
14. Bradley, A.R.: SAT-based model checking without unrolling. In: Jhala, R., Schmidt, D. (eds.) VMCAI 2011. LNCS, vol. 6538, pp. 70–87. Springer, Heidelberg (2011). https://doi.org/10.1007/978-3-642-18275-4_7
15. Adleman, L.M.: Molecular computation of solutions to combinatorial problems. Science **266**, 1021–1024 (1994)
16. Braich, R.S., Chelyapov, N., Johnson, C.P., Rothemund, W.K., Adleman, L.M.: Solution of a 20-variable 3-SAT problem on a DNA computer. Science **296**, 499–502 (2002)
17. Jonoska, N., Sa-Ardyen, P., Seeman, N.C.: Computation by self-assembly of DNA graphs. Genet. Program Evolvable Mach. **4**(2), 123–137 (2003)
18. Yoshida, H.: Solution to 3-SAT by breadth first search. DIMACS Ser. Discrete Math. Theor. Comput. Sci. **54**, 9–20 (2000)
19. Yang, C.N., Yang, C.B.: A DNA solution of SAT problem by a modified sticker model. BioSystems **81**(1), 1–9 (2005)
20. Wang, X.L., Bao, Z.M., Hu, J.J., Wang, S., Zhan, A.B.: Solving the SAT problem using a DNA computing algorithm based on ligase chain reaction. BioSystems **91**(1), 117–125 (2008)
21. Song, B.S., Prez-Jimnez, M.J., Pan, L.Q.: An efficient time-free solution to SAT problem by P systems with proteins on membranes. J. Comput. Syst. Sci. **82**(6), 1090–1099 (2016)

22. Marques-Silva, J.P., Sakallah, K.: GRASP: a search algorithm for propositional satisfiability. IEEE Trans. Comput. **48**(5), 506C521 (1999)
23. Sinz, C., Iser, M.: Problem-sensitive restart heuristics for the DPLL procedure. In: Kullmann, O. (ed.) SAT 2009. LNCS, vol. 5584, pp. 356–362. Springer, Heidelberg (2009). https://doi.org/10.1007/978-3-642-02777-2_33
24. Dubois, O., Dequen, G.: A backbone-search heuristic for efficient solving of hard 3-SAT formulae. In: Proceedings of IJCAI, pp. 248–253 (2001)
25. Vizel, Y., Weissenbacher, G., Malik, S.: Boolean satisfiability solvers and their applications in model checking. Proc. IEEE **103**(11), 2021–2035 (2015)
26. Dal Palu, A., Dovier, A., Formisano, A., Pontelli, E.: CUD@SAT: SAT solving on GPUS. J. Exp. Theor. Artif. Intell. **27**(3), 293–316 (2015)
27. Lamya, A,G., Aziza, I.H., Hanafy, M.A.: Parallelization of unit propagation algorithm for SAT-based ATPG of digital circuits. In: Proceedings of ICM, pp. 184–188 (2016)
28. Sohanghpurwala, A.A., Hassan, M.W., Athanas, P.: Hardware accelerated SAT solvers—a survey. J. Parallel Distrib. Comput. **106**, 170–184 (2017)
29. Kautz, H., Selman, B.: The state of SAT. Discrete Appl. Math. **155**(12), 1514–1524 (2007)
30. Monasson, R., Zecchina, R., Kirkpatrick, S., Selman, B., Troyansky, L.: Determining computational complexity from characteristic 'phase transitions'. Nature **400**, 133–137 (1999)
31. Arita, M., Kobayashi, S.: DNA sequence design using templates. New Gener. Comput. **20**(3), 263–277 (2002). https://doi.org/10.1007/BF03037360

Base Conversion Model Based on DNA Strand Displacement

Zhao Chen[1], Zhixiang Yin[2(✉)], Jianzhong Cui[3,4], Zhen Tang[1],
and Qiang Zhang[5]

[1] School of Mathematics and Big Data, Anhui University of Science and Technology,
Huainan 232001, Anhui, China
[2] School of Mathematics, Physics and Statistics,
Shanghai University of Engineering Science,
Shanghai 201620, People's Republic of China
zxyin66@163.com
[3] School of Electronic and Information Engineering,
Anhui University of Science and Technology, Huainan 232001, Anhui, China
[4] Department of Computer, Huainan Union University,
Huainan 232038, Anhui, China
[5] School of Computer Science, Dalian University of Technology,
Dalian 116024, Liaoning, China

Abstract. DNA computing has the advantages of high parallelism and large storage. In this paper, DNA strand displacement techniques are applied to binary to decimal system. In the field of DNA calculation, binary numerical calculation is relatively mature, but it is difficult to implement decimal calculation. So it is very necessary to study the conversion of binary system to decimal system. In this paper, DNA strand displacement are used to construct a logic model that converts binary system into decimal system. The output strand is obtained through DNA strand displacement reaction in the logical device, and the decimal result is judged by the number of hairpin structures formed. The model has good operability, flexibility.

Keywords: DNA strand displacement · Hairpin structure · Conversion of number systems

1 Introduction

The electronic computer is one of the greatest technological inventions in the 20th century, which has played a very important role in the development of human beings. However, with the development of science and technology, problems such as slow computing speed, high energy consumption and small storage capacity of electronic computers are gradually revealed. Therefore, scientists are gradually trying to dig deeper into the computer power. From then on, DNA computers come into people's vision. In 1994, Adleman innovatively transformed

© Springer Nature Singapore Pte Ltd. 2020
L. Pan et al. (Eds.): BIC-TA 2019, CCIS 1160, pp. 93–102, 2020.
https://doi.org/10.1007/978-981-15-3415-7_8

DNA computing from theory to reality, and realized solving the 7-point Hamilton problem with DNA coding [1]. In 1996, Guarnieri et al. used serial algorithm to realize binary addition calculation by DNA coding [2,3]. In the same year, Olilver designed a DNA computing model to solve the multiplication of special Boolean matrices and special real matrix [4]. In 1999, Labean used the tile-based self-assembly DNA model to accumulate allogeneic or parallel operations [5,6]. In 2000, Wasiewicz established a parallel computing model of binary addition [7].

The advantages of simple conditions, strong operability and high yield of DNA strand displacement reaction make it be an important application method in the field of nanoscience. In 2000, Bernard and Andrew et al. designed a molecular machine composed of three single strands of DNA, whose "opening" and "closing" are achieved by adding two other strands respectively. When "opening" the closed molecular machine, the biological manipulation technique used is the DNA strand displacement reaction [8]. In 2006, Rothemund first used DNA origami to construct patterns. In this technique, A long single-stranded DNA molecule (scaffold) extracted from the M13mp18 phage is folded and assembled into squares, triangles and pentagons, supported by short, single-stranded DNA oligonucleotides (staples). The invention of scaffold DNA origami quickly attracted the research focus of many disciplines and became one of the hot topics [9]. In the same year, Qian et al. used DNA origami technology to build an asymmetric simulated map of China with a diameter of about 150 nm [10]. In 2008, Andersen designed the structural mark of dolphins based on DNA origami, which can control the tail movement of dolphins [11]. In 2011, Winfree and Qian constructed a biochemical circuit that could solve the square root of four binary Numbers based on strand displacement reaction and assembled four artificial neurons with artificial intelligence neural network [12,13]. In 2012, the first autonomous nano-device based on DNA origami was reported [14]. In 2013, bipedal DNA walkers took a crucial step on the DNA origami track [15]. In 2014, DNA origami robot was used for routine calculation [16]. In 2017, Nature published four papers on DNA origami. Tikhomirov et al. used square DNA origami tiles with patterns on the surface as the basic construction unit to construct a two-dimensional DNA origami lattice with a width of 0.5 μ [17]. Ong et al. designed a micron scale self-assembly method for three-dimensional DNA structure [18]. Wagenbaue realized the self-assembly of 3d DNA origami nanostructures at the micron level by adopting the multi-layer self-assembly method [19]. In 2018, Bui found that the dynamics of most DNA walkers were not hindered by basic principles [20].

It is a very important branch of DNA compute to study DNA computer to solve algebra operation. At present, in the field of DNA calculation, binary numerical calculation is relatively mature, but it is difficult to implement decimal calculation, so it is very necessary to study the conversion of binary system to decimal system. This paper constructs a model to convert binary to decimal by DNA strand displacement technique. The model has the advantages of simple operation and easy implementation. The model is parallel computing, which can fully reflect the advantages of DNA computer.

2 Methodology

2.1 DNA Strand Displacement and DNA Origami

2.1.1 DNA Strand Displacement

DNA strand displacement technology has attracted more and more attention in the field of biological computing because of its programmable reaction process and predictable DNA strand dynamics. It has been widely used in the construction of DNA logic gates, biochemical logic circuits and nanomachines.

DNA strand displacement reaction is a spontaneous reaction between DNA molecules, which is completely driven by intermolecular forces. DNA strand displacement reaction refers to the reaction between a single strand of DNA and a partially complementary double strand structure to replace and release the single strand in the original structure and generate a new double strand structure. The principle of strand displacement reaction is that the binding force of different single DNA strands is different, and in the molecular hybridization system, the free energy tends to be stable, so that the input strand with strong binding force replaces the DNA strand with weak binding force in some complementary structures. In simple terms, the longer DNA strand replaces the shorter DNA strand, and the replaced strand serves as an output signal to perform molecular logical operations. The basic process of DNA strand displacement is shown in Fig. 1. As the input signal, the single strand ab and part of the double strand structure undergo strand displacement reaction. Firstly, the region a and region a^* form complementary double strands through certain binding force. Then, the identification area of input single strand b will gradually replace the original binding single strand b until it is completely replaced and the single strand b is released, that is, the output signal is released to achieve stability and complete the strand displacement reaction.

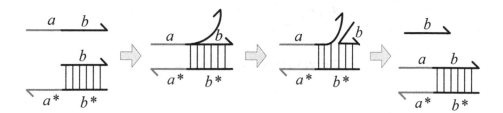

Fig. 1. Basic reaction principle of DNA strand displacement

2.1.2 Binary Conversion to Decimal Calculation Problem

Let the binary representation of the number n be $n(2)$ and the decimal $n(10)$ if $n(2) = a_k a_{k-1} \cdots a_i \cdots a_1 a_0$, then

$$n(10) = a_k \times 2^k + a_{k-1} \times 2^{k-1} + \cdots + a_i \times 2^i + \cdots + a_1 \times 2^1 + a_0 \times 2^0 \quad (2.1)$$

$i \in [0, k], i \in N$, the value of a_i is only 0 or 1, $a_k \neq 0$. The following is a detailed calculation of the conversion from binary to decimal. (1) Represents the value (0 or 1) corresponding to a_i in $n(2)$ and the bitmark i corresponding to the value; (2) The value of a_i in $n(2)$ times 2^i, such as $a_i \times 2^i$; (3) Sum all the results of the second step, that is, calculate the value of $n(10)$,

$$n(10) = a_k \times 2^k + a_{k-1} \times 2^{k-1} + \cdots + a_i \times 2^i + \cdots + a_1 \times 2^1 + a_0 \times 2^0. \quad (2.2)$$

2.2   Calculation Model

The designed logic diagram is divided into two parts. The first part is reaction logic unit a_i. There are two kinds of DNA strands in reaction logic a_i, namely gate:output and fuel. When an input strand enters the reaction logic, no matter how many input strand there are, the reaction logic can generate a predetermined number of output strands. The goal is to find the decimal value of each bit in binary separately. The second part is read logic $\sum I$, in which only one inspection strand exists, and the inspection strand will form a hairpin structure with the output strands. This is partly for easy reading. The specific structure is shown in Fig. 2.

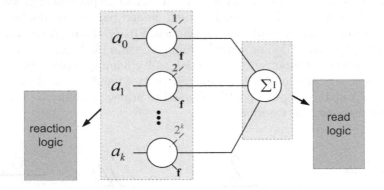

Fig. 2. Complete logic diagram

2.2.1   Design of DNA Strands in Reaction Logic and the Output Strand Structure

The reaction logic contains DNA strands with two structures, gate:output and fuel. Gate:output is a partially complementary double strand, consisting of 5 regions, in which S and S^* are complementary, and T and T^* are complementary. The structure is shown in Fig. 3(a). The fuel strand is a short strand of DNA composed of three regions, as shown in Fig. 3(b). Input strand a_i is composed of three regions. When it is the a_i reaction logic, the structure is shown in Fig. 3(c).

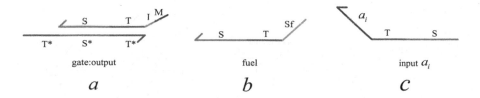

Fig. 3. Design of DNA strands

The output strand generated by each reaction logic is identical and consists of three regions. The structure is shown in Fig. 4. The output strand is all generated by gate:output strand and signal strand through DNA strand displacement reaction. Therefore, the same structure of gate:output strand in all reaction logics can realize the same output strand generated by different reaction logics.

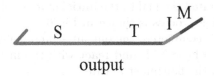

output

Fig. 4. Output strand structure

2.2.2 Reaction Logic, Read Logic and Complete Logic Diagram

The reaction logic model consists of a specific number of gate:output strand and a sufficient amount of fuel strand. In order to ensure the displacement of the input strand after the reaction with gate:output, the fuel strand can continue the reaction. The number of gate:output strands depend on the location of the reaction logic. In Fig. 5(b), the red number represents the amount of gate:output in the corresponding reaction logic, and the number of gate:output in reaction logic a_i is 2^i. Figure 5(a) is the model of reaction logic a_i. For example, in reaction logic a_i, gate:output quantity is 2. There is no requirement for the number of input strands. As long as there is an input strand, the reaction will last until all gate:output strand finishes the reaction. When $a = 0$, reaction logic a_i does not join the input strand, and strand replacement reaction will not occur in the logic. That is, no output strand is generated.

When $a = 1$, the input strand is added to reaction logic a_i. At this point, the logic contains 1 input strand, 2^i gate:output strand and sufficient fuel strand. Strand displacement reaction occurs in the logic. Firstly, input strand and gate:output strand are replaced, and output strand and process strand are generated. Then the process strand and fuel strand undergo strand displacement reaction to generate an input strand and a waste strand. The input strand will have strand displacement reaction with another gate:output strand, and repeat the above process until all gate:output strands have finished the reaction with

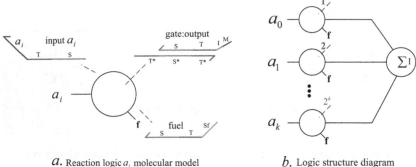

$a.$ Reaction logic a_i molecular model $b.$ Logic structure diagram

Fig. 5. Logic model

the input strand, and the reaction can be terminated. At this point, the number of output strands generated is equal to the initial number of gate:output strands, which is 2^i. The reaction process is shown in Fig. 6.

After the reaction is completed, the output strand generated in the reaction logic will enter the read logic $\sum I$, and react with the inspection strand in the read logic $\sum I$ to form the hairpin structure.

There is a inspection strand in the read logic $\sum I$, which has 3^{k+1} identical parts, and each part is composed of 3 regions. Part of the structure is shown in Fig. 7(a). When the reaction of the reaction logic is completed, all the output strand obtained and other strand enter into the logic $\sum I$, at this time, the IM part of the output strand will be complementary to the I^*, M^* part of the test strand and the test strand to form a hairpin structure, the reaction process is shown in Fig. 7(b). After the reaction is complete, the longest strand is extracted by gel electrophoresis, which is the test strand completed by the reaction. Under the observation of electron microscope, the output number of hairpin structures on the inspection strand is converted into decimal value. For example: checking the hairpin structure on the inspection strand with 5, so the final value converted to decimal is 5.

2.3 Inspection Strand the Process of Generating a Hairpin Structure

2.3.1 Biological Algorithm

Step 1: find the corresponding logic for each a_i in $n(2)$. The input strands of the construction are added separately to all reaction logics of $a_1 = 1$. That is to add nucleotide strand a_iTS to the reaction logic a_i. The different number of output strands generated in each reaction logic is the corresponding value of $n(2)$ in a_i times 2^i.

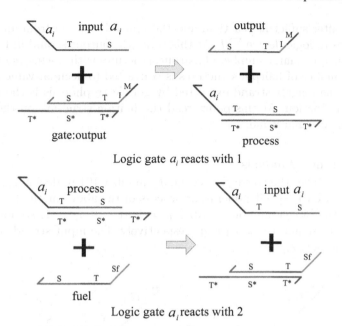

Fig. 6. Initial molecular structure and final state of reaction logic a_i

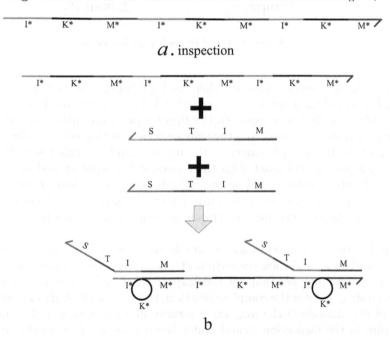

Fig. 7. Inspection strand the process of generating a hairpin structure

Step 2: after sufficient reaction, add the reacted solution from all the above logic devices to logic device $\sum I$. At this time, the output strand in logic device $\sum I$ will generate a large number of hairpin structures with the inspection strand. The total number of hairpin structures is converted to decimal values.

Step 3: the longest strand extracted by gel electrophoresis is the inspection strand after reaction. At this point, read the hairpin structure on the count, is converted to decimal value.

2.3.2   Instance Analysis

The following takes the conversion of binary number 101 to decimal as an example to illustrate the specific operation process of the above method.

Step 1: the binary number is 101, $a_0 = 1, a_1 = 0, a_2 = 1$, so add the constructed input strand to a_0, a_2 logic respectively. The input strand structure is shown in Fig. 8(1, 2).

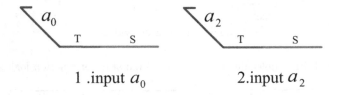

1 .input a_0 2.input a_2

Fig. 8. Input strand in each reaction logic

Step 2: in the reaction logic a_0, there is 1 gate:output strand, which reacts with $a_0 TS$ strand to generate 1 output strand. In a_2, there are 4 gates:output strand. After the first reaction with the input strand, an output strand and a process strand are generated, and the displacement reaction between the process strand and the fuel strand generates the input strand. At this point, the generated input strand will react with the second gate:output strand to generate the second output strand, and so on. a_2 will generate a total of four output strands. As there are too many reaction processes, only the reaction processes of generating the first two output strands in logic a_2 are shown here, as shown in Fig. 9.

Step 3: after the reaction logic is completed, all output strands are entered into the read logic, and then the output strand will generate hairpin structures at the corresponding positions of the inspection strand. Figure 10 shows the inspection strand. After the complete reaction, the longest DNA strand extracted by gel electrophoresis is the inspection strand after the reaction. The number of hairpins in the inspection strand under the microscope is 5, so the result of conversion to decimal system is 5.

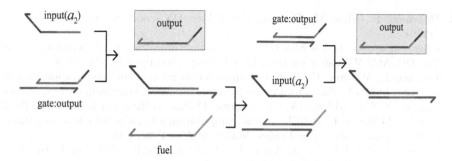

Fig. 9. Reaction logic a_2 reaction process

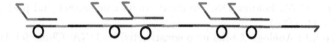

Fig. 10. The reaction completes the inspection strand

3 Conclusion

This paper constructs a binary to decimal model based on DNA origami and DNA strand displacement, and determines the model results by the number of hairpin structures formed by DNA origami. The model has simple implementation conditions and accurate response. The model uses parallel computation, so the solution time can be greatly improved. With the increase of the value of the obtained results, the strand index required by the calculation model increases, the number of hairpin structures increases rapidly, and the reading is relatively tedious. Therefore, the improvement of this aspect will be the focus of the next stage of research.

Acknowledgments. The authors would like to thank every author appeared in the references. This work was supported by National Natural Science Foundation of China [grant number 61672001, 61702008] and [18-163-ZT-005-009-01].

Conflicts of Interest. The authors declare that there are no conflicts of interest regarding the publication of this paper.

References

1. Adleman, L.: Molecular computation of solutions to combinatorial problems. Science **266**, 1021–1024 (1994)
2. Guarnieri, F., Bancroft, C.: Use of horizontal strand reaction for DNA-based addition. In: Proceedings of the 2nd DIMACS Workshop on Computers DNA-Based, pp. 159–249 (1996)

3. Duarnieri, F., Fliss, M., Bancrof, C.: Making DNA add. Science **273**, 220–223 (1996)
4. Oliver, J.: Computation with DNA: matrix multiplication. In: Proccedings of the 2nd DIMACS Workshop on DNA-Based Computers, pp. 236–248 (1996)
5. LaBean, T., Winfree, E., Reif, J.: Experimental progress in computation by self-assembly of DNA tilings. In: Proceedings 5th DIMACS Workshop on DNA-Based Computers, Cambridge, MA, USA 14 June–15 June, Institute of Technology (1999)
6. Mao, C., Leabean, T., Reif, J.: Logical computation using algorithmic self-assembly of DNA triple-crossover molecules. Nature **407**, 493–496 (2000)
7. Wasiewicz, P., Mulawka, J., Rudnicki, W., et al.: Adding Number With DNA. Prog. Nat. Sci.: Engl. 705–709 (2004)
8. Yurke, B., Turberfield, A., Mills Jr., A.P., et al.: A DNA-fuelled molecular machine made of DNA. Nature **406**, 605–608 (2000)
9. Rothemund, P.W.: Folding DNA to create nanoscale shapes and patterns. Nature **440**, 297–302 (2006)
10. Qian, L., et al.: Analogic China map constructed by DNA. Chin. Sci. Bull. **51**(24), 2973–2976 (2006)
11. Andersen, E., Dong, M., Nielsen, M., et al.: DNA origami design of dolphin-shaped structures with flexible tails. ACS Nano **2**(6), 1213–1218 (2008)
12. Qian, L., Winfree, E.: Scaling up digital circuit computation with DNA strand displacement cascades. Science **332**(6034), 1196–1201 (2011)
13. Qian, L., Winfree, E., Bruck, J.: Neural network computation with DNA strand displacement cascades. Nature **475**, 368–372 (2011)
14. Douglas, S., Bachelet, I., Church, G.M.: A logic-gated nanorobot for targeted transport of molecular payloads. Science **335**, 831–834 (2012)
15. Tomov, T.E., Tsukanov, R., Liber, M., Masoud, R., Plavner, N., Nir, E.: Rational design of DNA motors: fuel optimization through single-molecule fluorescence. J. Am. Chem. Soc. **135**, 11935–11941 (2013)
16. Amir, Y., Ben-Ishay, E., Levner, D., Ittah, S., Abu-Horowitz, A., Bachelet, I.: Universal computing by DNA origami robots in a living animal. Nat. Nanotechnol. **9**, 353–357 (2014)
17. Tikhomirov, G., et al.: Fractal assembly of micrometre-scale DNA origami arrays with arbitrary patterns. Nature **552**, 67–71 (2017)
18. Ong, L., et al.: Programmable self-assembly of three-dimensional nanostructures from 10,000 unique components. Nature **552**, 72–77 (2017)
19. Wagenbauer, K., et al.: Gigadalton-scale shape-programmable DNA assemblies. Nature **552**, 78–83 (2017)
20. Bui, H., Shah, S., Mokhtar, R., Song, T., Garg, S., Reif, J.: Localized DNA hybridization strand reactions on DNA origami. ACS Nano **12**, 1146–1155 (2018)

Simulating Collective Behavior in the Movement of Immigrants by Using a Spatial Prisoner's Dilemma with Move Option

Bingchen Lin, Can Zhou, Zhe Hua, Guangdi Hu, Ruxin Ding, Qiyuan Zeng, and Jiawei Li[✉]

University of Nottingham Ningbo China, Ningbo, China
jiawei.li@nottingham.edu.cn

Abstract. The movement of immigrants is simulated by using a spatial Prisoner's Dilemma (PD) with move option. We explore the effect of collective behavior in an evolutionary migrating dynamics. Simulation results show that immigrants adopting collective strategy perform better and thus gain higher survival rate than those not. This research suggests that the clustering of immigrants promotes cooperation.

Keywords: Simulation · Collective behavior · Spatial Prisoner's Dilemma · Evolutionary game theory

1 Introduction

Competition and cooperation are two ubiquitous and inalienable actions in nature [1]. How cooperation evolves in a population of self-interest individuals is an important longstanding scientific question in both biology and social sciences. A powerful framework which has been used frequently to investigate these problems is evolutionary game theory, which includes the changing strategy adoption of a single agent, leading to the concept of rounds [2].

The most relevant game models are two-player games where two players choose to either cooperate (C) or defect (D). The payoff of a player depends on the choices of both sides. There are four possible payoffs in total. Both two players will get a reward (R) if they both choose to cooperate while a punishment (P) will be given if they both choose to defect. In the situation where one player chooses to cooperate whereas the other player chooses to defect. A temptation reward (T) will be given to the defector, whereas the cooperator will get the suckers punishment (S). Different ranking about the above four values determines what game they are playing. In PD game, four values are set as $T > R > P > S$. In terms of these inequalities, it is clear that defection is dominant: every player is better off to choose defect whatever the other player chooses. This makes mutual defection to be a stable strategy in a well-mixed population. There are two other

© Springer Nature Singapore Pte Ltd. 2020
L. Pan et al. (Eds.): BIC-TA 2019, CCIS 1160, pp. 103–114, 2020.
https://doi.org/10.1007/978-981-15-3415-7_9

relevant social dilemma games with PD [3]. The first game is called Chicken [4] or Snowdrift (SD) with a rank as $T > R > S > P$. A reasonable explanation is that in most animal contests, mutual defection will lead to the worst consequence for both players where the loss even exceeds the cost of being exploited. Another game is called Stag Hunt (SH) [5]. The rank set in this game is $R > T > P > S$, which means the reward of cooperation surpasses the temptation. An example of this situation is that when disasters or powerful enemies occur, it is meaningless to still defect other self-interested agents. In the next part of the introduction, evolutionary games with spatial structure will be discussed.

Unlike classical evolutionary game theory which considers the population of players as randomly mixed, a spatial game model puts all players into a two-dimensional lattice, allowing them to continuously interact with their direct neighbors. This spatial structure simulates strategic interactions in the real world, where individuals are placed in certain locations, not move far away from their birthplace and only interact with their relatively fixed neighbors [6]. In each generation of a spatial PD game, every player will play a series of two-player PD games with all its neighbors (Moore neighborhood is commonly used here, which indicates the players in eight direct-linked sites) and accumulate the payoff. At the end of each generation, each player will compare its payoff with all its neighbors and change its strategy, following the one with the highest payoff among them. Compared with the classical evolutionary game, the competition in the spatial game is more complex, and it may lead to different but more useful results. While defection dominates cooperation in PD game, the existence of spatial structure gives cooperation the possibility of survival and grow into clusters [7,8]. Inside a cluster, the cooperation behavior provides its members with enough payoff to overweight the exploitation from outside defectors [9]. In a scale-free model where strong correlations exist between individuals, cooperation may even dominate over defection [10].

In 2007, Vainstein et al. introduced a random-walk mechanism to spatial game model, in which each player can move to a randomly chosen empty neighboring site with a probability [9]. Later, more directional migration mechanisms were introduced. Meloni et al. studied the case where the direction of migration is payoff biased and the movement is with a certain velocity [11]. Helbing et al. introduced the famous success-driven migration model, where players move to the empty neighboring site with the highest estimated payoff. In neighbor-considered migration model [12], a new concept called fairness payoff is introduced. In this model, while players still tend to move to the sites with a higher payoff, reducing the disparity of payoff among their neighbors becomes another target.

To date, there are many studies focused on the effects of mobility on the evolution of cooperation of two or more groups(strategies) in a spatial version of the N-player PD game. However, group cooperation in humans and animals is common [5,13], collective strategies also have great potential research value.

The collective strategy is a summary or a systematic name of collective behaviors. The concept of collective behavior traced back to Park and Burgess [14] and Blumer [15]. It is a special kind of social interaction and refers to a group of unorganized people who are stimulated and influenced by a certain factor. This expression has been expanded to other creatures, such as cells, social animals like birds and fish, and insects including ants [16].

Immigration is the international migration of people into a destination country in order to settle or reside there. Typical immigrants include permanent residents, naturalized citizens or those who expect to be employed as a temporary foreign worker or as a migrant worker [17]. Immense studies about influences brought by immigrants has been done. In terms of economic effects, research suggests that both sending and receiving countries benefit from migration [18,19]. Furthermore, some researches drew a conclusion that immigration on average has positive economic influences towards the native population. However, it is unclear whether low-skilled immigrants have negative effects on low-skilled natives [20,21]. In conclusion, immigration is a ubiquitous phenomenon in human societies, and it influences many aspects of social development. Therefore, immigration contains a significant value to be studied and immigration model is a useful one chosen by many other researches. Collective behavior brings advantage to the group of players who implements it. However, the performance of collective strategy on immigrants still needs more studies. Therefore, this paper studies if collective behavior is still powerful and how powerful it is for immigrants to survive in a new environment.

Though there has been a number of studies about migration and collective strategy separately, there are few studies combine migration with collective behavior. In this paper, we will continue this procession of research by studying the performance of collective strategies as immigrants in a spatial N-player PD game with random mobility.

2   Model: A Spatial Prisoners Dilemma (SPD) with Move Option

In the game model of this paper, some modification is done to the migration method in order to emphasize the existence of strategic interaction. The migration method is similar with that in Schelling model, which is a long studied mathematical model in social sciences. In Schelling model [22], a map is initialized with randomly distributed agents from several groups where each group represents a race. Each agent has an indicator of satisfaction, which value will increase as the number of its neighbors of the same race increases. Agents are stimulated to move if they are unsatisfied with their own satisfaction level. The studies on Schelling model came out with that even a slight homophyllic bias will lead to significant segregation of the different groups of agents. Since parts of the Schelling model corresponds to the spatial migrating PD game (for example, the spatial structure, the interaction with the neighbors, and the migrating behavior), some ideas from Schelling model is applied into our game model. More specifically, similar with Schelling model, in our game model player with collective strategy will stop moving only if the number of its neighbors of same group is equal or larger than a predefined mobility.

The game model introduced in this paper divides the player into two groups, natives and immigrants. The numbers of players in the two groups are not equal, while the population of natives is at least ten times more than the population of immigrants. The tag of the two groups is visible to all players. The players strategy may depend on its opponents group. According to Chiong and Kirley [23], proper move action can enhance the cooperation between immigrants when compared with the natives who always stay at the same place. In this paper, the effect of collective behaviour will be observed with movement, which might come to a more conspicuous and observable result. In the game model, immigrants with collective strategy will invade a larger group of native inhabitants. The collective behavior is reflected in two aspects. Firstly, the player with collective strategy could identify its opponents identity; it will always cooperate with the player from the same group and defect with the player from different groups. Secondly, the player with collective strategy tends to settle down in environments where its group dominates. A real-world example of this game model is the immigration in human society, where the native people is hostile toward the immigrant. To survive in the new unfriendly environment, isolated immigrants may assemble and develop as a group. Through this study, a new game model about collective behavior and migration is generated, which may fill some vacancy in the existing studies.

In this section, three perspectives will be introduced and expounded: the static rule of the game, the method under this set of rules and the experiment settings. Key terms are explained below:

group a collection of all players using the same strategy.

native a group having a larger quantity and have a low tendency of movement.

immigrant a group having a less quantity and have a comparably higher tendency of movement.

neighbor a player located near the current player. A player can have at most eight neighbors (Moore Neighborhood).

mobility a value indicating tendency of movement of a collective strategy, in this paper, referring to the minimum number of neighbors of same strategy in neighborhood to activate movement. Instances of strategies with different mobility can be seen in Appendix.

2.1   Rule

This model implements a spatial PD game; therefore, previous traditional studies are referenced for the rules.

- Every player locates in a cell of the 2D map.
- There are 50% players on the original map.
- Two adjacent players play a single PD game in an iteration.
- The result of each PD game is a pair of payoff value adding to scores of two players. The score is determined according to the payoff matrix in Fig. 1. After playing with each neighbor, the score (total payoff in this iteration) of

central player will be determined. At the start of each iteration, scores of all players will be reset to 0.
- The game ends when no change can be made or one of the two groups disappears.

Player II

		Cooperate	Defect
Player I	Cooperate	$R=3$ $R=3$	$T=5$ $S=0$
	Defect	$S=0$ $T=5$	$P=1$ $P=1$

Fig. 1. The payoff matrix

In this game, several rules are set in order to evaluate and simulate the relationship transition of two groups.

Players are allowed to move to a random nearby empty cell. The move decision is generated based on strategies.

At the end of every iteration, a player adopts the strategy of the neighbour who has received the highest score. If two neighbours gain equal highest score, one strategy will be chosen randomly.

The group adopts an efficient strategy dominates other groups with weak strategy over time.

Therefore, efficiency and strength of each strategy can be estimated according to the number of players remained in evolution.

Six strategies are tested in simulations. Strategies for native players include AllD, AllC and CS. Strategies for immigrants are AllDM, AllCM and CSM. Details of these strategies are given in the Appendix. Different distribution of initial population leads to different result which could be an equilibrium state where native strategies and immigrant strategies coexist, or a homogeneous population In this game, the natives are supposed to stay at the location they are and the immigrants choose whether to move based on their payoffs. In detail, AllD shows the exclusiveness of the native and AllC shows the hospitality of the native. In addition, CS probably dominate in most PD competitions that allows players to cooperate with kin members and defects against non-kin members. This is a strong simulation of group behaviour, which reflects the characteristic of immigrants as well. Besides, AllDM, AllCM and CSM are the same as the three strategies above, except the movement.

To simulate the typical scenarios in the movement of immigrants, it is distinct that immigrants represent less part of the population comparing to the native. In addition, it is possible to develop a different ratio of natives and immigrants. Therefore, 20% and 10% are applied in the study to restore society. The reason for implementing different ratio is to observe the influence of the number of players on the survival rate in spatial PD. Accordingly, a more convincing conclusion could be drawn even though with a different ratio of two groups.

In our first series of experiments, two percentages of immigrants are simulated: 20% and 10%. In those experiments, different ratios of players and different strategies are adopted. There are 18 experiments in total. Different strategy combinations are listed below.

1. Native strategy is AllD, immigrant strategies are AllDM, AllCM and CSM respectively.
2. Native strategy is AllC, immigrant strategies are AllDM, AllCM and CSM respectively.
3. Native strategy is CS, immigrant strategies are AllDM, AllCM and CSM respectively.

In another series of experiments, different mobility levels of CSM are tested.

1. Native strategy is AllD, immigrant strategies are CSM, CSM2, CSM3 and CSM4 respectively.
2. Native strategy is AllC, immigrant strategies are CSM, CSM2, CSM3 and CSM4 respectively.
3. Native strategy is CS, immigrant strategies are CSM, CSM2, CSM3 and CSM4 respectively.

3   Simulation Results Analysis

In the first series of experiments, we focus on whether collective strategy helps immigrants survive among natives. Firstly, three groups of tests were made, in each group the same native strategy played against three immigrant strategies ALLCM, ALLDM and CSM in separated simulations.

Figure 2 shows that CSM strategy could survive in the environments full of one of the three strategies (ALLD, ALLC, CS). In Fig. 2(a), where the native inhabitants take ALLD strategy, CSM not only becomes the only survival immigrant strategy, but also reaches a high percentage in the population (about 94%). In Fig. 2(b), among natives of ALLC strategy, the increasing rate of CSM is as fast as ALLDM's. The difference between results of CSM and ALLDM is that, while the immigrants with ALLDM strategy take the place of all native inhabitants, the percentage of CSM strategy increased to about 90% and reached an equilibrium. In the games of Fig. 2(c), the native inhabitants with collective strategy is a threat to all immigrant strategies. At the initial iterations, the percentages of population of all immigrant strategies plummeted. However, the immigrants with ALLDM strategy has a high rate may disappear during the plummet stage. The CSM and ALLCM immigrants have a high probability of maintaining a low population and surviving through the game.

To see if the initial percentage of immigrants will affect the process of game, another group of games are played (the result is shown in Fig. 3), where the initial percentage of immigrant increased from 10% to 20%. The trends of data

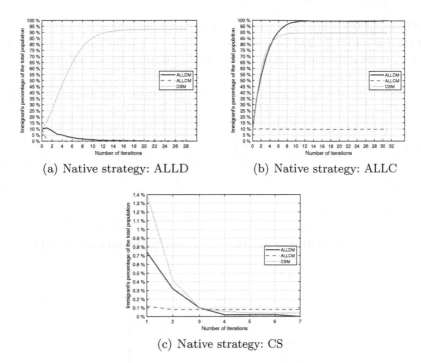

(a) Native strategy: ALLD   (b) Native strategy: ALLC

(c) Native strategy: CS

Fig. 2. Changes in immigrant population with the initial population containing 10% immigrants.

in Fig. 3 is generally same as that in Fig. 2, which indicates that the change of initial population not significantly affect the process of game.

Through these two groups of games, three discoveries are drawn. Firstly, while ALLDM and ALLCM both failed to survive in all games, CSM successfully survived in all games. Though in some games, CSM could only maintain a minimal subsistence, in majority of the games CSM led the immigrants to prosperity. In summary, CSM has proven its ability of surviving in SPD games with migration. Collective strategy is still in advantage in this SPD game simulating migration.

The second discovery is that, under most situations, though CSM could rapidly reach a high population, it will get into equilibrium at that population, instead of totally take place of the native inhabitants. This phenomenon was especially significant in Figs. 2(b) and 3(b), where the ALLDM soon deracinated the native inhabitants, CSM still left some living space for the native inhabitants. Though ALLDM and CSM both take defection strategy toward the native inhabitants, the invasion of CSM is more moderate than ALLDM. A probable reason for this phenomenon is that at the later iterations of the game the CSM immigrants has reached a big population and settled down among clusters of immigrants. According to CSM strategy, when CSM immigrants have neighbors from the same group, they will not move around anymore. As the population and density of CSM increases, the movability of CSM immigrants will decrease. After the CSM immigrants formed a stable cluster, they had no more motivation to move across some empty cells to invade the residual native inhabitants.

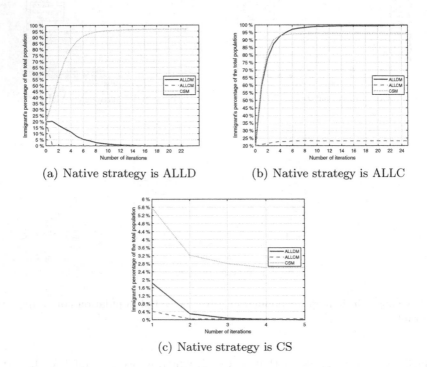

(a) Native strategy is ALLD   (b) Native strategy is ALLC

(c) Native strategy is CS

Fig. 3. Changes in immigrant population with the initial population containing 20% immigrants.

Thirdly, as Fig. 4(d) shows, the empty spaces between native inhabitants and immigrants will segregate the two groups, preventing the residual native inhabitants from extinction. The small cluster of native inhabitants will not be invaded by immigrants, so the native inhabitants could still survive in an environment where CSM immigrants prospered.

Moreover, there is a hypothesis that the move condition may affect the experimental results in CSM. As shown in Fig. 5, CSM with different move conditions are compared at the same initial ratio of immigrants. The move conditions have a limited impact on the final result. They influence the final ratio of immigrants in equilibrium population. However, they do not determine whether an immigrant strategy survive or not.

There are four kinds of move conditions for CS, which are when the mobility of move is 1, 2, 3 and 4. The higher the mobility means the higher probability of move.

In the case where natives are ALLD (Fig. 5(a)) and ALLC (Fig. 5(b)), all CSM immigrants will eventually become dominant populations. From the overall

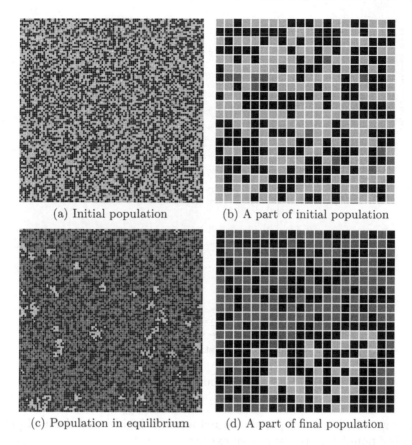

(a) Initial population   (b) A part of initial population

(c) Population in equilibrium   (d) A part of final population

Fig. 4. Simulating migration process. Green (light) cells are natives with ALLC strategy. Red (dark) cells are CSM immigrants. Black cells are empty spaces. The initial population contains 10% immigrants and 90% natives. (Color figure line)

trend, the higher the mobility, the greater the probability of movement, and the higher the percentage of immigrant in the total population at equilibrium. It is worth noting that when the mobility is 4–the most active one, the immigrant has a great probability of accounting for 100% of the total population.

When the native strategy is CS and the immigrant strategy is CSM (Fig. 5(c)), the survival rate of immigrant is sharply reduced. With the increase in mobility, the proportion of immigrant to the total population becomes lower at equilibrium. When the mobility is the highest (4), the immigrant has a high probability of dying

To sum up, this experiment compares several collective strategies as the survivability of immigrant. When natives use ALLD or ALLC strategy, the higher the mobility, the better the growth rate and the higher the final population ratio. However, in the unfavourable case – the natives use CS strategy, when the

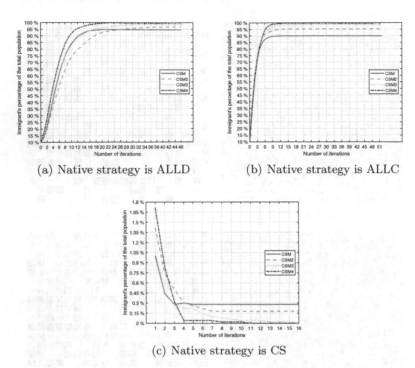

(a) Native strategy is ALLD   (b) Native strategy is ALLC

(c) Native strategy is CS

Fig. 5. Changes in immigrant population.

mobility grows, the rate of population reduction is higher, and the final proportion of the population is lower, even extinct.

4   Conclusion

This paper has discussed the effects of collective strategy in spatial PD via two series of experiments. We have analysed different results when adopting diverse strategies. The investigation shows that, comparing with immigrants without collective behaviour, those with collective strategy have better survival rate and a higher proportion of population in equilibrium. We have also studied collective behaviours with different levels of mobility. The result suggests that the mobility of CS has an influence on the survival rate of immigrants. AllC and AllD strategy of the native tends to accelerate the growth rate of immigrants, while the population of immigrants declines and even extinct when the native strategy is CS.

Although the current study is based on a small sample of experiments, it shows the significance of collective behaviour in strategic interactions. Different settings of the game may have an influence on the final result. For example, the initial population are set to 50% of the lattice, which might lead to a different result if the value is changed. Further research will be conducted in the future with different settings.

5   Appendix

Always Defect (AllD) Defect on every move.

Always Cooperate (AllC) Cooperate on every move.

Collective Strategy (CS) Cooperate with kin members and defect against non-kin members.

Always Cooperate with Move (AllCM) Cooperate on every move. If there exist kin members in neighborhood, then stay, else randomly move to one of eight neighboring empty places.

Always Defect with Move (AllDM) Defect on every move. If there exist kin members in neighborhood, then stay, else randomly move to one of eight neighboring empty places.

Collective Strategy with Move (CSM) Cooperate with kin members and defect against non-kin members. If there exist kin members in neighborhood, then stay, else randomly move to one of eight neighboring empty places.

Collective Strategy with Move (CSM2) Cooperate with kin members and defect against non-kin members. If there exist more than 1 kin members in neighborhood, then stay, else randomly move to one of eight neighboring empty places.

Collective Strategy with Move (CSM3) Cooperate with kin members and defect against non-kin members. If there exist more than 2 kin members in neighborhood, then stay, else randomly move to one of eight neighboring empty places.

Collective Strategy with Move (CSM4) Cooperate with kin members and defect against non-kin members. If there exist more than 3 kin members in neighborhood, then stay, else randomly move to one of eight neighboring empty places.

References

1. Smith, J.M., Szathmary, E.: The Major Transitions in Evolution. Oxford University Press, Oxford (1997)
2. Smith, J.M.: Evolution and the Theory of Games. Cambridge University Press, Cambridge (1982)
3. Liebrand, W.B.G.: A classification of social dilemma games. Simul. Games **14**(2), 123–138 (1983)
4. Rapoport, A.: Two-Person Game Theory. Courier Corporation, Chelmsford (2013)
5. Boyd, R., Richerson, P.J.: The evolution of reciprocity in sizable groups. J. Theoret. Biol. **132**(3), 337–356 (1988)
6. Hamilton, W.D.: The genetical evolution of social behaviour. (ii). J. Theoret. Biol. **7**(1), 17–52 (1964)
7. Nowak, M.A.: Evolutionary Dynamics. Harvard University Press, Cambridge (2006)
8. Szabó, G., Fath, G.: Evolutionary games on graphs. Phys. Rep. **446**(4–6), 97–216 (2007)
9. Vainstein, M.H., Silva, A.T.C., Arenzon, J.J.: Does mobility decrease cooperation? J. Theoret. Biol. **244**(4), 722–728 (2007)

10. Santos, F.C., Pacheco, J.M.: Scale-free networks provide a unifying framework for the emergence of cooperation. Phys. Rev. Lett. **95**(9), 098104 (2005)
11. Meloni, S., et al.: Effects of mobility in a population of prisoner's dilemma players. Phys. Rev. E **79**(6), 067101 (2009)
12. Ren, Y., et al.: Neighbor-considered migration facilitates cooperation in prisoner's dilemma games. Appl. Math. Comput. **323**, 95–105 (2018)
13. Suzuki, S., Akiyama, E.: Reputation and the evolution of cooperation in sizable groups. Proc. R. Soc. B: Biol. Sci. **272**(1570), 1373–1377 (2005)
14. Park, R., Burgess, E.: Introduction to the Science of Sociology. The University of Chicago, Chicago (1921)
15. Blumer, H.: New Outline of the Principles of Sociology, p. 67121. Barnes and Nobel, New York (1951)
16. Gordon, D.M.: The ecology of collective behavior. PLoS Biol. **12**(3) (2014)
17. England: Refugee Council London. Who's who: Definitions, 9 (2015). An optional note
18. Di Giovanni, J., Levchenko, A.A., Ortega, F.: A global view of cross-border migration. J. Eur. Econ. Assoc. **13**(1), 168–202 (2015)
19. Ahmed, S.A., Go Delfin S., Dirk, W.: Global migration revisited: short-term pains, long-term gains, and the potential of south-south migration. The World Bank (2016)
20. Card, D., Dustmann, C., Preston, I.: Immigration, wages, and compositional amenities. J. Eur. Econ. Assoc. **10**(1), 78–119 (2012)
21. Van den Berg, H., Bodvarsson, Ö.B.: The Economics of Immigration: Theory and Policy. Springer, Heidelberg (2009). https://doi.org/10.1007/978-3-540-77796-0
22. Rogers, T., McKane, A.J.: A unified framework for Schelling's model of segregation. J. Stat. Mech.: Theory Exp. 2011(07) (2011). Article no. P07006
23. Chiong, R., Kirley, M.: Random mobility and the evolution of cooperation in spatial n-player iterated prisoner's dilemma games. Phys. A **391**(15), 3915–3923 (2012)

Exploring Computation Tree Logic with Past-Time Operator Model Checking Using DNA Computing

Ying-Jie Han, Xiao-Fei Nan, Shao-Huan Ban, and Qing-Lei Zhou[✉]

School of Information Engineering, Zhengzhou University, Zhengzhou 450001, China
ieqlzhou@zzu.edu.cn

Abstract. Computation tree logic (CTL) model checking is a verification technique that is important to safety-critical systems. DNA computing provides new ideas for improving the efficiency and solving the state space explosion problem of CTL model checking. However, existing research mainly focuses on DNA computing methods for checking CTL with future-time operators and has not addressed CTL with past-time operators (CTLP). In this paper, we propose a DNA computing method for CTLP model checking. First, a system to be checked and a CTLP formula are encoded by DNA strands. Next, all the strands are mixed into a test tube. Then, the complementary strands in the test tube hybridize and form fully or partially double-stranded DNA molecules. Finally, a series of biochemical operations are performed to detect the double-stranded DNA molecules, and whether the system satisfies the CTLP formula is determined. Simulations show the validity and effectiveness of the method. Our new method enhances the power and lays the foundation for the completeness of CTL model checking.

Keywords: Computation tree logic · DNA computing · Model checking · Past-time operator

1 Introduction

Deoxyribonucleic acid (DNA) computing is a new computational paradigm in which DNA molecules and biological enzymes are materials and biochemical reactions are a means of computation [1]. In the study of DNA computing, models are one of the main research focuses. The DNA computing models proposed so far can be divided into two generations. The first generation of DNA computing models consists of laboratory-scale, human-operated models that basically aim to solve complex computational problems [2–8]. The second generation of

This work was supported by the National Natural Science Fund of China under Grant 61572444 and Science and Technology Research Plan of Henan province (International Scientific Cooperation Projects) under Grant 172102410065.

© Springer Nature Singapore Pte Ltd. 2020
L. Pan et al. (Eds.): BIC-TA 2019, CCIS 1160, pp. 115–133, 2020.
https://doi.org/10.1007/978-981-15-3415-7_10

DNA computing models consists of molecular-scale, autonomous, partially programmable models that essentially target complex computational problems [9–12], theoretical model [13–15], nanometer materials [16], DNA logic circuits [17–22], information security [23] and medical diagnosis and treatment [24–28].

CTL model checking is a formal verification method that can answer questions such as whether a system automatically satisfies a given property specified by a CTL formula [29]. It is widely applied in hardware verification [30,31], software verification [32,33], and biological systems [34]. However, severe spatiotemporal complexity problems restrict its application in industry. The large storage capacity and parallelism of DNA molecules provide new ideas for resolving this issue. Emerson first designed an algorithm to check EFp, which is a CTL formula with future-time operators [35]. Zhu et al. proposed model-checking algorithms for CTL with future-time operators [36]. Nevertheless, previous research did not address the model-checking method for CTL with past-time operators (CTLP). In view of this, Han et al. proposed a DNA computing method for checking CTLP formulas based on the memoryless filtering model [37]. However, this method can check only four basic CTLP formulas, and it was time-consuming and error-prone due to human-operated computations in the laboratory.

To address the above problems, we propose a model-checking method based on the sticker automata, an autonomous DNA computing model, and verify the validity of the new method via simulation. The new method overcomes the shortcomings related to time consumption and error proneness. Furthermore, it can theoretically check all the CTL and CTLP formulas, thereby greatly improving the power of the DNA computing method for CTL model checking.

The remainder of this paper is organized as follows. First, we give a brief introduction of CTLP, model checking and the sticker automata in Sect. 2. Then, we elaborate on the new CTLP model checking method in Sect. 3 and describe the simulations in Sect. 4. Finally, conclusions and future work are discussed in Sect. 5.

2   Background

2.1   CTL with Past-Time Operators and Model Checking

CTLP extends CTL by allowing past-time operators with branching-time semantics. It is well known that allowing past-time operators makes temporal specification easier to write, more natural to understand [38], and more expressive [39]. Moreover, adding past-time operators does not increase the complexity of model-checking problems [38].

The definition of CTLP is as follows.

Definition 1. Let $AP = \{p, q...\}$ be a nonempty finite set of atomic propositions. The CTLP formulas φ and ψ are defined as follows [40]:

$\varphi, \psi ::= p|q|\neg\varphi|\varphi \vee \psi|EH\varphi|AH\varphi|EO\varphi|AO\varphi|E(\varphi S\psi)|A(\varphi S\psi)|EY\varphi|AY\varphi...$

where E (for "there exists a path") and A (for "every path") are the path quantifiers, and S (for "since"), Y (for "previous"), H (for "historically") and

O (for "once") are the past-time operators. The corresponding basic CTLP formulas are AHp, EHp, AOp, EOp, $A(pSq)$, $E(pSq)$, AYp and EYp. Supposing that M is a system model, the above formulas are explained below, and their intuitive meanings are shown in Fig. 1 (p is valid on the current state except AOp and EOp).

- AHp: for all paths in M, p held at all times in the past.
- EHp: for some paths in M, p held at all times in the past.
- AOp: for all paths in M, p held at some times in the past.
- EOp: for some paths in M, p held at some times in the past.
- $A(pSq)$: for all paths in M, q held at some times in the past, and p has been holding ever since.
- $E(pSq)$: for some paths in M, q held at some times in the past, and p has been holding ever since.
- AYp: for all paths in M, p held on the previous state starting from the current state.
- EYp: for some paths in M, p held on the previous state starting from the current state.

We model the basic CTLP formulas by finite state automata (FSA) and the to-be-checked system by labeled finite state automata (LFSA). Their definitions are as follows.

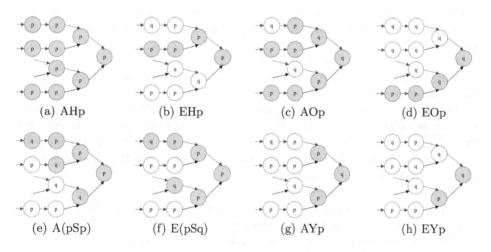

(a) AHp   (b) EHp   (c) AOp   (d) EOp

(e) A(pSp)   (f) E(pSq)   (g) AYp   (h) EYp

Fig. 1. The basic CTLP formulas and their intuitive meanings

Definition 2. An FSA is a five-tuple $(\sum, S, \delta, s_0, F)$, where $\sum$ is a finite alphabet, S is a finite set of states, $\delta : S \times \sum \to S$ is a finite set of transitions, $s_0 \in S$ is an initial state, and $F \subseteq S$ is a set of acceptance states.

Definition 3. An LFSA is a six-tuple $(\sum, S, \delta, s_0, F, L)$, where $\sum$, S, δ, s_0 and F share the same meanings given in Definition 2. $L : S \rightarrow 2^{AP}$ assigns each state $s \in S$ to an atomic proposition $L(s)$ that is valid in s.

The difference between FSA and LFSA is as follows: the atomic propositions are satisfied on the transitions in FSA, whereas the atomic propositions are satisfied on the states in LFSA.

Definition 4. A reverse path of an LFSA is a finite sequence of states $s_0 s_1 s_2 ... s_n$, such that $(s_{i+1}, s_i) \in \delta$ holds for all $i \in \{1, 2, ..., n-1\}$, where δ is the finite set of transitions and $(s, s_n) \notin \delta$ for all $s \in S$.

The CTLP model-checking problem is as follows: given a system model and a CTLP formula, determine whether the system model satisfies the formula.

2.2   The Sticker Automaton and Its Principle

The sticker automaton [14] is an autonomous DNA computing model for FSA. Given a single-stranded DNA molecule representing an input string and an FSA, the sticker automaton of the FSA can be used to determine whether the input string is accepted by the FSA. The sticker automaton operates in three steps: data preprocessing, computation and detection. Data preprocessing is performed by the following steps:

Step 1: Encode an input string of an FSA with state set $S = \{s_0, ..., s_{m-1}\}$, where m represents the number of states. Each symbol a in $\sum$ is encoded by a DNA strand $5'C(a)3'$. In this way, each input string $x = a_1 a_2 ... a_n$ over $\sum$ will be encoded by a DNA strand consisting of alternating symbol encodings and spacers,

$C(x) = 5'I_1 X_0 ... X_m C(a_1) ... X_0 ... X_m C(a_n) X_0 ... X_m I_2 3'$

where $X_0 ... X_m$ is the spacer sequence, I_1 is the initiator sequence, and I_2 the terminator sequence.

Step 2: Construct the sticker automaton of the FSA. The initial state $s_i \in S$ is encoded by $3'\overline{I_1 X_0 ... X_i}5'$, and each terminal state s_j is encoded by $3'\overline{X_{j+1} ... X_m I_2}5'$. The transition $\delta(s_i, a) = s_j (a \in \sum)$ is encoded by $3'\overline{X_{i+1} ... X_m C(a) X_0 ... X_j}5'$, where $\overline{X}$ and $\overline{C(a_i)}$ denote the Watson-Crick complements of X and $C(a_i)$, respectively.

Step 3: All the DNA strands are placed in a test tube, and complementary strands are allowed to hybridize. Then, ligase is added to the test tube to obtain (partially) double-stranded DNA molecules.

After preprocessing, an accepted input string corresponds to a complete, double-stranded DNA molecule, whereas a nonaccepted input string corresponds to a partially double-stranded DNA molecule.

The computation is carried out by mung bean nuclease. Mung bean nuclease has lower intrinsic activity on duplex DNA and is able to degrade the single-stranded region in a nonaccepted input string. As a consequence, the double-stranded DNA molecules corresponding to the accepted input strings will remain intact after degradation.

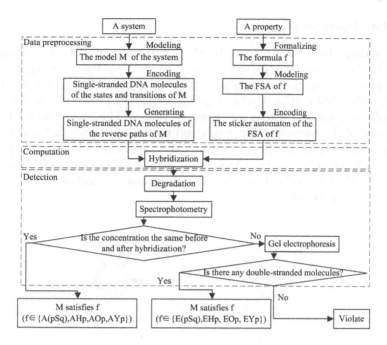

Fig. 2. Overview of the new CTLP model-checking method

Detection is performed by gel electrophoresis or polymerase chain reaction (PCR). When the lengths of the molecules are known, gel electrophoresis can be employed to separate the DNA molecules by size. Hence, the accepted input strings can be detected. When the length of the DNA molecules is unknown, PCR can be used to detect the molecules corresponding to the accepted input strings because the DNA molecule encoding an accepted input string has both an initiator and a terminator.

3   The Proposed Method

We proposing a CTLP model-checking method based on sticker automata in this section. Suppose that the LFSA of a to-be checked system is M, and the basic CTLP formula is f, where $f \in \{AHp, EHp, AOp, EOp, A(pSq), E(pSq), AYp, EYp\}$. Figure 2 presents an overview of the method, which is divided into three phases: data preprocessing, computation and detection. In the data preprocessing phase, the sticker automaton of the FSA of f is encoded by specific DNA strands, and the reverse paths of M are generated as the input strings (see Sect. 3.1). In the computation phase, the computation is performed by the hybridization of the above two types of DNA strands (see Sect. 3.2). In the detection phase, after degrading the product, spectrophotometry and gel electrophoresis are employed to determine whether the reverse paths are accepted by the FSA of f so that the result of whether M satisfies f can be deduced (see Sect. 3.3).

3.1   Data Preprocessing

Constructing the FSA and the Sticker Automaton of f. The union FSA $\{\{p, q\}, \{s_0, s_1, s_2\}, s_0, \{s_2\}, \{\delta(s_0, p) = s_0, \delta(s_0, q) = s_0, \delta(s_0, p) = s_1, \delta(s_0, q) = s_1, \delta(s_0, p) = s_2, \delta(s_0, q) = s_2, \delta(s_2, p) = s_2, \delta(s_2, q) = s_2, \delta(s_1, p) = s_2\}$, which is of $\{EHp, AHp, EOp, AOp, A(pSq), E(pSq), AYp, EYp\}$ is shown in Fig. 3, and its sets of states and transitions are shown in Table 1. The FSA of each formula f is composed of the corresponding states and transitions shown in Table 1, where "1" indicates being selected, and "0" indicates not being selected. As shown in Table 1, each pair of formulas $E(pSq)$ and $A(pSq)$, EHp and AHp, EOp and AOp, and EYp and AYp share the same automaton, which we call pSq, Hp, Op, and Yp, respectively.

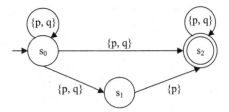

Fig. 3. The union FSA of f

Table 1. The states and transitions of the FSA of f

Object		The FSA of f			
		$E(pSq)$	EHp	EOp	EYp
		$A(pSq)$	AHp	AOp	AYp
The initial state	s_0	1	1	1	1
	s_1	0	0	0	1
The terminal state	s_2	1	1	1	1
The transitions	$\delta(s_0, q) = s_2$	1	0	0	0
	$\delta(s_0, p) = s_2$	0	1	1	0
	$\delta(s_0, q) = s_1$	0	0	0	1
	$\delta(s_0, p) = s_1$	0	0	0	1
	$\delta(s_1, p) = s_2$	0	0	0	1
	$\delta(s_2, p) = s_2$	1	0	1	1
	$\delta(s_2, q) = s_2$	1	0	1	1
	$\delta(s_0, q) = s_0$	0	0	1	0
	$\delta(s_0, p) = s_0$	1	1	1	0

The sticker automaton of f is constructed by encoding the initial state s_0, the accepting state s_2, and the transitions δ into DNA single strands, as described in Sect. 2.2.

Generating the Reverse Paths of M. To obtain the reverse paths of M, the states and transitions are encoded by single-stranded DNA molecules first, as described below.

1. Each state $s_i(s_i \neq s_{ini}, and\, s_i \notin F)$ is encoded by $5'X_0...X_mC(L(s_i))3'$, where $L(s_i)$ represents the atomic proposition valid in s_i, and $C(L(s_i))$ is the DNA sequence of $L(s_i)$. To meet the input requirements of the sticker automaton, the initial state s_{ini} is encoded by $5'X_0...X_mC(L(s_{ini}))X_0...X_mI_23'$, and the accepting state s_{acc} is encoded by $5'I_1X_0...X_mC(L(s_{acc}))3'$. We assume that the DNA sequence of s_i is $5'H(s_i)T(s_i)3'$, where the formal part is $5'H(s_i)3'$, and the latter part is $5'T(s_i)3'$.
2. The transition $(s_i, s_j)(s_i \neq s_{ini}, and\, s_j \notin F)$ is encoded by $5'T(s_j) H(s_i)3'$, the transition $(s_i, s_j)(s_i = s_{ini}, and\, s_j \notin F)$ is encoded by $5'T(s_j) H(s_i)T(s_i)3'$, the transition $(s_i, s_j)(s_i \neq s_{ini}, and\, s_j \in F)$ is encoded by $5'H(s_j)T(s_j)H(s_i)3'$, and the transition $(s_i, s_j)(s_i = s_{ini}, and\, s_j \in F)$ is encoded by $5'H(s_j)T(s_j)H(s_i)T(s_i)3'$.

After synthesizing the DNA strands for the states and transitions, the DNA strands are mixed together; the complementary DNA strands are annealed and form short-double DNA molecules, and then, they are concatenated into long-double DNA molecules by ligase. The DNA strands of the reverse paths starting with s_{acc} and ending with s_{ini} are extracted by amplifying the product via PCR. The double-stranded DNA molecules are forced to be separated into single-stranded DNA molecules by increasing the temperature. Each single-stranded DNA molecule corresponds to a reverse path with s_{acc} as the start and s_{ini} as the end. In this way, the single-stranded DNA molecules of the reverse paths of M are generated.

3.2 Computation

The computation is carried out by the following steps:

Step 1: The single-stranded DNA molecules of the reverse paths of M and the sticker automaton of f are mixed together.
Step 2: The complementary strands hybridize and generate double-stranded molecules through specific hybridization or partially double-stranded molecules through nonspecific hybridization. A double-stranded molecule indicates that a reverse path of M is accepted by the FSA of f, whereas a partially double-stranded molecule indicates that a reverse path of M is not accepted by the FSA of f.

3.3 Detection

Detection is performed by the following steps:

Step 1: An appropriate amount of mung bean nuclease is added into the product of the hybridization to degrade the single-stranded fragment of the partially double-stranded DNA molecules.

Step 2: Spectrophotometry is used to observe the concentration before and after hybridization to determine whether the DNA molecules generated are completely double-stranded. If the concentration of the DNA molecules is the same before and after hybridization, M satisfies f, where $f = \{A(pSq), AHp, AOp, AYp\}$; otherwise, Step 3 is performed.

Step 3: Gel electrophoresis is employed to separate the double-stranded DNA molecules since their lengths are known. If the electrophoresis bands corresponding to the lengths of the reverse paths exist, M satisfies f, where $f = \{E(pSq), EHp, EOp, EYp\}$; otherwise, M does not satisfies f.

4 Simulations

We carry out simulations with a system model shown in Fig. 4 and the above eight CTLP formulas. In the simulations, we first use NUPACK [41] to design and evaluate the DNA sequences because they directly affect the speed, efficiency and reliability of the biochemical reactions. Then, we use NUPACK to simulate and verify the specific or nonspecific hybridization of the single DNA strands of the reverse paths of the system model and the sticker automaton of f since this is the key to our method.

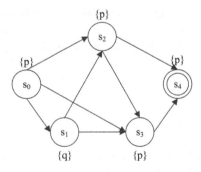

Fig. 4. A system model M_1 used in the simulations

4.1 The Design and Evaluation of DNA Sequences

Figure 5 shows the designed sequences used in the simulation and their thermo-mechanical analysis. Figure 5(a) shows the designed DNA sequence with 50 base pairs and its Watson-Crick complement at 23 °C. Figure 5(b) shows the structural properties of the designed sequences. As shown in Fig. 5(b), the normalized ensemble defect (NED) of the sequences is 0.3% (0% is best and 100% is worst). NED refers to the proportion of nucleotides that are not correctly paired when the biochemical reaction reaches equilibrium. Figure 5(c) shows the minimum free energy (MFE) of the designed DNA sequences. The MFE is the Gibbs free energy of the system when equilibrium is reached. As shown in Fig. 5(c), the dark red line suggests that the free energy converges to the MFE. Figure 5(d) shows the pairing probabilities of the designed DNA sequence ("strand1") with its Watson-Crick complement ("strand2"). The position of the red line suggests that all the bases of the two sequences are paired exactly in the corresponding order, and the color of the red line indicates that the pairing probabilities of all the bases are approximately equal to 100%. In other words, the designed DNA sequence and its complement satisfy the MFE constraint and have the same melting temperature. Therefore, the results obtained from the sequences are biologically effective and reliable.

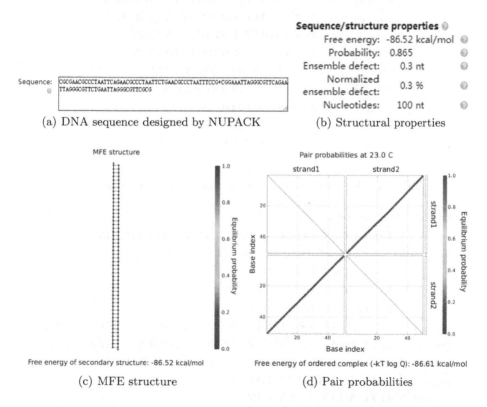

(a) DNA sequence designed by NUPACK

(b) Structural properties

(c) MFE structure

(d) Pair probabilities

Fig. 5. The DNA sequence designed by NUPACK and its thermodynamic analysis

The DNA sequences encoding the states and transitions of M_1 are shown in Table 2 according to Fig. 5(a). The DNA sequences for the initial state, the

Table 2. The DNA sequences for encoding the states and transitions of M_1

Object	DNA sequence
Initiator I_1	$5'CGCG3'$
Terminator I_2	$5'TCCG3'$
Spacer X_i	$X_0 = 5'AAC3', X_1 = 5'GCC3', X_2 = 5'CTA3', X_3 = 5'ATT3'$
p	$5'CAG3'$
q	$5'CTG3'$

Table 3. The DNA sequences of the union FSA in Fig. 3

Object	DNA sequence
s_0	$3'GCGCTTG5'$
s_2	$3'TAAAGGC5'$
$\delta(s_0, q) = s_2$	$3'CGGGATTAAGACTTGCGGGAT5'$
$\delta(s_0, p) = s_2$	$3'CGGGATTAAGTCTTGCGGGAT5'$
$\delta(s_0, q) = s_1$	$3'CGGGATTAAGACTTGCGG5'$
$\delta(s_0, p) = s_1$	$3'CGGGATTAAGTCTTGCGG5'$
$\delta(s_1, p) = s_2$	$3'GATTAAGTCTTGCGGGAT5'$
$\delta(s_2, p) = s_2$	$3'TAAGTCTTGCGGGAT5'$
$\delta(s_2, q) = s_2$	$3'TAAGACTTGCGGGAT5'$
$\delta(s_0, q) = s_0$	$3'CGGGATTAAGACTTG5'$
$\delta(s_0, p) = s_0$	$3'CGGGATTAAGTCTTG5'$

Table 4. The DNA sequences of the states and transitions of M_1

Object	DNA sequence
s_0	$5'AACGCCCTAATTCAGAACGCCCTAATTTCCG3'$
s_4	$5'CGCGAACGCCCTAATTCAG3'$
s_1	$5'AACGCCCTAATTCTG3'$
s_2	$5'AACGCCCTAATTCAG3'$
s_3	$5'AACGCCCTAATTCAG3'$
(s_0, s_1)	$5'ATTCTGAACGCCCTAATTCAGAACGCCCTAATTTCCG3'$
(s_0, s_2)	$5'ATTCAGAACGCCCTAATTCAGAACGCCCTAATTTCCG3'$
(s_0, s_3)	$5'ATTCAGAACGCCCTAATTCAGAACGCCCTAATTTCCG3'$
(s_1, s_2)	$5'ATTCAGAACGCCCTA3'$
(s_1, s_3)	$5'ATTCAGAACGCCCTA3'$
(s_2, s_3)	$5'ATTCAGAACGCCCTA3'$
(s_2, s_4)	$5'CGCGAACGCCCTAATTCAGAACGCCCTA3'$
(s_3, s_4)	$5'CGCGAACGCCCTAATTCAGAACGCCCTA3'$

Table 5. The DNA sequences of the reverse paths of M_1

No	Path	DNA sequence
1	$s_4 s_2 s_0$	5′$CGCGAACGCCCTAATTCAGAACGCCCTAATTCAGAACG$ $CCCTAATTCAGAACGCCCTAATTTCCG$3′
2	$s_4 s_3 s_0$	5′$CGCGAACGCCCTAATTCAGAACGCCCTAATTCAGAACG$ $CCCTAATTCAGAACGCCCTAATTTCCG$3′
3	$s_4 s_3 s_1 s_0$	5′$CGCGAACGCCCTAATTCAGAACGCCCTAATTCAGAACG$ $CCCTAATTCTGAACGCCCTAATTCAGAACGCCCTAATTTCCG$3′
4	$s_4 s_2 s_1 s_0$	5′$CGCGAACGCCCTAATTCAGAACGCCCTAATTCAGAACG$ $CCCTAATTCTGAACGCCCTAATTCAGAACGCCCTAATTTCCG$3′
5	$s_4 s_3 s_2 s_0$	5′$CGCGAACGCCCTAATTCAGAACGCCCTAATTCAGAACG$ $CCCTAATTCAGAACGCCCTAATTCAGAACGCCCTAATTTCCG$3′
6	$s_4 s_3 s_2 s_1 s_0$	5′$CGCGAACGCCCTAATTCAGAACGCCCTAATTCAGAACG$ $CCCTAATTCAGAACGCCCTAATTCTGAACGCCC$ $TAATTCAGAACGCCCTAATTTCCG$3′

accepting state and the transitions of the union FSA of f in Fig. 3 are shown in Table 3. The DNA sequences of the states and transitions of M_1 are obtained from Table 2 and shown in Table 4. Therefore, the DNA sequences of the six reverse paths of M_1 are generated and shown in Table 5. As shown in Table 5, the DNA sequence of *path 1* is the same as that of *path 2*, and the DNA sequence of *path 3* is the same as that of *path 4*.

4.2 Verification

The key to our method is the hybridization of the single-stranded DNA molecules of the reverse paths of M_1 with the sticker automaton of f. In our simulations, we verify only the hybridization of a reverse path molecule such as 5′$X_1...X_m C(p)...X_0...X_m C(q) X_0...X_{m-1}$3′ and the transitions of the FSA of f because the two ends (5′$I_1 X_0$3′ and 5′$X_m I_2$3′) of the reverse paths are completely complementary to the initial state and the accepting state of the FSA of f, respectively.

The Verification of EHp and AHp. Hybridization of the DNA strands of the reverse paths of M_1 with the transitions of the FSA of Hp is shown in Fig. 6, where strand1 represents the DNA strands of *path i*, and strand2 and strand3 represent the DNA strands of $\delta(s_0, p) = s_0$ and $\delta(s_0, p) = s_2$, respectively. As shown in Fig. 6(a), strand1 denotes the DNA strands of *path 1* (*path 2*). The three red lines in the upper right corner indicate that the single strands of *path 1* (*path 2*) in the 5′ $-$ 3′ direction pair completely with the single strands of $\delta(s_0, p) = s_0, \delta(s_0, p) = s_0$ and $\delta(s_0, q) = s_2$ from the 1st to the 15th base, from the 16th to the 30th base, and from the 31st to the 51st base in the 3′$-$5′ direction, respectively. According to the legend on the right, the pairing probability is nearly 100%, which suggests specific hybridization. This also indicates that *path 1* (*path 2*) is accepted by the FSA of Hp. The same situation exists for *path 5*, as shown in Fig. 6(c). Figure 6(b) and (d) show the opposite cases. Taking Fig. 6(b) as an example, strand1 represents the DNA strands of *path 3* (*path 4*). The red line in the upper right corner indicates that strand1 pairs with strand2 from the 1st to the 15th base. The green lines indicate that from the 16th base to the 30th base, some of strand1 pair with strand2, while some pair with strand3; from the 46th base to the 60th base, some of strand1 pair with strand2, while some pair with strand3; and from the 37th base to the 45th base, there are no DNA strands paired with strand1. According to the legend on the right, the pairing probability is less than 60%, which suggests nonspecific hybridization. This also indicates that *path 3* (*path 4*) is not accepted by the FSA of Hp. In other words, paths *1*, *2*, and *5* are accepted by the FSA of Hp, whereas paths *3*, *4*, and *6* are not accepted by the FSA of Hp, which indicates that M_1 satisfies EHp but not AHp.

The Verification of EOp and AOp. Hybridization of the DNA strands of the reverse paths of M_1 with the transitions of the FSA of Op is shown in Fig. 7, where strand1 represents the DNA strands of *path i*, and strand2, strand3 and strand4 represent the DNA strands of $\delta(s_0, (p|q)) = s_0, \delta(s_0, p) = s_2$ and $\delta(s_2, (p|q)) = s_2$, respectively. Figure 7(a) to (c) shows three cases of base pairing of *path 1* (*path 2*) with the transitions of the FSA of Op. Taking Fig. 7(a) as an example, strand1 represents the DNA strands of *path 1* (*path 2*). The three red lines in the upper right corner indicate that the single strands of *path 1* (*path 2*) in the $5' - 3'$ direction pair completely with the single-stranded molecules of strand2, strand2 and strand3 from the 1st base to the 15th base, from the 16th base to the 30th base, and from the 31st base to the 51st base in the $3' - 5'$ direction, respectively. According to the legend on the right, the pairing probability is nearly 100%, which suggests specific hybridization. This also indicates that *path 1* (*path 2*) is accepted by the FSA of Op. Figure 7(d) shows the details of the three cases. Figure 7(e) to (f) show that paths *3*, *4*, *5* and *6* are accepted by the FSA of Op. In other words, all the reverse paths are accepted by the FSA of Op, which indicates that M_1 satisfies both EOp and AOp.

The Verification of E(pSq) and A(pSq). Hybridization of the DNA strands of the reverse paths of M_1 with the transitions of the FSA of pSq is shown in Fig. 8, where strand1 represents the DNA strands of *path i*, and strand2, strand3, and strand4 represent the DNA strands of $\delta(s_0, p) = s_0, \delta(s_0, q) = s_2$, and $\delta(s_2, (p|q)) = s_2$, respectively. As shown in Fig. 8(a), strand1 denotes the DNA strands of *path 1* (*path 2*). The yellow and blue lines in the upper right corner indicate that from the 1st base to the 30th base, some of strand1 pair with strand2, while some pair with strand3; from the 31st base to the 51st base, some of strand1 pair with strand3, while some pair with strand4. The breakpoints in the blue lines represent the unpaired bases. According to the legend on the right, the pairing probability is less than 60%, which suggests nonspecific hybridization.

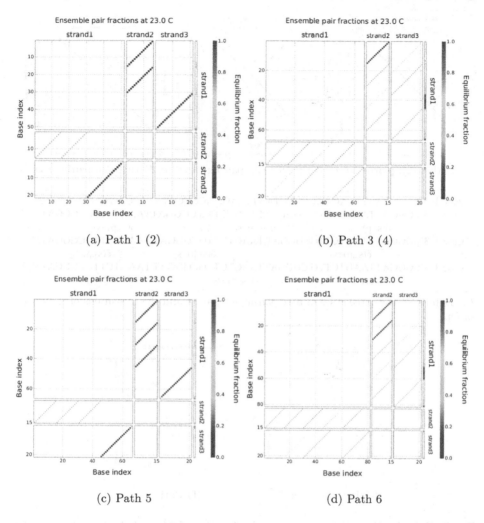

Fig. 6. Hybridization of the single strands of path i of M_1 and the transitions of the FSA of Hp: pairing positions and probabilities

This also indicates that *path 1* (*path 2*) is not accepted by the FSA of *pSq*. The same situation exists for *path 5*, as shown in Fig. 8(c). Figure 8(b) and (d) show the opposite cases. Taking Fig. 8(b) as an example, strand1 represents the DNA strand of *path 3* (*path 4*). The four red lines in the upper right corner indicate that the DNA strands of *path 3* (*path 4*) in the $5' - 3'$ direction pair completely with the single-stranded molecules of $\delta(s_0, p) = s_0, \delta(s_0, p) = s_0, \delta(s_0, q) = s_2$ and $\delta(s_2, p) = s_2$ from the 1st base to the 15th base, from the 16th base to the 30th base, from the 31st base to the 51st base, and from the 52nd base to the 66th base in the $3' - 5'$ direction, respectively. According to the legend on the right, the pairing probability is close to 100%, which suggests specific hybridization.

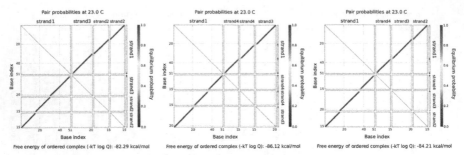

(a) Path 1 (2): pairing case 1(b) Path 1 (2): pairing case 2(c) Path 1 (2): pairing case 3

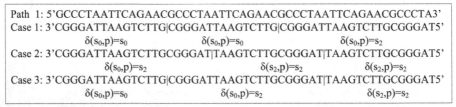

(d) Details for the three cases of base pairing for path 1 and the transitions of the FSA of Op

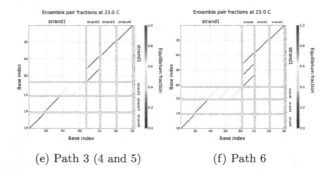

(e) Path 3 (4 and 5)   (f) Path 6

Fig. 7. Hybridization of the single strands of path i of M_1 and the transitions of the FSA of Op: pairing positions and probabilities

This also indicates that *path 3 (path 4)* is accepted by the FSA of pSq. In other words, paths *1, 2,* and *5* are not accepted by the FSA of pSq, whereas paths *3, 4,* and *6* are accepted by the FSA of pSq, which indicates that M_1 satisfies $E(pSq)$ but not $A(pSq)$.

The Verification of EYp and AYp. Hybridization of the DNA strands of the reverse paths of M_1 with the transitions of the FSA of Yp is shown in Fig. 9, where strand1 represents the DNA strand of *path i*, and strand2, strand3 and strand4 represent the DNA strand of $\delta(s_2, (p|q)) = s_2, \delta(s_1, p) = s_2$ and $\delta(s_0, (p|q)) = s_1$, respectively. As shown in Fig. 9(a), strand1 denotes the DNA

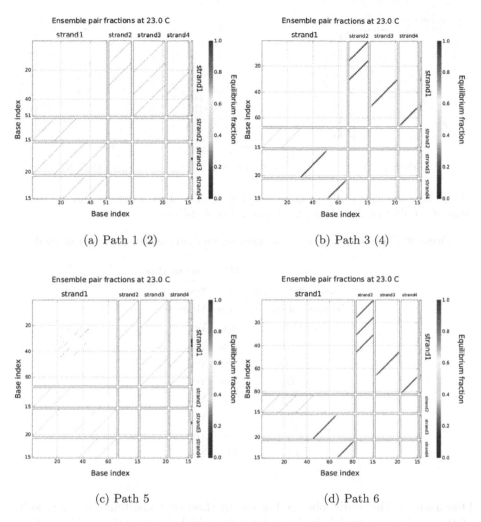

Fig. 8. Hybridization of the single-stranded molecules of path i of M_1 with the transitions of the FSA of pSq: pairing positions and probabilities

strands of *path 1* (*path 2*). The three red lines in the upper right corner indicate that the single-stranded molecules of *path 1* (*path 2*) in the $5' - 3'$ direction pair completely with the single-stranded molecules of $\delta(s_0, p) = s_1$, $\delta(s_1, p) = s_2$ and $\delta(s_2, p) = s_2$ from the 1st base to the 18th base, from the 19th base to the 36th base, and from the 37th base to the 51st base in the $3' - 5'$ direction, respectively. According to the legend on the right, the pairing probability is close to 100%, which suggests specific hybridization. This indicates that *path 1* (*path 2*) is accepted by the FSA of Yp. The same situation exists for paths *3*, *4*, *5* and *6*, as shown in Fig. 9(b) and (c). In other words, all the reverse paths are accepted by the FSA of Yp, which indicates that M_1 satisfies both EYp and AYp.

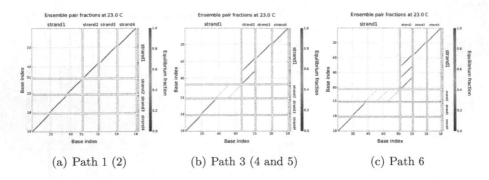

(a) Path 1 (2)   (b) Path 3 (4 and 5)   (c) Path 6

Fig. 9. Hybridization of the single-stranded molecules of path i of M_1 and the transitions of the FSA of Yp: pairing positions and probabilities

Table 6. Comparison between the existing method and the proposed method

		[37]	Our method
Autonomy or not		No	Yes
Number of species		Less	More
Power	EOp	Yes	Yes
	AOp	Yes	Yes
	EHp	Yes	Yes
	AHp	Yes	Yes
	EYp	No	Yes
	AYp	No	Yes
	$E(pSq)$	No	Yes
	$A(pSq)$	No	Yes

Discussion. The results obtained by our method are consistent with the results of the traditional model-checking methods, which confirm the validity of our method. In this experiment, we do not verify the detection phase because as long as the DNA coding scheme is valid and the hybridization is correct, the corresponding product can be correctly detected by available detection means.

The comparison between the existing method [37] and our method is shown in Table 6. Unlike [37], our method is autonomous. Given the coding scheme of a system model and an FSA of a CTLP formula, the computation is spontaneously completed without manual operations, thus solving the problems related to time consumption and error proneness. From the perspective of the number of species, the fewer the encoding species involved in the biochemical reactions, the lower the complexity of biochemical reactions, and the higher the efficiency of biochemical detection will be. In [37], only the system model needs to be encoded. In our method, the FSA of the CTLP formula needs to be encoded except for the system model. Therefore, the DNA encoding species of our method is more than the method of [37]. Nonetheless, the checking capability of our method is more powerful than that of [37]. The method of [37] can check only four basic CTLP formulas, while our method can check all eight basic CTLP formulas.

5 Conclusions

We propose a DNA model-checking method for CTLP and verify its validity through simulations. The checking power of our method is by far the strongest, but at the expense of higher biochemical reaction complexity. Despite this, improving checking power is the key. In addition, our method can theoretically check all the CTLP formulas given the coding scheme of a system model and an FSA of the CTLP formula, which greatly improve the power of the existing CTL model-checking methods.

References

1. Xu, J., Tan, G., Fan, Y., Guo, Y.: DNA computer principle, advances and difficulties (IV): on the models of DNA computer. Chin. J. Comput. **30**(6), 881–893 (2007). (In Chinese)
2. Adleman, L.: Molecular computation of solutions to combinatorial problems. Science **266**(5187), 1021–1023 (1994)
3. Lipton, R.: DNA solution of hard computational problems. Science **268**(5210), 542–545 (1995)
4. Roweis, S., et al.: A sticker-based model for DNA computation. J. Comput. Biol. **5**(4), 615–629 (1998)
5. Ouyang, Q., Peter, D., Liu, S., Libchaber, A.: DNA solution of the maximal clique problem. Science **278**(5337), 446–449 (1997)
6. Pan, L., Xu, J., Liu, Y.: A surface-based DNA algorithm for the minimal vertex cover problem. Prog. Nat. Sci. **12**(1), 78–80 (2003)
7. Li, K., Yao, F., Xu, J., Li, R.: An O(1.414n) volume molecular solutions for the subset-sum problem on DNA-based supercomputing. Chin. J. Comput. **30**(11), 1947–1953 (2007). (in Chinese)
8. Xu, J., Qiang, X., Yang, Y., Wang, B.: An unenumerative DNA computing model for vertex coloring problem. IEEE Trans. Nanobiosci. **10**(2), 94–98 (2011)
9. Sakamoto, K., Gouzu, H., Komiya, K., Kiga, D., Yokoyama, S., Yokomori, T., Hagiya, M.: Molecular computation by DNA hairpin formation. Science **288**(5469), 1223–1226 (2000)
10. Xu, J., Qiang, X., Zhang, K., Zhang, C., Yang, J.: A DNA computing model for the graph vertex coloring problem based on a probe graph. Engineering **4**(1), 61–77 (2018)
11. Yang, J., Yin, Z., Huang, K., Cui, J.: The maximum matching problem based on self-assembly model of molecular beacon. Nanosci. Nanotechnol. Lett. **10**, 213–218 (2018)
12. Yin, P., Turberfield, A.J., Sahu, S., Reif, J.H.: Design of an autonomous DNA nanomechanical device capable of universal computation and universal translational motion. In: Ferretti, C., Mauri, G., Zandron, C. (eds.) DNA 2004. LNCS, vol. 3384, pp. 426–444. Springer, Heidelberg (2005). https://doi.org/10.1007/11493785_37

13. Winfree, E., Liu, F., Wenzler, L., Seeman, N.: Design and self-assembly of two-dimensional DNA crystals. Nature **394**(6693), 539–544 (1998)
14. Martínez-Pérez, I., Zimmermann, K., Ignatova, Z.: An autonomous DNA model for finite state automata. Int. J. Bioinform. Res. Appl. **5**(1), 81–96 (2009)
15. Xu, J.: Probe machine. IEEE Trans. Neural Netw. Learn. Syst. **27**(7), 1405–1416 (2016)
16. Shi, X., Wu, X., Song, T., Li, X.: Construction of DNA nanotubes with controllable diameters and patterns by using hierarchical DNA sub-tiles. Nanoscale **8**(31), 14785–14792 (2016)
17. Pan, L., Wang, Z., Li, Y., Zhang, C.: Nicking enzyme-controlled toehold regulation for DNA logic circuits. Nanoscale **9**(46), 18223–18228 (2017)
18. Yang, J., et al.: Entropy-driven DNA logic circuits regulated by DNAzyme. Nucleic Acids Res. **46**(16), 8532–8541 (2018)
19. Yang, J., Jiang, S., Liu, X., Pan, L., Zhang, C.: Aptamer-binding directed DNA origami pattern for logic gates. ACS Appl. Mater. Interfaces **8**(49), 34054–34060 (2016)
20. Yang, J., Song, Z., Liu, S., Zhang, Q., Zhang, C.: Dynamically arranging gold nanoparticles on DNA origami for molecular logic gates. ACS Appl. Mater. Interfaces **8**(34), 22451–22456 (2016)
21. Zhang, C., Yang, J., Jiang, S., Liu, Y., Yan, H.: DNAzyme-mediated DNA origami pattern for logic gates. Nano Lett. **16**(1), 736–741 (2016)
22. Zhang, C., Shen, L., Liang, C., Dong, Y., Yang, J., Xu, J.: DNA sequential logic gate using two-ring DNA. ACS Appl. Mater. Interfaces **8**(14), 9370–9376 (2016)
23. Wang, Y., Han, Q., Cui, G., Sun, J.: Hiding messages based on DNA sequence and recombinant DNA technique. IEEE Trans. Nanotechnol. **18**, 299–307 (2019)
24. Benenson, Y., Gil, B., Ben-Dor, U., Adar, R., Shapiro, E.: An autonomous molecular computer for logical control of gene expression. Nature **429**(27), 1–6 (2004)
25. Nakagawa, H., Sakamoto, K., Sakakibara, Y.: Development of an *In Vivo* computer based on *Escherichia coli*. In: Carbone, A., Pierce, N.A. (eds.) DNA 2005. LNCS, vol. 3892, pp. 203–212. Springer, Heidelberg (2006). https://doi.org/10.1007/11753681_16
26. Martínez-Pérez, I.M., Zhang, G., Ignatova, Z., Zimmermann, K.H.: Computational genes: a tool for molecular diagnosis and therapy of aberrant mutational phenotype. BMC Bioinform. **8**(1), 365–365 (2007)
27. Rinaudo, K., Bleris, L., Maddamsetti, R., Subramanian, S., Weiss, R., Benenson, Y.: A universal RNAi-based logic evaluator that operates in mammalian cells. Nat. Biotechnol. **25**(7), 795–801 (2007)
28. Xie, Z., Wroblewska, L., Prochazka, L., Weiss, R., Benenson, Y.: Multi-input RNAi-based logic circuit for identification of specific cancer cells. Science **333**(6047), 1307–1311 (2011)
29. Clarke, E., Grumberg, O., Peled, D.A.: Model Checking. The MIT Press, Cambridge (1999)
30. Burch, J.R., Clarke, E.M., Long, D.E., McMillan, K.L., David, L.D.: Symbolic model checking for sequential circuit verification. IEEE Trans. Comput. Aided Des. Integr. Circuits Syst. **13**(4), 401–424 (1994)
31. Clarke, E.M., et al.: Verification of the futurebus+ cache coherence protocol. Formal Methods Syst. Des. **6**(2), 217–232 (1995). https://doi.org/10.1007/BF01383968
32. Chan, W., et al.: Model checking large software specifications. IEEE Trans. Softw. Eng. **24**(7), 498–520 (1998)

33. Fu, S., Tayssir, T.: Efficient CTL model-checking for pushdown systems. Theoret. Comput. Sci. **549**(3), 127–145 (2014)
34. Brim, L., Češka, M., Šafránek, D.: Model checking of biological systems. In: Bernardo, M., de Vink, E., Di Pierro, A., Wiklicky, H. (eds.) SFM 2013. LNCS, vol. 7938, pp. 63–112. Springer, Heidelberg (2013). https://doi.org/10.1007/978-3-642-38874-3_3
35. Emerson, E.A., Hager, K.D., Konieczka, J.H.: Molecular model checking. Int. J. Found. Comput. Sci. **17**(4), 733–742 (2006)
36. Zhu, W., Wang, Y., Zhou, Q., Nie, K.: Model checking computational tree logic using sticker automata. In: Gong, M., Pan, L., Song, T., Zhang, G. (eds.) BIC-TA 2016. CCIS, vol. 681, pp. 12–20. Springer, Singapore (2016). https://doi.org/10.1007/978-981-10-3611-8_2
37. Han, Y., Zhou, Q., Jiao, L., Nie, K., Zhang, C., Zhu, W.: Model checking for computation tree logic with past based on DNA computing. In: He, C., Mo, H., Pan, L., Zhao, Y. (eds.) BIC-TA 2017. CCIS, vol. 791, pp. 131–147. Springer, Singapore (2017). https://doi.org/10.1007/978-981-10-7179-9_11
38. Lichtenstein, O., Pnueli, A., Zuck, L.: The glory of the past. In: Parikh, R. (ed.) Logic of Programs 1985. LNCS, vol. 193, pp. 196–218. Springer, Heidelberg (1985). https://doi.org/10.1007/3-540-15648-8_16
39. Kupferman, O., Pnueli, A.: Once and for all. J. Comput. Syst. Sci. **78**(2012), 981–996 (1995)
40. Laroussinie, F., Schnoebelen, P.: Specification in CTL+Past for verification in CTL. Inf. Comput. **156**(1–2), 236–263 (2000)
41. NUPACK. http://www.nupack.org. Accessed 9 Aug 2019

Review on DNA Cryptography

Ying Niu[1(✉)], Kai Zhao[2], Xuncai Zhang[2], and Guangzhao Cui[2]

[1] School of Architecture Environment Engineering,
Zhengzhou University of Light Industry, Zhengzhou 450002, China
niuying@zzuli.edu.cn
[2] School of Electrical and Information Engineering,
Zhengzhou University of Light Industry, Zhengzhou 450002, China
zhangxuncai@163.com

Abstract. As a new encryption method, the DNA cryptography takes DNA as the information carrier, and makes full use of the advantages of DNA molecules, such as ultra-high storage density, ultra-low energy consumption, and the potential of ultra-large-scale parallel computing to realize the cryptographic functions of information encryption, authentication, and signature. In this paper, the current research status of DNA cryptography is reviewed, the encryption and authentication based on the DNA molecule and the analysis of traditional cryptography based on the DNA computing are introduced, and the DNA encryption algorithm based on the DNA origami is discussed. The future development of the DNA cryptography is also prospected.

Keywords: DNA molecule · DNA cryptography · DNA computing · Information security technology

1 Introduction

As one of the most important parts of information security technology, cryptography is an active security defense strategy, which provides protection for information storage and transmission. At the same time, cryptography is the basis of other security technologies, such as digital signature and key management. With the development of computing technology and mathematical theory, cryptography has been threatened. For instance, the 56bit block cipher system DES has been cracked, and the security of the famous public key system RSA has been threatened [1]. The two cryptographic algorithms MD5 and SHA-1, which have been widely used in the world, were cracked by Professor Wang of China in 2004 and 2005, respectively. This achievement has aroused great repercussions in the international community, especially in the field of international cryptography [2]. Then, the international cryptographer Lenstra used the MD5 collision provided by Professor Wang to forge a digital certificate meeting the X.509 standard, which shows that the deciphering of the MD5 is not only a theoretical deciphering result, but can lead to an actual attack, and the

© Springer Nature Singapore Pte Ltd. 2020
L. Pan et al. (Eds.): BIC-TA 2019, CCIS 1160, pp. 134–148, 2020.
https://doi.org/10.1007/978-981-15-3415-7_11

withdrawal of the MD5 is imminent. In 2015, Wang said that the currently SHA-1 has been theoretically deciphered and is not far from practical application [3]. In February 2017, Google announced that its official SHA-1 encryption algorithm had been cracked. With the continuous improvement of cryptanalysis and attack methods, the increasing speed of computer operations, and the growing demand for cryptographic applications, it is urgent to develop cryptographic theory and innovative cryptographic algorithms. People want to seek more secure and simpler cryptosystem. At the same time, biotechnology such as genetic engineering has developed rapidly, especially after the completion of the human genome sequencing project, a huge number of DNA sequences have been generated, which induces the idea of using biotechnology methods to encrypt information.

DNA is being exploited for molecular computing, data storage, and cryptography because of its ultra-large-scale parallelism, ultra-high storage density, and ultra-low energy consumption. In 1994, Professor Adleman successfully solved the Hamilton problem of seven vertices with the aid of DNA molecular computing [4]. Subsequently, the DNA computing is used to solve many complex decision-making problems. For example, Ouyang et al. gave the method of using DNA computing to solve the maximum clique problem of graphs [5]. Lipton et al. transformed the satisfiability problem into the directed Hamilton problem [6] and gave the DNA computing model to solve the satisfiability problem [7 9]. Meanwhile, the feasibility of using DNA system to construct the Turing machine has been proved [10].

DNA molecular computing has a unique data storage and computing mechanism. It solves the traditional difficult problems from a new perspective and successfully solves the NP complete problem. It also opens up a new field for cryptography and brings new opportunities for information security. With the increasing maturity of DNA computing methods, the DNA cryptography theory emerges as the times require. Its main principle is to use modern biotechnology as a tool and DNA as a data carrier to realize cryptographic functions such as encryption, steganography, authentication, and signature by mining the characteristics of high parallelism and high storage density of DNA molecule itself. The development and application potential of the DNA computing and DNA cryptography itself will trigger a new technological revolution in the field of information security.

2   DNA Computing and Traditional Cryptanalysis

2.1   DNA Computing

Computational problems have always been an important issue in the field of science and technology [11], and the development of cryptography is closely related to the development of computing technology. Now, cryptography or traditional cryptography, is based on mathematical cryptography and relies on mathematical difficulties. With the continuous improvement of computing ability, the cracking of mathematical difficulties has also been challenged. High-performance computing is widely used in the fields of economy, aerospace, cryptography, and biological information processing. On the one hand, it has played an important role in national economic and social life, on the other hand, it also faces major problems. For example, as the problems to be solved become more and more complex, new high-performance computing methods, technologies, and equipment are urgently needed. As the limit of the Moore's law approaches, the physical limit and high economic investment become the obstacles that high-performance computing cannot overcome. Therefore, as a new computing model, biological computing has attracted much attention. The development of the DNA computing in the field of biological computing has also aroused the interest of more researchers. It introduces new data structures and computing methods, and shows unique application prospects in cryptography, steganography and other fields.

Professor Feynman, who won the Nobel Prize in physics in 1959, put forward the idea of building a molecular computer [12], which marked the beginning of biological computing. However, due to the limitations of experimental conditions and technologies at that time, these ideas could not be verified by experiments, so the period of biological computing was only the beginning of the theoretical prototype.

In 1994, Professor Adleman of the University of California, USA, first used the DNA computing method to solve the "seven-vertex Hamilton path" problem and realized the DNA molecular computation (as shown in Fig. 1(a)) [4]. The DNA molecular computation uses DNA double helix structure and complementary base-pairing principle to encode information and map the object to be calculated into DNA molecular strand. Various data pools are generated by the action of biological enzymes. Specific enzymes act as "software" to perform the various information processing tasks required. Then, according to certain rules, the data operations of the original problem are mapped highly parallel to the controllable biochemical reaction process of the DNA molecular strand. Finally, molecular biological technologies such as (polymerization chain reaction PCR, polymerization overlap amplification POA), ultrasonic degradation, affinity chromatography, cloning, mutagenesis, molecular purification, electrophoresis, magnetic bead separation, etc. were used to read the calculation results.

In 1995, scientists from many countries and regions discussed the feasibility of the DNA computing at the First International DNA Conference, and they generally believed that the DNA computing is a field of great development value. In the same year, Professor Lipton from Princeton university in the United States provided a DNA computing model for the problem of satisfiability by coding DNA sequences, which can simulate the logic gate circuit of digital electronics to make the judgment of "yes" and "no" [13]. Five years later, his thoughts were validated by Faulhammer in the lab [14].

In 1995, Boneh et al. proposed the splicing model DNA algorithm [15], Adleman et al. proposed the sticking model DNA algorithm [16]. These two DNA algorithm models can crack DES cryptosystem. Boneh et al. used four months to crack the 56-bit key, which is the first time to use the DNA computing to

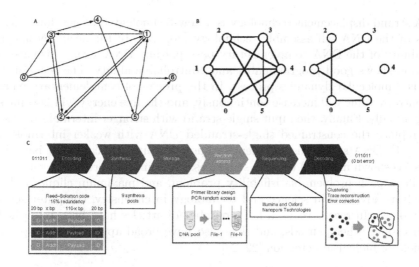

Fig. 1. DNA computing systems with different functions.

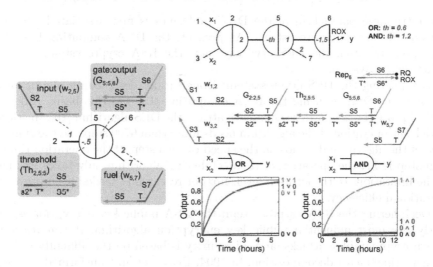

Fig. 2. Computing model of DNA strand displacement reaction.

crack the traditional encryption standard DES cryptosystem [15]. In 1996, Oliver proposed a DNA computing method for multiplying the Boolean matrix and positive real matrix [17]. Eng put forward a living DNA computing method for SAT problem [18]. Guarnieri et al. realized the access and carry of binary number by introducing the concepts of position operator, position shift operator and bit node symbol, which are similar to the method of digit exchange in electronic computer [19]. In 1997, Ouyang presented a DNA computing model, which can be used to solve the maximum clique problem of graphs [5].

DNA strand displacement technology is a new technology that developed on the basis of the DNA self-assembly technology [20]. Due to the accuracy and predictability of the DNA complementary base pairing. DNA strand displacement reaction shows good programmable and controllable ability. The reaction process is a molecular dynamics process. In the process of complementary pairing, the entropy value will increase continuously, and the free energy will become stable gradually. Finally, the input single-strand with stronger intermolecular force will replace the constrained single-stranded DNA with weaker intermolecular force [21]. DNA strand displacement technology is used to construct basic logic gates (as shown in Fig. 2) to solve some classical computer problems, such as the Hamilton path problem, satisfiability problem, and maximum clique problem, as well as advances in science and technology in chemistry, computer science, and biology. In recent years, a large number of articles have been published in Sciences and other journals, and they also have broad application prospects in the field of biological detection [22–24].

2.2 Traditional Cryptanalysis Based on the DNA Computing

The ultra-large parallelism of the DNA molecules is first associated with the analysis of current cryptosystems. At present, the DNA computing has been attempted to crack the DES cryptosystem, the RSA cryptosystem, and the NTRU cryptosystem.

Deciphering the DES cryptosystem: Boneh et al. designed an algorithm to decipher the DES by using the DNA computing [15]. On this basis, Adleman et al. [14] use the DNA sticker model to attack the DES algorithm [16]. In sticker model, the operation of the algorithm needs more than 6700 steps, so the running time of the algorithm depends on the speed of each step. If the operation time of each step is 1 s, the operation of the deciphering algorithm only takes 2 h. Due to the limitations of the existing level, further research is needed to improve the operational efficiency.

Deciphering the RSA cryptosystem: the RSA public key encryption is currently the most influential public key encryption algorithm. It can resist all known cryptographic attacks, and its security is based on the difficulty of large integer prime factor decomposition. In 1994, Beaver et al. transformed the large number decomposition problem into the Hamilton path problem with the idea of Adleman. The molecular calculation was used to solve the problem. It was analyzed that when decomposing 1000 digits, at least 10^6 vertices were needed. However, because the volume of the DNA solution needed exceeded 10 million liters, it was not feasible in practice. In 2007, Yuriy Brun proposed an addition and multiplication model based on the self-assembly model of the DNA tiles, which can be distributed through molecules or large computer networks. Subsequently, Yuriy proposed to decompose non-deterministic integers using the DNA tile self-assembly model [25,26], as shown in Fig. 3. From a theoretical point of view, the feasible solutions can be obtained by executing each non-deterministic path. The model is intended to find successful solutions in a variety of non-deterministic paths, as well as to find constraints on non-deterministic systems.

The key of the model lies in the establishment of the non-deterministic guessing factor system. The system randomly selects two numbers, and the product of the two numbers is obtained by the system, and the product is compared with the input number. If they are equal, two factors of the input number are found. Although the number of the DNA tiles and the error precision required in the experiment exceed the current technical level, the self-assembly model of the DNA tiles provides a powerful theoretical support for the cracking of the RSA algorithm.

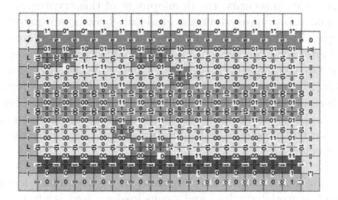

Fig. 3. Nondeterministic polynomial time factoring in the tile assembly model.

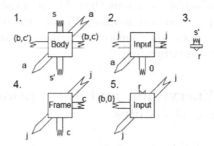

Fig. 4. Structure of DNA tile.

Deciphering the NTRU cryptosystem: the NTRU is considered as the most promising public key cryptosystem in the 21*st* century. It is fast and secure, and is widely used in data encryption, digital signature and other fields. Using the idea of the self-assembly, Pelletier realized the convolution calculation required in NTRU by defining the corresponding three-dimensional molecular tile structure (as shown in Fig. 4). By means of brute force cracking, all possible keys are convoluted and the key is found according to the characteristics of the NTRU.

But the scheme can only prove its feasibility in theory. A non-deterministic algorithm for deciphering the NTRU public key cryptosystem using the DNA self-assembly is presented [27].

Deciphering Diffie-Hellman key exchange algorithm of the elliptic curve: In 1985, Koblitz and Millef proposed the elliptic curve cryptosystem problems in their respective public key encryption algorithms. The system is a kind of cryptosystem which uses elliptic curve finite group instead of finite cyclic group in public key cryptosystem based on the discrete logarithm problem over finite field. Based on the advantages of the elliptic curve cryptosystem, especially in mobile communication security, the development of this cryptosystem has been accelerated, and this cryptosystem has gradually become an important branch of cryptography. Its commercial and military value is attracting more and more attention. Li gave the algorithm of solving the elliptic curve discrete logarithm based on the DNA computing model [28], Chen gave the algorithm of solving finite field $GF(2^n)$ multiplication inverse element and division operation using the DNA self-assembly model [29]. With the increasing amount of computation, the complexity of the DNA computing model in space increases significantly. At present, the method proposed by Koblitz et al. can only break the symmetric cryptosystem below 64 bits.

Deciphering knapsack cryptosystem: Knapsack cryptosystem are a typical optimization problem in operations research. It has important applications in budget control, material cutting, and cargo loading, and is often studied as a sub-problem of other issues. Darehmiraki et al. used the high parallelism of the DNA computing to solve the 0–1 knapsack problem in test tube [30]. By encoding technology, the problem to be solved is mapped to a set of DNA sequences, and the initial spatial solution is formed in the test tube. Then the DNA strands corresponding to the infeasible solutions that do not satisfy the constraints are deleted by means of separation and merging. Finally, the objective function values corresponding to each feasible solution strand are obtained, and the optimal solution is obtained by comparison.

3   Information Encryption and Decryption Based on DNA Molecules

The development of the DNA cryptography benefits from the research progress of the DNA computing (also known as molecular computing or biological computing). On the one hand, cryptosystems are always more or less related to the corresponding computing models. On the other hand, some biotechnologies used in the DNA computing and DNA cryptosystem have also been applied to some extent. With the development of the DNA nanotechnology, the biology cryptography has become a new field of cryptography. Compared with traditional cryptography based on mathematical problems, the DNA cryptography is not only based on mathematical problems, but also depends on biotechnology, which makes it more difficult to decipher DNA cryptography and makes the DNA cryptography more secure.

With the development of science and technology, cryptanalysis is becoming more and more advanced. It is imperative to develop new encryption methods. As an emerging field of information security, the DNA cryptography is expected to become one of the three branches of cryptography along with traditional cryptography and quantum cryptography. Here are some typical encryption schemes.

3.1 DNA Encryption and Decryption of the One-Time Pad

One-time pad refers to the use of random keys of the same length as messages in stream ciphers, and the key itself is used only once. Its security mainly depends on the random generation and non-reuse of keys. If the attacker does not have a password at a time, even with great computing ability, it cannot decipher a password at a time. This algorithm is absolutely safe in theory.

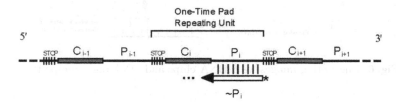

(a) One-time-pad Codebook DNA Sequences.

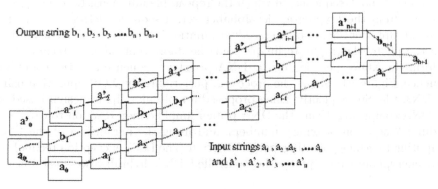

(b) XOR calculation using DNA tile.

Fig. 5. DNA cryptosystems using random one-time-pads.

DNA has the characteristics of small size, and it has the ability of information storage which is beyond the reach of traditional information storage media. As an information carrier, it can solve the huge cryptographic generation and storage problems well. In 1999, Gehani et al. of Duke University in the United States used the DNA to design a map substitution method and a DNA chip XOR method to realize a one-time pad encryption method [31]. The map alternative method replaces the fixed length DNA plaintext sequence unit with the corresponding

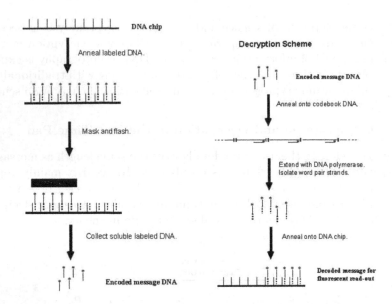

Fig. 6. Encrypting images with DNA chips and DNA one-time-pads.

DNA ciphertext sequence according to the defined map table. Figure 5(a) shows a one-time codebook sequence, in which the repeatable unit consists of a sequence letter C_i from the cryptographic alphabet set, a sequence letter P_i from the plaintext alphabet and a polymerase "termination" sequence. The XOR method uses the photolithography technique and the fluorescent labeling technique to perform the XOR operation of the DNA plaintext sequence and the codebook sequence. Figure 5(b) shows the process of performing the XOR operation using the DNA tile. Subsequently, Chen proposed a molecular cipher design based on the DNA computing. Using the DNA primer amplification reaction to carry out modular 2 addition of binary numbers, and using the parallelism of the DNA computing to achieve one-time pad encryption. The DNA encryption technology is to encrypt and decrypt by the controlled DNA hybridization reaction (as shown in Fig. 6).

In addition, the DNA hybridization reactions premised on the denaturation and renaturation of the DNA molecules include specific hybridization reactions and non-specific hybridization reactions. Among them, the specific hybridization reaction does not consider various constraints, the randomly generated DNA single strands constitute one DNA microdot, and each DNA microdot contains all DNA single strands. Each DNA microdot is a one-time pad, and once the password is generated, it can be encrypted once. The main processes of the encryption include data processing, key distribution, XOR operation, and information transmission. This encryption method can accurately obtain the results of encryption and decryption. However, in the actual biochemical operation process, how to quickly separate the required key, correct the errors of the DNA codebook and long-term preservation need further research.

3.2   DNA Cryptosystem Based on the Nanotechnology

DNA computing is a computational problem solved by the DNA technology. In DNA cryptography, various biological problems have been studied and used as the security basis of DNA cryptosystem. The process of encryption and decryption of the DNA codes is regarded as a computational process, and not all DNA computations are related to confidentiality.

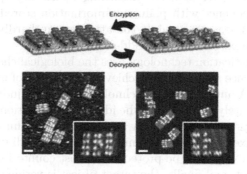

Fig. 7. Design of the encrypted nano-Morse code tile with corresponding code translation.

DNA nanotechnology is an interdisciplinary research field. It mainly uses the characteristics of nano-size DNA, rigid structure and strong coding to construct various nanostructures, which can be applied in biomedicine, chemistry, materials, and other fields. It can not only be used as a carrier of genetic information, but also as a tool to construct nanostructures. The excellent surface addressability of the DNA origami makes it possible to arrange nanoparticles or molecules in an orderly manner, which has great application potential in many fields due to the special properties of nanoparticles or molecules themselves. The nanoparticles or molecules that have been arranged in origami include nucleic acids, proteins, metal nanoparticles, quantum dots, and fluorescent molecules.

In 2013, Wong et al. demonstrated the directional reversible selection of the DNA origami modified with streptavidin by using different binding relationships among biotin, desulfurization biotin and streptavidin (Fig. 7). They presented an encrypted Morse code "NANO" and a reversible transformation of the letters "I" and lower-case "i", which realized the encryption and decryption of information on the DNA origami [32].

3.3   DNA Hiding and Authentication

The DNA hiding is to use biotechnology to process the encrypted information and hide plaintext information into the DNA vector, so as to realize information hiding and achieve the purpose of transmitting information. The DNA steganography has two layers of security: mathematical security and biological security. The traditional steganography technology has low security and is easy to decipher, while the DNA steganography technology has higher security and operability than traditional steganography technology. The researchers mixed the DNA sequences containing plaintext information with a large number of redundant DNA sequences without plaintext information to achieve the purpose of hiding. The mixed DNA sequences exist in the form of the DNA microdots, one of which contains tens of billions of DNA sequences, an the DNA sequences containing plaintext information are specially labeled. The correct receiver can easily find the specially labeled DNA sequence according to the key (primer), and then extract the DNA sequence with plaintext information stored from the numerous redundant DNA sequences, while the attacker can hardly obtain the DNA sequence with plaintext information hidden.

The DNA authentication technology uses the biological characteristics of the DNA, namely sequence specificity, to achieve the purpose of anti-counterfeiting. At present, the DNA authentication technology is used in judicial, financial and other fields to accurately authenticate the identity information of biological individuals. The basic process of the DNA authentication technology is as follows: first, the DNA is mixed with the medium, and then the medium with the DNA is coated on the target object to be protected against counterfeiting or infiltrated into the target object, and finally, the target object is verified to be the original marker by examining the DNA on the target object.

In 1999, Clelland et al. [33] succeeded in DNA steganography through the encryption of the DNA base triad through the famous message "June 6 Invasion: Normandy" during World War II. Due to the reliability of the complementary DNA bases, the information encrypted in DNA in this way is amplified by PCR and read by the DNA sequencing instrument. Figure 8 is the basic flow of this information hiding method. In 2000, the DNA Technology company in Canada successfully applied the information hiding method of Clelland in the trademark certification of goods for the Sydney Olympic Games. Everything from T-shirts to coffee cups is marked with a special ink containing the DNA of an unknown athlete. Using a portable scanner, the appraiser can identify the authenticity of the souvenir by the DNA information in the mark. Not only is the security tag cheaper than a standard trademark, but also very difficult to forge such randomly selected DNA information. In 2011, Mousa et al. successfully recovered the secret information by hiding secret information into the DNA sequence using reversible contrast map technology [34].

The DNA-based watermarking technology using the DNA encryption algorithm can not only print the DNA watermarking on objects, but also implant the DNA watermarking in living bodies to verify user identity or copyright information by identifying the DNA authentication information [35]. At present, the DNA authentication technology has been developed quite mature and widely used in accounting-based information security. If the DNA steganography is

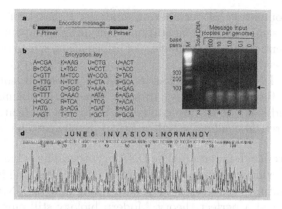

Fig. 8. Celland et al.'s approach to information hiding.

used in accounting-based identification or identification, and the information authentication in a broader sense can be carried out.

3.4 A Cryptographic Method Based on the Pseudo-DNA Computing

The previous encryption methods all require complex biotechnology and harsh biological experimental environment. The method of pseudo-DNA computing mainly adopts the idea of the DNA computing, including the DNA coding, DNA computing, and so on. In 2009, Ning et al. proposed to use the basic idea of the central law of biomolecules to achieve the encryption of information [36]. However, the security of single use still has some deficiencies, so it is usually used together with other information encryption theories to expand a new direction for the development of information encryption. The usual method is to establish a special connection between the four bases A, T, C, and G and the binary in the computer, and convert the binary number into A, T, C, and G, and encode the binary. Then use the idea of the DNA operation, namely the DNA XOR, DNA addition, DNA complementation and other rules to transform the transformed DNA sequence, complete the diffusion operations, and then combine with some mature encryption methods, such as chaos theory, to achieve image encryption [37,38].

4 Conclusions and Prospects

With the increasing computing ability of computers, the security of the traditional cryptography based on mathematical cryptography has been seriously flawed. The DNA cryptography, as an interdisciplinary subject of life science and cryptography, has advantages that traditional cryptography does not have. Its tremendous development potential breaks through the data complexity and computational complexity of deciphering analysis. It has been applied to the cryptanalysis, design and implementation of cryptographic functions such as encryption, authentication, and signature of information, which greatly improves the speed and encryption security.

Although the DNA computing has its own advantages in solving computational problems, there are few ways to solve practical problems by the DNA cryptography, and it is not mature enough in biological applications. There are still a lot of theoretical challenges and practical problems to be solved urgently:

(1) Weak theoretical foundation. The DNA cryptography is based on the DNA computing, which was proposed only in the mid-1990s, and its development is still immature. Neither the DNA computing nor the DNA cryptography has established a perfect theory. Modern biology still emphasizes experiment rather than theory. So that the operable cryptographic model cannot be formed. How to store the intermediate results of the DNA computing, and how to reduce the errors in the process of information processing and the scope of application remain to be further studied.

(2) Implementation is difficult and application is costly. Throughout the whole process of the cipher process, from the generation and distribution of the keys, to the artificial synthesis sequence of information coding, the transmission of messages, the amplification and decryption of sequences, the sequencing of sequences, and the restoration of plaintext information, are all artificial biochemical processes realized in the laboratory. Under the current biotechnology level, these steps can only be achieved in a fully equipped and perfect sterile room, which restricts the application of the cryptography in practice because of many restrictions, difficult operation and high accuracy requirements.

(3) The DNA cryptosystem needs to be improved urgently. In most DNA cryptosystems, the structure is based on "primer-coding rules". The primers and encoding methods act as the key together. In these systems based on "primer-coding rules", the primers play a core role, while coding rules that transform data into DNA sequences are mostly within the scope of classical cryptography. If the opponent obtains primers, these schemes which depend on coding rules may be cracked by classical cryptanalysis methods such as alphabetic frequency analysis to get the message content. Therefore, the research on the DNA cryptography needs to explore new methods that do not rely on coding rules, so that opponents cannot use classical cryptographic analysis to obtain secret information.

With the rapid development of the DNA nanotechnology, the potential of the DNA molecule in the field of information science will be constantly exploited, especially the emergence of the DNA origami, which makes the DNA nanostructures more and more diverse and complex. These nanostructures can be used as a code to provide security for information transmission, and has more advantages in security and efficiency, and has a broader development prospects in DNA cryptography research.

Acknowledgments. The work for this paper was supported by the Key Scientific Research Projects of Henan High Educational Institution (18A510020). National Natural Science Foundation of China (Grant nos. 61572446, U1804262, and 61602424), and Key Scientific and Technological Project of Henan Province (Grant nos. 174100510009, 192102210134).

References

1. Coppersmith, D.: The data encryption standard (DES) and its strength against attacks. IBM J. Res. Dev. **38**(3), 243–250 (1994)
2. Wang, X., Yu, H.: How to break MD5 and other hash functions. In: Cramer, R. (ed.) EUROCRYPT 2005. LNCS, vol. 3494, pp. 19–35. Springer, Heidelberg (2005). https://doi.org/10.1007/11426639_2
3. Stallings, W., Brown, L., Bauer, M.D., Bhattacharjee, A.K.: Computer Security: Principles and Practice. Pearson Education, Upper Saddle River (2012)
4. Adleman, L.M.: Molecular computation of solutions to combinatorial problems. Science **266**, 1021–1024 (1994)
5. Ouyang, Q., Kaplan, P.D., Liu, S., Libchaber, A.: DNA solution of the maximal clique problem. Science **278**(5337), 446–449 (1997)
6. Lipton, R.-J.: Using DNA to solve NP-complete problems. Science **268**(4), 542–545 (1995)
7. Braich, R.S., Chelyapov, N., Johnson, C., Rothemund, P.W., Adleman, L.: Solution of a 20-variable 3-SAT problem on a DNA computer. Science **296**(5567), 499–502 (2002)
8. Gifford, D.K.: On the path to computation with DNA. Science **266**(5187), 993–995 (1994)
9. Sakamoto, K., et al.: Molecular computation by DNA hairpin formation. Science **288**(5469), 1223–1226 (2000)
10. Calude, C.S., Păun, G.: Bio-steps beyond turing. BioSystems **77**(1–3), 175–194 (2004)
11. Reif, J.H.: Successes and challenges. Science **296**(5567), 478–479 (2002)
12. Fu, B., Beigel, R.: Length bounded molecular computing. BioSystems **52**(1–3), 155–163 (1999)
13. Lipton, R.J.: DNA solution of hard computational problems. Science **268**(5210), 542–545 (1995)
14. Faulhammer, D., Cukras, A.R., Lipton, R.J., Landweber, L.F.: Molecular computation: RNA solutions to chess problems. Proc. Natl. Acad. Sci. **97**(4), 1385–1389 (2000)
15. Lipton, R.J., Boneh, D., Dimworth, C.: Breaking DES using a molecular computer. In: DNA Based Computers, vol. 27, p. 37 (1996)
16. Adleman, L.M., Rothemund, P.W., Roweis, S., Winfree, E.: On applying molecular computation to the data encryption standard. J. Comput. Biol. **6**(1), 53–63 (1999)
17. Oliver, J.S., et al.: Computation with DNA: matrix multiplication. In: DNA Based Computers 2, pp. 113–122 (1996)
18. Eng, T.L., Serridge, B.M.: A surface-based DNA algorithm for minimal set cover. In: DNA Based Computers, pp. 185–192. Citeseer (1997)
19. Guarnieri, F., Fliss, M., Bancroft, C.: Making DNA add. Science **273**(5272), 220–223 (1996)
20. Fern, J., Schulman, R.: Modular DNA strand-displacement controllers for directing material expansion. Nat. Commun. **9**(1), 3766 (2018)

21. Hu, P., et al.: Cooperative toehold: a mechanism to activate DNA strand displacement and construct biosensors. Anal. Chem. **90**(16), 9751–9760 (2018)
22. Qian, L., Winfree, E.: Scaling up digital circuit computation with DNA strand displacement cascades. Science **332**(6034), 1196–1201 (2011)
23. Paulino, N.M., Foo, M., Kim, J., Bates, D.G.: PID and state feedback controllers using DNA strand displacement reactions. IEEE Control Syst. Lett. **3**, 805–810 (2019)
24. Srinivas, N., Parkin, J., Seelig, G., Winfree, E., Soloveichik, D.: Enzyme-free nucleic acid dynamical systems. Science **358**(6369), 1–11 (2017). eaal2052
25. Brun, Y.: Arithmetic computation in the tile assembly model: addition and multiplication. Theoret. Comput. Sci. **378**(1), 17–31 (2007)
26. Brun, Y.: Nondeterministic polynomial time factoring in the tile assembly model. Theoret. Comput. Sci. **395**(1), 3–23 (2008)
27. Pelletier, O., Weimerskirch, A.: Algorithmic self-assembly of DNA tiles and its application to cryptanalysis. In: Proceedings of the 4th Annual Conference on Genetic and Evolutionary Computation, pp. 139–146. Morgan Kaufmann Publishers Inc. (2002)
28. Li, K., Zou, S., Xv, J.: Fast parallel molecular algorithms for DNA-based computation: solving the elliptic curve discrete logarithm problem over GF (2^n). BioMed Res. Int. **2008**, 10 (2008)
29. Cheng, Z.: Computation of multiplicative inversion and division in GF (2^n) by self-assembly of DNA tiles. J. Comput. Theoret. Nanosci. **9**(3), 336–346 (2012)
30. Darehmiraki, M., Nehi, H.M.: Molecular solution to the 0–1 knapsack problem based on DNA computing. Appl. Math. Comput. **187**(2), 1033–1037 (2007)
31. Gehani, A., LaBean, T., Reif, J.: DNA-based cryptography. In: Jonoska, N., Păun, G., Rozenberg, G. (eds.) Aspects of Molecular Computing. LNCS, vol. 2950, pp. 167–188. Springer, Heidelberg (2003). https://doi.org/10.1007/978-3-540-24635-0_12
32. Wong, N.Y., Xing, H., Tan, L.H., Lu, Y.: Nano-encrypted morse code: a versatile approach to programmable and reversible nanoscale assembly and disassembly. J. Am. Chem. Soc. **135**(8), 2931–2934 (2013)
33. Clelland, C.T., Risca, V., Bancroft, C.: Hiding messages in DNA microdots. Nature **399**(6736), 533 (1999)
34. Mousa, H., Moustafa, K., Abdel-Wahed, W., Hadhoud, M.M.: Data hiding based on contrast mapping using DNA medium. Int. Arab J. Inf. Technol. **8**(2), 147–154 (2011)
35. Heider, D., Barnekow, A.: DNA-based watermarks using the DNA-crypt algorithm. BMC Bioinform. **8**(1), 176 (2007)
36. Ning, K.: A pseudo DNA cryptography method. arXiv preprint arXiv:0903.2693 (2009)
37. Zhang, X., Wang, L., Zhou, Z., Niu, Y.: A chaos-based image encryption technique utilizing hilbert curves and H-fractals. IEEE Access **7**, 74734–74746 (2019)
38. Zhang, X., Zhou, Z., Niu, Y.: An image encryption method based on the Feistel network and dynamic DNA encoding. IEEE Photonics J. **10**(4), 1–14 (2018)

Design of a Four-Person Voter Circuit Based on Memristor Logic

Qinfei Yang[1,2], Junwei Sun[1,2], and Yanfeng Wang[1,2(✉)]

[1] Henan Key Lab of Information-Based Electrical Appliances,
Zhengzhou University of Light Industry, Zhengzhou 450002, China
yanfengwang@yeah.net
[2] School of Electrical and Information Engineering,
Zhengzhou University of Light Industry, Zhengzhou 450002, China

Abstract. The development of traditional CMOS-based logic circuits in terms of speed and energy consumption is approaching the limit. Memristor is a kind of bio-inspired hardware with special structure, which has the advantages of simple structure, low power consumption and easy integration. It has a good application prospect in high performance memory and neural networks. The invention of memristors provides a new way to develop more efficient logic circuits. In this paper, the memristor-based logic gates are employed to implement complex logic functions. By changing the polarity of two parallel memristors, the OR logic and AND logic can be implemented separately. By using these basic logics, adders and comparators are performed, and further, a four-person voter is designed. The feasibility of a four-person voter based on memristor logic is verified by theoretical analysis and Pspice simulation. The memristor logic circuit provides the basis for the building of more complex circuits in the future. It also provides support for the development and application of bio-inspired hardware.

Keywords: Memristor · Logic circuit · Adder · Comparator · Four-person voter

1 Introduction

In 1971, based on the principle of symmetry, Chua speculated the existence of the fourth basic electronic element, which he called a memristor [1]. Then the development of the memristor had stalled for a long time, until 2008, it was fabricated by Hewlett-Packard in the lab [2]. A memristor is a resistor with memory property. Memristor can change its own structure independently and dynamically according to the change of external environment, which is consistent with the concept of bionic hardware proposed by Swiss federal institute of technology [3]. The memristance of a memristor changes with the amount of charge flowing through it. If the voltage applied on a memristor is removed, then the memristance remains its value. The development of memristor combines nanometrics, quantum electronics, molecular and biological electronics technology,

© Springer Nature Singapore Pte Ltd. 2020
L. Pan et al. (Eds.): BIC-TA 2019, CCIS 1160, pp. 149–162, 2020.
https://doi.org/10.1007/978-981-15-3415-7_12

which makes a contribution to the exploration of new mechanism of bio-inspired hardware. [4,5] Based on the special characteristics, many possible applications of memristors have been presented. Memristors are used to mimic synapses in [6–13,15]. The functions of storing data in memristors are realized in [14,16–22]. In addition, the memristor-based logic is another interesting application of memristors [23–32].

The high and low impedance states of the memristor can represent two logic states. So one approach for realising logic operations is to treat memristance as the logical states. Stateful logic operations were implemented based on memristive switches via material implication (IMPLY logic gate) [28]. The IMPLY logic gates can be integrated in an crossbar array together with a complete logic family. Due to the non-volatility of the memristance, the operation and storage of the memristive logic can be performed simultaneously. This provides a way for designing a new generation of computes with non-Von Neumann architecture. However, this method requires sequential active voltage at different locations in the circuit. And the IMPLY logic also needs additional circuit components such as a controller and an additional resistors. In [24], memristor-aid logic (MAGIC) is presented, which does not require a complicated structure. Unlike the IMPLY logic gates, the input and output in MAGIC are separated. The MAGIC needs only one applied control voltage to complete logic operation.

Another approach for logic with memristors is to treat voltages as the logical states. Practical memristors are compatible with standard CMOS technology [23]. Memristors were integrated with CMOS to perform the logic operations in [32]. Since memristors are smaller than transistors, the memristor-based circuits are smaller than transistor-only circuits. The read and write speed of memristor-based logic operations can reach nanosecond. In [25], memristor ratioed logic (MRL) for integration with CMOS is described. This MRL logic family uses the memristance for computation of Boolean AND/OR functions with voltage as the logic state. For MRL, the CMOS transistors perform logic inversion and amplification of the logic voltage signals. The MRL logic family contains complete logic gates.

Many complex functions can be realized through memristor-based logics. Adder is implemented by material implication logic in [33]. Memristor based carry lookahead adder architectures are designed in [34]. Linear feedback shift register with memristor-based logic is realized in [35]. Up-down counter is designed by using material implication logic in [36].

In this paper, the memristor-based logic gates that use voltages to represent logic values are adopted to construct adders and comparators. And the adders and comparators are employed to design a four-person voter circuit. The designed voter circuit consists of complex structure and a large number of basic logic units, which proves the feasibility of constructing large-scale circuits based on memristor logics. Pspice is utilized to simulate and verify its validity of the designed circuit.

The paper is organized as follows: Sect. 2 describes a mathematical model of a memristor with Biolek window function. Followed by, a schematic of design principle of AND and OR logic are given in Sect. 2. A full adder circuit and a comparator circuit are presented in Sect. 3. The design of a four-person voter circuit and SPICE simulation results are described in Sect. 4. Finally, Sect. 5 summarizes the content of the paper.

2 A Mathematical Model of Memristor and Basic Logic Gate

A memristor is a basic passive two-terminal element, which is defined by the relation between magnetic flux and charge. One of the popular mathematic model of memristor is the linear ion drift model that is based on the characteristics fabricated by Hewlett-Packard. As shown in Fig. 1(a), a memristor contains a doped region and an undoped region.

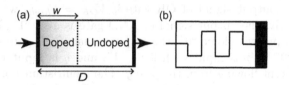

Fig. 1. (a) Schematic of a memristor. w represents the width of doped region, and D represents the width of a memristor. (b) Symbol of a memristor.

The memristance is

$$M = R_{on}\frac{w(t)}{D} + R_{off}(1 - \frac{w(t)}{D}).(R_{off} \gg R_{on}). \tag{1}$$

where M represents memristance, and R_{on} represents the memristance when a memristor is completely doped, R_{off} denotes the memristance when a memristor is not doped at all. The width of the doped and undoped regions changes with the amount of charge flowing through the memristor. The width w of the doped region is described as

$$w(t) = \mu_v\frac{R_{on}}{D}q(t). \tag{2}$$

where μ_v represents ion mobilitand q indicates the charge flowing through a memristor. As shown in Fig. 1(a), if voltage is applied to the memristor, then the current flows from left to right, w increases and memristance decreases. Conversely, when the current flows from right to left, w decreases and the memristance increases.

Based on this principle, a voltage divider circuit can be implemented. By changing the polarity of the memristors, the correct OR and AND logic values can be obtained.

As shown in the Fig. 2, the memristor-based OR gate and AND gate are described. High level and low level indicate logic "1" and logic "0", respectively.

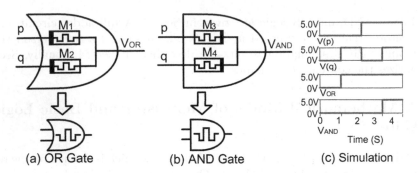

Fig. 2. (a) OR gate. (b) AND gate. (c) Simulation results. M_1, M_2, M_3 and M_4 are memristors. p and q are input signals, V_{out} is a output signal.

If p and q are simultaneously input with logic "1" Since no current flows through M_1 and M_2, the output signal of OR gate is $V_{OR} = V_{high}$, which represents logic "1". Similarly, the output signal of AND gate is also "1". If p and q are simultaneously input with logic "0", the output signal of OR gate and AND gate are both logic "0". If p is input with logic "1", and q is input with logic "0", there will be current flowing from the p to q. The memristance of M_1 decreases to R_{on} and the memristance of M_2 increases to R_{off} in OR gate circuit; the memristance of M_3 increases to R_{off} and the memristance of M_4 decreases to R_{on} in AND gate circuit. The output signal of OR gate is

$$V_{OR} = \frac{R_{off}}{R_{off} + R_{on}} * V_{high} \approx V_{high}. \tag{3}$$

The output signal of AND gate is

$$V_{AND} = \frac{R_{on}}{R_{off} + R_{on}} * V_{high} \approx 0v. \tag{4}$$

Then, p is given with logic "0", and q is given with logic "1". Since the circuit structures of OR gate and AND gate are respectively symmetrical, the output signal of OR gate is still logic "1", and the output signal of AND gate is still logic "0". The simulation results of OR gate and AND gate are shown in Fig. 2(c), which are consistent with the theoretical analysis.

Since the memristance takes a certain amount of time to switch between R_{on} and R_{off}, the circuit will have a delay in performing the AND and OR operations. When memristor-based logic gates are used to form a cascade logic circuit, the voltage value is gradually reduced, so that the subsequent voltage amplitude cannot guarantee normal logic operation. Attaching BUFFER gates to the memristor-based logic gates can effectively improve this situation.

3   Full Adder and Comparator Circuits

Using the basic memristor-based logic gates, some combinational logic circuits can be obtained. A full adder circuit is an important component for the designed

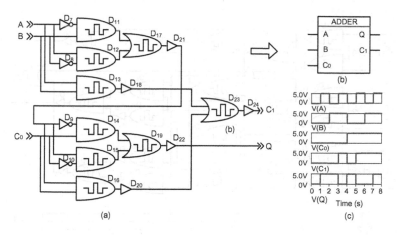

Fig. 3. (a) Full adder circuit. (b) Symbol of full adder. (c) Simulation result. A and B are the augend and addend, respectively. C_0 and C_1 represent the low carry and high carry, respectively. Q stands for the total sum. D_7–D_{10} are inverters, D_{11}–D_{16} are memristor-based AND gates, D_{17}, D_{19} and D_{23} are memristor-base OR gates, D_{18}, D_{20}, D_{21}, D_{22} and D_{24} are BUFFER gates.

complex circuit in the paper. The memristor-based full adder circuit is designed as Fig. 3(a), and its logical expression is

$$Q = A \oplus B \oplus C_0$$
$$C_1 = AB + (A \oplus B)C_0. \tag{5}$$

A and B are the augend and addend, respectively. C_0 and C_1 represent the low carry and high carry, respectively. Q stands for the total sum. If $A = B = C_0 = 0$, then $Q = C_1 = 0$. If $A = 0$, $B = 1$ and $C_0 = 0$, then the output signals of D_{21}, D_{18}, D_{22} and D_{20} are "1", "0", "1" and "0", respectively; the total sum Q is "1", and the high carry C_1 is "0". If $A = 1$, $B = 0$ and $C_0 = 0$, then the output signals of D_{21}, D_{18}, D_{22} and D_{20} are still "1", "0", "1" and "0", respectively; the total sum Q is "1", and the high carry C_1 is "0". If $A = B = 1$ and $C_0 = 0$, then the output signals of D_{21}, D_{18}, D_{22} and D_{20} are "0", "1", "0" and "0", respectively; the total sum Q is "0", and the high carry C_1 is "1". If $A = B = 0$ and $C_0 = 1$, then the output signals of D_{21}, D_{18}, D_{22} and D_{20} are "0", "0", "1" and "0", respectively; the total sum Q is "1", and the high carry C_1 is "0". If $A = 0$, $B = 1$ and $C_0 = 1$, then the output signals of D_{21}, D_{18}, D_{22} and D_{20} are "1", "0", "0" and "1", respectively; the total sum Q is "0", and the high carry C_1 is "1". If $A = 1$, $B = 0$ and $C_0 = 1$, then the output signals of D_{21}, D_{18}, D_{22} and D_{20} are still "1", "0", "0" and "1", respectively; the total sum Q is "0", and the high carry C_1 is "1". If $A = B = 1$ and $C_0 = 1$, then the output signals of D_{21}, D_{18}, D_{22} and D_{20} are "0", "1", "1" and "0", respectively; the total sum Q is "1", and the high carry C_1 is also "1". The simulation result of the memristor-based full adder is shown in Fig. 3(c) which is consistent with Table 1. The memristor-based full adder is able to perform the full addition operation

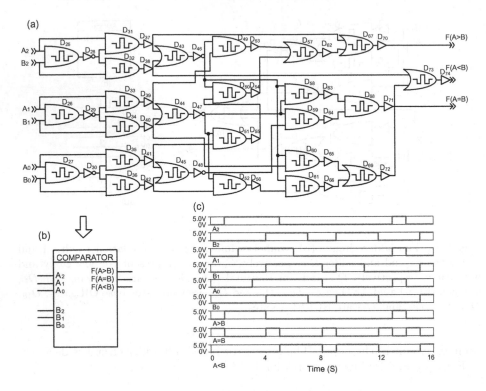

Fig. 4. (a) Comparator circuit. (b) Symbol of comparator. (c) Simulation result. $A_2A_1A_0$ and $B_2B_1B_0$ are the numbers to be compared. $F(A > B)$, $F(A < B)$ and $F(A = B)$ are the output signals.

correctly. In order to simplify the subsequent designed circuit, the full adder is integrated into a module which is represented as Fig. 3(b).

Table 1. Truth table of full adder.

A	B	C_0	Q	C1
0	0	0	0	0
0	0	1	1	0
0	1	0	1	0
0	1	1	0	1
1	0	0	1	0
1	0	1	0	1
1	1	0	0	1
1	1	1	1	1

Table 2. Truth table of comparator.

Input			Output		
A_2 vs B_2	A_1 vs B_1	A_0 vs B_0	$F(A > B)$	$F(A < B)$	$F(A = B)$
$A_2 > B_2$	X	X	1	0	0
$A_2 < B_2$	X	X	0	1	0
$A_2 = B_2$	$A_1 > B_1$	X	1	0	0
$A_2 = B_2$	$A_1 < B_1$	X	0	1	0
$A_2 = B_2$	$A_1 = B_1$	$A_0 > B_0$	1	0	0
$A_2 = B_2$	$A_1 = B_1$	$A_0 < B_0$	0	1	0
$A_2 = B_2$	$A_1 = B_1$	$A_0 = B_0$	0	0	1

A comparator is a device that compares two binary numbers. As shown in Fig. 4, the 3-bit comparator circuit is designed with memristor-based logic gates, and its logical expression is

$$F(A > B) = A_2\overline{B_2} + (\overline{A_2 \oplus B_2})A_1\overline{B_1} + (\overline{A_2 \oplus B_2})(\overline{A_1 \oplus B_1})A_0\overline{B_0}$$
$$F(A < B) = \overline{A_2}B_2 + (\overline{A_2 \oplus B_2})\overline{A_1}B_1 + (\overline{A_2 \oplus B_2})(\overline{A_1 \oplus B_1})\overline{A_0}B_0 \qquad (6)$$
$$F(A = B) = (\overline{A_2 \oplus B_2})(\overline{A_1 \oplus B_1})(\overline{A_0 \oplus B_0})$$

The 3-bit comparator is constructed based on a 1-bit comparator. D_{25}, D_{28}, D_{31}, D_{32}, D_{37}, D_{38}, D_{43} and D_{46} form a 1-bit comparator. The output signals of D_{37}, D_{46} and D_{38} indicate $A_2 > B_2$, $A_2 = B_2$ and $A_2 < B_2$, respectively. In the process of comparing numerical values, if the high order (A_2 and B_2) are not equal, then there is no need to compare the low order (A_1, A_0 and B_1, B_0). The results of the comparison between A_2 and B_2 represent the result of the comparison between the two numbers. The OR gate D_{67} generates an output signal when it receives a signal from D_{37}, the comparison result is $A > B$. The OR gate D_{73} produces an output signal when it receives a signal from D_{38}, the comparison result is $A < B$. If the higher order is equal, then the comparison result is determined by the comparison between lower orders. If $A_2 = B_2$ and $A_1 > B_1$, then the output signals of D_{46} and D_{39} are both "1", the output signals of D_{49}, D_{53}, D_{57}, D_{62}, D_{67} and D_{70} are all "1", the comparison result is $A > B$. If $A_2 = B_2$ and $A_1 < B_1$, then the output signals of D_{46} and D_{40} are both "1", the output signals of D_{60}, D_{65}, D_{69}, D_{72}, D_{73} and D_{74} are all "1", the comparison result is $A < B$. When the first two bits are equal, the last bit is compared. If $A_2 = B_2$, $A_1 = B_1$ and $A_0 > B_0$, then the output signals of D_{46}, D_{47} $D_{41}$are "1", and the output signals of D_{51},D_{55},D_{50}, D_{54}, D_{57}, D_{62}, D_{67} and D_{70} are all "1", the comparison result is $A > B$. If $A_2 = B_2$, $A_1 = B_1$ and $A_0 < B_0$, then the output signals of D_{46}, D_{47} D_{42} are "1", and the output signals of D_{52}, D_{56}, D_{61}, D_{66}, D_{69}, D_{72}, D_{73} and D_{74} are all "1", the comparison result is $A < B$. If $A_2A_1A_0 = B_2B_1B_0$, then the signals produce by D_{46}, D_{47}, D_{48} are all equal to "1", and the output signals of D_{58}, D_{59}, D_{63}, D_{64}, D_{68} and D_{71} are all "1", the comparison result is $A = B$. The simulation result of the memristor-based comparator is consistent with the truth table that is shown in Table 2, which verifies the correctness of the designed circuit. The comparator circuit is integrated into a module which is represented as Fig. 4(b).

4   Design of the Four-Person Voter Circuit

In the four-person voter circuit, four people express their own will to vote on a proposal. Each person controls two buttons, the button A indicates whether to abstain and the button B indicates whether to agree. Once the abstention button is selected, the second button is no longer valid. In order to ensure that the priority of abstention is higher than whether or not to agree, the input signals are processed by the signal conversion circuit as shown in Fig. 5(a). In the circuit, A' and B' represent the original input signals, A and B denote the signals which will be processed by the voter circuit. If $A' = 0$ and $B' = 0$, then $A = 0$ and $B = 0$. If $A' = 0$ and $B' = 1$, then $A = 0$ and $B = 1$. If $A' = 1$ and $B' = 0$, then $A = 1$ and $B = 0$. If $A' = 1$ and $B' = 1$, then $A = 1$ and $B = 0$, which indicates that once A' is "1", B' no longer works. These four conditions correspond to the

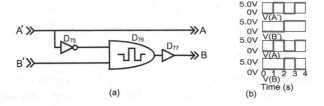

Fig. 5. (a) Signal conversion circuit. A' and B' represent the original input signals, A and B denote the signals which will be processed by the voter circuit. D_{75} is a inverter, D_{76} is a AND gate, D_{77} is a BUFFER. (b) Simulation result.

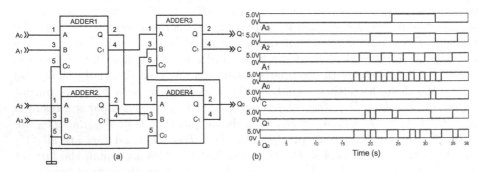

Fig. 6. (a) Counter circuit. A_0, A_1, A_2 and A_3 are the numbers to be summed. CQ_1Q_0 is the final result. (b) Simulation result.

signals of 1 s, 3 s, 2 s and 4 s in Fig. 5(b), respectively. In the voter circuit, if three or more people abstain from the proposal, the voting result is invalid. If more than half of the people agree, then the resolution is passed. Otherwise, the resolution would not be adopted. The voting result of each person is represented by A_iB_i, and the final result of the vote are indicated by X, Y and Z. $A_0 = 0$ and $B_0 = 0$ represent that the first person disagrees with the proposal; $A_0 = 0$ and $B_0 = 1$ denote that the first person agrees with the proposal; $A_0 = 1$ and $B_0 = *$ indicate that the first person stated to abstain, where the symbol $*$ represents an arbitrary value. $X = 1$ indicates the voting result is invalid, $Y = 1$ indicates the resolution is passed, and $Z = 1$ represents the resolution is not adopted. The voting result Q can be expressed by the following equation

$$Q = \begin{cases} X = 1 & \text{if } A_0 + A_1 + A_2 + A_3 \geq 3, \\ Y = 1 & \text{if } B_0 + B_1 + B_2 + B_3 \geq 3, \\ Z = 1 & \text{if other cases,} \end{cases} \tag{7}$$

In this paper, the above algorithm mainly contains two steps. The first step is to get the sum value $\sum_{i=0}^{3} A_i$ and $\sum_{i=0}^{3} B_i$. The second step is to compare, and then generates the results of logical operation.

4.1 Design of the Counter Circuit

The purpose of designing the counter circuit is to count the number of approvals and the number of abstentions in the voter. As shown in Fig. 6, the counter circuit is composed by four full adders. ADDER1 obtains the sum value from A_0 and A_1, ADDER2 gets the sum value from A_2 and A_3. The high order values and low order values are summed in ADDER3 and ADDER4, respectively. CQ_1Q_0 is the final result of the circuit. Figure 6(a) depicts the counter circuit of the abstain signals, which is similar to the counter circuit of the signals that represent whether or not to agree to the proposal.

The simulation results of the counting circuit are shown in Fig. 6(b). At 20 s, the input signals A_3 A_2 A_1 and A_0 are "0" "0" "1" "1", and the output signals C Q_1 Q_0 are "0" "1" "0", respectively. It can be seen that there are two abstentions at this time. At 32 s, the input signals A_3 A_2 A_1 and A_0 are "1" "1" "1" "1", and the output signals C Q_1 Q_0 are "1" "0" "0", respectively. It means that everyone chose to abstain at this time. At 38 s, the input signals A_3 A_2 A_1 and A_0 are "0" "1" "1" "1", and the output signals C Q_1 Q_0 are "0" "1" "1", respectively. It indicates that three people chose to abstain in 38 s.

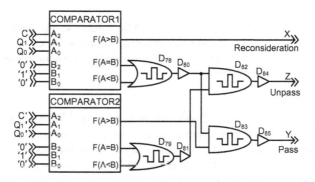

Fig. 7. Resolution circuit. B_2–B_0 indicate the number to be compared, which is set to 010. X, Y and Z are output signals which represent the resolution is reconsidered, passed, and unpassed, respectively.

4.2 Design of the Resolution Circuit

Using the data from the counter circuit, 3-bit comparators can be used in the resolution circuit to perform comparison according to the requirements, thereby, classify the voting situations, and prepare for the resolution result. As shown in Fig. 7, the input ports A_2–A_0 of the comparator obtain signals from the counting circuit. The output signals X, Y and Z represent reconsidered, passed, and unpassed, respectively. B_2–B_0 indicate the number to be compared. Since the formation of the resolution requires more than half of the people to agree or

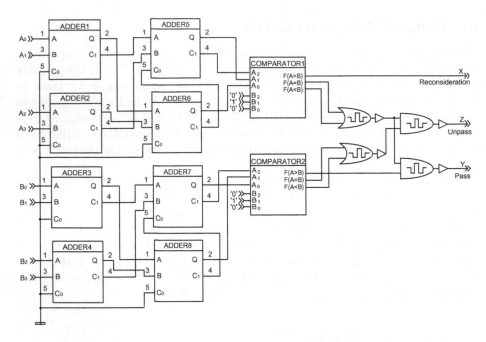

Fig. 8. Complete circuit of four-person voter. The values A_0–A_3 indicate whether to abstain or not, the values B_0–B_3 indicate whether to agree to a proposal or not. X, Y and Z are the final output signals which represent the resolution is reconsidered, passed, and unpassed, respectively.

abstain, $B_2B_1B_0$ is set to 010. Classification is obtained by comparing the input data with the stored data 010. Comparator 1 compares the number of abstainers to the stored data. If the number of abstainers is greater than 2, comparator 1 produces an output signal with $A > B$, which indicates the proposal should be reconsidered. If the number of abstainers is equal to 2, comparator 1 generates an output signal which indicates $A = B$. If the number of abstainers is less to 2, comparator 1 produces an output signal which indicates $A < B$. When the number of abstainers is less than 3, comparator 2 compares the number of people who agree with the proposal to the stored data. If the number of approvers is greater than 2, comparator 2 produces an output signal with $A > B$. D_{78}, D_{80}, D_{83} and D_{85} is "1", the final result indicates the proposal is passed. If the number of approvers is less than 3, comparator 2 generates an output signal with $A = B$ or $A < B$. D_{79}, D_{81}, D_{82} and D_{84} is "1", the final result indicates the proposal is unpassed.

4.3   Complete Circuit of the Four-Person Voter

The complete circuit of the designed Four-Person Voter is shown in Fig. 8. The counter circuit is connected to the resolution circuit. A_3B_3 to A_0B_0 respectively represent the voting results of four people; and X, Y, Z represent the final results.

The circuit first counts the number of abstainer and the number of people who agree to the proposal, and then compares the results with the stored value to obtain the final results.

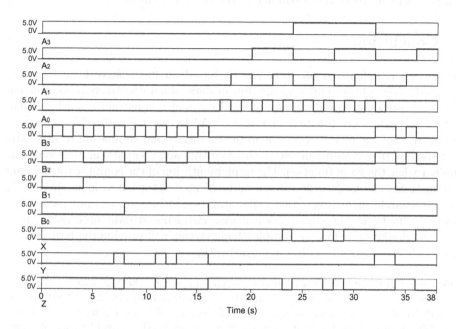

Fig. 9. Simulation results of the designed four-person voter circuit. A_0B_0–A_3B_3 represent the voting status of four people. For each status, "00" denotes the person disagrees to a proposal, "01" indicates the person agrees to the proposal, "10" indicate that the person abstains from the vote. X, Y and Z are the final output signals which represent the resolution is reconsidered, passed, and unpassed, respectively.

4.4   Simulation Results

The simulation results of the designed four-person voter circuit are shown in Fig. 9. A_0B_0–A_3B_3 are input signals which represent the voting status of four people. $A_i = 0$ and $B_i = 0$ represent that the person disagrees with the proposal; $A_i = 0$ and $B_i = 1$ denote that the person agrees with the proposal; $A_i = 1$ and $B_i = 0$ represent that the person state to abstain. X, Y and Z are the final output signals which represent the resolution is reconsidered, passed, and unpassed, respectively. At 0 s, everyone disagree to the proposal, $Y = 1$, indicating the resolution is unpassed. At 10 s, A_3B_3 is "00", A_2B_2 is "01", A_1B_1 is "00", A_0B_0 is "01", and the result is $Z = 1$. Only two people agree to the proposal and the result of the vote is still not passed. At 11 s, A_3B_3 is "01", A_2B_2 is "01", A_1B_1 is "00", A_0B_0 is "01", and the result is $Y = 1$. Three people agree to the proposal, the result of the vote is passed. At 37 s, A_3B_3 is "00", A_2B_2 is "10", A_1B_1 is

"10", $A_0 B_0$ is "10", and the result is $X = 1$. Three people state to abstain and one people disagrees to the proposal, the proposal need to be reconsidered. The simulation results of the rest of the time are still consistent with the theoretical analysis, which proves the feasibility of the designed circuit.

5   Conclusion

In the paper, a four-person voter circuit based on memristive logic has been designed. The voter circuit contains full adders which constitute a counter circuit, and contains comparator circuit modules which construct a resolution circuit. The basic logic function of OR and AND can be realized by changing the polarity of the memristors. Theoretical analysis and simulation results show that the designed circuit can realize the counting and comparison functions, and finally realize the voter function. The implementation of the complex voter circuit demonstrates the large-scale cascadability of memristor-based logic circuits. In future research, memristor-based logic circuits will be used to implement more complex functions, thereby, develop a memristor-based data processor. Moreover, the design of memristor complex logic circuit will provide the basis for the further development of bio-inspired hardware.

Acknowledgements. This work was supported in part by the National Key R and D Program of China for International S and T Cooperation Projects (2017YFE0103900), in part by the Joint Funds of the National Natural Science Foundation of China (U1804262), in part by the State Key Program of National Natural Science of China under Grant 61632002, in part by the National Natural Science of China under Grant 61603348, Grant 61775198, Grant 61603347, and Grant 61572446, in part by the Foundation of Young Key Teachers from University of Henan Province (2018GGJS092), and in part by the Youth Talent Lifting Project of Henan Province and Henan Province University Science and Technology Innovation Talent Support Plan under Grant 20HASTIT027.

References

1. Chua, L.O.: Memristor-the missing circuit element. IEEE Trans. Circ. Theory **18**(5), 507–519 (1971)
2. Strukov, D.B., Snider, G.S., Stewart, D.R., Williams, R.S.: The missing memristor found. Nature **453**(7191), 80 (2008)
3. Wang, Y., Cui, J., You, X., Huang, S., Yao, R.: Theory and technology development of bio-inspired hardware. Chin. Space Sci. Technol. **24**(6), 32–42 (2004)
4. Sipper, M., Sanchez, E., Mange, D., Tomassini, M.: A phylogenetic, ontogenetic, and epigenetic view of bio-inspired hardware systems. IEEE Trans. Evol. Comput. **1**(1), 83–97 (1997)
5. Duan, H., Shao, S., Su, B., Zhang, L.: New development thoughts on the bio-inspired intelligence based control for unmanned combat aerial vehicle. Sci. China Technol. Sci. **53**(8), 2025–2031 (2010). https://doi.org/10.1007/s11431-010-3160-z

6. Jo, S.H., Chang, T., Ebong, I., Bhadviya, B.B., Mazumder, P., Lu, W.: Nanoscale memristor device as synapse in neuromorphic systems. Nano Lett. **10**(4), 1297–1301 (2010)
7. Kim, H., Sah, M.P., Yang, C., Roska, T., Chua, L.O.: Memristor bridge synapses. Proc. IEEE **100**(6), 2061–2070 (2011)
8. Wang, Z., et al.: Memristors with diffusive dynamics as synaptic emulators for neuromorphic computing. Nat. Mater. **16**(1), 101–108 (2017)
9. Prezioso, M., Merrikh-Bayat, F., Hoskins, B., Adam, G.C., Likharev, K.K., Strukov, D.B.: Training and operation of an integrated neuromorphic network based on metal-oxide memristors. Nature **521**(7550), 61–64 (2015)
10. Choi, S., Ham, S., Wang, G.: Memristor synapses for neuromorphic computing. In: Memristors-Circuits and Applications of Memristor Devices. IntechOpen (2019)
11. Dang, B., et al.: Physically transient memristor synapse based on embedding magnesium nanolayer in oxide for security neuromorphic electronics. IEEE Electron Device Lett. **80**(8), 1265–1268 (2019)
12. Adnan, M.M., Sayyaparaju, S., Rose, G.S., Schuman, C.D., Ku, B.W., Lim, S.K.: A twin memristor synapse for spike timing dependent learning in neuromorphic systems. In: 31st IEEE International System-on-Chip Conference (SOCC), pp. 37–42. IEEE (2018)
13. Hong, Q., Zhao, L., Wang, X.: Novel circuit designs of memristor synapse and neuron. Neurocomputing **330**, 11–16 (2019)
14. Liu, C., Liu, F., Li, H.H.: Beyond CMOS: memristor and its application for next generation storage and computing. ECS Trans. **85**(6), 115–125 (2018)
15. Sun, J., Zhao, X., Fang, J., Wang, Y.: Autonomous memristor chaotic systems of infinite chaotic attractors and circuitry realization. Nonlinear Dyn. **94**(4), 2879–2887 (2018). https://doi.org/10.1007/s11071-018-4531-4
16. Chen, Y., Liu, G., Wang, C., Zhang, W., Li, R.W., Wang, L.: Polymer memristor for information storage and neuromorphic applications. Mater. Horiz. **1**(5), 489–506 (2014)
17. Liu, G., et al.: Organic biomimicking memristor for information storage and processing applications. Adv. Electron. Mater. **2**(2) (2016). https://doi.org/10.1002/aelm.201500298
18. Duan, S., Hu, X., Wang, L., Li, C., Mazumder, P.: Memristor-based RRAM with applications. Sci. China Inf. Sci. **55**(6), 1446–1460 (2012). https://doi.org/10.1007/s11432-012-4572-0
19. Xu, C., Dong, X., Jouppi, N.P., Xie, Y.: Design implications of memristor-based RRAM cross-point structures. In: Design, Automation & Test in Europe, pp. 1–6. IEEE (2011)
20. Shaarawy, N., Emara, A., El-Naggar, A.M., Elbtity, M.E., Ghoneima, M., Radwan, A.G.: Design and analysis of 2T2M hybrid CMOS-memristor based RRAM. Microelectron. J. **73**, 75–85 (2018)
21. Majumder, M.B., Hasan, M.S., Uddin, M., Rose, G.S.: A secure integrity checking system for nanoelectronic resistive RAM. IEEE Trans. Very Large Scale Integr. (VLSI) Syst. **27**(2), 416–429 (2018)
22. Zhang, X., et al.: Novel hybrid computing architecture with memristor-based processing-in-memory for data-intensive applications. In: 14th IEEE International Conference on Solid-State and Integrated Circuit Technology (ICSICT), pp. 1–3. IEEE (2018)
23. Xia, Q., et al.: Memristor-CMOS hybrid integrated circuits for reconfigurable logic. Nano Lett. **9**(10), 3640–3645 (2009)

24. Kvatinsky, S., et al.: MAGIC-memristor-aided logic. IEEE Trans. Circ. Syst. II Express Briefs **61**(11), 895–899 (2014)
25. Kvatinsky, S., Wald, N., Satat, G., Kolodny, A., Weiser, U.C., Friedman, E.G.: MRL-memristor ratioed logic. In: 13th International Workshop on Cellular Nanoscale Networks and their Applications, pp. 1–6. IEEE (2012)
26. Kvatinsky, S., Satat, G., Wald, N., Friedman, E.G., Kolodny, A., Weiser, U.C.: Memristor-based material implication (IMPLY) logic: design principles and methodologies. IEEE Trans. Very Large Scale Integr. (VLSI) Syst. **22**(10), 2054–2066 (2013)
27. Guckert, L., Swartzlander, E.E.: MAD gates-memristor logic design using driver circuitry. IEEE Trans. Circ. Syst. II Express Briefs **64**(2), 171–175 (2016)
28. Borghetti, J., Snider, G.S., Kuekes, P.J., Yang, J.J., Stewart, D.R., Williams, R.S.: 'Memristive' switches enable 'stateful' logic operations via material implication. Nature **464**(7290), 873–876 (2010)
29. Hu, X., Schultis, M.J., Kramer, M., Bagla, A., Shetty, A., Friedman, J.S.: Overhead requirements for stateful memristor logic. IEEE Trans. Circ. Syst. I Regul. Pap. **66**(1), 263–273 (2018)
30. Pershin, Y.V.: A demonstration of implication logic based on volatile (diffusive) memristors. IEEE Trans. Circ. Syst. II Express Briefs **66**(6), 1033–1037 (2018)
31. Danaboina, Y.K.Y., Samanta, P., Datta, K., Chakrabarti, I., Sengupta, I.: Design and implementation of threshold logic functions using memristors. In: 32nd International Conference on VLSI Design and 18th International Conference on Embedded Systems (VLSID), pp. 518–519. IEEE (2019)
32. Liu, G., Zheng, L., Wang, G., Shen, Y., Liang, Y.: A carry lookahead adder based on hybrid CMOS-memristor logic circuit. IEEE Access **7**, 43691–43696 (2019)
33. Teimoory, M., Amirsoleimani, A., Shamsi, J., Ahmadi, A., Alirezaee, S., Ahmadi, M.: Optimized implementation of memristor-based full adder by material implication logic. In: 21st IEEE International Conference on Electronics, Circuits and Systems (ICECS), pp. 562–565. IEEE (2014)
34. Shaltoot, A., Madian, A.: Memristor based carry lookahead adder architectures. In: IEEE 55th International Midwest Symposium on Circuits and Systems (MWSCAS), pp. 298–301. IEEE (2012)
35. Teimoory, M., Amirsoleimani, A., Ahmadi, A., Alirezaee, S., Salimpour, S., Ahmadi, M.: Memristor-based linear feedback shift register based on material implication logic. In: European Conference on Circuit Theory and Design (ECCTD), pp. 1–4. IEEE (2015)
36. Chakraborty, A., Dhara, A., Rahaman, H.: Design of memristor-based up-down counter using material implication logic. In: International Conference on Advances in Computing, Communications and Informatics (ICACCI), pp. 269–274. IEEE (2016)

Building of Chemical Reaction Modules and Design of Chaotic Oscillatory System Based on DNA Strand Displacement

Zhi Li[1,2], Yanfeng Wang[1,2(✉)], and Junwei Sun[1,2(✉)]

[1] Henan Key Lab of Information-Based Electrical Appliances,
Zhengzhou University of Light Industry, Zhenzhou 450002, China
junweisun@yeah.net
[2] School of Electrical and Information Engineering,
Zhengzhou University of Light Industry, Zhenzhou 450002, China

Abstract. DNA strand displacement as a new type of technology has provided different ways to build complex circuits. Research on chemical kinetics is helpful for exploiting the inherent potential property of biomolecular systems. It is common practice to use fluorophore and dark quencher to detect the luminous intensity of DNA chain, so as to determine the concentration of DNA chain. The luminous intensity of fluorophore and dark quencher is positively correlated with the concentration of DNA chain. In this study, six different chemical reaction modules have been designed and been demonstrated validity. The classical theory of chemical reaction networks can be used to describe the biological processes by mathematical modeling. Based on that, we have proposed a 3-variable chaotic oscillatory system and simulated by matlab. The result of simulation is convincing. A 3-variable chaotic oscillatory system as a bridge paves way to make some connection between chaotic oscillatory system and synchronized system in the future study.

Keywords: DNA strand displacement · Chaotic oscillatory system · Synchronized system

1 Introduction

DNA strand displacement is a potential dynamic DNA nanotechnology developed on the DNA self-assembly technology. It uses the bonding between the single chains of DNA molecules, which can be reacted with triggered chain to release the single chain products of DNA. This technique has the characteristics of self-induction, sensitivity and accuracy [1,2]. In recent years, DNA strands displacement technology has developed rapidly, and chain displacement cascade reaction has realized the dynamic connection between adjacent logic modules, which makes it possible for researchers to construct nanoscale large-scale complex logic circuits. The concentration of DNA chain represents the output of DNA circuit. Because it is difficult to detect the concentration of DNA, it is

© Springer Nature Singapore Pte Ltd. 2020
L. Pan et al. (Eds.): BIC-TA 2019, CCIS 1160, pp. 163–177, 2020.
https://doi.org/10.1007/978-981-15-3415-7_13

common practice to use fluorophore and dark quencher to detect the luminous intensity of DNA chain, so as to determine the concentration of DNA chain [3,4]. The luminous intensity of fluorophore and dark quencher is positively correlated with the concentration of DNA chain. This characteristic of the fluorophore and the dark quencher is particularly useful in the study of the cascade of DNA digital circuits, because the output of the digital circuit is a binary. The high luminous intensity is expressed as "1", and the low luminous intensity is expressed as "0" [5–10]. On this basis, with the help of the advantages of large capacity of information storage, high speed parallel computing and programmable simulation of biomolecule devices, it has become the focus for researchers. It has been deeply studied in the fields of molecular computing, nanomachines, disease diagnosis and treatment [11–13].

Because of its diverse structure and complex forms of motion, there is no unified and standard definition of chaotic system so far. In general, chaotic systems are described by differential equations, which are generally three-dimensional or four-dimensional, and some hyperchaotic systems may be composed of higher-dimensional equations [14,15]. The construction of a new type of chaotic system and the analysis of the basic dynamic characteristics of the system not only promote the progress of chaotic system in theory, but also make great progress in practical application [16,17].

Chaotic oscillatory system can be used in synchronization system [18–20]. Synchronization refers to the phenomenon to coordinate events so that they can get consistency in time of the same system. Synchronization is an important dynamic behavior, which is of great significance in the construction of complex networks, communication security and other fields [21–23]. Chaotic synchronization is a natural phenomenon, which usually refers to the phenomenon of coordination and consistency between the phases of at least two vibration systems. In recent years, the methods of chaotic synchronization have emerged continuously, and its application field has been rapidly extended from physics to biology, chemistry, medicine, electronics, information science and secure communications [24–26]. Because of its important value and broad application prospect in engineering technology, chaotic synchronization has always been one of the hot topics in the field of nonlinear science.

In this paper, we have designed a chaotic oscillatory system and verified the validity by Matlab. Then, reaction modules of catalysis, adjustment, and degradation have been designed to match the chaotic system. Given that chaotic synchronization is a significant dynamical behaviour, we have proposed a synchronization system, and linked the chaotic oscillatory system with the synchronization system. Different DNA strands have been marked with different colors of fluorophores. From the simulation of these modules, it has demonstrated that the DNA implementation of CRN can imitate the dynamical of the target CRN, and there is some relationship between DNA CRNs and target CRN. We used visual DSD software to design the DNA strands and test the feasibility of the chemical reaction with DNA strand displacement [27].

2   Modules Construction

DNA strand displacement as a type of chemical reaction has some connection between DNA CRNs and target CRN. CRN is an abbreviation of the chemical reaction network. CRNs is an abbreviation of the chemical reaction networks. Target CRN is a state of chemical reaction without auxiliary species, and provides basement of the establishment of chemical reaction modules. In order to get convincing conclusion, six different types of reaction modules are designed based on DNA CRNs and target CRN, where $[X]_0$ and $[X_t]_0$ represent the initial concentration of species X and X_t in DNA CRNs and target CRN, $[X]_0 = [X_t]_0$; $[Y]_0$ and $[Y_t]_0$ represent the initial concentration of species Y and Y_t in DNA CRNs and target CRN, $[Y]_0 = [Y_t]_0$; $[Z]_0$ and $[Z_t]_0$ represent the initial concentration of species Z and Z_t in DNA CRNs and target CRN, $[Z]_0 = [Z_t]_0$.

In the catalysis reaction module (1), as shown in Fig. 1, there are reactant X, and auxiliary species A, C. In the reactions of DNA CRNs, $\frac{d[X]_t}{dt} = -q_iCm[X]_t$, according to Eq. (1); $\frac{d[X]_t}{dt} = q_mCm[X]_t$, according to Eq. (2), where $q_i \ll q_m$. In the reaction of target CRN, $\frac{d[X_t]_t}{dt} = -k_1[X_t]_t$, according to Eq. (3). When DNA CRNs are states of balance, we can obtain approximately the mathematical expression, $q_iCm = k_1$. Initial concentration of the auxiliary species A and C is set to Cm, where $[X]_0 \ll Cm$, and $[X]_0$ is the initial concentration of species X. As shown in Fig. 2, the evolutions of X in DNA CRNs and target CRN are coincident to some extent, but there are errors between DNA CRNs and target CRN.

The DNA CRNs are:

$$A + X \xrightarrow{q_i} sp4 + sp5 \tag{1}$$

$$C + sp5 \xrightarrow{q_m} waste + 2X \tag{2}$$

The target CRN is:

$$X_t \xrightarrow{k_1} 2X_t \tag{3}$$

The Eqs. (1), (2) are DNA CRNs, Eq. (3) is target CRN. Where $[X]_0 = [X_t]_0 = 5nM$, $q_i = 10^{-7}nMs^{-1}$, $q_m = 10^{-3}nMs^{-1}$. A and C are auxiliary species. A and C are set to Cm, $A = C = 10^4nM$. $q_i \ll q_m$, $[X]_0 \ll Cm$, $q_iCm = k_1$.

In the catalysis reaction module (2), as shown in Fig. 3, there are reactants X and Y, auxiliary species A, B, D, F, and G. In the reactions of DNA CRNs, $\frac{d[X]_t}{dt} = -q_iCm[X]_t$, according to Eq. (4); $\frac{d[X]_t}{dt} = -q_iCm[X]_t$, according to Eq. (5); $\frac{d[Y]_t}{dt} = q_mCm[Y]_t$, according to Eq. (6); $\frac{d[Y]_t}{dt} = -q_sCm[Y]_t$, according to Eq. (7); $\frac{d[Y]_t}{dt} = q_sCm[Y]_t$, according to Eq. (8), where $q_i \ll q_m$, $q_i \ll q_s$, $q_m \ll q_s$. In the reaction of target CRN, $\frac{d[X_t]_t}{dt} = -k_2[X_t]_t$, according to Eq. (9); $\frac{d[Y_t]_t}{dt} = k_2[Y_t]_t$, according to Eq. (9). When DNA CRNs reach to stabilization, we can obtain approximately the conclusion, $q_iCm = k_2$. Initial concentration

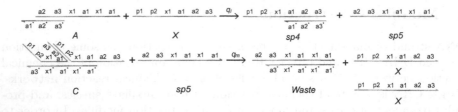

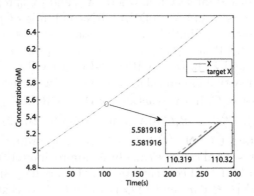

Fig. 1. Catalysis reaction module (1) $X \to 2X$.

Fig. 2. Simulation of species X and X_t.

of the auxiliary species A, B, D, F, and G is set to Cm, where $[X]_0 = [Y]_0$, $[X]_0 \ll Cm$, $[Y]_0 \ll Cm$, and $[X]_0$, $[Y]_0$ is the initial concentration of species X, Y respectively. As shown in Fig. 4, the evolutions of X, Y in DNA CRNs and target CRN are coincident to some extent, but there are nuances between DNA CRNs and target CRN.

The DNA CRNs are:

$$A + X \xrightarrow{q_i} sp8 + sp9 \tag{4}$$

$$B + X \xrightarrow{q_i} sp8 + sp9 \tag{5}$$

$$D + sp9 \xrightarrow{q_m} sp10 + Y \tag{6}$$

$$F + Y \xrightarrow{q_s} sp11 + sp12 \tag{7}$$

$$G + sp12 \xrightarrow{q_s} sp13 + Y \tag{8}$$

The target CRN is:

$$X_t + Y_t \xrightarrow{k_2} 2Y_t \tag{9}$$

The Eqs. (4), (5), (6), (7) and (8) are DNA CRNs, Eq. (9) is target CRN. Where $[X]_0 = [X_t]_0 = 1nM$, $[Y]_0 = [Y_t]_0 = 1nM$, $q_i = 10^{-6}nMs^{-1}$, $q_m = 10^{-3}nMs^{-1}$, $q_s = 7nMs^{-1}$. A, B, D, F and G are auxiliary species. A, B, D,

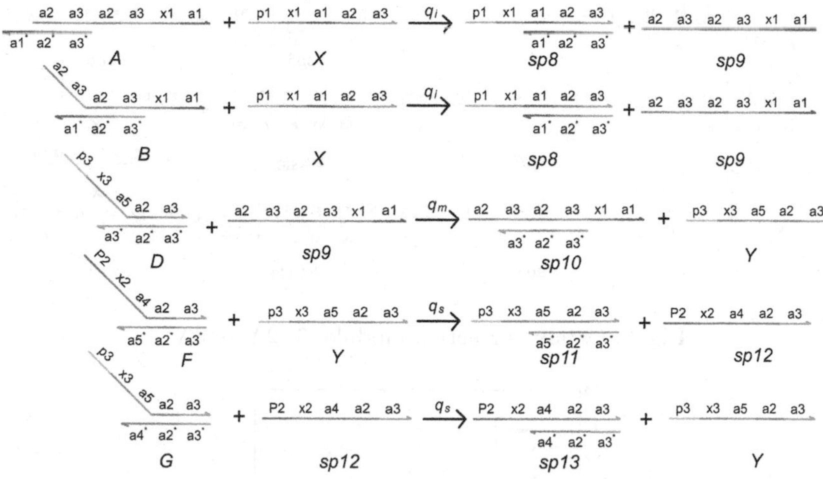

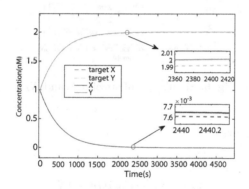

Fig. 3. Catalysis reaction module (2) $X + Y \rightarrow 2Y$.

Fig. 4. Simulation of species X, Y, X_t, and Y_t.

F and G are set to Cm, $A = B = D = F = G = 10^3 nM$. $q_i \ll q_m$, $q_m \ll q_s$, $[X]_0 \ll Cm$, $q_i Cm = k_2$.

In the catalysis reaction module (3), as shown in Fig. 5, there are reactant X, auxiliary species A, C, and D. In the reactions of DNA CRNs, $\frac{d[X]_t}{dt} = -q_i Cm[X]_t$, according to Eq. (10); $\frac{d[X]_t}{dt} = q_m Cm[X]_t$, according to Eq. (11); $\frac{d[X]_t}{dt} = q_m Cm[X]_t$, according to Eq. (12), where $q_i \ll q_m$. In the reaction of target CRN, $\frac{d[X_t]_t}{dt} = -k_3[X_t]_t$, according to Eq. (13). When DNA CRNs are states of balance, that can be thought $q_i Cm = k_3$. Initial concentration of the auxiliary species A, C and D is set to Cm, where $[X]_0 \ll Cm$, and $[X]_0$ is the initial concentration of species X. As shown in Fig. 6, the evolutions of X in DNA CRNs and target CRN are coincident to some extent, but there are errors between DNA CRNs and target CRN.

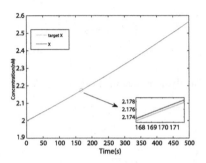

Fig. 5. Catalysis reaction module (3) $2X \to 3X$.

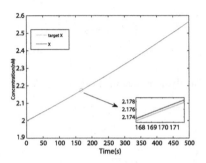

Fig. 6. Simulation of species X and X_t.

The DNA CRNs are:

$$A + X \xrightarrow{q_i} sp5 + sp6 \tag{10}$$

$$C + sp6 \xrightarrow{q_m} waste + 2X \tag{11}$$

$$D + sp6 \xrightarrow{q_m} waste + X \tag{12}$$

The target CRN is:

$$2X_t \xrightarrow{k_3} 3X_t \tag{13}$$

The Eqs. (10), (11) and (12) are DNA CRNs, Eq. (13) is target CRN. Where $[X]_0 = [X_t]_0 = 2nM$, $q_i = 10^{-6}nMs^{-1}$, $q_m = 10^{-3}nMs^{-1}$. A, C, and D are auxiliary species. A, C, and D are set to Cm, $A = C = D = 10^3 nM$. $q_i \ll q_m$, $[X]_0 \ll Cm$, $q_i Cm = k_3$.

In the adjustment reaction module (4), as shown in Fig. 7, there are reactants X and Z, auxiliary species A, B, D, F, and G. In the reactions of DNA CRNs, $\frac{d[X]_t}{dt} = -q_i Cm[X]_t$, according to Eq. (14); $\frac{d[X]_t}{dt} = -q_i Cm[X]_t$, according to Eq. (15); $\frac{d[Z]_t}{dt} = q_m Cm[Z]_t$, according to Eq. (16); $\frac{d[Z]_t}{dt} = -q_s Cm[Z]_t$, according to Eq. (17); $\frac{d[Z]_t}{dt} = q_s Cm[Z]_t$, according to Eq. (18), where $q_i \ll q_m$, $q_i \ll q_s$, $q_m \ll q_s$. In the reaction of target CRN, $\frac{d[X_t]_t}{dt} = -k_4[X_t]_t$, according to

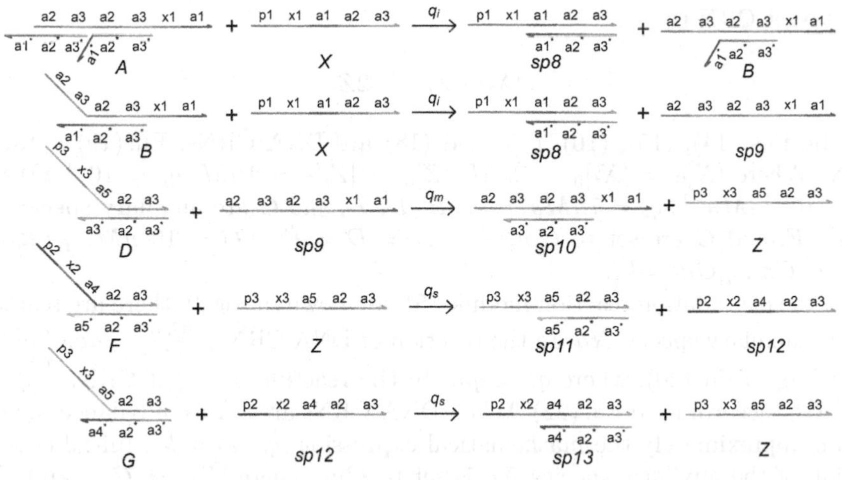

Fig. 7. Adjustment reaction module (4) $2X + Z \rightarrow 2Z$.

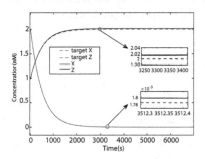

Fig. 8. Simulation of species X, Z, X_t and Z_t.

Eq. (19); $\frac{d[Z_t]_t}{dt} = k_4[Z_t]_t$, according to Eq. (19). When DNA CRNs are states of balance, we can obtain approximately the mathematical expression, $q_i Cm = k_4$. Initial concentration of the auxiliary species A, B, D, F, and G is set to Cm, where $[X]_0 = 2[Z]_0$, $[X]_0 \ll Cm$, and $[Z]_0 \ll Cm$, and $[X]_0$, $[Z]_0$ is the initial concentration of species X, Z respectively. As shown in Fig. 8, the evolutions of X and Z in DNA CRNs and target CRN are coincident to some extent, but there are errors between DNA CRNs and target CRN.

The DNA CRNs are:

$$A + X \xrightarrow{q_i} sp8 + B \tag{14}$$

$$B + X \xrightarrow{q_i} sp8 + sp9 \tag{15}$$

$$D + sp9 \xrightarrow{q_m} sp10 + Z \tag{16}$$

$$F + Z \xrightarrow{q_s} sp11 + sp12 \tag{17}$$

$$G + sp12 \xrightarrow{q_s} sp13 + Z \tag{18}$$

The target CRN is:

$$2X_t + Z_t \xrightarrow{k_4} 2Z_t \tag{19}$$

The Eqs. (14), (15), (16), (17) and (18) are DNA CRNs, Eq. (19) is target CRN. Where $[X]_0 = [X_t]_0 = 2nM$, $[Z]_0 = [Z_t]_0 = 1nM$, $q_i = 10^{-6}nMs^{-1}$, $q_m = 10^{-3}nMs^{-1}$, $q_s = 7nMs^{-1}$. A, B, D, F, and G are auxiliary species. A, B, D, F, and G are set to Cm, $A = B = D = F = G = 10^3 nM$. $q_i \ll q_m$, $[X]_0 \ll Cm$, $q_i Cm = k_4$.

In the degradation reaction module (5), as shown in Fig. 9, there are reactant X and auxiliary species Na. In the reaction of DNA CRNs, $\frac{d[X]_t}{dt} = -q_i Cm[X]_t$, according to Eq. (20), where $qi \ll qm$. In the reaction of target CRN, $\frac{d[X_t]_t}{dt} = -k_5[X_t]_t$, according to Eq. (22). When DNA CRNs are states of balance, we can obtain approximately the mathematical expression, $q_i Cm = k_5$. Initial concentration of the auxiliary species Na is set to Cm, where $[X]_0 \ll Cm$, and $[X]_0$ is the initial concentration of species X. As shown in Fig. 10, the evolutions of X in DNA CRNs and target CRN are coincident to some extent, but there are nuances between DNA CRNs and target CRN.

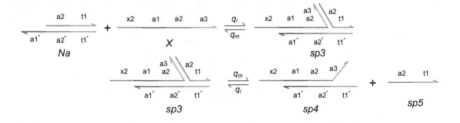

Fig. 9. Degradation reaction module (5) $X \to \phi$.

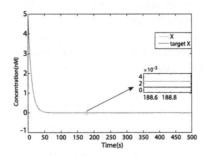

Fig. 10. Simulation of species X, Z, X_t and Z_t.

The DNA CRNs are:

$$Na + X \underset{q_m}{\overset{q_i}{\rightleftharpoons}} sp3 \tag{20}$$

$$sp3 \underset{q_i}{\overset{q_m}{\rightleftharpoons}} sp4 + sp5 \tag{21}$$

The target CRN is:

$$X_t \xrightarrow{k_5} \phi \tag{22}$$

The Eqs. (20) and (21) are DNA CRNs, Eq. (21) is target CRN. Where $[X]_0 = [X_t]_0 = 5nM$, $q_i = 3 \times 10^{-4} nMs^{-1}$, $q_m = 1.126 \times 10^{-1} nMs^{-1}$. Na is auxiliary species. Na is set to Cm, $Na = 10^2 nM$. $q_i \ll q_m$, $[X]_0 \ll Cm$, $q_iCm = k_5$.

In the synchronization reaction module (6), as shown in Fig. 11, there are reactants X and Y, auxiliary species A, B, C, D, E, F, H and G. In the reactions of DNA CRNs, $\frac{d[X]_t}{dt} = -q_iCm[X]_t$, according to Eq. (23); $\frac{d[sp12]_t}{dt} = -q_mCm[sp12]_t$, according to Eq. (23); when the reaction (23) reach the state of balance, $\frac{d[X]_t}{dt} = \frac{d[sp12]_t}{dt}$, in other words, $\frac{[X]_t}{[sp12]_t} = \frac{q_m}{q_i}$. According to Eqs. (23), (24), (25) and (29), $[X_t]_t = [X]_t + [sp12]_t$, we can get the equation $[X_t]_t = \frac{(q_m+q_i)[sp12]_t}{q_i}$ by "$\frac{[X]_t}{[sp12]_t} = \frac{q_m}{q_i}$". According to Eq. (24), $\frac{d[sp12]_t}{dt} = -q_sCm[sp12]_t$. According to Eq. (29), $\frac{d[X_t]_t}{d_t} = -k_6[X_t]_t$. When DNA CRNs reach to stability, we can get the equation $\frac{d[sp12]_t}{dt} = \frac{d[X_t]_t}{d_t}$, where $qs \ll qi$, $qs \ll qm$. In the end, we can calculate the value $k_6 = \frac{q_sq_iCm}{q_m+q_i}$. Initial concentration of the auxiliary species A, B, C, D, E, F, H and G is set to Cm, where $[X]_0, [Y]_0 \ll Cm$, and $[X]_0, [Y]_0$ is the initial concentration of species X, Y respectively. As shown in Fig. 12, X and Y have the same dynamic behavior in DNA CRNs and target CRN, but the synchronization values are different.

The DNA CRNs are:

$$A + X \underset{q_m}{\overset{q_i}{\rightleftharpoons}} B + sp12 \tag{23}$$

$$C + sp12 \xrightarrow{q_s} waste + sp14 \tag{24}$$

$$D + sp14 \xrightarrow{q_m} sp15 + Y \tag{25}$$

$$E + Y \underset{q_m}{\overset{q_i}{\rightleftharpoons}} F + sp17 \tag{26}$$

$$H + sp17 \xrightarrow{q_s} waste + sp19 \tag{27}$$

$$G + sp19 \xrightarrow{q_m} sp20 + X \tag{28}$$

The target CRNs are:

$$X_t \xrightarrow{k_6} Y_t \tag{29}$$

$$Y_t \xrightarrow{k_6} X_t \tag{30}$$

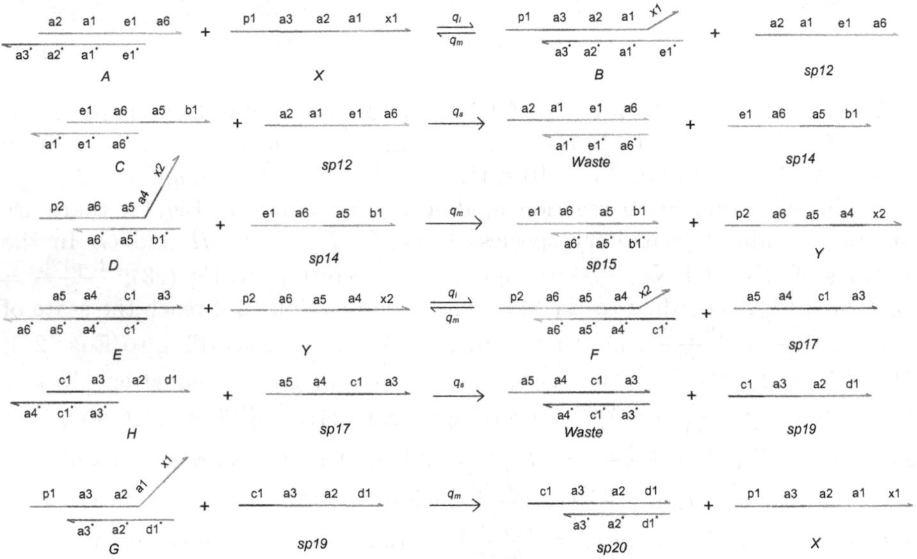

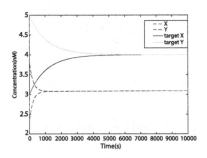

Fig. 11. Synchronization reaction module (6).

Fig. 12. Simulation of species X, Y, X_t and Y_t.

The Eqs. (23), (24), (25), (26), (27) and (28) are DNA CRNs, Eqs. (29) and (30) are target CRNs. Where $[X]_0 = [X_t]_0 = 3nM$, $[Y]_0 = [Y_t]_0 = 5nM$,

$q_i = 3 \times 10^{-4} nMs^{-1}$, $q_m = 10^{-3} nMs^{-1}$, $q_s = 10^{-6} nMs^{-1}$. A, B, C, D, E, F, H and G are auxiliary species. A, B, C, D, E, F, H and G are set to Cm, $A = B = C = D = E = F = H = G = 2 \times 10^3 nM$. $q_s \ll q_m$, $q_s \ll q_i$, $[X]_0, [Y]_0 \ll Cm$, $k_6 = \frac{q_s q_i Cm}{q_m + q_i}$.

As shown in Fig. 12, X and Y have the same dynamic behavior in DNA CRNs and target CRN, but the synchronization values are different. The reasons are as follows:

In DNA CRNs:

$$[X]_0 + [Y]_0 = [X]_t + [Y]_t + [sp12]_t + [sp14]_t + [sp17]_t + [sp19]_t \qquad (31)$$

When DNA CRNs are states of balance, the following can be obtained:

$$[X]_t = [Y]_t \qquad (32)$$

$$[sp12]_t = [sp17]_t \qquad (33)$$

$$\frac{[X]_t}{[sp12]_t} = \frac{q_m}{q_i} \qquad (34)$$

$$[sp14]_t = [sp19]_t \approx 0 \qquad (35)$$

Base on Eqs. (31–35), we can obtain the synchronization values:

$$[X]_t = [Y]_t = \frac{q_m([X]_0 + [Y]_0)}{2(q_i + q_m)} \qquad (36)$$

In target CRNs:

$$[X_t]_0 + [Y_t]_0 = [X_t]_t + [Y_t]_t \qquad (37)$$

When target CRNs approach stability:

$$[X_t]_t = [Y_t]_t \qquad (38)$$

Therefore, the synchronization values X_t and Y_t are obtained as follows:

$$[X]_t = [Y]_t = \frac{[X]_0 + [Y]_0}{2} \qquad (39)$$

3   Chaotic Oscillatory System and Its Dynamic Analysis

3.1   Chaotic Oscillatory System

A 3-variable chaotic oscillatory model can be described as follows:

$$X_1 \xrightarrow{r_1} 2X_1 \qquad (40)$$

$$X_1 + X_2 \xrightarrow{r_2} 2X_2 \qquad (41)$$

$$X_1 + X_3 \xrightarrow{r_3} 2X_3 \qquad (42)$$

$$X_2 \xrightarrow{r_4} \phi \qquad (43)$$

$$X_3 \xrightarrow{r_5} \phi \qquad (44)$$

CRNs can be represented by the following set of ODEs

$$\dot{X}_1 = r_1 X_1 - r_2 X_1 X_2 - r_3 X_1 X_3 \tag{45}$$
$$\dot{X}_2 = r_2 X_1 X_2 - r_4 X_2 \tag{46}$$
$$\dot{X}_3 = r_1 X_1 X_3 - r_5 X_3 \tag{47}$$

Based on the above conclusion, we have made some adjustments, we added two new gates to link some connection with chaotic oscillatory system as follows:

$$X_1 \xrightarrow{r_1} 2X_1 \tag{48}$$
$$X_1 + X_2 \xrightarrow{r_2} 2X_2 \tag{49}$$
$$2X_1 + X_3 \xrightarrow{r_3} 2X_3 \tag{50}$$
$$2X_1 \xrightarrow{r_4} 3X_1 \tag{51}$$
$$X_2 \xrightarrow{r_5} \phi \tag{52}$$
$$X_3 \xrightarrow{r_6} \phi \tag{53}$$

The new CRNs can be represented by the following set of ODEs

$$\dot{X}_1 = r_1 X_1 - r_2 X_1 X_2 - r_3 X_1^2 X_3 + r_4 X_1^2 \tag{54}$$
$$\dot{X}_2 = r_2 X_1 X_2 - r_5 X_2 \tag{55}$$
$$\dot{X}_3 = r_3 X_1^2 X_3 - r_6 X_3 \tag{56}$$

3.2   Dynamic Analysis

The system is represented by the following third-order equation:

$$\begin{cases} \dot{x} = ax + bx^2 - cx^2 z - mxy \\ \dot{y} = -ny + gxy \\ \dot{z} = -dz + hx^2 z \end{cases} \tag{57}$$

Among them, a, b, c, m, n, g, d, h are the parameters of the system, x, y z are the state variables of the system. When $a = 1$, $b = 2$, $c = 2.9851$, $d = 3$, $m = 1$, $n = 1$, $g = 1$, $h = 2.9851$, the behaviour of the system represent the state of chaos. The Lyapunov exponent is defined as follows:

$$\lambda = \frac{1}{t_M - t_0} \sum_{k=1}^{M} \ln \frac{L(t_k)}{L(t_{k-1})} \tag{58}$$

$L(t_k)$ represents the distance from the t_k^{th} point to the origin, and M is the number of steps in the iteration. t_0 and t_k represent the initial time and the iteration time of M step, respectively.

When $x(0) = 1.2$, $y(0) = 1.2$, $z(0) = 1.2$, in initial condition, we make a particular study in stability and the behaviour characteristic of the system

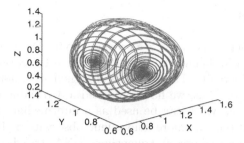

Fig. 13. Chaotic attractors of the system in different directions.

using the Cybernetics theory and simulation software. The chaotic attractors of the system in different directions are shown in the Fig. 13:

The influence of parameter g. The influence of the parameters on the dynamic behavior of the system is discussed in detail by means of Lyapunov exponential diagram, bifurcation diagram and stability criterion, which are shown in Figs. 14 and 15 respectively.

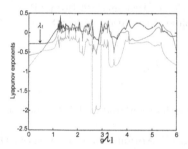

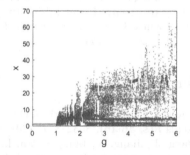

Fig. 14. Lyapunov exponential diagram of parameter g.

Fig. 15. Bifurcation diagram of parameter g.

When $g \in [0, 6]$, $a = 1$, $b = 2$, $c = 2.9851$, $d = 3$, $m = 1$, $n = 1$, $h = 2.9851$, the initial condition $x(0) = 1.2$, $y(0) = 1.2$, $z(0) = 1.2$, the dynamic behavior trajectory of the system changes with the adjustment of the parameter g. The analysis of the motion characteristics of the system is helpful to better understand and apply the chaotic characteristics of the system.

First, the Lyapunov exponent of the system is calculated by using the selected region $[0, 6]$ as the variation interval of the parameter g, and the dynamic behavior of the system with the parameter g is analyzed by using the Lyapunov exponential stability criterion. At the same time, the bifurcation diagram of the system is compared. When $g \in [0, 1]$, the largest Lyapunov exponent $\lambda_1 < 0$, and λ_1 is increasing from negative to zero. When $g \in [1, 2] \cup [4, 5]$, the chaotic behavior of the system is obvious. Meanwhile, the bifurcation behavior of the system also mainly occurs in the $g \in [1, 2] \cup [4, 5]$.

4   Conclusions

Building complex chemical circuits is a crucial step for DNA computing. Achieving the building of basic chemical reaction networks is the cornerstone of building of complex chemical circuits. Thus, we have designed six different chemical reaction modules. Basis on that, we have proposed a 3-variable chaotic oscillatory system as an example, which can be used as a intermediate link to implement chaotic synchronization. For more convenient observation, DNA strands have been designed with fluorophores and quenchers. DNA strands have been designed by DSD software. We have analyzed the differences between DNA CRNs and target CRN, and the results have been shown in corresponding figures. The method of using mathematic ODEs to represent chaotic oscillatory system is available.

Acknowledgements. This work was supported in part by the National Key R and D Program of China for International S and T Cooperation Projects (2017YFE0103900), in part by the Joint Funds of the National Natural Science Foundation of China (U1804262), in part by the State Key Program of National Natural Science of China under Grant 61632002, in part by the National Natural Science of China under Grant 61603348, Grant 61775198, Grant 61603347, and Grant 61572446, in part by the Foundation of Young Key Teachers from University of Henan Province (2018GGJS092), and in part by the Youth Talent Lifting Project of Henan Province (2018HYTP016) and Henan Province University Science and Technology Innovation Talent Support Plan under Grant 20HASTIT027.

References

1. Yang, J., et al.: Entropy-driven DNA logic circuits regulated by DNAzyme. Nucleic Acids Res. **46**(16), 8532–8541 (2018)
2. Yang, J., Jiang, S., Liu, X., Pan, L., Zhang, C.: Aptamer-binding directed DNA origami pattern for logic gates. ACS Appl. Mater. Interfaces **8**(49), 34054–34060 (2016)
3. Yang, J., Song, Z., Liu, S., Zhang, Q., Zhang, C.: Dynamically arranging gold nanoparticles on DNA origami for molecular logic gates. ACS Appl. Mater. Interfaces **8**(34), 22451–22456 (2016)
4. Zhang, C., Yang, J., Jiang, S., Liu, Y., Yan, H.: DNAzyme-mediated DNA origami pattern for logic gates. Nano Lett. **16**(1), 736–741 (2016)
5. Zhang, C., Shen, L., Liang, C., Dong, Y., Yang, J., Xu, J.: DNA sequential logic gate using two-ring DNA. ACS Appl. Mater. Interfaces **8**(14), 9370–9376 (2016)
6. Zou, C., Wei, X., Zhang, Q., Liu, C., Zhou, C., Liu, Y.: Four-analog computation based on DNA strand displacement. ACS Omega **2**(8), 4143–4160 (2017)
7. Sun, J., Li, X., Cui, G., Wang, Y.: One-bit half adder-half subtractor logical operation based on the DNA strand displacement. J. Nanoelectron. Optoelectron. **12**(4), 375–380 (2017)
8. Li, W., Zhang, F., Yan, H., Liu, Y.: DNA based arithmetic function: a half adder based on DNA strand displacement. Nanoscale **8**(6), 3775–3784 (2016)
9. Sun, J., Zhao, X., Fang, J., Wang, Y.: Autonomous memristor chaotic systems of infinite chaotic attractors and circuitry realization. Nonlinear Dyn. **94**(4), 2879–2887 (2018). https://doi.org/10.1007/s11071-018-4531-4

10. Cui, G., Zhang, J., Cui, Y., Zhao, T., Wang, Y.: DNA strand-displacement digital logic circuit with fluorescence resonance energy transfer detection. J. Comput. Theor. Nanosci. **12**(9), 2095–2100 (2015)

11. Engelen, W., Wijnands, S.P., Merkx, M.: Accelerating DNA-based computing on a supramolecular polymer. J. Am. Chem. Soc. **140**(30), 9758–9767 (2018)

12. Ito, K., Murayama, Y., Takahashi, M., Iwasaki, H.: Two three-strand intermediates are processed during Rad51-driven DNA strand exchange. Nat. Struct. Mol. Biol. **25**(1), 29 (2018)

13. Guo, Y., et al.: DNA and DNA computation based on toehold-mediated strand displacement reactions. Int. J. Mod. Phys. B **32**(18), 1840014 (2018)

14. Sawlekar, R., Montefusco, F., Kulkarni, V.V., Bates, D.G.: Implementing nonlinear feedback controllers using DNA strand displacement reactions. IEEE Trans. Nanobiosci. **15**(5), 443–454 (2016)

15. Zhang, Z., Fan, T.W., Hsing, I.M.: Integrating DNA strand displacement circuitry to the nonlinear hybridization chain reaction. Nanoscale **9**(8), 2748–2754 (2017)

16. Barati, K., Jafari, S., Sprott, J.C., Pham, V.T.: Simple chaotic flows with a curve of equilibria. Int. J. Bifurc. Chaos **26**(12) (2016). https://doi.org/10.1142/S0218127416300342

17. Li, C., Sprott, J.C., Xing, H.: Constructing chaotic systems with conditional symmetry. Nonlinear Dyn. **87**(2), 1351–1358 (2016). https://doi.org/10.1007/s11071-016-3118-1

18. Zou, C., Wei, X., Zhang, Q.: Visual synchronization of two 3-variable Lotka-Volterra oscillators based on DNA strand displacement. RSC Adv. 8(37), 20941–20951 (2018)

19. Li, Y., Kuang, Y.: Periodic solutions of periodic delay Lotka-Volterra equations and systems. J. Math. Anal. Appl. **255**(1), 260–280 (2001)

20. Dunbar, S.R.: Traveling wave solutions of diffusive Lotka-Volterra equations: a heteroclinic connection in R^4. Trans. Am. Math. Soc. **286**(2), 557–594 (1984)

21. Zou, C., Wei, X., Zhang, Q., Liu, Y.: Synchronization of chemical reaction networks based on DNA strand displacement circuits. IEEE Access **6**, 20584–20595 (2018)

22. Apraiz, A., Mitxelena, J., Zubiaga, A.: Studying cell cycle-regulated gene expression by two complementary cell synchronization protocols. JoVE (J. Vis. Exp.) (124), e55745 (2017)

23. Fries, P., Reynolds, J.H., Rorie, A.E., Desimone, R.: Modulation of oscillatory neuronal synchronization by selective visual attention. Science **291**(5508), 1560–1563 (2001)

24. Zhang, Q., Wang, X., Wang, X., Zhou, C.: Solving probability reasoning based on DNA strand displacement and probability modules. Comput. Biol. Chem. **71**, 274–279 (2017)

25. Olson, X., Kotani, S., Yurke, B., Graugnard, E., Hughes, W.L.: Kinetics of DNA strand displacement systems with locked nucleic acids. J. Phys. Chem. B **121**(12), 2594–2602 (2017)

26. Lakin, M.R., Stefanovic, D.: Supervised learning in adaptive DNA strand displacement networks. ACS Synth. Biol. **5**(8), 885–897 (2016)

27. Lakin, M.R., Youssef, S., Polo, F., Emmott, S., Phillips, A.: Visual DSD: a design and analysis tool for DNA strand displacement systems. Bioinformatics **27**(22), 3211–3213 (2011)

Reference Point Based Multi-objective Evolutionary Algorithm for DNA Sequence Design

Haozhi Zhao[1](✉), Zhiwei Xu[1](✉) (iD), and Kai Zhang[1,2](✉)

[1] School of Computer Science and Technology, Wuhan University of Science
and Technology, Wuhan 430065, Hubei, China
873674474@qq.com, {xuzhiwei,zhangkai}@wust.edu.cn
[2] Hubei Province Key Laboratory of Intelligent Information Processing
and Real-Time Industrial System, Wuhan 430065, Hubei, China

Abstract. DNA computing is a parallel computing model based on DNA molecules. High-quality DNA sequences can prevent unwanted hybridization errors in the computation process. The design of DNA molecules can be regarded as a multi-objective optimization problem, which needs to satisfy a variety of conflicting DNA encoding constraints and objectives. In this paper, a novel reference point based multi-objective optimization algorithm is proposed for designing reliable DNA sequences. In order to obtain balance Similarity and H-measure objective values, the reference point strategy is adapted to searching for idea solutions. Firstly, every individual should be assigned a rank value by the non-dominated sort algorithm. Secondly, the crowding distance is replaced by the distance to the reference point for each individual. Lastly, the proposed algorithm is compared with some state-of-the-art DNA sequence design algorithms. The experimental results show our algorithm can provide more reliability DNA sequences than existing sequence design techniques.

Keywords: DNA encoding · Multi-objective optimization · Reference point

1 Introduction

DNA computing is a new computational paradigm, which has shown great potential to solve NP-complete problems, such as Hamiltonian path problem (HPP) [1], satisfaction problem (SAT) [2], traveling salesman problem (TSP) [3] and graph coloring problem (GCP) [4]. High-quality DNA sequences can improve the efficiency and reliability. Therefore, the design of DNA molecules should be carefully designed to prevent unwanted hybridization errors. The design of DNA

Supported by the National Natural Science Foundation of China (Grant Nos. U1803262, 61702383, 61602350).

© Springer Nature Singapore Pte Ltd. 2020
L. Pan et al. (Eds.): BIC-TA 2019, CCIS 1160, pp. 178–188, 2020.
https://doi.org/10.1007/978-981-15-3415-7_14

molecules can be regarded as a multi-objective optimization [22–25] problem, which need satisfy a variety of conflicting DNA encoding constraints and objectives [5].

In the past few decades, a lot of efficient algorithms have been proposed to solve DNA sequences design problem. Frutos et al. [6] proposed the Template-Map method, but it is difficult to derive templates and mappings that satisfy the combined constraints when there are many constraints. Hartemink et al. [7] implemented an exhaustive search algorithm to design DNA sequences, which satisfy the constraints, but the algorithm has a high time complexity. Feldkamp [8] uses directed trees to design DNA encoding that the fixed length subsequences are only allowed to appear once, but the length of the sub-sequences requires a lot of testing to determine in the actual design. Recent years, evolutionary algorithms are widely adapted for DNA encoding design, such as genetic algorithm [9,11,14,20], particle swarm optimization [11,15,17,18], ant colony algorithm [12], simulated annealing [16], and multi-objective evolutionary algorithms [10,13,19]. However, existing algorithms often obtain the DNA sequences set with bias Similarity or H-measure values, which is easy to introduce errors during the DNA computing process.

In this paper, a reference point based multi-objective optimization evolutionary algorithm is proposed for designing DNA sequences. Firstly, the non-dominated sort algorithm is adapted to select convergent solutions rank by rank. Secondly, the crowding distance sort algorithm is adapted to choose the solutions that are closer to the reference point. The algorithm can provide a set of DNA sequences near idea point which are more reliable and efficient for DNA computing. Finally, to validate the proposed algorithm, we compare our algorithm with some state-of-art techniques. The experimental results confirm the performance of our algorithm in designing high-quality DNA sequences set efficiently.

In the following chapters, Sect. 2 introduces the relevant basis of DNA sequence design problem. The proposed reference-based evolution algorithm for designing DNA sequences is then detailed in Sect. 3. In Sect. 4, we present the experiment results of the proposed algorithm and compare them with the other literatures. Finally, conclusions are drawn in Sect. 5 along with pertinent observations identified.

2 Problem Formulation

During the process of DNA computing, single-strand DNA molecules dismissed randomly in the vitro, therefore four kinds of molecules exist simultaneity, including DNA molecular X, corresponding complementary sequence X C, reverse sequence X R and reverse complementary sequence X RC. Most existing DNA computing models are based on the specific hybridization between a given molecular X and it's unique Watson-Crick complement X C. In fact, the non-specific hybridization often occurs because unwanted mismatches maybe take place between random molecules, as shown in Fig. 1.

However, the non-specific hybridization could introduce errors, such as false positives and negatives, and degrade efficiency. Obviously, it is important to

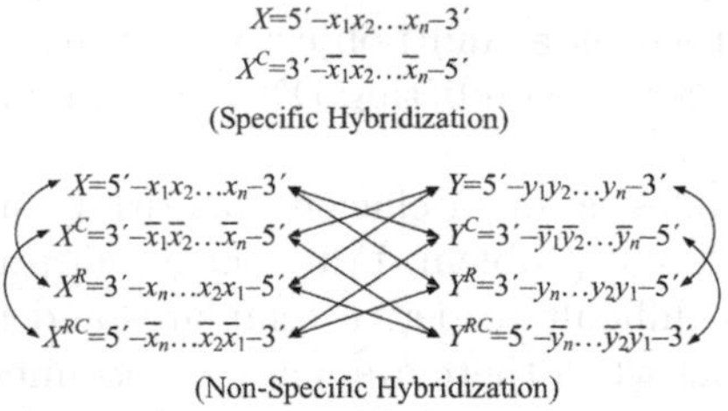

$$X{=}5'{-}x_1x_2...x_n{-}3'$$
$$X^C{=}3'{-}\bar{x}_1\bar{x}_2...\bar{x}_n{-}5'$$

(Specific Hybridization)

(Non-Specific Hybridization)

Fig. 1. Specific and non-specific hybridizations.

design reliable DNA sequences for DNA computing, and the key is to avoid the non-specific hybridization. The design of reliable DNA sequences involves several conflicting design constraints which have to be considered simultaneously. In mathematical terms, DNA sequence design problem can be formulated as a multi-objective optimization problem as Eq. (1).

$$\min f(X) = \min\left[f_1(X), f_2(X), \cdots, f_M(X)\right]^T, f(X) \in R^M \qquad (1)$$

where DNA sequence $X = [x_1, x_2, \cdots, x_N]^T \in \Omega$ consists of N bases $x_i \in \{A, C, G, T\}$, and the search space Ω is 4^N. $f(X)$ consists of M objective functions $f_m(x)$, $m = 1, \ldots, M$. R^M denotes the objective space. Several typical biochemical design criteria are chosen that other relevant authors use to evaluate and generate reliable DNA libraries. The formal definition for each design criteria is provided in the following subsections.

2.1 Similarity Criterion

Let X_i and X_j be two different DNA sequences, the similarity criterion refers to the degree of similarity in base composition between X_i and X_j. By controlling the similarity, non-specific hybridization between X_i and the complementary of X_j, (i.e. X_j^C). The calculation of Similarity is shown in Eq. (2).

$$\begin{aligned}
f_{\text{Similariy}}(X) &= \sum_{i=1}^{n}\sum_{j=1}^{n} \text{Similarity}\,(X_i, X_j) \\
&= \sum_{i=1}^{n}\sum_{j=1}^{n} \text{Max}_{g,i}\left(Si_{dis}\,(X_i, X_j, s) + Si_{con}\,(X_i, X_j, s)\right)
\end{aligned} \qquad (2)$$

where the function $Max_{g,i}$ represents traversing all possible values of g and i and taking the maximum value as a result. The function $Si_{dis}\,(X_i, X_j, s)$ represents the number of identical bases in which the DNA sequence X_i is shifted to the right by the s position compared with the sequence X_j. The function $Si_{con}\,(X_i, X_j, s)$ represents that the DNA sequence X_i shifts to the right by the s bit and the base compared with X_j is continuously the same penalty value.

2.2   H-Measure Criterion

For DNA sequences X and Y, the H-measure constraint is to limit non-specific hybridization between X and reverse Y. The calculation of H-measure is as shown in Eq. (3).

$$
\begin{aligned}
f_{H\text{-measioe}}(X) &= \sum_{i=1}^{n} \sum_{j=1}^{n} H\text{-measure}\,(X_i, X_j) \\
&= \sum_{i=1}^{n} \sum_{j=1}^{n} \text{Max}_{g,i}\left(h_{dis}\left(X_i, X_j^{\text{R}}, s\right) + h_{con}\left(X_i, X_j^{\text{R}}, s\right)\right)
\end{aligned}
\tag{3}
$$

where the function $Max_{g,i}$ represents traversing all possible values of g and i and taking the maximum value as a result. The function $h_{ds}\left(X_i, X_j^{\text{R}}, s\right)$ represents the number of base complements in which the DNA sequence X_i shifts to the right by the s-bit compared with the sequence X_j. The function $h_{con}\left(X_i, X_j^{\text{R}}, s\right)$ represents a base continuous pairing penalty value in which the DNA sequence X_i is shifted to the right by the s-bit compared with X_j.

Similarity Criterion describes the degree of similarity between DNA sequences, and H-measure Criterion describes the degree of complementary hybridization between DNA sequences. Similarity Criterion and H-measure Criterion are two conflicting objectives, and they are difficult to optimize at the same time. Shin et al. [10] had proved that Similarity Criterion and H-measure Criterion are conflicting, and they are both discontinuous functions and have many locally optimal solutions.

2.3   Continuity Criterion

Continuity constraint means that in the single strand of DNA, the same base appears continuously, and an undesired secondary structure occurs under the hydrogen bonding force of the base molecule. The calculation of Continuity is as shown in Eq. (4).

$$
\begin{aligned}
f_{\text{Continuity}}(X) &= \sum_{i=1}^{n} \text{Continuity}\,(X_i) \\
&= \sum_{i=1}^{n} \sum_{i=1}^{l-t+1} T\left(c_a(x,i), t^2\right)
\end{aligned}
\tag{4}
$$

2.4   Hairpin Structure Criterion

The hairpin structure constraint refers to a single-stranded DNA molecule formed by reverse folding of itself, resulting in a secondary structure of a hairpin shape. The calculation of Hairpin is as shown in Eq. (5).

$$f_{\text{Hairpin}}(X) = \sum_{i=1}^{n} \text{Hairpin }(X_i)$$

$$= \sum_{i=1}^{n} \sum_{s=s_{mn}}^{(l/R_{mn})/2} \sum_{r=R_{mn}}^{l-2s} \sum_{i=1}^{l-2s-r} T \left(\sum_{j=1}^{s} bp\,(x_{s+i-j}, x_{s+i+r+j}), \frac{s}{2} \right) \tag{5}$$

2.5   GC Content Criterion

DNA computing prefer the DNA molecules with uniform GC content. The GC content refers to the number or percentage of bases G and bases C in the DNA sequence. The calculation of GC% is as shown in Eq. (6).

$$f_{GC}(X) = \max_i \{ GC\,(X_i) \} - \min_j \{ GC\,(X_j) \}$$
$$GC = \sum_{i=1}^{n} \sum_{i=1}^{l} gc\,(x_i),\, gc\,(x_i) = \begin{cases} 1, x_i = G \text{ or } x_i = C \\ 0, x_i = A \text{ or } x_i = T \end{cases} \tag{6}$$

2.6   Melting Temperature Criterion

The melting temperature is the temperature at which 50% of the DNA molecules open the double strand into a single strand during the warming denaturation of the double-stranded DNA molecule. The melting temperature is an important parameter for evaluating the thermodynamic stability of DNA molecules. The calculation of Tm is as shown in Eq. (7).

$$f_{Tm}(X) = \max_i \{ \text{Tm}\,(X_i) \} - \min_j \{ \text{Tm}\,(X_j) \}$$
$$\text{Tm}\,(X_i) = \sum_{i=1}^{n} \frac{\Delta H^{\circ}}{\Delta S^{\circ} + R \ln(|C_T|/4)} \tag{7}$$

ΔH° is the total enthalpy of the adjacent base, ΔS° is the total entropy of the adjacent base, R is the gas constant (1.987 cal/Kmol), and C_T is the DNA molecule concentration.

3   Problem Formulation

Because Similarity and H-measure are two conflict objectives, we would obtain a set of non-dominated solutions using MOEA. However, the DNA sequences which have high Similarity values will lead to non-specific hybridization between X and the complementary strand Y^C. Moreover, the DNA sequences which have high H-measure values will lead to non-specific hybridization between X and the reverse strand Y^R. Among the whole PF, only the nonbiased point is the idea solutions for DNA sequences design problem, as shown in Fig. 2.

In response to the above problems, we adopt R-NSGA-II [21] to search for DNA sequences, in which the crowding distance is replaced by the distance to the reference point. In our algorithm, the reference distance (RD) can be calculated as shown in Eq. (8).

$$RD = \sqrt{\sum_{i=1}^{m} \left(\frac{f_i(x) - R_i}{f_i^{max} - f_i^{min}} \right)^2} \tag{8}$$

where f_i^{max} and f_i^{min} are the global maximum and minimum function values of the i-th objective function, and R_i is the reference value of the i-th objective. In our algorithm, the reference point is set to $R = (f_1, f_2, f_3, f_4, f_5, f_6) = (0, 0, 0, 0, \frac{N}{2}, 50)$.

Most of the algorithms calculate the objective functions on the entire population P_t. However, two objective functions Similarity an H-measure are the full correlation with all the individuals. If k DNA sequences with best fitness values are selected for DNA computing, they may not remain optimal. In our algorithm, we re-evaluate the individuals with the population P_{t+1}, and update the fitness values in P_t one by one. Three main procedures are iteratively run in our algorithm, specifically the non-dominated sort, the reference crowding distance sort, and full correlation fitness update. The algorithm procedure is also shown in Fig. 3.

Firstly, tournament selection is adapted on Pt, and the winner of two randomly selected individuals should be added into mating pool Qt. Then, crossover and mutation operators are adapted to generate new offspring, and replace the individuals in mating pool Qt. Secondly, the non-dominated sorting is applied to the union set PtQt, and the non-dominated fronts are copied to parent population rank by rank. Thirdly, the reference distance should be calculated for every individual, and the individual with minimum reference distance could be added into the new population Pt + 1 until the population size N. Moreover, in order to select the individuals with optimal full correlation objective values, we re-evaluate the population Pt when individual is selected and added into Pt + 1. The pseudocode is shown as Algorithm 1.

Algorithm 1. *Proposed Algorithm*

1: Initialization P_0
2: **while** (stopping criterion is not satisfied) **do**
3:      $Q_t = $ Tournament Selection (P_t)
4:      $Q_t = $ Crossover and Mutation (P_t)
5:      $R_t = P_t \cup Q_t$
6:      EvaluatePopulation on R_t
7:      Non-dominated Sort (R_t)
8:      **for** $i = 0$ to P_t **do**
9:          Reference Distance calculate (R_t)
10:          $P_{t+1} = P_{t+1} + $ Nearest Individual with min(R_t)
11:          Re-EvaluatePopulation on P_{t+1}
12:      **end for**
13: **end while**

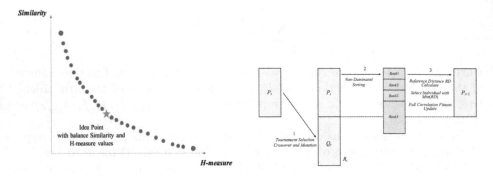

Fig. 2. Idea solution within the non-dominated solutions.

Fig. 3. The procedure of our algorithm.

4   Result and Discussion

In order to verify the effectiveness of the proposed algorithm, we compare the obtained results with various known algorithms. In our comparison, the population size is set to 200, the DNA length is set to 20, and the maximum number of iterations is set to 1000. The algorithm is implemented in Eclipse Java and tested on a PC (running environment intel® $Core^{TM}$ i5-8400 CPU @ 2.802 GHz, 8G RAM, Windows 10).

Table 1 shows the obtained sequences generated by MGA [20], NACST/Seq [10], and our algorithm. As can be seen from Table 1, all the algorithms obtain same Hairpin and GC content values. The sequences of MGA and our algorithm have same Continuity values, which are better than the sequences of NACST/Seq. The MGA has most uniform melting temperature values fluctuated within one degree Celsius. The temperature fluctuation range of our sequences is ±1.2526 °C, which is better than NACST/Seq.

The Similarity value of our algorithm is 290, which is much smaller than MGA(444) and NACST/Seq(374). In addition, the H-measure value of our algorithm is 284, which is also much smaller than MGA(438) and NACST/Seq(338). Moreover, the balanced Similarity and H-measure values imply that our sequences are more reliable and have a lower probability of unwanted non-hybridization.

Table 2 shows obtained larger group of DNA sequences by three compared algorithms. As can be seen from Table 2, all the algorithms obtain same Continuity and GC content values. Our algorithm obtains best Hairpin objective value, however, ten sequences in MGA set have poor Hairpin objective values. The sequences of our algorithm have most uniform melting temperature values fluctuated within ±0.9002 °C. The melting temperature of MGA and NACST/Seq fluctuate in range ±1.7500 and ±2.9574 respectively. The H-measure and Similarity values of the sequences designed by our algorithm are balance and much lower than compared algorithms, which means the mismatch and non-hybridization between the coding sequences can be greatly reduced.

Table 1. Comparison results of the obtained seven sequences with 20 bases.

Sequence	Continuity	Hairpin	H-measure	Similarity	Tm	GC%
MGA [20]						
TAGACCACTGTTGCACATGG	0	0	58	52	56.0900	50
ATTCGGTCAGACTTGCTGTG	0	0	64	52	56.2400	50
ATAGTGCGGACAGTAGTTCC	0	0	66	59	54.9200	50
AATACGCGGAACGTAACCTC	0	0	61	85	55.8300	50
AATACGCGGAACGTAACCTC	0	0	61	85	55.4000	50
ACAGCCTTAAGCCTAACTCC	0	0	65	54	56.0641	50
ATGCTTCCGACATGGAATGG	0	0	63	57	55.8500	50
Objective values	0	0	438	444	55.5800 (±0.6600)	50 (±0)
NACST/Seq [10]						
CTCTTCATCCACCTCTTCTC	0	0	43	58	46.6803	50
CTCTCATCTCTCCGTTCTTC	0	0	37	58	46.9393	50
TATCCTGTGGTGTCCTTCCT	0	0	45	57	49.1066	50
ATTCTGTTCCGTTGCGTGTC	0	0	52	56	51.1380	50
TCTCTTACGTTGGTTGGCTG	0	0	51	53	49.9252	50
GTATTCCAAGCGTCCGTGTT	0	0	55	49	50.7224	50
AAACCTCCACCAACACACCA	9	0	55	43	51.4735	50
Objective values	9	0	338	374	49.0769 (±2.3966)	50 (±0)
Our algorithm						
ACAACAACCACCACCACCAA	0	0	37	45	50.2236	50
CCAAGGAAGGAAGGAAGGAA	0	0	54	33	49.0486	50
CCTCTCCTCTTCTTATCTCC	0	0	34	49	49.6556	50
GTGTGTGTGTGTGTGTGTGT	0	0	48	25	50.9244	50
CCAACCAACCAACCAACCAA	0	0	34	45	51.3054	50
CTTCTTCCTCCTTCTTCTCC	0	0	36	45	48.8003	50
CTCTCGCTCTATATCTCTCC	0	0	41	48	49.4115	50
Objective values	0	0	284	290	50.0529 (±1.2526)	50 (±0)

Table 2. Comparison results of the obtained fourteen sequences with 20 bases.

Sequence	Continuity	Hairpin	H-measure	Similarity	Tm	GC%
MGA [20]						
CTCATCTAATCAGCCTCGCA	0	0	135	114	55.2900	50
CTAATAGTGACAGCTGCGTG	0	3	131	119	53.9200	50
GCATCGTTAGAGACACCTAC	0	3	134	124	53.1000	50
GCATCAATATGCGCGACTAC	0	0	131	125	54.8700	50
CATTAAGTAGACGCTGTCGG	0	3	132	114	53.6100	50
TATGGATGAGGAGGACCTAG	0	3	133	117	53.2300	50
CAGAGATGTTCTGTACCACC	0	3	128	117	53.2000	50
CGTCGAGAATTCGTAGCTCA	0	0	137	119	55.1300	50
TCTGTTACCGTATCGGATCG	0	3	129	115	54.4900	50
AGAAGAGTTCGACTTGCTGG	0	3	134	121	55.6300	50
GCAAGGAATTCACCGTCTGT	0	3	133	129	56.6000	50
CGTGTGAAGAGAGTGGTTCA	0	0	127	123	55.5000	50
CGACTGAATCATGGACCTGT	0	3	134	126	55.5300	50
TACCGAGAAGTAGGACTGCA	0	3	134	124	56.0100	50
Objective values	0	30	1852	1687	54.8500 (±1.7500)	50 (±0)
NACST/Seq [10]						
GTGACTTGAGGTAGGTAGGA	0	3	129	115	47.2490	50
ATCATACTCCGGAGACTACC	0	3	132	121	47.2304	50
CACGTCCTACTACCTTCAAC	0	0	128	121	47.4589	50
ACACGCGTGCATATAGGCAA	0	3	141	117	52.5401	50
AAGTCTGCACGGATTCCTGA	0	3	132	115	50.5497	50
AGGCCGAAGTTGACGTAAGA	0	0	132	116	51.0482	50
CGACACTTGTAGCACACCTT	0	0	132	123	50.2683	50
TGGCGCTCTACCGTTGAATT	0	0	135	116	52.0565	50
CTAGAAGGATAGGCGATACG	0	0	134	117	46.6253	50
CTTGGTGCGTTCTGTGTACA	0	0	140	116	50.5774	50
TGCCAACGGTCTCAACATGA	0	0	132	121	51.8587	50
TTATCTCCATAGCTCCAGGC	0	0	136	117	48.1017	50
TGAACGAGCATCACCAACTC	0	0	121	121	50.3351	50
CTAGATTAGCGGCCATAACC	0	0	127	119	47.6383	50
Objective values	0	12	1851	1655	49.2420 (±2.9574)	50 (±0)
Our algorithm						
GAGAATAGAGAAGGAGGAGG	0	0	84	115	49.6556	50
TGTTGTGGTGTGGTGTGGTT	0	0	124	80	50.1562	50
GAAGGAAGGAAGGAAGGAAG	0	0	77	106	49.4336	50
GAGAGTGAGAGGATAAGAGG	0	0	91	112	49.5929	50
TTGTTCTGGTGGTGGTGGTT	0	0	116	82	49.6702	50
GTTGGTTGGTTGGCTTGGTT	0	0	113	84	50.1442	50
CACACGCACAGACATACACA	0	0	99	98	50.2702	50
GGAAGAGCAATAGCAGAAGG	0	0	88	116	49.0941	50
CAACGACCAAGAACGACCAA	0	0	95	109	49.6784	50
AACACATCACACAGCACACC	0	0	103	102	49.9036	50
ACACACCTCACACTCAACAC	0	0	105	97	49.9141	50
CCACACGACACACTACACAA	0	0	102	104	50.8945	50
AACCAGCAACTACCAGCAAC	0	0	103	104	49.2441	50
AATGGAATGGAATGGCGAGG	0	0	100	111	49.8795	50
Objective values	0	0	1400	1420	49.9943 (±0.9002)	50 (±0)

5   Conclusion

In this study, a multi-objective DNA sequence design algorithm had been successfully implemented for reliable DNA computation. The algorithm was based on ideal reference point, which could guide the population to search for balance Similarity and H-measure objective values efficiently. The algorithm was compared with some state-of-the-art approaches. The experimental results showed our algorithm can generate high quality DNA sequences set which satisfied various conflict DNA encoding criterions.

References

1. Yang, R., Zhang, C., Gao, R.: A new bionic method inspired by DNA computation to solve the hamiltonian path problem. In: IEEE International Conference on Information and Automation (ICIA), pp. 219–225. IEEE (2017)
2. Song, B., Pérez-Jiménez, M.J., Pan, L.: An efficient time-free solution to SAT problem by P systems with proteins on membranes. J. Comput. Syst. Sci. **82**(6), 1090–1099 (2016)
3. Wang, X.: Research on solution of TSP based on improved genetic algorithm. In: International Conference on Engineering Simulation and Intelligent Control (ESAIC), pp. 78–82. IEEE (2018)
4. Jafarzadeh, N., Iranmanesh, A.: A new graph theoretical method for analyzing DNA sequences based on genetic codes. MATCH-Commun. Math. Comput. Chem. **75**(3), 731–742 (2016)
5. Chaves-González, J.M., Vega-Rodrgíuez, M.A.: A multiobjective approach based on the behavior of fireflies to generate reliable DNA sequences for molecular computing. Appl. Math. Comput. **227**, 291–308 (2014)
6. Frutos, A.G., Liu, Q., Thiel, A.J., et al.: Demonstration of a word design strategy for DNA computing on surfaces. Nucleic Acids Res. **25**(23), 4748–4757 (1997)
7. Hartemink, A.J., Gifford, D.K., Khodor, J.: Automated constraint-based nucleotide sequence selection for DNA computation. Biosystems **52**(1–3), 227–235 (1999)
8. Feldkamp, U., Saghafi, S., Banzhaf, W., Rauhe, H.: DNASequenceGenerator: a program for the construction of DNA sequences. In: Jonoska, N., Seeman, N.C. (eds.) DNA 2001. LNCS, vol. 2340, pp. 23–32. Springer, Heidelberg (2002). https://doi.org/10.1007/3-540-48017-X_3
9. Arita, M., Nishikawa, A., Hagiya, M., et al.: Improving sequence design for DNA computing. In: Conference on Genetic and Evolutionary Computation, pp. 875–882. Morgan Kaufmann Publishers Inc. (2000)
10. Shin, S.Y., Kim, D.M., Lee, I.H., et al.: Evolutionary sequence generation for reliable DNA computing. In: 2002 Proceedings of the 2002 Congress on Evolutionary Computation, CEC 2002, pp. 79–84. IEEE (2002)
11. Xu, C., Zhang, Q., Wang, B., et al.: Research on the DNA sequence design based on GA/PSO algorithms. In: The International Conference on Bioinformatics and Biomedical Engineering, pp. 816–819. IEEE (2008)
12. Kurniawan, T.B., Khalid, N.K., Ibrahim, Z., et al.: Sequence design for direct-proportional length-based DNA computing using population-based ant colony optimization. In: ICCAS-SICE, pp. 1486–1491. IEEE (2009)

13. Wang, Y., Shen, Y., Zhang, X., et al.: An improved non-dominated sorting genetic algorithm-II (INSGA-II) applied to the design of DNA codewords. Math. Comput. Simul. **151**, 131–139 (2018)
14. Zhang, Q., Wang, B., Wei, X., et al.: DNA word set design based on minimum free energy. IEEE Trans. Nanobioscience **9**(4), 273–277 (2010)
15. Muhammad, M.S., Selvan, K.V., Masra, S.M.W., et al.: An improved binary particle swarm optimization algorithm for DNA encoding enhancement. In: Swarm Intelligence, pp. 1–8. IEEE (2011)
16. Mantha, A., Purdy, G., Purdy, C.: Improving reliability in DNA-based computations. In: IEEE International Midwest Symposium on Circuits and Systems, pp. 1047–1050. IEEE (2013)
17. Ibrahim, Z., Khalid, N.K., Lim, K.S., et al.: A binary vector evaluated particle swarm optimization based method for DNA sequence design problem. In: Research and Development, pp. 160–164. IEEE (2012)
18. Kurniawan, T.B., Khalid, N.K., Ibrahim, Z., et al.: Evaluation of ordering methods for DNA sequence design based on ant colony system. In: Second Asia International Conference on Modelling and Simulation, pp. 905–910. IEEE Computer Society (2008)
19. Jeong, K.S., Kim, M.H., Jo, H., et al.: Search of optimal locations for species- or group-specific primer design in DNA sequences: non-dominated sorting genetic algorithm II (NSGA-II). Ecol. Inform. **29**, 214–220 (2015)
20. Peng, X., Zheng, X., Wang, B., et al.: A micro-genetic algorithm for DNA encoding sequences design. In: International Conference on Control Science and Systems Engineering, pp. 10–14. IEEE (2016)
21. Deb, K., Sundar, J.: Reference point based multi-objective optimization using evolutionary algorithms. In: Proceedings of the 8th Annual Conference on Genetic and Evolutionary Computation, pp. 635–642. ACM (2006)
22. Pan, L., He, C., Tian, Y., Su, Y., Zhang, X.: A region division based diversity maintaining approach for many-objective optimization. Integr. Comput.-Aided Eng. **24**(3), 279–296 (2017)
23. He, C., Tian, Y., Jin, Y., Zhang, X., Pan, L.: A radial space division based evolutionary algorithm for many-objective optimization. Appl. Soft Comput. **61**, 603–621 (2017)
24. Pan, L., He, C., Tian, Y., Wang, H., Zhang, X., Jin, Y.: A classification-based surrogate-assisted evolutionary algorithm for expensive many-objective optimization. IEEE Trans. Evol. Comput. **23**(1), 74–88 (2018)
25. Pan, L., Li, L., He, C., Tan, K.C.: A subregion division-based evolutionary algorithm with effective mating selection for many-objective optimization. IEEE Trans. Cybern. (2019). https://doi.org/10.1109/TCYB.2019.2906679

Research on DNA Cryptosystem Based on DNA Computing

Shuang Cui, Weiping Peng$^{(\boxtimes)}$, and Cheng Song

School of Computer Science and Technology, Henan Polytechnic University,
Jiaozuo 454002, China
pwphpu@163.com

Abstract. As a new type of cryptography, DNA cryptography generally uses DNA molecule as the information carrier and biological technology as the implementation tool. Due to its prominent advantages such as large storage capacity, high parallel computing, low energy consumption and abundant resources in nature, DNA cryptography has attracted wide attention. DNA cryptography involves biology, computer, mathematics and other disciplines. On the basis of the traditional cryptosystems, DNA molecular computing methods are combined to form a more reliable and stable new cryptosystems, which brings opportunities and challenges to the modern cryptosystems. This paper depicts several DNA cryptosystems, further analyzes the security and performance of these schemes, summarizes the shortcomings of current DNA cryptography research, and looks forward to its development prospect in the field of information security.

Keywords: DNA cryptosystems · DNA molecular computing · Information security

1 Introduction

As the genetic material of organism, DNA molecule plays a key role in the reproduction of organism. Since the discovery of the computational ability of DNA molecule, the research on the computational function of DNA molecule has become the focus of many scientists. In 1994, the United States at the university of California, Dr Adleman [1] have discovered similarities between computer code and the molecular structure of DNA. The four bases A, G, C, T of DNA molecules and computer code can be fixed in accordance with the rules together. Therefore, academic operations of addition, subtraction, multiplication, and division of informatics and logical operation can be implemented using DNA molecules [2]. Based on the large-scale parallelism inherent in DNA computing, scientists have designed many models to crack the traditional encryption algorithms of DES (Data Encryption Standard), RSA (Rivest-Shamir-Adleman), NTRU (Number Theory Research Unit) and so on [3, 4] and [5] proposed the algorithm of DNA computation to crack DES, and [6] and [7] proposed the computational model of decoding RSA based on DNA computation. [8] designed a method based on self-assembly model to decipher Diffie-Hellman key exchange algorithm, which could threaten the security of Diffie-Hellman key exchange in 2012.

© Springer Nature Singapore Pte Ltd. 2020
L. Pan et al. (Eds.): BIC-TA 2019, CCIS 1160, pp. 189–197, 2020.
https://doi.org/10.1007/978-981-15-3415-7_15

In reference [9], an algorithm based on the DNA computation model to solve the discrete logarithm of elliptic curve was presented. The above articles are only based on theoretical research, and the current DNA calculation does not pose a threat to traditional encryption algorithms [10].

With the rapid development of modern technology, DNA and cryptography begin to be combined in recent years. As a new type of cryptography, DNA cryptography is quite different from traditional cryptography [11]. Traditional cryptography using computer chip as storage medium and DNA cryptography using DNA strand or DNA chip as storage medium. Traditional cryptography are mainly serial operation and DNA cryptography have the ability of parallel operation under the concurrent reaction of DNA. Traditional cryptography is based on difficult mathematical problems, while the security of DNA codes depends on difficult biological problems. In 2004, Gehani et al. [12] designed two kinds of one-time-pad DNA cryptography methods, which include substitution and XOR operation methods. DNA cryptography also includes symmetric and asymmetric cryptography based on DNA technology, modern DNA biotechnology and microarray technology are applied in cryptography. In 2007, [13] designed a symmetric encryption system combining modern genetic engineering technology and cryptography technology, in which encryption and decryption keys are DNA probes and ciphertext is a specially designed DNA chip. In 2010, [14] designed an asymmetric encryption and signature system, which is similar to the traditional public key cryptography. Furthermore, DNA cryptography also contains DNA steganography and DNA authentication encryption schemes. The safety of DNA steganography is that in the DNA mixture, the plaintext DNA strands are hidden in a large number of similar DNA strands. In the process of decryption, PCR (Polymerase Chain Reaction) technology was used to amplify the plaintext DNA strands determined by primers, and the plaintext information was obtained by sequencing technique.

In the second chapter, this paper introduces several classical DNA cryptosystems, and analyzes the advantages and disadvantages of these schemes. In the third chapter, the safety and performance of these schemes are analyzed and compared. The fourth chapter summarize the shortcomings of current DNA cryptography research and look forward to its development prospects in the field of information security.

2 DNA Cryptography

2.1 DNA Cryptography Based on DNA Molecule

In 2008, Chen et al. [15] designed a one-time-pad encryption algorithm based on DNA tile structure. This tile self-assembly is different from common DNA molecule self-assembly. These tiles have sticky ends called pads that can match other DNA tiles at the corresponding sticky ends to form larger structures for DNA splicing. In this paper, four kinds of tile systems are proposed, which respectively implement encryption, ciphertext extraction, key extraction and decryption. The implementation of DNA one-step encryption extends the self-assembling tile model and achieves true randomness in DNA one-step encryption. As can be seen above, the security of this scheme is mainly based on biosafety. Although the algorithm is complex to operate, the coding rules used are relatively simple. If an attacker obtains the algorithm operation flow, the security of this scheme will be greatly reduced.

In 2014, Yang et al. [16] proposed a new one-time-pad cryptography scheme based on DNA self-assembly structure, which realized one-time-pad mainly by using DNA toehold sequence and DNA strand replacement technology. Encryption and decryption algorithms use DNA self-assembly structures to xor binary plaintext and key. Finally, the fluorescence intensity spectrum corresponding to the ciphertext binary results can be transmitted to the receiver through the public channel. After receiving the ciphertext fluorescence intensity spectrum, the receiver can decrypt the ciphertext with different self-assembly structures. The flow chart of encryption and decryption of the scheme is shown in Fig. 1. The security analysis of this algorithm shows that the security mainly depends on biological operations such as strand replacement by self-assembling structure. Without knowing these biological reactions, even if the attacker obtains ciphertext spectrum, it is impossible to crack the corresponding plaintext information. However, the plaintext conversion of this algorithm is based on the same simple coding rules as the above references. The key book has the characteristic of one encryption at a time, but there is no corresponding encryption measure for the key book. If the attacker gets the key book and the structure construction method, it is easy to break the ciphertext and get the plaintext information.

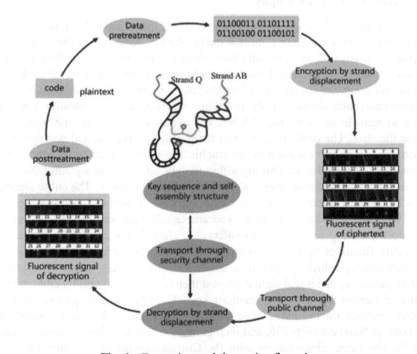

Fig. 1. Encryption and decryption flow chart

At the same time, Wang et al. [17] put forward another one-time-pad cryptography algorithm based on DNA coding in 2014. In the algorithm, the key is generated randomly, and one-time-pad key book is composed of synthetic single strands. The PCR technology is used for key distribution. The algorithm can ensure the one-time-pad, basically satisfy the random property of key, and effectively eliminate the influence of non-specific hybridization reaction between DNA molecules on the algorithm. The algorithm mainly adopts the way of triplet coding. The combination of three bases corresponds to letters, Numbers and symbols, and the corresponding relationship are only C_{64}^{40}, However, the coding security is relatively low. The triplet coding is shown in Table 1.

Table 1. Triplet coding

"A"=CGA	"I"=ATG	"Q"=AAC	"Y"=AAA	"6"=TTA
"B"=CCA	"J"=AGT	"R"=TCA	"Z"=CTT	"7"=ACA
"C"=GTT	"K"=AAG	"S"=ACG	"0"=ACT	"8"=AGG
"D"=TTG	"L"=TGC	"T"=TTC	"1"=ACC	"9"=GCG
"E"=GGC	"M"=TCC	"U"=CTG	"2"=TAG	""=ATA
"F"=GGT	"N"=TCT	"V"=CCT	"3"=GCA	","=TCG
"G"=TTT	"O"=GGA	"W"=CCG	"4"=GAG	"."=GAT
"H"=CGC	"P"=GTG	"X"=CTA	"5"=AGA	":"=GCT

2.2 Pseudo DNA Cryptography

In 2014, Wan et al. [18] proposed a one-time-pad encryption algorithm based on hyper-chaos mapping DNA computing. In this paper, the key sequence in the original one-time-pad algorithm was analyzed and three defects were obtained: limited use times, excessively long occupation and the true randomness is unpredictable. Based on the defect optimization, the encryption scheme is proposed. Finally, scheme was applied to the image encryption simulation, the resulting ciphertext image correlation coefficient is close to zero. In addition, the 256 bits key are used to simulate the images with different file sizes. The analysis shows that the average encryption and decryption time is much lower than the common cryptographic algorithm, which proves the security and feasibility of the scheme. This algorithm mainly using the binary and DNA single strand mutual transformation, does not provide biological security. The entire algorithm process is the replacement and joint encryption process. Its security mainly depends on the key sequence generated according to parameters. If parameters are leaked, without the support of coding security and biosafety, the algorithm is easy to be cracked.

In 2016, Bonny et al. [19] proposed a symmetric encryption algorithm based on DNA technology, which proposed a secure symmetric key generation process, including generating the initial ciphertext, and then converting the initial ciphertext into the final ciphertext by using the random key. The random key generation in the algorithm depends on the index table, whose size is 256 and 4 base combinations correspond to Numbers of 1–256, and then generates a random key Pk with a range of 1 to 256. The index table changes with Pk. Compared with the traditional DES and other DNA-based encryption algorithms, the coding method proposed in this paper has better and faster performance. However, this algorithm does not provide biological security, and its security mainly depends on the random key. Due to the simple encoding and small key space, the random key is easy to be cracked.

In 2018, Thangavel et al. [20] proposed an enhanced DNA and ElGamal cryptosystems. In this paper, a new DNA cryptosystem adopted symmetric cryptography to encrypt data transmitted between data owners and data users in the cloud. The enhanced ElGamal cryptography system uses asymmetric cryptography to solve key management problems in the cloud by securely transferring key files between data owners and data users. Compared with ElGamal cryptosystems, this system has higher security and is difficult to implement brute force attacks and cryptanalytic attack. The dynamic generation of coding tables and introns reduces the possibility of cryptanalysis and improves the security of data. The biology of DNA makes the system more random. This work improves the randomization of key generation, encryption and decryption from ElGamal cryptography system, but the proposed algorithm key generation is a time-consuming process, in addition, the system does not provide biological security guarantee like the above schemes.

3 Performance Analysis

In order to qualitatively compare the performance of DNA cryptosystems, Ubaidurrahman et al. [21] proposed the performance of six efficient DNA cryptosystems in 2015, Peng et al. [22] defined five performance parameters in 2018. We select some key parameters from them, and the definitions are shown in Table 2.

Table 2. Definition of performance parameters

Parameters	Definition
DNA coding integrity	The DNA coding table should provide DNA encoding sequences for the complete character set
Dynamic coding table	To ensure an increased level of security, the encoding table should be changed at periodic intervals or for every interaction session between the sender and receiver
Structure randomness	In order to improve the security, the encryption structure should be unique and different every time
Biological process simulation	The biological process of all proposed DNA encryption and decryption algorithms should be simulated to adapt to the digital computing environment
Key randomness	To ensure the security of encryption algorithm, it is necessary to provide randomness of key generation, storage and distribution

An effective DNA cryptosystem requires the combination of DNA cryptography and modern cryptography to effectively guarantee biosecurity and computational security. The integrity and dynamics of the DNA encoding table ensure that all plaintext sequences can be randomly converted into different DNA sequences. Structural randomness provides the biosecurity of schemes, and the simulation of biological processes based on some biological characteristics of DNA computing makes the algorithm more randomized and ensures the computational security. Random key generation, storage and distribution are the most important parts of traditional cryptosystems. Qualitative comparison on the performance of the DNA cryptosystems mentioned in this paper is shown in Table 3.

Table 3. Comparisons of the performance

	DNA coding integrity	Dynamic coding table	Structure randomness	Biological process simulation	Key randomness
Lu et al. [13]	×	×	×	×	*
Lai et al. [14]	×	×	×	×	*
Chen et al. [15]	×	×	*	×	√
Yang et al. [16]	√	×	×	√	×
Wang et al. [17]	√	×	×	√	*
Wan et al. [18]	√	×	×	×	*
Bonny et al. [19]	√	√	×	×	×
Thangavel et al. [20]	√	√	×	√	×
Ubaidurrahman et al. [21]	√	√	×	√	×
Peng et al. [22]	√	√	√	√	*

×-Indication of minimum level of support.
√-Indication of acceptable level of support.
*-Partial fulfillment.

From the performance comparison of the above DNA cryptosystems, it can be clearly seen that the DNA coding table of most schemes is integrated, some schemes realize biological process simulation, and a few schemes complete the Dynamic coding table. In addition, fewer schemes can achieve the randomness of DNA structure and true key randomness.

4 Security Analysis

DNA cryptography is usually based on DNA computing, and high security encryption and decryption operations are realized by means of DNA coding, base operation and confusing coding table. The security strength of the scheme is analyzed by the probability value selected by each random factor. We assume that the length of the primer is m, the encoding selection probability is P_1, the ciphertext combination probability is P_2, and then the total ciphertext cracking probability P is:

$$P = P_1 \times P_2 \tag{1}$$

Without considering the selection probability of DNA structure, the methods proposed in some schemes were compared. The probability statistics of each scheme were shown in Table 4.

Table 4. Comparisons of the probability of selection and cracking

	P_1	P_2	P
Chen et al. [15]	/	$\frac{1}{C_{4m}^2}$	$\frac{1}{C_{4m}^2}$
Yang et al. [16]	/	$\frac{1}{24 \times 24}$	$\frac{1}{24 \times 24}$
Wang et al. [17]	$\frac{1}{C_{64}^{40} \times 24}$	$\frac{1}{C_{4m}^2}$	$\frac{1}{C_{64}^{40} \times 24 \times C_{4m}^2}$
Bonny et al. [19]	$\frac{1}{24 \times 256}$	/	$\frac{1}{24 \times 256}$
Thangavel et al. [20]	$\frac{1}{24 \times 24^2 \times 256^2}$	/	$\frac{1}{24^3 \times 256^2}$
Peng et al. [22]	$\frac{1}{24}$	$\frac{1}{24 \times C_{4m}^2 \times 24}$	$\frac{1}{C_{4m}^2 \times 24^3}$

According to the comparison of cracking probability of schemes, it can be seen that some of the DNA cryptosystems can also provide high computational security by means of DNA encoding, encoding table and the conversion and transmission of biological signals when the key is allowed to leak, but the computational security that can be provided is not enough to resist exhaustive attacks.

5 Prospect

At present, the difficulty in the field of DNA cryptography is that there is no complete security theory and convenient and feasible implementation conditions [23]. Therefore, the main goal of DNA cryptography research should be to explore the characteristics and reactions of DNA molecules, create new technology and algorithm theory, and explore various possible research paths, so as to make the implementation of DNA cryptography simple and easy to operate and lay a foundation for its future development. The research directions in this field are supplemented as follows:

- From the perspective of DNA self-assembly structure, many scholars in the field of biology have proposed various models of DNA self-assembly structure. DNA molecules can be used to make a variety of graphics display, using biotechnology in this field, combining DNA computing and traditional cryptography to study DNA cryptosystems.
- From the perspective of DNA coding, most of the DNA cryptosystems used in this paper have weak coding security and do not consider biological difficulties. The computational security provided by most schemes is not enough to resist exhaustive attacks. Therefore, designing a feasible complex and regular DNA coding will be a new opportunity for the development of DNA cryptography.
- At present, most researches on DNA cryptography lack theoretical support. Although DNA cryptography has developed a lot so far, the establishment of the theoretical system of DNA cryptography is still the main work of current scholars in the field of DNA cryptography.

More and more scholars are studying and exploring the DNA cryptography in depth, hoping to explore more perfect and safer cryptography schemes with the characteristics of DNA molecules in the future.

6 Conclusion

In this paper, through a detailed analysis of the DNA cryptosystems, we can know that the DNA code is not only based on the mathematical difficulties of the traditional code scheme, but also depends on the biological difficulties. The DNA cryptography based on DNA molecule mainly uses related biological technologies such as DNA hybridization technology, DNA amplification technology and DNA self-assembly structure to encrypt the information, providing biological security such as cracking the self-assembly structure, the order of the DNA bases and physical isolation. The Pseudo DNA cryptography mainly uses the methods of DNA coding, base operation and coding table for information confusion, which provides high computational security. If the advantages of both can be combined, it will provide a more secure DNA cryptosystems. By analyzing the performance and security of DNA cryptosystems, we can design more effective DNA cryptosystems according to the defined performance parameters and crack probability formulas.

References

1. Chen, H., Huo, J., Xu, B., Zhang, W.: New Directions in Cryptography: From Quantum No Cloning to DNA's Perfect Reproducting. National Defense Industry Press, Beijing (2015)
2. Adleman, L.: Molecular computation of solutions to combinatorial problems. Science **266** (5187), 1021–1024 (1994)
3. Chao, L., Jing, Y., Cheng, Z.: Research progress for DNA cryptography. Netinfo Security (2015)
4. Boneh, D., Dunworth, C., Lipton, R.J.: Breaking DES using a molecular computer. DNA Based Comput. **27**, 37 (1996)
5. Adleman, L.M., Rothemund, P.W.K., Roweis, S.: On applying molecular computation to the data encryption standard. J. Comput. Biol. **6**(1), 53–63 (1999)
6. Beaver, D.: Factoring: the DNA solution. In: Pieprzyk, J., Safavi-Naini, R. (eds.) ASIACRYPT 1994. LNCS, vol. 917, pp. 419–423. Springer, Heidelberg (1995). https://doi.org/10.1007/BFb0000453
7. Brun, Y.: Arithmetic computation in the tile assembly model: addition and multiplication. Theor. Comput. Sci. **378**(1), 17–31 (2007)
8. Cheng, Z.: Nondeterministic algorithm for breaking Diffie-Hellman key exchange using self-assembly of DNA tiles. Int. J. Comput. Commun. Control **7**, 616–630 (2012)
9. Li, K., Zou, S., Xu, J.: Fast parallel molecular algorithms for DNA-based computation: solving the elliptic curve discrete logarithm problem over GF2. J. Biomed. Biotechnol. **2008** (1), 518093 (2014)
10. Chen, Z., Shi, X., Cheng, Z.: Impact and application of DNA nanotechnology in information security. Bull. Chin. Acad. Sci. **29**(01), 70–82 (2014)
11. Xiao, G., Lu, M.: DNA computation and DNA cryptography. Chin. J. Eng. Math. **23**(1), 1–6 (2006)
12. Gehani, A., LaBean, T., Reif, J.: DNA-based cryptography. In: Jonoska, N., Păun, G., Rozenberg, G. (eds.) Aspects of Molecular Computing. LNCS, vol. 2950, pp. 167–188. Springer, Heidelberg (2003). https://doi.org/10.1007/978-3-540-24635-0_12
13. Lu, M., Lai, X., Xiao, G., Qin, L.: Symmetric-key cryptosystem with DNA technology. Sci. China Ser. F: Inf. Sci. **50**(3), 324–333 (2007)

14. Lai, X., Lu, M., Qin, L.: Asymmetric-key cryptosystem and signature with DNA technology. Sci. Sin.: Inf. **40**(02), 240–248 (2010)
15. Chen, Z., Xu, J.: One-time-pads encryption in the tile assembly model. In: International Conference on Bio-Inspired Computing: Theories and Applications, pp. 23–30. IEEE (2010)
16. Yang, J., Ma, J., Liu, S., Zhang, C.: A molecular cryptography model based on structures of DNA self-assembly. Chin. Sci. Bull. **59**(11), 1192–1198 (2014)
17. Wang, Z., Zhao, X., Wang, H.: One-time-pad cryptography algorithm based on DNA cryptography. Comput. Eng. Appl. **50**(15), 97–100 (2014)
18. Wan, R., Mo, H., Yu, S.: Document and image encryption based on OTP optimized by hyper-chaos mapping DNA computing. Comput. Measur. Control **22**(10), 3278–3281 (2014)
19. Bonny, B.R., Vijay, J.F., Mahalakshmi, T.: Secure data transfer through DNA cryptography using symmetric algorithm. Int. J. Comput. Appl. **133**(2), 19–23 (2016)
20. Thangavel, M., Varalakshmi, P.: Enhanced DNA and ElGamal cryptosystem for secure data storage and retrieval in cloud. Cluster Comput. **21**(2), 1411–1437 (2017)
21. Ubaidurrahman, N.H., Balamurugan, C., Mariappan, R.: A novel DNA computing based encryption and decryption algorithm. Proc. Comput. Sci. **46**, 463–475 (2015)
22. Peng, W., Cheng, D., Song, C.: One time-pad cryptography scheme based on a three-dimensional DNA self-assembly pyramid structure. PLoS ONE **13**(11), e0206612 (2018)
23. Xiao, G., Lu, M., Qin, L., Lai, X.: New field of cryptography: DNA cryptography. Chin. Sci. Bull. **51**(12), 1413–1420 (2006)

Performing DNA Strand Displacement with DNA Polymerase

Zhiyu Wang[1], Yingxin Hu[1], Zhekun Chen[1], Sulin Liao[2], and Yabing Huang[3(✉)]

[1] Key Laboratory of Image Information Processing
and Intelligent Control of Education Ministry of China,
School of Artificial Intelligence and Automation,
Huazhong University of Science and Technology, Wuhan 730074, China
{wangzhiyu0471,yingxinhu,zkchen}@hust.edu.cn
[2] School of Computer Science and Technology,
Huazhong University of Science and Technology, Wuhan 730074, China
liaosulin_2030@qq.com
[3] Department of Pathology,
Renmin Hospital of Wuhan University, Wuhan 430060, China
drybhuang@gmail.com

Abstract. Significant improvements to the dynamic DNA nano science have emerged in recent years, primarily due to the elaborate design of toehold-mediated DNA strand displacement (TMSD). However, it remains an ongoing challenge to design huge-scaled TMSD based sophisticated dynamic DNA structures without base-pair mismatching or unwanted secondary structures, because the sequence design complexity of TMSD will dramatically increase as the scale increases. Here we report a new polymerase triggered strand displacement (PTSD) mechanism to realize DNA strand displacement with the potential of constructing complicated dynamic DNA nanostructures. By employing proper polymerase with strand displacement activity, the target strand in a duplex will be displaced and freed during the polymerization. In this manner, DNA strand displacement takes place without considering the design complexity of branch migration domain. The mechanism was successfully applied in constructing one-layer and two-layer circuits in which the parameters involved were explored, suggesting that PTSD is feasible and able to be harnessed to construct cascading DNA circuits with high complexity.

Keywords: Nucleic acids · Molecular circuit · DNA computing · DNA strand displacement · Polymerase

1 Introduction

The excellent programmability and specific binding offered by the Watson-Crick base pairing principle make DNA an ideal material for building nanoscale structures and devices with varying complexity and functionality [1–3]. The dynamic interaction of most DNA structures and devices, including sensors

© Springer Nature Singapore Pte Ltd. 2020
L. Pan et al. (Eds.): BIC-TA 2019, CCIS 1160, pp. 198–208, 2020.
https://doi.org/10.1007/978-981-15-3415-7_16

[4–6], machines [7–10], circuits [11–17] and reconfigurable structures [18–20], rely on strand-exchange reactions in which an invading strand replaces an origin strand of a duplex. However, in strand-exchange reactions between stable duplexes and identical invading strands, the reaction rates are quite slow [21]. To overcome this kinetic barrier, the concept of toehold-mediated DNA strand displacement (TMSD) was introduced by Yurke et al. [22]. In their design, a short sticky end domain called toehold was introduced to the duplex while a corresponding complementary domain to the toehold was introduced to the invading DNA strands (Fig. 1A). The introduction of toehold dramatically accelerates the strand-exchange rates at the branch migration (BM) domain to over 10^6 fold, leading to the flourishing of dynamic DNA nanoscience [1–20,23–26].

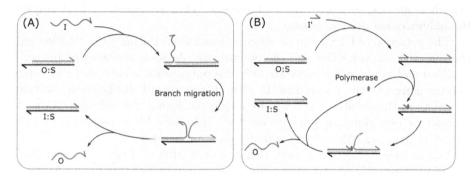

Fig. 1. Principles of toehold-mediated DNA strand displacement reaction $O:S+I\xrightarrow{BM}I:S+O$ (A) and polymerase triggered DNA strand displacement reaction $O:S+I'\xrightarrow{Pol.}I:S+O$ (B).

Traditional toehold-mediated strand displacement exhibits strong powers on constructing complex networks. However, this is based on the complete knowledge on the sequence and elaborate sequence design which get much harder as the scale of the networks dramatically increases. Inspired by the organic lives in nature, enzymes based methods are proposed and may offer a feasible approach to address this problem. Enzymatically driven reactions exhibit greater nonlinear kinetics, versatility and irreversibility, thus being widely used in constructing complicated DNA networks. For instance, networks based on DNA replication, nicking, and degradation have shown to be highly modular and have been engineered to display stable oscillations [27], multistability [28] and chaotic dynamics [29].

In this work, we introduce a polymerase triggered strand displacement (PTSD) mechanism that may hopefully resolve the problem of constructing huge-scale network by simplifying the sequence designing. Phi29, a kind of DNA polymerase with strand displacement activity, was employed to perform strand displacement. During the polymerization of phi29, the target strand will be displaced and released in the presence of input strand, and the released target strand can play the role of input to the downstream reaction, making PTSD cascadable. Parameters such as the length of input strand were explored to optimize the performance. A two-layer PTSD circuit was further designed to test the expansibility. Via fluorescence tracking, PTSD was proven to be feasible, reliable and eligible for constructing more complex DNA dynamic systems.

2   Results and Discussion

The Principle of Polymerase Triggered Strand Displacement

The principle of polymerase triggered strand displacement is shown in Fig. 1B. Unlike the traditional toehold-mediated DNA strand displacement relying on branch migration to realize the replacement between strands, PTSD employs the polymerizing process instead.

The principle of PTSD can be described as follow. Initially, the solution only contains the substrate O:S and the polymerase. As soon as the short primer I' is added and binds to the substrate O:S, the polymerase is activated and begins polymerizing. As the polymerase is with strong strand displacement activity, strand O in the substrate O:S will be displaced during the polymerization. As a result, a new completely complementary substrate I:S is produced and the strand O that can trigger a downstream reaction is released.

Polymerase triggered strand displacement differs from the traditional toehold-mediated DNA strand displacement mainly in that it relies on the strong strand displacement activity during the polymerizing by the polymerase, rather than the branch migration process. Therefore, the key property of the polymerase used in PTSD is good strand displacement activity.

The Characteristics of Phi29 DNA Polymerase

Phi29 DNA polymerase is the replicative polymerase from the Bacillus subtilis phage phi29. The polymerase is characterized by strong strand displacement activity and widely used in DNA amplification such as rolling circle amplification (RCA) [30] and multiple displacement amplification (MDA) [31]. Such exceptional strand displacement activity makes phi29 DNA polymerase a promising candidate for polymerase triggered strand displacement reaction.

Three DNA structures (Str1, Str2 and Str3 in Fig. 2) were designed to explore the properties of phi29 DNA polymerase. Str1 is a very simple linear structure. After being catalyzed for 12 h by phi29 DNA polymerase, the vast majority of Str1 were transformed into the completely complementary duplex (Lane 2 in Fig. 2). Based on structure Str1, Str2 was designed to contain a multiple T tail hanging. After catalyzed by phi29 DNA polymerase, about one half of Str2 were transformed into the completely complementary duplex (Lane 4 in Fig. 2). This can be explained by the exonuclease activity of phi29 DNA polymerase. To suppress the exonuclease activity, we added a Phosphorothioate modification on the 3' end of strand Str2u (the red strand of Str2) and named it as Str2(p). After being catalyzed for 12 h, Str2(p) remained the initial state (Lane 6 in Fig. 2). To test the ability of strand displacement, a target strand (the green strand on Str3) was added. After the catalyzation, the completely complementary duplex was formed and the target strand was displaced (Lane 8 in Fig. 2).

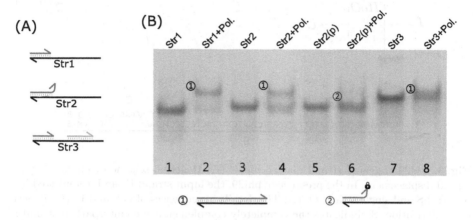

Fig. 2. The characteristics of phi29 DNA polymerase. (A) Three DNA structures designed to characterize phi29 DNA polymerase. (B) Non-denaturing gel electrophoresis for each structure in (A). Lane 1: Str1, lane 2: Str1+phi29, lane3: Str2, lane 4: Str2+phi29, lane 5: Str2 (phosphorothioate-modified at 3' end), lane 6: Str2 (phosphorothioate-modified at 3' end)+phi29, lane 7: Str3, lane 8: Str3+phi29. (Color figure online)

In conclusion, phi29 DNA polymerase is with enough strand displacement activity and strong exonuclease activity. Meanwhile, the exonuclease activity can be inhibited by the Phosphorothioate modifications. Such properties make phi29 DNA polymerase a suitable candidate for PTSD.

Phi29 Triggered Strand Displacement

According to the principle of polymerase triggered strand displacement, the reaction scheme was designed (Fig. 3A). In the presence of phi29, the input strand I' can be converted into duplex I:S and single strand O. A fluorophore and a quencher were modified at the 5' end of S and 3' end of O, respectively. Thus if strand O was displaced and released, the increase of fluorescence will appear due to the separation of the fluorophore and the quencher. The process of the reaction can be abstracted as a converter performing *ItoO* function.

The detailed sequences of O, S and I' are illustrated in Fig. 3B. I' is designed with a length of 24 nucleotides (nt) to ensure that it can bind to strand S firmly while there is a 20-nt single strand domain (SS domain) between I' and O when binding. Strand O and S are modified with BHQ quencher and FAM fluorophore, respectively. It is worth mentioning that strand I' and S are with phosphorothioate modifications at 3' end owing to the exonuclease activity of phi29.

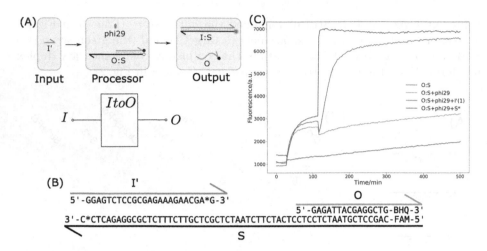

Fig. 3. Phi29 triggered strand displacement. (A) Reaction scheme of phi29 triggered strand displacement. In the presence of phi29, the input strand I' can be converted into duplex I:S and single strand O. (B) The details in sequences of O:S and I'. (C) Kinetic characterization. S* indicates the completely complementary counterpart of strand S. Obvious fluorescence increase can be observed after phi29 and I' were added (green curve). (Color figure online)

To monitor the PTSD in detail, we used a qPCR instrument to track the fluorescent signal (Fig. 3C). Reactions of O:S with no input (the blue curve), phi29 (the orange curve) and both phi29 and I' (the green curve) were tested, and a reaction of traditional toehold-mediated DNA strand displacement (the red curve) was performed as a control trial. It can be illustrated from the results that in the presence of primer I', phi29 can displace and release strand O during the polymerization (the green curve).

To better explore the mechanism of PTSD, an experiment varying the length of I' and another varying the concentration of I' were performed. In the experiment varying the length of I', another three variations of I' with different lengths, I'(14), I'(34) and I'(44), was designed (sequence details can be find in Table 1). The results showed that the length of center single strand domain (the SS domain between I' and O when binding in Fig. 3B) plays a significant role in PTSD (Fig. 4A). The PTSD takes place normally when the length of center single strand domain varying from 30 nt (the orange curve) to 10 nt (the red curve) whereas it can hardly take place with a center single strand domain length of 0 (the purple curve).

In the other experiment shown in Fig. 4B, the concentrations of I' vary from $0.05\,\mu M$ ([I']:[O:S] = 0.25, the green curve) to $0.8\,\mu M$ ([I']:[O:S] = 4, the pink curve). As the concentration of I' increases, the displacement performance improves. At relatively high concentrations ([I']:[O:S] $\geq$ 1), phi29 triggered strand displacement performs as well as toehold-mediated strand displacement while at relatively low concentrations ([I']:[O:S] $\leq$ 0.5), PTSD does not perform that well.

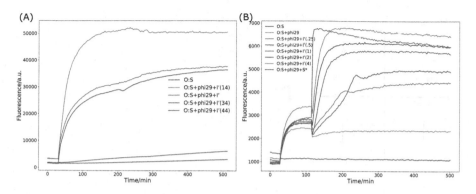

Fig. 4. Kinetic characterization under (A) varying lengths and (B) varying concentration of I'. (Color figure online)

The experiments above demonstrate that phi29 based polymerase triggered strand displacement is feasible and in optimized conditions it can perform as well as toehold-mediated strand displacement. There are also some moderate leakages in the reactions when only phi29 is added (the orange curve in Fig. 3C). This can be explained by the unwanted structures formed by mismatching and the leakage could diminish via elaborate sequence design. In the design of PTSD, the length and the concentration of input strand are two key parameters. In a sense, the polymerization of phi29 is with "inertance". Hence a too long input strand, which means no enough space to target strand, may lead to poor performance (the purple curve in Fig. 4A). On the other hand, the performance is also influenced by the concentration of input strand. About a same amount of input strand and substrate would be reasonable for the strand displacement occurring normally. It is interesting to note that, in the several experiments, the PTSD generally occurs more slowly than toehold-mediated strand displacement. In this experiment, as the reaction rate depends on the concentration of phi29, a higher concentration of phi29 may address this problem.

Cascading Phi29 Triggered Strand Displacement

To expand the mechanism to larger system, we designed a two-layer phi29 triggered strand displacement circuit (Fig. 5A). The circuit contains two layers (Layer 1 and Layer 2) and an input (I2'). The output strand of Layer 2 (I1') can trigger the reaction of Layer 1 as the input strand, producing a noticeable fluorescence increase. In the circuit, the strand displacement occurred on both I1':S1 and O:S with the help of phi29.

Figure 5B depicts the fluorescence kinetics of the two-layer phi29 triggered strand displacement circuit. In the presence of phi29, the addition of the input strand (I2') leads to a distinct increase on fluorescence intensity.

The results above demonstrate that the phi29 triggered strand displacement is of feasibility and can be utilized to construct cascadable DNA circuits.

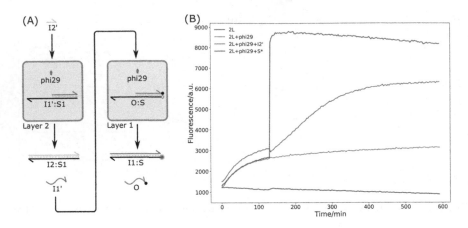

Fig. 5. Cascading phi29 triggered strand displacement. (A) Reaction scheme. The output of layer 2, I1', is the input to layer 1. (B) Kinetic characterization. "2L" indicates the mixture of I1':S1 and O:S. In the presence of phi29, as the input strand I2' was added, output strand O was released, leading to the increase of fluorescence (the green curve). (Color figure online)

Compared to TMSD, PTSD relies on the polymerization process instead of branch migration to realize the strand displacement. With proper parameters (length, concentration of the input strand and concentration of phi29), PTSD can perform as well as the toehold-mediated strand displacement. The greatest advantage PTSD possesses over the toehold-mediated strand displacement is no requirement of knowing the sequence of branch migration. In toehold-mediated strand displacement, one needs to know the whole sequence of the substrate to design the invading strand whereas he only needs to know the last several nt of the 3' end on the substrate in PTSD. In other words, one input strand can trigger strand displacement on several different substrates varying the branch migration domain sequences.

3   Materials and Method

Materials and Reagents

Ammonium persulfate (APS), N,N,N',N'-tetramethylethylenediamine (TEM-ED), 40% acrylamide/bis-acrylamide solution, and DNA loading buffer were purchased from Sangon Biotech Co., Ltd. (Shanghai, China). Phi29 polymerase, phi29 reaction buffer were purchased from Sangon Biotech Co., Ltd. All DNA samples were purchased from Sangon Biotech Co., Ltd. and purified by HPLC for fluorophore/quencher modified DNA and ULTRAPAGE for DNA without modification.

Table 1. DNA sequences and modifications.

Strands	Sequences(5'→3')
Str1u	CCCGTCTCGGCG
Str1d	CTTTCGTCTTCTCCCTTCTTAATTTCCTCTCCATGTCCGCCGAGAC GGG
Str2u	CCCGTCTCGGCGTTT
Str2u(p)	CCCGTCTCGGCGTT-Phosphorothioate-T
Str3u	AGAGGAAATTAAGAAGGGAG
O	GAGATTACGAGGCTG-BHQ
S	FAM-CAGCCTCGTAATCTCCTCCTCATCTTCTAATCTCGCTCGTTCT TTCTCGCGGAGACTC-Phosphorothioate-C
I'	GGAGTCTCCGCGAGAAAGAACGA-Phosphorothioate-G
S*	GGAGTCTCCGCGAGAAAGAACGAGCGAGATTAGAAGATGAGGAGG AGATTACGAGGCTG
I'(14)	GGAGTCTCCGCGA-Phosphorothioate-G
I'(34)	GGAGTCTCCGCGAGAAAGAACGAGCGAGATTAG-Phosphorothioate-A
I'(44)	GGAGTCTCCGCGAGAAAGAACGAGCGAGATTAGAAGATGAGGA-Phosphorothioate-G
I2'	TGAATGAGGTGGAT-Phosphorothioate-G
I1'	GGAGTCTCCGCGAGAAAG-Phosphorothioate-A
S1	TCTTTCTCGCGGAGACTCCCATTTAATCTGCCCACTCGTCTCTCCA TCCACCTCATTC-Phosphorothioate-A

DNA Sequences and Design

DNA sequences and modifications are listed in Table 1. The components of each structure are listed in Table 2. Sequences were designed using NUPACK design function and then analyzed using NUPACK to ensure minimal crosstalk between unrelated domains.

Annealing

All annealing processes were performed using a LongGene Multi-Block thermo-cycler. The samples (typically at a final duplex concentration of 1 μM) were heated to 95 °C for 5 min and then gradually cooled to room temperature at a constant rate over a period of 2 h.

Table 2. The components of each structure.

Structures	Components
Str1	Str1u, Str1d
Str2	Str2u, Str1d
Str2(p)	Str2u(p), Str1d
Str3	Str1u, Str1d, Str3u
O:S	O, S
I1':S1	I1', S1

Native Polyacrylamide Gel Electrophoresis

Samples were run on 12% native polyacrylamide gel in $1 \times$ TAE buffer at 85 V for 2 h at 4 °C. Gels were scanned with a Fluorchem FC2 gel scanner.

Quantitative Fluorescence Tracking

The fluorescent experiments were implemented using real-time PCR (Tianlong, TL988) equipped with a 48-well fluorescence plate reader. In a typical 25-μL reaction volume, the $1\times$ reaction concentration was 0.4 μM. The reactions were performed in $1 \times$ NEB Cutsmart buffer. Fluorescence intensity was measured every 3 min for 600 min.

4   Conclusion

In this work, we have established a polymerase triggered strand displacement mechanism where the phi29 DNA polymerase was employed to perform strand displacement during its polymerization. The PTSD mechanism is proven to be feasible and the effects of several parameters on the performance of PTSD were explored. Furthermore, a two-layer PTSD circuit was constructed by cascading two one-layer PTSD circuits. It is worth mentioning that the reactions of both the two layers were driven by phi29, indicating the PTSD mechanism is cascadable and of potential for constructing complicated dynamic DNA structures.

The greatest advantage PTSD possesses over the traditional toehold-mediated strand displacement is no requirement of knowing the sequence of branch migration domain, which makes it possible for one input strand to trigger strand displacement on different substrates varying the branch migration domain sequences. Considering the characteristics, the PTSD has a great many potential applications in constructing dynamic DNA nanodevices which used to employ toehold-mediated strand displacement. Furthermore, as the strand displacement in PTSD is triggered by polymerases, some unwanted structures may also lead to the occurrence of strand displacement and the leakages take place in this way. Thus finding ways to lower the leakages and improve the performance of PTSD may be a promising future direction.

Acknowledgments. We would like to acknowledge Cheng Zhang for the constructive suggestions at the early time of the experiment. This work was supported by National Key R&D Program of China for International S&T Cooperation Projects (No. 2017YFE0103900) and National Natural Science Foundation of China (No. 61772214).

References

1. Aldaye, F.A., Palmer, A.L., Sleiman, H.F.: Assembling materials with DNA as the guide. Science **321**(5897), 1795–1799 (2008)
2. Jones, M.R., Seeman, N.C., Mirkin, C.A.: Programmable materials and the nature of the DNA bond. Science **347**(6224), 1260901–1260901 (2015)

3. Zhang, D.Y., Seelig, G.: Dynamic DNA nanotechnology using strand-displacement reactions. Nat. Chem. **3**(2), 103–113 (2011)
4. Li, B., Jiang, Y., Chen, X., Ellington, A.D.: Probing spatial organization of DNA strands using enzyme-free hairpin assembly circuits. J. Am. Chem. Soc. **134**(34), 13918–13921 (2012)
5. Yang, X., Tang, Y., Mason, S.D., Chen, J., Li, E.: Enzyme-powered three-dimensional DNA nanomachine for DNA walking, payload release, and biosensing. ACS Nano **10**(2), 2324–2330 (2016)
6. You, M., Zhu, G., Chen, T., Donovan, M.J., Tan, W.: Programmable and multi-parameter DNA-based logic platform for cancer recognition and targeted therapy. J. Am. Chem. Soc. **137**(2), 667–674 (2015)
7. Grosso, E.D., Dallaire, A.-M., Vallée-Bélisle, A., Ricci, F.: Enzyme-operated DNA-based nanodevices. Nano Lett. **15**(12), 8407–8411 (2015)
8. Jung, C., Allen, P.B., Ellington, A.D.: A stochastic DNA walker that traverses a microparticle surface. Nat. Nanotechnol. **11**(2), 157–163 (2015)
9. Liu, M., et al.: A DNA tweezer-actuated enzyme nanoreactor. Nat. Commun. **4**, 2127 (2013)
10. Peng, H., Li, X.-F., Zhang, H., Le, X.C.: A microRNA-initiated DNAzyme motor operating in living cells. Nat. Commun. **8**, 14378 (2017)
11. Chen, Y.-J., et al.: Programmable chemical controllers made from DNA. Nat. Nanotechnol. **8**(10), 755–762 (2013)
12. Cherry, K.M., Qian, L.: Scaling up molecular pattern recognition with DNA-based winner-take-all neural networks. Nature **559**(7714), 370–376 (2018)
13. Qian, L., Winfree, E.: Scaling up digital circuit computation with DNA strand displacement cascades. Science **332**(6034), 1196–1201 (2011)
14. Qian, L., Winfree, E., Bruck, J.: Neural network computation with DNA strand displacement cascades. Nature **475**(7356), 368–372 (2011)
15. Seelig, G., Soloveichik, D., Zhang, D.Y., Winfree, E.: Enzyme-free nucleic acid logic circuits. Science **314**(5805), 1585–1588 (2006)
16. Soloveichik, D., Seelig, G., Winfree, E.: DNA as a universal substrate for chemical kinetics. Proc. Nat. Acad. Sci. **107**(12), 5393–5398 (2010)
17. Yin, P., Choi, H.M.T., Calvert, C.R., Pierce, N.A.: Programming biomolecular self-assembly pathways. Nature **451**(7176), 318–322 (2008)
18. Ke, Y., Meyer, T., Shih, W.M., Bellot, G.: Regulation at a distance of biomolecular interactions using a DNA origami nanoactuator. Nat. Commun. **7**, 10935 (2016)
19. Saccà, B., et al.: Reversible reconfiguration of DNA origami nanochambers monitored by single-molecule FRET. Angewandte Chemie Int. Ed. **54**(12), 3592–3597 (2015)
20. Zhang, F., Nangreave, J., Liu, Y., Yan, H.: Reconfigurable DNA origami to generate quasifractal patterns. Nano Lett. **12**(6), 3290–3295 (2012)
21. Reynaldo, L.P., Vologodskii, A.V., Neri, B.P., Lyamichev, V.I.: The kinetics of oligonucleotide replacements. J. Mol. Biol. **297**(2), 511–520 (2000)
22. Yurke, B., Turberfield, A.J., Mills Jr., A., Simmel, F.C., Neumann, J.L.: A DNA-fuelled molecular machine made of DNA. Nature **406**(6796), 605–608 (2000)
23. Yang, J., Dong, C., Dong, Y., Liu, S., Pan, L., Zhang, C.: Logic nanoparticle beacon triggered by the binding-induced effect of multiple inputs. ACS Appl. Mater. Interfaces **6**, 14486–14492 (2014)
24. Yang, J., Jiang, S., Liu, X., Pan, L., Zhang, C.: Aptamer-binding directed DNA origami pattern for logic gates. ACS Appl. Mater. Interfaces **8**, 34054–34060 (2016)
25. Yang, J., et al.: Entropy-driven DNA logic circuits regulated by DNAzyme. Nucleic Acids Res. **46**, 8532–8541 (2018)

26. Pan, L., et al.: Aptamer-based regulation of transcription circuits. Chem. Commun. **55**(51), 7378–7381 (2019)
27. Montagne, K., Plasson, R., Sakai, Y., Fujii, T., Rondelez, Y.: Programming an in vitro DNA oscillator using a molecular networking strategy. Mol. Syst. Biol. **7**(1), 466–466 (2010)
28. Padirac, A., Fujii, T., Rondelez, Y.: Bottom-up construction of in vitro switchable memories. Proc. Nat. Acad. Sci. **109**(47), E3212–E3220 (2012)
29. Fujii, T., Rondelez, Y.: Predator–prey molecular ecosystems. ACS Nano **7**(1), 27–34 (2013)
30. Ali, M.M., et al.: Rolling circle amplification: a versatile tool for chemical biology, materials science and medicine. Chem. Soc. Rev. **43**(10), 3324 (2014)
31. Sato, M., Ohtsuka, M., Ohmi, Y.: Usefulness of repeated GenomiPhi, a phi29 DNA polymerase-based rolling circle amplification kit, for generation of large amounts of plasmid DNA. Biomol. Eng. **22**(4), 129–132 (2005)

The Design of Logic Gate Based on Triplex Structures

Yingxin Hu[1,2], Zhiyu Wang[1], Zhekun Chen[1], Sulin Liao[3],
and Yabing Huang[4(✉)]

[1] Key Laboratory of Image Information Processing and Intelligent Control of
Education Ministry of China, School of Artificial Intelligence and Automation,
Huazhong University of Science and Technology, Wuhan 430074, Hubei, China
{yingxinhu,wangzhiyu0471,zkchen}@hust.edu.cn
[2] College of Information Science and Technology,
Shijiazhuang Tiedao University, Shijiazhuang 050043, Hebei, China
[3] School of Computer Science and Technology,
Huazhong University of Science and Technology, Wuhan 430074, Hubei, China
liaosulin_2030@qq.com
[4] Department of Pathology,
Renmin Hospital of Wuhan University, Wuhan 430060, Hubei, China
drybhuang@gmail.com

Abstract. Molecular carch hotspots in the area of DNA nanotechnology due to its inherent large storage capacity, high parallelism, and energy efficiency. The assembled structures in the existing computing units mainly adopt a double-helix structure. However, besides duplex, DNA is known to adopt other helical forms such as triplex form. The triplex structure has introduced a novel paradigm into the DNA engineer owing to their unique characteristic sequence, structural and assembly requirements. Based on the assembly and disassembly of the triplex structure, a YES logic gate was constructed via strand displacement reaction. The experiment results showed that GC content and stem length would influence the displacement reaction. Then AND logic gate was also established through cooperative effect by two DNA strands. The construction of the cascading circuit indicated triplex-based logic gate could be integrated into larger circuits and enrich the toolbox of the computing devices.

Keywords: DNA computing · Logic gate · Molecule circuit · Triplex structure · Strand displacement

1 Introduction

DNA is not only the carrier of genetic information but also an excellent material for building nanostructures due to its low cost and high programmability according to Waston-Crick base paring rule [1–8]. The self-assembled structures have been applied in biocomputing [9–21], biosensing [22–25], stimulus-responsive

© Springer Nature Singapore Pte Ltd. 2020
L. Pan et al. (Eds.): BIC-TA 2019, CCIS 1160, pp. 209–220, 2020.
https://doi.org/10.1007/978-981-15-3415-7_17

nanodevices and nanorobotics [26–31]. Among which, molecular computing has great advantages over traditional computing due to its inherent large storage capacity, high parallelism, and energy efficiency [32]. However, most of the components in molecular computing are based on double-stranded structure. In addition to the classical double-helix, a variety of other helical forms such as triplex or tetraplex were also observed [33–35]. Among the non-canonical helical forms, triplex complexes have been used as part of the rich toolbox of structures owing to their unique characteristic sequence, structural and assembly requirements. Moreover, their potential application in regulation of gene regulation and therapeutic are attracting broad interest [36]. As a result, it is essential to construct computing units based on triplex structures which would enrich the toolbox of molecule computing. Moreover, toehold-mediated strand displacement reaction is a powerful and frequently used mechanism to construct logic circuits [9–12], catalytic amplifiers and switches [37–39]. This mechanism could also apply to triplex structures by adding toehold to the end of a strand.

In this work, based on different triplex structure, YES gate, AND gate, and cascading circuit were constructed utilizing strand displacement reaction. First, a YES gate based on the basic hairpin triplex structure was designed and the factors such as length of the stem, GC content, and toehold that influenced the gate were also investigated. Then, single-stranded arms and double-stranded arms triplex structure were also used to build the YES gate. Based on double-stranded arms triplex structure, an AND gate was constructed. Finally, to verify the extensibility of the scheme, a cascading circuit was also established. The native PAGE results confirmed the feasibility and validity of using triplex structures as the computing components.

2 Experimental

2.1 Material

All DNA strands were purified by polyacrylamide gel electrophoresis (PAGE) and purchased from Sangon Biotech Co., Ltd. (Shanghai, China). The DNA oligonucleotides were dissolved in water as the stock solution and quantified using Nanodrop 2000, and absorption intensities were recorded at $\lambda = 260\,\text{nm}$. $50 \times$ TAE buffer, acrylamide/bis-acrylamide (29:1) 40% (w/v) stock solution, ammonium persulphate, TEMED, DNA ladder were all purchased from Sangon Biotech Co., Ltd. (Shanghai, China). All other chemicals were of analytical grade and used without further purification. The DNA sequences in this study were designed by NUPACK software to minimize unnecessary crosstalk and they are listed in Table 1 below.

2.2 Method

The Formation of the Gate Structure. The DNA complexes were formed by mixing corresponding single strands with equal concentrations in reaction

Table 1. DNA sequences

Strands	Sequences (5'→3')
H1	TTTTCTTTTCTTCAATGTACAGTATTGTTCTTTTCTTTT
M1	TTAACCAGACACAAAAGAAAAGAA
M1*	TTCTTTTCTTTTGTGTCTGGTTAA
H2	TTCTTTTCTTTTCTTTTCTTCAATGTACAGTATTGTT CTTTTCTTTTCTTTTCTTC
M2	AGACACAAGAAAAGAAAAGAAAAGAA
M2*	TTCTTTTCTTTTCTTTTCTTGTGTCT
M23	AAGAAAAGAAAAGAAAAGAAAGACAC
M23*	GTGTCTTTCTTTTCTTTTCTTTTCTT
M24	TTAACCAGACACAAGAAAAGAAAAGAAAAGAA
M24*	TTCTTTTCTTTTCTTTTCTTGTGTCTGGTTAA
H3	TCTCTCCTTTCTCAATGTACAGTATTGTCTCTCCTTTCT
M3	TTAACCAGACACAGAAAGGAGAGA
M3*	TCTCTCCTTTCTGTGTCTGGTTAA
H4	TTCTTCTTCTCTCAATGTACAGTATTGTCTCTTCTTCTT
M4	TTAACCAGACACAGAGAAGAAGAA
M4*	TTCTTCTTCTCTGTGTCTGGTTAA
H5	ATGGCATTAACCTTGCTTCTCTCCTTTCTCAATGTACAGTATT GTCTTTCCTCTCTAGGTTCATCATCAACTAG
L	AGCAAGGTTAATGCCAT
R	CTAGTTGATGATGAACCT
M5	AGAAAGGAGAGAAAGGAAAGAGGA
M5s	AGAAAGGAGAGA
M5*	TCCTCTTTCCTTTCTCTCCTTTCT
Ls	AGCAAGGTTAAT
Rs	TGATGATGAACCT
Lr1	GAGAAGCAAGGTTAATGCCAT
Rr1	CTAGTTGATGATGAACCTAGAG
Lr2	GAGAGAAGCAAGGTTAATGCCAT
Rr2	CTAGTTGATGATGAACCTAGAGAG
R1	TTTTCTTCCTCTTTCCTTTCTCTCCTTTCT
R2	AGAAAGGAGAGAAAGGAAAGAGGA

buffer (24 mM Tris-HCl (pH = 7.9), 3.6 mM $MgCl_2$). The mixture was annealed in a polymerase chain reaction (PCR) thermal cycler at the reaction condition of 85 °C for 5 min, 65 °C for 30 min, 50 °C for 30 min, 37 °C for 30 min, 25 °C for 30 min, and finally kept at 25 °C.

The Strand Displacement Reaction. The reaction was conducted at 25 °C for 2 h.

PAGE Experiments. 5.92 mL H_2O, 3 mL 40% (w/v) acrylamide stock solution, 1 mL 10 × TAE buffer, 75 µL 10% ammonium persulfate solution (w/v), 7.5 µL TEMED were added to a beaker sequentially and shaken gently. The solution mixture (total volume: 10 mL) was slowly added to the gap of glass plates and kept at room temperature for 30 min. Then the samples were mixed with 36% glycerin solution and subjected to electrophoresis analysis. The analysis was carried out in 1X TAE buffer (40 mM Tris, 20 mM acetic acid, 2 mM EDTA, pH 8.0) supplemented with 12.5 mM $MgCl_2$ at 90 V for 90 min at 4 °C. After stains all (Sigma-Aldrich) or EB staining, Gels were imaged using the scanner or Gel imager (Bio-Rad).

DNA Sequences and Design. The DNA sequences in each gate were designed and analyzed by NUPACK software.

3   Results and Discussion

The triplex structure is composed of a normal DNA duplex and an extra specific single-stranded DNA strand. The normal DNA duplex consists of a homopurine (A/G) DNA strand and a homopyrimidine (T/C) DNA strand and they hybridize via Watson-Crick base-pairings. The extra homopyrimidine (T/C) strand binds to the duplex in the major groove via Hoogsteenor or reverse Hoogsteen interactions. As shown in Fig. 1, the triplex structure consisted of two DNA strands H1 and M1. One segment of strand M1 (A/G) was designed to bind with the H1 stems (T/C) through Watson-Crick and Hoogsteen base pairing. The other segment of M1 was a toehold far away from the loop of the triplex structure for strand displacement. Here, the stem of the triplex structure H1/M1 was 12bp. Upon addition of strand M1* which was fully complementary

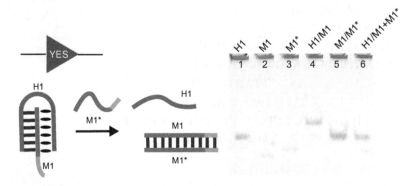

Fig. 1. The YES logic gate based on the triplex structure and the native PAGE analysis of YES gate products. The DNA strands and the assembled complex involved were labeled above the lane. Lane 1, H1; lane 2, M1; lane 3, M1*; lane 4, H1/M1; lane 5, M1/M1*; lane 6, H1/M1+M1*. The concentration of each sample was 1 µM.

with M1, the triplex structure was disrupted and generated single-stranded H1 and two double-stranded M1/M1*. Thus, a YES gate was established which took triplex structure H1/M1 as the gate, M1* and M1/M1* as the input and output respectively.

The gate was verified by native PAGE gel (Fig. 1). A band with a lower migration than the one corresponding to H1 was observed in lane 4 which demonstrated the well-formed triplex structure. When input strand M1* was added to the gate complex, the band corresponding to H1/M1 disappeared and H1 and M1/M1* appeared in lane 6. The gel results confirmed the feasibility of the YES gate based on the triplex structure.

To achieve the strand displacement based on the triplex structure under neutral PH, it is needed to find an optimal thermodynamic trade-off that requires to satisfy the following conditions. First, the triplex conformation with both Watson-Crick and Hoogsteen interactions should well-formed. Second, the triplex structure should not be too stable so that it's disruption by the strand displacement would be allowed.

We have studied the stem of triplex structures with different length and GC content (leading to complexes of different stabilities), the toehold with different length, the toehold at different positions. The DNA sequences used in this part are illustrated in Table 1.

First, the stem length of the triplex structure was fixed at 20bp. As can be seen in Fig. 2a, when the toehold at 5 end is 6nt, two bands ran lower than the one of H2 in lane 2 which indicated other secondary structures were also formed except the triplex structure. In addition, the bands of H2 and M2/M2* were not observed upon addition of M2* in lane 4 demonstrated that no displacement occurred under this condition. Then, the position of toehold was altered from '5' end to '3' end, from Fig. 2b, only one band ran slower than H2 was observed in lane 2 which accounted for the well-formed triplex conformation. However, the bands of H2 and M23/M23* was invisible upon addition of M23* in lane 4 and this result indicated that the displacement still not occurred. This is still the case in lane 7 even the toehold was further lengthened to 12nt. The experimental results showed that the triplex structure with 20bp stem was too stable and shorter stem was needed to facilitate the strand displacement reaction. Therefore, the stem was shortened to 12bp in following experiments. Moreover, considering that GC content would influence the formation of the triplex conformation, the stem with different GC contents was also tested. As expected, in Fig. 2c and d, the bands corresponding to strand H3(H4) and M3/M3*(M4/M4*) were visible upon addition of M3* (M4*) which indicated the displacement indeed occurred. Nevertheless, high GC content such as 41.7% in Fig. 2c lane 4 or 33.4% in Fig. 2d lane 5 would lead to a poor yield of triplex structure . While low GC content such as 20% (Fig. 1) could achieve not only well-formed triplex conformation but also fast-displacement reaction. In our next experiments, we employed a stem of 12bp with 20% GC content as the basic gate structure.

To further verify the extensibility of the strand displacement based on triplex structure, the other two triplex conformations were also studied. First, as

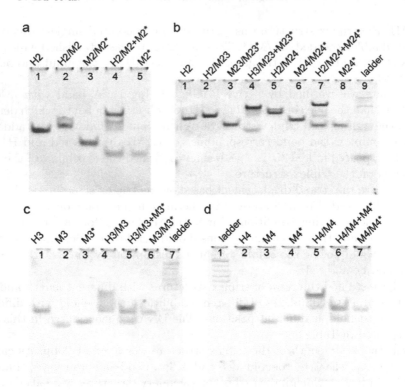

Fig. 2. Native PAGE analysis of strand displacement effects based on the triplex structure under various conditions. (a) The stem length of the triplex is 20bp and the toehold at '5' end is 6nt. Lane 1, H2; lane 2, H2/M2; lane 3, M2/M2*; lane 4, H2/M2+M2*; lane 5, M2*. (b) The stem length of the triplex is 20bp and the toehold at '3' end is 6nt (Lane 1–4) or the toehold at 5' end is 12nt (lane 5–8). Lane 1, H2; lane 2, H2/M23; lane 3, M23/M23*; lane 4, H2/M23+M23*; lane 5, H2/M24; lane 6, M24/M24*; lane 7, H2/M24+M24*; lane 8, M24*; lane 9, 20bp DNA ladder. (c) The stem length of the triplex is 12bp with 41.7% GC content. Lane 1, H3; lane 2, M3; lane 3, M3*; lane 4, H3/M3; lane 5, H3/M3+M3*; lane 6, M3/M3*; lane 7, 20bp DNA ladder. (d) The stem length of the triplex is 12bp with 33.4% GC content. ; lane 1, DNA ladder; Lane 2, H4; lane 3, M4; lane 4, M4*; lane 5, H4/M4; lane 6, H4/M4+M4*; lane 7, M4/M4*. The concentration of each sample was $1\,\mu$M and the input strand M2*, M23*, M24*, M3*, and M4* was $2\,\mu$M.

illustrated in Fig. 3a, two single-stranded arms were involved in the gate structure. In the gel results (Fig. 3b), only one lower migration band corresponding to the gate structure was observed and when the input strand M5* was added, the bands of H5 and M5/M5* appeared. The experimental results indicated the well-formed triplex structure and good displacement ability. Further, two double-stranded arms were involved into the gate structure. As shown in Fig. 3c, the gate structure consisted of four strands H5, L, R, M5(M5s). Strands L and R hybridized with H5 at two ends respectively, and M5(M5s) was designed to bind

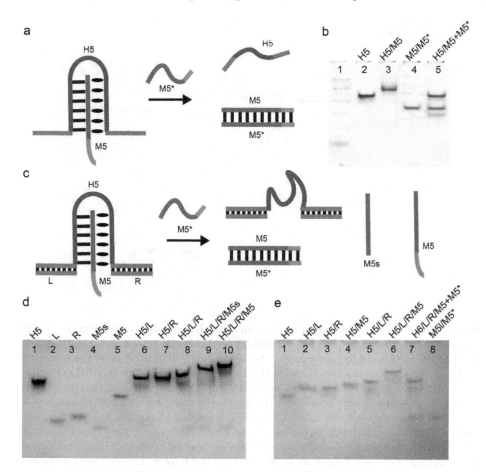

Fig. 3. The strand displacement reaction based on different triplex structures and the native PAGE analysis of the reaction. The DNA strands and the assembled complex involved were labeled above the lane. (a) The schematic illustration of strand displacement based on the triplex structure with left and right single-stranded arms. (b) The native PAGE analysis of the reaction products. Lane 1, 20 bp DNA ladder; Lane 2, H5; lane 3, H5/M5; lane 4, M5/M5*; lane 5, H5/M5+M5*. (c) The schematic illustration of strand displacement based on the triplex structure with left and right double-stranded arms. (d) The native PAGE analysis of the formation of the triplex structure. Lane 1, H5; Lane 2, L; lane 3, R; lane 4, M5s; lane 5, M5; lane 6, H5/L; lane 7, H5/R; lane 8, H5/L/R; lane 9, H5/L/R/Ms; lane 10, H5/L/R/M5. (e) The native PAGE analysis of the reaction products. Lane 1, H5; Lane 2, H5/L; lane 3, H5/R; lane 4, H5/M5; lane 5, H5/L/R; lane 6, H5/L/R/M5; lane 7, H5/L/R/M5+M5*; lane 8, M5/M5*. The concentration of each sample was 1 μM and the input strand M5* was 2 μM.

with H5 stems through Watson-Crick and Hoogsteen base pairing with overhang (without overhang). Before testing the displacement reaction, the assembled structures were first confirmed. In Fig. 3d, the bands ran slower with the addition

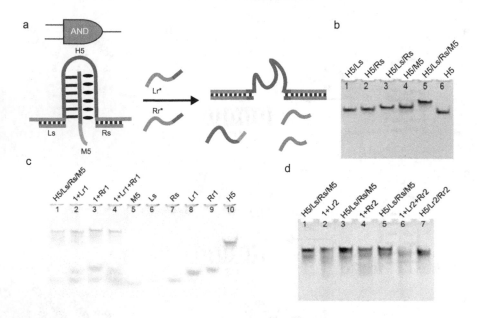

Fig. 4. The AND logic gate based on the armed-triplex structure and the native PAGE analysis of gate products. (a) The schematic illustration of cooperative strand displacement based on armed-triplex structure. (b) The native PAGE analysis of the formation of the armed-triplex structure. Lane 1, H5/Ls; Lane 2, H5/Rs; lane 3, H5/Ls/Rs; lane 4, H5/M5; lane 5, H5/Ls/Rs/M5; lane 6, H5. (c) The native PAGE analysis of the strand displacement reaction in which the displacing stand had 4nt complementary to the stem of the triplex. Lane 1, H5/Ls/Rs/M5; Lane 2, 1+Lr1; lane 3, 1+Rr1; lane 4, 1+Lr1+Rr1; lane 5, M5; lane 6, Ls; lane 7, Rs; lane 8, Lr1; lane9, Rr1; lane 10, H6. (d) The native PAGE analysis of the strand displacement reaction in which the displacing stand had 6nt complementary to the stem of the triplex. Lane 1, H5/Ls/Rs/M5; Lane 2, 1+Lr2; Lane 3, H5/Ls/Rs/M5; lane 4, 1+Rr2; Lane 5, H5/Ls/Rs/M5; lane 6, 1+Lr2+Rr2; lane 7, H5/Lr2/Rr2. The concentration of each sample was 1 μM.

of L, R, M5s, and M5 respectively. The gel results demonstrated that both blunt-ended M5s and elongated M5 could form the triplex structure. In Fig. 3e, upon addition of input strand M5* in lane 7, the band of gate structure H5/L/R/M5 disappeared and the bands corresponding to H5/L/R and M5/M5* appeared, indicating the successful displacement reaction based on the triplex structure with two rigid arms. The DNA sequences used in this part:

To achieve the universality of the triplex structure, the AND gate was also established. In YES gate, the input strand was fully complementary with one of the strands in the gate structure. In addition, the construction of logic gates whose input had less identical sequences as the strand displaced was vital.

As shown in Fig. 4a, the gate structure H5/Ls/Rs/M5 had left a single-stranded segment at left and right arm respectively. Therefore, the cooperative action of two strands from two ends could disassemble the gate structure. In Fig. 4b lane 5, the band corresponding to the assembled structure H5/Ls/Rs/M5 ran slower than that of H5, H5/Ls, H5/Rs, H5/M5, indicating the well-formed gate structure. The length of the displacing strand corresponding to the stem of the M5 would influence the displacement reaction. Upon addition of Lr1, Rr1, or both Lr1 Fig. 4c lane 2, 3, 4, which illustrated that the 4nt-length was not enough for disassembly of the triplex structure. Further lengthening to 6nt, only when both Lr2 and Rr2 were added, the band of complex H5/Ls/Rs/M5 disappeared and complex H5/Lr2/Rr2 was observed. While when only one of them was added respectively, the band of gate complex H5/Ls/Rs/M5 still could be observed. The experimental results indicated that the logic gate could operate successfully. The DNA sequences used in this part are listed in Table 1.

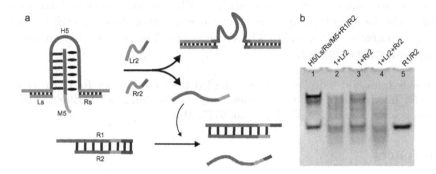

Fig. 5. The cascading circuit based on the armed-triplex structure and the native PAGE analysis of gate products. (a) The schematic illustration of cascading circuit based on armed-triplex structure. (b) The native PAGE analysis of the reaction products. Lane 1, H5/Ls/Rs/M5+R1/R2; Lane 2, 1+Lr2; lane 3, 1+Rr2; lane 4, 1+Lr2+Rr2; lane 5, R1/R2. The concentration of each sample was $1\,\mu M$.

To demonstrate the capability of constructing the molecular circuit with the triplex structure, a cascading circuit was designed. As shown in Fig. 5a, the components of the circuit were the triplex complex H5/Ls/Rs/M5, duplex complex R1/R2, and single-stranded Lr2, Rr2. In the initial state, the segment of strand M5 which could react with R1/R2 was encapsulated in the preformed triplex structure, thus, the two complexes coexisted and no reaction occurred. On addition of the two input strands Lr2 and Rr2, the strand M5 was released and the strand displacement reaction could proceed. In Fig. 5b, the bands of H5/Ls/Rs/M5 and R1/R2 were clearly observed in lane 1 and upon addition of Lr2 or Rr2 alone in lane 2 and lane 3 the complex R1/R2 still could be observed, while both Lr2 and Rr2 were added, the complex H5/Ls/Rs/M5 and R1/R2 disappeared in lane 4. The experimental results validated the effectiveness of the cascading circuit. The DNA sequences used in this part used in this part are listed in Table 1.

4   Conclusion

To sum up, different from the strand displacement based on duplex structure, a series of logic gates were established which employed triplex conformation as the gate structure and strand displacement reaction as the means to induce the disassembly of the gate. The factors such as the stem length, GC content, toehold position and length that influenced the validity of the gate were studied. In addition, different triplex structures, for example, no-armed complex, single-stranded arms complex were tested to achieve the extensibility. The experimental results demonstrated the effectiveness of using triplex structures as the building block to construct molecule circuits.

Due to its unique characteristic such as PH-responsiveness and adjustable stability via changing the sequence, the proposed scheme based on triplex structures has potential not only for DNA computing but also for biosensing and programming nanomachine.

Acknowledgments. This research was funded by National Key R&D Program of China for International S&T Cooperation Projects (No. 2017YFE0103900), National Natural Science Foundation of China (No. 61772214).

References

1. Seeman, N.C.: DNA in a material world. Nature **421**, 427 (2003)
2. Yan, H., Park, S.H., Finkelstein, G., Reif, J.H., LaBean, T.H.: DNA-templated self-assembly of protein arrays and highly conductive nanowires'. Science **301**(5641), 1882–1884 (2003)
3. Liu, D., Park, S.H., Reif, J.H., LaBean, H.: DNA nanotubes self-assembled from triple-crossover tiles as templates for conductive nanowires. Proc. Nat. Acad. Sci. USA **101**(3), 717–722 (2004)
4. He, Y., Chen, Y., Liu, H., Ribbe, A.E., Mao, C.: Self-assembly of hexagonal DNA two-dimensional (2D) arrays. J. Am. Chem. Soc. **127**(35), 12202–12203 (2005)
5. Rothemund, P.W.K.: Folding DNA to create nanoscale shapes and patterns. Nature **440**(7082), 297–302 (2006)
6. Yin, P., Choi, H.M., Calvert, C.R., Pierce, N.A.: Programming biomolecular self-assembly pathways. Nature **451**(7176), 318–322 (2008)
7. Seelig, G., Soloveichik, D., Zhang, D.Y., Winfree, E.: Enzyme-free nucleic acid logic circuit. Science **314**(5805), 1585–1588 (2006)
8. Xu, F., Wu, X.T.F., Shi, L., Pan, L.Q.: A study on a special DNA nanotube assembled from two single-stranded tiles. Nanotechnology 30 (2019)
9. Zhang, D.Y., Turberfield, A.J., Yurke, B., Winfree, E.: Engineering entropy-driven reactions and networks catalyzed by DNA. Science **318**(5853), 1121–1125 (2007)
10. Qian, L., Winfree, E.: Scaling up digital circuit computation with DNA strand displacement cascades. Science **332**(6034), 1196–1201 (2011)
11. Qian, L., Winfree, E., Bruck, J.: Neural network computation with DNA strand displacement cascades. Nature **475**(7356), 368–372 (2011)
12. Li, W., Yang, Y., Yan, H., Liu, Y.: Three-input majority logic gate and multiple input logic circuit based on DNA strand displacement. Nano Lett. **13**(6), 2980–2988 (2013)

13. Du, Y., Peng, P., Li, T.: Logic circuit controlled multi-responsive branched DNA scaffolds. Chem. Commun. **54**(48), 6132–6135 (2018)
14. Erbas-Cakmak, S., et al.: Molecular logic gates: the past, present and future. Chem. Soc. Rev. **47**(7), 2228–2248 (2018)
15. Fern, J., Schulman, R.: Modular DNA strand-displacement controllers for directing material expansion. Nat. Commun. **9**, 1–8 (2018)
16. Quan, K., et al.: Dual-microRNA-controlled double-amplified cascaded logic DNA circuits for accurate discrimination of cell subtypes. Chem. Sci. **10**(5), 1442–1449 (2019)
17. Zhong, W.Y., et al.: A DNA arithmetic logic unit for implementing data backtracking operations. Chem. Commun. **55**(6), 842–845 (2019)
18. Yang, J., Dong, C., Dong, Y., Liu, S., Pan, L., Zhang, C.: Logic nanoparticle beacon triggered by the binding-induced effect of multiple inputs. ACS Appl. Mater. Interfaces **6**, 14486–92 (2014)
19. Yang, J., Jiang, S.X., Liu, X.R., Pan, L.Q., Zhang, C.: Aptamer-binding directed DNA origami pattern for logic gates. ACS Appl. Mater. Interfaces **8**, 34054–34060 (2016)
20. Pan, L.Q., Wang, Z.Y., Li, Y.F., Xu, F., Zhang, Q., Zhang, C.: Nicking enzyme-controlled toehold regulation for DNA logic circuits. Nanoscale **9**, 18223–18228 (2017)
21. Pan, L., Hu, Y., Ding, T., Xie, C., Wang, Z., Chen, Z., et al.: Aptamer-based regulation of transcription circuits. Chem. Commun. **55**, 7378–7381 (2019)
22. Nutiu, R., Li, Y.: Structure-switching signaling aptamers. J. Am. Chem. Soc. **125**(16), 4771 (2003)
23. Ono, A., Togashi, H.: Highly selective oligonucleotide-based sensor for mercury(II) in aqueous solutions. Angew. Chem. **43**(33), 4300–4302 (2004)
24. Liu, J., Lu, Y.: Smart nanomaterials responsive to multiple chemical stimuli with controllable cooperativity. Adv. Mater. **18**(13), 1667–1671 (2010)
25. Wang, R., Zhou, X., Shi, H., Luo, Y.: T-T mismatch-driven biosensor using triple functional DNA-protein conjugates for facile detection of Hg 2+. Biosensors Bioelectron. **78**, 418–422 (2016)
26. Douglas, S.M., Ido, B., Church, G.M.: A logic-gated nanorobot for targeted transport of molecular payloads. Science **335**(1), 831–834 (2012)
27. Fu, J.L., Yan, H.: Controlled drug release by a nanorobot. Nat. Biotechnol. **30**(5), 407–408 (2012)
28. Amir, Y., Ben-Ishay, E., Levner, D., Ittah, S., Abu-Horowitz, A., Bachelet, I.: Universal computing by DNA origami robots in a living animal. Nat. Nanotechnol. **9**(5), 353–357 (2014)
29. Torelli, E., et al.: A DNA origami nanorobot controlled by nucleic acid hybridization. Small **10**(14), 2918–2926 (2014)
30. Thubagere, A.J., et al.: A cargo-sorting DNA robot. Science **357**(6356), 1–11 (2017)
31. Li, S.P., et al.: A DNA nanorobot functions as a cancer therapeutic in response to a molecular trigger in vivo. Nat. Biotechnol. **36**(3), 258 (2018)
32. Braich, R.S., Chelyapov, N., Johnson, C., Rothemund, P.W., Adleman, L.: Solution of a 20-variable 3-SAT problem on a DNA computer. Science **296**(5567), 499–502 (2002)
33. Mirkin, S.M., Lyamichev, V.I., Drushlyak, K.N., Dobrynin, V.N., Filippov, S.A., Frank-Kamenetskii, M.D.: DNA H form requires a homopurine-homopyrimidine mirror repeat. Nature **330**(6147), 495–497 (1987)
34. Haider, S.M., Parkinson, G.N., Neidle, S.: Structure of a G-quadruplex-ligand complex. J. mol. Biol. **326**(1), 117–125 (2003)

35. Gehring, K., Leroy, J.L., Gueron, M.: A tetrameric DNA structure with protonated cytosine-cytosine base pairs. Nature **363**(6429), 561–565 (1993)
36. Chandrasekaran, A.R., Rusling, D.A.: Triplex-forming oligonucleotides: a third strand for DNA nanotechnology. Nucleic Acids Res. **46**(3), 1021–1037 (2018)
37. Yurke, B., Turberfield, A.J., Mills, A.P., Simmel, F.C., Neumann, J.L.: A DNA-fuelled molecular machine made of DNA. Nature **406**(6796), 605–608 (2000)
38. Kim, J., White, K.S., Winfree, E.: Construction of an in vitro bistable circuit from synthetic transcriptional switches. Mol. Syst. Biol. **2**, 68 (2006)
39. Yang, J., Wu, R.F., Li, Y.F., Wang, Z.Y., Pan, L.Q., Zhang, Q., et al.: Entropy-driven DNA logic circuits regulated by DNAzyme. Nucleic Acids Res. **46**, 8532–8541 (2008)

Research on the Application of DNA Cryptography in Electronic Bidding System

Jianxia Liu[1], Yangyang Jiao[2(✉)], Yibo Wang[2], Hongxuan Li[2], Xuncai Zhang[3], and Guangzhao Cui[3(✉)]

[1] School of Information Management,
Nanjing University, Nanjing 210000, China
jianxialiuedu@163.com
[2] School of Economics and Management,
Wuhan University, Wuhan 430072, China
yangyangjiao@whu.edu.cn
[3] School of Electrical and Information Engineering,
Zhengzhou University of Light Industry, Zhengzhou 450002, China
cgzh@zzuli.edu.cn

Abstract. With the rapid development of bidding market, a new type of bidding method–electronic bidding method has been widely used in the bidding market. This new-type of method is a program that the bidding is completed by graphics and textures. Generally, the electronic bidding provides doubles bidders with the data information of the transaction through the Internet. In this study we propose an information encryption and digital signature method based on DNA cryptography which is combined with HASH function, symmetric encryption and asymmetric encryption comprehensively for the purpose of the information resources' integrality, usability, security and reliability in the electronic bidding system. The results of the study show that our new method provides double guarantee for the bidding system under the role of symmetric encryption and asymmetric encryption used DNA cryptography, and DNA molecular encryption has the characteristics of electronic immunity, concealment, independent of mathematical problems, massive parallelism, etc. It provides a double guarantee for the bidding system. The security analysis results show that the algorithm can effectively protect the security and integrity of the bidder and bidder's data information.

Keywords: Electronic bidding · Data and information · DNA cryptography · Digital signature

1 Introduction

With the rapid development of information technology, it has provided great convenience for the rapid development of enterprises. The rapid development

© Springer Nature Singapore Pte Ltd. 2020
L. Pan et al. (Eds.): BIC-TA 2019, CCIS 1160, pp. 221–230, 2020.
https://doi.org/10.1007/978-981-15-3415-7_18

of enterprises requires mutual cooperation. One way of cooperation is through bidding. The original method of bidding and tendering is through the paper materials, and the transmission process is usually proved cumbersome and complex, and the efficiency is relatively low [1]. The emergence of the Internet brings a novel way for bidding which is electronic bidding mode. The electronic bidding system is a set of whole process of online bidding through paperless way. The electronic bidding system has greatly simplified the bidding process, improving the efficiency of the bidder and the bid inviter [2].

A complete set of electronic bidding process includes following stages. Firstly, the bidding information is formulated and published online by bidding organizers; Secondly, the bidders collect information from the tenderer according to their business; Thirdly, the bidders bid on the electronic bidding system according to the information provided by the tenderer; Fourthly, the tenderer conducts electronic online selection in the electronic bidding system; Fifthly, the tenderer chooses the winning bidder(s) according to its own needs [3]. The electronic bidding system can effectively solve the problem of human operation, which can better reflect the fairness, fairness and openness of the bidding process. Moreover, in the electronic bidding system, the tenderer publishes the bidding information according to its own needs, so that the bidder can obtain the information in time, save the time of the tenderer and the bidder, and improve the efficiency. Lastly, the new method makes contribution to promote the better development of enterprises, facilitates the development of biddings internationally, and makes enterprises move towards international development [4].

The emergence of the electronic bidding mode brings great convenience to the bidding work. Because all the data and information transmission is stored and transmitted on the Internet, these resources are vulnerable to be stolen, even the bidding information can be modified, deleted by hackers simultaneously. Any leaked information may lead to tens of thousands of losses of bidders and the whole bidding process will be failed with the lost biding information data. Therefore, the data and information of electronic bidding system must be protected effectively [5]. There are already some cryptosystems in the face of strong cracking caused by hacker attackers, such as: DES, AES and RSA cryptosystems based on traditional mathematical computing. However, these traditional cryptosystems cannot satisfy the encryption of data and information completely because of the low security. In 2014, Montenegro and Lopez [6] proposed a sealed bidding and multi-currency auction method based on secure multi-party computation. These functions are used without revealing any information, or even reveal to the trusted third parties (such as auctioneers). In 2018, Jin et al. [7] proposed an efficient certificateless aggregated deniable authentication protocol that based on certificateless public key cryptography. It is neither related to public key certificate management in traditional public key infrastructure (PKI) cryptography nor key escrow in identity-based cryptography. It can be used in e-voting system, e-tender system and secure network negotiation.

Because DNA molecules are in very large parallelism inherently with low energy consumption and high storage density make the encryption algorithms have unique advantages that traditional cryptographic algorithms do not have based on DNA calculation [8]. DNA cryptogram is the storage or hiding of data and information carried by DNA molecules, using molecular biology technology such as molecular information synthesis technology, DNA chain amplification technology and DNA chain coding technology to process, store and hide data and realize encryption system for information data. In 2018 Ben Slimane et al. [9] designed an information encryption system for pictures based on DNA sequence operation, single neuron model and chaotic mapping. The initial conditions and system parameters of the dynamic system are generated by using 512-bit hash values highly dependent on the flat image. In the same year, a new image encryption algorithm was proposed by combining chaotic system with DNA coding, this encryption scheme has large key space and high security. In 2019 Enayatifar [10] proposed a fast and secure multi-image encryption (MIE) algorithm based on DNA sequence and image matrix index which can resist common attacks sufficiently.

In order to protect the data and information which are consisted in the electronic bidding system more effectively. In this article, we proposed a new method of information encryption and digital signature combining with DNA cryptography, it proved that the new method can protect the data and information in the electronic bidding system better and provide higher security for the whole bidding process.

2 Design of Bidding System Model

A complete online bidding system can program the bidding process effectively, improve the efficiency of tenderer and bidders, and reduce their costs, and the whole successful bidding process depends on the security of bidding data. Bidders must ensure the information security and integrity in the process of transmission of bidding data, which are mostly commercial confidential and extremely important privacy information. Hence, it is indispensable to design a system to encrypt bidding data and information in order to avoid bidding information leakage and ensure the authenticity and integrity of bidding information effectively. And make it not be tampered with in the process of dissemination; avoid the problems of information being destroyed in the process of dissemination [11].

Decryption of data and signature verification are also described in Fig. 1. Detailed steps are omitted there.

Security and integrity of the data are two key elements of the research on information security technology. Security requires that secret data shouldn't be acquired or destroyed by third parties in the process of storage and exchange. Integrity ensures the consistency of data in the process of communication and other attributes of data are accurate and complete. Data encryption can be used to realize the confidentiality of data, and digital signature can protect the integrity of data in the process of communication.

Data encryption scrambles and replaces the data by encryption algorithm, so that unaccredited users can not get the original data because they do not know the encryption algorithm or the keys of the encryption algorithm. Generally, encryption algorithms include symmetrical encryption and asymmetrical

224   J. Liu et al.

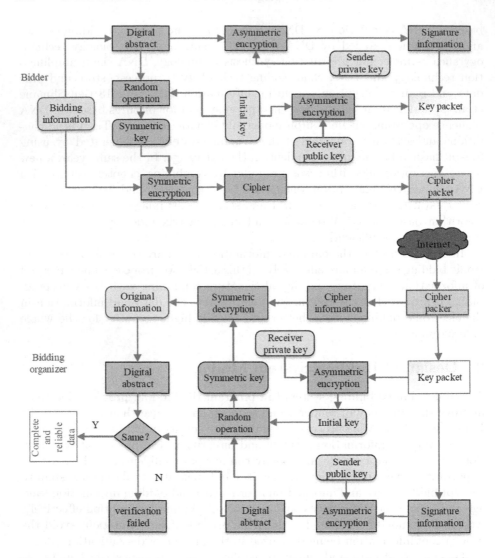

Fig. 1. Flow chart of information encryption and decryption in electronic bidding system.

encryption. In symmetric encryption, the same key to encrypt and decrypt data is used. And velocity of the encryption and decryption is faster in symmetrical encryption, however, symmetrical encryption algorithm need control the key distribution strictly because the same key is used in encryption and decryption. Asymmetrical encryption includes a pair of keys for encryption and decryption of data which are called public key and private key. It improves the security of the data for the different keys of the encryption and decryption and the keys can be used for a long run in this method. Velocity of this method is slow in encrypting large amounts of data. Integrate data ensures that the data can't be tampered in the transmission process which realized by using digital signature with the help of Hash function.

In these circumstances, a data encryption and digital signature method is designed with high practicability and security under the application of symmetrical encryption algorithm, asymmetrical encryption algorithm and Hash algorithm comprehensively. For the purpose of the security and integrity of the data, data encryption and digital signature are operated before the sender sends the data or when they are sending the data. The model of electronic bidding system designed in this paper is shown in Fig. 1.

The procedures of the system are executed as follows:

Step one: HASH arithmetic. The unique digital abstract information H of the original data can be obtained using HASH function to perform HASH operation on the original data;

Step two: Digital signature. The digital signature message S is obtained by using the sender's private key Sa to sign the original data abstract information by asymmetrical encryption;

Step three: Generation of the symmetrical key. The symmetric key K is obtained by calculating the digital abstract and the initial key, which is used for symmetrical encryption;

Step four: Symmetrical encryption. The ciphertext information C1 is obtained based on the symmetrical encryption of bidding information by using the symmetrical key which is generated in step three;

Step five: Encryption of symmetrical key K. The symmetrical key K generated in step three is encrypted using the receiver's public key Ra and combining with signature information S to obtain the key information package;

Step six: Combination of the ciphertext. Combine the ciphertext information generated in step 4 with the key information package generated in step five to get the ciphertext combination package, and then transmit it on the Internet.

3   Selection of Encryption Methods

Modern cryptography is based on mathematical cryptography mostly, so the security of most cryptosystems depends on difficult mathematical problems entirely except one secret at a time. These cryptosystems can be easily deciphered if an attacker has enough computational ability because these cryptosystems only have computational security. With the emergence and maturity of new computing methods in recently years, the attack methods of cryptographic attackers are becoming more and more advanced, and the existing cryptographic technology is more difficult to meet the growing demand of information security. As the degree of informatization in human society is becoming higher and higher, and the means of cryptographic attack are becoming more and more advanced, the existing cryptographic technology is difficult to meet the growing needs of information security. Therefore, it is imperative to explore and develop new cryptography theory and technology while strengthening the research of traditional cryptography theory and technology.

Cryptography always changes with the progress of computing technology from Caesar cryptography to modern cryptography. As it is known to all that the application of electronic computer improves people's computing speed greatly which plays a great role in promoting the development of human society. However, electronic circuits have entered the era of super-large-scale integration circuits and ultra large scale integrated circuits with the development of modern industrial technology, the design of transistors is approaching the physical limit, and the electronic computing technology is getting to a bottleneck. In order to meet the demand of calculation, researchers from all walks of life are seeking a new generation of computing technology and computer architecture to improve the calculation speed and information storage capacity constantly. Therefore, unique advantages and great potential have emerged on optical computers, quantum computers and molecular biological computers one after another.

DNA calculation has great advantages in computing parallelism, information storage capacity and energy consumption which uses DNA molecule as information carrier. DNA computer has higher velocity far ahead of traditional computer because of the DNA molecules have very large-scale parallelism [12]. It is estimated that the total number of operations of DNA computers in a week is equivalent to the total number of operations of all modern computers in the world since their birth. The double helix structure of DNA molecule makes it have powerful storage space. The data shows that DNA solution can store at least 1 trillion bits of modern computer data in per cubic meter. But the energy consumption of DNA is less than one-billionth of the energy which is consumed by modern computers. And DNA resources are very abundant and the price is generally cheap makes a strong economical type. In 1994, professor Adleman worked out the Hamilton problem of seven Vertices successfully with the help of DNA molecule calculation [13]. And then DNA calculation is used to solve a variety of complex decision-making problems, such as: Ouyang et al. proposed a method for solving the maximum clique problem of graphs on the basic of DNA calculation [14]. Lipton et al. [15] made a DNA computing model for solving the satisfiability problems which is presented by transforming the satisfiability problems into a directed Hamilton problem, so DNA calculation methods are widely discussed and used [16–18]. At the same time, it has been proved that the feasibility of using DNA system to construct Turing machine.

Application of molecular biological calculation not only provides a new type method to solve NP complete problems, but also lays a foundation for opening up a new field of cryptography by the new data structure and computational method. This poses a challenge to the traditional cryptosystem based on mathematical difficulties, but also brings new opportunities for information security technology.

DNA cryptography is a new field of cryptography with the research of DNA calculation in recent years, which includes three research directions: DNA encryption, DNA steganography and DNA authentication. The development of

DNA cryptography benefits from the research progress of DNA computing (also known as molecular calculation or biological calculation). On the one hand, cryptosystems are always related to the corresponding computing models more or less. On the other hand, DNA computing used in some biotechnologies which have also been used in DNA cryptosystem. Since Professor Adleman solved NP complete problems by DNA calculation, DNA cryptography derived from DNA computing has made rapid progress both in theory and in experiment. In 1999, Clelland et al. [19] a well-known message called "June 6 Invasion": Normandy in world war II was steganized by encrypting DNA base triplets successfully. Because of the great reliability of DNA base complementarity, the information encrypted in DNA can be amplified by PCR and read by DNA sequencer. In the same year, Gehani et al. [20] proposed two types of methods one-time cryptographic schemes includes substitution method and XOR method using DNA sequences as carriers. In 2000 Leier realized two cryptographic schemes in information hiding using DNA binary sequence to encrypt information [21]. In 2003a coding system based on DNA molecular sequence was constructed by Chen et al. [22]. Lu proposed a symmetric encryption system-DNASC which is designed by combining modern genetic engineering technology with cryptography technology [23]. In 2013, Le et al. realized a particle array encryption model by combining DNA microparticle technology with thermal shrinkage [24]. In 2017, Zhang designed a DNA encrypting algorithm by using DNA chip and polymerase chain reaction (PCR). The algorithm is mainly based on two points: it is difficult to decipher unknown information without knowing the correct primers and probes and break probes when DNA sequencing is in difficulty [25]. In 2018, Zhang presented an encryption scheme based on DNA chip [26].

However, it is difficult to complete the corresponding experiments in general laboratories that mentioned above DNA encryption is based on DNA biological operation mostly which requires rigorous experimental environment and expensive experimental equipments. So many scholars just use the idea of DNA calculation and combine it with traditional encryption methods to encrypt information, which is the mainstream pseudo-DNA encryption algorithm at this stage.

Usually, the image or text is converted into a DNA sequence by means of DNA coding, and then the converted DNA sequence is transformed by the idea of DNA operation, that is, DNA addition, DNA complementation, etc., to complete the information diffusion operation. However, the image encryption method using purely DNA coding is not flexible, so the idea of DNA coding is combined with some mature image encryption methods. In 2009, Ning et al. [27] proposed a pseudo-DNA encryption method, which uses the basic idea of the central principle of biomolecular to realize the encryption of text information. In 2017, we proposed an image encryption algorithm based on chaotic systems and DNA sequence operations [28].

The algorithm first uses Keccak to calculate the hash value of a given DNA sequence as the initial value of the chaotic map, and generates a chaotic map index to scramble the pixel position of the image, and then combines the butterfly network to scramble the bit to achieve bit level scrambling [29,30]. The dynamic security of the encryption is further improved by performing dynamic DNA coding on the image and performing an exclusive OR operation with the generated hash value to implement pixel replacement [31,32]. To this end, both symmetric encryption and asymmetric encryption are implemented using DNA cryptography to improve the security and integrity of bidding data [33].

4   Security Analysis

The electronic bidding system model designed in this paper has strong security. Firstly, the sender package the ciphertext, signature messages and encrypted initial key and send it to the receiver. The symmetric key is the result of the sender's public key operation. The symmetric key participating in the encryption will not be the same because of the original data sent by the sender are different. The whole process not only guarantees the confidentiality and integrity of the original data, but also ensures the randomness of the symmetric key due to the association between the plaintext and the symmetric key, which can effectively resist the explicit text attack, and reduces the danger of the symmetric key being cracked. Therefore, compared with the traditional data encryption and digital signature methods, this method improves the security [30]. Secondly, due to the ultra-large-scale parallelism inherent in DNA molecules, ultra-low energy consumption and ultra-high storage density, DNA-based encryption algorithms have unique advantages comparing with traditional cryptographic algorithms. DNA encryption has two layers of security. The first layer of security comes from biotechnology and is the most important security protection in the program; the second layer is the security layer from the computational problem. The massive parallelism of DNA molecules is difficult for attackers to break [34].

5   Conclusion

Based on the theory of symmetric and asymmetric encryption algorithms, this paper designs and implements a data encryption and digital signature method, and applies it to the electronic bidding system. The symmetric encryption and digital signature in the scheme are all realized by DNA cryptography, which effectively ensures the confidentiality and integrity of data in electronic bidding system, and has high practical application value.

Acknowledgments. The work is supported by the National Natural Science Foundation of China (Grant Nos. 61572446, 61602424, and U1804262).

References

1. Zhang, W.P.: Information security risk analysis in the implementation of electronic bidding. Mod. Bus. **5**(9), 261–262 (2010)
2. Liao, B.X.: Application of CA multiple encryption technology in construction engineering electronic bidding platform. Low-Carbon World **6**(32), 252–253 (2016)

3. Zhang, G.J., Zhang, J.F.: The application of electronic signature technology in online bidding system. J. Changzhou Vocat. Coll. Inf. Technol. **10**(4), 22–24 (2011)
4. Xu, L.X., Wu, X.L., Zhang, X.C.: Application of encryption and digital signature based on online bidding. Comput. Eng. Des. **26**(6), 1431–1433 (2005)
5. Sun, X.: Analyses PKI technology in the application of the electronic bidding. China New Commun. **3**(10), 17–18 (2012)
6. Montenegro, J.A., Lopez, J.: A practical solution for sealed bid and multi currency auctions. Comput. Secur. **45**, 186–198 (2014)
7. Jin, C., Zhao, J.: Certificateless aggregate deniable authentication protocol for ad hoc networks. Int. J. Electron. Secur. Digit. Forensics **10**(2), 168–187 (2018)
8. Xiao, G.Z., Lu, M.X.: DNA computation and DNA coding. J. Eng. Math. **23**(1), 1–6 (2006)
9. Ben Slimane, N., Aouf, N., Bouallegue, K., Machhout, M.: A novel chaotic image cryptosystem based on DNA sequence operations and single neuron model. Multimed. Tools Appl. **77**(23), 30993–31019 (2018). https://doi.org/10.1007/s11042-018-6145-8
10. Enayatifar, R., Guimaraes, F.G., Siarry, P., et al.: Index-based permutation-diffusion in multiple-image encryption using DNA sequence. Opt. Lasers Eng. **115**, 131–140 (2019)
11. Xia, S.Y., Xu, D.: Security requirements and solutions of online bidding system. Netw. Secur. Technol. Appl. (7), 34–38 (2003)
12. Elbaz, J., Lioubashevski, O., Wang, F., et al.: DNA computing circuits using libraries of DNAzyme subunits. Nat. Nanotechnol. **5**(6), 417 (2010)
13. Adleman, L.M.: Molecular computation of solutions to combinatorial problems. Science **266**, 1021–1024 (1994)
14. Ouyang, Q., Kaplan, P., Liu, S., et al.: DNA solution of the maximal clique problem. Science **278**(5337), 446–449 (1997)
15. Lipton, R.: Using DNA to solve NP-complete problems. Science **268**(4), 542–545 (1995)
16. Braich, R.S., Chelyapov, N., Johnson, C., et al.: Solution of a 20-variable 3-SAT problem on a DNA computer. Science **296**(5567), 499–502 (2002)
17. Gifford, D.K.: On the path to computation with DNA. Science **266**(5187), 993–994 (1994)
18. Sakamoto, K., Gouzu, H., Komiya, K., et al.: Molecular computation by DNA hairpin formation. Science **288**(5469), 1223–1226 (2000)
19. Clelland, C.T., Risca, V., Bancroft, C.: Hiding messages in DNA microdots. Nature **399**(6736), 533 (1999)
20. Gehani, A., Labean, T., Reif, J.: DNA based cryptography. In: Proceedings of the 5th Annual DIMACS Meeting on DNA Based Computers (DNA 5). MIT, Cambridge (1999)
21. Leier, A., Richter, C., Banzhaf, W., et al.: Cryptography with DNA binary strands. Biosystems **57**(1), 13 (2000)
22. Chen, J.: A DNA based biomolecular cryptography design. In: Proceedings of the 2003 International Symposium on Circuits and Systems, vol. 3, p. 822. IEEE, Piscataway (2003)
23. Lu, M.X., Lai, X.J., et al.: Symmetric encryption based on DNA technology. Sci. China **37**(2), 175–182 (2007)
24. Le, G.G.C., Blum, L.J., Marquette, C.A.: Shrinking hydrogel-DNA spots generates 3D microdots arrays. Macromol. Biosci. **13**(2), 227 (2013)

25. Zhang, Y., Wang, Z., Wang, Z., et al.: A DNA based encryption method based on two biological axioms of DNA chip and Polymerase Chain Reaction (PCR) amplification techniques. Chem. CA Eur. J. **23**(54), 13387–13403 (2017)
26. Zhang, X.C., et al.: A visual cryptography scheme-based DNA microarrays. Int. J. Perform. Eng. **14**(2), 334–340 (2018)
27. Ning, K.: A pseudo DNA cryptography method. ArXiv preprint arXiv:0903.2693 (2009)
28. Zhang, X., Han, F., Niu, Y.: Chaotic image encryption algorithm based on bit permutation and dynamic DNA encoding. Comput. Intell. Neurosci. **2017**, 11 (2017)
29. Lai, X.J., Lu, M.X., Qin, L., et al.: Asymmetric encryption and signature method with DNA technology. Sci. China Ser. F (Inf. Sci.) **53**(3), 506–514 (2010)
30. Dagadu, J.C., Li, J.-P., Aboagye, E.O.: Medical image encryption based on hybrid chaotic DNA diffusion. Wirel. Pers. Commun. **108**(1), 591–612 (2019). https://doi.org/10.1007/s11277-019-06420-z
31. Feng, W., He, Y.G.: Cryptanalysis and improvement of the hyper-chaotic image encryption scheme based on DNA encoding and scrambling. IEEE Photonics J. **10**(6), 1–15 (2019)
32. Zhang, X., Wang, X.: Multiple image encryption algorithm based on DNA encoding and chaotic system. Multimed. Tools Appl. **78**(6), 7841–7869 (2019)
33. Fu, X.Q., Liu, B.C., Xie, Y.Y., et al.: Image encryption-then-transmission using DNA encryption algorithm and the double chaos. IEEE Photonics J. **PP**(99), 1–1 (2018)
34. Liu, P., Zhang, T., Li, X.: A new color image encryption algorithm based on DNA and spatial chaotic map. Multimed. Tools Appl. **78**(11), 14823–14835 (2018). https://doi.org/10.1007/s11042-018-6758-y

Neural Networks and Artificial Intelligence

Optimal-Operation Model and Optimization Method for Hybrid Energy System on Large Ship

Xi Chen[✉] and Qinqi Wei

China Ship Development and Design Centre, Wuhan, China
cx040504@yeah.net

Abstract. For the large ocean-going ship installed with hybrid energy system (HES), the ship's electric power is supplied by the cooperation of onboard PV system, battery, diesel and the onshore-power. In this study, a novel optimal-operation model of the maritime HES is developed, and the evolutionary optimization method of the aforementioned model is also presented. The proposed model and optimization method is tested by numerical experiments when the ship stops in port. Experimental results show that the proposed methodology can optimally dispatch the power flows of the maritime HES, and obtains significant electricity cost savings.

Keywords: Hybrid energy system · Intelligent ship · Swarm intelligence · Evolutionary computing

1 Introduction

Nowadays, the green energy techniques are widely developed for energy conservation and emission reduction. In the maritime transportation, the concept of green ship and the corresponding control and communications technologies attract worldwide attentions in recent years [1–4]. For the large ocean-going ship, the photovoltaic system (PV), battery banks, and the fuel cells can be introduced for reducing the fuel consumption. In addition, when the ship stops in port, the onshore-power can be also connected with the ship's power-grid and used to supply power for the electric loads of the ship. As a result, the typical structure of a maritime hybrid energy system (HES) mainly consists of the PV system, battery banks, onboard diesel and the onshore-power.

Due to the large amount of exhaust emissions, the ports local air quality always suffers from serious pollutions. In order to protect the port's local environment, a lot of strict regulations are gradually published, e.g., restricting the use of onboard diesel when the ship stops in port [5]. That is to say, the ship should use the paid onshore-power service instead of turning on the onborad diesel in the entire hoteling period, or use the onboard green energy system.

© Springer Nature Singapore Pte Ltd. 2020
L. Pan et al. (Eds.): BIC-TA 2019, CCIS 1160, pp. 233–242, 2020.
https://doi.org/10.1007/978-981-15-3415-7_19

As a result, in order to make full use of the renewable PV power, minimize the electricity cost and also satisfy the regulations from the port, the optimal-operation model of maritime HES and its optimization algorithm is worthy to be studied. In this study, the optimal-operation of maritime HES is modelled as a large-scale optimization problem. Then, the decentralizing and coevolving differential evolution (DCDE) algorithm is employed for solving this problem. Experimental results show that the evaluated HES can optimally dispatch the power flows of the maritime HES, and obtains significant electricity cost savings.

2   Related Works

In maritime transportation, the fuel consumption and greenhouse gas emission per year are extremely large, and the utilization of clean energy is urgently required for energy-saving and environment protection. In recent years, the concept of green ship attracts the attentions of world-wide scholars, and different kinds of green techniques, e.g., the photovoltaic system, have been applied in ships and obtained good performance. For example, the Solar Sailor with solar/wind HES was manufactured in Australia in 2000. In 2012, a solar energy system with 768 panels was installed on the Emerald Ace ship in Japan. However, the maritime PV system differs from the land-based system in many aspects, so a lot of open issues with respect to the operational safety and efficiency of maritime HES is still worthy to be studied in detail. Tang et al. studied the topological structure of large-scale PV system, and the maximum power point tracking (MPPT) method based on evolutionary algorithm is also presented [6]. In a recent study of Tang et al., the optimal-operation of HES is modelled as an optimization problem subject to a number of constraints, and the optimal control method and model predictive control method are developed and compared to solve the aforementioned problem [7]. Lee et al. proposed a hybrid photovoltaic/diesel ship, in which the distributed power system can be connected to the smart grid and micro-grid [2].

In the maritime HES, the battery bank is required for power storage and voltage stabilization. In addition, as restricted by the port's regulations, some hoteling loads need to be satisfied by the paid onshore-power service. As a result, the power flow dispatching is needed for optimizing the output of PV array, battery banks, onboard diesels, and also the onshore-power, in order to maximize the cost-saving when the ship stops in port. This study develops a novel optimal-operation model of the maritime HES, and an evolutionary algorithm called the DCDE is employed for optimizing the aforementioned model. In addition, some simulation experiments and analysis are also presented in this study.

3   Optimal-Operation Model of Maritime HES

3.1   Electricity Cost Model of HES

Assume the maritime HES consists of the PV system, battery bank, onboard diesel and the onshore-power. The schematic of the evaluated HES is shown in Fig. 1, in which the directional power flows are denoted by arrows.

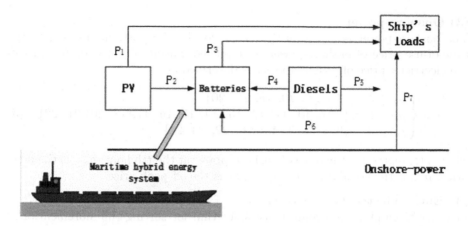

Fig. 1. Schematic of the PV/battery/diesel/onshore-power maritime HES

(1) Battery bank

As shown in Fig. 1, the batteries can be charged by the PV output, onboard diesel, and the on-shore power. The output of batteries can be only employed for feeding the ship's electricity demand. As the batteries are always in charging and discharging processes, the battery's state of charge (SOC) can be described as

$$\begin{cases} S_{bat}(t) = S_{bat}(0) + \eta_{cha} \sum_{i=1}^{t} P_{cha}(t) - \frac{1}{\eta_{dis}} \sum_{i=1}^{t} P_3(i) \\ P_{cha}(t) = P_2(t) + P_4(t) + P_6(t) \end{cases} \tag{1}$$

where $S_{bat}(t)$ denotes the SOC of battery bank at the tth hour; η_{cha} and η_{dis} are constants within the interval $[0, 1]$; $P_{cha}(t)$, which is equal to the sum of $P_1(t)$, $P_4(t)$ and $P_6(t)$, denotes the charging power flows at the tth hour; $P_3(t)$ denotes the corresponding discharging power flow.

(2) Onboard diesel

In the traditional power system of a ship, diesel is the unique power supplier for feeding all the ship's electric loads. The diesel generates electric power by burning the fuel, and the efficiency of a diesel is related to the relationship of the rated power and real output power. In this study, the cost of onboard diesel is modelled as

$$\begin{cases} V_{die}(t) = n_a \cdot P_{rat} + n_b \cdot P_{die}(t) \\ C_{dis}(t) = \frac{P_{fue} \cdot V_{die}(t)}{P_{die}(t)} \\ P_{die}(t) = P_4(t) + P_5(t) \end{cases} \tag{2}$$

where P_{fue} denotes the fuel price per liter; P_{rat} and P_{die} denote the rated power and output power of the diesel generator, respectively; n_a and n_b are the fuel intercept coefficient and the fuel slop coefficient [8]; C_{dis} denotes the cost of diesel-generated power per kWh.

(3) Onshore-power

The onshore-power is a paid service provided by the port. In this study, the time-of-use price of onshore-power is employed. Specifically, the daily time-of-use electricity price of onshore-power is described as

$$\rho(t) = \begin{cases} \rho_{pea}, & t = \{7,\ 8,\ 9,\ 10,\ 18,\ 19,\ 20\} \\ \rho_{sta}, & t = \{6,\ 7,\ 10,\ 11,\ 12,\ 13,\ 14,\ 15,\ 16,\ 17,\ 18,\ 20,\ 21,\ 22\} \\ \rho_{off}, & t = \{0,\ 1,\ 2,\ 3,\ 4,\ 5,\ 6,\ 22,\ 23,\ 24\} \end{cases} \quad (3)$$

where $\rho(t)$ represents the price of onshore-power at the tth hour; ρ_{pea} is the peak price, ρ_{sta} is the standard price, and ρ_{off} is the off-peak price.

(4) Entire electricity cost of HES

In order to employ the evolutionary algorithm for solving the aforementioned problem, the optimizing vector is required. In this study, the optimizing vector is modelled as the power flows of each distributed source in the dispatching period, that is to say, for a dispatching period with N hours, the optimizing vector is modelled as

$$\begin{aligned} \boldsymbol{x} = [&P_2(1), ..., P_2(N), P_3(1), ..., P_3(N), P_4(1), \\ &..., P_4(N), P_5(1), ..., P_5(N), P_6(1), ..., P_6(N)] \end{aligned} \quad (4)$$

After the optimal power flow x^* is worked out, the power flows $P_1(t)$ and $P_7(t)$ $(t = 1, 2, ..., N)$ can be obtained by using the follow equations

$$\begin{cases} P_1(t) = P_{pv}(t) - P_2(t) & (t = 1, 2, \dots, N) \\ P_7(t) = P_L(t) - P_2(t) - P_3(t) - P_5(t) & (t = 1, 2, \dots, N) \end{cases} \quad (5)$$

where $P_{pv}(t)$ denotes the PV output at the tth hour; $P_L(t)$ denotes the ship's electric load-demand at the tth hour.

As discussed above, optimal operation of the evaluated HES aims to minimize the electricity cost when the ship stops in port. The total cost (C_{tot}) contains three parts: first of all, the fuel consumption cost (C_{fue}); secondly, the battery's charging and discharging costs (C_{bat}); finally, the onshore-power service cost (C_{ser}). For a dispatching period with N hours, C_{tot} can be expressed as

$$C_{tot} = C_{fue} + C_{bat} + C_{ser} \quad (6)$$

$$\begin{cases} C_{fue} = \sum_{t=1}^{N} \frac{P_{fue} \cdot \{n_a \cdot P_{rat} + n_b \cdot [P_4(t) + P_5(t)]\}}{P_4(t) + P_5(t)} \\ C_{bat} = \sum_{t=1}^{N} w_a \cdot P_3(t) + N \cdot w_b \\ C_{ser} = \sum_{t=1}^{N} \rho(t) \cdot [P_6(t) + P_7(t)] \end{cases} \quad (7)$$

where w_a is the coefficient of battery wearing cost and w_b is the hourly wearing cost of other components.

3.2 Penalty Functions of the Optimal-Operation Model

For optimally dispatching the power flows, many constraints must be satisfied, e.g., SOC boundary, diesel output, port's regulation, power flow boundary and so on. In this study, in order to employ the evolutionary algorithms, all the constraints are described as penalty functions.

(1) SOC boundary constraint: Due to the limited capacity, SOC of the battery bank should be constrained within a certain interval. The penalty function of SOC lower bound constraint is formulized as

$$P_{S-l}(t) = \begin{cases} |P_3(t) - P_1(t) - P_4(t) - P_6(t)| + Const & if \ S_b(t) < S_l \\ 0 & else \end{cases} \quad (8)$$

where S_l denote the lower bound of SOC; Const is a positive constant.

The penalty function of upper bound constraint is formulized as

$$P_{S-u}(t) = \begin{cases} |P_1(t) + P_4(t) + P_6(t) - P_3(t)| + Const & if \ S_b(t) > S_u \\ 0 & else \end{cases} \quad (9)$$

where S_u denote the upper bound of SOC.

(2) Diesel output constraint: The output of onboard diesel P_{die} should not exceed the diesel's rated power P_{rat}. The corresponding penalty function is formulized as

$$P_{die}(t) = \begin{cases} P_4(t) + P_5(t) + Const & if \ P_4(t) + P_5(t) > P_{rat} \\ 0 & else \end{cases} \quad (10)$$

(3) Port's regulation constraint: As restricted by the port's environment protecting regulations, the use of onboard diesel is strictly constrained, that is, the power generated by the diesel should less than a certain value. The penalty function of port's regulation constraint is formulized as

$$P_{por} = \begin{cases} \{\sum_{t=1}^{N} [P_4(t) + P_5(t)]\} + Const & if \ \sum_{t=1}^{N} [P_4(t) + P_5(t)] \leq P_{con} \\ 0 & else \end{cases} \quad (11)$$

where P_{con} denotes the constrained value of the power generated by onboard diesel.

(4) Power flow boundary constraint: The power flows should be non-negative and less than the maximum allowable value. The penalty function of power flow boundary constraint is formulized as

$$P_{bou}(t) = P_2(t) + P_3(t) + P_5(t) + Const \quad (12)$$

3.3 HES Optimal-Operation Model and the Optimization Algorithm

As discussed above, the optimal-operation model of the evaluated HES is formulized as

$$\min f(x) = C_{tot} + P_{tot} \quad (13)$$

where C_{tot} denotes the total cost in Eq. (6); P_{tot} denotes the entire penalty function, which is formulized as

$$P_{tot} = \sum_{t=1}^{N} [P_{S-l}(t) + P_{S-u}(t) + P_{die}(t) + P_{bou}(t)] + P_{por} \tag{14}$$

In this study, the DCDE algorithm which has been proved to be effective for solving 1000-dimensional problem [9], is employed to solve the aforementioned dispatching problem. The individuals in DCDE are represented by n-dimensional vectors x_i, $(i = 1, 2, NP; n = 6 \times N)$, where NP denotes the population size. As the traditional DE algorithm, the DCDE contains three basic operations, that is, the mutation, crossover and selection. However, different from the traditional DE, the mutation operation in DCDE is improved as

$$v_i = x_i + F \cdot (SP\text{-}best\text{-}ring_i - x_i) + F \cdot (x_{r1} - x_{rsp}) \tag{15}$$

where, v_i denotes the ith newly created mutant individual after the mutation operation. F is a positive coefficient used to control the differential variation. r_1 is a distinct integer uniformly randomized from the set $1, 2, ..., NP$. The $SP\text{-}best\text{-}ring_i$ and x_{rsp} are two individuals generated with some special mechanism proposed in the DCDE. In addition, the famous cooperative coevolution (CC) framework is also employed in the DCDE. The detailed principles of DCDE can be found in its original papers [9].

4   Simulation Results and Analysis

4.1   Parameters Settings

In this section, the DCDE algorithm is employed for optimizing the optimal-operation model of maritime HES, and some other state-of-the-art evolutionary algorithms are also employed for comparison. The sizing of PV system and battery bank is set to 210 kW and 432 kWh, respectively. The other parameters settings of the evaluated HES are listed in Table 1.

The ship's electric load-demand (P_L) and PV output (P_{pv}) used for dispatching are based on the forecasting values. The profiles of daily P_L and P_{pv} employed in the following experiments are listed in Table 2.

In the following experiments, the performance of DCDE is compared with some state-of-the-art evolutionary algorithms, including the CCPSO2 [10], SaDE [11], JADE [12] and AMCCDE [13]. The population size of all the compared algorithms are set to 50, and the maximum number of fitness evaluations is set to 2×10^7. The dynamic group sizes of DCDE and CCPSO2 are set as $S = 2, 4, 6$. The other parameters settings of the compared algorithms are the same as their original papers. The dispatching period is set to 48 h.

Table 1. Parameters settings of the evaluated HES

Parameter	Value	Parameter	Value
$S_{bat}(0)$	240 kWh	η_{cha}	85%
P_{rat}	350 kW	η_{dis}	100%
S_u	432 kWh	ρ_{pea}	0.31538$/kWh
S_l	129.6 kWh	ρ_{sta}	0.15948$/kWh
W_a	0.001	ρ_{off}	0.06558$/kWh
W_b	0.002	ρ_{fue}	0.78571$/kWh
n_a	0.01609 L/kWh	n_b	0.2486 L/kWh

Table 2. Profiles of daily load demand and PV output

Time	P_L(kW)	P_{pv}(kW)	Time	P_L(kW)	P_{pv}(kW)
0:30	72	0	12:30	103.2	159.34
1:30	72	0	13:30	103.2	139.5
2:30	72	0	14:30	103.2	110.36
3:30	72	0	15:30	103.2	72.23
4:30	72	0	16:30	103.2	34.41
5:30	93.6	0	17:30	86.4	4.03
6:30	93.6	5.89	18:30	110.88	0
7:30	79.2	37.51	19:30	182.88	0
8:30	64.8	82.46	20:30	110.88	0
9:30	156	122.45	21:30	110.88	0
10:30	156	151.59	22:30	110.88	0
11:30	103.2	162.75	23:30	64.8	0

Table 3. Comparison of total costs for different algorithms

DCDE ($)	CCPSO2 ($)	SaDE ($)	JADE ($)	AMCCDE ($)
560.5085	589.6587	622.5862	582.2658	566.2153

4.2   Experimental Results and Analysis

For the 48 h dispatching period, output of the compared algorithms are listed in Table 3. As shown in the table, DCDE obtains the best performance of 560.5085 $, and AMCCDE is the second-best performer (obtains 566.2153 $). Then, the JADE obtains the electricity cost of 582.2658 $, which is followed by 589.6587 $ obtained by the CCPSO2. Finally, the SaDE obtains the worst performance, which should pay 622.5862 $ to feed the entire hoteling loads.

The optimal power flows obtained by DCDE are plotted in Fig. 2 in detail. As shown in the figures, the onboard diesel (P_5) is rarely turned on in the daytime

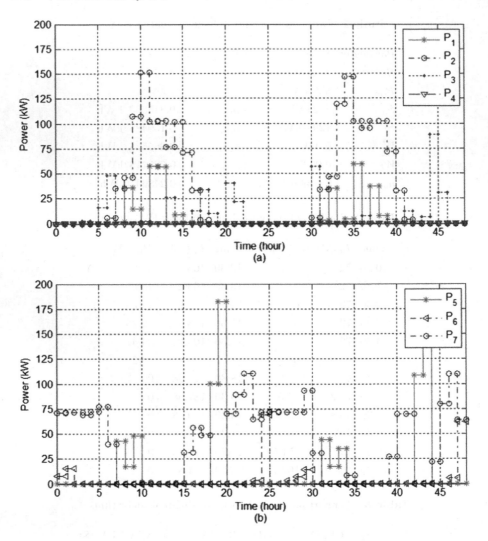

Fig. 2. Optimal power flows obtained by DCDE

(from $t = 10$ to $t = 15$ for the first day, and from $t = 34$ to $t = 39$ for the second day), so the ship's load demand is mostly fed by the PV power (P_2) during this period. In addition, the surplus PV power is also used to charge the battery in the daytime (P_1). When the PV output is not enough to satisfy all the electricity demand in the morning and afternoon ($t = 7$ to $t = 10$ and $t = 18$ to $t = 20$, $t = 31$ to $t = 34$ and $t = 42$ to $t = 44$), the onboard diesel will be turned on. Then, in the night ($t = 1$ to $t = 7$, $t = 17$ to $t = 31$, and $t = 42$ to $t = 48$), the ship's load is mostly fed by onshore-power (P_7) and the onboard diesel (P_5), that is because the PV system cannot work during these periods.

5   Conclusion

Due to the increasingly stringent emission restriction of maritime transportation, the green energy systems like PV array, battery banks, and also the onshore-power are widely introduced into large ships, so the optimal energy management of maritime HES is of great importance for ship's operational safety and efficiency. This paper studies a novel optimal-operation model of maritime HES, in which the power flow dispatching is modelled as an optimization problem with different penalty functions. Then, the DCDE algorithm is employed for optimally dispatching all the power flows of the HES, and some state-of-the-art evolutionary algorithms are also employed for comparison. Experimental results have shown that the evolutionary-algorithm-based dispatching model can be well optimized, and the DCDE obtains the best performance compared with the other evolutionary algorithms.

References

1. Zhang, H.: Towards global green shipping: the development of international regulations on reduction of GHG emissions from ships. Int. Environ. Agreem. Polit. Law Econ. **16**(4), 561–577 (2014). https://doi.org/10.1007/s10784-014-9270-5
2. Lee, K.J., Shin, D., Yoo, D.W., Choi, H.K., Kim, H.J.: Hybrid photovoltaic/diesel green ship operating in standalone and grid-connected mode – experimental investigation. Energy **49**, 475–483 (2013)
3. Kong, Z., Yang, S., Wang, D., Hanzo, L.: Robust beamforming and jamming for enhancing the physical layer security of full duplex radios. IEEE Trans. Inf. Forensics Secur. **14**(12), 3151–3159 (2019)
4. Tang, R., Wu, Z., Fang, Y.: Configuration of marine photovoltaic system and its MPPT using model predictive control. Sol. Energy **158**, 995–1005 (2017)
5. Vaishnav, P., Fischbeck, P.S., Morgan, M.G., Corbett, J.J.: Shore power for vessels calling at US ports: benefits and costs. Environ. Sci. Technol. **50**(3), 1102 1110 (2016)
6. Tang, R., Wu, Z., Fang, Y.: Maximum power point tracking of large-scale photovoltaic array. Sol. Energy **2016**(134), 503–514 (2016)
7. Tang, R., Wu, Z., Li, X.: Optimal operation of photovoltaic/battery/diesel/cold-ironing hybrid energy system for maritime application. Energy **2018**(162), 697–714 (2018)
8. Ramli, M.A.M., Hiendro, A., Twaha, S.: Economic analysis of PV/diesel hybrid system with flywheel energy storage. Renew. Energy **78**, 398–405 (2015)
9. Tang, R.: Decentralizing and coevolving differential evolution for large-scale global optimization problems. Appl. Intell. **47**(4), 1208–1223 (2017). https://doi.org/10.1007/s10489-017-0953-9
10. Li, X.D., Yao, X.: Cooperatively coevolving particle swarms for large-scale optimization. IEEE Trans. Evol. Comput. **16**(2), 210–224 (2012)

11. Qin, A.K., Suganthan, P.N.: Self-adaptive differential evolution algorithm for numerical optimization. In: Proceedings of the IEEE Congress on Evolutionary Computation, pp. 1785–1791 (2005)
12. Zhang, J.Q., Sanderson, A.C.: JADE: adaptive differential evolution with optional external archive. IEEE Trans. Evol. Comput. **13**(5), 945–958 (2009)
13. Tang, R., Li, X.: Adaptive multi-context cooperatively coevolving in differential evolution. Appl. Intell. **48**(9), 2719–2729 (2017). https://doi.org/10.1007/s10489-017-1113-y

Dual-Graph Regularized Sparse Low-Rank Matrix Recovery for Tag Refinement

Dengdi Sun[1], Yuanyuan Bao[1], Meiling Ge[1], Zhuanlian Ding[2(✉)], and Bin Luo[1]

[1] Key Lab of Intelligent Computing and Signal Processing
of Ministry of Education, School of Computer Science and Technology,
Anhui University, Hefei 230601, China
[2] School of Internet, Anhui University, Hefei 230039, China
dingzhuanlian@163.com

Abstract. In recent years, extremely large amounts of images with manual tags are easily available in many social websites such as Twitter, Flickr, and Instagram. However, these user-provided tags are often imprecise and incomplete, which inevitably limits the performances of image retrieval and other related applications. To this end, tag refitment technology aims at improving the quality of images tags automatically, and has been a fundamental challenge in Internet era. In this paper, we propose a novel dual-graph regularized sparse low-rank matrix recovery method, referred to as DGSLR briefly, to infer and improve the manual tags. Specifically, our DGSLR model first suppress the sparse noisy tags by ℓ_1 norm, and decompose simultaneously the residual low-rank matrix to learning the low-dimensional vector representations of images and tags. Moreover, the visual similarities and the tag pairwise correlation are fully exploited to smooth the decomposition results by using the dual lapalcian graph regularization terms. For optimization, an improved alternative iteration strategy is designed to solve the resulting objective function. Extensive experiments on the tasks of tag refinement demonstrate the superior performance of the proposed algorithms over several representative methods, especially when the manual tag data is highly noisy and sparse.

Keywords: Tag refinement · Matrix recovery · Graph regularization

1 Introduction

With the emergency and rapid development of novel web social media such as Twitter, Flickr, and Instagram, people are allowed to upload images freely on the websites and annotate the images they are interested in [17–19], leading to the volume of images and tags in the websites are growing explosively. However, the quality of tags provided by users is far from satisfactory, which limits the performance seriously of following related works such as image browse, manage and retrieve [4,5,15]. Therefore, image tag refinement currently has been a

© Springer Nature Singapore Pte Ltd. 2020
L. Pan et al. (Eds.): BIC-TA 2019, CCIS 1160, pp. 243–258, 2020.
https://doi.org/10.1007/978-981-15-3415-7_20

fundamental and active research direction in computer vision and data mining [2,11,13]. It aims at inferring and improving the quality of images tags automatically, and has received increasing attentions due to its wide range of applications in Internet and social media.

Generally, in websites, the noisy tags can be divided into three types: wrong tags, missing tags, and imprecise tags. In this paper, to address the noisy annotations, we propose a novel image tag refinement approach aiming to improve the quality of tags provided by Internet users. Our approach is motivated by the following observations from real-world social images: First, visually similar images often have similar tags. Image visual similarities play a critical role in tag annotation and other related works, this observation has been demonstrated in several previous works [6,25,27]. Secondly, tags are always related to each other in terms of semantic or content correlations, such as tag 'fish' often appear with tag 'animal' (semantic) and 'water' (content). Thus, if a tag is annotated to one image, it's highly correlated tags are usually annotated to that image. Based on this observation, finding the latent correlation between tags is also an important factor for tag refinement. Finally, although user-provided tags are imprecise and incomplete, to a certain extent, they are still reliable because users often have common knowledge. Studies [1] have shown that user-provided tags have about 50 precision rate, so these tags can reveal the basic content of images. Thus, the residue matrix between original tag matrix and the refined result should be very sparse, which is called as error sparsity.

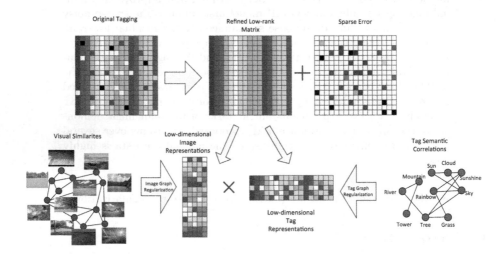

Fig. 1. The overview of the proposed tag refinement approach

Based on above observations, we solve image tag refinement problem from an optimization perspective which takes consideration of visual similarities and tag correlations simultaneously. In addition, considering the sparsity of the noise tags, we will also use the ℓ_1 norm-based sparse coding to suppress noises and minimize the sparse residue matrix. Sparse coding is an important brain-inspired cognitive technology that has been validated in the human visual system and widely used in computer vision. With these three factors incorporated, a dual-graph regularized sparse low-rank matrix recovery method, named as DGSLR is proposed in this paper, which is illustrated in Fig. 1. Here, the original image-tag matrix is split into a low rank component L and a sparse residue component E, and we further decompose A into two low-dimensional matrices for representing the latent features of images and tags. Two graphs that characterize image visual similarities and tag correlations respectively are incorporated to regularize the low-dimensional representations. Finally, re-multiply the two representation matrices and superimpose the sparse residual matrix as the recovery result. To solve the proposed model, we also design an efficient algorithm to speed up the optimization, which facilitates the proposed model can work on large-scale dataset.

2   Related Works

Numerous algorithms have been proposed for social image tag refinement, they can roughly be grouped into two major categories, model-free and model-based approaches. As a pioneer work of model-free work, Jin et al. [12] estimate the semantic correlations among the annotated tags and remove weakly correlated ones by utilizing WordNet. Later, content-based image refinement is proposed by Wang [24]. In this method, existing image annotation method is firstly employed to obtain candidate tags. Then, the candidate annotations are re-ranked and only the top ones are reserved as the final annotations. Compared with method firstly proposed in [24], Cheng [25] proposed a novel approach to automatically refine the image tags incorporating tag semantic and visual similarity, and re-rank the tags using a fast random walk with restart to identify the top tags reserved as the final results. Recently, Fu [6] proposed that a visual concept should be presented by two components: generic view and a specific view, so they first find candidate tags for an input image, and then guided this view-dependent concept representation to eliminate irrelevant tags. Although these model-free methods have made some progress, they are not suitable for large scale datasets.

For model-based methods, [20] and [22] are both based on nonnegative tensor factorization to learning relationship among users, images and tags. [23] addressed the image tag refinement based on visual synset which is an organization of images. Specifically, linear SVM is utilized to predict the visual synset membership, and a weighted voting rule is used to construct a ranked list of predicted annotations. [14] is one of the representative work which is proposed by Liu. They formulated tag refinement as a multiple graph-based multi-label learning problem, and proposed the concept of tag-specific graph to calculate image visual similarities. Generally, the number of tags collected from open websites is extremely large and the user-provided tags are noisy, which is a big challenge for this method.

3 Methodology

3.1 Sparse Low-Rank Matrix Recovery

Suppose we have m images and n tags, $X \in \mathbb{R}^{n \times m}$ denotes a image-tag matrix. If j-th tag is related to the visual content of i-th image, $X_{ij} = 1$, otherwise $X_{ij} = 0$. Considering the possible corruption and sparse noise in tagging matrix X, it is necessary to investigate how to accurately recover the desired low-rank matrix from a corrupted data. To this end, Wright et al. [26] propose the following convex optimization problem:

$$\min_{A,E} \|E\|_1 + \eta \|A\|_*, s.t. \ rank(A) \leq r, X = A + E, \tag{1}$$

where $\| \cdot \|_*$ denotes the nuclear norm of a matrix (i.e., the sum of its singular values), $\| \cdot \|_1$ denotes ℓ_1-norm, that is the sum of the absolute values of matrix entries, and η is a positive weighting parameter. However, the nuclear norm minimization requires to execute expensive singular value decompositions on the whole data. To mitigate the computational pressure, Theorem 1 provides a bridge between Nuclear norm minimization and bilinear factorization.

Theorem 1. *For any matrix* $A \in \mathbb{R}^{n \times m}$, *the following relationship holds [16]:*

$$\|A\|_* = \min_{U,V} \frac{1}{2}\|U\|_F^2 + \frac{1}{2}\|V\|_F^2, s.t. \ A = UV,$$

If $rank(L) = rmin(m,n)$, *then the minimum solution above is attained at a factor decomposition* $A = UV$, *where* $U \in \mathbb{R}^{n \times r}$ *and* $V \in \mathbb{R}^{r \times m}$. □

In the sequel, applying Theorem 1 on Eq. 1 reads:

$$\min_{U,V} \|X - UV\|_1 + \frac{\eta}{2}(\|U\|_F^2 + \|V\|_F^2). \tag{2}$$

Compared with directly restrict $X = UV + E$, the model 2 inherits the advantage of Eq. 1, which avoids overfitting when r is larger than the intrinsic rank, and $\|X - UV\|$ describe the error sparsity.

3.2 Dual-Graph Regularization

Based on matrix completion strategy, we aim to seek two low-rank latent matrices $U \in \mathbb{R}^{n \times r}$ and $V \in \mathbb{R}^{r \times m}$ as the low-dimensional representations of images and tags respectively, and construct the visual similarity graph $G^U \in \mathbb{R}^{m \times m}$ and tag correlation graph $G^V \in \mathbb{R}^{n \times n}$ to regularize them. For convenience, i-th row vector $u_{i.}, 1 \leq i \leq n$ in U denotes the low-dimensional representation of image i, and j-th column vector $v_{.j}, 1 \leq j \leq m$ denotes the low-dimensional representation of tag j.

Visual Similarity Graph

In order to measure the visual similarity between images, we first extract crucial features for images [7,21]. In our framework, each image is extracted a 428-dimensional feature vector x as the content representation, including 128-dimensional wavelet texture features, 75-dimensional edge distribution histogram features, and 225-dimensional blockwise color moment features generated from 5-by-5 fixed partition on each image [22,27]. Then the image similarity matrix W^U is defined as the following formulation:

$$W_{ij}^U = \begin{cases} e^{-||x_i - x_j||^2 / \delta_I{}^2} & if\ j \in N_k(i)\ or\ i \in N_k(j) \\ 0 & otherwise \end{cases} \tag{3}$$

where $N_k(.)$ presents the index set of the k nearest neighbors of a image, here Euclidean distance is utilized to find the k nearest neighbors for each image, and k is set as $0.001n$ (where n is the size of image dataset), δ_I is set as the average of distance.

Depending on similarity matrix W^U, we can construct an undirect weighted visual similarity graph G^U, and then define the G^U graph regularization as follows,

$$\frac{1}{2} \sum_{ij} ||u_{i\cdot} - u_{j\cdot}||_F^2 W_{ij}^U = tr(U^T L_U U), \tag{4}$$

where $tr(.)$ denotes the trace of a matrix, and $L_U = D^U - W^U$ is the Laplacian of G^U graph, D^U is diagonal degree matrix with $D_{ii}^U = \sum_j W_{ij}^U$. This formulation is a smoothness constraints with image internal information, that is, the larger visual similarity W_{ij}^U between image i and image j, the smaller distance $||u_{i\cdot} - u_{j\cdot}||_F^2$, indicating that these two images should be assigned similar tags.

Tag Correlation Graph

We can also construct the tag correlation graph and laplacian regularization in the same way. Due to the observation of tag correlation mentioned in Sect. 1, we collect intra-relations among tags. If tag i and tag j have highly correlation, they might have high frequency of co-occurrence in the same image. To better utilize the intra-relation between tags, we use the cosine similarity [8,9] to measure correlations. Let z_i and z_j are annotation assignment indication vectors for tag i and tag j (the i-th and j-th columns of original tagging matrix X), thus we define tag similarity matrix W^Y as follows:

$$W_{ij}^Y = \frac{<z_i, z_j>}{||z_i|| ||z_j||} \tag{5}$$

where $<.>$ indicates inner product, $||z_i||$ represents the total number of times of tag i appeared in dataset. Obviously, if tag i and tag j have high co-occurrence, the value of W_{ij}^V is large, otherwise, W_{ij}^V is small.

Depend on correlation matrix W^V, we can construct an undirect weighted tag correlation graph G^U, and then define the G^V graph regularization as follows,

$$\frac{1}{2} \sum_{ij} ||v_{\cdot i} - v_{\cdot j}||_F^2 W_{ij}^V = tr(V L_V V^T), \tag{6}$$

where, $L_V = D^V - W^V$, that is the Laplacian for W^V, and D^V is diagonal degree matrix with $D_{ii}^V = \sum_j W_{ij}^V$.

3.3 Objective Function

By integrating the sparse low-rank matrix recovery (Eq. 2) and Dual-graph regularization (Eqs. 4 and 6) together, we define the unified objective function for our DGSLR model as follows.

$$\min_{U,V} \|X - UV\|_1 + \frac{\eta}{2}(\|U\|_F^2 + \|V\|_F^2) + \lambda_1 tr(U^T L_U U) + \lambda_2 tr(V L_V V^T) \quad (7)$$

where η, λ_1 and λ_2 are positive parameters. In this function, the first term requires the model to approximate the given matrix (bipartite) X and suppress the noisy entries. The second term can be regarded as the Tikhonov regularization which minimizes the norms of both U and V to ensure the low-rank and smoothness. The last two terms are respectively for image graph and tag graph regularization via minimizing the distance between latent feature vectors of two neighboring nodes. These regularization terms prevent overfitting and increase generalization capability.

4 Optimization

In this section, we propose an efficient algorithm to optimize the proposed DGSLR model, Eq. 7 employing Augmented Lagrange Multiplier (ALM) scheme [3]. The algorithm iteratively solves three sub-problems: one is a simplified LASSO defined in Eq. 9 with simple exact solution; the other two are the matrix equation from Eqs. 11 and 15. The augmented Lagrangian function of Eq. 7 is defined as:

$$\begin{aligned} \mathcal{L}_{E,U,V} = \|E\|_1 + <\Lambda, X - UV - E> + \frac{\mu}{2}\|X - UV - E\|_F^2 \\ + \frac{\eta}{2}(\|U\|_F^2 + \|V\|_F^2) + \lambda_1 tr(U^T L_U U) + \lambda_2 tr(V L_V V^T) \end{aligned} \quad (8)$$

Here we introduce an auxiliary variable E to replace $X - UV$ in first term. Accordingly, we use the ALM approach by enforcing an additional constraint $E = X - UV$ using Lagrange multipliers (matrix Λ) and quadratic penalty parameter μ.

4.1 Solving the Subproblems

The key step of the algorithm is solving the two subprograms of Eq. 8 for each set of parameter values of Λ, μ. Fortunately, this can be solved in closed form solutions for E and group of (U, V).

A. E-subproblem

First, when updating E with U and V fixed, we need to solve the following problem:

$$\min_{E} \mathcal{L}(E) = \|E\|_1 + \frac{\mu}{2}\|E - P\|_F^2, \tag{9}$$

where $P = X - UV + \frac{\Lambda}{\mu}$ is a constant matrix independent of E, and the Λ means the Lagrange multiplier. So, this problem has closed form solution:

$$E_{ij}^* = sign(P_{ij}) \cdot max(|P_{ij}| - \frac{1}{\mu}, 0), \tag{10}$$

where, $sign(\cdot)$ means the sign function that extracts the sign of a real number.

B. U-subproblem

In the next, we solve U while fixing E and V. The objective function becomes:

$$\min_{U} \mathcal{L}(U) = \frac{\mu}{2}\|Q - UV\|_F^2 + \lambda_1 tr(U^T \widetilde{L}_U U). \tag{11}$$

where, $\widetilde{L}_U = L_U + \frac{\eta}{2\lambda_1}I$, I is the identity matrix, and the $Q = X - E + \frac{\Lambda}{\mu}$ is a constant matrix independent of U and V as follow. Set derivative of $\mathcal{L}(U)$ with respect to U to 0, we have:

$$\frac{\partial \mathcal{L}}{\partial U} = -\mu Q V^T + \mu U V V^T + 2\lambda_1 \widetilde{L}_U U = 0. \tag{12}$$

The formula of Eq. 12 is exactly same to the matrix equation $AX + XB = C$, which is solved by the known Sylvester equation. Therefore, the Eq. 12 will be formulated as:

$$[I \otimes (2\lambda_1 \widetilde{L}_U) + (\mu V V^T) \otimes]U(:) = \mu Q V^T(:) \tag{13}$$

In the equation we formulated above, $\otimes$ represents the Kronecker tensor product, and $U(:)$ and $\mu Q V^T(:)$ denotes the matrices U and $\mu Q V^T$ as long vectors respectively. In order to avoid errors in Sylvester's equation in the iterative process, for example, if the rank does not meet the condition for the equality, we add disturbance σI ($\sigma \to 0$) to one of the terms. At last, we get the iteration:

$$U \leftarrow Sylvester(2\lambda_1 \widetilde{L}_U, \mu V V^T + \sigma I, \mu Q V^T). \tag{14}$$

C. V-subproblem

Based on the above description, we obtain the equation with respect to V while fixing E and U in the same way,

$$\min_{V} \mathcal{L}(V) = \frac{\mu}{2}\|Q - UV\|_F^2 + \lambda_2 tr(V \widetilde{L}_V V^T), \tag{15}$$

where $\widetilde{L}_V = L_V + \frac{\eta}{2\lambda_2}I$. The derivative of $\mathcal{L}(V)$ with respect to V is:

$$\frac{\partial \mathcal{L}(V)}{\partial V} = -\mu U^T Q + \mu U^T U V + (2\lambda_2 V \widetilde{L}_V) = 0 \tag{16}$$

Likewise, the formula of Eq. 16 matches the Sylvester equation well. So we obtain the close-form solution of V:

$$[I \otimes (\mu U^\mathrm{T} U) + (2\lambda_2 \tilde{L}_V) \otimes] U(:) = \mu U^\mathrm{T} Q(:) \tag{17}$$

where $V(:)$ and $\mu U^\mathrm{T} Q(:)$ denotes the matrices V and $\mu U^\mathrm{T} Q$ as long vectors respectively. As well as the Eq. 17, in order to avoid errors in Sylvester's equation in the iterative process, we add disturbance σI (with very small constant σ) to one of the terms. Then, the formula is:

$$V \leftarrow Sylvester(\mu U^\mathrm{T} U + \sigma I, 2\lambda_2 \tilde{L}_V, \mu U^\mathrm{T} Q) \tag{18}$$

4.2 Updating Parameters

In each iteration of ALM, after obtaining consistent E, U and V, the parameters Λ and μ are updated as the following

$$\Lambda \Leftarrow \Lambda + \mu(X - D - E), \tag{19}$$

$$\mu \Leftarrow \mu\rho. \tag{20}$$

where $\rho > 1$ is a constant and the complete algorithm is described in Algorithm 1. In Algorithm 1, we initialize (U_0, V_0) by SVD solution. We know that when the objective function is convex, ALM algorithm usually converges. After extensive experiments, we find that the convergent solutions obtained by different initializations are very close to each other [23], and there is no obvious performance difference for tag inference and noisy tags suppression. Moreover, since the result of standard PCA may be close to the optimal solution of DGSLR, the initial values of U and V using the decomposition factor of standard PCA may significantly accelerate the convergence rate. Once the U and V in algorithm 1 are calculated, the product of U and V is the best low-rank approximation of the tagging matrix under ℓ1-norm. Then the original tags could be updated via the desired low-rank recovery matrix with threshold t, as shown in Algorithm 2 in detail.

5 Experiments

In this section, we first introduce the large-scale social image dataset MIRFlickr-25k which is widely used. Then, we conduct several experiments to demonstrate the performance of our proposed method, meanwhile the comparison methods are illustrated in the same subsection. Finally, we report and discuss the refinement results.

Algorithm 1. Dual-graph regularized sparse low-rank matrix decomposition

---

Require: Tagging matrix X; graph laplacians L_U, L_V; parameters: d, λ_1 and λ_2.
1: Initialize $\mu = ||X||_F$, $\rho = 1.01$, U_0, V_0, $E_0 = X - U_0 V_0$
2: **repeat**
3:    Computing E_k using Eq.(12)
4:    Computing U_k, V_k using Eq.(17) and Eq.(21)
5:    $\Lambda = \Lambda + \mu(X - U_k V_k - E_k)$
6:    $\mu = min(\mu\rho, 10^{10})$
7:    $J_k = ||E_k||_1 + \lambda_1 tr(U^T \widetilde{L}_U U) + \lambda_2 tr(V \widetilde{L}_V L^T)$
8:    $k = k + 1$
9: **until** $|J_k - J_{k-1}| \leq 10^{-2}$
Ensure: Low-dimensional image representation $U \in \mathbb{R}^{n \times d}$; Low-dimensional tag representation $V \in \mathbb{R}^{d \times m}$.

Algorithm 2. Tag refinement with DGSLR

Require: X, G^U, G^V, parameters: d, λ_1 and λ_2
1: Computing the adjacency matrix W_U of graph G^U using Eq.(4)
2: Computing the adjacency matrix W_V of graph G^V using Eq.(6)
3: Computing graph Laplacian L^U and L^V using W^U and W^V respectively
4: Solve the following problem using Algorithm 1,

$$\min_{U,V} ||X - UV||_1 + \frac{\eta}{2}(||U||_F^2 + ||V||_F^2) + \lambda_1 tr(U^T L_U U) + \lambda_2 tr(V L_V V^T)$$

and obtain an optimal solution (U^*, V^*)
5: Construct the recovered low-rank taging matrix $X^* = U^* V^*$
Ensure: Identify tags from the recovery matrix X^*.

5.1   DataSet

Our experiments were performed on the MIRFlickr-25k [10], which contains 25000 images which is all download form Flickr, and 1386 unique tags. The ground-truth for several tags are offered by the LIACS Medialab at Leiden University. Before experiments, we perform a pre-process to discard those unpopular tags, correct the misspellings and compound nouns, and finally capture totally 259 unique meaningful tags images. The following experiments are all based on these pre-processed tags.

5.2   Experiment Settings

To confirm the effectiveness of our proposed DGSLR method, we compare it with the following algorithms, the parameter settings are also illustrated:

- PCA: $\min_{U,V} ||A - UV||_F^2, s.t.\ U^T U = I$. The dimensionality d is tuned by the grid 5, 10, 15.
- $\ell 1$-PCA: $\min_{U,V} ||A - UV||_1$. The dimensionality d is also tuned by the grid 5, 10, 15.

- SLR+Image Graph: $\min_{U,V} \|X - UV\|_1 + \frac{\eta}{2}(\|U\|_F^2 + \|V\|_F^2) + \lambda_1 tr(U^\mathrm{T} L_U U)$, The dimensionality d is tuned by the grid 5, 10, 15, and the parameter λ_1 is tuned by the grid of $\{0.1, 1, 10, 100\}$.
- SLR+Tag Graph: $\min_{U,V} \|X - UV\|_1 + \frac{\eta}{2}(\|U\|_F^2 + \|V\|_F^2) + \lambda_2 tr(V L_V V^\mathrm{T})$, The dimensionality d is tuned by the grid 5, 10, 15, and the parameter λ_2 is tuned by the grid of $\{0.1, 1, 10, 100\}$.
- DGSLR: $\min_{U,V} \|X - UV\|_1 + \frac{\eta}{2}(\|U\|_F^2 + \|V\|_F^2) + \lambda_1 tr(U^\mathrm{T} L_U U) + \lambda_2 tr(V L_V V^\mathrm{T})$, The dimensionality d is tuned by the grid 5, 10, 15, and both parameters λ_1 and λ_2 are tuned by the grid of $\{0.1, 1, 10, 100\}$.

Finally, to investigate the performance of the refined result, we adopted F-score for a detailed evaluation on those tags which have been provided the ground-truth annotations. The F-score was widely used in previous works, which is defined as:

$$F\text{-}score = \frac{2 \times Precision \times Recall}{Precision + Recall} \tag{21}$$

Here, 'Precision' and 'Recall' is defined as follows respectively:

$$Precision = \frac{TP}{TP + FP} \qquad Recall = \frac{TP}{TP + FN} \tag{22}$$

where TP (True Positive) is the number of tags which correctly annotated to the corresponding images; FP (False Positive) is the number of tags we annotated to image but have no relation with the image's content; FN (False Negative) is the number of tags which can describe visual content of corresponding image but we do not annotate.

5.3   Tag Refinement Result

Convergence of the Optimization Process
Before going any further, the convergence property of the optimization process is illustrated firstly. In the experiments, we observed that our proposed objective function converge to the minimum after about 15 iterations. Figure 2(a) shows the change of objective function values in the convergence process. Moreover, the refinement result against iteration times is illustrated in Figure 2(b). It is clearly shown that the refinement result(calculated by F-Score) increases steadily and finally reaches a satisfactory performance. It is necessary to know that, here the dimensionality of the low-dimensional is set to 20, and the parameter λ_1 and λ_2 are set with the value of 1 and 0.1 respectively.

Refinement Performance Against Noise Rate
To adequately evaluate the tag refinement capability of our proposed approach, we will recover the tagging matrix at different synthetic noise tag levels. Firstly, we assume the original tagging matrix X consist of N_1 entries with value 1 (annotated) and N_0 entries with value 0 (unannotated). Then, we simulate the annotated noises at different rates as following steps: For a given rate α of noise

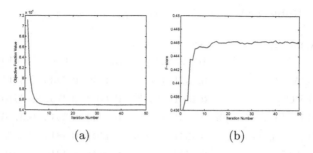

(a)   (b)

Fig. 2. The behavior of objective function and performance during iterations of our proposed algorithm. (a) Convergence of the objective function value, (b) Refinement result (F-score) vs. iteration times

$(0 \leq \alpha \leq 1)$, we first randomly select $N_{1\rightarrow0} = \alpha N_1$ images of the values 1 are changed from 1 to 0. Likewise, $N_{0\rightarrow1}\beta N_0$ images with value 0 are randomly selected and adjusted as 1, where $\beta = \frac{N_1}{N_0}\alpha$. In this experiment, $\frac{N_1}{N_0} \approx 5$, so the noise rate α is set by the grid 0.1, 0.2, 0.3, 0.4, 0.5, and the noise rate β is set by the grid 0.02, 0.04, 0.06, 0.08, 0.10 respectively. Each experiment was carried out 10 times from different initial value of U and V independently. The average performance of each dimensionality is shown in Tables 1, 2 and 3 respectively.

As shown in the three tables, we observe our SLR+Image Graph and SLR+Tag Graph approaches, which are two special cases of proposed DGSLR model with $\lambda_1 = 0$ and $\lambda_2 = 0$ respectively, outperform the first two baseline methods, which indicates that preserving the graph structure is very important to refine the noisy tags. Especially, our DGSLR model incorporating the image visual similarities and tag correlations simultaneously achieves better performance than all the other methods. Although the original tagging matrix is sparse, the graphs of visual similarities and tag correlations can alleviate the problem of sparsity, and aid the matrix factorization to obtain more interpretable low-dimensional representations of images and tag. Therefore, our proposed method is more suitable for refining the noisy user-provided tags associated with the social images.

In addition, with increasing the noise rate, the tag refinement performance degrades gradually in all approaches, but our method still achieves consistent better than other methods under different noise rate, this gives another evidence to support the effectiveness of our proposed methods. Moreover, the higher the dimensionality of the representation is, the better the performance is. However, we find that when the dimensionality increases to 20, there is no significant improvement through our experiments.

Effectiveness for Individual Tag Category

We further analyze the detailed effectiveness of using DGSLR in tag refinement over image tags independently. The F1-score of the five comparison methods for 16 tags are shown in Fig. 3. Due to space limitations in the article, here we list the 16 most common/representative tag categories. For the same reason, we just show the class-wise F1-score of tag refinement under the dimension $d = 15$ and

Table 1. Average F1-score comparison under d = 5

Algorithm	Noise rate (α%, β%)				
	10+2	20+4	30+6	40+8	50+10
PCA	0.4521	0.4131	0.3756	0.3436	0.3214
ℓ1-PCA	0.4801	0.4435	0.4100	0.3752	0.3410
SLR+Image Graph	0.5231	0.4789	0.4366	0.3913	0.3740
SLR+Tag Graph	0.5341	0.4860	0.4401	0.3814	0.3670
DGSLR	0.5428	0.5162	0.4831	0.4310	0.4010

Table 2. Average F1-score comparison under d = 10

Algorithm	Noise rate (α%, β%)				
	10+2	20+4	30+6	40+8	50+10
PCA	0.4810	0.4581	0.4361	0.4010	0.3751
ℓ1-PCA	0.5132	0.4801	0.4510	0.4278	0.3946
SLR+Image Graph	0.5410	0.5107	0.4731	0.4413	0.4176
SLR+Tag Graph	0.5501	0.5206	0.4697	0.4307	0.4297
DGSLR	0.5796	0.5431	0.5115	0.4672	0.4241

Table 3. Average F1-score comparison under d = 15

Algorithm	Noise rate (α%, β%)				
	10+2	20+4	30+6	40+8	50+10
PCA	0.4967	0.4601	0.4478	0.4206	0.3971
ℓ1-PCA	0.5297	0.4950	0.4610	0.4278	0.3946
SLR+Image Graph	0.5671	0.5317	0.4891	0.4510	0.4371
SLR+Tag Graph	0.5704	0.5476	0.4970	0.4479	0.4301
DGSLR	0.6103	0.5890	0.5431	0.5271	0.4610

noise rate $\alpha = 40\%$ and $\beta = 8\%$. Apparently, it's a very difficult and challenging situation, where almost half of entries in input tag matrix X are imprecise and incomplete. As shown in Fig. 3, the F1-score of DGSLR are consistently superior to the other comparison methods in most of the listed tags, which demonstrates the detailed effectiveness of our proposed approach.

Case Study of Tag Refinement

We further illustrate some case studies to demonstrate the effectiveness of our proposed method. Figure 4 shows some exemplary images and their corresponding refined tags. Let's take Fig. 4(c) and (f) as example. In Fig. 4(c), 'sole' and 'romagna' are noisy tags which can't describe the image visual content. Although the image content is very challenging for refinement task, our pro-

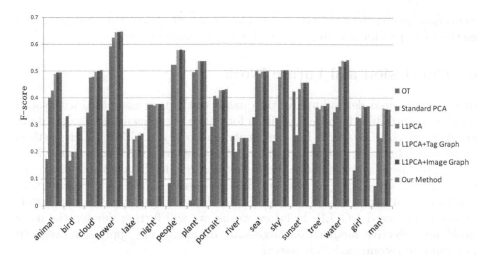

Fig. 3. Detailed performances for each tag of different methods on MIRFlickr-25K dataset

Original Tag:amyn fullmoon beach sand Parasol longexplosure	Refined Tag: beach sand sea ocean people art	Original Tag: sudan fly bird nature insect	Refined Tag:bird animal nature outdoor landscape	Original Tag: sunset sole sun romagna	Refined Tag: sunset sun cloud silhouette orange
(a)		(b)		(c)	

Original Tag: girl portrait taiwan	Refined Tag: portrait girl people face self	Original Tag: architecture	Refined Tag: architecture building street city urban	Original Tag:nubes clelo agua water lake azul blue landscape	Refined Tag: water blue lake reflection nature landscape sky
(d)		(e)		(f)	

Fig. 4. Example of tag refinement results obtained by our proposed method

posed method removes successfully the subjective tags, and enriches the tags 'cloud', 'silhouette', and 'orange', which are high correspondence with the visual content. It is a reasonable result because we incorporate image visual graph and tag correlations to the basic model of error sparsity. In Fig. 4(f), the refined tags are better to describe image content compared with original tags, as 'sky' 'reflection' are enriched. These experimental results show that our proposed method can provide a better strategy for tag refinement task.

6   Conclusion and Future Work

In this paper, in order to improve the quality of the tags associated with social images, we introduced a novel algorithm for image tag refinement based on dual-graph regularization sparse low-rank matrix recovery (DGSLR). This method formulates refinement task as an optimization problem, which incorporates visual content of images and correlations between tags provided by users, and maintains the error sparsity in tagging recovery simultaneously. Extensively experimental results on real image collections from Flickr demonstrated the superior performance of proposed DGSLR. In future works, we will further develop the more rapid and effective methods on large-scale database and apply our proposed method into recommendation system.

Acknowledgements. This work was supported by the Key Natural Science Project of Anhui Provincial Education Department (KJ2018A0023), the Guangdong Province Science and Technology Plan Projects (2017B010110011), the Anhui Key Research and Development Plan (1804a09020101), the National Basic Research Program (973 Program) of China (2015CB351705) and the National Natural Science Foundation of China (61906002, 61402002, 61876002 and 61860206004).

References

1. Ames, M., et al.: Why we tag: motivations for annotation in mobile and online media. In: CHI 2007: Proceeding of the SIGHI Conference on Human Factors in Computing Systems, pp. 971–980. ACM Press (2007)
2. Arulmozhi, K., Perumal, S.A., Priyadarsini, C.S.T., Nallaperumal, K.: Image refinement using skew angle detection and correction for Indian license plates. In: 2012 IEEE International Conference on Computational Intelligence and Computing Research, pp. 1–4, December 2012. https://doi.org/10.1109/ICCIC.2012.6510316
3. Chen, J., Yang, J.: Robust subspace segmentation via low-rank representation. IEEE Trans. Cybern. **44**(8), 1432–1445 (2014). https://doi.org/10.1109/TCYB.2013.2286106
4. Cheng, Z., Shen, J., Miao, H.: The effects of multiple query evidences on social image retrieval. Multimed. Syst. **22**(4), 509–523 (2014). https://doi.org/10.1007/s00530-014-0432-7
5. Chua, T.S., Tang, J., Hong, R., Li, H., Luo, Z., Zheng, Y.: NUS-WIDE: a real-world web image database from National University of Singapore. In: CIVR (2009)
6. Fu, J., Wang, J., Rui, Y., Wang, X., Mei, T., Lu, H.: Image tag refinement with view-dependent concept representations. IEEE Trans. Circuits Syst. Video Technol. **25**(8), 1409–1422 (2015). https://doi.org/10.1109/TCSVT.2014.2380211
7. Garcia, D.H., Mitchell, J.: Feature-extraction-based image scoring (2015)
8. Wang, H., Ding, C., Huang, H.: Multi-label linear discriminant analysis. In: Daniilidis, K., Maragos, P., Paragios, N. (eds.) ECCV 2010. LNCS, vol. 6316, pp. 126–139. Springer, Heidelberg (2010). https://doi.org/10.1007/978-3-642-15567-3_10

9. Wang, H., Huang, H., Ding, C.: Multi-label feature transform for image classifications. In: Daniilidis, K., Maragos, P., Paragios, N. (eds.) ECCV 2010. LNCS, vol. 6314, pp. 793–806. Springer, Heidelberg (2010). https://doi.org/10.1007/978-3-642-15561-1_57

10. Huiskes, M.J., Lew, M.S.: The MIR Flickr retrieval evaluation. In: Proceedings of the 1st ACM International Conference on Multimedia Information Retrieval, pp. 39–43. ACM (2008)

11. Jin, Y., Khan, L., Prabhakaran, B.: Knowledge based image annotation refinement. J. Sig. Process. Syst. **58**(3), 387–406 (2010)

12. Jin, Y., Khan, L., Wang, L., Awad, M.: Image annotations by combining multiple evidence and word-net. In: Proceedings of the ACM MM, pp. 706–715 (2005)

13. Wang, L., Zhou, T.H., Lee, Y.K., Cheoi, K.J., Ryu, K.H.: An efficient refinement algorithm for multi-label image annotation with correlation model. Telecommun. Syst. **60**(2), 285–301 (2015). https://doi.org/10.1007/s11235-015-0030-9

14. Liu, D., Yan, S., Hua, X., Zhang, H.: Image retagging using collaborative tag propagation. IEEE Trans. Multimed. **13**(4), 702–712 (2011). https://doi.org/10.1109/TMM.2011.2134078

15. Liu, D., Hua, X.S., Zhang, H.J.: Content-based tag processing for Internet social images. Multimed. Tools Appl. **51**(2), 723–738 (2011). https://doi.org/10.1007/s11042-010-0647-3

16. Mazumder, R., Hastie, T., Tibshirani, R.: Spectral regularization algorithms for learning large incomplete matrices. J. Mach. Learn. Res. **11**, 2287–2322 (2010)

17. Mislove, A., Druschel, P., Bhattacharjee, B., Gummadi, K.P.: Growth of the Flickr social network. In: WOSN (2008)

18. Nov, O., Chen, Y.: Why do people tag? Motivations for photo tagging. Commun. ACM **53**(7), 128–131 (2010). https://doi.org/10.1145/1785414.1785450

19. Pan, X., He, S., Zhu, X., Fu, Q.: How users employ various popular tags to annotate resources in social tagging: an empirical study. J. Assoc. Inf. Sci. Technol. **67**(5), 1121–1137 (2016)

20. Qian, Z., Zhong, P., Wang, R.: Tag refinement for user-contributed images via graph learning and nonnegative tensor factorization. IEEE Sig. Process. Lett. **22**(9), 1302–1305 (2015). https://doi.org/10.1109/LSP.2015.2399915

21. Ran, X., Chen, J.: Feature extraction for rescue target detection based on multispectral image analysis. In: 2015 International Conference on Transportation Information and Safety (ICTIS), pp. 579–582, June 2015. https://doi.org/10.1109/ICTIS.2015.7232204

22. Sang, J., Xu, C., Liu, J.: User-aware image tag refinement via ternary semantic analysis. IEEE Trans. Multimed. **14**(3), 883–895 (2012). https://doi.org/10.1109/TMM.2012.2188782

23. Tsai, D., Jing, Y., Liu, Y., Rowley, H.A., Ioffe, S., Rehg, J.M.: Large-scale image annotation using visual Synset. In: 2011 International Conference on Computer Vision, pp. 611–618, November 2011. https://doi.org/10.1109/ICCV.2011.6126295

24. Wang, C., Jing, F., Zhang, L., Zhang, H.: Content-based image annotation refinement. In: 2007 IEEE Conference on Computer Vision and Pattern Recognition, pp. 1–8, June 2007. https://doi.org/10.1109/CVPR.2007.383221

25. Cheng, W., Wang, X.: Image tag refinement using tag semantic and visual similarity. In: Proceedings of 2011 International Conference on Computer Science and Network Technology, vol. 4, pp. 2146–2149, December 2011. https://doi.org/10.1109/ICCSNT.2011.6182401

26. Wright, J., Ganesh, A., Rao, S., Peng, Y., Ma, Y.: Robust principal component analysis: exact recovery of corrupted low-rank matrices via convex optimization. In: Bengio, Y., Schuurmans, D., Lafferty, J.D., Williams, C.K.I., Culotta, A. (eds.) Advances in Neural Information Processing Systems 22, pp. 2080–2088. Curran Associates, Inc. (2009). http://papers.nips.cc/paper/3704-robust-principal-component-analysis-exact-recovery-of-corrupted-low-rank-matrices-via-convex-optimization.pdf
27. Zhu, G., Yan, S., Ma, Y.: Image tag refinement towards low-rank, content-tag prior and error sparsity. In: Proceedings of the 18th ACM international conference on Multimedia, pp. 461–470 (2010)

Optimal Quasi-PR Control of Grid-Connected Inverter with Selective Harmonic Compensation

Shuai Zhang[1], Gaifeng Lu[1], Shuai Du[2(✉)], Haideng Zhang[3], and Yuanhui Ge[4]

[1] School of Electric Power, North China University of Water Resources
and Electric Power, Zhengzhou, China
[2] School of Electric Engineering, Zhengzhou University, Zhengzhou, China
shuai.du@me.com
[3] State Grid Shanghai Electric Power Supply Company, Zhumadian, China
[4] Henan Senyuan Electric Co., Ltd., Xuchang, China

Abstract. This paper aims at analyzing and compensating the harmonic current effects in the Grid-Connected Photovoltaic (PV). Traditional harmonic compensation methods extract all harmonic current, so that the compensator of harmonic current tends to be complex and less flexible. Therefore, the design of the quasi-PR control with selective harmonic compensation which is based on BP neural network is applied to compensate selected harmonic currents. The strategy can not only reduce the distortion of the grid current, but also improve the adaptive ability of the current controller. The design of the PR current control and the selective harmonic compensator will be carried out using Matlab. Matlab simulation results show that compare with traditional quasi-PR control, the quasi-PR control with selective harmonic compensation which is based on BP neural network decreases the total harmonic distortion rate of tracking current, increases the dynamic response performance, and improve the stability of the controller.

Keywords: Grid-connected inverter · Harmonic compensator · PR control · BP neural network

1 Introduction

Harmonics generated by Distributed Power Generation Systems is a major power quality issue, especially considering the increasing speed of these systems connected to the grid is always fast [1]. This means that it is extremely urgent to control the harmonics generated by these inverters to limit their adverse effects on the grid power quality. Different areas in the world have made their own standards about harmonic control [2]. IEEE Std. 519-1992 titled "IEEE Recommended Practices and Requirements for Harmonic Control in Electric Power

Tenth Innovative Topic of North China University of Water Resources and Electric Power (YK2018-01).

© Springer Nature Singapore Pte Ltd. 2020
L. Pan et al. (Eds.): BIC-TA 2019, CCIS 1160, pp. 259–270, 2020.
https://doi.org/10.1007/978-981-15-3415-7_21

System", is the main document for harmonic limitations in North America [3], and European IEC standards (IEEE 929, IEEE 1547 and IEC 61727) suggest the following aspects: harmonic limits generated by Photovoltaic (PV) Systems, Distributed Power Resources for the current total harmonic distortion (THD) factor and also for the magnitude of each harmonic.

Furthermore, experts and academics have made comprehensive researches. A PR controller based on second order generalized integrator has been proposed [4], this method can enhance the capability of locking voltage phase and reduce the contend of current harmonics. However, the response circle is too long to get the exact signal at times. Some scholars propose a PR current controller with harmonic compensation [5]. Although the controller can effectively eliminate low-frequency odd harmonics, it can not adaptively adjust the gain of each harmonic compensation. Others believe that proportional integral control combing the voltage outer loop and the grid-connected current inner loop can realize the tracking of specific frequency and phase without static error [6], but it also has poor dynamic characteristics, and the ability of restraining harmonics is limited. A adaptive filter based on variable step size algorithm is proposed to replace the low-pass filter [7], this method changes the step factor of the filter according to the input signal, but a complete control system is not proposed. The paper [8] applies the neural network to the PR controller using the parameter self-tuning, it reduces the total harmonic distortion and improves the dynamic response speed, but the compensation of each harmonic is not obvious. In view of the characteristics of poor anti-disturbance ability, poor control dynamic performance, limited ability to suppress harmonic interference and difficult to realize, this paper adopts the quasi-PR control with selective harmonic compensation which is based on BP neural network method to solve this problem.

This paper is divided into four sections. Section 2 present the comprehensive introduction of the single-phase photovoltaic inverter. The principle and mathematical model of PR control system are also in this section. Section 3 covers the design of selective harmonic compensation strategy and the theory for self-adaptive harmonic compensation adjustment, while Sect. 4 presents the simulations and the results respectively, it also covers the comparison of results of the PR current control alone with the PR current control including the selective harmonic compensation. This paper concludes with final comments in Sect. 5. For the quasi-PR control strategy, the compensation gain is adjusted by improved BP neural network. Compared with the traditional quasi-PR control strategy, the total distortion rate of grid-connected current (THD) is reduced from 3.42% to 2.06%, while the THD is reduced from 5.32% to 3.10% when the load suddenly changes which improves the stability of the controller.

2   Grid-Connected Inverter Principle and Mathematical Model of Control System

In this paper, a single-phase grid-connected photovoltaic system is adopted. The Boost circuit raised the DC voltage at the pre-stage, and the grid-connected inverter realized the conversion from DC voltage to AC voltage at the last stage.

Figure 1 shows the structure of system. The system contains photovoltaic array, DC boost converter, single-phase inverter, LCL filter, inverter controller, etc. [9]. In this figure, U_{pv} is the output voltage of PV array; U_{dc} is the input of the inverter; U_g is the single-phase grid voltage; I_{pv} is the output current of the PV array; i_g is the real grid-connected current; i_{ref}^* is the reference value of the grid-connected current. L_1 is the inverter side inductor, and L_2 is the grid side inductor. C is the filter capacitor.

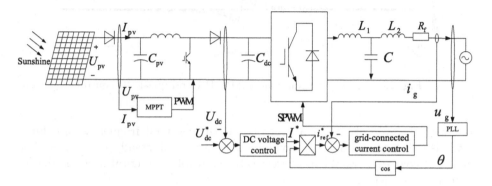

Fig. 1. Single-phase grid-connected photovoltaic system

2.1   PR Control

The transfer function of the grid-connected inverter in the system is [10,11]:

$$G_{\text{PWM}}(s) - \frac{K_{\text{PWM}}}{T_s s + 1} \tag{1}$$

In Eq. (1), K_{PWM} is the gain of the output voltage; the delay time of the inverter is Ts, which is usually taken as one sampling period.

The transfer function of the LCL filter is:

$$G_{\text{LCL}}(s) = \frac{1}{L_f s + R_f} \tag{2}$$

Where, R_f is the resistance of the connected reactor, and L_f is the sum of the inductor L_1 and L_2.

As shown in Fig. 2, the PI controller is a first-order controller, and the gain of the fundamental frequency is limited, so the capacity of tracking current amplitude and phase would exist some errors. Although the static error can be reduced by increasing the magnification, it cannot be completely eliminated. The ideal resonant term on its own PR controller provides an infinite gain at the fundamental frequency and no phase shift, but the ideal PR controller should be made non-ideal called quasi-PR controller by introducing damping in (3). So

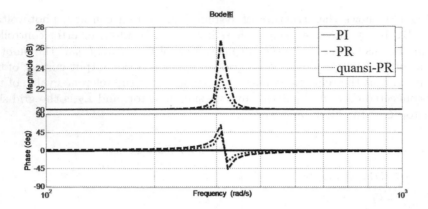

Fig. 2. Frequency response among PI, PR and quasi-PR control

that with the gain of quasi-PR controller at the resonant frequency ω_0 is finite but it is still large enough to provide only a very small steady state error. The quasi-PR controller would be more easily realizable in digital systems due to their finite precision.

The quasi-PR controller transfer function in Fig. 2 should be expressed as:

$$G_{PR}(s) = K_P + K_R \frac{2\omega_c s}{s^2 + 2\omega_c s + \omega_0{}^2} \tag{3}$$

In Eq. (3), K_P is the proportional gain term, K_P is the resonant gain term; the resonant frequency is ω_0, and ω_c is the bandwidth around the resonant frequency of ω_0.

2.2   Control System Design

The single-phase grid-connected photovoltaic system generally consists of three control loops, which are the pre-stage DC MPPT loop, the DC voltage loop and the grid-connected current loop. The pre-stage MPPT loop boosts voltage by modulating the duty cycle of the boost converter. The DC voltage loop and the grid-connected current loop operate independently in last stage.

Figure 3 shows the structure of inverter control system. The DC voltage U_{dc} and the reference DC voltage U_{dc}^* are the DC outer voltage loop reference values. Through the PI controller, it gives a reference amplitude I^* of the current inner loop. Meanwhile, the grid side voltage extracts the phase θ through the phase lock system. The control system using I^* and θ to merge into the reference current inner loop value i_{ref}^*. The inner current loop uses the quasi-PR principle to convert the current i_α to the voltage u_α, and the inverter is modulated by

PWM. The control function should be expressed as:

$$\begin{cases} I^* = (K_{\text{op}} + \frac{K_{oi}}{s})(U_{\text{dc}}^* - U_{\text{dc}}) \\[2mm] i_\alpha = i_{\text{ref}}^* - i_{\text{g}} \\[2mm] u_\alpha = K_{\text{P}} + K_{\text{R}}\frac{2\omega_c s}{s^2 + 2\omega_c s + \omega_0})i_\alpha \end{cases} \qquad (4)$$

In (4), K_{op} is the proportional gain term of the DC voltage outer loop, K_{oi} is the integral gain term of the DC voltage outer loop, and in Fig. 3, Δu is the output value of the harmonic compensation system. The meanings of the other variables are not repeated here.

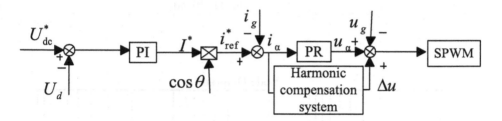

Fig. 3. Structure of inverter control system

3   Adaptive Current Harmonic Control Strategy

3.1   Quasi-PR Control of Selective Harmonic Compensation

In this section, the detection method of harmonic signals is selective harmonic detection. Because of the limitations of the inverter and the accuracy of the detection system, the 3rd, 5th, and 7th harmonics with large content and low order are selected [10]. The detection system adopts a method based on phase-locked loop. Firstly, the grid-connected current is extracted by FFT transform, so the amplitude and phase of the selective harmonics are obtained, then the amplitude and phase are reconstructed to the selective harmonics.

For the selective harmonics, the harmonic compensation transfer function is representing as shown in (5). K_{hR} (h $= 0$, 1, 2, ...) is the resonant term at the particular harmonic, and $h_\omega 0$ is the particular harmonic frequency. ω_c is the bandwidth around the $h_{\omega 0}$.

$$G_{\text{h}}(s) = K_{\text{hR}}\frac{2\omega_c s}{s^2 + 2\omega_c s + (h\omega_0)^2} \qquad (5)$$

Figure 4 shows the bode diagram with the fundamental PR controller and the harmonic compensation system. In this paper, the harmonic compensators are designed for the 3rd, 5th and 7th harmonics, and the harmonic compensators change the transfer function. Each different harmonic number compensator is designed on its own and then combines together with the fundamental quasi-PR controller in Matlab. Ultimately fine tuning of the compensator is performed to obtain the optimum operation of the compensators by varying ω_c and K_{hR}. The parameters of harmonic compensator are shown in Table 1.

Table 1. Parameter of harmonic compensators

Harmonic order	K_{hR}	ω_c
3rd	211.208	2.5 rad/s
5th	83.867	4.5 rad/s
7th	40.834	10 rad/s

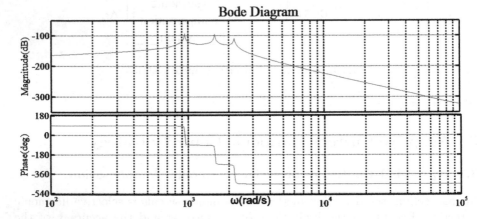

Fig. 4. Open loop bode diagram of the system with harmonic fundamental PR controller and the harmonic compensators

3.2   Self-adaptive Harmonic Compensation Adjustment

The harmonic detection system can obtain harmonic current $i_{\alpha l}$ by detecting the grid current i_α. Then, the compensation voltage signal is generated by the quasi-PR control with the selective harmonic compensator, and then reconstructed the PWM modulated. In order to improve the dynamic ability of the inverter controller, the BP neural network with variable learning rate (VLBP) [11] is adopted. Figure 5 shows the adaptive compensation gain adjustment principle.

The VLBP neural network is based on the variable learning rate, which shortens the training time and overcomes the problem that usual BP neural network

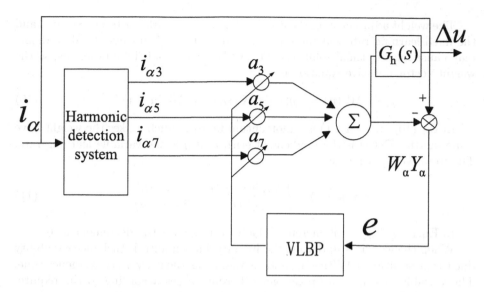

Fig. 5. Adaptive compensation weight adjustment principle

is easy to get the local optimal solution [12]. When RMSE is minimal, the process to adjust the compensation gain is ending, and the current compensation weight is the best. In this paper, the 3, 5, and 7th current harmonics are taken as the input of VLBP neural network, and the actual error current is the target output. The VLBP neural network can dynamically adjust the compensation weights, and the iterative equation are converged when the compensation effect is optimal.

All the selective harmonics are used as the column vector of the matrix Y_α; the weight vector is W_α, the initial weight value is 1; and the target output is i_α.

$$\begin{cases} Y_\alpha = [i_{\alpha 3}, i_{\alpha 5}, i_{\alpha 7}]^{\mathrm{T}} \\ W_\alpha = [a_3, a_5, a_7] \end{cases} \tag{6}$$

In Eq. (7), n represents the number of iteration, the nth error signal should be defined as:

$$en = i_\alpha(n) - W_\alpha(n)Y_\alpha(n) \tag{7}$$

K is the length of the output layer neurons, and the error energy is defined as:

$$\varepsilon(n) = \frac{1}{2} \sum_{k=1}^{K} e_k^2(n) \tag{8}$$

In the process of error back propagation, the weight adjustment ΔW can follow the following general formula (9):

$$\Delta W = LearningRate(\eta) \times LocalGradient(\delta) \times UpperLayerOutputSignal(\nu) \tag{9}$$

The weight adjustment of the usual BP neural network exists some errors, and the fixed learning rate will cause long learning time. Meanwhile, local gradients may cause local optimal solutions. The VLBP can avoid these problems, so the weight vector iterative equation is:

$$\Delta W(n) = -\eta(1-a)\nabla e(n) + a\Delta W(n-1) \tag{10}$$

In Eq. (10), $a(0 < a < 1)$ is a momentum factor, which makes the weight have some inertia, $\nabla e(n)$ is the gradient of e(n), and η is the variable learning rate. The iterative equation is:

$$\eta(n+1) = \begin{cases} k_{\text{inc}}\eta(n) & e(n+1) < e(n) \\ k_{\text{dec}}\eta(n) & e(n+1) > e(n) \end{cases} \tag{11}$$

In Eq. (11), k_{inc} is the increment factor and k_{dec} is the decrement factor.

When the weight is updating, the Eq. (10) has an anti-disturbance capability due to the addition $a\Delta W(n-1)$, and also shortens the network convergence time. The variable learning rate η can get different values according to the requirements in different training stages. The specific performance of η is:

(1) If the pre-iteration and post-iteration have the same updating direction, the algorithm will increase the updating step size, accelerate the convergent speed, and reduce the time in a single position during the iterative process.
(2) If the pre-iteration and post-iteration have the contrary updating direction, which indicates that there is a minimum point of η, the algorithm will decrease the amplitude of $\Delta W(n)$ by $a\Delta W(n-1)$, and update in a smaller step size to make η easier get the minimum point.

4   Simulations

The Grid-connected PV Inverter is modeled and simulated in MATLAB/ Simulink. The DC-link voltage is set to 400 V, the grid voltage is set to 220 V AC, the Boost converter side inductor L_1 is 1.3 mH, the grid side filter capacitor C is 8 μF, the grid side inductor L_2 is 1 mH, and the fundamental frequency is 50 Hz.

K_{op} is the proportional gain of the PI controller, which is used for the voltage outer loop and K_{oi} is the integral. The parameters of the quasi-PR controller are $K_P = 0.8$, $K_R = 100$, and $\omega_c = 6.28$.

4.1   Simulation Results When the Load Changes

For the situation of grid-connected current when the load changes, the traditional quasi-PR control is compared with the quasi-PR control which uses the selective harmonic compensation, and the results are shown in Figs. 6 and 7:

Compare Fig. 6 with Fig. 7, the traditional quasi-PR control and the quasi-PR control with selective harmonic compensation both can output the grid-connected current, Under the traditional quasi-PR control strategy, the total

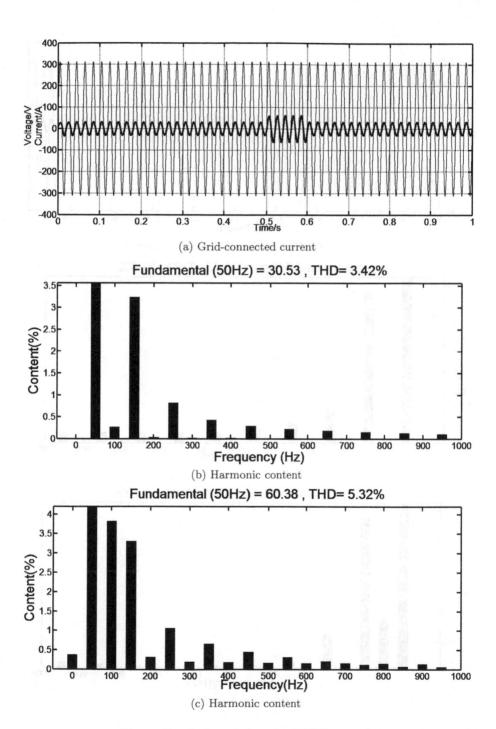

(a) Grid-connected current

(b) Harmonic content

(c) Harmonic content

Fig. 6. Simulation results of quasi-PR control

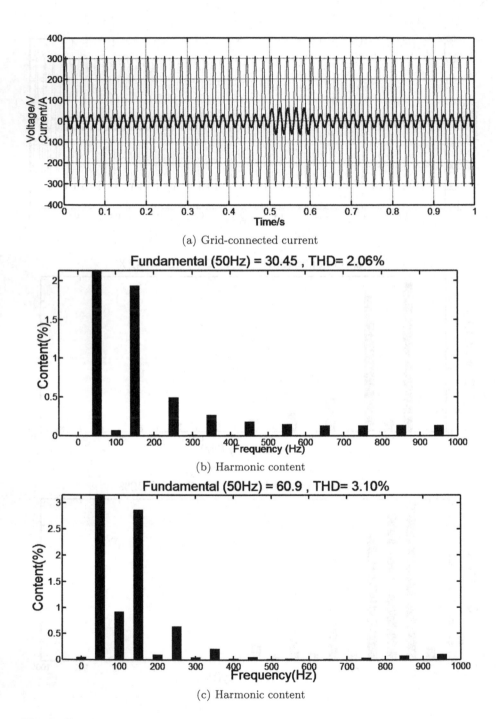

(a) Grid-connected current

(b) Harmonic content

(c) Harmonic content

Fig. 7. Simulation results of quasi-PR control with selective harmonic compensation

distortion rate of grid-connected current (THD) is 3.42%. When the compensation gain is adjusted by the improved BP neural network, the THD decreases to 2.06%. It can be seen that the quasi-PR control with selective harmonic compensation optimizes current more efficient, and 3rd, 5th, 7th harmonics are the main reduction. However, when a sudden decrease in the load at 0.5 s in simulation, the grid-connected current of traditional quasi-PR control is highly affected by the load changes, and the THD is 5.32%. The quasi-PR control with selective harmonic compensation improves the grid-connected current, and the THD is 3.10%.

5   Conclusion

This paper presents a procedure to design a quasi-PR current control with additional selective harmonic compensation for Grid Connected Photovoltaic (PV) Inverters. A grid connected PV inverter is designed and built for this research in MATLAB/Simulink. This paper covers the design of the PR control and also the design of the selective harmonic compensators for the 3rd, 5th and 7th harmonics. Results from simulations and analysis of the quasi-PR current control with selective harmonic compensation are presented. Simulation results show the effectiveness of the compensator to reduce the harmonics in the grid-connected current. When the load suddenly changes. The total distortion rate of grid-connected current (THD) is reduced from about 5.32% to about 3.10%. This reduction in harmonics make the grid-connected inverter compliant to the standard regulations.

References

1. Chatterjee, A., Mohanty, K.B.: Current control strategies for single phase grid integrated inverters for photovoltaic applications-a review. Renew. Sustain. Energy Rev. **92**, 554–569 (2018)
2. Mousavi, S.Y.M., Jalilian, A., Savaghebi, M., et al.: Coordinated control of multifunctional inverters for voltage support and harmonic compensation in a grid-connected microgrid. Electr. Power Syst. Res. **155**, 254–264 (2018)
3. Hwang, D.H., Lee, J.Y., Cho, Y.: Single-phase single-stage dual-buck photovoltaic inverter with active power decoupling strategy. Renew. Energy **126**, 454–464 (2018)
4. Xavier, L.S., Cupertino, A.F., Mendes, V.F., et al.: Adaptive current control strategy for harmonic compensation in single-phase solar inventers. Electr. Power Syst. Res. **142**, 84–95 (2017)
5. Zammit, D., Staines, C.S., Apap, M., et al.: Design of PR current control with selective harmonic compensators using Matlab. J. Electr. Syst. Inf. Technol. **4**, 347–358 (2017)
6. Meng, J., Shi, X., Fu, C., et al.: Optimal control of photovoltaic grid-connected current based on PR control. Electr. Power Autom. Equip. **34**(02), 42–47 (2014)
7. Peng, Y., Zhang, K., Li, Y., et al.: Research on harmonic detection method based on adaptive algorithm. Electr. Meas. Instrum. **55**(09), 6–9 (2018)

8. Zhou, Z., Shi, S.: Research on quasi-PR control based on drnn of self-tuning for photovoltaic grid-connected system. Acta Energiae Solaris Sinica **38**(11), 2932–2940 (2017)
9. Chen, Y., Xie, Z., Zhou, L., et al.: Optimized design method for grid-current-feedback active damping to improve dynamic characteristic of LCL-type grid-connected inverter. Electr. Power Energy Syst. **100**, 19–28 (2018)
10. Savaghebi, M., Vasquez, J.C., Jalilian, A., et al.: Selective compensation of voltage harmonics in grid-connected microgrids. Math. Comput. Simul. **91**, 211–228 (2013)
11. Dang, C., Tong, X., Yin, J., et al.: The neutral point-potential and current model predictive control method for Vienna rectifier. J. Frankl. Inst. **354**, 7605–7623 (2017)
12. Fan, B., Luo, X., Liao, Z., et al.: Quasi PR photovoltaic grid-connected inverter control method based on BP neural network. Proc. CSU-EPSA **28**(03), 30–34 (2016)

Recognition Method of Mature Strawberry Based on Improved SSD Deep Convolution Neural Network

Zhongchao Liu[1,2(✉)] and Dongyue Xiao[1]

[1] School of Electronic and Electrical Engineering,
Nanyang Institute of Technology, Nanyang 473000, Henan, China
liuzhongchao2008@sina.com
[2] College of Mechanical and Electronic Engineering,
Northwest A&F University, Yangling 712100, Shaanxi, China

Abstract. Recognition and localization of ripe strawberries is the basis of strawberry automatic picking system. Aiming at the problems of outdoor illumination change, uneven brightness, mutual occlusion of fruits and leaves, low recognition rate and poor generalization ability of traditional recognition methods, this paper proposes a mature strawberry target detection method based on improved single shot multibox detector (SSD) in-depth learning. The lightweight network MobileNet V2 was used as the basic network in SSD model to reduce the time spent in extracting image features and the amount of computation. The information loss caused by downsampling operation was avoided while synthesizing multi-scale features. The target of strawberry image recognition was established through the TensorFlow depth neural network framework. The results showed that the mAP of the model test set was 82.38%, the recognition accuracy of the improved SSD model was 97.4%, the recall rate was 94.5%, and the average recognition time of single frame image was 125 ms. This method can effectively recognize mature strawberries in natural environment and provide technical support for automatic production of strawberry harvesting.

Keywords: Strawberry target recognition · Deep learning · SSD · MobileNet

1 Introduction

In recent years, strawberry planting has been widely promoted in the world, and the area of strawberry planting in China is increasing. However, there is a shortage of labor force in strawberry harvesting, which requires at least two harvesting times a day during the harvesting period. The labor intensity and workload of strawberry harvesting are very large, which seriously restricts the development of strawberry cultivation [1]. Therefore, the accurate recognition of ripe strawberry targets in natural growth environment is of great significance to the automation and intellectualization of strawberry harvesting.

Fruit recognition has been extensively studied at home and abroad [2, 3]. The main methods of fruit target recognition are color difference method [4, 5], K-means

© Springer Nature Singapore Pte Ltd. 2020
L. Pan et al. (Eds.): BIC-TA 2019, CCIS 1160, pp. 271–281, 2020.
https://doi.org/10.1007/978-981-15-3415-7_22

clustering method [6], fuzzy C-means method [7], K-nearest neighbor (KNN) method [8], artificial neural network (ANN) [9] and support vector machine (SVM) [10, 11], although the above methods can do well. Fruit target recognition in image is based on fruit color, shape or texture characteristics. When the fruit surface is not uniform, shaded or occluded due to light or natural environment factors, the recognition accuracy will decrease significantly.

In recent years, deep convolution neural network (DCNN) has developed rapidly. It directly drives feature abstraction to recognize targets by data itself, instead of using manual feature extraction like traditional methods. Its accuracy and speed are much faster than traditional methods based on traditional artificial design features such as HOG and SIFT [12]. At present, target detection schemes based on in-depth learning mainly include two kinds: R-CNN (Region Based Convolutional Neural Network) proposed by Girshick [13], improved Fast R-CNN [14] (Fast Region-based Convolutional Network) and Faster R-CNN (Faster Region-based Convolution Network) proposed by Ren [15].

Some scholars have applied the target detection framework based on convolution neural network to agricultural and animal husbandry production, and achieved good recognition results. The YOLO V2 model was used to identify green citrus and mango in natural environment [16, 17]. In the field of animal target recognition, based on the YOLO V3 model target detection framework, the accuracy of underwater river crab target recognition can reach 96.65% [18]. Recognition of dairy cow's individual identity based on convolution neural network, the recognition rate of single frame image is 90.55% [19]. Based on faster R-CNN network, the accuracy of identification of feeding behavior of herd pigs was 99.6%, and the recall rate was 86.93% [20].

Therefore, based on the above analysis, in order to achieve real-time monitoring of ripe strawberry fruits in natural environment, this paper proposes an improved SSD, which takes ripe strawberries in natural planting environment as recognition object, effectively and accurately identifies the target of strawberry in the image, and lays the foundation for realizing the automation of strawberry picking.

2   Test Data

2.1   Image Acquisition

For the strawberry planting base in Jiangdong District, Jinhua City, Zhejiang Province, as shown in Fig. 1. The shooting date is February. The image acquisition device is the iPhone 7 plus. The image resolution is 3 024 * 3 024 pixels and the format is JPEG.

Fig. 1. Strawberry planting base

2.2   Sample Data Set

In the experiment, 8000 mature strawberry images were collected. In order to reduce the running time of the follow-up experiment, the image resolution was reduced to 500 × 500 pixels, and then the image was enhanced. Data enhancement is a general method to improve the robustness of the algorithm without reducing the detection accuracy. In specific scenarios and practical applications, it is often difficult to collect data satisfying various conditions for training and testing. Therefore, many target detection algorithms use data enhancement to evaluate the generalization of the designed algorithm.

In this paper, rotate, noise enhancement, mirror and random brightness are used to enhance the image data collected in natural environment, which makes the model have better generalization ability in complex light environment and different angles. The image of strawberry data set is increased. The strong effect is shown in Fig. 2. After enhancement, 16 000 samples were obtained, including 12 800 training sets and 3 200 test sets. The proportion of training sets and test sets was 80% and 20%.

2.3   Image Annotation

In addition to a large number of sample images, the data set also needs to annotate the detected objects in the images. The labelImg tool is used to mark the image in the experiment. When the label tool is used, the user-defined targets are only marked in the picture, and the tool can automatically generate the corresponding configuration files. In image annotation, the smallest outer rectangle of each mature strawberry is used to ensure that there is only one strawberry target in each rectangle annotation frame and as few background pixels as possible.

| (a) Original graph | (b) Increased brightness | (c) Brightness reduction |

| (d) Mirror flip | （e）Rotate | (f) Rotation plus noise |

Fig. 2. Strawberry data enhancement image

2.4   Data Set Preparation

After labeling the data set photos manually, the minimum outer rectangular box information of ripe strawberries (width, height and central point coordinates) in each photo is labeled, and the labeled files are converted into tfrecord format data by python program, which is put into the eyes together with the original photos. The structure catalogue of the application programming interface is checked and the preparation of the data set is completed.

3   Construction of Deep Convolution Neural Network

3.1   SSD Network Framework

The SSD network structure is divided into two parts: base network and auxiliary network. The basic network is a network with high classification accuracy in the field of image classification and its classification layer is removed. The auxiliary network is a convolution network structure added to the basic network for target detection, and the size of these layers is gradually increasing. It can be reduced so that multi-scale prediction can be carried out. Each additional auxiliary network layer generates a fixed prediction set through a series of convolution kernels. For a feature layer of m × n × p (p is the number of channels, m, n is the size), each auxiliary layer will use the convolution kernels of 3 × 3 × p to predict and generate a category of score values, or the object relative to silence. The position offset of the boundary box is recognized, and the corresponding values are predicted at m × n positions respectively.

SSD uses VGG-16 network as its basic network and adds feature extraction layer after the network. These increased convolution layers decrease step by step, and feature maps of different sizes can be extracted for detection. Large feature maps are used to detect objects with smaller physical volume, and smaller feature maps are used to detect objects with larger physical volume. In each feature map of different sizes, smaller convolution kernels are directly used to extract the detection results. In order to reduce the difficulty of training, reduce the operation time and required hardware performance. SSD refers to the concept of anchor in Faster R-CNN, and sets default bounding boxes with different aspect ratios in each partitioned cell. When the convolution core passes through these partitioned units, each pre-set priori box outputs a set of detection values. This set of detection values contains two parts of information: confidence of each category and location information of boundary boxes.

3.2   MobileNet_SSD Algorithm

The VGG network model used in SSD target detection model has many parameters, which occupies most of the running time in the process of feature extraction, and the information loss in the process of transformation is caused by the existence of non-linear transformation in the forward propagation process. The classical basic network VGG-16 is replaced by the improved MobileNet in this paper. The improved MobileNet model has the advantages of compact size, less operation parameters and much less computation than the traditional convolution neural network, which reduces the training time of the recognition model and reduces the hardware requirements.

MobileNet is a lightweight network model proposed by the Google team. It uses deep separable convolution instead of traditional convolution to reduce model parameters and improve computing speed. The network consists of 28 layers, except the standard convolution layer with convolution core size of 3 × 3 in the first layer, and the following one. Each separable convolution module consists of two layers. First, a single filter is applied to each input channel by using depth convolution, and then the output of different depth convolution is combined by using point convolution (1 × 1) to achieve the same output effect as standard convolution, but the computational complexity is the same. In addition, MobileNet abandons the maximum pooling down-sampling method and sets the stride of partially separable convolution layer to 2 to achieve the purpose of down-sampling. This paper uses MobileNet as the skeleton network to construct a lightweight single-step detection model based on it.

MobileNet-SSD trains the model by calculating the loss functions of confidence for label and prior box for ground truth at different scales. The loss functions of SSD are weighted by location error (LOC) and confidence loss (CONF). And, as shown in Formula (1).

$$L(x, c, l, g) = \frac{1}{N} \left(L_{conf}(x, c) + \alpha L_{loc}(x, l, g) \right) \tag{1}$$

Formula (1): N is the positive sample number; c is the predicted value of class confidence; l is the predicted value of the corresponding boundary box of a priori box; g is the location parameter of the real target.

3.3   MobileNet_SSD Network Model

The model structure of SSD-Mobile Net algorithm is shown in Fig. 3. The model is mainly divided into two parts, one is the front-end MobileNet network, whose function is to extract the initial features of the target; the other is the back-end multi-scale feature detection network, whose main work is to carry out the characteristics of the feature layer generated by the front-end network under different scales. Extraction. As can be seen from Fig. 3, six scales of information point to the final detection module, which can better predict the location and classification of the target. At the end of the model, the non-maximum suppression (NMS) module is used to filter out the targets of repeated prediction. The model can effectively extract the information of the target to be identified, accurately identify the mature strawberry, and has the advantages of displacement invariance and fast detection speed. It has good robustness to the changing target.

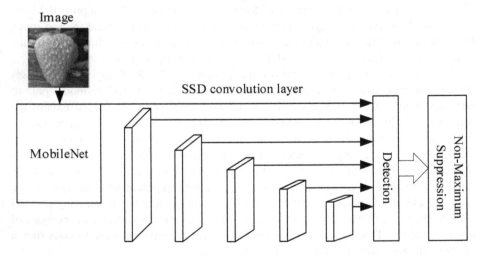

Fig. 3. Structure of SSD-MobileNet model

3.4   Model Fine-Tuning

The network model of the system design mainly carries on the fine-tuning of the model from three aspects.

(1) Deepwise convolution is used to replace the 3×3 standard convolution in the auxiliary feature extraction network, which further reduces the network parameters and computational complexity.

(2) In this paper, Atrous convolution is used to reduce the noise caused by padding and the extraction of redundant features.

(3) By adjusting the convolution step or feature dimension of some auxiliary feature extraction convolution network, the size of the auxiliary feature network decreases while the number of dimensions decreases, which reduces the influence of redundant features on training and recognition results.

4   Test Results and Analysis

4.1   Test Platform

The experimental processing platform is a desktop computer, the processor is Intel Core i7-8700k, the main frequency is 3.7 GHz, 32 GB memory, 500 GB hard disk, and the operating environment is Windows 10 (64-bit) system. The TensorFlow deep learning framework was built, and the training and testing of the mature strawberry target recognition network model were realized by Python language programming.

4.2   Training Parameter Setting

In order to save training time and speed up convergence, transfer learning is used to train in-depth learning model. First load the trained. The parameters of good Mobi-leNetV2 classification network are assigned to the corresponding parameters in SSD model except the final classification layer. The other parameters of each layer are randomly initialized by Gaussian distribution with 0 as the mean and 0.01 as the standard deviation.

In this paper, the batch random gradient descent algorithm is used. The training times are 20000, batchsize is 128, impulse is 0.9, weight attenuation coefficient is 2×10^{-3}, maximum iteration times are 2×10^5, initial learning rate is 0.005, attenuation rate is 0.9, attenuation once every 1000 iterations, after every 10,000 iterations, guarantee Save the model once, and finally select the model with the highest accuracy. The change of learning rate during training is shown in Fig. 4.

learning_rate

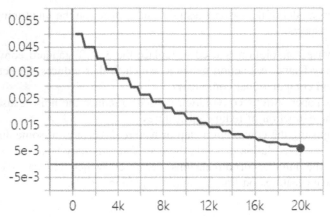

Fig. 4. Changes in learning rate during training

4.3   Network Model Training

Mean average precision represents the average AP value of many kinds of objects. It is a standard for measuring the sensitivity of network to target objects. The higher the mAP value, the higher the recognition accuracy of convolutional neural network, but the cost is that the detection speed will be slower. Figure 5 shows the change of mAP and Loss with the number of iterations during the training process. It can be seen that the final mAP reaches 82.38%. In the training of the model before 6000 steps, the state of the model is under-fitting due to the small amount of training, so the mAP is almost zero in the evaluation process, and the model is trying to approach the real results continuously. The gradient of loss function of model optimizer decreases, the Loss value decreases, and the average accuracy of model evaluation mAP increases rapidly. After 10000 iterations, with the decrease of learning rate, the index of the model will tend to fluctuate steadily and slightly.

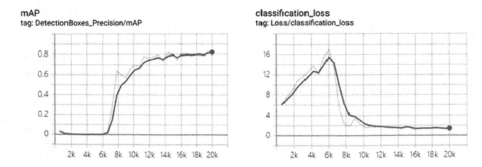

Fig. 5. Changes in mAP and loss

4.4   Result Analysis

In this study, the accuracy and recall rate are used to evaluate the recognition results of mature strawberry target recognition.

$$P = \frac{T_P}{T_P + F_P} \times 100\% \tag{2}$$

$$R = \frac{T_P}{T_P + F_N} \times 100\% \tag{3}$$

P denotes accuracy, R denotes recall, T_P denotes the number of ripe strawberries recognized by the algorithm, F_P denotes the number of ripe strawberries recognized by background or immature strawberries, and F_N denotes the number of ripe strawberries not recognized.

According to the designed network model, the test set pictures are tested, and the results of network recognition analysis are given in Table 1 T_P and F_N indicated the percentage of correct and wrong judgments in cow crawling cross-estrus test samples respectively, while F_P and T_N indicated the percentage of wrong judgments and correct judgments in non-estrus test samples.

Table 1. Network test recognition results

Convolutional neural network structure	Test category	Sample size	T_P	F_N	F_P	T_N
SSD+MobileNet	Ripe strawberry	3200	3027	173		
	Immature strawberry	1000			82	918

Table 1 shows that the recognition accuracy of ripe strawberries in the test set is 97.4%, the recall rate is 94.5%, and the average recognition time of single frame image is 125 ms, which can meet the requirements of automatic picking of strawberries.

The trained model is used to detect the images collected under the conditions of sunshine (full light), cloudy (dark light), shadows (occlusion, poor light angle), as shown in Figs. 6 and 7, which shows that the designed network structure can better recognize mature strawberries. From the actual test results, it can be seen that the improved SSD network structure has better confidence and location regression effect on mature strawberry.

Fig. 6. Test results under weak and strong light

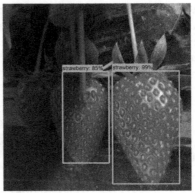

Fig. 7. Test results under different shadows and light angles

5 Conclusion

In this paper, a method of target recognition for mature strawberry in natural environment based on SSD convolution neural network is proposed. The characteristics of mature strawberry can be learned from complex data and the robustness of the model can be enhanced by using the convolution neural network to express the characteristics of the data itself in a non-linear way. Meanwhile, based on the target detection framework of MobileNet and SSD deep convolution neural network, a network structure suitable for strawberry recognition is constructed by modifying the input and output of the network, and a data set is constructed to train and test the model. The average accuracy of the output model reaches 82.38% s. The recognition accuracy of mature strawberry is 97.4%, the recall rate is 94.5%, and the average recognition time of single frame image is 125 ms, which can meet the requirements of automatic real-time picking.

Acknowledgments. This work is supported by the National Natural Science Foundation of China No. 61504072.

References

1. Qiu, Q., He, C.: Design of strawberry picking car based on TCS3200 color recognition. Mech. Res. Appl. **32**(3), 104–106 (2019)
2. Gongal, A., Amatya, S., Karkee, M., et al.: Sensors and systems for fruit detection and localization: a review. Comput. Electron. Agric. **116**, 8–19 (2015)
3. Lu, J., Sang, N.: Detecting citrus fruits and occlusion recovery under natural illumination conditions. Comput. Electron. Agric. **110**, 121–130 (2015)
4. Arefi, A., Motlagh, A., Mollazade, K., et al.: Recognition and localization of ripen tomato based on machine vision. Aust. J. Crop Sci. **5**(10), 1144 (2011)
5. Zhou, R., Damerow, L., Sun, Y., et al.: Using colour features of cv. 'Gala' apple fruits in an orchard in image processing to predict yield. Precis. Agric. **13**(5), 568–580 (2012)
6. Wachs, J., Stern, H., Burks, T., et al.: Low and high-level visual feature-based apple detection from multi-modal images. Precis. Agric. **11**(6), 717–735 (2010)
7. Zhu, A., Yang, L.: An improved FCM algorithm for ripe fruit image segmentation. In: IEEE International Conference on Information and Automation, pp. 436–441. IEEE (2014)
8. Linker, R., Cohen, O., Naor, A.: Determination of the number of green apples in RGB images recorded in orchards. Comput. Electron. Agric. **81**(1), 45–57 (2012)
9. Arefi, A., Motlagh, M.: Development of an expert system based on wavelet transform and artificial neural networks for the ripe tomato harvesting robot. Aust. J. Crop Sci. **7**(5), 699–705 (2013)
10. Qiang, L., Jianrong, C., Bin, L., et al.: Identification of fruit and branch in natural scenes for citrus harvesting robot using machine vision and support vector machine. Int. J. Agric. Biol. Eng. **7**(2), 115–121 (2014)
11. Zhao, Y., Lee, W., He, D.: Immature green citrus detection based on colour feature and sum of absolute transformed difference (SATD) using colour images in the citrus grove. Comput. Electron. Agric. **124**, 243–253 (2016)
12. Hu, Q., Wang, P., Shen, C., et al.: Pushing the limits of deep CNNs for pedestrian detection. IEEE Trans. Circ. Syst. Video Technol. **28**(6), 1358–1368 (2017)
13. Girshick, R., Donahue, J., Darrell, T., et al.: Rich feature hierarchies for accurate object detection and semantic segmentation. In: Proceedings of the IEEE Conference on Computer Vision and Pattern Recognition, pp. 580–587 (2014)
14. Girshick, R.: Fast R-CNN. In: Proceedings of the IEEE International Conference on Computer Vision, pp. 1440–1448 (2015)
15. Ren, S., He, K., Girshick, R., et al.: Faster R-CNN: towards real-time object detection with region proposal networks. In: Proceedings of 29th Annual Conference on Neural Information Processing Systems (2015)
16. Xiong, J., Liu, Z., Lin, R., et al.: Unmanned aerial vehicle vision detection technology of green mango on tree in natural environment. Trans. Chin. Soc. Agric. Mach. **49**(11), 23–29 (2018)
17. Xue, Y., Huang, N., Tu, S., et al.: Immature mango detection based on improved YOLOv2. Trans. CSAE. **34**(7), 173–179 (2018)
18. Zhao, D., Liu, X., Sun, Y., et al.: Detection of underwater crabs based on machine vision. Trans. Chin. Soc. Agric. Mach. **50**(3), 151–158 (2019)
19. Zhao, K., He, D.: Recognition of individual dairy cattle based on convolutional neural networks. Trans. CSAE **31**(5), 181–187 (2015)
20. Yang, Q., Xiao, D., Lin, S.: Feeding behavior recognition for group-housed pigs with the faster R-CNN. Comput. Electron. Agric. **155**, 453–460 (2018)

Analysis of Switching Network Based on Fourth-Order Chua's Circuit with a Memristor

Wudai Liao[1], Xiaosong Liang[1(✉)], Jinhuan Chen[2], Jun Zhou[1],
and Zongsheng Liu[1]

[1] College of Electric and Information, Zhongyuan University of Technology,
Zhengzhou 450007, China
liangxiaosong0909@qq.com
[2] College of Science, Zhongyuan University of Technology, Zhengzhou 450007, China

Abstract. Fourth-order Chua's circuit based on a memristor is investigated in this paper. Based on the physical properties of memristor, the mathematical model of fourth-order Chua's circuit is obtained by Kirchhoff laws. Then some verifiable stability conditions are constructed for the circuit by applying the Lyapunov method. Furthermore, due to the segmentation characteristics of the memristor, the existence theory of Filippov solutions is also discussed in this paper. Finally, the simulation of the system is presented to demonstrate the effectiveness of the theoretical results.

Keywords: Memristors · Chaotic circuit · Lyapunov stability

1 Introduction

In a seminal paper [1], the existence of the memristor was first predicted in 1971, and considered the fourth fundamental passive circuit element beside the resistance, capacitance and inductance. However, it was ever just a theoretical conception, no one could ever build one, therefore no much attention was paid to memristor. After 37 years, Hewlett Packard Labs (HP) announced that they had discovered the memristor by switching networks in Nature [2,3] in 2008. In their study, the memristor is nano-scale two terminal element. Due to its characteristic, which remember the voltage before turned off even after the circuit turned off, it provides more opportunities for future research. And its value depends on the magnitude and polarity of the voltage applied to it and length of time that the voltage has been applied [4]. Evidence suggests that memristor is among the most important alternative device for synaptic of recurrent neural networks [5]. After that, extensive research has been carried out on memristor. The existing body of research on memristor suggests that memristor-based switching network can be

This work is supported by National Natural Science Foundation (NNSF) of China under Grant 6187020178.

© Springer Nature Singapore Pte Ltd. 2020
L. Pan et al. (Eds.): BIC-TA 2019, CCIS 1160, pp. 282–292, 2020.
https://doi.org/10.1007/978-981-15-3415-7_23

used to solve efficiently classes of optimization problems arising in practical engineering application [6]. Non-volatile memory can be stored in memristors and it can be applied in the fields of associative memory and neural networks [7], but opportunity always be side with challenge. Memristor has nonlinear characteristics, which is difficult to construct a model to express its electric property [8,9]. There are two kinds of memristors, one is charge-controlled resistor, the other is flux-controlled resistor, and the value of memristor is related by the charge and magnetic flux, which is represented by $M(\Phi, q)$. Where q and Φ denote the charge and magnetic flux [10]. Both have its advantages. The flux-controlled memristor is easier to realize because it is easier to operate the voltage, and the charge-controlled memristor is more closely related to the physical device. Now, more and more researchers are focus on circuit with memristor because of its potential applications in brain-like "neural" computer. In [11,12], the iris classification method based on memristor neural network circuit is studied. In [13], the synchronous adjustment of weights of multi-layer neural network circuits based on memristor is studied, which improves the training speed of weights in neural networks. Therefore, the stability of memristor-based Chua's circuit has been receiving increasing attention recently. However, no previous study has investigated the stability of the fourth-order Chua's circuit with a memristor. It is believed that the paper provides sufficient conditions can be used to solve more complex optimization problems.

The essay has been organised in the following way. The paper first gives a brief overview of the recent history of memristor. The characteristic of memristor will be analyzed in Sect. 2. The Sect. 3 is concerned with the stability of a memristor-based fourth-order circuit, the parameter range satisfying the condition is obtained. The fourth section presents three simulation examples of the research and the simulation results also verify the correctness of the conclusions. In Sect. 5, the conclusion of this paper is given.

2   The Characteristic of Memristor

As the fourth fundamental element, memristor has nonlinear characteristics and its value is not constant. According to Itoh and Chua's research, the memristor is defined as follows [14]:

$$M(q) = \frac{d\Phi(q)}{dq} \quad or \quad W(\Phi) = \frac{dq(\Phi)}{d\Phi}. \tag{1}$$

where $M(q)$ denotes the memristance of the memristor, $W(\Phi)$ denotes the value of the memristor, q denotes the charge flowing through the memristor, and $\Phi(q)$ denotes the magnetic flux. In this paper, the model of memristor charge can be rewritten as follows:

$$W(\Phi(t)) = \frac{dq(\Phi(t))}{d\Phi(t)} = \begin{cases} \alpha, & |\Phi(t)| < 1, \\ \beta, & |\Phi(t)| > 1. \end{cases} \tag{2}$$

where $\alpha, \beta > 0$. In this paper, the systems' memductance can be given by $W(\Phi)$.

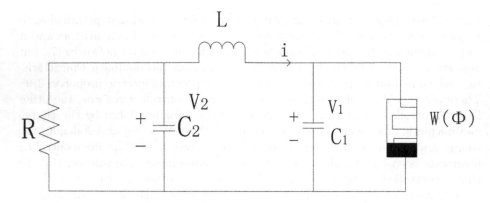

Fig. 1. The fourth-order Chua's circuit with a memristor.

3   Fourth-Order Chua's Circuit Based on a Memristor

In this paper, the circuit diagram, a memristor based fourth-order Chua's circuit, is shown in Fig. 1, which is deformed from Chua's circuit by replacing Chua's diode with a memristor [14]. In Fig. 1, L stand for inductor; i denote the current through the inductor; R stand for resistor; $W(\Phi)$ represent memristor; C_1, C_2 are two capacitors; V_1, V_2 denote the voltage at both ends of C_1, C_2. According to Kirchhoff laws and the charge-controlled model of memeristor, the following differential equations are obtained:

$$\begin{cases} \frac{dV_1}{dt} = \frac{1}{C_1}[i - W(\Phi)V_1] \\ \frac{di}{dt} = \frac{1}{L}(V_2 - V_1) \\ \frac{dV_2}{dt} = \frac{1}{C_2}\left(i - \frac{V_2}{R}\right) \\ \frac{d\Phi}{dt} = V_1. \end{cases} \tag{3}$$

where V_1, V_2, i are describe in Fig. 1, and the charge-controlled memristor $W(\Phi)$ satisfied (2), Choose the following state variables: $x_1 = V_1$, $x_2 = V_2$, $x_3 = i$, $x_4 = \Phi$, and let $\frac{1}{C_1} = a$, $\frac{1}{C_2} = b$, $\frac{1}{C_2 R} = c$, $\frac{1}{L_1} = d$, the differential equations of the circuit can be transformed into state-pace Eq. (4), which can be expressed as follows:

$$\begin{cases} \dot{x}_1 = -aW(x_4)x_1 + ax_3 \\ \dot{x}_2 = -cx_2 + bx_3 \\ \dot{x}_3 = -dx_1 + dx_2 \\ \dot{x}_4 = x_1 \end{cases} \tag{4}$$

To obtain the equilibrium point of the system (4), let $\dot{x} = 0$, then the equilibrium set can be obtain by $\Omega = \{(x_1, x_2, x_3, x_4) | x_1 = x_2 = x_3 = 0, x_4 = c\}$, where c denote a real constant. When appropriate parameters are selected, the circuit exhibits chaotic characteristics [15].

By analyzing the characteristics of the memristor, the state matrix for system (4) can be written as follows:

(1) If $|x_4(t)| < 1$, then

$$A = \begin{pmatrix} -a\alpha & 0 & a & 0 \\ 0 & -c & b & 0 \\ -d & d & 0 & 0 \\ 1 & 0 & 0 & 0 \end{pmatrix}.$$

(2) If $|x_4(t)| > 1$, then

$$A = \begin{pmatrix} -a\beta & 0 & a & 0 \\ 0 & -c & b & 0 \\ -d & d & 0 & 0 \\ 1 & 0 & 0 & 0 \end{pmatrix}.$$

In case of $|x_4(t)| < 1$, $A = A_1$, and in case of $|x_4(t)| > 1$, $A = A_2$. Therefore the state-pace of system (4) can be written as follow:

$$\dot{x}(t) = \sum_{i=1}^{2} m_i(t) A_i f(x(t)) \tag{5}$$

here $m_i(t)$ are piecewise membership function:

$$m_1(t) = \begin{cases} 1, & |\Phi(t)| < 1, \\ 0, & |\Phi(t)| > 1 \end{cases}$$

$$m_2(t) = \begin{cases} 0, & |\Phi(t)| < 1, \\ 1, & |\Phi(t)| > 1 \end{cases}$$

where $f(x(t)) = (f_1(x_1), f_2(x_2), f_3(x_3), f_4(x_4))^T$ with $f(0) = 0$, which is the input-output activation function, and the function is bounded non-decreasing piecewise continuous. The initial condition of system (5) can be written $x(0) = x_0$.

Here, we define the following symbols as follows: the Euclidean norm of the column vector x can be written as $\|x\|_2$. The maximum real part of the eigenvalue a real matric A can be written $Re\{\lambda_{max}(A)\}$. $K[U]$ denote the convex set closure of U, i.e., $K[U] = \bar{co}(U)$. $\mu(U)$ represents the Lebesgue measure of set U. $B(x, r) := \{y \in R^4 \mid \|y - x\|_2 < r\}$ denote a sphere of radius r with center $x \in R^4$. For set-valued map $\phi : x \rightarrow \phi(x)$, and remove the set of zero measure. Therefore the map can be written as follow:

$$\phi(x) = \bigcap_{\epsilon > 0} \bigcap_{\mu(N)=0} K[h(B(x, \epsilon) \setminus N)] \tag{6}$$

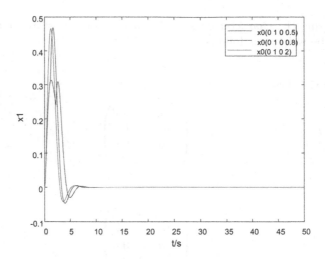

Fig. 2. The state trajectories of x_1. This shows a figure consisting of the state trajectories of x_1 which are under three given initial states. The trajectory starting from the initial position is finally stabilized at the equilibrium position after oscillation.

Here $h(x) := \sum\limits_{i=1}^{2} m_i(t) A_i f(x(t))$, N is an arbitrary set with measure zero. For any solution of switching network (5), for all $t \in [0, t_n]$, $0 \le t_n \le +\infty$ with initial condition $x(0) = x_0$, which is absolutely continuous function, and satisfies the differential inclusion $\dot{x}(t) \in \phi(x(t))$. According to the theorem of the Filippov solution local existence, moreover consider the set-value mapping, the following conclusions can be given:

$$0 \in \sum_{i=1}^{2} m_i(t) A_i K\left[f(0)\right] := \sum_{i=1}^{2} m_i(t) A_i K\left[f(0)\right] = \phi(c) \tag{7}$$

Therefore $c = 0$ is the an equilibrium of the network (5).

To find the range of parameters that make switching network (5) stable, an impulsive control sequence, $\{t_k, C_k x(t_k)\}$ is introduced here. Hence, the following impulsive differential equation can be obtained:

$$\begin{cases} \dot{x}(t) = \sum\limits_{i=1}^{2} m_i(t) A_i f(x(t)), \forall t \ge 0, t \ne t_k \\ \Delta x(t) = C_k x(t), t = t_k, k = 1, 2, \cdots, \end{cases} \tag{8}$$

where $0 < t_1 < t_2 < \cdots < t_k < \cdots$, $t_k \to +\infty$, $C_k = diag(c_{1k}, c_{2k}, c_{3k}, c_{4k})$, are constant matrices and $\Delta x(t) = x(t_k^+) - x(t_k)$, $x(t_k^+) = \lim\limits_{t \to t_k^+} x(t_k)$.

The equilibrium point of system (4) is uniformly stably only if Lyapunovs' uniform stability theorem is satisfied.

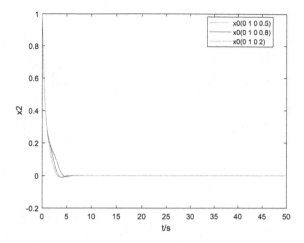

Fig. 3. The state trajectories of x_2. This shows a figure consisting of the state trajectories of x_2 which are under three given initial states. The trajectory starting from the initial position is finally stabilized at the equilibrium position after oscillation.

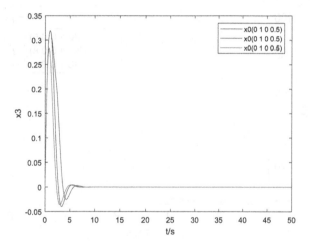

Fig. 4. The state trajectories of x_3. This shows a figure consisting of the state trajectories of x_3 which are under three given initial states. The trajectory starting from the initial position is finally stabilized at the equilibrium position after oscillation.

Theorem 1. *The equilibrium point of the system (8), $c = 0$, is uniformly stable if and only if the system satisfies the following two conditions:*

(1) $max\{Re\{\lambda_{max}(A_1)\}, Re\{\lambda_{max}(A_2)\}\} < 0$

(2) $0 < c_{ik} < 1,\ i = 1, 2, \cdots,\ k = 1, 2, \cdots.$

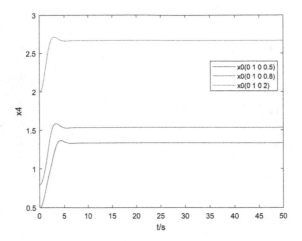

Fig. 5. The state trajectories of x_4. This shows a figure consisting of the state trajectories of x_4 which are under three given initial states. The trajectory starting from the initial position is finally stabilized at the equilibrium position after oscillation.

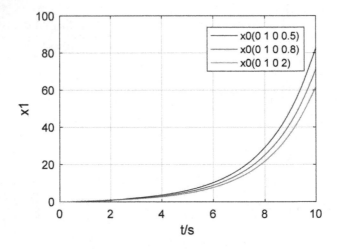

Fig. 6. The state trajectories of x_1. This shows a figure consisting of the state trajectories of x_1 which are under three given initial states.

Proof. Choose the following function as the Lyapunov function:

$$V(t) = \sum_{j=1}^{4} \int_{0}^{x_j(t)} f_j(\theta)d\theta$$

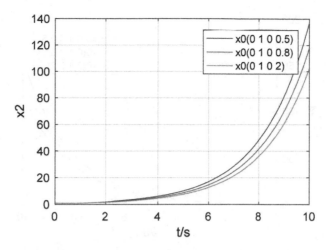

Fig. 7. The state trajectories of x_2. This shows a figure consisting of the state trajectories of x_2 which are under three given initial states.

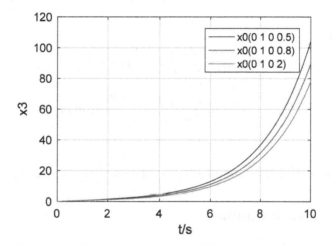

Fig. 8. The state trajectories of x_3. This shows a figure consisting of the state trajectories of x_3 which are under three given initial states.

When $t \neq t_k$, the upper right Dini derivative of $V(t)$ as follow:

$$D^+V(t) = (f(x(t,x_0)))^T \left[\sum_{i=1}^{2} m_i(t) A_i \right] f(x(t,x_0))$$
$$\leq (f(x(t,x_0)))^T \left[max\{Re\{\lambda_{max}(A_1)\}, Re\{\lambda_{max}(A_2)\}\} \right] f(x(t,x_0))$$
$$\leq 0$$

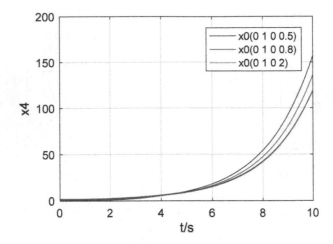

Fig. 9. The state trajectories of x_4. This shows a figure consisting of the state trajectories of x_4 which are under three given initial states.

When $t = t_k$,

$$V(t)\mid_{t=t_k^+} = \sum_{j=1}^{4} \int_0^{x_j(t_k^+)} f_j(\theta)d\theta$$

$$= \sum_{j=1}^{4} \int_0^{(1-c_{jk})x_j(t)} f_j(\theta)d\theta$$

$$\leq V(t).$$

In conclusion, the equilibrium point $c = 0$ of system (8) is uniformly stable.

4   Illustrative Examples

In order to verify that the equilibrium point $c = 0$ of system (5) is uniformly stable when the parameters meet the above conditions, the system is simulated as follows. Firstly, select the following parameters for system (5): $\alpha = 1$, $\beta = 2$, $a = 2$, $b = 1$, $c = 2$, $d = 1$, let the activation function $f(x) = (f_1(x_1), f_2(x_2), f_3(x_3), f_4(x_4))^T = ((exp\{x_1\} - exp\{x_1\})/(exp\{x_1\} + exp\{x_1\}),$ $(exp\{x_2\} - exp\{x_2\})/(exp\{x_2\} + exp\{x_2\}), 1/2(|x_3 + 1| - |x_3 + 1|), (exp\{x_4\} - exp\{x_4\})/(exp\{x_4\} + exp\{x_4\}))^T$. For generality, this paper take three initial states are $x_0 = (0 \quad 1 \quad 0 \quad 0.5)^T$, $x_0 = (0 \quad 1 \quad 0 \quad 0.8)^T$, $x_0 = (0 \quad 1 \quad 0 \quad 2)^T$.

Simulation was used to simulate the system and the graphs of the changes of each state with time were obtained Figs. 2, 3, 4, 5. It is not difficult to see from these graphs that the equilibrium point Ω of the system is uniformly stable.

The correctness of the conclusion can be verified again by contrastive experiments. The parameters are as follows: $a = 2$, $b = 2$, $c = 1$, $d = 1$. At this time, the maximum value of the real part of the eigenvalue of the system is 0.7321.

This does not meet the conditions of stability in the conclusion. The system is unstable. In Figs. 6, 7, 8, 9, the state variables of the system start from the initial point and eventually tend to infinity, which is not close to the equilibrium point. This is an unstable situation. This experiment verifies the correctness of the conclusion in this paper.

5 Conclusions

This paper mainly studies the stability of memristor based fourth-order Chua's circuit. As we all know, Chua's circuit is a special chaotic circuit. When appropriate circuit parameters are selected, the circuit shows chaotic characteristics. However, when a resistance element in the circuit is replaced with a memristor, the stability of the circuit on the same initial condition are changes, and even when choose the appropriate parameters, the circuit is uniform stability. We prove this conclusion not only by theory, but also by simulation. In this paper, we use Lyapunov stability theorem, switching network theory and Filipov's solution theorem. The conclusions of this paper can be extended to the study of memristor-based recurrent neural networks.

References

1. Chua, L.: Memristor-the missing circuit element. IEEE Trans. Circ. Theory **18**(5), 507–519 (1971)
2. Yang, J.J.: Memristive switching mechanism for metal/oxide/metal nanodevices. Nat. Nanotechnol. **3**(7), 429–433 (2008)
3. Strukov, D.B.: The Missing Memristor Found. Nature **453**(7191), 80–83 (2008)
4. Hu, J., Wang, J.: Global uniform asymptotic stability of memristor-based recurrent neural networks with time delays. In: The 2010 International Joint Conference on Neural Networks (IJCNN) (2010)
5. Wu, A., Zeng, Z.: Dynamic behaviors of memristor-based recurrent neural networks with time-varying delays. Neural Netw.: Off. J. Int. Neural Netw. Soc. **36**(C), 1–10 (2012)
6. Wu, A., Shen, Y., Zeng, Z., Zhang, J.: Analysis of a memristor-based switching network. In: Proceedings of 10th International Conference on Information Science and Technology. IEEE, Nanjing (2011)
7. Guo, Y., Wang, X., Zeng, Z.: A Compact memristor-CMOS hybrid look-up-table design and potential application in FPGA. IEEE Trans. Comput.-Aided Design Integr. Circ. Syst. 1 (2017)
8. Da, S., Sun, Y., Yang, H.: Analysis on influence parameter of memristance. Transd. Microsyst. Technol. **36**, 43–49 (2017)
9. Bao, B., Wang, C., Wu, G., et al.: Dimensionality reduction modeling and characteristic analysis of memristive circuit. Acta Phys. Sinica **63**(2), 257–264 (2014)
10. Xiao, J., Shen, Y., Zeng, Z.: Analysis of a third-order circuit based on a memristor. In: Proceedings of 10th International Conference on Information Science and Technology. IEEE, Nanjing (2011)
11. Xiao, S., Xie, X., Wen, S., et al.: GST-memristor-based online learning neural networks. Neurocomputing **272**, 677–682 (2018)

12. Wen, S., Xiao, S., Yan, Z., et al.: Adjusting learning rate of memristor-based multi-layer neural networks via fuzzy method. IEEE Trans. Comput. Aided Des. Integr. Circ. Syst. **38**(6), 1084–1094 (2019)
13. Yang, L., Zeng, Z., Shi, X.: A memristor-based neural network circuit with synchronous weight adjustment. Neurocomputing **363**, 114–124 (2019)
14. Itoh, M., Chua, L.: Memristor oscillators. Int. J. Bifurc. Chaos. **18**(11), 3183–3206 (2008)
15. Shen, W., Zeng, Z., Wang, G.: Feedback stabilization of memristor-based hyperchaotic systems. In: Proceedings of 12th International Conference on Information Science and Technology. IEEE, Yangzhou (2013)

Software Design of Online Monitoring System of Large Linear Vibrating Screen Fault Diagnosis

Qinghui Zhu[1(✉)] and Zhikui Wang[2]

[1] School of Electronic and Electrical Engineering,
Nanyang Institute of Technology, Nanyang 473000, Henan, China
0zhu@163.com
[2] School of Mechanical and Automotive Engineering,
Nanyang Institute of Technology, Nanyang 473000, Henan, China

Abstract. In this paper, the online monitoring system software of side cracks of large linear vibrating screen is designed on the basis of the system hardware platform. This software is constituted by six functional modules which are separately data acquisition, storage, communication, analysis, display and user management ones. Based on wavelet analysis, neural network theory and genetic algorithms, the main online fault diagnosis interface in PC with friendly and easy operation is designed, where the database, Visual Basic and Matlab software are comprehensively used for processing and analyzing the fault data. The system can be used for early warning of the early fatigue side cracks of large linear vibrating screen and has achieved the expected goal in real application.

Keywords: Fault diagnosis · Vibrating screen · Online detection · Software design

1 Introduction

Large linear vibrating screen is widely used in coal, ore and other levels of screening or separation of impurities in the mine screening equipment [1]. Because the long-term dynamic stress of the two sides of the screen is relatively concentrated, it is particularly easy to produce cracks and even cause side fracture. If these early cracks are not found in time, it will bring potentially great harm to production.

In this paper, we take the equipment of large linear vibrating screen DZK2466 in Tianzhuang Coal Preparation Plant of Pingdingshan Coal Group as the research objects, establishing the side crack online monitoring and fault diagnosis system, detecting the side cracks and making the early warning through the comprehensive control of hardware and software. The system design includes two parts: hardware and software. The hardware part mainly completes the collection and storage of the fault data and the software part mainly makes the data conversion, processing and analysis. In the independent development of online monitoring and control system hardware platform, we focus on the analysis of the computer diagnosis system, including the

© Springer Nature Singapore Pte Ltd. 2020
L. Pan et al. (Eds.): BIC-TA 2019, CCIS 1160, pp. 293–300, 2020.
https://doi.org/10.1007/978-981-15-3415-7_24

design of diagnosis interface based on VB and database, and also the background Matlab program design and call of wavelet analysis and neural network theory [2, 3] in order to complete the online identification and early warning for crack levels on the sides of the screen. This system has achieved satisfactory results of fault diagnosis through the debugging, simulation and field test.

2 System Hardware Composition

The hardware of this on-line monitoring system is mainly composed of MCU, A/D converter, sensor, charge amplifier, LCD and upper monitor. The sensor uses three axis acceleration sensor BZ1114 and the charge amplifier is designed whose function is to convert the voltage and charge standard. A built-in 16 bit microcontroller with 2 channels of 10/8 ADC named MC9S12XS128 is adopted as the system MCU. The hardware of the system is shown in Fig. 1.

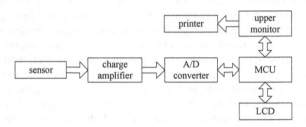

Fig. 1. System hardware composition

3 System Software Design Idea

System software design includes two parts: the MCU program and PC program and interface. The software part can be divided into six modules according to functions which are data acquisition, storage, communication, analysis, display and user management modules, as shown in Fig. 2.

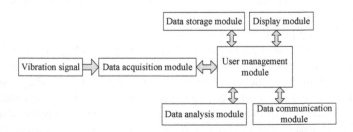

Fig. 2. The constitution of system software function modules

3.1 Data Acquisition and Storage Modules

The data acquisition module and storage module mainly complete the signal acquisition, conversion and storage from BZ1114 to SCM as well as the setting and monitoring of the parameters of the data acquisition process.

3.2 Data Analysis Module

The main function of data analysis module is that the host computer analyzes processed the original discrete signal collected and stored in real time. These ways mainly include wavelet noise reduction, wavelet packet fault feature extraction, genetic neural network optimization and fault identification and training etc. Matlab, Visual Basic and database are taken as the main analysis tools.

3.3 Communication Module

Communication module mainly completes the data exchange between SCM and PC through writing communication protocols to the host computer and debugging procedures. Single chip microcomputer can pre process, pre store and display fault for collected data, while the host computer is enable to analyze and process the data from SCM and is enable to send the results of the diagnosis to SCM to display.

3.4 User Management Module

The function of user management module is to monitor and process the data in real time. The user interface is written in the upper computer, and the system parameters and the diagnosis results are set up and searched in real time by the background program.

3.5 Data Display Module

The main task of the data display module is to display the fault diagnosis information of the equipment by LCD and PC respectively. The design of upper computer display interface will be introduced in Sect. 3.4.

4 Design and Implementation of Computer Monitoring Software

PC monitoring software is mainly to complete the online detection, analysis and treatment for side panel cracks of vibration screen. The development of diagnosis interface on PC is performed by using comprehensively wavelet analysis theory, network algorithm, software such as Visual Basic, Access database and ADO database connection technology etc. The fault diagnosis result is obtained through signal wavelet noise reduction and signal energy feature extraction of wavelet packet which can be completed by calling the Matlab calculation program and then by calling the neural network of database [4–6].

4.1 PC Software Programming Flow

The software design flow of the upper computer is signal timing acquisition, manual acquisition, signal wavelet noise reduction, energy characteristic calculation and fault diagnosis. The system main program flow of upper computer is shown in Fig. 3.

4.2 Data Tables Relationship Analysis

In order to facilitate the retrieval and view of the collected signals in the specific time period, the data tables and their relationship in the Access database are designed, as shown in Fig. 4.

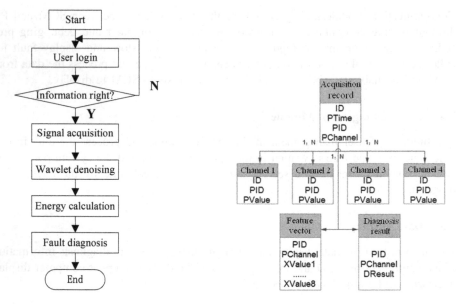

Fig. 3. Main program flow of upper computer

Fig. 4. Data tables in Access database

There are 7 data tables in the picture, which are the acquisition record table, the four channels sensor data tables, the characteristic vector table and the diagnosis result table. In the process of on-line fault diagnosis of the vibration screen, these tables are used to store the collected sensor signals, the noise data, the fault features and the diagnosis results.

4.3   Design of Main Function Modules of Upper Computer

Signal Acquisition Module. The signal acquisition module calls the acquisition card interface function GetSignal () to obtain the signals and stores them into corresponding signal tables according to the channel number, while the signal waveform is displayed in the acquisition window in real-time. The main program code is as follows:

```
Dim Plv As Single
Dim i As Integer
Dim Ylv As Single
Dim GetSig as Single
Private Sub ReadSignal (PID as Long, Channel as Byte)
   GetSig = GetSignal(Channel) 'read acquisition signal
   sql = "Insert Into signal table" & Channel & "(PID, PValue) Values (" & PID
& "," & GetSig & ")" 'Building storage database SQL statement
   Conn.Execute sql 'Save the signal to the database
   ' Display signal waveform to the foreground in real-time
   Plv = Me.Pic1(0).ScaleWidth /2048
   Ylv = Me.Pic1(0).ScaleHeight /26
   Me.Pic1(0).DrawStyle = vbSolid
   Me.Pic1(0).CurrentX = 0
   Me.Pic1(0).CurrentY = 0
   Me.Pic1(0).Line Step(0, 0)-(j, (13 - GetSig) * Ylv)
   Plv = Plv + Plv
   i = i+1
End Sub
```

Signal Noise Reduction Module. Signal noise reduction module calls M file ReduceNoise (Signal as Single) of Matlab, the real-time acquisition of the signal wavelet denoising, and the signal waveform after noise reduction to the foreground. The main program code is as follows:

```
Private Sub RedNoise(Signal As Single)
   RedNoise = ReduceNoise(Signal)
   ' Display the waveform of the signal after noise reduction to the fore-
ground in real time
   Plv = Me.Pic1(1).ScaleWidth /2048
   Ylv = Me.Pic1(1).ScaleHeight /26
   Me.Pic1(1).DrawStyle = vbSolid
   Me.Pic1(1).CurrentX = 0
   Me.Pic1(1).CurrentY = 0
   Me.Pic1(1).Line Step(0, 0)-(j, (13 - RedNoise) * Ylv)
   Plv = Plv + Plv
   i = i+1
End Sub
```

Energy Fault Feature Extraction Module. Energy fault feature extraction module calls the packet energy calculation file of CalEnergy (Signal as Single) in Matlab which can calculate the 8 energy feature values after denoising the signal, and displays to the foreground front in bar graph. At the same time, the calculation results are stored in the feature vector data tables, as the input signal of the genetic neural network. Here the program is omitted.

4.4   Implementation of Online Monitoring Interface

The upper computer diagnosis process of on-line monitoring system of large scale linear vibration screen is as follow:

First, Enter the user login interface. By inputting the account number and password, the main interface of the on-line monitoring system for the side crack of the vibrating screen is introduced. The main interface of the system is composed of four parts: signal acquisition, wavelet analysis, neural network and control center. The first three interfaces are introduced here.

Signal Acquisition. The interface of signal acquisition is shown in Fig. 5. Data from four sensors on the two sides of vibrating screen can be collected online at the same time through operation in this interface. From the figure, we can see that the time-domain waveform of data can also be dynamically displayed in the corresponding graph area where the acquisition frequency and group numbers of sampling can be set manually and also the sampling interval time and date of expiration date.

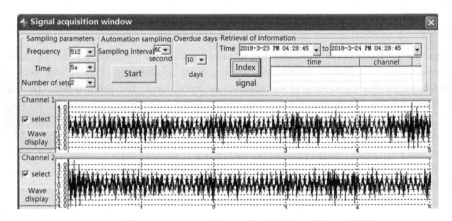

Fig. 5. The interface of signal acquisition

Wavelet Analysis. The wavelet analysis interface mainly completes the wavelet denoising processing and wavelet packet energy feature calculation.

First, you should check the corresponding channels to collect signals. Click the "noise" button to call Matlab wavelet noise reduction program through background and to the process current channel signal noise reduction. The denoised signal is stored in the corresponding relational tables in the database. At the same time, the time domain waveform of the denoised signal is displayed in the area of the corresponding signal. Wavelet de-noising decomposition layers and each layer of noise threshold can be separately set through this interface and can be read to the noise reduction program by Matlab automatically, achieving the intelligent function of soft threshold noise reduction processing.

Next, click on the "energy calculation" button, wavelet packet energy fault feature extraction program of Matlab is called through background. This program can realize three-layer decomposition of wavelet packet. Then, it calculates the signal energy of eight frequency bands and normalizes them. Meanwhile, this data is stored in the corresponding relation tables of database and the fault feature vector in the form of a bar chart is displayed in the interface below.

Neural Network. The neural network diagnostic interface realized the fault monitoring by selecting any one of the four input signal channels which mean that the detection signals of four channels can be diagnosed separately. The input signals of the neural network are the wavelet packet energy characteristic signals stored in the database while the outputs of the neural network are four kind fault levels of vibrating screen side cracks which are normal, early fatigue crack, formation crack and formed crack [7, 8]. The diagnosis result of channel 1 is displayed in Fig. 6 which shows the diagnosis result of channel 1 is "Early fatigue crack!".

After selecting the data channel to diagnose, click the "fault diagnosis" button. The software can also automatically diagnose the side crack grade and just we can do is selecting "automatically diagnose" button.

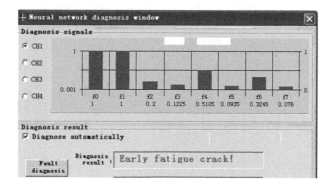

Fig. 6. Diagnosis interface of neural network

5   Conclusions

Because this online diagnosis system used VB programming, database storage, wavelet analysis and Matlab in integration, it has multiple functions and friendly interface and it is easy to be operated. The software design of large linear vibrating screen DZK2466 side crack fault characteristics implements the fault diagnosis of semi-automatic or fully automatic online processing function. Through actual application, it has been proved that this system has the advantages of classification accuracy and faster diagnostic speed, especially for fatigue crack of early warning of large scale linear vibrating screen. This software of online monitoring system has a very positive practical significance for the maintenance of the equipment of vibrating screens.

References

1. Ren, Z., Li, H., Jin, T., Zhang, C.: Vibration test and fault analysis on shaker of large-scale screen. Coal Mine Mach. **29**(11), 184–187 (2008)
2. Hua, Y.: Wavelet Transform and Engineering Application. Science Press, Beijing (1999)
3. Han, L.: Artificial Neural Network Theory, Design and Application. Chemical Industry Press, Beijing (2002)
4. Liu, H., Cai, Z., Wang, Y., et al.: Wavelet packet analysis and fault diagnosis of the rolling bearigns. Mech. Sci. Technol. **18**(2), 301–303 (1999)
5. Hou, Z., Gupta, M.M., et al.: A recurrent neural network for hierarchical control of interconnected dynamic systems. IEEE Trans. Neural Networks **18**(2), 466–481 (2007)
6. Xue, D.: Computer aided design of control system - MATLAB language and application, pp. 51–63. Tsinghua University Press (2006)
7. Chiang, H., Russell, L., Braatz, D.: Fault Detection and Diagnosis in Industrial Systems. Springer, Hong Kong (2001)
8. Yang, L., Li, X., Li, G.: Development of image displaying and processing software based on image grabbing card. J. Liquid Cryst. Disp. **25**(6), 909–913 (2010)

Research and Design of Lubrication System Based on J1939 Protocol

Wudai Liao⑩, Zongsheng Liu$^{(\boxtimes)}$⑩, Xiangyang Lu⑩, Jun Zhou⑩,
and Xiaosong Liang⑩

College of Electronic Information, Zhongyuan Institute of Technology,
Zhengzhou 45007, China
{wdliao,lzs919816}@sina.com

Abstract. In this paper, a lubrication system based on J1939 protocol is designed, and a method of detecting lubrication system based on fractional Fourier transform is proposed. According to the Raman effect detection in fractional order domain, a photosensitive sensor is used to accurately detect the lubricating oil, and a pressure sensor is used to monitor the oil pressure. When the refueling machine works, the system will automatically detect the refueling situation, lubricating oil pressure, oil condition and refueling times, which will be displayed on the OLED display screen through the CAN bus. Fractional Fourier analysis is applied to the collected data, and the lubrication system based on J1939 protocol realizes closed-loop control, which not only achieves the centralized lubrication effect, but also enables real-time monitoring of the oil supply system.

Keywords: J1939 protocol · CAN bus · Fractional order domain · Raman effect

1 Introduction

In the concentrated lubrication of the car, the current main use of manual refueling, manual lubricating oil refueling time and labor, refueling too little amount will lead to mechanical wear when the lubricating oil is exhausted or excessive unnecessary waste. The design of lubricating oil filling system will effectively solve these problems, and the use of fractional order domain Raman effect closed-loop detection technology for real-time monitoring.

In order to make the system work stably under the environment of high interference and noise. CAN bus is adopted for data transmission of the system. Due to its short frame number, fast data transmission and CRC check, it has excellent anti-interference ability and a complex dual-line structure, and CAN resist noise [1]. As the world's unified application-layer protocol standard [2], J1939

Supported by Zsupported by National Natural Science Foundation (NNSF) of China under Grant 6187020178.

© Springer Nature Singapore Pte Ltd. 2020
L. Pan et al. (Eds.): BIC-TA 2019, CCIS 1160, pp. 301–309, 2020.
https://doi.org/10.1007/978-981-15-3415-7_25

protocol is widely used in vehicles and large engine systems. J1939 is a closed-loop control network capable of supporting high-speed communication between multiple communication nodes. The 1939 protocol on CAN bus is selected as the application layer protocol of the system, which takes CAN bus as the core of communication network [1].

The lubrication system adopts multiple pipelines. In order to detect the oil output of each pipeline, the fractional order domain Raman effect terminal detection is designed, and the detection results are displayed in OLED screen. The CAN signal is tested and analyzed after the lubrication system based on J1939 protocol is connected to the automobile system.

2   Fractional Fourier Transform

Fourier analysis is widely used in signal processing and other branches of science and engineering. The fractional Fourier transform is the generalized Fourier transform, also known as the Angle Fourier transform or the rotational Fourier transform [3].

Fractional-order Fourier transform(FRFT) has been applied in wave propagation, optics and optical signal processing, time and space frequency analysis, pattern recognition, digital signal processing, image processing, optical communication and other fields [4]. Most applications are based on replacing the ordinary Fourier transform with the fractional Fourier transform. Because FRFT has additional degrees of freedom (order a), it is usually possible to generalize and improve the previous results.

2.1   Fractional Fourier Transform Algorithm

The fractional Fourier transform, the Angle α of $f(t)$ function is defined as

$$f_\alpha(u) = F^\alpha[f(t)](u) = \int_{-\infty}^{+\infty} K_\alpha(u,t)f(t)dt \tag{1}$$

Where the f ordinary Fourier transform of f_α. Integral kernel is

$$K_\alpha(u,t) = \begin{cases} A_\alpha e^{j\frac{u^2+t^2}{2}\cot\alpha - jut\csc\alpha} & \alpha \neq k\pi \\ \delta(u-t) & \alpha = 2k\pi \\ \delta(u+t) & \alpha = (2k+1)\pi \end{cases} \tag{2}$$

among $\alpha = a\pi/2$ is the angle of time-frequency rotation, k is integer,

$$A_\alpha = \sqrt{(1 - j\cot\alpha)/2\pi} \tag{3}$$

Therefore, the classical Fourier transform of signal can be regarded as the projection of its time-frequency distribution on the frequency axis. The spectrum of chirp signals in the Fourier domain will expand, as shown in Fig. 1. We can also say that the inverse transformation of an angular FRFT is an angular FRFT.

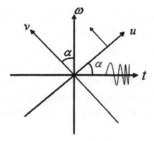

Fig. 1. Projection of LFM signal on the Fourier domain

2.2 Fractional Order Domain Detection Principle

The sensor adopts an optical system to detect the grease flow at any time, and transmit the detection signal to the internal MCU controller [5]. After the controller receives and processes the signal, it is sent into the bus by interruption or polling, and the monitoring situation is displayed on the display screen in real time according to the J1939 protocol. The Raman effect in fractional order domain is adopted for detection. The specific method is shown in Fig. 2.

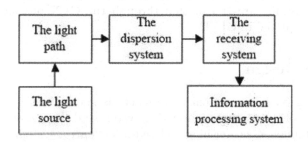

Fig. 2. Block diagram of fractional Raman detection

The collected data will be processed by the following equation.

$$I_0(x) = \frac{(x - x_1)(x - x_2)\Lambda(x - x_n)}{(x_0 - x_1)(x_0 - x_2)\Lambda(x_0 - x_n)} = \prod_{1 \le j \le n} \frac{x - x_j}{x_0 - x_j} \tag{4}$$

$$I_1(x) = \frac{(x - x_1)(x - x_2)\Lambda(x - x_n)}{(x_1 - x_0)(x_1 - x_2)\Lambda(x_1 - x_n)} = \prod_{\substack{1 \le j \le n \\ j \ne 1}} \frac{x - x_j}{x_1 - x_j} \tag{5}$$

$$I_n(x) = \frac{(x - x_1)(x - x_2)\Lambda(x - x_n)}{(x_n - x_1)(x_n - x_2)\Lambda(x_n - x_n)} = \prod_{1 \le j \le n-1} \frac{x - x_j}{x_n - x_j} \tag{6}$$

After the value is obtained, frit-order FFT operation is carried out to convert to the frequency domain, and the frit-order domain matrix after the transformation is obtained. Here is the change of order P and the result of transformation. The change of order describes the range of signal frequency change. The result of transformation is the magnitude of energy. By setting the corresponding threshold matrix $X_p(u_0)$, judge the frequency change degree, and judge the test result according to the frequency change. Specifically, Fractional Fourier Transform of $x(t)$ Signal:

$$X_p(u) = \int_{-\infty}^{+\infty} x(t)K_p(t,u)dt \tag{7}$$

After obtaining the calculated result of frequency variation and comparing it with the corresponding initial value, the equation is as follows:

$$p_1(x) = X_p(u) - X_p(u_0) \tag{8}$$

$$p_2(x) = X_p(u)/X_p(u_0) \tag{9}$$

Where $P_1(x)$ is the result of the change, $P_2(x)$ is the speed of the change in frequency.

The frequency changes of $P_1(x)$ and $P_2(x)$ are used to detect the lubrication system end. Judge whether the pipeline has lubricating oil and filling amount. The detection results are transmitted through the CAN bus based on J1939 protocol and displayed in OLED.

3   System Design

The lubricating oil filling system is designed as shown in Fig. 3. The system mainly consists of the design of lubricating system and the end detection of fractional order Raman effect, forming a closed-loop control system. The hardware system is as follows: the main control circuit USES SWM220 as the main control chip, and its peripheral circuits include power supply circuit, SWD download debugging circuit, CAN controller circuit, reset circuit, clock circuit and indicator circuit. The main control chip of SWM220C8T7 is the control and processing center of the whole system, which is responsible for driving each module, collecting data of each module, or sending data to each module [6]. MCU is used to form a central control mechanism, and each module can work cooperatively. The function of CAN controller circuit is to send CAN signal of MCU. The reset circuit will generate a reset after receiving the signal. The clock circuit provides a stable clock frequency for the chip. Finally, automatic lubricating oil filling is realized.

3.1   CAN Transceiver Circuit Design

CAN transceiver is used to communicate between lubrication system and automobile system network [7]. As shown in Fig. 4, CAN signal receiving pin

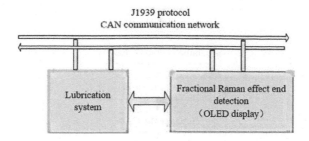

Fig. 3. General block diagram of system design

CAN_RX and sending pin CAN_TX are directly connected with the single chip microcomputer. CAN controller adopts 5 V power supply. In order to prevent over current impact, CANH by 120 Ω resistance and CANL pin connection. The stability of CAN bus transmission is improved. CAN controller module is used to connect the chip to the upper computer and send the data to the upper computer. Complete the communication between lubrication system and automobile network.

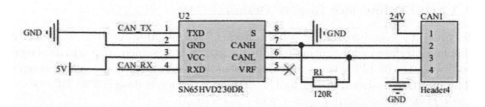

Fig. 4. CAN transceiver circuit

3.2 Automatic Transmission Program Design Based on J1939 Protocol CAN Bus

CAN bus automatically sends software unit program flow chart, as shown in Fig. 5. Firstly, the related pins are initialized. Then carry out OLED initialization, timer initialization, CAN bus initialization. Sets a loop in the main function and interrupts it in the loop body according to the timer. It detects whether the CAN signal of the upper computer is received or not every second. If it is not received for a long time, the OLED CAN bus connection fails. If it is successfully received, the single chip microcomputer will send the state of the filling machine, liquid level,pressure,motor state and other information every second.

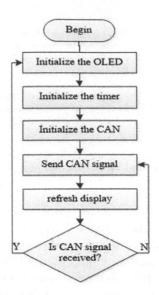

Fig. 5. CAN bus program flow chart

3.3 OLED Interface Display Design

In order to visualize the detection effect of fractional-order Raman effect, a display interface of lubricating oil filling status was designed. Use OLED screen to display the oil filling times, working status, oil level status and pressure status. The interface is shown in Fig. 6. The display interface is designed as follows:

$OLEDShowChinese(0, Line0, 4, ZhuYouCiShu)$;
$OLEDShowChinese(0, Line1, 4, GongZuoZhuangTai)$;
$OLEDShowChinese(0, Line2, 4, YouWeiZhuangTai)$;
$OLEDShowChinese(0, Line3, 4, YaLiZhuangTai)$;

After the interface is planned, programs are written to implement subsequent displays, displaying values and states, and refreshing every second. The filling interface realizes real-time monitoring of lubrication filling state.

Fig. 6. OLED display interface

4   CAN Signal Test Analysis

CAN bus system integration test. Bus system integration test is to connect each node to form a complete CAN bus system. The system is tested to verify the integrity and correctness of the system operation, the communication robustness of the system, the electrical robustness, and the fault-tolerant self-recovery function of the system. Build the test environment, implement the test, and finally analyze the test data.

4.1   Lubrication System and Automobile System CAN Signal Detection

Firstly, the CAN signal sent by the lubricating oil filling system based on J1939 protocol and the internal CAN signal are analyzed. Waveform was detected by oscilloscope. From Fig. 7, CAN signal amplitude is 1.50 V, basic voltage is 2.51 V to 3.23 V, and the differential pressure range is 0.72 V. According to the test voltage, the waveform of lubricating oil filling system based on J1939 protocol is in normal state. Figure 8 shows the internal CAN signal waveform of the automobile. The waveform is in a normal state after being tested by the detection tool.

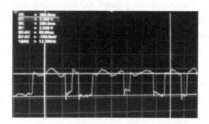

Fig. 7. Waveform of lubricating oil filling system

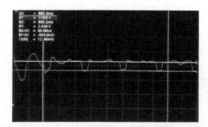

Fig. 8. CAN signal waveform inside the car

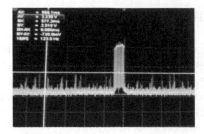

Fig. 9. Waveform of vehicle control system connected to CAN communication of automobile system by lubrication system

4.2   CAN Signal Test and Analysis of Lubrication System Connected to Internal Ssystem of Automobile

When the lubricating oil filling system based on J1939 protocol is connected to the automobile system, CAN communication appears abnormal. As can be observed in Fig. 9, the basic voltage is consistent, and the signal waveform presents a double-layer state. After analysis and detection, the CAN signal sent by the lubricating oil filling system affects the CAN signal of the vehicle itself. Part of the function prompts abnormal communication. The upper waveform is the CAN signal sent by the control panel of lubricating oil filling system, while the lower waveform is the CAN signal of the vehicle itself. The receiving end can receive normally. According to the common mode interference [8], the hardware part of lubricating oil filling system based on J1939 protocol is optimized. Stable LM2596 voltage stabilizer chip is used to ensure the reliability of waveform [9]. After adding the power regulator module and modifying the baud rate to 250 KHz, it is found that the interference has disappeared, the detection circuit waveform is normal, and the communication is normal. It CAN be seen from Fig. 10 that the lubricating oil filling system CAN communicate with the meter normally through the CAN network. The ground waveform detected by CAN logic analyzer is shown in Fig. 11. The waveform is normal, and the lubrication system connected to the internal system of the automobile CAN communication returns to normal.

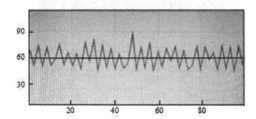

Fig. 10. Ground waveform of lubricating oil filling system after CAN bus communication system is connected

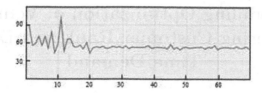

Fig. 11. Ground alignment results read by CAN logic analyzer

5 Conclusion

This paper designs a lubricating oil filling system based on J1939 protocol. The lubrication system and end detection transmit data through CAN bus based on J1939 protocol, which realizes the closed-loop control of the system and makes the system more stable. The fractional-order Raman effect is used for terminal detection and the results are displayed intuitively. Finally, the system CAN signal detection analysis and the lubrication system based on J1939 protocol access to the car system CAN bus communication analysis. The lubrication system and automobile system realize communication control and verify the feasibility of the design system.

References

1. Wu, J., Li, Y., Li, J., Li, Y., Yu, H., Song, L.: Vehicle drive control CAN bus based on SAEJ1939 protocol. J. Jilin Univ. (Eng. Sci. Edn.) **39**(04), 855–858 (2009)
2. Vehicle Application Layer-Diagnostics. SAE Standards. SAE J1939/73 (2005)
3. Tao, R., Deng, B., Wang, Y.: Fractional Fourier Transform and Its Applications, pp. 13–15. Tsinghua University Press, Beijing (2009)
4. Ma, J., Miao, H., Su, X., Gao, C., Kang, X., Tao, R.: Research progress on fractional Fourier transform theory and its application. Optoelectron. Eng. **45**(06), 5–28 (2012)
5. Zhou, L., Liu, X.: Detection of tunnel low contrast cracks based on frying-order Fourier transform. Comput. Sci. **46**(S1), 208–210 (2019)
6. Yu, Z., Zhang, X.: Design and implementation of battery management system based on J1939 protocol. Power Technol. **40**(10), 1950–1952 (2016)
7. Xu, B., Pan, M.: Design of vehicle communication system based on J1939 protocol. Sci. Technol. Eng. **12**(33), 9114–9117 (2012)
8. Yi, W., Luo, S., Peng, Y.: Simulation and test of all-electric vehicle drive control system based on CAN bus. China Test **40**(03), 105–108 (2014)
9. Cui, S., Gong, Y.: Modeling and fault diagnosis simulation of electric vehicle pedal and shift system based on CAN bus. J. Chongqing Univ. Technol. (Nat. Sci.) **33**(03):15–20+85 (2019)

Scheduling Optimization of Vehicles Considering Customer Rank and Delivery Time Demand

Wenqiang Yang[1,2(✉)], Hao Guo[1,2], and Jianxiu Su[1]

[1] School of Mechanical and Electrical Engineering,
Henan Institute of Science and Technology, Xinxiang, China
yangwqjsj@163.com
[2] Postdoctoral Research Base, Henan Institute of Science and Technology,
Xinxiang, China

Abstract. Customer satisfaction is an important factor to evaluate the service quality of the distribution of the vehicles, which is mainly reflected in the vehicles delivering goods to customers on time, which usually has certain correlation with customer rank for maintaining the customer relationships. So the scheduling optimization problem of the vehicles is modeled with the minimum transportation cost, and the earliness and tardiness penalty regarded as the optimization goal, which is solved by an artificial fish swarm algorithm (AFS). And yet, AFS has low optimization precision and low convergence speed in the later period of the optimization. To overcome such shortcomings, this paper proposes an improved artificial fish swarm algorithm (IAFS) based on elitist guiding evolution strategy, crossover operator with cyclic misalignment and heuristic mutation strategy. Finally, simulation examples show that the validity and effectiveness of the IAFS.

Keywords: Scheduling optimization of vehicles · Artificial fish swarm algorithm · Customer rank · Elitist guiding evolution strategy · Crossover operator with cyclic misalignment · Heuristic mutation strategy

1 Introduction

Distribution plays an important role in the logistics system, while vehicle routing problem (VRP) is of vital importance for logistics distribution management. The relevant data shows that the distribution cost accounts for approximately 54.5% in all kinds of logistics cost [1]. So taking the scheduling optimization of vehicle as the subject of this paper has the theoretical and practical meaning for reducing distribution costs, improving the operational earnings of enterprise and achieving scientific logistics. In past years, much research has been devoted

Supported by the National Nature Science Foundation of China (No. 61773156).

© Springer Nature Singapore Pte Ltd. 2020
L. Pan et al. (Eds.): BIC-TA 2019, CCIS 1160, pp. 310–325, 2020.
https://doi.org/10.1007/978-981-15-3415-7_26

to methods reducing logistics distribution time or distance. For solving the periodic vehicle routing problem, Tenahua et al. [2] present an iterated local search metaheuristic, the problem is solved in two phases: the first step is to assign days of visit to each customer, and in the second step to determine the routes that each vehicle must perform each day. Elgesem et al. [3] introduce a single-ship routing problem with stochastic travel times that is faced by a chemical shipping company, which is modeled as a stochastic traveling salesman problem with pickups and deliveries, in which the goal is to find the route within the port with maximized probability that its total length does not exceed a threshold. Ho et al. [4] consider multi-depot vehicle routing problem as a complex combination optimization, and propose a hybrid genetic algorithm to solve it. Kalayci et al. [5] extend the basic vehicle routing problem to a typical vehicle routing problem where pickup and delivery operations are simultaneously taken into account. Meanwhile, a hybrid metaheuristic algorithm based on an ant colony system and a variable neighborhood search is developed for its solution. Salhi et al. [6] establish the mathematic model for the multi-depot fleet size and mix vehicle routing problem, which is solved by using the variable neighborhood search method. Morais et al. [7] address the vehicle routing problem with cross-docking where the vehicles leave a single cross-dock towards the suppliers, pick up products and return to the cross-dock, and make use of iterated local search heuristics to get the objective amount. Song et al. [8] analyze the characters of the capacitated vehicle routing problem with recycling in reverse logistics and formulate it as a combinatorial optimization problem, which a food chain algorithm is designed for. Silvestrin et al. [9] deem that it is essential for multiple compartments frequently arises in practical applications when there are several products of different quality or type, that must be kept or handled separately. It is treated as a variant of the vehicle routing problem that allows vehicles with multiple compartments and solved with an iterated search. Liu et al. [10] develop two mixed-integer programming model for the vehicle scheduling problem with simultaneous delivery and pickup. And correspondingly, propose two heuristic algorithms to solve the two models. Defryn et al. [11] research the clustered vehicle routing problem in which customers are grouped into predefined clusters, and all customers in a cluster must be served consecutively by the same vehicle, and obtain the solution with the fast two-level variable neighborhood search method. Vaz et al. [12] model a complex vehicle routing problem applied to the response phase after a natural disaster, which involves a heterogeneous fleet of vehicles, multiple trips, multiple depots, and vehicle-site dependencies, and is solved through the proposed generic hybrid heuristic.

Vehicle routing problem have received quite some attention in the above literatures. However, the customer being the important resources of the enterprises, with which the relationships must be maintained positively. Therefore, the rank of the customer which reflects the degree of each customer importance can be used as a reference during routing vehicles. In this way, it can enhance customer satisfaction and boost profits for enterprises. For this reason, the problem of vehicle routing while respecting the rank of the customers is researched in this study.

The remainder of this paper is organized as follows: Sect. 2 describes the problem concerned in this study, introduces the model assumptions and models the scheduling of vehicles using mixed integer programming approach. Section 3 proposes an improved artificial fish swarm algorithm, and Sect. 4 derives a theoretical analysis for convergence and stability of the proposed algorithm. In Sect. 5, the computational experiments and comparison results are discussed. Finally, Sect. 6 concludes and suggests some directions for further research.

2 Problem Description and Modeling

2.1 Problem Description

Here, we describe the problem of vehicle scheduling optimization considering the rank of the customers as a variant of the standard vehicle routing problem. The process that we describe is as follows. A given vehicle picks the orders. Vehicle starts completely empty and then receives a set of pre-specified orders to be picked. The weights of received orders the vehicle do not exceed its maximum load. Once all the orders have been picked, subsequently, the vehicle distributes the orders based on the optimal delivery sequence.

2.2 Model Assumptions

The following assumptions are made in this study:

(I) Single distribution center is assumed.
(II) The vehicle drives at a constant speed.
(III) The order demands of customers are certain in advance.
(IV) The vehicle faults during the transportation are ignored.
(V) Running out of inventory in the distribution center is not considered.

2.3 Notation

O: The set of customer and distribution center.
W_i, V_i: The weight and the volume of customer order i, respectively. $i \in O \backslash \{0\}$.
Q_{max}, V_{max}: The maximum load and the maximum volume of the vehicle, respectively.
R: The set of the subtours of vehicle completing all the customer orders.
O_f, O_l: The set of first and last customer order in the subtours of vehicle.
v: The average speed, which the vehicle drives at.
S_{ir}: The binary decision variable, indicating whether or not order i is distributed in subtour r, $i \in O, r \in R$.
H_{ijr}: The binary decision variable, indicating whether the vehicle goes from customer i to customer j directly in subtour r, $i, j \in O, r \in R$.
d_{ij}: Travel distance between customer and distribution center, or between customer and customer, $i, j \in O$, while i or j equal 0, it denoting the distribution center accordingly.

2.4   Problem Modeling

Zero inventory service is provided to customs is a goal for which distribution center strives. As it helps the customs reduce costs and increase profits. Therefore, the cost generated by earliness and tardiness should be considered. Meanwhile, the customer, which is the important resources of the enterprises. The key customer, especially, is the enterprise's main benefit origin. In order to maintain good relationship with customers, it is necessary to distribute orders according to customer rank. Thus, the mathematical model in this paper including the total travel cost (TC), earliness and tardiness penalty based on customer rank (ETP), can be minimized.

TC can be formulated as:

$$TC = \sum_{i \in O} \sum_{j \in O \setminus \{i\}} \frac{d_{ij}}{v} \sum_{r \in R} H_{ijr} S_{ir} S_{jr} \qquad (1)$$

Besides, ETP can be formulated as:

$$ETP = \sum_{i \in O} (\lambda_i E_i + \delta_i T_i) \qquad (2)$$

where λ_i and δ_i which are relative to the rank of customer i, are the earliness penalty per unit time and the tardiness penalty per unit time, respectively. E_i, the earliness time of customer i. T_i, the tardiness time of customer i. They can be expressed as respectively:

$$E_i = max\{0, l_i - c_i\} \qquad (3)$$

$$T_i = max\{0, c_i - u_i\} \qquad (4)$$

Where l_i, the lower bound of the due time window for customer i. In the same way, u_i, the upper bound of the due time window. c_i, the completion time customer order i.

Therefore, the problem in this study can be modeled as:

$$min(TC + ETP) = min(\sum_{i \in O} \sum_{j \in O \setminus \{i\}} \frac{d_{ij}}{v} \sum_{r \in R} H_{ijr} S_{ir} S_{jr} + \sum_{i \in O} (\lambda_i E_i + \delta_i T_i)) \quad (5)$$

Subject to

$$\sum_{r \in R} S_{ir} = 1 \qquad \forall i \in O \qquad (6)$$

$$\sum_{j \in O \setminus \{i\}} H_{ijr} = \sum_{i \in O \setminus \{j\}} H_{ijr} = 1 \, \forall i \in O \setminus \{O_l\}, \forall j \in O \setminus \{O_f\}, \forall r \in R \qquad (7)$$

$$\sum_{i,j \in O} H_{ijr}(W_i + W_j) \le Q_{max} \qquad \forall r \in R \qquad (8)$$

$$\sum_{i,j \in O} H_{ijr}(V_i + V_j) \leq V_{max} \qquad \forall r \in R \tag{9}$$

$$S_{ir} \in \{0, 1\} \qquad \forall i \in O, \forall r \in R \tag{10}$$

$$H_{ijr} \in \{0, 1\} \qquad \forall i, j \in O, i \neq j; \forall r \in R \tag{11}$$

The Eq. (5) is the problem objective function. Constraint (6) ensures that only one customer order is allocated to each vehicle. Constraint (7) guarantees that there is no circuit between the customer i and the customer j. Constraint (8) and (9) limits the total weight (volume) in a sub-tour to less than the maximum load (volume) of the vehicle. Constraints (10) and (11) are binary decision variables, which only are either 0 or 1.

3 Improved Artificial Fish Swarm Algorithm (IAFS)

In order to solve the problem of this paper better, a so-called IAFS algorithm is proposed, in which elitist guiding evolution strategy, crossover operator with cyclic misalignment and heuristic mutation strategy are introduced, as discussed below in detail.

3.1 Basic Principle of AFS

Artificial fish swarm algorithm is a biological modeling algorithm and meta-heuristic searching algorithm firstly developed by Li [13] in 2002, which is inspired by the foraging behavior of the fish swarm and has been applied to solve many optimization problems [14–19]. Generally speaking, a certain water area with the greater number of the fish, where the more nutrition there is. Here, the water area with more nutrition can be seen as the optimal or suboptimal solution. Therefore, such characteristic of AFS can be used to realize the global optimization.

3.2 Fundamental Behaviors of AFS

Given N artificial fish, the position of each artificial fish X_i ($i \in \{1, 2, ..., N\}$) can be represented as $X_i = (X_{i1}, X_{i2}, ..., X_{in})$, where n is the dimensionality of the problem to optimize. Furthermore, the food concentration being in the position X_i is evaluated by the function $Y_i = f(X_i)$. To mimic each artificial fish being in finding food, three main behaviors are defined as follows:

Foraging Behavior. The artificial fish looks for food by mean of vision. Once the area with more food is discovered, which the artificial fish will swim toward. i.e., a position X_j that is within the vision scope of X_i is selected randomly by Eq. (12). If $Y_j > Y_i$, the artificial fish swims in the direction of X_j by Eq. (13); Otherwise, if fails after try-number attempts, the artificial fish swims randomly by Eq. (14).

$$X_j = X_i + Visual \cdot rand() \tag{12}$$

$$X_{i\_next} = X_i + rand() \cdot step \cdot \frac{X_j - X_i}{\|X_j - X_i\|} \tag{13}$$

$$X_{i\_next} = X_i + rand() \cdot step \tag{14}$$

where $Visual$ is the vision scope of the artificial fish, $rand()$ is a random number ($rand() \in [0, 1]$), step denoting the maximum distance that the artificial fish swims in forage behavior.

Swarming Behavior. To avoid dangers, the artificial fish tend to cluster. i.e., within the vision scope of X_i, the number of the artificial fish is n_f, the central position is X_c. If $\frac{Y_c}{n_f} > \delta Y_i$, meaning more food and less crowded in X_c relative to X_i, the artificial fish swims to the central position X_c by Eq. (15); otherwise performs foraging behavior.

$$X_{i\_next} = X_i + rand() \cdot step \cdot \frac{X_c - X_i}{\|X_c - X_i\|} \tag{15}$$

where δ is the crowded-factor.

Rearing Behavior. In order to acquire more food rapidly, the artificial fish tries to follow the fish positioned at a place with more food. i.e., within the vision scope of X_i, the number of the artificial fish is n_f, the position with more food is X_{more}. If $\frac{Y_{more}}{n_f} > \delta Y_i$, reflecting more food concentration X_j relative to X_i, the artificial fish swims to the position X_{more} by Eq. (16); otherwise performs foraging behavior.

$$X_{i\_next} = X_i + rand() \cdot step \cdot \frac{X_{more} - X_i}{\|X_{more} - X_i\|} \tag{16}$$

In this section, we mainly introduce the characteristics and basic idea of AFS, the pseudocode of AFS can be shown in Algorithm 1.

Algorithm 1. The pseudo code of AFS
1. Initializing a population of N artificial fish with random positions
2. Initialization of parameters
3. Evaluate each artificial fish in the population using the fitness function, update the bulletin board with the best individual
4. Define the maximal number of iterations $MaxGen$, and set $t=1$
5. $while(t < MaxGen)$ do
6. for $i=1:N$
7. if $\frac{Y_{more}}{n_f} > \delta Y_i$ $then$
8.   Performing rearing behavior
9. $elseif$ $\frac{Y_c}{n_f} > \delta Y_i$ $then$
10.   Performing swarming behavior
11. $else$
12.   Performing foraging behavior
13. $endif$
14. $endfor$
15.Compare the position of each artificial fish with bulletin board. If superior to that of the bulletin board, then update the bulletin board
16.Set $t=t+1$
17.$endwhile$
18.output the best solution

3.3   Improvements to AFS

Elitist Guiding Evolution Strategy. In AFS, the positions of the fittest artificial fish in each generation are preserved in the bulletin board, but not involved in the subsequent evolutions. i.e., such elitists dont make use of their own advantages to guide the evolutionary search so as to improve the convergence efficiency. However, guiding excessively that makes it easy to lead the AFS to be trapped in local optima. To overcome such drawbacks, the elitist guiding evolution strategy is only valid for such artificial fish, the objective function value of which has not been improved after try_number attempts during foraging behavior. Then Eq. (13) is modified as follow:

$$X_{i\_next} = X_i + rand() \cdot step \cdot \frac{X_j - X_i}{\|X_j - X_i\|} \tag{17}$$

where X_i and X_j represents the current position of artificial fish i and the position of artificial fish with the highest food concentration in the bulletin board, respectively. Where X_i and X_j represents the current position of artificial fish i and the position of artificial fish with the highest food concentration in the bulletin board, respectively.

Crossover Operator with Cyclic Misalignment. Crossover operation is performed to generate better offspring by trading genes between two chosen parents [20]. To fully exploit the space around the suboptimal solutions, the two chosen parents, of which one comes from the bulletin board, the other is the current individual. Furthermore, a two-point crossover operation is employed during which two gene segments are randomly chosen and the two segments of parent chromosomes are swapped to form offspring chromosomes. However, if the two chosen parents are identical, the children produced are also identical. Such cases make the population be trapped into local optimum obviously, so that the optimum cannot be obtained. To overcome such shortcoming, a cyclic misalignment crossover, which selects the two gene segments of the same size located in different places in their respective parents, is proposed. e.g., as shown in Fig. 1. Finally, the offspring satisfying constraints that has better fitness than the current individual, and replace it; Otherwise, discarded.

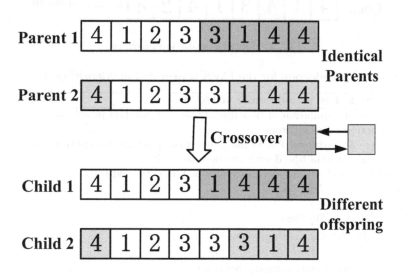

Fig. 1. Crossover operator with cyclic misalignment.

Heuristic Mutation Strategy. From the above analysis of AFS, swarming behavior and rearing behavior during search process lower population diversity, which make it very difficult for AFS to effectively explore the solution space, and even easily trapped into local optimum. To date, mutation operator has been regarded as an effective method, which is used to solve the premature convergence problem. Likewise, if mutation operation is performed randomly, it will enhance the stochastic behavior and blindness of AFS. At this point, it is necessary to study the mechanism of the local optimum to propose a mutation operation to suit AFS. As far as we know, whether swarming behavior or rearing behavior, both are equally subject to the scope of vision during evolution. Inspired by this, in this paper, mutation operator should be limited to such individuals within the scope of vision. Meanwhile, in order to enlarge the solution space and jump out of local optimum by a greater probability, a novel heuristic mutation strategy is presented based on customer rank and customer order cost, i.e., the two orders representing the largest portion and the smallest portion of the costs of all customer orders, respectively, are selected to exchange for each other, and thus it will have a very significant effect on the distribution cost of all customer orders, e.g., a concrete process of how to implement the method is described through Fig. 2.

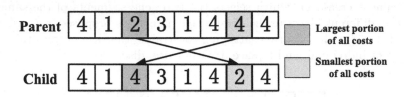

Fig. 2. Heuristic mutation operator.

A detailed pseudocode for the IAFS is provided in Algorithm 2.

Algorithm 2. The pseudo code of IAFS
1. Initializing a population of N artificial fish with random positions
2. Initialization of parameters
3. Evaluate each artificial fish in the population using the fitness function, update the bulletin board with the best individual
4. Define the maximal number of iterations $MaxGen$, and set $t=1$
5. $while(t < MaxGen)\ do$
6. $for\ i=1{:}N$
7. $if\ \frac{Y_{more}}{n_f} > \delta Y_i\ then$
8.   Performing rearing behavior
9. $elseif\ \frac{Y_c}{n_f} > \delta Y_i\ then$
10.   Performing swarming behavior
11. $else$
12.   Performing foraging behavior with elitist guiding evolution strategy
13. $endif$
14. Compare the position of the current artificial fish with bulletin board. If superior to that of the bulletin board, then update the bulletin board
15. $endfor$
16. $if\ rand() < $ crossover probability $p_c\ then$
17.   Performing crossover operation with cyclic misalignment
18. $endif$
19. $if\ rand() < $ mutation probability $p_m\ then$
20.   Performing heuristic mutation operation
21. $endif$
22. Compare the position of each artificial fish with bulletin board. If superior to that of the bulletin board, then update the bulletin board
23. Set $t=t+1$
24. $endwhile$
25. output the best solution

4   Convergence Analysis of IAFS

Definition 1. *Let* $\{X_k, k = 0, 1, ...\}$ *be a discrete sequence, where* X_k *takes values in a finite state space S. For* $\forall k \geq 0$ *and* $\gamma_q \in S$ $(q \in \{1, 2, ..., k+1\})$, *if the Eq. (18) is always valid, then is called as a finite state Markov chain.*

$$P\{X_{k+1} = \gamma_{k+1}|X_0 = \gamma_0, X_1 = \gamma_1, \cdots, X_k = \gamma_k\}$$
$$= P\{X_{k+1} = \gamma_{k+1}|X_k = \gamma_k\} \tag{18}$$

Definition 2. *The conditional probability* $P\{X_{k+1} = \gamma_j|X_k = \gamma_i\}$ *that represents the probability for the Markov chain* $\{X_k, k = 0, 1, \cdots\}$ *changing from* γ_i *at the moment k to* γ_j *at the moment* $k+1$, *is noted as* $p_{ij}(k)$. *Furthermore, if* $p_{ij}(k)$ *is irrelevant to the moment k,* $\{X_k, k = 0, 1, \cdots\}$ *is known as a finite homogeneous Markov chain.*

Proposition 1. *The state sequences of the population of IAFS is a finite homogeneous Markov chain.*

Proof. Here, the set of all possible artificial fish positions denoted as S is called the artificial fish state space. Since the scale of problem to be solved is limited, thus S is finite. Based on the principle of IAFS, we can conclude that transition probability $p_{ij}(k)$ corresponding to the artificial fish changing from the state γ_i at the moment k to γ_j at the next moment by means of foraging, rearing, swarming, crossover or mutation, only has something to do with γ_i and is independent of the moment k. Consequently, the state sequences of the population of IAFS is a finite homogeneous Markov chain according to Definition 2.

Definition 3. *Suppose* $Q_k = \min\{F(x_i)|i \in (1, 2, \cdots, U)\}$ *includes all the best feasible solutions up to iteration k, and if satisfies*

$$\lim_{k \to \infty} P\{Q_k \cap Q^* \neq \emptyset\} = 1 \tag{19}$$

where Q^* *are global optimal solutions, then the corresponding algorithm is global convergent.*

Proposition 2. *The IAFS presented in this paper possesses global convergence.*

Proof.
$$P\{Q_k \cap Q^* \neq \emptyset\} = 1 - P\{Q_k \cap Q^* = \emptyset\} \tag{20}$$

As the strategy that the fittest individual is held is employed in this study, in view of Bayes formula, the Eq. (20) can be also represented as follows.

$$P\{Q_k \cap Q^* \neq \emptyset\} = 1 - P\{Q_1 \cap Q^* = \emptyset, Q_2 \cap Q^* = \emptyset, \cdots, Q_k \cap Q^* = \emptyset\}$$
$$= 1 - P\{Q_1 \cap Q^* = \emptyset\} \times P\{Q_2 \cap Q^* = \emptyset|Q_1 \cap Q^* = \emptyset\} \tag{21}$$
$$\times \cdots \times P\{Q_k \cap Q^* = \emptyset|Q_{k-1} \cap Q^* = \emptyset\}$$

Assume

$$\varsigma = \max\{P\{Q_1 \cap Q^* = \emptyset, P\{Q_k \cap Q^* = \emptyset | Q_{k-1} \cap Q^* = \emptyset\}\}, k = 2, 3, \cdots \quad (22)$$

Then the following can be deduced refer to Eqs. (21) and (22).

$$P\{Q_k \cap Q^* \neq \emptyset\} \geq 1 - \varsigma^k \quad (23)$$

Furthermore, due to $\varsigma \in (0, 1)$ Hence

$$\lim_{k \to \infty} P\{Q_k \cap Q^* \neq \emptyset\} \geq 1 - \lim_{k \to \infty} \varsigma^k = 1 - 0 = 1 \quad (24)$$

and because

$$0 \leq \lim_{k \to \infty} P\{Q_k \cap Q^* \neq \emptyset\} \leq 1 \quad (25)$$

With Eqs. (24) and (25), the following equation is derived.

$$\lim_{k \to \infty} P\{Q_k \cap Q^* \neq \emptyset\} = 1 \quad (26)$$

In conclusion, the IAFS presented in this paper has global convergence.

5   Experimental Results and Analysis

To test the performance of the proposed IAFS method, the experiments are carried out in a PC with Windows 10 system, 3.7 GHz Intel Core, 4 GB RAM, and MATLAB R2014b. Here, a set of customer orders are generated randomly, which is summarized in Table 1. Furthermore, the detailed information of each item is illustrated in Table 2. The earliness penalty per unit time λ and the tardiness penalty per unit time ρ both relevant to customer rank are given in Table 3. Other parameters are set as $v = 50$ km/h, $Q_{max} = 8{,}000$ kg, $V_{max} = 900$ m^3. Here, the distances between any two locations being included in the locations of distribution center and customers, which are uniformly distributed between 20 km and 250 km.

5.1   Analysis of Parameters Influence on the IAFS

Different setting of control parameters may bring some effect on the evolution performance of IAFS, such as $Visual$, $Try - number$, $Step$ and $Crowded - factor$. In order to find the best parameter combinations, it is necessary to study the parameter setting. Figure 3 depicts the impact of such parameters on the performance of the proposed algorithm, which indicates that the best setting for $Visual$ is 11, $Try - number$ is 50, $Crowded - factor$ is 0.7 and $Step$ is 6. Furthermore with Fig. 3, we conclude that the larger $Visual$ may cause the blindness of search; the larger $Try_n umber$ makes it easy to trap in the local minimum; the fixed $step$, no matter what the $step$ is, which will adversely affect the performances of IAFS; the larger the $Crowded_f actor$ that represents the degree of crowd of the artificial fish, the easier to escape from the local minimum, but it's convergence is slower.

Table 1. The information of customer orders.

No_C	Rank_C	Description of orders	Due time
C1	1	(A, 2), (C, 6), (D, 3), (E, 6), (F, 6)	(8:20, 8:50)
C2	2	(A, 3), (B, 8), (C, 7), (E, 7), (F, 6), (G, 10)	(9:10, 9:25)
C3	4	(B, 5), (C, 6), (D, 7), (G, 8)	(8:10, 8:32)
C4	3	(A, 10), (B, 15), (D, 3), (G, 9)	(11:20, 12:10)
C5	5	(B, 12), (C, 6), (D, 9), (E, 4), (F, 13), (G, 8)	(10:05, 10:17)
C6	4	(B, 4), (C, 10), (D, 4), (G, 12)	(8:20, 8:50)
C7	4	(A, 11), (B, 4), (C, 14), (D, 3), (E, 12), (F, 4), (G, 10)	(9:14, 9:59)
C8	4	(B, 5), (C, 6), (D, 7), (E, 12), (G, 8)	(12:10, 12:25)
C9	1	(A, 13), (B, 15), (C, 12), (D, 13), (E, 2), (F, 9), (G, 14)	(9:28, 9:37)
C10	1	(B, 12), (C, 3), (D, 8), (G, 6)	(9:07, 9:46)
C11	2	(A, 5), (B, 10), (C, 8), (D, 13), (E, 4), (G, 13)	(11:41, 12:04)
C12	1	(B, 7), (C, 16), (D, 13), (F, 14), (G, 11)	(13:01, 13:18)
C13	5	(A, 13), (B, 14), (C, 13), (D, 4)	(8:32, 8:47)
C14	5	(A, 9), (C, 10), (D, 13), (E, 10), (F, 16), (G, 11)	(9:21, 9:24)
C15	3	(B, 7), (C, 10), (D, 14), (G, 20)	(11:09, 11:18)
C16	2	(B, 12), (C, 10), (E, 12), (F, 10), (G, 13)	(13:24, 13:34)
C17	4	(A, 6), (C, 13), (E, 10), (F, 16), (G, 18)	(8:37, 9:15)
C18	5	(A, 2), (C, 18), (E, 17), (G, 18)	(12:08, 12:34)
C19	1	(B, 12), (C, 13), (G, 16)	(11:51, 12:15)
C20	1	(A, 6), (B, 7), (C, 3), (D, 2), (E, 13), (G, 16)	(11:21, 12:34)
C21	2	(A, 14), (B, 10), (C, 3), (D, 5), (G, 12)	(9:26, 9:59)
C22	5	(A, 9), (B, 17), (C, 6), (D, 7), (E, 8), (F, 17)	(10:37, 11:01)
C23	3	(B, 16), (C, 13), (D, 12), (E, 10), (F, 9), (G, 18)	(11:41, 12:04)
C24	2	(A, 9), (C, 18), (E, 13), (F, 7), (G, 5)	(11:11, 11:46)
C25	3	(B, 21), (D, 5), (E, 7), (F, 16)	(12:07, 12:45)
C26	1	(A, 13), (B, 12), (C, 20), (D, 12), (G, 13)	(9:23, 9:45)
C27	1	(B, 12), (C, 9), (D, 15), (G, 22)	(9:26, 9:54)
C28	4	(B, 13), (C, 15), (D, 11), (E, 10), (F, 14), (G, 16)	(13:17, 13:44)
C29	2	(A, 12), (C, 11), (D, 19), (G, 13)	(9:46, 9:58)
C30	1	(B, 13), (C, 15), (D, 14), (F, 10), (G, 19)	(9:54, 10:24)

Table 2. The information of items.

Item name	A	B	C	D	E	F	G
Unit weight (kg)	10	5	31	20	21	28	7
Unit volume (m³)	7	3	7	6	5	7	5

Table 3. λ and ρ corresponding to each customer rank.

Rank	1	2	3	4	5
λ	1.1	1.6	2	2.3	2.9
ρ	2.1	2.5	3.1	3.5	4.3

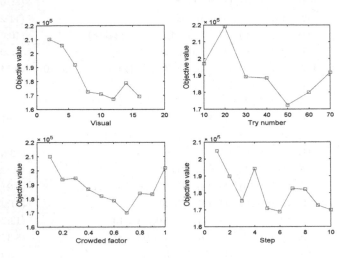

Fig. 3. Crossover operator with cyclic misalignment.

5.2   Compared with Other Evolutionary Algorithms

In order to validate the performance of the proposed algorithm in this paper, compared with standard genetic algorithm (GA), standard particle swarm optimization algorithm (PSO) and standard artificial fish swarm algorithm (AFS) based on the order information in Tables 1, 2 and 3. As for all the algorithms, the common control parameters are set as follows: the population size is set to 50, the maximum iteration number is 600. Other special parameters of each algorithm that include crossover probability 0.8 and mutation probability 0.05 being from GA. inertia weight 1.3, both learning factor are 2 being from PSO. *Step* 5, *visual* 14, *trynumber* 40 and *crowdedfactor* 0.5 being from AFS. For IAFS, basic parameter settings has been described in Sect. 5.1, other parameters, such as crossover probability 0.8 and mutation probability 0.05. Finally, the simulation results are compared, as shown in Figs. 4 and 5.

Without loss of generality, simulation experiments of different-scaled problems are conducted. Table 4 records the experimental results. All the test problems are evaluated by 30 independent runs. The best and the standard deviation denoted as std are computed to evaluate the performance of such algorithms.

Through Fig. 4, we can intuitively see that IAFS is obviously superior to the other three algorithms on convergence rate and the solution quality. As for IAFS, furthermore, according to whether considering customer rank or not, the differences about distribution on time rate between the two conditions are com-

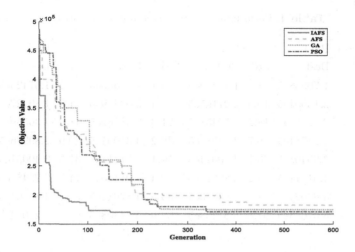

Fig. 4. Convergence curves of four algorithms.

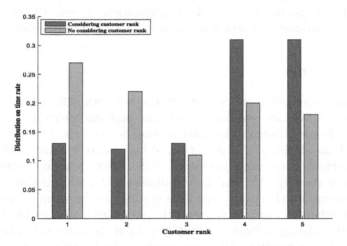

Fig. 5. Distribution on time rate comparison.

pared in Fig. 5, which shows that customer rank under consideration can ensure distribution on time. It is conducive to maintain good cooperation relations with customers and avoid customers lost, especially VIP customers. In other words, it indirectly increases the economic efficiency of enterprises. To further evaluate the performance of IAFS, numerical simulation experiments of orders of different-scale are conducted among IAFS, AFS, PSO and GA in Table 4. In contrast to other three algorithms, Table 4 tells that IAFS could not only prevent premature and guarantee convergence to global optimum, but also improve convergence rate and the solution quality, especially in resolving large-scale problems. The reason why IAFS presents such advantages, is that elitist guiding evolution strategy, crossover operator with cyclic misalignment and heuristic mutation strategy are

Table 4. Performance comparison among algorithms.

Order size	PSO		GA		AFS		IAFS	
	Best	Std	Best	Std	Best	Std	Best	Std
30	1.70988e5	1.7	1.74649e5	3.1	1.82565e5	4.5	1.67423e5	0
50	2.13001e5	10.5	2.21165e5	7.4	2.03459e5	11.8	1.89927e5	3.3
80	2.6107e5	18.2	2.50451e5	21.3	2.58862e5	18.6	2.36559e5	5.9
100	3.22781e5	16.7	3.24011e5	30.2	3.31001e5	20.9	2.89976e5	4.1
120	3.69408e5	40.1	3.70014e5	50.1	3.7131e5	47.5	3.40051e5	10.6
150	4.01811e5	44.2	3.89211e5	60.4	4.0116e5	71.2	3.73487e5	9.7
200	4.89022e5	61.5	4.84356e5	49.5	4.87011e5	80.1	4.65501e5	16.8
230	5.51643e5	91.8	5.54338e5	102.9	5.44308e5	98.6	5.21054e5	19.4

employed. It does lead IAFS to find the best result effectively, improve the quality of the offspring and the diversity of the population. Thanks to these improvements, the exploration and exploitation ability of IAFS can be enhanced greatly.

6   Conclusions

The IAFS proposed in this paper offers an effective method for solving scheduling optimization of vehicles considering customer rank and delivery time demand, which can effectively balance the contradiction between the key customer satisfaction and the efficiency of distribution. Whats more, it achieves the win-win situation between corporate and customers. So it will bring corporate achievements to grow steadily. IAFS mainly introduces elitist guiding evolution strategy, crossover operator and heuristic mutation strategy to the standard artificial fish swarm algorithm. Not only it can lead the population to the global optimum and make the offspring inherit excellent gene from the parents, but improve the diversity of the population. All these causes make it not easy to trap in the local extreme points and the rapidity of convergence accelerate consequently. Finally, by some practical instances, the proposed method is verified that it is superior to GA, PSO and AFS. Based on this, IAFS can be extensively used in relevant fields of the logistics in future research, such as shipping, intelligent transportation, production planning, and so on.

References

1. http://www.chinawuliu.com.cn/lhhkx/201802/06/328520.shtml. [EB/OL]
2. Tenahua, A., Olivares, B.E., Diana, S.: ILS metaheuristic to solve the periodic vehicle routing problem. Int. J. Comb. Optim. Probl. Inform. **9**(3), 55–63 (2018)
3. Elgesem, A.S., Skogen, E.S., Wang, X.: A traveling salesman problem with pickups and deliveries and stochastic travel times: an application from chemical shipping. Eur. J. Oper. Res. **269**(3), 844–859 (2018)

4. Ho, W., Ho, G.T.S., Ji, P., et al.: A hybrid genetic algorithm for the multi-depot vehicle routing problem. Eng. Appl. Artif. Intell. **21**(4), 548–557 (2008)
5. Kalayci, C.B., Kaya, C.: An ant colony system empowered variable neighborhood search algorithm for the vehicle routing problem with simultaneous pickup and delivery. Expert Syst. Appl. **66**, 163–175 (2016)
6. Salhi, S., Imaran, A., Wassan, N.A.: The multi-depot vehicle routing problem with heterogeneous vehicle fleet: formulation and a variable neighborhood search implementation. Comput. Oper. Res. **52**, 315–325 (2014)
7. Morais, V.W.C., Mateus, G.R., Noronha, T.F.: Iterated local search heuristics for the vehicle routing problem with cross-docking. Expert Syst. Appl. **41**(16), 7495–7506 (2014)
8. Song, Q., Gao, X., Santos, E.T.: A food chain algorithm for capacitated vehicle routing problem with recycling in reverse logistics. Int. J. Bifurcat. Chaos **25**(14), 1540031 (2015)
9. Silvestrin, P.V., Ritt, M.: An iterated tabu search for the multi-compartment vehicle routing problem. Comput. Oper. Res. **81**, 192–202 (2017)
10. Liu, R., Xie, X., Augusto, V., et al.: Heuristic algorithms for a vehicle routing problem with simultaneous delivery and pickup and time windows in home health care. Eur. J. Oper. Res. **230**(3), 475–486 (2013)
11. Defryn, C., Sorensen, K.: A fast two-level variable neighborhood search for the clustered vehicle routing problem. Comput. Oper. Res. **83**, 78–94 (2017)
12. Vaz, P., Puca, H., Santos, A.C., et al.: Vehicle routing problems for last mile distribution after major disaster. J. Oper. Res. Soc. **69**(8), 1254–1268 (2018)
13. Li, X., Shao, Z., Qian, J.: An optimizing method based on autonomous animats: fish swarm algorithm. Syst. Eng. Theory Pract. **22**(11), 32–38 (2002)
14. Zhao, W., Du, C., Jiang, S.: An adaptive multiscale approach for identifying multiple flaws based on XFEM and a discrete artificial fish swarm algorithm. Comput. Methods Appl. Mech. Eng. **339**, 341–357 (2018)
15. Tsai, H.C., Lin, Y.: Modification of the fish swarm algorithm with particle swarm optimization formulation and communication behavior. Appl. Soft Comput. **11**(8), 5367–5374 (2011)
16. Zhang, Z., Wang, K., Zhu, L., et al.: A pareto improved artificial fish swarm algorithm for solving a multi-objective fuzzy disassembly line balancing problem. Expert Syst. Appl. **86**, 165–176 (2017)
17. Sengottuvelan, P., Prasath, N.: BAFSA: breeding artificial fish swarm algorithm for optimal cluster head selection in wireless sensor networks. Wirel. Pers. Commun. **94**(4), 1979–1991 (2017)
18. Azad, M., Rocha, A., Fernandes, E.: Improved binary artificial fish swarm algorithm for the 0–1 multidimensional knapsack problems. Swarm Evol. Comput. **14**, 66–75 (2014)
19. Xian, S., Zhang, J., Xiao, Y., Pang, J.: A novel fuzzy time series forecasting method based on the improved artificial fish swarm optimization algorithm. Soft. Comput. **22**(12), 3907–3917 (2017). https://doi.org/10.1007/s00500-017-2601-z
20. Zhang, Y., Huang, G.: Traffic flow prediction model based on deep belief network and genetic algorithm. IET Intel. Transp. Syst. **12**(6), 533–541 (2018)

Classification of Tongue Color Based on Convolution Neural Network

Yifan Shang[1](✉), Xiaobo Mao[1](✉), Yuping Zhao[2], Nan Li[1], and Yang Wang[1]

[1] School of Electrical Engineering, Zhengzhou University, Zhengzhou 450001, China
13197155876@163.com, 594473748@qq.com
[2] China Academy of Chinese Medical Sciences, Beijing 100020, China

Abstract. The color of the tongue is closely related to the patient's physical condition and pathological condition, and it plays a very important position in the treatment and diagnosis of traditional Chinese medicine (TCM) clinical medicine. Convolutional neural network (CNN) has achieved fruitful results in image classification. Therefore, the method combining CNN with tongue color classification is proposed. Frist, initial data set is obtained by standardizing tongue image acquisition and tongue image preprocessing. Then, the Otsu method is applied on the image multi-channel to remove the background of the tongue accurately as the input of the model. At the same time, data augmentation is applied to avoid over-fitting of the model and improve the accuracy of model classification. The accuracy of the trained tongue color classifier based on CNN is 90.5% in the clinically collected data set, which is better than the traditional machine learning methods for tongue color classification.

Keywords: Image classification · Tongue color · CNN

1 Introduction

The color discrimination of the tongue image is the most important part of the tongue diagnosis process of Chinese medicine [1]. It largely reflects the physical condition of a person. In recent years, there are mainly the following methods for tongue diagnosis [2]: K-Nearest Neighbor (KNN), Support Vector Machine (SVM), AdaBoost and other algorithms. However, due to the diversity and complexity of tongue color, there is no significant local or global visual feature. Therefore, the above technique is applied to the classification of tongue image color recognition with limited accuracy. Different from the above traditional methods, the application of deep learning to tongue color classification can directly input the image of multidimensional vector into the network, thus avoiding the complexity of data reconstruction in feature extraction and classification process. The application of deep learning can extract more abundant image information

Supported by key project at central government level (2060302).

© Springer Nature Singapore Pte Ltd. 2020
L. Pan et al. (Eds.): BIC-TA 2019, CCIS 1160, pp. 326–335, 2020.
https://doi.org/10.1007/978-981-15-3415-7_27

to make the classifier have a better recognition effect. In 1998, LeCnn designed the convolutional neural network LeNet-5 to realize the recognition of handwritten numbers. Subsequently, with the rise of Internet big data, convolutional neural networks have been widely used in the field of image recognition [3–5], and have become an important tool for image classification problems. Based on the above research results, this paper selects the convolutional neural network based on Tensorflow to complete the recognition and classification of tongue image color. The overall research process is shown in Fig. 1.

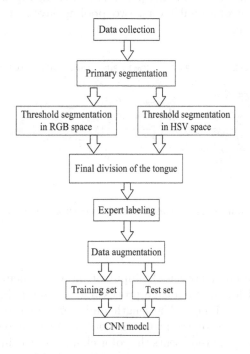

Fig. 1. Tongue image color classification framework flow chart

This article mainly divides the color of the tongue into the following four categories: Light red tongue, Red tongue, Dark red tongue, and Green and Purple tongue. In the diagnosis of Chinese medicine, the color of the tongue is closely related to the body mechanism, and different colors represent different physical symptoms of the human body. From the perspective of Chinese medicine, the light red tongue represents the human body, and the tongues of other colors have their respective symptoms.

2   The Structure of CNN for Tongue Color Recognition

CNN are feedforward neural networks with convolutional structures that are close to the actual neural network. In recent years, it has developed rapidly and has a unique advantage in speech recognition and image processing with its special structure of local connection and weight sharing.

2.1 Network Model of CNN

A convolutional neural network model has a multi-layer structure, which is mainly divided into the following five parts: input layer, convolution layer (Con), pooling layer (Pool), fully connected layer (FC), and Softmax layer. The CNN model structure constructed in this paper is shown in Fig. 2. It consists of an input layer, two convolutional layers, two pooling layers, two fully connected layers, and a Softmax layer.

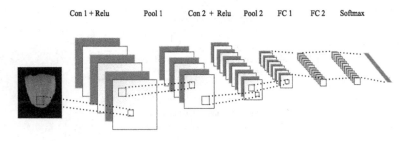

Fig. 2. Convolutional neural network model structure constructed in this paper

Input Layer. The input layer is the input of the whole neural network model. In the CNN model for tongue image processing, the input layer represents the pixel matrix of tongue image. The length and width of the three-dimensional matrix in the input layer represent the size of the image, while the depth of the three-dimensional matrix represents the color channel of the image. For example, the depth of black-and-white image is 1, while in RGB color mode, the depth of the image is 3. In this paper, the dimension of the input tongue image is 64*64*3.

Convolution Layer. Convolution layer is the most important part of the whole network. Its function is to extract the features of each small part of the picture. For example, there is a 6*6 image in which the information of the image is stored in each pixel. Then a convolution kernel is defined, which is weighted enough to extract features from the image. The output of convolution layer is obtained by multiplying and adding the corresponding phases of convolution core and digital matrix. The machine can get different output values to determine which convolution cores can best represent the characteristics of the image by comparing with different convolution cores. The higher the output value of the convolution layer, the better the convolution core can represent the characteristics of the convolution core. The convolution process is shown in Fig. 3.

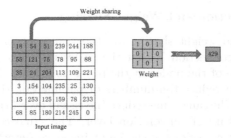

Fig. 3. The convolution layer

Pooling Layer. The input of the pooling layer is the output matrix multiplied by the original image matrix and the corresponding convolution kernel. The function of convolution layer is to reduce the number of training parameters and the dimension of the output eigenvector of convolution layer, while avoiding over-fitting of the model, so as to extract the image features with high correlation and reduce the transmission of noise. As shown in Fig. 4, the common pooling methods are max-pooling and mean-pooling. The maximum pooling is to select the maximum number in the designated area of the output matrix of the pooling layer to represent the whole area, while the average pooling is to select the numerical average value in the designated area to represent the whole area.

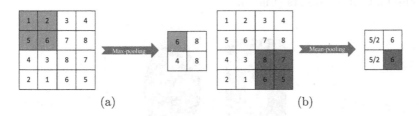

Fig. 4. Two most commonly used pooling methods (a) max-pooling (b) mean-pooling

Fully Connected Layer. In the full connection layer, the output tensor of the pooling layer is re-cut into some vectors, then multiplied by the weight matrix and added the bias value, and then a classifier is generated by using the Relu activation function [6].

Softmax Layer. The Softmax layer is mainly used for classification problems. Through the Softmax layer, we can get the probabilities of different classes in the current samples.

2.2 Local Connection and Weight Sharing

Local connection and weight sharing are the two most prominent features of CNN model. Local connection is that the nodes in convolution layer are only connected with some of the nodes in the previous layer to learn local features. Local connection can reduce the numbers of parameters, thus speeding up the learning rate, and at the same time reduce the risk of over-fitting. Weight sharing is an important feature of convolution layer. The same convolution core can convolute with different regions of the input image to detect the same features, which greatly reduces the parameters of the network, reduces the complexity of the network, and speeds up the training speed of the network [7].

3 Pretreatment of Tongue Image

3.1 Data Acquisition

In order to avoid environmental interference, the acquisition environment is a closed space. The tongue image is taken under the background of the standard light source and the same position. The color temperature of the light source of the acquisition device is 5000K, which conforms to the relevant regulations of the International Organization for Standardization on the standard observation environment and the standard light source [8] (ISO3664:2000). The acquisition equipment is Nikon 700d phase. Machine. Data were collected from college students and hospital patients (Fig. 5).

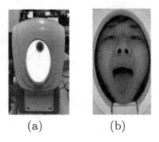

(a)   (b)

Fig. 5. (a) Acquisition device (b) Standard tongue image

After screening, 197 cases of data were obtained, and the data distribution was shown in Table 1.

Table 1. Categories and quantities of collected data.

Tongue color	Light red tongue	Red tongue	Dark red tongue	Green and Purple tongue
Quantity	75	50	40	32

3.2   Tongue Segmentation

The photographs taken by the camera include useless backgrounds such as lips and faces, which have a high impact on accuracy. Therefore, it is necessary to segment the tongue and then train the model. Firstly, rough segmentation is carried out, because the tongue position is fixed, the face and background can be removed by extracting the corresponding size of the image center. Then we need to remove the lips and other useless parts to further extract the tongue. The steps of tongue segmentation are shown in Fig. 6. Two color spaces RGB and HSV are used to extract the tongue body. A large number of literatures and experiments show the following results [9–11]. In RGB space, the RGB mixed channel is obtained by subtracting the pixel values of B channel and G channel. Then the threshold of the tongue body region can be automatically determined by Otsu in RGB mixed channel. It is difficult to distinguish between mouth and lip, but V channel in HSV color space can remove the shadows and lips better by Otsu method. Therefore, the combination of RGB spatial pixel channel and HSV spatial pixel channel can segment better tongue photographs.

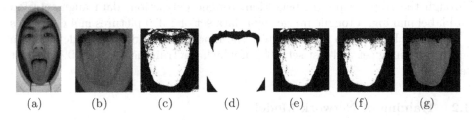

(a)   (b)   (c)   (d)   (e)   (f)   (g)

Fig. 6. Tongue segmentation process (a) Original image (b) Pre-cutting (c) Binary image obtained by threshold segmentation of B/G mixed channel (d) Binary image obtained by threshold segmentation of V-channel (e) Binary image obtained by threshold segmentation of RGB/HSV combination space (f) Binary map after morphological processing (g) Final extraction of tongue image

3.3   Data Augmentation

Because deep learning model has a large number of free parameters, the training of convolutional neural network model needs a lot of data to improve the generalization ability of the model. However, data acquisition and labeling is a huge project in practice, but data augmentation is one of the solutions to such problems. Because the collected sample data is limited, it is necessary to increase the original data set several times without changing the tongue image category, and to ensure good training effect [12]. As shown in Fig. 7, the common methods include horizontal flip, vertical flip, a certain degree of displacement, tailoring, rotation and so on. Data augmentation can prevent small sample set from overfitting during training, so it can improve the accuracy and generalization ability of the model.

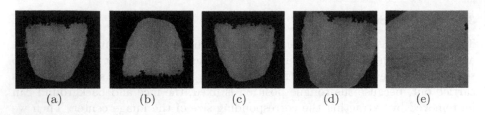

(a)   (b)   (c)   (d)   (e)

Fig. 7. (a) Original image (b) Vertical flip (c) Horizontal flip (d) Enlargement (e) Amplify and rotate

4 The Results and Analysis of the Experiment

4.1 Data Set

The data set is composed of 267 samples of tongue images, which are calibrated by five experts from Chinese Academy of Medical Sciences. For the tongue color of the same picture, only when all five experts have the same viewpoint can they be included in the sample set. In the end, 197 samples meet the standard. Then through the steps of pre-segmentation, tongue extraction, data augmentation and label making, a tongue image color data set of 1 350 pictures in 4 categories is obtained. In order to train and test the model, we randomly divide the data set into training set and test set, 80% of which are training set and 20% are test set.

4.2 Training of Network Model

The hardware environment of the machine used for training model is CPU I5-4200U@1.60 GHz with 8 G memory. The system environment is Elementory OS 5.0 Juno. Detailed parameter settings of the model are shown in Table 2 (f: the size of convolution or pooling nuclei, s: step size, d: the depth of convolution nuclei).

Table 2. Detailed parameter configuration of the model

Sequence	Type of operation	Information	Latitude (input)	Latitude (output)
0	Data input	–	64*64*3	64*64*3
1	Conv 1	d = 64 f = 3*3 s = 1	64*64*3	64*64*64
2	Activation function	Relu	64*64*64	64*64*64
3	Pool 1	f = 3*3 s = 2	64*64*64	32*32*64
4	Conv 2	d = 16 f = 3*3 s = 1	32*32*64*	32*32*16
5	Activation function	Relu	32*32*16	32*32*16
6	Pool 2	F = 3*3 d = 2	32*32*16	16*16*16
7	FC 1	128 Neurons	16*16*16	1*128
8	FC 2	128 Neurons	1*128	1*128
9	Softmax	–	1*128	1*4

4.3   Result Analysis

The following results are obtained from the training and testing of the CNN model based on Tensorflow. The comparison of the accuracy of tongue color 2-classification between CNN and two traditional methods (Knn and SVM) is shown in Table 3. It can be found that the accuracy of CNN in tongue image 2-classification is 98%, which is far beyond the accuracy of the traditional two machine learning methods.

Table 3. The results of color classification of tongue image

Methods	SVM [13]	KNN [14]	PCA+ Bayes [15]	PCA-AdaBoost [15]	CNN
2-classification	0.797	0.738	—	—	0.968
4-classification	—	—	0.675	0.851	0.905

Table 3 also describes the accuracy of four classification methods in tongue image color. The average accuracy of the four classifications obtained by our method is 0.95, which is much higher than that of PCA+Bayes and a little higher than that of PCA-AdaBoost.

Furthermore, as shown in Fig. 8, PCA+ Bayes is far behind PCA-AdaBoost and CNN in the accuracy of 4-classification of tongue color. As shown in Fig. 8, PCA-AdaBoost is slightly higher in the recognition accuracy of light red tongue, red tongue and crimson tongue than this method, but far lower in the recognition of blue-purple tongue than this method, which shows the universality of the model in this paper. The convolutional neural network model created in this paper has only seven layers, so the structure of the model is relatively simple. Because of the small scale of existing data, the network model with deeper

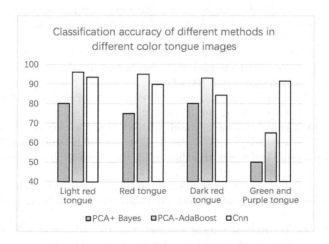

Fig. 8. Classification accuracy of different methods in different color tongue images

structure can easily lead to over-fitting [16]. With the increase of data and the deepening of convolution neural network model structure, the accuracy will be improved.

5   Conclusion

This paper chooses convolution neural network model instead of traditional methods of Manually extracting the color of the tongue. It can learn the color features of tongue image independently and accurately recognize the color of tongue image. After standardized image acquisition, preprocessing and tongue extraction, the accuracy of the 4-classifications reaches 0.905, which greatly exceeds the traditional manual feature extraction method, and proves the validity of convolutional neural network in tongue image recognition. However, the accuracy of the model can be improved greatly. The next step is to optimize the performance of the CNN model and increase the size of the data set, so as to obtain more accurate tongue color classification accuracy.

References

1. Hinton, G.E., Salakhutdinov, R.R.: Reducing the dimensionality of data with neural networks. Science **313**(5786), 504–507 (2006)
2. Zhou, L., Zhang, P., Cheng, B.: Automatic tongue color analysis of traditional Chinese medicine based on image retrieval. In: 13th International Conference on Control Automation Robotics Vision (ICARCV), pp. 637–641. IEEE, Singapore (2015)
3. Tao, X.: Tongue image classification based on rough set theory. Comput. Eng. Appl. **43**(27), 216–19 (2007)
4. Krizhevsky, A., Sutskever, I., Hinton, G.E.: ImageNet classification with deep convolutional neural networks. Commun. ACM **60**(6), 84–90 (2017)
5. Szegedy, C., Liu, W., Jia, Y., Sermanet, P.: Going deeper with convolutions. In: IEEE Conference on Computer Vision and Pattern Recognition (CVPR), pp. 1–9. IEEE, Boston (2015)
6. Jiang, S., Hu, J., Xia, C., Qi, J., Peng, Y.: A tongue image separation method based on Otsu threshold method and morphological adaptive correct. Comput. Intell. Neurosci. **10**(1155), 102–106 (2016)
7. Harangi, B.: Skin lesion classification with ensembles of deep convolutional neural networks. J. Biomed. Inform. **86**, 25–32 (2018)
8. ISO 36642000 Graphic technology and photography–Viewing conditions
9. Sladojevic, S., Arsenovic, M., Anderla, A.: Deep neural networks based recognition of plant diseases by leaf image classification. Chin. High Technol. Lett. **27**(2), 150–155 (2017)
10. Zhang, L., Qin, J., Zeng, Y.: Tongue-coating image segmentation based on combination of morphological gradient and watershed algorithms. Imaging Sci. J. **59**(6), 311–316 (2011)
11. Wu, K., Zhang, D.: Robust tongue segmentation by fusing region-based and edge-based approaches. Expert Syst. Appl. **42**(21), 8027–8038 (2015)

12. Pham, T.-C., Luong, C.-M., Visani, M., Hoang, V.-D.: Deep CNN and data augmentation for skin lesion classification. In: Nguyen, N.T., Hoang, D.H., Hong, T.-P., Pham, H., Trawiński, B. (eds.) ACIIDS 2018. LNCS (LNAI), vol. 10752, pp. 573–582. Springer, Cham (2018). https://doi.org/10.1007/978-3-319-75420-8_54
13. Zhang, X., Shen, L.: Application of weighted SVM on the classification and recognition of tongue images. Chin. J. Biomed. Eng. **2006**(02), 230–233 (2006)
14. Liang, J., Yang, H., Zhang, H.: The classification of common tongue body and tongue coating based on the feature of color. Microcomput. Appl. **36**(17), 102–105 (2017)
15. Hui, K., Li, W., Shi, G.: Color classification of tongue based on PCA-AdaBoost in traditional Chinese medicine. J. Guangxi Normal Univ. **27**(3), 158–161 (2009)
16. Smirnov, E.A., Timoshenko, D.M., Andrianov, S.N.: Comparison of regularization methods for imagenet classification with deep convolutional neural networks. In: 2nd AASRI Conference on Computational Intelligence and Bioinformatics (CIB), pp. 89–94. Elsevier Science, South Korea (2014)

Multiple Classifiers Combination Hyperspectral Classification Method Based on C5.0 Decision Tree

Dongyue Xiao[✉] and Xiaoyan Tang

School of Electronics and Electrical Engineering,
Nanyang Institute of Technology, Nanyang 473004, China
568044943@qq.com

Abstract. Hyperspectral image classification is nowadays an essential method for remote sensing data analysis and information extraction. In order to improve the classification accuracy of hyperspectral image, this paper introduces a novel method which is based on multi-classifier combination of C5.0 decision tree. This method selects the three methods of minimum distance, maximum likelihood, spectral angle mapping (SAM) and support vector machine (SVM) for classifying images, which have better classification results. The C5.0 decision tree is used to complete the classification. The experimental results show that the classification accuracy of the combined classifier is significantly higher than that of the single sub-classifier, which is an effective method to improve the classification accuracy of remote sensing images.

Keywords: Hyperspectral classification · Classifier combination · Classification accuracy · C5.0 decision tree

1 Instruction

Hyperspectral remote sensing is a technique for remote sensing imaging of features using a narrow and continuous spectrum of electromagnetic spectrum [1]. Hyperspectral image classification is the division of each pixel or region in an image into different physical categories based on its spectral and image characteristics [2]. Moreover, it is an important method for remote sensing image analysis and information extraction, and is an important research direction in the field of remote sensing [3].

Compared with conventional remote sensing, hyperspectral remote sensing data have several difficulties in the classification and identification of ground objects. Firstly, it is difficult to obtain sufficient and uniform training samples, which makes the prior knowledge estimation required for supervised classification inaccurate. Secondly, Hyperspectral imagery has a large amount of data due to the high spectral dimension, so the performance of the computer and the processing efficiency of the algorithm are high, and the high-dimensional is prone to Hughes phenomenon. Thirdly, due to the low spatial resolution of hyperspectral, there is a problem of mixed pixels occurred; Lastly, because some conventional remote sensing classification methods are not

© Springer Nature Singapore Pte Ltd. 2020
L. Pan et al. (Eds.): BIC-TA 2019, CCIS 1160, pp. 336–344, 2020.
https://doi.org/10.1007/978-981-15-3415-7_28

suitable for hyperspectral images, a new hyperspectral classification algorithm is needed.

Remote sensing image classification usually chooses a single classifier as the final result. Commonly used classifiers are: minimum distance method, maximum likelihood method, spectral angle mapping method (SAM), support vector machine (SVM) [4] and decision tree [5], etc. However, the theory and practice of pattern recognition show that various classifiers have different characteristics and different application scopes. The selection of optimal classifiers will be affected by many factors. In order to improve the classification accuracy of remote sensing images, the combination and fusion of multi-classifiers has been introduced into the field of hyperspectral remote sensing image classification, which has been widely used [6–8] and has become an important direction to improve classification accuracy [9, 10]. In recent years, through the comparative study of multiple classifiers, it has been found that for a sample that is misclassified by a classifier, it is possible to obtain the correct category label by using other classifiers, that is, different classifiers have complementary information for the classification mode.

In this paper, C5.0 decision tree algorithm is used as a combination algorithm of multiple classifiers. Maximum likelihood classification, SAM and SVM are used as member classifiers to generate classification rules for sample training and learning, and to achieve fine classification of features. This method can effectively improve the classification accuracy of hyperspectral images.

2 Research Method

2.1 SVM Algorithm

Set in the D-dimensional space there is a training set $\{x_i, y_i\}$ ($i = 1, 2, \ldots, k$, k is the number of samples; $x_i \in R^d, y_i \in \{-1, 1\}$), that is, if x_i belongs to category 1, then $y_i = 1$, if x_i belongs to category 2, then $y_i = -1$. The support vector machine is to find a segmentation plane that meets the requirement $\omega \cdot x + b = 0$, where ω is the weight vector, b is the offset of the segmentation plane, so that the data of the training samples can be classified correctly and the interval of the classified data is maximized. The equation $(\omega \cdot x_i) + b = \pm 1$ is called the classification boundary [11]. The interval between the sample points of the $y = 1$ category and the normative hyperplane meets the formula as follows:

$$\min_{y_i = 1} \frac{|(\omega \cdot x_i) + b|}{\|\omega\|} = \frac{1}{\|\omega\|} \tag{1}$$

Optimal hyperplane means parameter $\frac{2}{\|\omega\|}$ maximized.

In order to find the optimal hyperplane, the problem can be transformed into a quadratic programming problem, which is described by the mathematical model as follows:

$$min(\frac{1}{2}\|\omega\|^2 + C\sum_{i=1}^{k}\xi_i) \tag{2}$$

$$y_i(\omega \cdot x_i + b) \geq 1 - \xi_i; i = 1, 2, \ldots k, \xi_i \geq 0 \tag{3}$$

Where ξ_i is a relaxation variable, indicating the degree of deviation between the model and the ideal linear case, and C is a penalty factor, which controls the degree of penalty for misclassified samples.

From this, the classification function can be obtained.

$$f(x) = sgn(\sum_{i=1}^{k}\alpha_i^* y_i(x_i \cdot x) + b^*) \tag{4}$$

2.2 Maximum Likelihood Classification Algorithm

The maximum likelihood method assumes that the spectral features of the ground objects in the training area approximately obey normal distribution, and chooses the training area, calculates the attribution probability of each sample area, and then classifies them [12].

Maximum likelihood classification algorithm generally assumes that each category is based on the Gaussian multivariate normal distribution, and its decision function is:

$$g_i(x) = -\ln\left|\sum_i\right| - (x - m_i)^T \sum_i^{-1} (x - m_i) \quad i = 1, \ldots, C \tag{5}$$

Where x is the pixel vector, m_i is the mean vector of category i, $\sum_i$ is the covariance matrix of category i, and C is the number of categories.

The decision rule is:

$$x \in \omega_i; \, if \, g_i(x) > g_j(x), \, for \, all \, j \neq i. \tag{6}$$

Maximum likelihood classification method is suitable for remote sensing image classification with few bands or features. It has the problems of long time and insufficient training samples when applied directly to hyperspectral images. For a classification problem with d-dimensional features, at least 10d training samples are required, while a reliable statistical estimation requires 100d training samples. It is difficult to estimate the statistical parameters if the number of training samples is not sufficient. The methods to solve the problem are block processing algorithm and band dimension reduction processing. After dimension reduction, the spectral dimension can be reduced and the training samples needed can be reduced. The other is to use block maximum likelihood classification method for hyperspectral images. Firstly, the bands are divided into subspaces, and the likelihood probability is estimated in each subspace. This can effectively reduce the time and accurately estimate the required parameters [13].

2.3   C5.0 Decision Tree Algorithm

C5.0 algorithm [14] is a decision tree algorithm with wide application and high efficiency. C5.0 algorithm uses information gain rate as the criterion to determine the
decision tree branching criteria, prunes or merges the cotyledons of the decision tree to
improve the classification accuracy, and finally determines the optimal threshold of
each leaf [15].

Assuming that the training sample set S can be divided into n subsets as
$[s_1, s_2, \cdots s_n]$ according to attribute X, $|S|$ is the total number of samples S. $freq(c_i, S)$ is
the number of samples belonging to the category $c_i (i = 1, 2, \cdots, N)$. in set S, then the
probability of one sample belonging to the category c_i is given as $freq(c_i, S)/|S|$.

The entropy of the training set S is expressed by following Eq. (7), which is also
the total amount of information needed to identify all samples in S in information
theory.

$$info(S) = \sum_{i=1}^{N} \frac{freq(c_i, S)}{|S|} log_2 \left(\frac{freq(c_i, S)}{|S|} \right) \tag{7}$$

The information entropy of each subset in n subsets of the training set is calculated,
and the expected information of the set is calculated according to Eq. (8).

$$info_x(S) = \sum_{j=1}^{n} \frac{|s_j|}{|S|} \times info(s_j) \tag{8}$$

Gain standard Gain (X) is used to measure the difference between the partitioned
information entropy and the information entropy of each subset, and each partition is
carried out according to the attribute of the highest information gain.

$$Gain(X) = info(S) - info_x(s_i) \tag{9}$$

The potential information after segmentation into n subsets can be calculated by
Eq. (10).

$$split\_Info(X) = -\sum_{i=1}^{n} \frac{|s_i|}{|S|} \times log_2 \left(\frac{|s_i|}{|S|} \right) \tag{10}$$

The information gain rate of S partitioned by attribute X can be calculated by
Eq. (11).

$$Gain\_ratio(X) = \frac{Gain(X)}{split\_info(X)} \tag{11}$$

Assuming that the classifier set is $C = \{c_1, c_2, \cdots, c_k\}$, each basic classifier in the
combined classifier set predicts a probability distribution for the subsample, then the
prediction vector generated by the sample s_i on C can be expressed as $P_{s_i}^C = \left(p_{s_i}^{c_1}, p_{s_i}^{c_2} \cdots p_{s_i}^{c_k} \right)$, where $p_{s_i}^{c_j}$ is the information gain rate of the sample s_i on the classifier

c_k. The classifier with the highest information gain rate is selected as the classifier for this leaf sample. If the classification result in the leaf satisfies the stop growth threshold, the leaf stops growing and outputs the sample classification result, otherwise it returns to the beginning as a sample to continue to classify the cotyledon sample.

2.4   Algorithm

In this paper, C5.0 decision tree algorithm is used as a combined algorithm of multiple classifiers to classify hyperspectral data. Compared with other algorithms (voting method, Bayesian method, evidential reasoning, artificial neural network, etc.), the decision tree algorithm has the following advantages: the algorithm has no input parameters, and the training sample points can be randomly selected; The classification and combination rules are easy to understand, and the classification accuracy is higher than or equal to other algorithms. The overall flow chart of the algorithm to classify hyperspectral data presented in this paper is shown in Fig. 1.

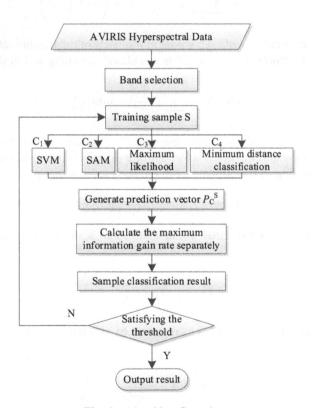

Fig. 1. Algorithm flow chart.

Specific steps are as follows:

(1) Firstly, unsupervised classification of hyperspectral images, and using the classi-fication results as a supplementary means of supervised classification, determining the selection and classification of training samples;

(2) For multi-classifier combination systems, the classification accuracy of member sub-classifiers should be high and the classification results should be diverse. For different classification images, select the member classifier: maximum likelihood classification, minimum distance, spectral angle mapping (SAM), support vector machine (SVM), three combinations of four supervised classifiers;

(3) Using the combined classifier as the characteristic attribute of C5.0 decision tree to carry out sample training and learning to generate classification rules, and then use the rules generated by training to supervise and classify the whole image;

(4) Accuracy evaluated and result analysis of classified images.

3 Experimental Results

The experimental images in this paper is AVIRIS hyperspectral data 92AV3C, which were taken in June 1992 at the Indian Pine Test Site in Indiana, USA, and has 145×145 pixels, spatial resolution of 20 m/pixel, 220 bands. 35 bands were selected for experiments by focusing on the best band in the reference [16] of this paper. There are 16 different categories of ground objects in this image, 6 of which are too few to be representative, so more ten categories are selected, as shown in Table 1.

Table 1. Number of training samples and test samples

Classification categories	Training sample	Test sample
C1 Stone-steel towers	45	19
C2 Corn-no till	96	174
C3 Corn-min	84	197
C4 Corn/Pasture	50	98
C5 Grass/Trees	48	162
C6 Hay-windrowed	51	206
C7 Soybean-no till	73	311
C8 Soybeans-min	148	626
C9 Soybeans-clean	161	110
C10 Woods	150	322
Total	873	2225

The four-member classifiers and the combined classifier classification results pre-sented in this paper are shown in Fig. 2. It can be qualitatively seen from the figure that the accuracy of the combined classifier classification result is higher than that of the other four classifiers.

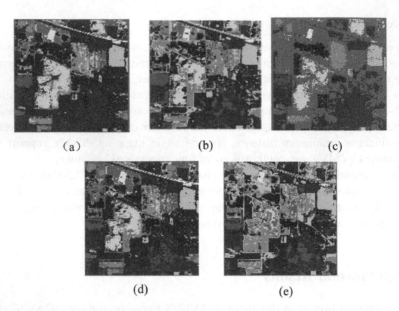

(a)   (b)   (c)

(d)   (e)

Fig. 2. Supervised classification results (a) Combined classifier; (b) SVM; (c) Maximum likelihood; (d) SAM; (e) Minimum distance.

Table 2 shows that the overall classification accuracy of the combined classifier is higher than that of a single classifier. This shows that the multi-classifier combination classification method based on C5.0 decision tree effectively improves the classification accuracy and greatly reduces the misclassification and misclassification of various categories.

Table 2. Comparison of classification accuracy of different classifiers

Classification accuracy [%]	Minimum distance	Maximum likelihood	SAM	SVM	Combined classifier
C1	89.5	100	94.7	63.1	100
C2	55.2	54.0	47.1	52.9	61.5
C3	55.8	50.8	62.9	78.7	78.2
C4	75.5	75.5	73.5	100	75.5
C5	85.8	74.1	86.4	80.8	91.3
C6	97.1	90.3	98.5	100	99.5
C7	36.0	44.4	57.9	77.5	73.9
C8	20.1	55.7	46.2	64.2	77.6
C9	1.8	100	53.6	21.8	37.3
C10	66.8	99.4	75.8	77.0	99.4
Overall	49.0	67.7	63.4	72.3	80.2
Kappa	42.7	63.5	58.2	68.1	76.7

From the perspective of classification accuracy, the overall classification accuracy of the image is not very high. There are several reasons for this: on the one hand, the accuracy of the original image capture is not high, and the spatial resolution is low, which lead to the existence of many mixed pixels; on the other hand, the uneven distribution of training samples affects the accuracy; The algorithm also needs to be further optimized. The combination classifier is not all class accuracy higher than that in the sub-classifier. For example, the classification accuracy of the combination classification category 9 in Table 2 is lower than that of the maximum likelihood classifier.

4 Conclusion

In order to improve the classification accuracy of hyperspectral image, this paper proposes to take maximum likelihood classification, minimum distance, spectral Angle mapping and support vector machine as member classifiers of C5.0 algorithm to conduct sample training and learn and generate classification rules to realize fine classification of ground objects. The experimental results of AVIRIS hyperspectral image show that the accuracy of the proposed algorithm is significantly higher than that of the classical single supervised classification algorithm: maximum likelihood classification, minimum distance, spectral Angle mapping, support vector machine.

References

1. Tong, Q., Zhang, B., Zheng, L.: Hyperspectral Remote Sensing: Principle, Technology and Application. Higher Education Press, Beijing (2006)
2. Zhang, B., Gao, L.: Hyperspectral Image Classification and Target Detection. Science Press, Beijing (2012)
3. Du, P., Xia, J., Xue, Z., et al.: Progress in classification of hyperspectral remote sensing images. J. Remote Sens. 20(2), 236–256 (2016)
4. Wang, S., Ai, Z., Du, W., et al.: Study on multi-source remote sensing image classification based on SVM different kernel functions. J. Henan Univ. Technol. (Nat. Sci.) 30(03), 304–309 (2011)
5. Wang, H., Tan, B., Fang, X., et al.: Precise classification of forest types uses hyperion image based on C5.0 decision tree algorithm. J. Zhejiang A&F Univ. 35(4), 724–734 (2018)
6. Zhang, L., Zhang, L., Tao, D., et al.: On combining multiple features for hyperspectral remote sensing image classification. IEEE Trans. Geosci. Remote Sens. 50(3), 879–893 (2012)
7. Liu, D., Yang, F., Wer, H., et al.: Vegetation classification method based on C5.0 decision tree of multiple classifiers. J. Graph. 38(5), 722–728 (2017)
8. Cui, Y., Xu, K., Lu, Z., et al.: Combination strategy of active learning for hyperspectral images classification. J. Commun. 39(4), 91–99 (2018)
9. Wang, Q., Fu, G., Wang, H., et al.: Fusion of multi-scale feature using multiple kernel learning for hyperspectral image land cover classification. Opt. Precis. Eng. 26(4), 980–988 (2018)
10. Pal, M., Foody, G.M.: Feature selection for classification of hyperspectral data by SVM. IEEE Trans. Geosci. Remote Sens. 48(5), 2297–2307 (2010)

11. Alam, F., Mehmood, R., Katib, I.: Comparison of decision trees and deep learning for object classification in autonomous driving. In: Mehmood, R., See, S., Katib, I., Chlamtac, I. (eds.) Smart Infrastructure and Applications. EICC, pp. 135–158. Springer, Cham (2020). https://doi.org/10.1007/978-3-030-13705-2_6
12. Jia, X., Richards, J.A.: Efficient maximum likelihood classification for imaging spectrometer data sets. IEEE Trans. Geosci. Remote Sens. **32**(2), 274–281 (1994)
13. Richards, J.A., Jia, X.: Using suitable neighbors to augment the training set in hyperspectral maximum likelihood classification. IEEE Geosci. Remote Sens. Lett. **5**(4), 774–777 (2008)
14. Wen, X., Hu, G., Yang, X.: Extraction of ETM+ image information based on C5.0 decision tree classification algorithm. Geogr. Geo-Inf. Sci. **23**(6), 26–29 (2007)
15. Wang, Y., Li, J.: Analysis of feature selection and its impact on hyperspectral data classification based on decision tree algorithm. J. Remote Sens. **11**(1), 69–76 (2007)
16. Meng, W., Ni, G., Gao, K., et al.: Spectral focusing of infrared hyperspectral spectroscopy. Infrared Laser Eng. **42**(3), 774–779 (2013)

Risk Prediction of Esophageal Cancer Using SOM Clustering, SVM and GA-SVM

Yuli Yang[1,2], Zhi Li[1,2], and Yanfeng Wang[1,2(✉)]

[1] Henan Key Lab of Information-Based Electrical Appliances,
Zhengzhou University of Light Industry, Zhengzhou 450002, China
[2] School of Electrical and Information Engineering,
Zhengzhou University of Light Industry, Zhengzhou 450002, China
yanfengwang@yeah.net

Abstract. In order to find the blood indicators significantly associated with the survival of esophageal cancer and predict the classification of patients' risk levels in an affordable, convenient, and accurate manner, a method based on self-organizing maps (SOM) neural network clustering and support vector machine prediction risk levels is proposed. Seventeen blood indicators of 501 esophageal cancer patients are pretreated. Nine factors related to patient survival are found by using SOM clustering method, and verified by using COX multi-factor risk regression model. Two critical thresholds for survival are found by plotting the ROC curve twice, and the lifetime are divided into three risk levels. The following is to select the data information of 9 blood indicators of 180 patients, including risk level 1, risk level 2, and risk level 3. Using the SVM method, patients' risk levels are predicted, the accuracy rate reached 91.11%. After the parameters optimization of genetic algorithm (GA), the accuracy rate reached 93.33%. Compared with BP neural network, it is concluded that SVM is superior to BP neural networks algorithm, and GA-SVM is better than SVM. This article provides a new method for early diagnosis and prediction of esophageal cancer.

Keywords: SOM · ROC curve · SVM · Genetic algorithm

1 Introduction

Tumors have always been one of the leading causes of death in both developed and developing countries, causing enormous social and economic burdens. Esophageal cancer is one of the highest morbidity and mortality types worldwide, ranking sixth in the cause of cancer-related deaths [1]. Esophageal cancer has become an important disease affecting the health of people [2].

With the advancement of science and technology and the innovation of medical technology, the treatment methods and treatment concepts of esophageal cancer have been continuously improved, but the comprehensive treatment of

© Springer Nature Singapore Pte Ltd. 2020
L. Pan et al. (Eds.): BIC-TA 2019, CCIS 1160, pp. 345–358, 2020.
https://doi.org/10.1007/978-981-15-3415-7_29

esophageal cancer models has certain limitations on the rapid recovery and long-term prognosis of patients [3]. And at present, due to the pathological complexity and early occultism of esophageal cancer, manual diagnosis will inevitably have some errors. The research on various pathological data of esophageal cancer patients is mostly based on traditional statistical analysis [4].

Therefore, the key to improving the survival rate of patients with esophageal cancer is early prediction [5]. By summarizing blood characteristics information of patients with esophageal cancer, a risk level prediction system can be established by using machine learning methods to make reasonable predictions and diagnoses of patients with unknown risk levels. Early prediction can effectively help people stay away from cancer [6].

The core content of this paper is to use self-organizing competitive maps (SOM) neural network and support vector machine algorithm (SVM) to find blood indicators thats are significantly related to patient survival, and to predict and classify patients' risk levels. Firstly, the self-organizing competitive neural network (SOM) is used to cluster the seventeen blood indicators of patients to find out the combination including nine indicators with high correlation. Through the COX regression method of MedCalc software, it is verified that the combination of these nine indicators has a significant correlation to the survival of patients. Then the survival periods are used as the test variable to draw the ROC curve. By plotting the ROC curve twice and calculating the Youden index, two critical thresholds for survival are obtained, dividing the patients into three risk levels, where the patients' nine indicators and three risk levels are obtained [7]. Through the algorithm of support vector machine, the patients' risk level is predicted and classified [8]. Optimized support vector machine with genetic algorithm, the prediction accuracy is improved. The method used in this paper accurately and effectively predicts the risk level of patients with esophageal cancer, and provides a method for early diagnosis of esophageal cancer.

2   Data Analysis and Processing of Blood Factors

2.1   SOM Cluster Analysis

Self-organizing maps (SOM) neural network is an unsupervised learning neural network with self-organizing functions. In the network structure, it is generally a two-layer network composed of an input layer and a competition layer, as shown in Fig. 1. The neurons between the two layers implement a two-way connection, and the network has no hidden layers [9]. SOM is adaptive and belongs to an unsupervised network, without distinguishing between training set and test set [10]. The "near excitation near suppression" function in the brain's nervous system is achieved through competition between neurons, with the ability to map high dimensional inputs to low dimensions. The SOM model is shown in Fig. 2.

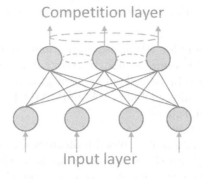

Fig. 1. SOM structure.

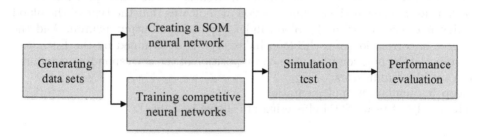

Fig. 2. SOM model building steps.

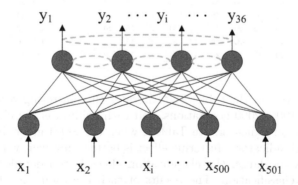

Fig. 3. SOM network structure diagram.

SOM algorithm clustering model is implementation as flowing:

Step 1. Data selection

Seventeen blood indicators for 180 patients' information are selected, including: WBC count, lymphocyte count, monocyte count, neutrophil count, eosinophil count, basophil count, red blood cell count, hemoglobin concentration, platelets count, total protein, albumin, globulin, PT, INR, APTT, TT, FIB, cluster analysis of these 17 blood indicators.

Step 2. Data normalization

The 180 patients' information of 17 blood indicators was normalized to [0, 1] by MATLAB, and brought into the SOM model for label clustering. The purpose of normalization is to make the algorithm converge quickly and reduce the error.

Step 3. Network establishment

As shown in Fig. 3, establish a self-organizing neural network of 180-36 structure.

Step 4. Network training

Train the data of each blood indicators and mark the indicator type neurons with the largest output. If the output neuron is at the same position as the output index type at the output position, it indicates that the type of the blood index is consistent with the blood index type of the output neuron. And the degree of correlation of each blood index type is determined by the Euclidean distance of the position from the neuron position of the corresponding maximum output blood indicators.

Step 5. Simulation implementation

The results of using SOM clustering are shown in Table 1.

Table 1. SOM clustering output.

Iteration steps	y_1	y_2	y_3	y_4	y_5	y_6	y_7	y_8	y_9	y_{10}	y_{11}	y_{12}	y_{13}	y_{14}	y_{15}	y_{16}	y_{17}
10	13	1	1	1	1	1	1	24	36	30	3	3	5	1	3	3	36
50	1	1	1	1	1	1	1	24	36	27	6	9	1	1	6	2	36
100	26	31	31	32	31	31	25	18	35	5	3	15	20	1	9	13	36
200	13	1	1	7	1	1	3	22	24	33	31	20	14	1	26	4	36
500	33	32	31	33	31	31	19	18	29	5	9	1	13	1	3	21	36
1000	20	33	32	27	32	32	25	17	24	5	9	1	36	1	3	13	36
2000	19	25	31	26	31	31	33	18	29	5	16	1	13	1	3	14	36

In the each row of the table, SOM iterates the clustering results of 10, 50, 100, 200, 500, 1000, 2000 generations, and each column is the clustering result of seventeen blood indicators. In Table 1, when the SOM algorithm is 50 iterations, the blood indicators clustering effect is better. However, when the number of iterations is too large, the classification effect is too detailed, resulting in unsatisfactory classification. The results of each iteration can also be verified by the subsequent COX regression. The smaller the P value, the greater the correlation. Therefore, the combination of blood indicators that has a greater correlation with the survival of the patients is 1, 2, 3, 4, 5, 6, 7, 13, 14 corresponding to: WBC count, lymphocyte count, monocyte count, neutrophil count, eosinophil count, basophil count, red blood cell count, PT, INR.

2.2 COX Regression Analysis

The COX regression model is also known as the "proportional hazards model" (Cox model), is a semiparametric regression model proposed by the British statistician D.R. Cox (1972). The model takes the survival outcome and survival time as the dependent variables, and can simultaneously analyze the influence of many factors on the survival period, analyze the data with truncated survival time. Due to the above-mentioned excellent properties, the model has been widely used in medical follow-up research since its inception, and it is the most widely used multi-factor analysis method in survival analysis [11].

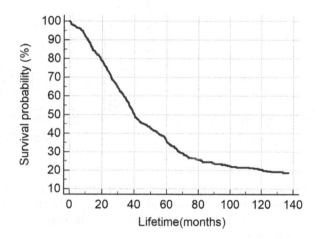

Fig. 4. Survival function at the mean of the covariate.

Using the software "MedCalc 18.2.1" as the operating platform, the survival time was taken as the time. The nine blood indicators obtained from the previous part of the cluster were used as covariates for COX regression analysis. The survival function at the mean of covariates is shown in Fig. 4. The results show that the P value of the overall score of the nine blood indicators is 0.0041 far less than 0.05, so the combination of these nine blood indicators is significantly related to the patients' survival [12].

2.3 ROC Curve Analysis

Receiver operating characteristic curve (ROC) is also known as sensitivity curve. The ROC curve is based on a series of different two-category methods (demarcation values or decision thresholds) [13]. The ROC curve combines sensitivity and specificity in a graphical manner, which accurately reflects the relationship between the specificity and sensitivity of an analytical method and is a comprehensive representation of the accuracy of the test [14]. The area under the curve can be used to evaluate the diagnostic accuracy [15].

The ROC curve can easily detect the ability to recognize performance at any threshold and select the best diagnostic threshold. The closer the ROC curve is to the upper left corner, the higher the accuracy of the test [16]. The point closest to the ROC curve in the upper left corner is the best threshold with the least error, with the fewest false positives and false negatives [17].

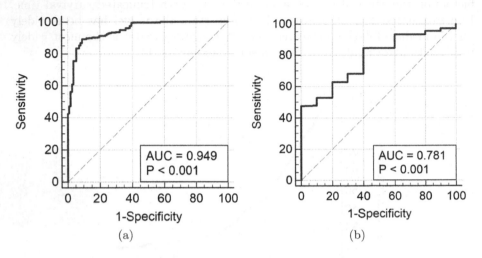

Fig. 5. The ROC curve analysis. (a) ROC curve of all samples; (b) ROC curve of samples with less than 27 months of survival. The ordinate is "Sensitivity" and the abscissa is "1-Specificity", the curve is clearly located at the upper left of the diagonal and has a good significance.

The ROC curve is plotted with the survival month of all samples as the variable, named "ROC for all samples" here, as shown in Fig. 5(a). The value of AUC (underline area) is 0.949, larger than 0.5, P < 0.0001. It's obvious that a good threshold can be found for the second classification of survival [18]. There is a better critical point which divides the lifetime into two levels. For the survival period, a critical threshold can be found to divide the survival period into two risk levels. The Youden index can be calculated as follows:

$$Youden\ Index = Sensitivity - (1 - Specificity) \qquad (1)$$

Finding the lifetime of the largest Youden index is the threshold, by calculating the lifetime threshold [19]. Here, the threshold for survival is 67.39 months. The Youden index is shown in Table 2.

Table 2. Youden index.

Project	ROC for all samples	ROC for low survival samples
Youden index J	≤67.39	≤27.38
Sensitivity	89.24	47.65
Specificity	90.99	100

Then, the sample information with a survival period of less than 67.39 months is summarized. Similarly, according to the above method, the ROC curve is drawn with the survival month as the variable, named "ROC for low survival samples", as shown in Fig. 5(b). The Youden index is shown in Table 2. The area under the ROC curve is 0.781. The threshold for survival is 27.38 months by calculating the Youden index.

Table 3. Risk classification of esophageal cancer patients.

Risk level	Lifetime (months)
Risk level 1	≤27.38
Risk level 2	27.38–67.39
Risk level 3	≥67.39

Therefore, the lifetime is divided into three risk levels by month. The survival period is not more than 27.38 months for "risk level 1", the survival period is 27.38 to 67.39 months for "risk level 2", and the survival period for more than 67.39 months is "risk level 3", as shown in Table 3. Thus, the patients' characteristic data set is obtained, which contains nine features and three levels, as shown in Table 4.

Table 4. Data of three risk levels of esophageal cancer patients.

WBC count	Lymphocyte count	Monocyte count	Neutrophil count	Eosinophil count	Basophil count	Red blood cell count	PT	INR	Life time (months)	Risk level
5.7	1.7	0.3	3.3	0.3	0.1	4.61	9.1	0.66	13.3	1
7	1	0	5	0	0	4.6	11.7	0.94	0.26	1
5.8	1.5	0.3	3.9	0	0.1	4.6	7.8	0.53	16	1
......										
6.6	1.3	0.4	4.7	0.1	0.1	4.91	8.1	0.56	62.6	2
7	2	0	4	0	0	4.9	8.7	0.62	28.7	2
6	2	0	3	0	0	4.9	11.2	0.89	65.2	2
......										
7.5	2.6	0.6	4.2	0.1	0	3.72	7.1	0.46	82.7	3
6.4	1.2	0.6	4.5	0.1	0	3.72	9.2	0.67	134	3
7.1	1.5	0.4	4.8	0.1	0.3	3.63	8.8	0.63	77.8	3

3 Risk Level Prediction

3.1 Support Vector Machine

Support Vector Machine (SVM) is derived from the theory of statistical learning. It is a set of machine learning theory established by Russian statistician and mathematician Vapnik. [20]. The basic idea of SVM is to use the principle of structural minimization to establish the optimal hyperplane in the attribute space, and to get the maximum classification interval, so that the classifier can get the global optimal in the whole sample space [21].

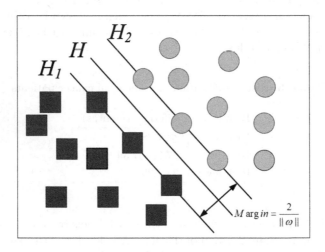

Fig. 6. Support vector machine optimal classification surface.

As shown in Fig. 6, square and circular points represent two samples. H is a classification line. H_1 and H_2 are the samples of each type that are closest to the classification line and parallel to the classification line, points on H_1 and H_2. The point (x_i, y_i) is called the support vector, and the classification interval is the distance between H_1 and H_2, which is Margin. When extended to a high-dimensional space, the optimal classification line becomes the optimal classification plane. Support vectors are the data points closest to the decision plane and are the most difficult to classify data points, so they are directly related to the optimal position of the decision plane [22].

3.2 SVM Kernel Function Selection

To find an effective classifier for risk levels prediction of esophageal cancer, the choice of SVM kernel function is a key step [23]. In order to solve the linear inseparability problem in the original space, it is transformed into a nonlinear separation problem by selecting the appropriate kernel function. Samples are projected from low-dimensional space to corresponding high-dimensional spaces. SVM kernel functions include linear kernel function, polynomial kernel function, Gaussian radial basis function (RBF) kernel function, and sigmoid kernel function. RBF kernel function has good application range and fitness, and low complexity. It is not only suitable for large samples, but also for small samples. Therefore, we choose the RBF kernel function here. The RBF is as follows:

$$K\left(x_i, y_i\right) = exp\left(-g\left\|x_i - x_j\right\|^2\right), g > 0 \qquad (2)$$

Parameter g represents the width of the kernel function, which plays an important role in improving the performance of SVM classification.

Fig. 7. SVM classification flowchart.

3.3   Support Vector Machine Classification Steps

The basic classification steps of the SVM algorithm are shown in Fig. 7.

Step 1. All 180 sets of data are loaded. Among them, there are 60 groups with risk level 1, 60 groups with risk level 2, and 60 groups with risk level 3. The data set contains three grades, nine blood indicators. Here, the first 45 sets of case samples of three risk levels were selected for training, as a training set. The last 15 case samples of each grade were tested as a test set. The training set had a total of 135 datasets, and the test set had a total of 45 datasets.

Step 2. The normalization function mapminmax is used to normalize the previously defined training sets and test sets. The normalization here is to map the data to the [0, 1] interval, prevent slow convergence of results and long training time.

Step 3. The RBF kernel function needs to set two parameters, c and g, which also have an important effect on the result, which are defaulted to 2 and 2. The default parameters g, c are brought in, and the training model is achieved.

Step 4. The risk level of 45 samples of the test sets are tested by the achieved model.

3.4   Support Vector Machine Prediction Results

LIB-SVM 2.83 and MATLAB R2016a software are used to predict the risk levels of esophageal cancer patients, and the running environment is Windows 10. All 180 esophageal cancer patients' information of three risk levels were investigated, and nine blood indicators were extracted from each patient's information using SOM neural network. The three risk levels of 135 cases of esophageal cancer cases were used as training samples, and 45 samples of each risk level were used for training; three risk level data from 45 cases were used as test samples, 15 sample data for each risk level for testing. The RBF parameters are $c = 2$ and $g = 2$. The actual classification and prediction classification results are shown in Fig. 8. The confusion matrix is shown in Fig. 9.

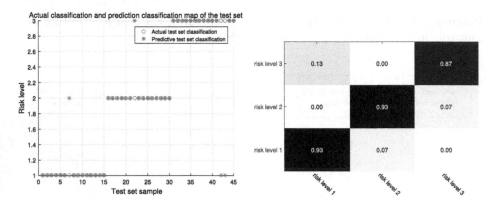

Fig. 8. SVM prediction results.   **Fig. 9.** Confusion matrix.

As it turns out, the SVM algorithm classifies the risk level of esophageal cancer very well. For the prediction results of 45 samples in the test set, the risk level 1 is correctly predicted in 14 cases with the accuracy rate of 0.93, the risk level 2 is correctly predicted in 14 cases with an accuracy rate of 0.93, and the risk level 3 is correctly predicted in 13 cases with an accuracy rate of 0.87. The test set predicts a total of 41 cases correctly with an accuracy rate of 91.11%.

Support vector machine has achieved good results in predicting the risk grade of esophageal cancer. It fully embodies the powerful classification of SVM algorithms and the unique advantages of classification and recognition in nonlinear and high dimensions.

3.5   Genetic Algorithm-Optimized Support Vectors Machines

In order to improve the performance of classifier, penalty parameter c and kernel parameter g need to be optimized [24]. The basic idea of GA is to simulate Darwin's natural evolution principle to search for the optimal solution, which can deal with many complex problems [25].

In this article, the accuracy of training set in the sense of cross validation is taken as the fitness function value in GA. The largest evolutionary algebra is 300, the maximum number of populations is 20, the range of the parameter c is [0, 100], the range of the parameter g is [0, 100], the cross validation parameter

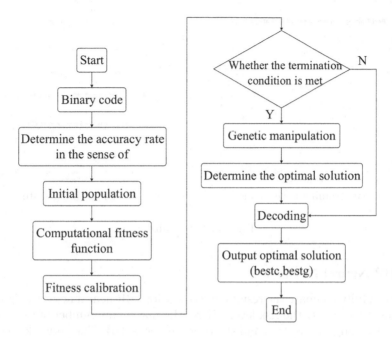

Fig. 10. GA-SVM flow chart.

is 5, the probability of crossover is 0.9, and the probability of variation is 0.01. The flow chart is shown in Fig. 10. The classification results are shown in Fig. 11, and the fitness curve is shown in Fig. 12. After the parameters optimization of GA, only three samples are not identified correctly, and its recognition accuracy rate rises to 93.33%.

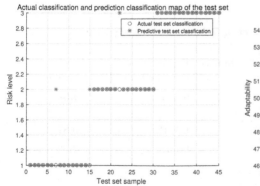

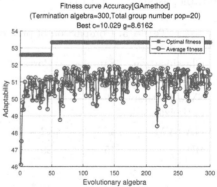

Fig. 11. Prediction based on SVM-GA.

Fig. 12. Fitness curve of GA optimization parameters.

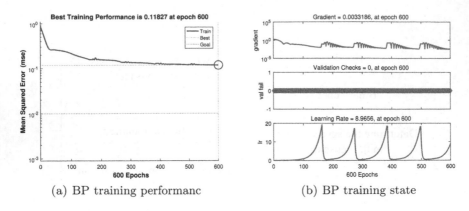

(a) BP training performanc   (b) BP training state

Fig. 13. BP results.

3.6 BP Neural Network

In order to fully compare the results of recognition, BP neural network algorithm is also used to classify the risk levels. Here, the maximum number of iterations is 600, the learning rate is 0.02, and the target error is 0.01. The network structure is 6-20-3-3, the algorithm uses traingdx, and the error change is mean square error (MSE).

Using the same characteristic data, the recognition rate of BP neural network is 86.67%. The training performance is shown in Fig. 13(a), and the training state is shown in Fig. 13(b).

3.7 Comparison of Prediction Algorithms

The comparison of SVM, GA-SVM and BP prediction results can be written in the table, as shown in Table 5.

Table 5. Comparison of prediction results of three algorithms.

Classification method	Samples number	Correctly identify number	Recognition accuracy rate (%)	Running times (S)
SVM (without optimization)	45	41	91.11	**0.56**
GA-SVM	45	42	**93.33**	19.87
BP	45	39	86.67	3.52

1. The SVM classification effect is better than that of BP neural network. The algorithms with higher accuracy rate are SVM classifiers with parameters optimization. Genetic algorithm-support vector machine (GA-SVM) have better classification results.
2. SVM has a good classification effect on small samples. Usually, the correct recognition rate is improved by optimizing the parameters c and g.

3. BP neural network has strong generalization ability and self-learning. The disadvantage is that the gradient descent method is slow, and sometimes, it may enter the local minimum and the training fails.

4 Conclusion

A method for prediction and evaluating risk levels of esophageal cancer was presented. The SOM neural network, ROC curve, SVM, GA-SVM and BP neural network were used in this method. The goal is to find more effective and accurate multiple blood indicators related to the survival of patients with esophageal cancer, and predictive classification of patient risk levels. In this paper, SOM network clustering and support vector machine-genetic algorithmlevel prediction have achieved good results and have a good effect on early risk prediction of esophageal cancer. The method for blood indicator analysis and risk prediction of esophageal cancer was provided. In the future, the optimization of the algorithm will make the predicted results more accurate.

Acknowledgements. This work was supported in part by the National Key R and D Program of China for International S and T Cooperation Projects (2017YFE0103900), in part by the Joint Funds of the National Natural Science Foundation of China (U1804262), in part by the State Key Program of National Natural Science of China under Grant 61632002, in part by the National Natural Science of China under Grant 61603348, Grant 61775198, Grant 61603347, and Grant 61572446, in part by the Foundation of Young Key Teachers from University of Henan Province (2018GGJS092), and in part by the Youth Talent Lifting Project of Henan Province (2018HYTP016) and Henan Province University Science and Technology Innovation Talent Support Plan under Grant 20HASTIT027.

References

1. Menya, D., Kigen, N., Oduor, M.: Traditional and commercial alcohols and esophageal cancer risk in Kenya. Int. J. Cancer **144**(3), 459–469 (2019)
2. Gillies, C., Farrukh, A., Abrams, R.: Risk of esophageal cancer in achalasia cardia: a meta-analysis. JGH Open **3**(3), 196–200 (2019)
3. Lin, S., Zhang, N.: Radiation modality use and cardiopulmonary mortality risk in elderly patients with esophageal cancer. Cancer **122**(6), 917–928 (2016)
4. Raymond, D., Seder, C., Wright, C.: Predictors of major morbidity or mortality after resection for esophageal cancer: a society of thoracic surgeon's general thoracic surgery database risk adjustment model. Ann. Thorac. Surg. **102**(1), 207–214 (2016)
5. Takeuchi, M., Suda, K., Hamamoto, Y.: Technical feasibility and oncologic safety of diagnostic endoscopic resection for superficial esophageal cancer. Gastrointest. Endosc. **88**(3), 456–46 (2018)
6. McCormack, V., Menya, D., Munishi, M.: Informing etiologic research priorities for squamous cell esophageal cancer in Africa: a review of setting-specific exposures to known and putative risk factors. Int. J. Cancer **140**(2), 259–271 (2017)

7. Miwata, T., et al.: Risk factors for esophageal stenosis after entire circumferential endoscopic submucosal dissection for superficial esophageal squamous cell carcinoma. Surg. Endosc. **30**(9), 4049–4056 (2015). https://doi.org/10.1007/s00464-015-4719-3

8. Omari, T., Szczesniak, M., Maclean, J.: Correlation of esophageal pressure-flow analysis findings with bolus transit patterns on video fluoroscopy. Dis. Esophagus **29**(2), 166–173 (2016)

9. Jin, C., Pok, G., Lee, Y.: A SOM clustering pattern sequence-based next symbol prediction method for day-ahead direct electricity load and price forecasting. Energy Convers. Manag. **90**, 84–92 (2015)

10. Delgado, S., Higuera, C., Calle-Espinosa, J.: A SOM prototype-based cluster analysis methodology. Expert Syst. Appl. **88**, 14–28 (2017)

11. El-Zimaity, H., Di, P., Novella, R.: Risk factors for esophageal cancer: emphasis on infectious agents. Ann. N. Y. Acad. Sci. **1434**(1), 319–332 (2018)

12. Ide, S., Toiyama, Y., Shimura, T.: Angiopoietin-like protein 2 acts as a novel biomarker for diagnosis and prognosis in patients with esophageal cancer. Ann. Surg. Oncol. **22**(8), 2585–2592 (2015)

13. Zeng, H., Zheng, R., Zhang, S.: Esophageal cancer statistics in China, 2011: estimates based on 177 cancer registries. Thorac. Cancer **7**(2), 232–237 (2016)

14. Kanzaki, N., Kataoka, T., Etani, R.: Analysis of liver damage from radon, X-ray, or alcohol treatments in mice using a self-organizing map. J. Radiat. Res. **58**(1), 33–40 (2017)

15. Roy, A., Bhattacharya, S., Guin, K.: Prediction of esophageal cancer using demographic, lifestyle, patient history, and basic clinical tests. Int. J. Adv. Eng. Sci. Appl. Math. **9**(4), 214–223 (2017). https://doi.org/10.1007/s12572-017-0199-0

16. Yerokun, B., Sun, Z., Yang, C.: Minimally invasive versus open esophagostomy for esophageal cancer: a population-based analysis. Ann. Thorac. Surg. **102**(2), 416–423 (2016)

17. Haisley, K.R., Hart, C.M., Kaempf, A.J., Dash, N.R., Dolan, J.P., Hunter, J.G.: Specific tumor characteristics predict upstaging in early-stage esophageal cancer. Ann. Surg. Oncol. **26**(2), 514–522 (2018). https://doi.org/10.1245/s10434-018-6804-z

18. Arnold, M., Laversanne, M., Brown, L.: Predicting the future burden of esophageal cancer by histological subtype: international trends in incidence up to 2030. Am. J. Gastroenterol. **112**(8), 1247 (2017)

19. Mora, A., Nakajima, Y., Okada, T.: Comparative study of predictive mortality scores in esophagostomy with three-field lymph node dissection in patients with esophageal cancer. Dig. Surg. **36**(1), 67–75 (2019)

20. Huang, S., Cai, N., Pacheco, P.P.: Applications of support vector machine (SVM) learning in cancer genomics. Cancer Genomics-Proteomics **15**(1), 41–51 (2018)

21. Kourou, K., Exarchos, T.P., Exarchos, K.P.: Machine learning applications in cancer prognosis and prediction. Comput. Struct. Biotechnol. J. **13**, 8–17 (2015)

22. Sun, J., Zhao, X., Fang, J., Wang, Y.: Autonomous memristor chaotic systems of infinite chaotic attractors and circuitry realization. Nonlinear Dyn. **94**(4), 2879–2887 (2018). https://doi.org/10.1007/s11071-018-4531-4

23. Huang, M.W., Chen, C.W., Lin, W.C.: SVM and SVM ensembles in breast cancer prediction. PLoS ONE **12**(1), e0161501 (2017)

24. Sukawattanavijit, C., Chen, J., Zhang, H.: GA-SVM algorithm for improving landcover classification using SAR and optical remote sensing data. Geosci. Remote Sens. Lett. **14**(3), 284–288 (2017)

25. Tao, Z., Huiling, L., Wenwen, W.: GA-SVM based feature selection and parameter optimization in hospitalization expense modeling. Appl. Soft Comput. **75**, 323–332 (2019)

Physical Constitution Discrimination Based on Pulse Characteristics

Nan Li[1], Yuping Zhao[2], Xiaobo Mao[1], Yang Wang[1], Yifan Shang[1],
and Luqi Huang[1,2(✉)]

[1] School of Electrical Engineering, Zhengzhou University, Zhengzhou 450001, China
[2] China Academy of Chinese Medical Sciences, Beijing 100020, China
810920179@qq.com

Abstract. Pulse palpation is an important diagnostic tool in Traditional Chinese Medicine (TCM) and related Oriental medicine systems. Pulse wave contains a lot of physiological and pathological information. How to effectively extract the information contained in pulse wave has been concerned at home and abroad. In this paper, a comprehensive introduction about the pulse wave characteristic is given. Furthermore, a new method of distinguishing students' physical constitution based on pulse characteristic information is proposed. First, pulse data were collected, preprocessed and pulse cycles were segmented. Second, time domain and pulse features coefficients of pulse wave were extracted. Finally, useful pulse wave features were evaluated and the features are classified to distinguish students' constitution by SVM classifier. Number experiments have proved the correctness and feasibility of the proposed theory.

Keywords: Traditional Chinese Medicine · Pulse characteristics · Physical constitution · SVM

1 Introduction

Traditional Chinese medicine (TCM) is a complete medical system that has widely application in disease diagnosis, treatment and prevention for over 30 centuries [1]. As one of the most important complementary and alternative medicines, TCM has been accepted and used increasingly in the world. Pulse Diagnosis (PD) is one of the most important techniques in Traditional Chinese Medicine [2]. In Oriental as well as Western medicine, the pulse is considered as a fundamental signal of life, carrying essential information about a person's health status [3]. It has now become a reality that modern medicine techniques can reveal well how human pulses are generated, so the pulse wave is widely used in aid of diagnosing cardiovascular disease, such as hypertension, heart disease, as well as other diseases, such as thyroid function abnormality, anemia, etc. [4,5].

Supported by key project at central government level (2060302).

© Springer Nature Singapore Pte Ltd. 2020
L. Pan et al. (Eds.): BIC-TA 2019, CCIS 1160, pp. 359–370, 2020.
https://doi.org/10.1007/978-981-15-3415-7_30

As we all known, the spectrum of arterial pulse signals exposes discrepancies between healthy people, sub-healthy people and patients with certain diseases. Experienced doctors can determine the cause of disease and other information through the pulse of patients, and give reasonable treatment. Leonard et al. [6] applied wavelet analysis to pulse waveforms of healthy and unwell children and found that it is possible to tell the different groups by using wavelet power features and wavelet entropy features at different frequency bands. Although the number of samples used by Leonard et al. is too small to reach statistical significance, it indicates that pulse features can be used to classify the patients into different groups based on their illness severity. By perceiving the pressure fluctuation of wrist pulse signal, the TCM doctors could tell the pathological changes of the patients and further classify the patient into particular groups defined by TCM theory for particular treatment [7]. The Chinese medical doctors divide the terminal region of the radial artery into three adjacent intervals called Chon, Gwan and Cheok, and use the three fingers of index, middle and ring fingers simultaneously or individually to determine various characteristic features of the pulse wave in pulse diagnosis [8,9]. A scientific way of studying the pulse, then, should be to analyze its frequency spectrum and to correlate the spectral features with health conditions [10,11]. By analyzing the pressure fluctuation of the pulse, we may obtain some diagnostic information that electrocardiograph cannot reflect. For example, the relationship between pulse wave and hemodynamic parameters in patients with hypertension was studied [12,13], and research has shown that blood pressure values can be predicted by pulse waves. In addition, compared with Western medicine, traditional Chinese medicine is a convenient, inexpensive, painless, bloodless, non-invasive and side effect free method to analyze the human body internal pathological changes [14].

In ancient Oriental medicine texts, various pulse qualities have been described [15–17]. The pulse modes cover the following 28 pulse qualities: floating, sunken, slow, rapid, surging, fine, vacuous, replete, long, short, slippery, rough, string-like, tight, soggy, moderate, faint, weak, dissipated, hollow, drumskin, firm, hidden, stirred, intermittent, bound, skipping, and racing [18]. TCM can distinguish everyone's physical constitution according to the corresponding pulse characteristics: Ping-He or Not Ping-He constitution [1,19]. However, the pulse assessment is a matter of technical skill and subjective experience in China. Therefore, a solid quantitative description of pulse diagnosis will pave the way for the modernization of pulse diagnosis in TCM [2,20].

At present, more and more attention has been paid to the diagnostic model of combining traditional Chinese medicine with artificial intelligence, such as deep learning [21] and Convolutional Neural Network [22]. With modern technology and means in medical diagnosis, AI assistant system can simulate the clinical diagnosis thinking and reasoning judgment process of medical experts, automatically analyze and calculate clinical data, provide clinical decision support, and has gradually become an effective assistant tool for disease prevention, management, diagnosis and treatment. In the classification module, the support vector machine (SVM) classifier is used to distinguish person constitution. SVM is related to the regularization networks [23] and offers an advancement over the ANN model. It is based on statistical learning theory that adopt least square methods to solve the problem [24] to reduces by least square solutions via a set of linear equations based on structural risk minimization. Consequently, the SVM model can avoid over fitting of the training data, does not require iterative tuning of model parameters [25], has better generalizability, requires few kernels and has good performance [26]. Yang et al. proposed a Prior knowledge Support Vector Machine (P-SVM) to build an Information Management System of TCM Syndrome [27]. Cho et al. [28] applied several machine learning techniques, such as support vector machine (SVM) classification and feature selection methods, to predict the onset of diabetic nephropathy using an irregular and unbalanced diabetes data set. It shows that SVM could be used for TCM clinical diagnosis study with high dimensions.

In this work, a comprehensive introduction about the pulse wave is reviewed and their characteristics are summarized in a centralized way. Furthermore, a new method of distinguishing students' constitution based on pulse characteristic information is proposed. Specifically, according to the collected pulse data of students, the pulse characteristics and numerical fitting method are extracted. Finally, the extracted pulse characteristics are used to distinguish students' constitution by SVM classifier. The experiment verifies the feasibility of the proposed method.

The rest of paper is organized as follows. In Sect. 2, a introduction about pulse prepare work is given. In Sect. 3, a new method of physique classification for different populations based on pulse image is proposed in detail. In Sect. 4, experimental result and analysis are given. Finally, some conclusions and suggestions for future work are given in Sect. 5.

2   Prepare Work

In this part, pulse data acquisition, preprocessing, pulse wave cycle segmentation, and feature extraction were performed.

2.1   Data Acquisition

Data were collected from students of Zhengzhou University by ZM-300 Intelligent Pulse Meter of Traditional Chinese Medicine. The subjects were allowed to rest for 3–5 min before data collection and were instructed to sit, breathe quietly, relax the upper arm, extend the forearm, and flex the shoulder and elbow to about 120°, with the left wrist on a pulse pillow. Then, the TCM pulse bracelet was placed over the Guan position in the left hand and right hand to capture the best pulse signals for 10 s. Subjects were excluded from analysis if they lacked complete data for control or outcome variables or had significant heart disease. Pulse wave cycle data were obtained for 10 s in all subjects during data preprocessing (Figs. 1 and 2).

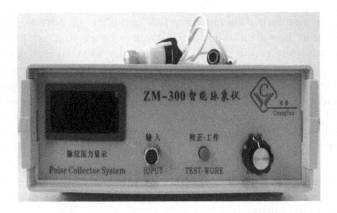

Fig. 1. ZM-300 intelligent pulse meter.

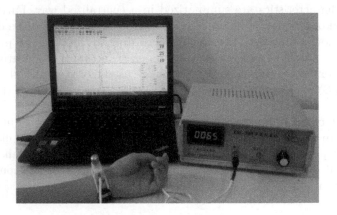

Fig. 2. Pulse acquisition.

2.2 Pulse Wave Analysis

Baseline wandering of original pulse data was removed with a high-pass filter in the sampling device. A pulse wave cycle was defined as the interval between two initial sets of pulse data. Pulse wave is the manifestation of various information carried by heart ejection activity and pulse wave propagating along all levels of arteries and vessels. Therefore, the rise, fall, wave and isthmus of pulse wave have corresponding physiological significance.

Figure 3, a typical cycle of a measured pulse signal is illustrated including main parameters that are often used to characterize the waveform. As shown in Fig. 3, the starting point of the pulse signal is point A, the three peaks are B, D and F, respectively, and the two valleys are C and E. B is the main pulse wave, AB and BC are its ascending and descending branches, and D is the front wave of heavy wave. Based on the above points and the related parameters formed by these points, some time-domain characteristic parameters of pulse can be

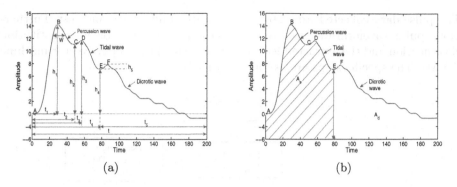

Fig. 3. Time domain parameters of the pulse wave signal.

extracted and a series of definitions are given in detail. One advantage of time domain parameters is that they are quite straightforward to comprehend and interpret and have some physiological significance.

t_1: time distance between the starting point of pulse chart and the main wave peak, which corresponds to the rapid ejection period of the left ventricle.

t_2: time value between the starting point of the pulse and the main wave isthmus.

t_3: time distance between the starting point of pulse and the tidal wave peak.

t_4: time distance between the starting point of pulse chart and dicrotic notch, which corresponds to the systole period of the left ventricle.

t_5: time distance between dicrotic notch and the ending point of pulse chart, which corresponds to the diastolic period of the left ventricle.

t: time distance between the starting point and the ending point.

h_1: The amplitude of the percussion wave.

h_2: The amplitude of the beginning of the tidal wave.

h_3 : The amplitude of the crest of the tidal wave.

h_4: The amplitude of the beginning of the dicrotic wave.

h_5: The amplitude of the crest of the dicrotic wave.

A_s: The systolic pulse wave area in t_4.

A_d: The diastolic pulse wave area between the end of t_4 and the end of t.

w_1: The width of main peak at $\frac{1}{3}$ of h_1, which means the duration of high intra-arterial pressure.

w_2: The width in $\frac{1}{5}$ amplitude of the percussion wave.

3   Constitution Analysis Based on Pulse Wave

In this section, the constitution of 100 students based on pules characters is analyzed in detail. Each person has 32 physical characteristics, of which the left hand and the right hand have 16 characteristics, respectively. People's constitution can be divided into Ping-He constitution and Non Ping-He constitution.

The pulse data collected from 100 students are train, study and test. Furthermore, pulse waveform was used to extract the relevant features for constitution classification and the extracted features are classified by support vector machine (SVM). The specific steps are as follows (Fig. 4).

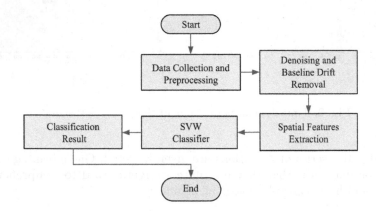

Fig. 4. Flow chart of physique constitution identification.

3.1 Pulse Processing

The data of left and right hand pulse of healthy college students were collected by pulse collector, and different pulse signals were obtained by each hand under 6 different pressures. Pulse signal is a kind of low-frequency signal, which mainly concentrates between 0.5 and 30 Hz. In the processing of original pulse waveform, it is necessary to remove the low-frequency noise of respiratory wave and baseline drift and other external high-frequency noise. The noise signals below 0.5 Hz and above 30 Hz are removed from each signal through a bandpass filter. The original pulse wave and filter pulse wave are shown in Figs. 5 and 6, respectively.

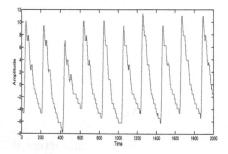

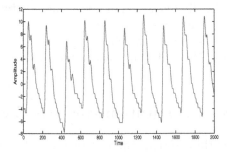

Fig. 5. Original pulse wave.   **Fig. 6.** Filter pulse wave.

3.2   Extraction of Time Domain Features

(1) A original pulse waveform based on the measured data are draw as shown in Fig. 5.

(2) The obtained waveform is filtered to remove irrelevant interference by band-pass filter, such as respiratory wave and baseline drift in Fig. 6.

(3) The peak point of pulse wave is found by extremum method, and the minimum point is found by minimum method and window function as the starting and ending point of pulse wave. The waveform within the starting and ending point is regarded as a complete pulse period. The complete waveforms and the extreme points during 10s are taken as the research objects.

(4) Extract a maximum point adjacent to the minimum point in step (3). The difference between the maximum value and the mean value of the two adjacent minimum values is h_1. Six main wave amplitudes h_1 can be obtained under six pulse pressures. The pressure at the maximum h_1 is taken as the optimal pulse pressure, and characteristics under the optimal pulse pressure are extracted. t_1 means the difference that is the abscissa corresponding to the maximum and the minimum divided by the sampling frequency. The amplitude and time characteristics of pulse waveform are extracted and recorded by this technique. In particular, $t = t_4 + t_5$ and $h_{5p} = h_4 + h_5$.

(5) The w_1 means the width of main wave in its 1/3 height position, The w_2 means the width of main wave in its 4/5 height position.

(6) Means of eigenvalues obtained in all periods are taken as the final time-domain eigenvalues (Fig. 7).

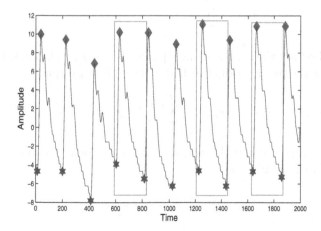

Fig. 7. Periodic pulse waveform.

3.3   Support Vector Machine (SVM)

Support Vector Machine (SVM) is a kind of generalized linear classifier which classifies data by supervised learning. Its decision boundary is the maximum margin hyperplane for solving learning samples. The classical SVM model belongs to a class of kernel-based ML methods that are used to solve the regression problems by a set of predictor variables (x) that are intrinsically related to the objective (forecasted) variable (y). A versatility of the SVM model is that the kernel function is able to implicitly convert the predictor data to a higher-dimensional feature space, where the results in the higher-dimensional feature space correspond to the results of the original, but lower-dimensional input space. To solve a nonlinear problem, an input variable that corresponds to predictor variable is non-linearly mapped into a high-dimension feature space, and is correlated linearly with objective variable in order to formalize the SVM model:

$$R_n = f(x) = \omega\phi(x) + b \qquad (1)$$

where ω is the weighted vector, b is the constant, $\phi(x)$ denotes a mapping function that is utilized in feature space. The coefficients ω and b are estimated by minimization as follows:

$$R_{reg}(f) = C\frac{1}{N}\sum_{i=1}^{N} L_\varepsilon f((x_i), y_i) + \frac{1}{2}\parallel \omega \parallel^2 \qquad (2)$$

$$L_\varepsilon(f(x) - y) = \begin{cases} |f(x) - y| - \omega & |f(x) - y| \geq \varepsilon \\ 0 & \text{otherwise} \end{cases}$$

where the C and ε are the model's prescribed parameters. After applying the Lagrangian and optimal conditions, a non-linear regression function is obtained:

$$f(x) = \sum_{i=1}^{l}(\alpha_i - \alpha_i^*)\kappa(x_i, x) + b \qquad (3)$$

where α_i and α_i^* are the Lagrange multipliers. The $\kappa(x_i, x)$ is the kernel function that describes the inner product in D-dimension feature space.

$$\kappa(x_i, x) = \sum_{i=1}^{D} \phi_j(x)\phi_i(y) \qquad (4)$$

Among the different kernel functions used, we have utilized radial basis function (RBF) in this study viz:

$$\kappa(x_i, x) = exp(\frac{- \parallel x_i - x_j \parallel^2}{2\sigma^2}) \qquad (5)$$

where σ is the width of the kernel and x_i and x_j are the inputs in the ith and jth dimensions, respectively.

4   Experimental Result and Discussion

In our experiments, we adopt support vector machine (SVM) with Gaussian RBF kernel for that it has good generalization on small dataset. The experiments were done under the MATLAB environment by using the SVM-KM toolbox. The students' constitutions are discriminated based on the 32 pulse features collected and 16 pulse features collected by right hand and left hand, respectively. 80% of the students were selected as training and the remaining 20% as testing. The average classification accuracy was obtained by 10 cross-validation. The SVW classification results are as follows (Table 1).

Table 1. Recognition rate of the SVW.

Parameter	Accuracy rate
Left and right hands	0.7450 ± 0.0858
Right hand only	0.7667 ± 0.1100
Left hand only	0.7210 ± 0.0720

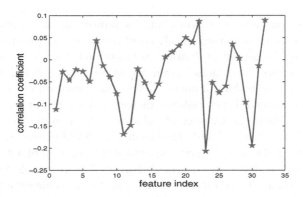

Fig. 8. Correlation coefficient between pulse characteristics and target tags.

The phenomenon can be explained by the fact that there exist some relevant attributes between human' pulse and constitution. In this work, 32 pulse characteristics were extracted and the absolute value of correlation coefficient was used to evaluate the contribution of classification accuracy. The greater absolute value of correlation coefficient is, the greater its influence on physical constitution classifies. The correlation coefficient between pulse characteristics and target tags is shown in Fig. 8.

The correlation between 32 features and target tags was analyzed and are classified by SVM classifier in Fig. 9. It can be seen that the classification results show a general upward trend and downward trend. The first seven pulse characteristics can make the discrimination accuracy exceed 85%, so multiple features have advantages in classification. However, with the increase of the number of features, feature redundancy will occur. It can be concluded that the increase of feature will affect the accuracy of classification.

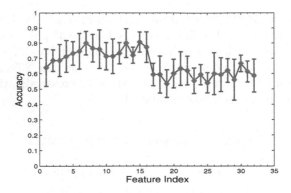

Fig. 9. Accuracy of constitution classification

5   Conclusion and Recommendations

The wrist pulse signal of a person contains important information about the pathologic changes of the person's body condition. Chinese pulse diagnosis plays an important role in patient evaluation in traditional Chinese medicine. In this paper, the characteristics of pulse signal are introduced in detail. In addition, a new method of distinguishing human constitution based on pulse characteristic information is proposed. First, pulse data were collected and pulse wave cycles were segmented. Then, the features of pulse wave were extracted. Finally, pulse wave features were evaluated and are classified by SVM classifier. Experimental results show that pulse signals carry important information of a person, and the proposed method achieves an accuracy of over 85% in classifying the healthy person's constitution. In the future, we will build large scale data set with more kinds of characteristics to further verify this new pulse diagnosis approach by different classifiers.

Acknowledgments. The work is supported by key project at central government level (Grant No. 2060302).

References

1. Wu, H.K., Ko, Y.S., Lin, Y.S., Wu, H.T., Tsai, T.H., Chang, H.H.: The correlation between pulse diagnosis and constitution identification in traditional Chinese medicine. Complement. Ther. Med. **30**, 107–112 (2017)
2. Velik, R.: An objective review of the technological developments for radial pulse diagnosis in traditional Chinese medicine. Eur. J. Integr. Med. **7**(4), 321–331 (2015)
3. De, M.N., Cordovil, I., De, S.F.A.: Traditional Chinese medicine wrist pulse-taking is associated with pulse waveform analysis and hemodynamics in hypertension. J. Integr. Med. **14**(2), 100–113 (2016)
4. Zhang, Z., Zhang, Y., Yao, L., Song, H., Kos, A.: A sensor-based wrist pulse signal processing and lung cancer recognition. J. Biomed. Inform. **79**, 107–116 (2018)

5. Qiao, L.J., et al.: The association of radial artery pulse wave variables with the pulse wave velocity and echocardiographic parameters in hypertension. Evid.-Based Complement. Altern. Med. **2018** (2018). Article ID 5291759
6. Leonard, P., Beattie, T.F., Addison, P.S., Watson, J.N.: Wavelet analysis of pulse oximeter waveform permits identification of unwell children. J Energ. Med. **21**, 59–60 (2004)
7. Xu, L., Wang, K., Li, Y.: Modern researches on traditional Chinese pulse diagnosis. Eur. J. Orient. Med. **8**(1), 56–63 (2004)
8. Jeon, Y.J., et al.: A clinical study of the pulse wave characteristics at the three pulse diagnosis positions of Chon, Gwan and Cheok. Evid.-Based Complement. Altern. Med. **2011** (2011). Article ID 904056
9. Bae, J.H., Jeon, Y.J., Kim, J.Y., Kim, J.U.: New assessment model of pulse depth based on sensor displacement in pulse diagnostic devices. Evid.-Based Complement. Altern. Med. **2013** (2013). Article ID 938641
10. Yallapragada, V.J., Rigneault, H., Oron, D.: Spectrally narrow features in a supercontinuum generated by shaped pulse trains. Opt. Express **26**(5), 5694–5700 (2018)
11. Khanna, A., Paul, M., Sandhu, J.S.: Efficacy of two relaxation techniques in reducing pulse rate among highly stressed females. Calicut Med. J. **5**(2), 23–25 (2007)
12. RibeirodeMoura, N.G., Cordovil, I., de Sá Ferreira, A.: Traditional Chinese medicine wrist pulse-taking is associated with pulse waveform analysis and hemodynamics in hypertension. J. Integr. Med. **14**, 100–113 (2016)
13. Moura, N.G.R., Ferreira, A.: Pulse waveform analysis of Chinese pulse images and its association with disability in hypertension. JAMS J. Acupunct. Meridian Stud. **9**, 93–98 (2016)
14. Xu, J., Yang, Y.: Traditional Chinese medicine in the Chinese health care system. Health Policy **90**(2–3), 133–139 (2009)
15. Nestler, G.: Traditional Chinese medicine. Med. Clin. **86**(1), 63–73 (2002)
16. Bilton, K., Zaslawski, C.: Reliability of manual pulse diagnosis methods in traditional East Asian medicine: a systematic narrative literature review. J. Altern. Complement. Med. **22**(8), 599–609 (2016)
17. Hajar, R.: The pulse in ancient medicine part 1. Heart Views Off. J. Gulf Heart Assoc. **19**(1), 36 (2018)
18. Tang, A.C.Y., Chung, J.W.Y., Wong, T.K.S.: Validation of a novel traditional chinese medicine pulse diagnostic model using an artificial neural network. Evid. Based Complement Altern. Med. **2012** (2012). Article ID 685094
19. Huan, E.Y., et al.: Deep convolutional neural networks for classifying body constitution based on face image. Comput. Math. Methods Med. **2017** (2017). Article ID 9846707
20. Li, X., et al.: Computerized wrist pulse signal diagnosis using gradient boosting decision tree. In: 2018 IEEE International Conference on Bioinformatics and Biomedicine, pp. 1941–1947. IEEE, Madrid, Spain (2018)
21. Cui, Z., Xue, F., Cai, X., Cao, Y., Wang, G., Chen, J.: Detection of malicious code variants based on deep learning. IEEE Trans. Industr. Inf. **14**(7), 3187–3196 (2018)
22. Cui, Z., Du, L., Wang, P., Cai, X., Zhang, W.: Malicious code detection based on CNNs and multi-objective algorithm. J. Parallel Distrib. Comput. **129**, 50–58 (2019)
23. Chen, W.H., Hsu, S.H., Shen, H.P.: Application of SVM and ANN for intrusion detection. Comput. Oper. Res. **32**(10), 2617–2634 (2005)
24. Lu, X., Fan, B., Huang, M.: A novel LS-SVM modeling method for a hydraulic press forging process with multiple localized solutions. IEEE Trans. Industr. Inf. **11**(3), 663–670 (2015)

25. Yu, P.S., Chen, S.T., Chang, I.F.: Support vector regression for real-time flood stage forecasting. J. Hydrol. **328**(3–4), 704–716 (2006)
26. Salcedo, S.S., Deo, R.C., Carro, C.L.: Monthly prediction of air temperature in Australia and New Zealand with machine learning algorithms. Theoret. Appl. Climatol. **125**(1–2), 13–25 (2016)
27. Yang, X.B., Liang, Z.H., Zhang, G.: A classification algorithm for TCM syndromes based on P-SVM. In: International Conference on Machine Learning and Cybernetics, pp. 3692–3697. IEEE, Guangzhou, China (2005)
28. Cho, B.H., Yu, H., Kim, K.W., Kim, T.H., Kim, I.Y., Kim, S.I.: Application of irregular and unbalanced data to predict diabetic nephropathy using visualization and feature selection methods. Artif. Intell. Med. **42**(1), 37–53 (2008)

Research on Emotional Classification of EEG Based on Convolutional Neural Network

Huiping Jiang$^{(\boxtimes)}$ ⓘ, Zequn Wang ⓘ, Rui Jiao ⓘ, and Mei Chen ⓘ

School of Information Engineering, Minzu University of China, Beijing 100081, China
jianghp@muc.edu.cn

Abstract. With the rapid development of machine learning technology, deep learning algorithms have been widely used in the processing of physiological signal. In this article, we used electroencephalography (EEG) signals based on Convolutional Neural Network (CNN) model in the deep learning algorithm to identify the positive and negative emotional states, and compare them with the support vector machine (SVM). By collecting the EEG signals of the subjects under different emotional stimuli states, the CNN and the SVM are used to identify the emotion data based on different feature transformations. The research results show that the average accuracy of using differential entropy (DE) features by SVM is 86.51%, which is better than the previous research on the same batch of datasets. At the same time, the classification effect of CNN is better than the traditional SVM (average classification accuracy is 86.90%), and its accuracy and stability have correspondingly better trends.

Keywords: Deep learning · Convolutional Neural Network · Electroencephalogram · Emotional classification

1 Introduction

Emotional recognition plays an increasingly important role in our lives and work, and also irreplaceable in human-computer interaction systems, such as autonomous driving technology [1], polygraph technology [2], medical applications [3], etc. It has become a significant research area of artificial intelligence. In recent years, researchers have tried many ways to identify the emotions of the subjects. Some of the main methods and techniques include facial expression recognition, gesture recognition, speech recognition, natural language processing, human physiological signal recognition (cortisol levels, heart rate, blood pressure, breathing, skin electrical activity, pupil diameter, EEG, etc.), multimodal emotion recognition [4] and so on. Although the non-physiological signal is simple and easy, it has low reliability and poor authenticity. Compared with other non-physiological signals, EEG signals not only have higher authenticity, but also apply to any population. EEG has been applied to health care, intelligent input, education and teaching, entertainment equipment and many other aspects.

© Springer Nature Singapore Pte Ltd. 2020
L. Pan et al. (Eds.): BIC-TA 2019, CCIS 1160, pp. 371–380, 2020.
https://doi.org/10.1007/978-981-15-3415-7_31

In the past, research on EEG emotion recognition mainly used traditional shallow machine learning algorithms to identify emotional state, including the following steps: emotional induction, EEG signal acquisition, EEG signal preprocessing, EEG feature extraction, emotion classification [5]. For example, Jiang et al. [6] pre-processed and extracted features of multiple physiological signals under different emotional states, and then used J48 decision tree algorithm to realize the recognition of four emotional states. Frantzidis et al. [7] divided the subjects' emotions into two types of comparisons: pleasure and low, high excitement and low excitement. The classification of emotions using the Mahalanobis distance classification method also achieves higher classification accuracy. The above-mentioned studies generally use shallow classifiers, which require manual feature selection and extraction. Shallow machine learning algorithms have limited ability to represent complex functions in the case of finite samples and computational units, and their generalization ability is constrained for complex classification problems [8]. Braverman [9] also pointed out that there is a large class of models that cannot be represented by shallow learning networks. These mathematical results point to the limitations of shallow learning networks. Therefore, the classification results of traditional shallow machine learning algorithms are limited in some complex models. This prompted researchers to turn to deep neural networks to model and classify complex problems.

A series of algorithms based on a deep learning framework is learned by a deep nonlinear network structure. The realization of complex function approximation can extract and select features that reflect the essence of the data set from a small number of sample sets [10]. Pan et al. [11] also proposed an algorithm mainly for solving expensive optimization problems where only a small number of real fitness evaluations are allowed. Since Krizhevsky et al. [12] used the Convolutional Neural Network (CNN) in the Computer Vision System Identification Challenge to win the championship with a rate of more than 10% in the second place. In the academic, the upsurge of deep learning algorithm research has been triggered.

In summary, the deep learning algorithm can mine deeper emotional features compared with the traditional machine learning algorithm, which is beneficial to the analysis and research of EEG signals. In this paper, the related applications of CNN based on EEG emotion recognition are studied. Convolutional neural networks are used to classify EEG features, and the classification effects are analyzed.

2   CNN Model

In 1998 LeCun [13] proposed the LeNet-5 model, which is the first formal CNN model. The key to the success of the LeNet-5-model is that it takes the approach of weight sharing and local feeling field. On the one hand, the reduced number of weights makes the network easy to optimize and on the other hand reduces the risk of overfitting. Since then, the CNN has not received much attention. Until 2006, Hinton [14] pointed out that multiple hidden layer neural networks have superior feature learning capabilities, and the training complexity can be effectively alleviated by layer-by-layer initialization. The CNN now enters people's field of vision again.

CNN have been widely used in the classification of speech and images and have achieved good results. However, there are few studies on the application of EEG, and there is very little in the emotional recognition based on EEG. In this paper, the CNN is introduced into EEG emotion recognition, and the application of convolutional neural network in EEG is explored. Because the EEG signal is weak and the features are not obvious, that is to say, the feature vector obtained by the feature extraction method is not necessarily used for emotion classification. Therefore, we introduce a CNN to process and classify the eigenvectors of EEG signals. Designed to improve the accuracy and robustness of classification. At the same time, we also directly used the convolutional neural network on the EEG data after dimension reduction, extracted the EEG features after dimension reduction, and evaluated the classification results.

In this paper, the input of the CNN model we used is normalized between samples. Since the input matrix is not a regular square matrix, in order to improve the accuracy of the classification, we also use a rectangular convolution kernel. We use $W_{x,y}$ for the weight of the representation filter, b for the bias term of the filter, and f for the activation function. Commonly used activation functions are Sigmoid functions, $tanh$ functions, and the more commonly used *ReLU* functions in recent years. The output of the filter is:

$$g = f(\sum_x \sum_y a_{x,y} \times W_{x,y} + b) \qquad (1)$$

The training of CNN is divided into two processes: forward propagation and error back propagation. First, the output value of the network is calculated by forward propagation. Then the error between the output value and the expectation is calculated by the loss function. Finally, the network is trained by error back propagation. Commonly used loss functions have mean square error and cross entropy. In this paper, we use the method of cross entropy verification. The cross entropy characterizes the distance between two probability distributions. Given two probability distributions p and q, the formula for the cross entropy of p by q is:

$$H(p,q) = -\sum_x p(x)logq(x) \qquad (2)$$

The error back propagation is based on the principle of gradient descent. We just need to update in the direction of the negative gradient. Let J be the cost function, then the iterative process of each $W_{i,j}$ and $b_{i,j}$ is:

$$W_{i,j}^{(l+1)} = W_{i,j}^{(l)} - \alpha \frac{\partial J}{W_{i,j}^{(l)}} \qquad (3)$$

$$b_{i,j}^{(l+1)} = b_{i,j}^{(l)} - \alpha \frac{\partial J}{b_{i,j}^{(l)}} \qquad (4)$$

Where α is the learning rate, $\frac{\partial J}{W_{i,j}^{(l)}}$ and $\frac{\partial J}{b_{i,j}^{(l)}}$ are the partial derivatives of the error.

3   Acquisition and Preprocessing of EEG Signals

This paper studies the classification of positive and negative emotions, so it is necessary to select pictures that can induce positive and negative emotions. The stimulating pictures are from IAPS, CAPS and images collected by the laboratory. The data is the EEG signals of 6 experimenters, and the subjects are students of Minzu University of China. Between 22 and 26 years of age, right-handed, healthy, sleep well, no brain damage and mental illness. And this study protocol was approved by the institutional review boards (ECMUC2019008CO) at Minzu University of China. All participants provided IRB-approved written informed consent after they were given an explanation about the experimental procedure.

Before the experiment begins, the participants should carefully read the instructions on the screen to fully understand the flow and details of the experiment. The specific flow chart of the experiment is shown in Fig. 1.

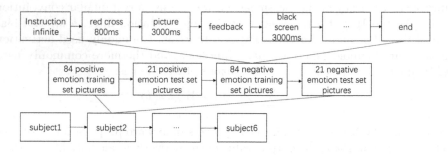

Fig. 1. Experimental flow chart.

The EEG signal is very weak, and it is very susceptible to changes from internal or external environment during the measurement process, so that the measured signal is unreliable and is doped with many electrical activity disturbances that are not caused by the brain. These disturbances are called artifacts or arti-facts [6]. In this paper, the pretreatment of EEG is done by Scan 4.5 software, which mainly works to remove bad areas, remove electro-optical artifacts, remove other artifacts, and digital filtering.

4   EEG Signal Feature Extraction

Feature extraction is to achieve the dimensionality reduction of the original EEG signal by extracting the method that can characterize the EEG signal parameters. In this paper, the method of feature extraction mainly includes wavelet transform. Previously, the laboratory did a lot of experiments. In the emotional classification experiment, the sym8 wavelet is better for reducing the original signal. In the paper [15], some comparisons were made on the mother wavelet. The effect of sym8 is better than other mother wavelets. Let's take a brief introduction to the symlet (symN) wavelet. The symlet wavelet function is a wavelet modified by IngridDaubechies for the db function. The dbN wavelet does not have symmetry, that is, the mother wavelet of the dbN wavelet and the scaling function are asymmetrical. symN has better symmetry than dbN wavelet, which can reduce the phase distortion when analyzing and reconstructing signals to a certain extent. Therefore, we choose sym8 wavelet as the mother wavelet. After wavelet transform, we extract four characteristics of band energy, band energy ratio, band energy ratio logarithm and differential entropy. Here is a brief introduction to these four characteristics.

(1) The band energy, that is, the energy of each frequency band after wavelet transform, is obtained by square summing the coefficients of each frequency band, as shown in Eq. (5):

$$E_i = \sum_{j=1}^{n_i} d_{ij}^2 \qquad (5)$$

Where d_{ij} is the j-th wavelet coefficient of the i-th frequency band, n_i is the number of wavelet coefficients of the i-th frequency band, and E_i is the energy of the i-th frequency band.

(2) The energy ratio REE_i of frequency bands is the ratio of each frequency band to the total energy, as shown in Eq. (6):

$$REE_i = \frac{E_i}{\sum_{j=1}^{n} E_j} \qquad (6)$$

Where E_i is the energy in the i-th frequency band, E_j is the energy in the j-th frequency band and n is the number of frequency bands.

(3) The logarithm of the energy ratio of the frequency bands is obtained by calculating the logarithm of the energy ratio of each frequency band, which was defined as Eq. (7):

$$LREE_i = log_{10}^{REE_i} \qquad (7)$$

(4) In the practical application, The Differential Entropy (DE_i) was instead of the logarithm of the band energy value commonly. The simplified formula of DE_i is shown in Eq. (8):

$$DE_i = log_{10}^{E_i} \qquad (8)$$

Where E_i is the energy in the i-th band.

5   Experimental Results and Analysis

5.1   WT-SVM

First, we studied the model that uses wavelet transform to decompose the signal and then classify it by support vector machine. There are many methods of feature extraction after wavelet transform. In this paper, four common and efficient methods are analyzed, namely band energy, band energy ratio, band energy ratio logarithm, and differential entropy. The evaluation of wavelet features uses a method of controlling variables. That is, the SVM classifier is selected as the criterion and only the wavelet features are changed to find the EEG features most relevant to the emotion classification. There are a lot of classifiers in the field of machine learning. The reason why this paper chooses SVM as the classifier is because the SVM is more mature in the field of EEG.

We use the SVM as the classifier, and select different WT feature extraction methods. The results of emotion recognition for aw of subjects are shown in Table 1.

Table 1. Classification results of WT-SVM model.

Subject	Wavelet	Feature	Accuracy
aw	sym8	E	54.76%
		REE	38.10%
		DE	78.57%
		LE	25.71%

We performed the averaging and variance calculations on the classification results of the four EEG features of the six subjects. The statistical results are shown in Table 2.

Table 2. Mean and variance of four feature classification results in WT-SVM model.

Feature	Average accuracy	The variance of accuracy
E	75.40%	0.036659108
REE	71.43%	0.036507937
DE	86.51%	0.008087680
LE	73.02%	0.054119426

In order to facilitate the intuitive observation, we made a line chart that marks the data according to the data in Table 2. In Fig. 2, the main ordinate axis represents the average accuracy of the classification, and the secondary ordinate axis represents the variance of the classification accuracy of the six subjects.

From Fig. 2, we can clearly see when differential entropy is used as the feature. Not only does the classification have the highest average accuracy and the variance is the smallest. This shows that the selection of differential entropy as a feature classification not only has higher accuracy when used for emotion

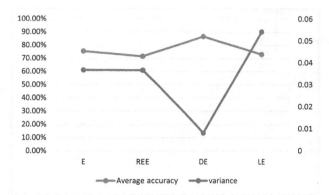

Fig. 2. Mean and variance of 4 feature classification results in WT-SVM model.

classification, but also has better stability between different subjects. From the data of aw in Table 1, it can also be found that the classification accuracy is very low or even less than 50% when replacing several other features. The replacement differential entropy as the feature classification accuracy rate still reached 78.57%.

Based on the above analysis results, it is easy to see that differential entropy is the best feature classification in WT-SVM model, and the average accuracy of classification is 86.51%.

5.2 WT-CNN

In Sect. 5.1, we selected the optimal wavelet feature, differential entropy, and its classification effect obviously has other wavelet features. In this section, we replace differential entropy as the EEG feature extraction method to establish the WT-CNN model.

When constructing a convolutional neural network model, different parameters determine different network structures. It is necessary to select the appropriate number of layers and determine the number and size of each layer of convolution kernels. These parameters directly determine the training speed of the network, the accuracy of the classification and the stability of the network. Moreover, over-fitting may occur when the number of hidden layer nodes is too large. This requires us to take the most compact structure possible while satisfying the accuracy requirements, that is, to take as few hidden layer nodes as possible. There is no clear definition of the choice of parameters, which needs to be determined based on the experimenter's experience and constant adjustments. After our continuous debugging, we finally established the CNN model for classifying differential entropy features as shown in Fig. 3.

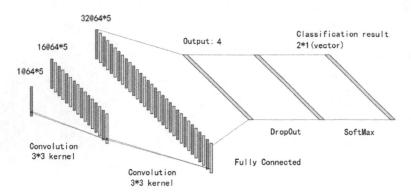

Fig. 3. CNN network model structure.

In the convolutional neural network model of Fig. 3, there are two connectional layers, one fully connected layer, one dropout layer, and one softmax layer. The size of the two connectional layers' kernel is 3 * 3, the first connectional layer has 16 convolution kernels, and the second has 32 convolution kernels. Fully connected layer has 4 outputs. Dropout layer is used to prevent overfitting and speed up the training of the network. Softmax layer is used to turn the raw output of the neural network into a probability distribution.

Since the size of the differential entropy feature matrix is 64 * 5, the sample feature dimension used for classification is not very large and the number of convolution kernels is small. In order to preserve more effective information and improve the accuracy of classification, we did not use the pooling layer for dimensionality reduction in this model. The main parameters of the convolutional neural network in the WT-CNN model are shown in Table 3.

Table 3. Main parameters of CNN in WT-CNN model.

Layer	OutPut	Param
Conv2d-1	(16, 64, 5)	160
Conv2d-2	(32, 64, 5)	4640
Dense-1	(4)	40964
Sum		45764

The classification results of the WT-CNN model for 6 subjects EEG data are shown in Table 4.

Table 4. Classification results of WT-CNN model.

Subject	Accuracy
aw	80.95%
ll	92.86%
sc	97.62%
xcl	73.81%
xtc	85.71%
dst	90.48%

Table 5 compares the classification results of the two classification models of WT-SVM and WT-CNN. As can be seen from Table 5, after our continuous adjustment, the classification result of the CNN has a slight improvement compared with the SVM. The average accuracy of the WT-CNN model is about 0.39% higher than that of the WT-SVM model. Moreover, the variance of the accuracy of the WT-CNN model has also decreased. This shows that the stability of classification between different subjects is also slightly better than the WT-SVM model. At present, based on EEG's emotion recognition, wavelet transform is used for feature extraction, and support vector machine is used for its classification are mature. A small improvement in the classification effect of the WT-CNN model indicates that it is practical for EEG-based emotion recognition.

Table 5. Comparison of results between WT-SVM and WT-CNN classification models.

Model	Feature	Average accuracy	The variance of accuracy
WT-SVM	DE	86.51%	0.00808768
WT-CNN		86.90%	0.00742630

6   Conclusion

In the WT-SVM model, the wavelet base selected for wavelet transform is sym8 wavelet. There are many methods for feature extraction after wavelet transform. In this paper, the four kinds of EEG characteristics of the band energy, the band energy ratio, the differential entropy and the band energy ratio are studied. The results show that the differential entropy is the highest classification and the stability of classification is the best, and the average accuracy is 86.51%. It is shown that differential entropy is more suitable for emotion recognition research than other wavelet features.

In order to facilitate the comparison between CNN and SVM. The paper uses differential entropy as the feature extraction method, and uses CNN for quadratic feature extraction and classification to establish the WT-CNN model. The classification effect of WT-CNN model and WT-SVM model is similar, and the accuracy of WT-CNN model has been slightly improved, and the average accuracy rate has reached 86.90%.

In addition, the experiment has the following areas that can be improved. For example, the sample size of the trial is small, and there are only 6 subjects, the amount of data is limited. Subsequent experiments we used more subjects and increased the number of trials per subject. On a more adequate data base, try a more sophisticated deep learning model that may have better results.

References

1. Sutherland, M.: Generating brain waves that pierce attention, pp. 1–4 (2005). http://www.sutherlandsurvey.com/
2. Krapohl, D., Mcmanus, B.: An objective method for manually scoring polygraph data. Polygraph **28**(3), 209–222 (1999)
3. De, R.S., He, X., Goldberg, A.P., et al.: Synaptic, transcriptional and chromatin genes disrupted in autism. Nature **515**(7526), 209–215 (2014)
4. Liu, J.-W.: Research and development on deep learning. Appl. Res. Comput. **31**(7), 1921–1930 (2014)
5. Niu, D.: A survey on EEG based emotion recognition. Chin. J. Biomed. Eng. **31**(4), 595–606 (2012)
6. Jiang, X.M.: Emotion recognition based on J48 decision tree classifier and results analysis. Comput. Eng. Des. **38**(3), 761–767 (2017)
7. Frantzidis, C.-A.: On the classification of emotional biosignals evoked while viewing affective pictures: an integrated data-mining-based approach for healthcare applications. IEEE Trans. Inf Technol. Biomed. **14**(2), 309–318 (2010)
8. Bengio, Y.: Learning deep architectures for AI. Found. Trends® Mach. Learn. **2**(1), 1–127 (2009)
9. Braverman, M.: Poly-logarithmic independence fools bounded-depth Boolean circuits. Commun. ACM **54**(4), 108 (2011)
10. Le, Q.V.: Building high-level features using large scale unsupervised learning. In: 2013 IEEE International Conference on Acoustics, Speech and Signal Processing (ICASSP), pp. 8595–8598 (2013)
11. Pan, L., He, C., Tian, Y., et al.: A classification based surrogate-assisted evolutionary algorithm for expensive many-objective optimization. IEEE Trans. Evol. Comput. **23**(1), 74–88 (2018)
12. Krizhevsky, A.: ImageNet classification with deep convolutional neural networks. NIPS, pp. 1097–1105. Curran Associates Inc. (2012)
13. Lecun, Y.: Gradient-based learning applied to document recognition. Proc. IEEE **86**(11), 2278–2324 (1998)
14. Hinton, G.E.: Reducing the dimensionality of data with neural networks. Science **313**(5786), 504–507 (2006)
15. Song, X.Y.: Research of EEG Based on Video Stimuli. Minzu University of China, Beijing (2016)

Heterogeneous Kernel Based Convolutional Neural Network for Face Liveness Detection

Xin Lu and Ying Tian[⊠]

School of Computer Science and Software Engineering,
University of Science and Technology Liaoning, Anshan 114051, China
astianying@126.com

Abstract. Liveness detection is a part of living biometric identification. While the face recognition system is promoted, it is also vulnerable to deceived and attacked from fake faces. Face liveness detection in traditional method needs network take long time to training and easy to appear over-fitting. Therefore, this paper proposed a Heterogeneous Kernel-Convolutional Neural Network (HK-CNN), the method replaces the standard convolutional kernel with heterogeneous convolutional kernel to detect the real face. In the two classic databases of NUAA and CASIA-FASD, the algorithm can improve accuracy and reduce the training cost of the model. Compared with traditional convolutional neural network methods, this algorithm has higher efficiency.

Keywords: Face recognition · Liveness detection · Heterogeneous kernel · Convolutional neural network

1 Introduction

With the continuous progress of information technology towards intelligence, face recognition technology has achieved good results, and its recognition rate is also increasing steadily. Compared with other biometric recognition methods, face recognition systems have the advantages of simple operation, hidden deeply, fine interface and non-contact collection. At the same time, the security problems caused by it are more and more important. But in general, face recognition algorithm cannot distinguish between real faces and fake faces. Face recognition system is vulnerable to spoofed and attacked from fake faces, in such scenarios as face payment, attendance, and remote identity verification. Therefore, in order to prevent such spoofing, the face recognition system needs to carry out liveness detection. Illegal attack method mainly include: printing photos, screen images or videos, realistic face masks or customers' 3D models authorized and so on [1].

Face liveness detection is a binary classification problem. Only when the real face is detected can the next step be carried out, otherwise, it is regarded as an illegal attack. Early face liveness detection was mainly based on traditional methods, which by extracting the different features between real face and non-living attacking face (fake face). Finally, the classifier makes the decision. Chakraborty et al. summarized common anti-spoofing methods: analysis based on frequency and texture [2], analysis based on variable focus, analysis based on eye movement, analysis based on optical flow,

© Springer Nature Singapore Pte Ltd. 2020
L. Pan et al. (Eds.): BIC-TA 2019, CCIS 1160, pp. 381–392, 2020.
https://doi.org/10.1007/978-981-15-3415-7_32

analysis based on lip movement, etc. Li et al. proposed a Fourier spectral analysis method based on a single face image to detect real faces [3], which according the fact that high-frequency components in the fake face images are less than the real face images; Määttä et al. proposed using multi-scale local binary pattern to analyze face images texture [4], then compared with the previous methods on the NUAA dataset, and achieved better results; Pereira et al. proposed an attack strategy based on the LBP-TOP algorithm to combine spatial and temporal information into a single multi-resolution texture descriptor [5].

Due to traditional feature extraction methods have some drawbacks when facing large data sets, so Yang et al. first proposed the application of AlexNet architecture for biometric identification in 2014 [6], Menotti et al. proposed a method of structure optimization and filter optimization, that to find an optimal filter weight setting for face liveness detection [7]. Alotaibi et al. used the nonlinear diffusion operator to preprocess the image and applied convolutional neural network (CNN) to detect the processed image [8]. Lucena et al. proposed a transfer learning method based on CNN for face anti-spoofing [9], which integrated FASNet and VGG (Visual Geometry Group), and fine-tuning them to obtain better experimental results. So far, it is still has great research value to introduce CNN algorithm into face anti-spoofing detection. However, there has problem such as high computation complexity, large amount of computation, the accuracy and computing cost cannot take into the same time for the deep CNN.

Heterogeneous kernel based CNN for face liveness detection, which replaces the traditional standard convolutional filters with heterogeneous convolutional filters. The experimental result show that the CNN with heterogeneous kernel can greatly reduce the network computing cost and improve the network efficiency while maintaining the accuracy of the model.

2   Related Work

The design of CNN is derived from the organism's visual perception mechanism, which is a deep feedforward neural network with local connection and shared weights. The difference from BP neural network is that the input format is two-dimensional matrix data, while the images expressed the pixel position information through the two-dimensional matrix, so CNN becomes the core of deep learning in the image processing. CNN has been in use since the 1990s, Hinton and his student Alex Krizhevsky designed the AlexNet which has won the first prize in the ImageNet competition until 2012. They reduced the error of image classification to 15%, which caused great attention in the industry. Then some classical and representative architectures appeared, such as ZFNet [10], VGGNet [11], ResNet [12] and so on.

2.1   Convolutional Layer

The purpose of convolution is feature extraction. As the number of convolutional layers increases, the network combines low-level features into high-level features and then classifies them. Assuming that a gray image's size is $M \times N$ and the size of filter W is $m \times n$, the convolutional formula can be defined as formula (1):

$$y_{ij} = \sum_{u=1}^{m} \sum_{v=1}^{n} w_{uv} \cdot x_{i+u-1,j+v-1} (1 \leq i \leq M - m + 1,\ 1 \leq j \leq N - n + 1) \quad (1)$$

where: w_{uv} is the value of the filter W which in the uth row and vth column, $x_{i+u-1,j+v-1}$ is the value of the filter W which in the $(i + u - 1)$th row and the $(j + v - 1)$th column on the image. In order to improve the convolutional neural network expression ability and represent image features better. In each layer can use multiple different feature maps. Add the convolutional results and sum with a bias named b, to get the convolutional layer net input Z^p, and then through the nonlinear activation function to get the output mapping Y^p:

$$Z^p = W^p \otimes X + b^p = \sum_{d=1}^{D} W^{p,d} \otimes X^d + b^p \quad (2)$$

$$Y^p = f(Z^p) \quad (3)$$

where: p is the pth convolution layer, P is the convolution neural network layer number, $W^{p,d}$ is the dth filter in the pth layer, $f(\cdot)$ is a nonlinear activation function. In the convolutional layer where the input is X and the output is Y, each input feature mapping needs D filters and a bias, the parameters numner is $P \times D(m \times n) + P$.

2.2 Pooling Layer

The purpose of the pooling layer is to reduce the dimension of the feature mapping input by the convolutional layer, which can reduce the data space, parameters number and calculation complexity, thereby improving the efficiency and controlling the over-fitting effectively. Mean-pooling and max-pooling are two common pooling functions. Max-pooling divides the input feature mapping into several sub regions, while preserving the maximum value of each sub region, reducing the phenomenon that the convolutional layer parameter error causes the estimated mean shift, and makes the image texture information better expressed. The algorithm in this paper adopts the max-pooling function, assumes that the input feature mapping group is $X \in R^{M \times N \times D}$. Dividing each feature mapping X^d into multiple $R_{m,n}^d$ region with size of $M' \times N'$, $1 \leq m \leq M'$, $1 \leq n \leq N'$, the pooling layer output mapping is as formulas (4) and (5):

$$Y_{m,n}^d = \max_{i \in R_{m,n}^d} x_i \quad (4)$$

$$Y^d = \left\{ Y_{m,n}^d \right\} \quad (5)$$

2.3   Activation Layer

Activation function introduces non-linear factors into the neural network, which can help the neural network to solve complex problems and to improve the network expression ability. Leaky ReLU is a variant of the nonsaturating activation functions Relu, which generally through back propagation algorithm to optimized parameters, multiply negative values by a non-zero slope named λ, such as formula (6):

$$f(x) = \begin{cases} \dfrac{x}{\lambda}, x \leq 0 \\ x, x > 0 \end{cases} (1 \leq \lambda \leq +\infty) \tag{6}$$

The negative input values in Relu cause neurons do not update parameters, Leaky ReLU solved this problem, it makes the network sparser and reducing the dependence between parameter, also trains the network in a supervised way without relying on unsupervised pre-training. The Leaky Relu function image is shown as Fig. 1:

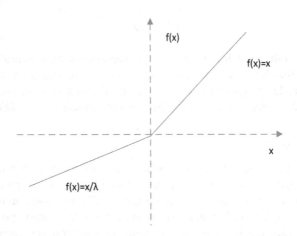

Fig. 1. The image of Leaky Relu function

2.4   Fully Connected Layer

The fully connected layer function is to integrate the output feature mapping from the pooling layer, transform the distributed feature space into the sample label space, and assume the role of classifier. The fully connected layer output is obtained by multiplying each node in the previous layer by the weight coefficient and plus the bias. After the neurons in the fully connected layer are activated, output the combined features, and get the conclusion finally.

3   The HK-CNN for Face Liveness Detection

Convolutional kernel, also known as filter, convolution operation according to convolutional kernel type can be divided into homogeneous kernels and heterogeneous convolution. Homogeneous kernel means that all convolutional kernels have the same size, such as deep convolution, pointwise convolution [13], groupwise convolution [14] are all belong to traditional filters. At present, most network architectures using combined filters to improve the model efficiency. Heterogeneous filters contain different sizes of convolutional kernels, which can reduce computing cost effectively and improved network efficiency under the condition of ensuring the accuracy [15].

3.1   Heterogeneous Convolutional Kernel

Suppose a image size is $M_i \times M_i \times S$, Where M_i is the width and height of the input image, S is the channel number. The output feature mapping is set to $N_i \times N_i \times T$, where N_i is the width and height of the output image, T is the channel number. T filters with size of $D \times D \times S$ is used to obtain the feature mapping, and the layer L computing cost FL_s is as formula (7):

$$FL_s = N_i \times N_i \times S \times T \times D \times D \tag{7}$$

As can be seen, the factors determine the computing cost size are the convolutional kernel size, the feature mapping size, the number of input and output channels, these make the computing cost high. But using the heterogeneous convolutional kernel, in order to control the different number of convolutional kernels types in the convolutional filter, the algorithm defines a parameter P, in S kernels, S/P kernels size is $D \times D$, the rest kernels size is 1×1. In the layer L, the computing cost of the kernel size $D \times D$ in the heterogeneous convolution filter with the P parameter as follows:

$$FL_D = (N_i \times N_i \times S \times T \times D \times D)/P \tag{8}$$

It can be seen that the computing cost of layer L is reduced by P times. The computing cost of the convolution kernel size 1×1 is as follow:

$$FL_1 = (N_i \times N_i \times T) \times (M - M/P) \tag{9}$$

The total computing cost FL_{HC} in layer L is the sum of formulas (8) and (9), to compared with standard convolutional filter, so the heterogeneous convolutional filter computing cost reduction rate R_{HC} is as follow:

$$R_{HC} = \frac{FL_D + FL_1}{FL_S} = \frac{1}{P} + \frac{(1 - 1/P)}{D^2} \tag{10}$$

Obviously, this algorithm reduces the spatial range of the filter and improves the R_{HC} effectively. The heterogeneous convolutional kernel structure of HK-CNN is shown as Fig. 2:

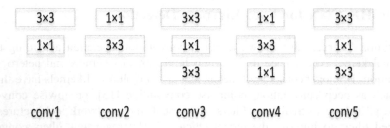

Fig. 2. The heterogeneous convolutional kernel structure in HK-CNN

3.2 Heterogeneous Convolutional Neural Network

The HK-CNN structure is shown as Fig. 3: there are five convolutional layers, five pooling layers, and two fully connected layers, each convolutional layer including the Leaky ReLU activation layer and batch normalization layer, and add the softmax layer after the fully connected layer.

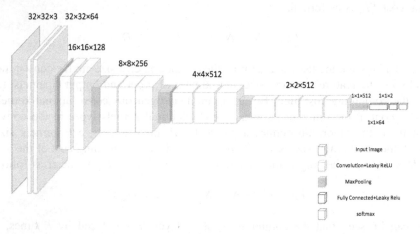

Fig. 3. The structure of HK-CNN

In order to facilitate the subsequent calculation, the algorithm preprocesses the image. The algorithm prevents network over-fitting by image enhancement of samples, and normalizes the positive and negative samples of face liveness detection to size of 32 × 32. For the five convolutional layers, each layer is connected with a max-pooling layer which with size of 2 × 2 and stride is 2.

The first convolutional layer extracts the preprocessed image feature, then through two convolutions of 64 convolutional kernels, the convolutional kernel size of the first convolution is 3 × 3, and the second convolutional kernel size is 1 × 1, finally get

$32 \times 32 \times 64$ feature images. The feature mapping that previous layer output through the first pooling layer to obtain the feature images which size of $16 \times 16 \times 64$.

The second convolutional layer designs 128 convolution kernels, like the first convolutional layer, with the size of 1×1, 3×3 convolution kernels to extract the image features, to get the $16 \times 16 \times 128$ feature images, and then sends to the second pooling layer to obtain $16 \times 16 \times 64$ feature images.

There are 256 convolution kernels in the third convolutional layer. The convolution kernels size are 3×3, 1×1, 3×3, and then get $8 \times 8 \times 128$ feature images. After the third pooling layer obtains $4 \times 4 \times 256$ feature images.

There are 512 convolutional kernels in the fourth convolutional layer. The convolutional kernels size are 1×1, 3×3, 1×1, and obtains $4 \times 4 \times 512$ feature images. After the fourth pooling layer to get $2 \times 2 \times 512$ feature mappings.

The structure of fifth convolutional layer is the same as the fourth convolutional layer, then the feature images though the fifth pooling layer into the fully connected layer, and finally get feature mappings of $1 \times 1 \times 512$.

The fully connected layer has 64 neurons, adding the softmax layer after the fully connected layer. As the output layer of the network, the softmax function performs classification operations, usually at the last layer of CNN. The output is between 0 and 1. This layer consists of two neurons, which are used to solve the binary classification problem. The model of the softmax function is as formula (11):

$$y_j = \frac{e^{y_j}}{\sum_j e^{y_j}} = \sum_{j=1}^{n} x_i \cdot w_{i,j} + b_j \tag{11}$$

where x_i is the output of the previous layer, $w_{i,j}$ is the weight of each layer corresponding to x_i, b_j is the bias. Due to the algorithm mainly considers the binary classification problem, so set j to 2.

4   Experimental Results and Analysis

The experimental part of the algorithm is based on deep learning framework Keras +Tensorflow, Environment Configuration: CentOS, CPU: Intel®Xeon(R) CPU E5-2650 v4 @ 2.20 GHz $\times$ 24, Memory: 94G, two graphics cards are GTX1080Ti, Memory Capacity: 11 GB, written in Python, to use two classic face anti-spoofing sample databases: NUAA and CASIA-FASD.

NUAA is a liveness detection sample for anti-spoofing face detection provided by Nanjing University of Aeronautics and Astronautics, it is the first one recognized as anti-spoofing face detection database. It contains 15 participants, divided into three collection time periods according to different lighting conditions. The image number of the three stages to the real face positive sample are: 4500, 3500 and 4500; the image numbers of the three stages to the fake face negative samples are: 4500, 18000 and 37500.

The experiment preprocessed the NUAA database by data enhanced and normalized, then put the data into HK-CNN to learn feature, and the results are made as binary classification problem. To verify the algorithm is valid, the experiment introduces accuracy, loss rate and time as the evaluation model indexes. Adjust the learning rate to 1×10^{-4}, and set the batch size to 8. A large number of experiments show that when epochs number reach 50, the model tends to be stable gradually, if epochs number more than 50, model begin to over-fitting. Therefore, the experiment compared the indexes when the epoch number was 50. Experiments using standard convolutional kernel and heterogeneous convolutional kernel on the NUAA database respectively, where 75% of the total samples were taken from the training set, and 25% of the total samples were taken from the test set. The experimental datas are shown in Table 1.

Table 1. Comparison of standard convolutional kernel and heterogeneous convolutional kernel in NUAA database

The algorithm	Epoch	Accuracy/%	Loss/%	Time/min
Standard convolutional kernel	50	99.71	0.16	30
Heterogeneous convolutional kernel	50	99.82	0.13	25

Table 1 shows that the accuracy of the heterogeneous kernel based CNN algorithm is 0.1% higher than using the standard convolutional kernel algorithm in the NUAA dataset, the loss rate reduced by 0.03%, the time reduced significantly, and the model efficiency improved greatly. Compared with other algorithms on the NUAA dataset, the results are shown in Table 2:

Table 2. Comparison of different algorithms accuracy in NUAA database

The algorithm	Accuracy/%
LTP [16]	91.10
LSP [17]	98.5
ND-CNN [8]	99.00
HK-CNN	99.82

From the data in Table 2, the convolutional neural network algorithm is better than traditional feature extraction method. On the basis of the convolutional neural network, the proposed algorithm which using heterogeneous convolutional kernel can improve accuracy. Figure 4 reflects with the increase of the Epoch value, the accuracy of the training set and the test set is also increased and the loss rate gradually decreases until tends to be stable.

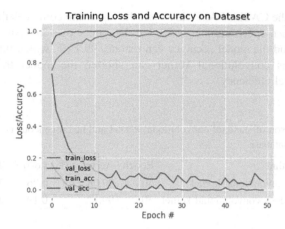

Fig. 4. The relationship between the accuracy, loss rate and epoch of training set and test set

The CASIA-FASD face anti-spoofing database contains 50 people face samples, 600 videos, the total sample is divided into two groups, 60% of which are used as training sets and 40% as test sets. Each group contains 3 real face videos and 9 fake face videos. Input the pre-processed face dataset samples into the model designed by the algorithm, and introduces the HTER as the standard of the evaluation model. HTER can be calculated by the false rejection rate (FRR) and the false acceptance rate (FAR).

$$HTER = \frac{1}{2}(FRR + FAR) \tag{12}$$

Table 3 compares different algorithms experimental data in the CASIA-FASD database, It can be seen that convolutional neural network has better performance than traditional methods in face liveness detection. In the experimental results, the HTER value of face liveness detection algorithm based on heterogeneous convolutional neural network is 1.03, the result is significantly lower than the HTER value of the VCST algorithm. Comparing with the DB-CNN algorithm, it is still has some improvement. At the same time, the model accuracy is also improved. It is further shown that the proposed algorithm has a certain anti-spoofing effect in face liveness detection.

Table 3. Comparison of different algorithms HTER in CASIA-FASD database

The algorithm	Accuracy/%	HTER
VCST [18]	93.25	14.26
LTP [16]	93.02	6.1
DLTP [16]	94.40	5.4
DB-CNN [8]	–	2.27
HK-CNN	99.17	1.03

Similarly, in the CASIA-FASD database, the algorithm compares the accuracy, loss rate, and time when epoch number is 50, the data are shown in Table 4. The learning rate unchanged and the model accuracy can reaches 99.17%, which further proves that the heterogeneous convolutional kernel can shorten the model training time and improve the model efficiency.

Table 4. Comparison of standard convolutional kernel and heterogeneous convolutional kernel in CASIA-FASD database

The algorithm	Epoch	Accuracy/%	Loss/%	Time/min
Standard convolutional kernel	50	99.08	0.26	36
Heterogeneous convolutional kernel	50	99.17	0.19	29

Figure 5 shows part of the experimental results tested on the NUAA dataset with the proposed algorithm. The first row is the positive samples of the test set, and the second row is the negative sample of the test set, the detection accuracy can reach 99.82%. Figure 6 shows the part of the experimental results of the algorithm tested on the CASIA-FASD dataset. The two images on the left are positive samples of the test set, and the two images on the right are negative samples of the test set, the can reach accuracy can reach 99.17%. So applying the proposed algorithm to the trained model can distinguish between real faces and the attacking faces.

Fig. 5. Some experimental results of NUAA dataset

Fig. 6. Some experimental results of CASIA-FASD dataset

5 Conclusion

HK-CNN uses heterogeneous convolutional neural network for face liveness detection by replacing the standard convolution kernel with heterogeneous convolution kernels, and to detect the real face and the fake face. Through the theoretical derivation and experiment, it shows that the accuracy of face liveness detection based on HK-CNN can reach 99.82%, at the same time it reducing the amount of network computation. The network designed by this algorithm has five convolutional layers, five max-pooling layers, one fully connected layer and one softmax layer. In the two classic databases, NUAA and CASIA-FASD, the experimental results prove that using heterogeneous convolutional kernel in convolutional neural network can improve the model efficiency and has certain feasibility and practicability. The next study will consider continuing to optimize the network to further enhance its security performance.

Acknowledgements. This work was supported by the National Natural Science Foundation of China (Project Number: 72472081).

References

1. Yang, J., Lei, Z., Liao, S., Li, S.Z.: Face liveness detection with component dependent descriptor. In: International Conference on Biometrics, pp. 1–6 (2013)
2. Chakraborty, S., Das, D.: An overview of face liveness detection. Int. J. Inf. Theory **3**(2), 11–25 (2014)
3. Li, J., Wang, Y., Tan, T., Jain, A.K.: Live face detection based on the analysis of fourier spectra. Proc. SPIE **5404**, 296–303 (2004)
4. Määttä, J., Hadid, A., Pietikäinen, M.: Face spoofing detection from single images using texture and local shape analysis. IET Biom. **1**(1), 3–10 (2012)
5. de Freitas Pereira, T., Anjos, A., De Martino, J.M., Marcel, S.: *LBP − TOP* based countermeasure against face spoofing attacks. In: Park, J.-I., Kim, J. (eds.) ACCV 2012. LNCS, vol. 7728, pp. 121–132. Springer, Heidelberg (2013). https://doi.org/10.1007/978-3-642-37410-4_11
6. Yang, J., Lei, Z., Li, S.Z.: Learn convolutional neural network for face anti-spoofing. Comput. Sci. **9218**, 373–384 (2014)

7. Menotti, D., et al.: Deep representations for iris, face, and fingerprint spoofing detection. IEEE Trans. Inf. Forensics Secur. **10**(4), 864–879 (2015)
8. Alotaibi, A., Mahmood, A.: Deep face liveness detection based on nonlinear diffusion using convolution neural network. SIViP **11**(4), 713–720 (2017)
9. Lucena, O., Junior, A., Moia, V., Souza, R., Valle, E., Lotufo, R.: Transfer learning using convolutional neural networks for face anti-spoofing. In: Karray, F., Campilho, A., Cheriet, F. (eds.) ICIAR 2017. LNCS, vol. 10317, pp. 27–34. Springer, Cham (2017). https://doi.org/10.1007/978-3-319-59876-5_4
10. Zeiler, M.D., Fergus, R.: Visualizing and understanding convolutional networks. In: Fleet, D., Pajdla, T., Schiele, B., Tuytelaars, T. (eds.) ECCV 2014. LNCS, vol. 8689, pp. 818–833. Springer, Cham (2014). https://doi.org/10.1007/978-3-319-10590-1_53
11. Simonyan, K., Zisserman, A.: Very deep convolutional networks for large-scale image recognition. Computer Science (2014)
12. He, K., Zhang, X., Ren, S., Sun, J.: Deep residual learning for image recognition. In: Computer Vision and Pattern Recognition, pp. 770–778 (2016)
13. Szegedy, C., et al.: Going deeper with convolutions. In: Computer Vision and Pattern Recognition, pp. 1–9 (2015)
14. Krizhevsky, A., Sutskever, I., Hinton, G.E.: ImageNet classification with deep convolutional neural networks. Neural Inf. Process. Syst. **141**(5), 1097–1105 (2012)
15. Singh, P., Verma, V.K., Rai, P., Namboodiri, V.P.: HetConv: heterogeneous kernel-based convolutions for deep CNNs. arXiv: Computer Vision and Pattern Recognition (2019)
16. Parveen, S., Ahmad, S.M., Abbas, N.H., Adnan, W.A., Hanafi, M., Naeem, N.: Face liveness detection using dynamic local ternary pattern (DLTP). First Comput. **5**(2), 10 (2016)
17. Kim, W., Suh, S., Han, J.: Face liveness detection from a single image via diffusion speed model. IEEE Trans. Image Process. **24**(8), 2456–2465 (2015)
18. Pinto, A.D., Pedrini, H., Schwartz, W.R., Rocha, A.: Face spoofing detection through visual codebooks of spectral temporal cubes. IEEE Trans. Image Process. **24**(12), 4726–4740 (2015)

Cell-like Spiking Neural P Systems with Anti-spikes and Membrane Division/Dissolution

Suxia Jiang[1], Jihui Fan[1], Dan Ling[1], Feifei Yang[1], Yanfeng Wang[1(✉)], and Tingfang Wu[2(✉)]

[1] School of Electrical and Information Engineering,
Zhengzhou University of Light Industry, Zhengzhou 450002, Henan, China
wangyanfeng@zzuli.edu.cn
[2] School of Computer Science and Technology, Soochow University,
Suzhou 215006, Jiangsu, China
tfwu@hust.edu.cn

Abstract. Cell-like spiking neural P systems (cSN P systems) are a class of membrane computing models, which have both the hierarchical structure of living cells and the way of communicating and processing information by means of neuron spikes. The universality of cSN P systems as number generating devices has been investigated, and these systems with request or evolution rules have been proved to be Turing universal. In this work, anti-spike is introduced into cSN P systems, we investigate the computational power of cSN P systems with anti-spikes by using membrane division and dissolution rules avoiding the use of producing more spikes and replicating spikes instructions. On the one hand, by adding anti-spikes to cSN P systems, we reduce the unnecessary spikes consumption of systems when simulating the subtraction instructions. On the other hand, by introducing membrane division and dissolution rules, the number of spikes used in neurons of cSN P systems with anti-spikes is limited without loss of computational power.

Keywords: Bio-inspired computing · Membrane computing · Spiking neural P system · Cell-like P system · Anti-spike · Universality

1 Introduction

Bio-inspired computing is an area of researching computational models and algorithms through taking ideas from the biological world [24]. Membrane computing is an important branch of Bio-inspired computing, inspired by living cell structure, organization and function [19,37]. Membrane systems (P systems) are a class of distributed and parallel computational models in membrane computing. According to the structure of membranes, P systems consists of the following three types: cell-like P systems [18], tissue-like systems [9], neural-like systems

© Springer Nature Singapore Pte Ltd. 2020
L. Pan et al. (Eds.): BIC-TA 2019, CCIS 1160, pp. 393–408, 2020.
https://doi.org/10.1007/978-981-15-3415-7_33

[7]. As is known to all, P systems have extensively applications in mathematical and theoretical computer science, for instance, solving NP-complete problems [20,25] and numerical calculation problems [15], investigating the computational power of various P systems [6,17,34]. Specifically, P systems can also be applied to solve real-life problems, such as knapsack problem [36], mobile robot controllers [2,11,35], trajectory tracking [33] and face recognition [1]. The reader can consult [21] for more information about membrane computing. Comprehensive research results and the most up-to-date references on membrane computing can be found at the website http://ppage.psystems.eu.

Spiking neural P systems (SN P systems, for short) were initially proposed in [7], which are a kind of computational models inspired by some biological phenomena of neurons processing and communicating information by means of electrical impulses (also called spikes). SN P systems can be used as computing devices for generating/accepting arbitrary natural numbers [10,16], language generators [4] and function computing devices [31]. Meanwhile, a lot of variants of SN P systems were proposed, and most of them have been proved to be computationally universal. For example, SN P systems with colored spikes [26], SN P systems with scheduled synapses [3], SN P systems with thresholds [39], SN P systems with rules on synapses [27] etc. Furthermore, many variants of SN P systems combined with successful applications were surveyed extensively and intensively, e.g. fuzzy reasoning SN P systems for fault diagnosis [22,32], SN P systems for image processing [5], fuzzy reasoning SN P systems for fault diagnosis of electric power systems [29], SN P systems for approximately solving combinatorial optimization problems [38].

Recently, a highly original variant of P systems were constructed, which combined the basic features of multiset rewriting P systems and spiking neural P systems, have the structure of neurons (cells) arranged hierarchically, and were called cell-like SN P systems (in short, cSN P systems) [30]. The P systems, having one type object, i.e. spikes a, using spiking rules that produce more spikes than the ones consumed or replicated, have been proved to be Turing universal. At the end of the reference [30], two open problems were provided: it is interesting to investigate the computational power of cSN P systems with anti-spikes; what is the computational power of cSN P systems avoid using target instructions in_{all} to generate more spikes. Later, the universality of cSN P systems were proved when using request rules or evolution rules [13,14]. In the following work, according to the open problems provided above, we introduce anti-spike into cSN P systems, propose cell-like SN P systems with anti-spikes (cSNa P systems, for short). Here, anti-spike is not detailed described, please consults the conference [12].

In this paper, anti-spike is considered and used to reduce the excess spikes consumed during the execution of the subtraction module in cSN P systems. Membrane division and dissolution rules (coming from the cell-like P systems) are applied to avoid using the target instruction in_{all}. Thus, cSNa P systems with membrane division and dissolution rules (referred to as cSNaDD P systems) are proposed, the computational power of cSNaDD P systems is investigated.

Specifically, we proved that cSNa P systems with membrane division and dissolution rules are Turing universal by simulating register machines.

The remainder of this paper is organized as follows: the related conception and definition of cSNa P systems with membrane division and dissolution rules are described in Sect. 2, cSNa P systems with membrane division and dissolution rules are constructed and the proof of universality is discussed in Sect. 3. Finally, conclusions and remarks for further work are outlined in Sect. 4.

2   Cell-like SN P Systems with Anti-spikes and Membrane Division/Dissolution

In this section, we mainly provide the definition of cell-like SN P systems with anti-spikes using membrane division and membrane dissolution rules (also called cSNaDD P systems), some basic concepts and notions of membrane computing and knowledge of formal language and automata theory are not detailed description here, please refer to [21, 23]. Besides, the formal definitions of SN P systems with anti-spikes can be found in [12], the reader also can consult [28] about the notions of membrane division and dissolution rules.

A cSN P system with anti-spikes and membrane division/dissolution rules of degree $m \geq 1$ is a construct of the form:

$$\Pi = (O, \mu, n_1, \cdots, n_m, R_1, \cdots, R_m, i_o),$$

Where

- $O = \{a, \bar{a}\}$ represents an alphabet where a is the spike and $\bar{a}$ is the anti-spike;
- μ represents a hierarchical membrane structure with m nested membranes;
- $n_i \geq 0$ $(1 \leq i \leq m)$ represents the initial number of spikes contained in membrane i;
- R_i $(1 \leq i \leq m)$ represents a finite set of rules placed in membrane i, which comes in the following forms:
  (1) spiking rule: $E/b^c \rightarrow (b^p, tar)$, where E is a regular expression over O, $b \in \{a, \bar{a}\}$, $c \geq p \geq 1$, and target indication $tar \in \{here, out, in, in_{all}\} \cup \{in_j, 1 \leq j \leq m\}$;
  (2) forgetting rule: $a^s \rightarrow \lambda$, where $s \geq 1$ and $a^s \notin L(E)$ for any spiking rule $E/b^c \rightarrow (b^p, tar)$ from R_i;
  (3) division rule: $[b^t]_i \rightarrow [d^f]_i[e^h]_i$, where $b, d, e \in (a, \bar{a})$, $t \geq f \geq 1$ and $t \geq h \geq 1$, indicates the membrane i is divided into two membranes with the same label;
  (4) dissolution rule: $[b^t]_i \rightarrow e^t$, where $b, e \in (a, \bar{a})$ and $t \geq 1$, indicates the membrane i is dissolved;
- $i_o \in \{1, \cdots, m\} \cup \{env\}$ indicates the output region of Π ($i_o = env$ represents the output region is environment).

In the set of rules R_i, $E/b^c \rightarrow (b^p, tar)$ is called extended spiking rules, where $b \in (a, \bar{a})$, c is the number of spikes consumed, p is the number of spikes produced, $c \geq p \geq 1$, that means if the neuron i contains k spikes, $b^k \in L(E)$ and $k \geq c$, the spike rule can be applied, then c spikes are consumed (only $k - c$ spikes remain in the neuron) and p spikes are produced (these spikes are sent to the corresponding neurons according to the target indication tar). Note that, unlike the classical cSN P systems, the number of spikes consumed is not less than the number of spikes produced in cSNaDD P systems. Note that two forms as follows: $E_1/\bar{a}^v \rightarrow (a^e, tar)$ and $E_2/a^v \rightarrow (\bar{a}^e, tar)$, where $v \geq e$, $L(E_1) \cap L(E_2) = \emptyset$. More broadly, if a neuron has a set of rules are of forms: $E_1/b^c \rightarrow (b^p, tar)$ and $E_2/b^c \rightarrow (b^p, tar)$, with $c \geq p$, $L(E_1) \cap L(E_2) \neq \emptyset$, in this case, it is possible that two or more spiking rules can be applied in the neuron, but only one of them is non-deterministically chosen and used.

A forgetting rule is denoted by $a^s \rightarrow \lambda$, it can be enabled and applied only if the neuron i contains exactly s spikes, and all of the s spikes are removed from the neuron immediately, but none are emitted.

A membrane division rule is described as $[b^t]_i \rightarrow [d^f]_i[e^h]_i$, which can be applied as follows: a membrane i contains t spikes b (either spikes a or anti-spike $\bar{a}$), which is neither the skin membrane nor the output region, the membrane can be divided into two membranes with the same label i, maybe of different type of spikes (d, e is either spikes a or anti-spike $\bar{a}$) and different number of spikes number (the number of spikes in each generated membrane is less than the number of spikes in the divided membrane, i.e. $f \leq t$ and $h \leq t$), t spikes are consumed in this moment. Meanwhile, each generated membrane contains developmental rules from R_i that the parent neuron already has.

A membrane dissolution rule is described as $[b^t]_i \rightarrow e^t$. If membrane i contains t spikes b (either spikes a or anti-spike $\bar{a}$), the membrane can be dissolved, all spikes inside are sent to the parent membrane, while the type of spikes specified in the rule can be modified.

In each time unit, rules of cSNaDD P systems are working in sequential manner at the level of each membrane, i.e. only one rule corresponds to each membrane can be used at one step, and in parallel at the level of the system, i.e. if a membrane has a rule that can be used at some step, the rule must be used.

In this system Π, a configuration is described by the number of spikes presented in each membrane, thus, the initial configuration is $<n_1, n_2, \cdots, n_m>$, where n_i is the initial number of spikes present in membrane i. The computation of a cSNaDD P system starts from the initial configuration and contains a sequence of transitions among consecutive configurations. A computation halts if it reaches a configuration where no rule in the system can be used. Note that there are two ways to define the computational result when a computation halts: (1) the number of spikes contained in the specified region is the result, denoted as $N_{all}(\Pi)$; (2) the time distance between the first two successive spikes sent to the environment is regarded as the result of a computation, denoted as $N_2(\Pi)$.

We denote by $N_{\alpha}^{DD}cSNaP_m(forg, here, in, in_t, in_{all}, out)$, $\alpha \in \{2, all\}$, the families of sets of numbers generated by cSNaDD P systems with at most m membranes, where DD indicates membrane division and dissolution rules, $forg$ indicates forgetting rules, $here, in, in_t, in_{all}, out$ indicates the types of target indications. As usual, when no bound is imposed on the number of membranes, the index m will be replaced with $*$. If the forgetting rule $forg$ or certain target indications are not used, the corresponding indications will be omitted.

3 Universality of cSNaDD P Systems

In this section, the computational power of cSNa P systems with membrane division and dissolution rules is investigated, using traditional spiking rules (the number of spikes consumed is not less than the number of spikes produced by a spiking rule) and forgetting rules, but excluding target indications $here$ and in_{all}, We prove that cSNa P systems with membrane division and dissolution rules are Turing universal as number generating devices in two ways of defining computation results mentioned above.

A register machine with three registers is denoted by $M = (3, H, l_1, \cdots, l_m, I)$, where H is the set of instruction labels, l_1 is the start label indicating an ADD instruction, l_m is the halt label assigned to the halt instruction. I is the set of instructions, every of which precisely corresponds to a unique label in H. The instructions are of the following forms:

- $l_i : (ADD(r), l_j, l_k)$ (ADD instruction: add 1 to register r, then indeterminately go to the instruction with label l_j or l_k);
- $l_i : (SUB(r), l_j, l_k)$ (SUB instruction: if register r is nonempty, then subtract 1 from it and go to the instruction with label l_j, otherwise go to the instruction with label l_k);
- $l_m : HALT$(halt instruction).

It is known that register machines with three registers can precisely generate a family of sets of recursively enumerable natural numbers, in other words, it can characterize NRE [8].

Theorem 1. $N_{all}^{DD}cSNaP_5(in, in_t, out) = NRE$.

Proof. Recalling the Turing-Church thesis, we only need to prove the inclusion $N_{all}^{DD}cSNaP_5(in, in_t, out) \supseteq NRE$. The process of proof is completed on the basis of simulating the register machines with three registers.

In what follows, a cSNa P system with membrane division and dissolution rules Π_1 is constructed to simulate a register machine $M = (3, H, l_1, \cdots, l_m, I)$, it is illustrated in a graphical way as shown in Fig. 1. In Theorem 1, when the computation halts, the number of spikes stored in register 1 is regarded as the computational result, other registers are empty. Besides, no SUB instructions are executed in register 1.

The system Π_1 consists of five membranes, represented by labels 0, 1, 2, 3, and 4, respectively. Membrane 0 is associated with the set of instructions of M. Membranes 1, 2, 3 correspond to registers 1, 2, 3 of M, respectively. The number of spikes contained in membrane 1 is equal to the number stored in register 1. Resisters 2 and 3 stored the number n correspond to $n(2m+1)$ copies of spikes in

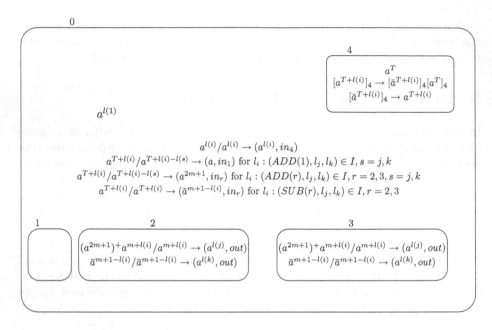

Fig. 1. System constructed for the proof of Theorem 1

membranes 2 and 3, respectively. For labels $l(i)$ correspond to each instruction of M, $1 \leq i \leq m$, is represented by $T + l(i)$ spikes, where $T = 3m + 1$ and $l(i) = i$. If and only if membrane 0 contains $T + l(i)$ spikes, then the rule associated with instruction $l(i)$ can be applied. In initial configuration, membrane 1 and 4, have one spike and T spikes, respectively. When $l(i)$ spikes are sent to membrane 4, afterwards, $a^{T+l(i)}$ are transmitted to membrane 0 by using membrane division and dissolving rules. Only membrane 0 contains $T + l(1)$ spikes, means that the simulation of the initial instruction l_1 of M. It is worth noting that after each instruction l_i is completed, the resulting spikes passes through membrane 0 and eventually enters membrane 4, and compensates for the loss of computing power caused by the execution of the previous instruction by performing membrane splitting and membrane dissolution operations. When $T + l(j)$ or $T + l(k)$ spikes are introduced in membrane 0, which means that systems Π_1 starts simulate instruction with label l_j or l_k. Besides, if membrane 0 has $T + l(m)$ spikes, the computation halt and no rule can be applied, M is completely simulated by system Π_1.

In what follows, we prove that system Π_1 can correctly simulate the register machine M, by showing that how system Π_1 simulates ADD and SUB instructions of M and outputs the computation result. Simulating an ADD instruction $l_i: (ADD(r), l_j, l_k)$. Initially, one spike in membrane 0, rule $a^{l(1)}/a^{l(1)} \rightarrow (a, in_4)$ can be enabled and a spike is sent to membrane 4. When $a^{T+l(1)}$ spikes in membrane 4, membrane division rule $[a^{T+l(1)}]_4 \rightarrow [a^T]_4[\bar{a}^{T+l(1)}]_4$ is fire, membrane 4 splits into two membranes, with same label. One of membranes 4 also maintain T spikes, but other membrane 4 have $T + l(1)$ anti-spikes. By using membrane dissolving rule $[\bar{a}^{T+l(1)}]_4 \rightarrow a^{T+l(1)}$, the corresponding membrane 4 is removed, $T + l(1)$ spikes in membrane 0. The instruction l_1 star simulation by rule $a^{T+l(1)}/a^{T+l(1)-l(s)} \rightarrow (a, in_1)$, the register 1 accumulates 1 and $l(s)$ spikes remain in the skin region, $s = j, k, 1 \leq j, k \leq m$. At the next step, $l(s)$ spikes repeat the above membrane splitting and dissolving operation, spikes $a^{T+l(s)}$ in skin membrane, is correspond to the two rules a $a^{T+l(j)}/a^{T+l(j)-l(s)} \rightarrow (a^{2m+1}, in_r)$ and $a^{T+l(k)}/a^{T+l(k)-l(s)} \rightarrow (a^{m+1}, in_r)$ in membrane 0 are enabled, $r = 2, 3$, non-deterministic choosing one of them to be applied. If $a^{T+l(j)}$ spikes in skin membrane, which indicates that system Π_1 starts to simulate an instruction with label l_j. Otherwise, system Π_1 starts to simulate instruction l_k. The simulation of an ADD instructions $l_i : (ADD(2), l_j, l_k)$ acting on register 2 and $l_i : (ADD(3), l_j, l_k)$ acting on register 3, is implemented in a similar way to the above process for register 1, we omit the description here.

Simulating SUB instruction $l_i: (SUB(r), l_j, l_k)$, because the SUB instruction is not executed on register 1, SUB instructions only act on registers 2 and 3. Take the example of executing the SUB instruction on register 2. Assume that at step t, the instruction $l_i : (SUB(2), l_j, l_k)$ are simulated, means that $a^{T+l(i)}$ spikes in membrane 0, rule $a^{T+l(i)}/a^{T+l(i)} \rightarrow (\bar{a}^{m+1-l(i)}, in_2)$ can be enabled. At the same moment, for membrane 2, it has the following two cases.

If membrane 2 is non-empty, received $\bar{a}^{m+1-l(i)}$ spikes, is corresponds to $(a^{2m+1})+a^{m+a^{l(i)}}$ spikes in membrane 2. Then, rule $(a^{2m+1})+a^{m+l(i)}/a^{m+l(i)} \rightarrow (a^{l(j)}, out)$ can be used, $a^{l(j)}$ spikes send to membrane 0, that is, the instruction l_j is correctly simulated. If membrane 2 is empty, received $\bar{a}^{m+1-l(i)}$ spikes in membrane 2. Then, rule $\bar{a}^{m+1-l(i)}/\bar{a}^{m+1-l(i)} \rightarrow (a^{l(k)}, out)$ fire, $a^{l(k)}$ spikes send to membrane 0, that is, the instruction l_k is correctly simulated.

Similarly, the SUB instructions $l_i : (SUB(3), l_j, l_k)$ acting on register 3 can be also correctly simulate, we omit the check here.

When $a^{T+l(m)}$ spikes in membrane 0, the register machine reaches the halt instruction l_m, there are no rules can be applied in the system. The number of spike in membrane 1 as a result, the system ends the computation. The proof is completed consequently.

Theorem 2. $N_{all}^{DD} cSNaP_9(in, out) = NRE$.

Proof. We consider an improvement in avoiding the indication in_t, that is, membrane 0 sends spikes to one of the inner membranes nondeterministically, which will be sent back to membrane 0 if the spikes arrive the wrong membrane. If only and if one membrane receives the corresponding spikes from other membrane, the computation continues. Here, we again consider register machine $M = (3, H, l_1, \cdots, l_m, I)$ as above, and we construct a system Π_2.

As shown in Figs. 2 and 3, there are nine membranes in the system. All instructions of registrar M are controlled in membrane 0. Registers $1, 2, 3$ are also corresponds to Membrane with labels $1, 2, 3$. In initially, one spike in membrane denote by $l(1)$, membrane $4'$ maintains T spikes. If the number contain in register $1, 2, 3$ of M is n, membranes $1, 2, 3$ will relatively contain nN copies of spikes

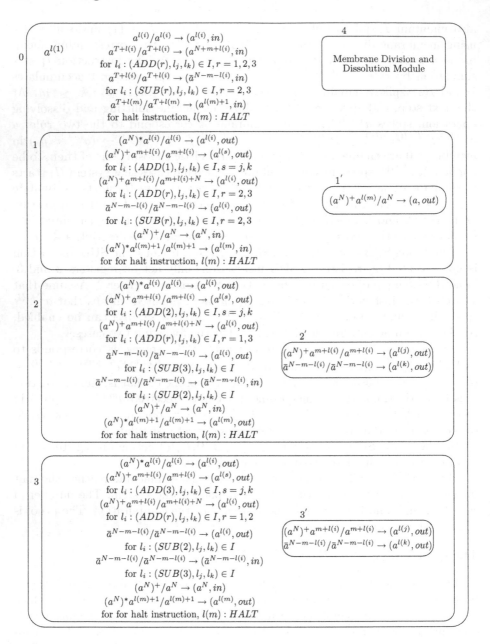

Fig. 2. System constructed for the proof of Theorem 2

inside, where $N = 2m + 1$. $T + l(i)$ is associated with instructions l_i of register machine M, where $T = 4m + 1$, $l(i) = i$, $1 \le i \le m$. It worth note that because of the uncertainty of the target instruction in, membrane $1^{'}$, $2^{'}$, $3^{'}$ and $4^{'}$ are

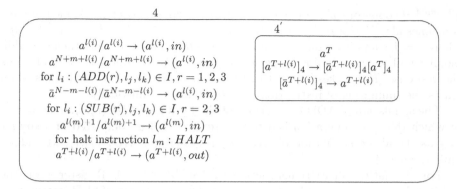

Fig. 3. Membrane division and dissolution Module

added to the membranes 1, 2, 3, and 4 to prevent the occurrence of anti-spikes entering in the membranes 1, 2, 3 and 4 to make misoperation. Meanwhile, the result of computation contain $N(2m + 1)$ spikes stored in membrane $1'$.

Simulating an ADD instruction l_i: $(ADD(r), l_j, l_k)$. There are $l(i)$ copies of spikes in membrane 0, $l(i)$ copies of spikes are sent to one of the inner membranes $1, 2, 3$, and 4 indeterminately. These spikes will send to the optimization module as show in Fig. 3, then, by using rules of the optimization module, $T + l(i)$ spikes are released into membrane 0, the systems Π_2 start simulates the instruction l_i : $(ADD(r), l_j, l_k), r = 1, 2, 3$. Similarly, when instruction l_i is executed, the spikes generated by the rule are sent indefinitely to either adjacent membrane until the correct membrane r is reached, means that the ADD instruction l_i is simulated.

Simulating SUB instruction l_i: $(SUB(r), l_j, l_k)$. Membrane 0 gets fired with $l(i)$ copies of spikes inside, the $l(i)$ copies of spikes are then sent to one of the three membranes $1, 2, 3, 4$ indeterminately. Only when the $T + l(i)$ copies of anti-spikes are received by the right membrane r, the system Π_2 start simulates the instruction l_i : $(SUB(r), l_j, l_k)$. Corresponding to the storage membrane r' of membrane r, there are two states of empty and non-empty, corresponding to $l(k)$ and $l(j)$ copies of spikes returned to membrane 0. The system thus passes to simulate instruction l_j or l_k.

Once $l(m)$ copies of spikes inside membrane 0, and after optimization module output $T + l(m)$ spikes to the membrane 0, that is, the HALT instruction can be enabled and membrane 1 is right. When there is no enabled rule can be applied, the system halts and the number of spikes in membrane 1 represents the output result (the result is n from membrane $1'$).

A more detailed description of the ADD, SUB, HALT instructions that simulate the registry is as follows.

For simulating an ADD instruction l_i: $(ADD(r), l_j, l_k)$. Membrane 0 have one spikes, and the spike is sent to one of the inner membranes $1, 2, 3$, and 4 indeterminately, which membrane 4 is right. When membrane 0 contain $a^{T+l(1)}$ copies of spikes, and rule $a^{T+l(1)}/a^{T+l(1)} \rightarrow (a^{N+m+l(1)}, in)$ can be fire, that is,

$a^{N+m+l(1)}$ copies of spikes are received by membrane 1. Afterwards, membrane 0 consumes $a^{m+l(1)}$ spikes by using the rule $(a^N)^+ a^{m+l(1)}/a^{m+l(1)} \to (a^{l(s)}, out)$, where $s = j, k$ and leave out N copies of spikes in membrane 1 and then stored in membrane $1'$, corresponding to register 1 accumulated 1. And membrane 1 sends the response ($l(s)$ spikes,) back to membrane 0. At the next step, the system starts the simulation of instruction l_j or l_k nondeterministically.

The simulation of ADD instructions of the form $l_i : (ADD(r), l_j, l_k), r = 2, 3$, in which the ADD operation is acted on register $2, 3$ is conducted in a similar way as the above simulation of ADD instructions $l_i : (ADD(1), l_j, l_k)$ and we omit the details.

For simulating an SUB instruction $l_i: (SUB(r), l_j, l_k)$. Register 1 on the register does not perform SUB instructions. The simulation of SUB instructions which act on register 2 of the form $l_i : (SUB(2), l_j, l_k)$ is as follows: membrane with $l(i)$ copies of spike inside, rule $a^{l(i)}/a^{l(i)} \to (a^{l(i)}, in)$ fires, and the spike is sent to one of the inner membranes $1, 2, 3$, and 4 indeterminately. Until to $a^{l(i)}$ copies of spikes in membrane 4, that is, membrane 0 will maintains $a^{l(i)+T}$ spikes. When membrane 0 fires, rule $a^{T+l(i)}/a^{T+l(i)} \to (\bar{a}^{N-m-l(i)}, in)$ is used, and $N - m - l(i)$ copies of anti-spikes are sent to one of the inner membranes $1, 2, 3$, and 4 indeterminately. Assume that at step t, only when the $\bar{a}^{N-m-l(i)}$ copies of spikes are correctly received by membrane 2, the computation can continue. There are two possible conditions to be considered:

If membrane 2 is non-empty, that is, membrane $2'$ contains $n(2m+1)$ spikes. Membrane 2 receives $\bar{a}^{N-m-l(i)}$, and then using rule $\bar{a}^{N-m-l(i)}/\bar{a}^{N-m-l(i)} \to (\bar{a}^{N-m-l(i)}, in)$ sends $\bar{a}^{N-m-l(i)}$ copies of anti-spikes to membrane $2'$. Afterwards, membrane $2'$ fires, using rule $(a^N)^+ a^{m+l(i)}/a^{m+l(i)} \to (a^{l(j)}, out)$, then, the $a^{l(j)}$ are sent into membrane 2. Membrane 2 fires by using rule $(a^N)^* a^{l(i)}/a^{l(i)} \to (a^{l(i)}, out)$, means that the system turns to the simulation of instruction l_j.

If membrane 2 is empty, that is, membrane $2'$ have no one spikes. Membrane 2 receives $\bar{a}^{N-m-l(i)}$, and then using rule $\bar{a}^{N-m-l(i)}/\bar{a}^{N-m-l(i)} \to (\bar{a}^{N-m-l(i)}, in)$ sends $N - m - l(i)$ copies of anti-spikes to membrane $2'$. Rule $\bar{a}^{N-m-l(i)}/\bar{a}^{N-m-l(i)} \to (a^{l(k)}, out)$ can be enabled, membrane $2'$ sends out the $l(k)$ copies of spikes to membrane 2. Then, by using rule $(a^N)^* a^{l(i)}/a^{l(i)} \to (a^{l(i)}, out)$, $a^{l(k)}$ copies of spikes are sent to membrane 0, that is, the system turns to the simulation of instruction l_k.

The simulation of SUB instructions of the form $l_i : (SUB(3), l_j, l_k)$ which act on register 3 is conducted in a similar way as the above simulation of SUB instructions $l_i : (SUB(2), l_j, l_k)$ and further details are omitted.

Assume that the computation in M halts, means that the halt instruction is reached. Membrane 0 contains $a^{T+l(m)}$ copies of spikes, rule $a^{T+l(m)}/a^{T+l(m)} \to (a^{l(m)+1}, in)$ is used, and $l(m) + 1$ spikes are sent to one of the inner membranes $1, 2, 3$, and 4 indeterminately. If these $l(m) + 1$ spikes are sent to membranes $2, 3$ and 4, by using rule $(a^N)^* a^{l(m)+1}/a^{l(m)+1} \to (a^{l(m)}, out)$ and then $l(m)$ spikes are sent back to the skin region, and the system continues to simulate the halt instruction. Once the $l(m) + 1$ spikes are sent to membrane 1, rule

$(a^N)^*a^{l(m)+1}/a^{l(m)+1} \rightarrow (a^{l(m)}, in)$ is applied, membrane $1'$ receives $l(m)$ spikes from membrane 1. Membrane $1'$ by using rule $(a^N)^+a^{l(m)}/a^N \rightarrow (a, out)$ sends n spikes into membrane 1, that is, the computation halts. It is concluded that the resister machine M is correctly simulated by the system, this completes the proof.

Theorem 3. $N_2^{DD}cSNaP_5(forg, in, in_t, out) = NRE$.

Proof. Based on the proof of Theorem 1, we present the improved system Π_1' to simulate the registration machine $M = (3, H, l_1, \cdots, l_m, I)$, as shown in Fig. 4. There are five membranes in the system Π_1'. Membrane 0 is associated with every instruction of register machine M, each instruction l_i is correspond to $l(i) + T$ spikes, where $l(i) = i$, $T = 4m + 1$. Membranes $1, 2, 3$ correspond to the three registers $1, 2, 3$ of M, respectively. If the number stored in registers $1, 2, 3$ is n, the corresponding membranes will relatively have $n(2m+1)$ copies of spikes inside. Initially, there are $l(1)$ copies of spikes in membrane 0 and membrane 4 contains T spikes, the system starts the simulation of instruction l_1.

System Π_1' simulates the ADD and SUB instructions are associated with the register $1, 2, 3$ of the register machine M, which is similar to the processing method of Theorem 1. Here, we will not make detailed introduction.

When the computation is reach halt at step t, the number of spikes in membrane 0 changes from $l(m)$ to $T+l(m)$. Rule $a^{T+l(m)}/a^{T+l(m)-m-2} \rightarrow (a^{l(m)}, in_1)$ can be enabled in membrane 0, sends $l(m)$ spikes to membrane 1. In next step, membrane 0 fires, rule $a^{m+2}/a \rightarrow (a, out)$ is applied, the first spike send into environment. Meanwhile, membrane 1 fires, send $2m+1$ copies of spikes to membrane 0 by using rule $a^m(a^{2m+1})^+/a^{2m+1} \rightarrow (a^{2m+1}, out)$ from step $t+1$. Then, rule $(a^{2m+1})^*a^{m+1}/a^{2m+1} \rightarrow \lambda$ can be used, $2m+1$ spikes are removed from membrane 0 at each step from step $t+1$ on. At step $t+n+1$, membrane 0 receives m spikes from membrane 1, rule $a^{2m+1}/a^{2m+1} \rightarrow (a, out)$ is used, sends the second spikes to the environment. The number of steps between first two spikes sent into the environment by system is $(t+n+1)-(t+1) = n$, is exactly the number n stored in register 1. Thus the proof is completed.

Theorem 4. $N_2^{DD}cSNaP_9(forg, in, out) = NRE$.

Proof. Similar to the proof of Theorem 2, system Π_2' is constructed to simulate the register machine M, where $M = (3, H, l_1, \cdots, l_m, I)$. System Π_2' is modified from system Π_2, as shown in Fig. 5.

In systems Π_2', there are nine membrane, among them, membrane $1, 2, 3$ are associated with register $1, 2, 3$ of M, respectively. If register has value $n(n \geq 0)$, is corresponded to the Nn copies of spikes in membrane, where $N = 2m + 1$. Each instructions $l(i)$ of register machine M is associated with $T + l(i)$ copies of spikes in membrane, where $T = 4m + 1, l(i) = i, 1 \leq i \leq m$.

Initially, $l(1)$ and T spikes contain in membrane 0 and $4'$, respectively. It worth note that the proof of system Π_2' uses the optimization module shown in Fig. 3. When $l(1)$ spike are sent to the optimization module, and then back

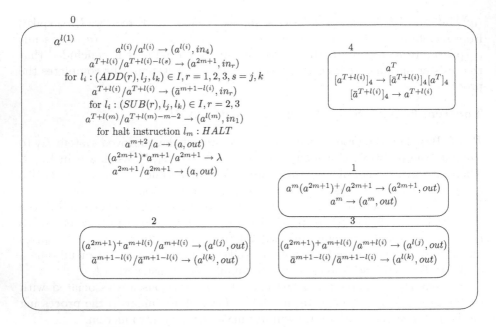

Fig. 4. System constructed for the proof of Theorem 3

$T + l(1)$ copies of spikes into membrane 0, which entails the simulation of the initial instruction with label l_1. The simulation of ADD instructions acting on register 1 is simulated in membrane 1. For other registers $2, 3$, the simulation of ADD and SUB instructions is simulated in membranes 2 and 3, respectively. The proof is similar to Theorem 2.

Assume that at step t, the computation in M halts, means that membrane 0 has $T + l(m)$ spikes is corresponding to the instructions l_m. Then rule $a^{T+l(m)}/a^{T+l(m)-m-2} \rightarrow (a^{l(m)+1}, in)$ can be enabled, leaving $m + 2$ spike in membrane 0 and sending $l(m) + 1$ spikes to one of inner membranes $1, 2, 3, 4$, nondeterministically chosen. At next step, rule $a^{m+2}/a \rightarrow (a, out)$ is applied in membrane 0, sending first spike to environment. Whichever of the $l(m) + 1$ spikes enters membrane $2, 3$, the $l(m)$ spikes is sent back to the skin membrane by using rule $(a^N)^* a^{l(m)+1}/a^{l(m)+1} \rightarrow (a^{l(m)}, out)$, and the system continues to simulate the halt instruction. Once the $l(m) + 1$ spikes are sent to membrane 1, rule $(a^N)^* a^{l(m)+1}/a^{l(m)+1} \rightarrow (a^{l(m)}, in)$ is applied, sending $l(m)$ spikes into membrane $1'$. From this step on, N spikes are removed from membrane $1'$ in each step. Until to membrane 0 received m spikes, $2m + 1$ spikes inside. Membrane 0 fires, and sends the second spike to environment by using rule $a^{2m+1}/a^{2m+1} \rightarrow (a, out)$. The time interval between first two spikes sent out by the system is exactly the value of register 1 of M in the moment when the computation in M halts, and this completes the proof.

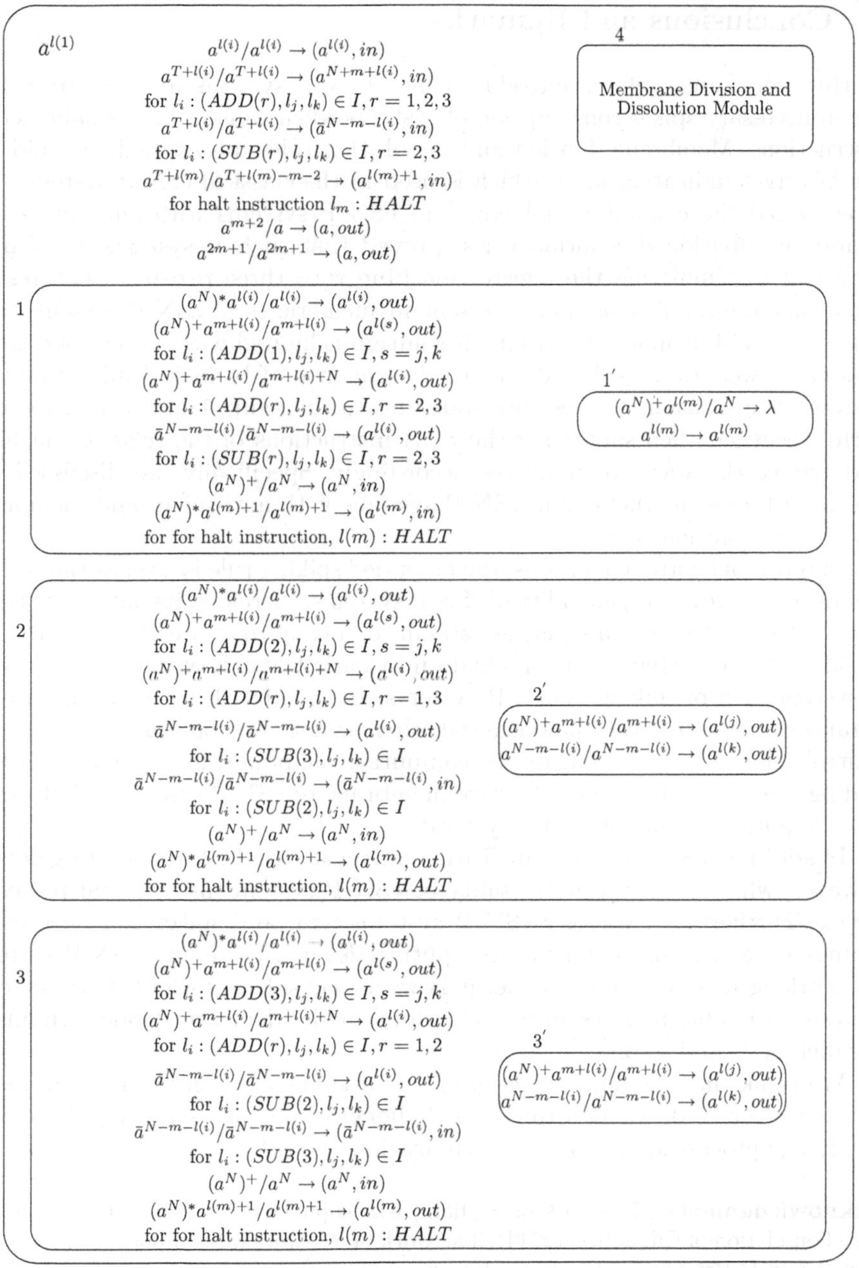

$a^{l(1)}$

0

$$a^{l(i)}/a^{l(i)} \rightarrow (a^{l(i)}, in)$$
$$a^{T+l(i)}/a^{T+l(i)} \rightarrow (a^{N+m+l(i)}, in)$$
for $l_i : (ADD(r), l_j, l_k) \in I, r = 1, 2, 3$
$$a^{T+l(i)}/a^{T+l(i)} \rightarrow (\bar{a}^{N-m-l(i)}, in)$$
for $l_i : (SUB(r), l_j, l_k) \in I, r = 2, 3$
$$a^{T+l(m)}/a^{T+l(m)-m-2} \rightarrow (a^{l(m)+1}, in)$$
for halt instruction $l_m : HALT$
$$a^{m+2}/a \rightarrow (a, out)$$
$$a^{2m+1}/a^{2m+1} \rightarrow (a, out)$$

4

Membrane Division and
Dissolution Module

1

$$(a^N)^*a^{l(i)}/a^{l(i)} \rightarrow (a^{l(i)}, out)$$
$$(a^N)^+a^{m+l(i)}/a^{m+l(i)} \rightarrow (a^{l(s)}, out)$$
for $l_i : (ADD(1), l_j, l_k) \in I, s = j, k$
$$(a^N)^+a^{m+l(i)}/a^{m+l(i)+N} \rightarrow (a^{l(i)}, out)$$
for $l_i : (ADD(r), l_j, l_k) \in I, r = 2, 3$
$$\bar{a}^{N-m-l(i)}/\bar{a}^{N-m-l(i)} \rightarrow (a^{l(i)}, out)$$
for $l_i : (SUB(r), l_j, l_k) \in I, r = 2, 3$
$$(a^N)^+/a^N \rightarrow (a^N, in)$$
$$(a^N)^*a^{l(m)+1}/a^{l(m)+1} \rightarrow (a^{l(m)}, in)$$
for for halt instruction, $l(m) : HALT$

$1'$

$$(a^N)^+a^{l(m)}/a^N \rightarrow \lambda$$
$$a^{l(m)} \rightarrow a^{l(m)}$$

2

$$(a^N)^*a^{l(i)}/a^{l(i)} \rightarrow (a^{l(i)}, out)$$
$$(a^N)^+a^{m+l(i)}/a^{m+l(i)} \rightarrow (a^{l(s)}, out)$$
for $l_i : (ADD(2), l_j, l_k) \in I, s = j, k$
$$(a^N)^+a^{m+l(i)}/a^{m+l(i)+N} \rightarrow (a^{l(i)}, out)$$
for $l_i : (ADD(r), l_j, l_k) \in I, r = 1, 3$
$$\bar{a}^{N-m-l(i)}/\bar{a}^{N-m-l(i)} \rightarrow (a^{l(i)}, out)$$
for $l_i : (SUB(3), l_j, l_k) \in I$
$$\bar{a}^{N-m-l(i)}/\bar{a}^{N-m-l(i)} \rightarrow (\bar{a}^{N-m-l(i)}, in)$$
for $l_i : (SUB(2), l_j, l_k) \in I$
$$(a^N)^+/a^N \rightarrow (a^N, in)$$
$$(a^N)^*a^{l(m)+1}/a^{l(m)+1} \rightarrow (a^{l(m)}, out)$$
for for halt instruction, $l(m) : HALT$

$2'$

$$(a^N)^+a^{m+l(i)}/a^{m+l(i)} \rightarrow (a^{l(j)}, out)$$
$$a^{N-m-l(i)}/a^{N-m-l(i)} \rightarrow (a^{l(k)}, out)$$

3

$$(a^N)^*a^{l(i)}/a^{l(i)} \rightarrow (a^{l(i)}, out)$$
$$(a^N)^+a^{m+l(i)}/a^{m+l(i)} \rightarrow (a^{l(s)}, out)$$
for $l_i : (ADD(3), l_j, l_k) \in I, s = j, k$
$$(a^N)^+a^{m+l(i)}/a^{m+l(i)+N} \rightarrow (a^{l(i)}, out)$$
for $l_i : (ADD(r), l_j, l_k) \in I, r = 1, 2$
$$\bar{a}^{N-m-l(i)}/\bar{a}^{N-m-l(i)} \rightarrow (a^{l(i)}, out)$$
for $l_i : (SUB(2), l_j, l_k) \in I$
$$\bar{a}^{N-m-l(i)}/\bar{a}^{N-m-l(i)} \rightarrow (\bar{a}^{N-m-l(i)}, in)$$
for $l_i : (SUB(3), l_j, l_k) \in I$
$$(a^N)^+/a^N \rightarrow (a^N, in)$$
$$(a^N)^*a^{l(m)+1}/a^{l(m)+1} \rightarrow (a^{l(m)}, out)$$
for for halt instruction, $l(m) : HALT$

$3'$

$$(a^N)^+a^{m+l(i)}/a^{m+l(i)} \rightarrow (a^{l(j)}, out)$$
$$a^{N-m-l(i)}/a^{N-m-l(i)} \rightarrow (a^{l(k)}, out)$$

Fig. 5. System constructed for the proof of Theorem 4

4   Conclusions and Remarks

In this work, anti-spike is introduced into cSN P systems in order to reduce
the unnecessary spikes consumption of systems when simulating the subtraction
instructions. Membrane division and dissolution rules are applied to avoid the
use of target indication in all which is used in the classical cSN P systems. We
investigated the computational power of cSN P systems with anti-spikes and
membrane division/dissolution rules, proved that such P systems are Turing
universal by simulating the register machine with three registers. The results
show that the number of spikes present in membranes of cSN P systems with
anti-spikes and membrane division/dissolution is limited without loss of computational power. Besides, based on the phenomenon of biological inhibition, the
systems only consume spikes corresponding to the value 1 in the register r in
each execution when simulating the SUB instructions of the register machine,
and reduce the waste of resources accordingly. Specifically, we discussed the
effects of target instructions in cSN P systems with anti-spikes and membrane
division/dissolution rules.

In terms of control conditions, the extended spiking rule is used as the technical means to prove the generality of cSN P systems, therefore, its natural to think
about whether the systems that use standard spiking rules are Turing universal.
Besides, cSN P systems is a combination of basic features of multiset-rewriting
P systems and of spiking neural P systems. In cSNaDD P systems, the way of
regular expression is very powerful, whether or not you could use some weaker
control conditions to investigate the computational power, for instance, consider
adding the electrical charges of active membrane to cSN P systems, what about
the computational power of this systems.

In addition, cSN P systems are Turing universal as a digital generating device,
however, what about the universality of cSN P systems as a digital receiving
device. Furthermore, whether cSN P systems with anti-spikes can be used as
a language generator or function computing devices. Generally, cSN P systems
are working in synchronous mode, it is worth considering whether or not cSN
P systems is able to work in asynchronous mode, depletion mode, minimally
parallel mode and so on.

We think that the application of cSN P systems with anti-spikes and membrane division or dissolution rules is a challenging area of researching, for example, image processing, parameter optimization and other aspects.

Acknowledgments. The work of S. Jiang was supported by National Natural Science Foundation of China (61772214). The work of Y. Wang was supported by National
Key R and D Program of China for International S and T Cooperation Projects 525
(2017YFE0103900), Joint Funds of the National Natural Science Foundation of China
(U1804262), and the State Key Program of National Natural Science Foundation of
China (61632002). The work of T. Wu was supported by the Priority Academic Program Development of Jiangsu Higher Education Institutions.

References

1. Alsalibi, B., Venkat, I., Al-Betar, M.A.: A membrane-inspired bat algorithm to recognize faces in unconstrained scenarios. Eng. Appl. Artif. Intel. **64**, 242–260 (2017)
2. Buiu, C., Vasile, C., Arsene, O.: Development of membrane controllers for mobile robots. Inf. Sci. **187**, 33–51 (2012)
3. Cabarle, F.G.C., Adorna, H.N., Jiang, M., Zeng, X.: Spiking neural P systems with scheduled synapses. IEEE Trans. Nanobiosci. **16**(8), 792–801 (2017)
4. Chen, H., Freund, R., Ionescu, M., Păun, G., Pérez-Jiménez, M.J.: On string languages generated by spiking neural P systems. Fundam. Inform. **75**(1–4), 141–162 (2007)
5. Díaz-Pernil, D., Gutiérrez-Naranjo, M.A., Peng, H.: Membrane computing and image processing: a short survey. JMC **1**(1), 58–73 (2019)
6. Leporati, A., Porreca, A.E., Zandron, C., Mauri, G.: Improved universality results for parallel enzymatic numerical P systems. Int. J. Unconv. Comput. **9**, 385–404 (2013)
7. Ionescu, M., Păun, G., Yokomori, T.: Spiking neural P systems. Fundam. Inform. **71**(2, 3), 279–308 (2006)
8. Korec, I.: Small universal register machines. Theor. Comput. Sci. **168**(2), 267–301 (1996)
9. Martin-Vide, C., Pazos, J., Păun, G., Rodríguez-Patón, A.: Tissue P systems. Theor. Comput. Sci. **296**(2), 295–326 (2003)
10. Neary, T.: Three small universal spiking neural P systems. Theor. Comput. Sci. **567**, 2–20 (2015)
11. Pavel, A.B., Buiu, C.: Using enzymatic numerical P systems for modeling mobile robot controllers. Nat. Comput. **11**(3), 387–393 (2012)
12. Pan, T., Păun, G.: Spiking neural P systems with anti-spikes. Int. J. Comput. Commun. **4**(3), 273–282 (2009)
13. Pan, T., Xu, J., Jiang, S., Xu, F.: Cell-like spiking neural P systems with evolution rules. Soft. Comput. **23**(14), 5401–5409 (2019)
14. Pan, L., Wu, T., Su, Y., Vasilakos, A.V.: Cell-like spiking neural P systems with request rules. IEEE Trans. Nanobiosci. **16**(6), 513–522 (2017)
15. Pan, L., Păun, G., Song, B.: Flat maximal parallelism in P systems with promoters. Theor. Comput. Sci. **623**, 83–91 (2016)
16. Păun, G., Păun, A.: Small universal spiking neural P systems. Biosystems **90**(1), 48–60 (2007)
17. Păun, G., Păun, R.: Membrane computing and economics: numerical P systems. Fundam. Inform. **73**(1, 2), 213–227 (2006)
18. Păun, G.: Computing with membranes. J. Comput. Syst. Sci. **61**(1), 108–143 (2000)
19. Păun, G.: Membrane Computing: An Introduction. Springer, Heidelberg (2002). https://doi.org/10.1007/978-3-642-56196-2
20. Păun, G.: P systems with active membranes attacking NP-complete problems. JALC **6**, 75–90 (2001)
21. Păun, G., Rozenberg, G., Salomaa, A. (eds.): The Oxford Handbook of Membrane Computing. Oxford University Press, New York (2010)
22. Peng, H., Wang, J., Wang, H., Shao, J., Wang, T., Pérez-Jiménez, M.J.: Fuzzy reasoning spiking neural P system for fault diagnosis. Inf. Sci. **235**, 106–116 (2013)
23. Rozenberg, G., Salomaa, A. (eds.): Handbook of Formal Languages. Springer, Berlin (1997). https://doi.org/10.1007/978-3-642-59136-5

24. Rozenberg, G., Back, T., Kok, J.N. (eds.): Handbook of Natural Computing. Springer, New York (2012). https://doi.org/10.1007/978-3-540-92910-9
25. Song, B., Kong, Y.: Solution to PSPACE-complete problem using P systems with active membranes with time-freeness. Math Probl. Eng. (2019). Article ID 5793234
26. Song, T., Rodríguez-Patón, A., Zheng, P., Zeng, X.: Spiking neural P systems with colored spikes. IEEE Trans. Cogn. Dev. Syst. **10**(4), 1106–1115 (2018)
27. Song, T., Pan, L., Păun, G.: Spiking neural P systems with rules on synapses. Theor. Comput. Sci. **529**, 82–95 (2014)
28. Woods, D., Murphy, N., Pérez-Jiménez, M.J., Riscos-Núñez, A.: Membrane dissolution and division in P. In: Calude, C.S., Costa, J.F., Dershowitz, N., Freire, E., Rozenberg, G. (eds.) UC 2009. LNCS, vol. 5715, pp. 262–276. Springer, Heidelberg (2009). https://doi.org/10.1007/978-3-642-03745-0_28
29. Wang, T., Zhang, G., Zhao, J., He, Z., Wang, J., Pérez-Jiménez, M.J.: Fault diagnosis of electric power systems based on fuzzy reasoning spiking neural P systems. IEEE Trans. Power Syst. **30**(3), 1182–1194 (2014)
30. Wu, T., Zhang, Z., Păun, G., Pan, L.: Cell-like spiking neural P systems. Theor. Comput. Sci. **623**, 180–189 (2016)
31. Wu, T., Zhang, Z., Pan, L.: On languages generated by cell-like spiking neural P systems. IEEE Trans. Nanobiosci. **15**(5), 455–467 (2016)
32. Wang, T., Zhang, G., Rong, H., Pérez-Jiménez, M.J.: Application of fuzzy reasoning spiking neural P systems to fault diagnosis. Int. J. Comput. Commun. **9**(6), 786–799 (2014)
33. Wang, X., et al.: Design and implementation of membrane controllers for trajectory tracking of nonholonomic wheeled mobile robots. Integr. Comput.-Aided Eng. **23**(1), 15–30 (2015)
34. Zhang, X., Liu, Y., Luo, B., Pan, L.: Computational power of tissue P systems for generating control languages. Inf. Sci. **278**, 285–297 (2014)
35. Zhang, G., Gheorghe, M., Pan, L., Pérez-Jiménez, M.J.: Evolutionary membrane computing: a comprehensive survey and new results. Inf. Sci. **279**, 528–551 (2014)
36. Zhang, G., Gheorghe, M., Wu, C.: A quantum-inspired evolutionary algorithm based on P systems for knapsack problem. Fundam. Inform. **87**(1), 93–116 (2008)
37. Zhang, G., Pérez-Jiménez, M.J., Gheorghe, M.: Real-life Applications with Membrane Computing. ECC, vol. 25. Springer, Cham (2017). https://doi.org/10.1007/978-3-319-55989-6
38. Zhang, G., Rong, H., Neri, F., Pérez-Jiménez, M.J.: An optimization spiking neural P system for approximately solving combinatorial optimization problems. Int. J. Neural Syst. **24**(05), 1–16 (2014)
39. Zeng, X., Zhang, X., Song, T., Pan, L.: Spiking neural P systems with thresholds. Nat. Comput. **26**(7), 1340–1361 (2014)

Research on EEG Emotional Recognition Based on LSTM

Huiping Jiang[(⊠)] and Junjia Jia[(⊠)]

School of Information Engineering, Minzu University of China,
Beijing 100081, China
jianghp@muc.edu.cn, 1634467879@qq.com

Abstract. Electroencephalogram (EEG) signals have been shown to provide insight into deeper emotional processes and responses. Emotion recognition based on EEG signals has been studied on a large scale. One of the research objectives is to find features suitable for EEG emotion recognition through various methods, then optimize model and improve accuracy of classification method. However, EEG signal is a random non-stationary time series signal, and the traditional classification method does not take into account its timing. The Long Short-Term Memory (LSTM) network in deep learning technology can solve this problem well due to its temporal recursive structure. However, a EEG sequence is generally long. If you use LSTM classification directly, it will lead to large computing resources and poor results. Therefore, in this paper, firstly the EEG signal is cut, and the differential entropy feature of each segment is extracted by wavelet transform. Finally, it is input into LSTM model. The experimental results show that the classification accuracy rate is about 0.89 when the 4 layers LSTM model is selected.

Keywords: EEG · LSTM · Emotion recognition

1 Introduction

The research on emotion recognition is of great significance both in theory and in practical application. Because human emotion is closely related to human cognition and behavior, recognition and research on emotion is also an important part of psychology and behaviour. Most researchers agree that 'emotions are generated by strong nerve impulses and are closely related to the cerebral cortex. They can make people produce positive or negative psychological reactions, so that the corresponding body organizations can act.' So many researchers begin to shift from expression and speech to emotion recognition based on EEG or multi-combination [1,2].

Lin et al. classified four emotions of happiness, anger, sadness and happiness by analyzing the subjects' EEG signals [3]. Nie, Lu and others used movie clips as inducement materials to classify positive and negative emotions by extracting the frequency domain characteristics of EEG signals [4]. Using Bayesian linear

© Springer Nature Singapore Pte Ltd. 2020
L. Pan et al. (Eds.): BIC-TA 2019, CCIS 1160, pp. 409–417, 2020.
https://doi.org/10.1007/978-981-15-3415-7_34

discriminant analysis, Yazdani et al. classified EEG features into six categories of emotions (joy, anger, aversion, sadness, surprise, fear), with an accuracy of more than 80% [5]. Petrantonakis and Hadjileontiadis used quadratic discriminant analysis (QDA) and SVM to classify emotions in six categories: pleasure, surprise, anger, fear, aversion and sadness. The accuracy rates were 62.3% and 83.33% respectively [6]. In the past, many researches on emotion recognition based on EEG used traditional machine learning methods such as KNN, LDA, SVM, RVM and so on. Traditional machine learning methods need to design and extract the features of EEG signals manually and the redundancy of features is very high. However dynamics of EEG signals, which are crucial for emotion recognition, is not taken into account. And EEG signals are often high-dimensional, which maybe is resolved by others algorithms if having no feature extraction [7]. LSTM in deep learning can solve the problem of long sequence dependence in neural networks by using its own gate structure and reduce dimension. EEG is a kind of time series signal. LSTM can be used to classify EEG emotions, which can better take into account the temporal dynamic information in EEG signals.

Therefore, an emotion recognition model based on Long Short-Term Memory (LSTM) network is used in this paper to perform EEG emotion recognition. First, laboratory personnel collect EEG data, and then EEG signals are divided into several segments. Five bands of differential entropy features are extracted from each signal to form a feature sequence. Then LSTM is used to learn time dynamic information from various feature sequences and makes final emotional prediction.

2   Related Work

The first step about emotion recognition emotion with EEG is to induce emotions and collect EEG signals. Many researchers choose different stimulus materials, such as visual, auditory, olfactory or multimedia, to induce emotions. Koelstra et al. combined pictures and music stimulus materials and used 120 music videos to stimulate subjects to produce different emotions [8]. Soleymani et al. constructed an emotional database. Which includes electroencephalogram and peripheral physiological signals [9]. The common method of EEG acquisition is to bond electrodes to scalp through gelatinous electroencephalogram paste, so as to reduce the resistance of electrodes to scalp and make EEG signal acquisition easier [10]. Then the original EEG signal is pretreated to remove noise and artifacts. Finally, the relevant features are extracted from the EEG signal and used to train the classifier.

In order to better express EEG signal characteristics, feature analysis has become an important part of EEG emotion recognition. The earliest feature analysis is mainly divided into time domain analysis and frequency domain analysis. Time domain analysis extracts useful waveform information directly from time domain, such as ANOVA, waveform parameter analysis and wave identification, Histogram, correlation analysis (CA), peak detection and so on. Because EEG signals are often simpler and more intuitive in frequency domain than in time domain, many researchers begin to analyze EEG signals in frequency domain by means of spectrum analysis and power spectrum analysis. However, simple time domain and frequency domain methods are mainly used for stationary signals. For EEG, which is a random and non-stationary signal, it is difficult to extract effective feature information [11]. Considering that time domain and frequency domain features can be combined together, researchers began to use time-frequency analysis method to combine the two methods, which makes up for the shortcomings of time domain and frequency domain analysis methods in the analysis of non-linear signals, and can accurately extract the features of non-linear signals changing with time. Wavelet transform is a typical time-frequency analysis method. Wavelet transform describes the signal by time and scale, fully considering the scalability and translation of the signal. Wavelet transform has many advantages. The time-frequency analysis results of signals can be obtained quickly by wavelet transform, and the time window and frequency window of signals can be changed, so they have strong multi-resolution characteristics [12].

At present, emotional recognition based on EEG has taken on a large scale. Yazdani, Lee and others used linear discriminant analysis to select EEG features, and then classified six kinds of emotions (pleasure, sadness, fear, disgust, anger, surprise) and achieved good results [5]. Nie, Wang et al. selected features by calculating the correlation coefficients between the features of each frequency band and emotional tags, so as to achieve the goal of feature dimensionality reduction. Finally, emotions were classified by using six-order linear support vector machine, and the accuracy was 87.53% [6]. Takahashi et al. used support vector machine and neural network to classify emotions, and compared classification results of them. The result proved that the classification effect of support vector machine was relatively good to a certain extent [13]. Murugappan et al. used unsupervised fuzzy C-means clustering method to cluster three kinds of emotions (digust, fear, pleasure). According to the inherent characteristics of the category itself, the best feature set for emotional classification was selected [14]. In this paper, we use wavelet transform to extract the features of five frequency bands, and finally feed them to LSTM for emotional classification.

3   LSTM-Based Classification Model

The LSTM algorithm is called Long short-term memory, first proposed by Sepp Hochreiter and Jürgen Schmidhuber in 1997, which is a specific form of RNN (Recurrent Neural Network), while RNN is a series of capable sequences. However, RNN encounters great difficulties in dealing with long-term dependencies (distant nodes in time series), because calculating the connections between distant nodes involves multiple multiplications of Jacobian matrices, which can lead to the disappearance of gradients (often occurring) or gradient expansion (rarely occurring). Then LSTM appears, which makes the weight of self-circulation change by adding input threshold, forget threshold and output threshold. In this way, the integral scale can be changed dynamically at different time when the model parameters are fixed, thus avoiding the problem of gradient disappearance or gradient expansion.

At the same time, the cell in LSTM will judge the information, and the information that conforms to the rules will be left behind, and the information that does not conform will be forgotten. Based on this principle, the problem of long sequence dependence in neural networks can be solved. EEG is a kind of time series signal. LSTM can be used to classify EEG emotions, which can better take into account the temporal dynamic information in EEG signals [2]. Therefore, LSTM is used to classify EEG signals after feature extraction.

The differential entropy features after wavelet transform are fed into LSTM classification model for training. The LSTM network structures used in this paper are as follows (see Fig. 1).

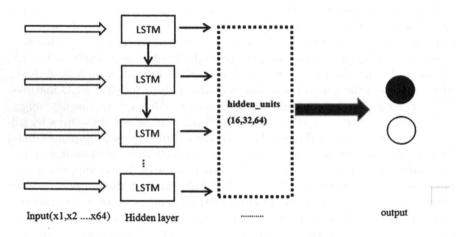

Fig. 1. LSTM struction.

4   Specific Implementation and Results

4.1   Data Set on Experiment

Data Collection. The EEG data sets in this paper are collected in laboratory. The subjects we collected were all college students, 8 of whom were aged between 22 and 26, right-handed, healthy, sleeping well, without brain damage and mental illness. Before the beginning of the experiment, we will tell the subjects the purpose, process and precautions of the experiment, and let them turn off their mobile phones in advance for a period of time to alleviate the mental stress of the subjects, so that the subjects can devote themselves to the experiment.

The stimulus files are selected from 8 videos lasting about 3 min, of which four were able to induce positive and negative emotions. Before the beginning of the experiment, the subjects should carefully read the instructions on the screen in order to fully understand the process and details of the experiment. In order to better induce the emotions of the subjects, we randomly present positive and negative videos and collect emotional samples. When the subjects watch the videos, we synchronously collect and save their EEG data. The sampling frequency of EEG is set to 1000 HZ. The experiment collects 64 EEG signals. The distribution of electrodes is generally adopted in the 10/20 system electrode placement method at home and abroad.

The flow chart of the formal emotional induction experiment is as follows (see Fig. 2).

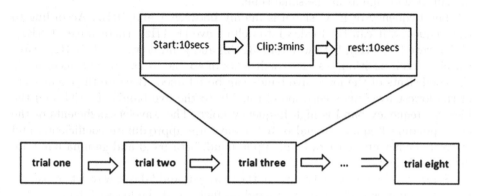

Fig. 2. Flow chart of experiment.

Total eight trials correspond to eight videos. Before the beginning of each video, subjects will have 10 s of attention-enhancing time. After the end of the video, they will have 10 s of feedback relaxation time, so as to obtain the behavior data of subjects, which is convenient to verify the validity and correctness of the experiment.

After the experiment, the EEG data of the subjects are saved.

Pretreatment. Firstly, the collected EEG data are preprocessed with NeuroScan's own data processing tools, mainly to remove the Eye Electricity and artifacts. Then normalize all EEG signals measured in the same channel to $(-1, 1)$ range, so that the difference in amplitude range of different channel signals can be eliminated. And the way can retain the difference in amplitude of the same channel under different stimuli and reduce the individual difference between subjects and others. In this way, more robust EEG data can be obtained, and the generalization ability of emotion recognition model can be improved.

Secondly, datas of 8 subjects after the above pretreatment are cut once every 4 s, and then partitioned by matlab. Finally, each subject corresponded to 480 segments of data. Eight subjects are divided into training set and test set according to 3:1 scale. In result, final data sets include 2880 training sets and 960 test sets.

4.2 Features Extraction About EEG

Wavelet transform is a kind of time-frequency analysis method. It can not only provide the whole information of the signal, but also the information of the drastic change of the signal in any part of the time [15]. It is the most common and effective method for feature extraction of EEG [16]. It can decompose signals into different frequency ranges according to different levels, and save time information of signals at the same time.

The frequency of EEG signal is mainly between 1 and 50 Hz. According to the frequency, it can be divided into delta wave (1–4 Hz), theta wave (4–8 Hz), alpha wave (8–14 Hz), beta wave (14–31 Hz) and gamma wave (31–50 Hz). After wavelet decomposition, wavelet coefficients of six bands can be obtained, while the coefficients of the lower five bands can be retained. Because the coefficients of the lowest five bands correspond roughly to the five bands of EEG, and the highest frequency band is high frequency noise. The wavelet coefficients of the corresponding frequency band of delta wave are approximate coefficients, and wavelet coefficients of theta band, alpha band, beta band and gamma band are detail coefficients.

Sym8 wavelet is selected as the mother wavelet, and 5-layer wavelet transform is made to preserve the frequency band coefficients of the lowest 5 bands. After wavelet transform of EEG signal, we get the wavelet coefficients of each frequency band, and then transform the wavelet coefficients according to the extracted feature types. This paper chooses the differential entropy feature. Compared with frequency band energy feature, differential entropy feature can balance the disparity of EEG energy in different frequency domains, reduce the error caused by the difference of energy value in subsequent calculation, and improve the discriminant ability of learning algorithm [17].

Energy of the frequency band, which is sum of squares of the wavelet coefficients of each frequency band, is solved by the following formula.

$$E_i = \sum_{j=1}^{n_i} d_{ij}^2 \tag{1}$$

In this equation, E_i is energy of the frequency band in i layer. n_i is the number of decomposition coefficients in i layer. d_{ij} is the wavelet coefficients of No.j in i layer.

By calculation, it can be seen that for EEG sequences of the same length, his differential entropy in a certain frequency band is equivalent to the logarithmic value of his energy in that frequency band, that is:

$$DE_i = \log_{10} E_i \tag{2}$$

After extracting features of EEG signal by wavelet transform, the data of single EEG signal with dimension 64*4000 is transformed to 64*5.

4.3   Analysis of Results

After feature extraction, EEG datas are fed into LSTM models with different structures. Table 1 gives classification results.

Table 1. Comparison of LSTM models with different structures.

Input dimension	Num units	Layers of LSTM	Accuracy
64*5	16	1	78
		2	79
		3	79
		4	80
64*5	32	1	78
		2	80
		3	82
		4	83
64*5	64	1	83
		2	85
		3	86
		4	89

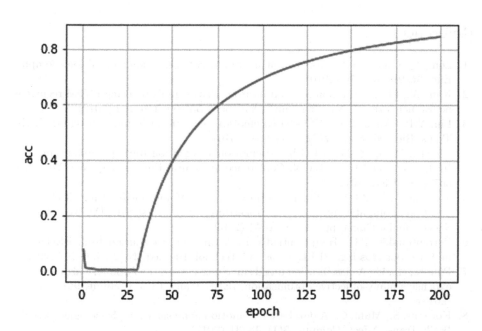

Fig. 3. Results of four-layer classification.

As can be seen from the above Fig. 3, the 4-layer LSTM model with 64 neurons in the hidden layer has the highest classification accuracy, reaching 89%. It proves the feasibility of LSTM in EEG emotion classification. The results of four-layer classification with 64 neurons in the hidden layer are as follows (see Fig. 3).

5 Conclusion and Prospect

In this paper, different LSTM models are used to classify extracted EEG features. The accuracy of last model reaches 89%. That is to say, multi-layer LSTM is more suitable for EEG classification, which can be considered in the classification model.

However, there are still some problems that need to be studied in the following stages in the process of emotion recognition system based on EEG. For example, in the aspect of feature extraction, whether the differential entropy feature extracted by wavelet will lose time series information or have better feature extraction methods. About EEG pretreatment, whether the denoising operation of EEG signal in the pretreatment process is the loss of effective EEG information or not is considered. In the emotional evoked place, how to select the stimulus material can better induce emotions to be conducive to validity of EEG data. Therefore, emotion recognition based on EEG still needs more in-depth exploration and research.

References

1. Zeng, C., Guang, Y.L.: Application of EEG signal in emotion recognition. Comput. Eng. **36**(9), 168–170 (2010)
2. Kan, W., Li, Y.: Emotion recognition from EEG signals by using LSTM recurrent neural networks. J. NanJing Univ. (Nat. Sci.) **55**(01), 116–122 (2019)
3. Lin, Y.P., Wang, C.H.: EEG-based emotion recognition in music listening. IEEE Trans. Biomed. Eng. **57**(7), 1798–806 (2010)
4. Nie, D., Wang X.-W.: EEG-based emotion recognition during watching movies. In: International IEEE/EMBS Conference on Neural Engineering, New York, pp. 667–668. IEEE (2011)
5. Yazdani, A., Lee J.-S.: Implicit emotional tagging of multimedia using EEG signals and brain computer interface. In: Proceedings of the First SIGMM Workshop on Social Media, China, pp. 81–88. ACM (2009)
6. Petrantonakis, P.C., Hadjileontiadis, L.J.: Emotion recognition from EEG using higher order crossings. IEEE Trans. Inf Technol. Biomed. **14**(2), 186–197 (2010)
7. Pan, L., He, C.: A classification based surrogate-assisted evolutionary algorithm for expensive many-objective optimization. IEEE Trans. Evol. Comput. **99**(1), 74–88 (2018)
8. Koelstar, S., Muhl, C.: A database for emotion analysis using physiological signals. IEEE Trans. Affect. Comput. **3**(1), 18–31 (2012)
9. Soleymani, M., Lichtenauer, J.: A multimodal database for affect recognition and implicit tagging. IEEE Trans. Affect. Comput. **3**(1), 42–55 (2012)

10. Ying, X., Lin, H.: Study on non-linear bistable dynamics model based EEG signal discrimination analysis method. Bioengineered **6**(5), 297–298 (2015)
11. Haiyu, W., Jianfeng, H., Yinglong, W.: A review of EEG signal processing methods. Comput. Era **2018**(1), 13–15 (2018)
12. Alotaiby, T., El-Samie, F.E.A., Alshebeili, S.A., Ahmad, I.: A review of channel selection algorithms for EEG signal processing. EURASIP J. Adv. Signal Process. **2015**(1), 1–21 (2015). https://doi.org/10.1186/s13634-015-0251-9
13. Takahashi, K.: Remarks on emotion recognition from multi-modal bio-potential signals. In: IEEE International Conference on Industrial Technology, Tunisia, pp. 1138–1143. IEEE (2004)
14. Murugappan, M., Rizon, M., Nagarajan, R.: Time-frequency analysis of EEG signals for human emotion detection. In: Abu Osman, N.A., Ibrahim, F., Wan Abas, W.A.B., Abdul Rahman, H.S., Ting, H.N. (eds.) 4th Kuala Lumpur International Conference on Biomedical Engineering 2008. Springer, Heidelberg (2008). https://doi.org/10.1007/978-3-540-69139-6_68
15. Wang, X.: Multi-focus fusion algorithm for noisy images. Opt. Precis. Eng. **19**(12), 2977–2984 (2012)
16. Bao-Guo, X.U., Ai-Guo, S.: Feature extraction and classification of EEG in online brain-computer interface. Acta Electron. Sinica **39**(5), 1025–1030 (2011)
17. Duan, R.N.: EEG-Based Emotion Recognition During Watching Videos. Jiao Tong University, Shanghai

Research on Two-Stage Path Planning Algorithms for Storage Multi-AGV

Tao Mu$^{(\boxtimes)}$, Jie Zhu, Xiaoling Li, and Juntao Li

School of Information, Beijing Wuzi University, Beijing 101149, China
349940276@qq.com

Abstract. In order to solve the problem of path conflict of multiple AGV (Automated Guided Vehicle) in warehousing environment during handling shelves, in this paper, a two-stage path planning algorithm is proposed. In the first stage, on the premise of ignoring the conflicts between robots, the optimal path of each AGV is obtained by using A* algorithm. In this paper, an improved A* algorithm with directional search is proposed, which can effectively reduce the search of unnecessary nodes in the path finding process. In the second stage, conflict is checked by time window, when collision conflicts occur in multi-robot system. When a post-conflict situation occurs, excessive energy consumption is caused by a AGV waiting for another AGV to pass, in order to solve the energy consumption caused by waiting, a path planning method of coupling conflict car is proposed to realize dynamic path planning of multi-AGV. The simulation results show that the proposed algorithm can effectively reduce the number of searching nodes and waiting times in the process of path finding and it can also improve the overall efficiency of the system under the condition of ensuring the optimal or sub-optimal path.

Keywords: AGV · Path planning · Improved A* algorithm · CBS algorithm

1 Introduction

AGV has been widely used as the main tool in intelligent warehousing and sorting work and brought many advantages such as saving labor costs and improving the accuracy of sorting. Path planning is a hot issue in the field of AGV. In recent years, more and more AGV path planning problems have attracted people's attention. As a classical heuristic search algorithm, A* algorithm has been widely used in AGV path planning [1]. But, many AGVs still have many problems in path planning under more complex logistics warehousing environment. The problems are as follows:

A* algorithm has some problems in path search, such as Lower search speed, many turning points and large search area. Many scholars have proposed the improved A* algorithm in the aspects of path smoothness, node expansion and operation efficiency. But these improved A* algorithms still have some shortcomings. For example, Li et al. [2] introduced the factors of AGV turning into A* algorithm and edge removal based on the improved A* algorithm is adopted to solve k shortest path problem for path planning. this method calculated a smoother path, but it does not improve the efficiency of A* algorithm. Chen et al. [3] introduced direction vector and parallel search into A*

© Springer Nature Singapore Pte Ltd. 2020
L. Pan et al. (Eds.): BIC-TA 2019, CCIS 1160, pp. 418–430, 2020.
https://doi.org/10.1007/978-981-15-3415-7_35

algorithm for path planning in, which reduced search time and search path points, but this method did not consider the smoothness of the path. Therefore, the improved A* algorithm can not fully meet the requirements of AGV path planning in complex environments.

Due to the emergence of dynamic obstacles in multi-AGV path planning, deadlock, collision, more start-stop times and more planning times are prone to occur. Hu et al. [4] proposed a dynamic path planning method based on time window. This method first searches for alternative paths, and then evades conflicts by calculating and scheduling time windows. It is a time window about edges, that is to say, a long lane can only be reserved by one AGV, edges time window is less efficient than nodes time window. Li and Yi [5] proposed a time-based multi-AGV coordinated collision avoidance planning method. Firstly, the static collision-free path of a single robot is planned one by one, and then the collision-free motion between multi-AGV is realized by staggering the time to modify the motion sequence. This method can effectively realize multi-AGV collision avoidance path planning.

In this paper, in order to solve a series of problems in multi-AGV dynamic path planning, A path planning method based on time window sorting is proposed. Using the directed A* algorithm, the paths are sequentially planned for AGV. The time window [6–8] is introduced by adding time series. Whether the time window overlaps to detect whether the car is in conflict or not. The collision car is planned in a Coupling-Based way. Under the condition of collision-free collision, the path time cost is lowest, and the overall efficiency of the system is improved.

2 Problem Description

2.1 Environmental Map

This paper studies the path under the application background of warehousing logistics [9–11]. Because the goods need to be transported to the designated location in warehousing logistics, it is necessary to set up location information related nodes in the environment, such as specifying loading and unloading points for AGV path planning. Therefore, this paper uses the grid model to construct the map, and makes the following assumptions:

(1) AGV can travel bi-directionally between adjacent nodes.
(2) Only one AGV is allowed in each node.
(3) AGV has four states: static state without task, unloaded driving task with task, load handling state and temporary waiting state.
(4) AGV runs at the same speed and remains constant under load and unload conditions.
(5) The mobile time of two adjacent nodes is 1 s.

2.2 Objective Function

The goal of multi-AGV path planning is to find a path with the least time cost. Specifically, it refers to the sum of walking time (t_{walk}^i), turning time (t_{turn}^i) and waiting time (t_{walk}^i) in all paths. Therefore, the sum of the time costs of all AGVs can be expressed by the following formula:

$$\min TimeCost = \sum_{i=1}^{k} (t_{walk}^i + t_{turn}^i + t_{wait}^i) x_{d_q}^i \tag{1}$$

Among them, K is the number of the picking task.

The task arrangement of each AGV is expressed as follows:

$$d_q = (a^i, y^i, p_s^i, p_g^i, d_q^{begin}, d_q^{end}, r_q^i)$$

y^i represents the priority of a^i, P_s^i represents the start node of a^i, p_g^i represents the goal node of a^i, d_q^{begin} represents the start execution time of task d_q, d_q^{end} represents the end time of task d_q, r_q^i represents a^i completes task d_q's driving route.

$$x_{d_q}^i = \begin{cases} 1, & \text{Let the car } a^i \text{ to perform the task } d_q \\ 0, & \text{The task } d_q \text{ is not assigned to the car } a^i \end{cases}$$

3 Two-Stage Coordinated Collision Avoidance Algorithm for Multi-robot

3.1 Phase 1: Initial Path Planning

In the first stage, on the premise of ignoring the conflicts between robots, the optimal path of each AGV is obtained by using A* algorithm.

Environmental Modeling. Assume that the map is a limited three-dimensional space environment. Each AGV occupies a grid in the map. Each grid contains data such as: X-axis, Y-axis and time window sets. The reserved time window of the current grid node can be expressed as: $node\_time = \{[t_{in}^1, t_{out}^1], [t_{in}^2, t_{out}^2], \ldots, [t_{in}^k, t_{out}^k]\}$ among them, t_{in}^i is the time when the AGV a^i enters the current node, t_{out}^i is the time when the AGV a^i leaves the current node.

Improved A* Algorithm. A* algorithm is a heuristic search algorithm and an effective method to solve the shortest path in static network. Traditional A* algorithm can search in eight directions in path finding. Considering the direction of AGV in warehousing environment, the adjacent reachable position of AGV is determined to be four directions. They are left and right front and back. The evaluation function of A* algorithm is as follows:

$$f(n) = g(n) + h(n) \tag{2}$$

$g(n)$ is the actual cost between the start node and the current node $h(n)$ is the estimated cost of the current node n and the goal node. *f(n) is* the evaluation function of the current node n. $h(n)$ plays a key role in the evaluation function, which determines the efficiency of A* algorithm. In this paper, Manhattan distance is used to evaluate the distance between two points. It's expression can be written as $h(n) = abs(x_g - x_s) + abs(y_g - y_s)$. (x_g, y_g) is the coordinate of the goal node g, (x_s, y_s) is the coordinate of the current node s. In Fig. 1, s represents the starting point, g represents the goal point, green grid is the shelf, and black connection represents the searched path. The searched path has five turning times and 44 searching nodes. It can be seen that the common A* algorithm has a large amount of computation, a large number of turning points and the path is not smooth enough.

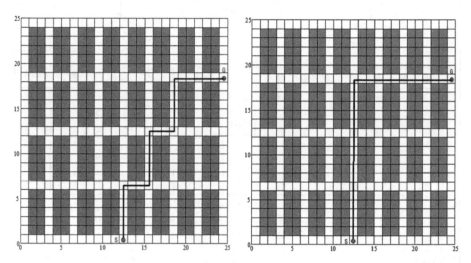

Fig. 1. Path planning for 1 A* algorithm. (Color figure online)

Fig. 2. Path planning for improved A* algorithm. (Color figure online)

In order to solve the problems existing in the above-mentioned paths, Therefore, an improved A* algorithm is proposed in this paper. By calculating the coordinate difference between the target node and the starting node, the direction of the extended node is determined, that is, if $x_g - x_s > 0$ and $y_g - y_s > 0$, extend the right and upper nodes of the current node. If $x_g - x_s > 0$ and $y_g - y_s < 0$, extend the right and bottom nodes of the current node. If $x_g - x_s < 0$ and $y_g - y_s > 0$, extend the left and upper nodes of the current node. If $x_g - x_s < 0$ and $y_g - y_s < 0$, extend the left and bottom nodes of the current node. If there is the same f value in the expansion process, the next node in the same direction with the current node has the right to choose first, so as to reduce the number of turns. The flow chart of the improved A* algorithm is shown in Fig. 3. The path obtained by the improved A* algorithm is shown in Fig. 2. It can be seen that the turning times are 1 and the path search nodes are 34. Compared with the ordinary A* algorithm, the performance has been improved.

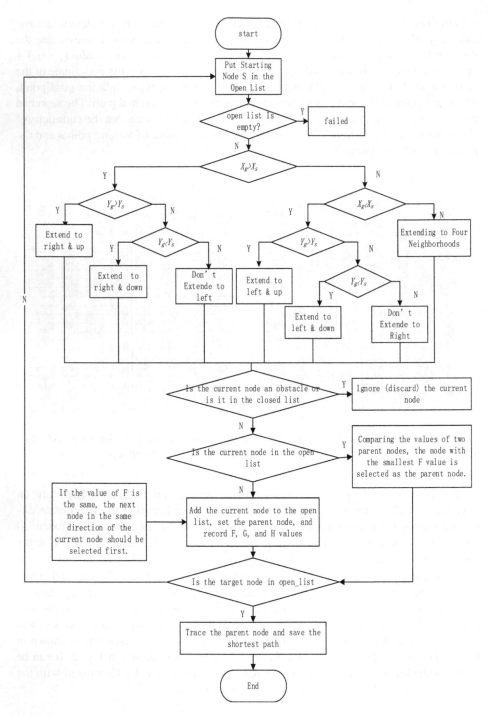

Fig. 3. Flow chart of improved A* algorithm.

3.2   Phase II: Multi-robot Coordination Obstacle Avoidance

Conflict Overview. In multi-AGV system, AGV is prone to collisions, blockages and deadlocks at intersections in bidirectional guidance paths. Figure 4 shows the two types of conflict encountered by the car during its journey, which are as follows:

Inter-directional conflict: The two AGVs are driving in opposite directions as shown in the figure. Two AGVs clashed because they occupied each other's path at the same time.

Vertical conflict: the two AGVs with vertical driving direction are conflicted because they compete for the same node.

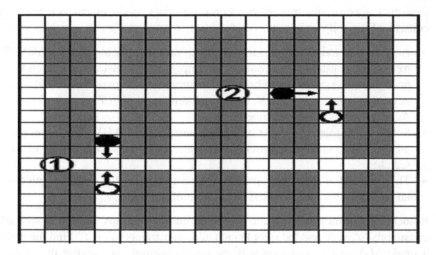

Fig. 4. Conflict types.

Conflict Resolution. In the first stage, the improved A* algorithm is used to calculate the optimal path to complete each task, and the corresponding time window of the path node is generated. By detecting the reserved time window of the grid node, we can determine whether the conflict has occurred and what type of conflict has occurred. There are two common conflict solutions. One is the waiting strategy, which allows the lower priority AGV to wait while the higher priority AGV passes through the inter- section in the face of vertical conflict; Another strategy is to re-plan the path. When the opposite conflict occurs, the current conflict node is set to the unavailable node, the previous path planning algorithm is re-invoked to regenerate the new path, and the conflict detection is carried out again until the new path does not conflict with the existing path. These two solutions only change the path of the current task, so it often lead to poor AGV path planning at this time, and it will cause multiple waiting. The start-stop action caused by waiting consumes too much time and energy, and the excessive number of re-planning paths also reduces the system efficiency.

In view of this shortcoming, the cost of waiting time is set to 3 s. In the face of conflict, this paper adopts the idea based on coupling, that is, when conflict occurs, CBS(Conflict-Based Search) algorithm is used to intercept part of the conflict section. This algorithm considers the cost of car routing in the conflict section as a whole. Then the overall optimal or sub-optimal driving path is redesigned. The number of waiting times for the planned path will be reduced, so the time cost of completing the order task will be effectively reduced.

CBS algorithm steps:

(a) Looking for the conflict node, the conflict is a tuple (a^i, a^j, v, t), which represents AGV^i and AGV^j jointly occupying the node v at time of t. Selecting the conflicting node and its front (back) nodes as a small path fragment, about this small path, its first node acts as the root node and the last node acts as the target node.

(b) Construct a constraint tree at the root node. Each node in the constraint tree consists of the following parts:

- A set of constraints (N. constraints). Each constraint belongs to an AGV. The root of the constraint tree contains an empty set of constraints. The child of a node in the constraint tree inherit the constraints of the parent node and add one new constraint for one AGV. The constraint is a tuple (a^i, v, t), indicates that AGV^i occupies node v at time t;

- Solution (N. solution). A set of k paths, one for each AGV. The target path of the AGV is to satisfy the current constraints, Such paths are found by low-level search;

- The total cost of the current solution (N. cost) (summation over the path cost of all AGVs). This cost is called the f value of the node N;

(c) Branch the constraint. Divide the root node into two groups, the left branch adds constraints (a^i, v, t), and the right branch adds constraints (a^j, v, t). Find the path of a single AGV that is consistent with the new constraint, then update the total cost of the current solution and the retention time window for each node;

(d) Determine whether the current solution is valid, that is to say, determine whether there is a conflict. If there is still has a conflict, then perform the constraint branch again; If the conflict disappears, that is to say, find the optimal solution.

3.3 Algorithmic Flow of Multi-AGV Conflict Resolution

See Fig. 5.

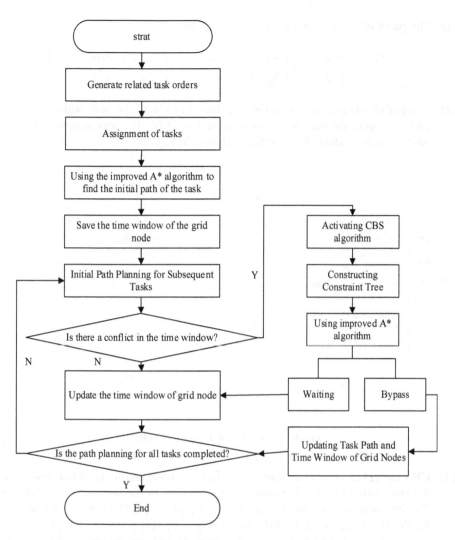

Fig. 5. The algorithm flow of two-stage path planning.

3.4   Example Illustration

Twenty picking order tasks are randomly generated and each task is assigned an idle AGV. We chose two of these conflicting tasks. They are q_1 and q_2.

$$q_1 = (a^1, 1, [19, 2], [3, 7], 0, 23, r_1^1) \text{ and } q_2 = (a^2, 2, [17, 2], [4, 7], 1, 21, r_2^2).$$

(1) The paths of tasks q_1 and q_2 are shown below.

$$r_1^1 = \{[19, 2], \ldots, [6, 4], \ [5, 4], \ [4, 4], \ [3, 4], \ [3, 5], \ [3, 6], \ [3, 7]\};$$
$$r_2^2 = \{[17, 2], \ldots, [5, 3], \ [4, 3], \ [4, 4], \ [4, 5], \ [4, 6], \ [4, 7]\};$$

(2) Conflict checking is performed on the time window of the path node. The overlap of the time windows of a^1 and a^2 at [4, 4] nodes indicates that two AGVs are in conflict. The conflict is shown in Fig. 6:

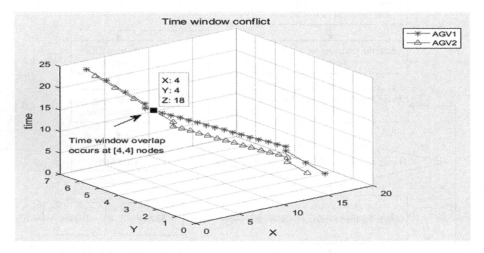

Fig. 6. Time window of conflict path.

(3) CBS algorithm to solve conflicts. The collision node and its front two nodes and the latter two nodes are extracted. The total length of the intercepted section is 5. The fragment paths of the two cars are {[4, 6], [4, 5], [4, 4], [3, 4], [3, 5]} and {[3, 5], [3, 4], [4, 4], [4, 5], [4, 6]} respectively. The first node and the last node of each section are regarded as the starting node and the goal node. The specific solutions are as follows:

① At the high level, the constraints tree is constructed as shown in Fig. 7. ① represents the root node, which contains an empty set. The time cost for two AGVs to complete the current path is 10, and the conflict nodes are [4, 4].

② The root node divides two leaf nodes ② and ③. Adds constraints (1, [4, 4], 2) and (2, [4, 4], 2) to each node respectively. (1, [4, 4], 2) means that AGV1 is prohibited from moving to the node of [4, 4] in the second step. (2, [4, 4], 2) means thatAGV2 is prohibited from moving to the node of [4, 4] in the second step. Then the improved A* algorithm is used to find the optimal path solution of each car and calculate the time cost it consumes under the current constraints.

③ Detecting whether the time window of the new path is conflicting. Finally, it is found that the path scheme of node ② can be used as a solution to the conflict, so stop the branch of node ②. However, it is found that the sub-node ③ has new conflict nodes, so we add constraint branches to the sub-tree node ③ and expand the sub-node ④ and sub-node ⑤ again. Sub-node ④ adds constraints (1, [4, 5], 1) on the basis of constraints (2, [4, 4], 2), and Sub-node ⑤ adds constraints (2, [4, 5], 1) on the basis of constraints (2, [4, 4], 2). Then, the A* algorithm is invoked to find the current path solution and the time cost.

④ The paths of ②, ④ and ⑤ can be used as appropriate solutions, because there is no conflict in their paths. Finally, the path of Leaf Node ② with the lowest cost is chosen as the optimal conflict solution. The corresponding optimal path is shown in Fig. 8. In order to confirm the validity of this method, the same conflict is solved by waiting strategy, and the final path scheme is shown in Fig. 9. By comparison, it is found that the route chosen in this paper reduces the waiting times of AGV due to the bypass strategy, and reduces the total path time cost by two time steps compared with the waiting strategy. This method reduces the energy consumption and time consumption of AGV in the waiting process, and effectively improves the efficiency of multi-AGV system in path planning.

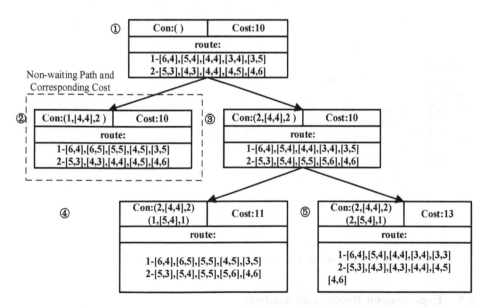

Fig. 7. Process display of conflict resolution using CBS algorithms.

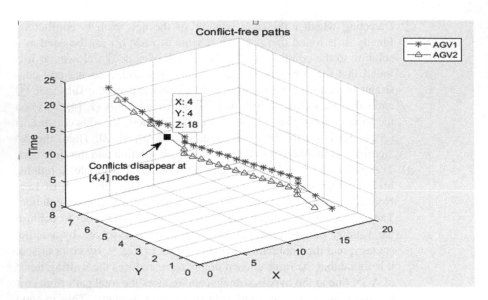

Fig. 8. Resolves conflict using CBS algorithm.

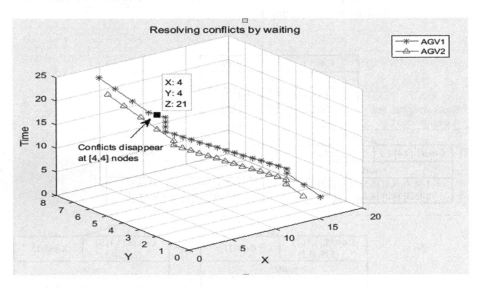

Fig. 9. Resolves conflict by waiting strategy.

3.5   Experimental Results and Analysis

In this paper, matlab 2016a is selected as the simulation tool. The running map is
25*25 grid. Random generation of 100, 200 task orders, Use 20 AGVs to complete
order picking task. Finally, the algorithms are analyzed and compared, which are the
traditional A* algorithm, the improved A* algorithm and the improved A*-CBS

algorithm. The first two algorithms rely on waiting or re-planning path to resolve conflicts. The latter one uses the coupling method to solve conflicts. The specific operation results are shown in Figs. 10 and 11.

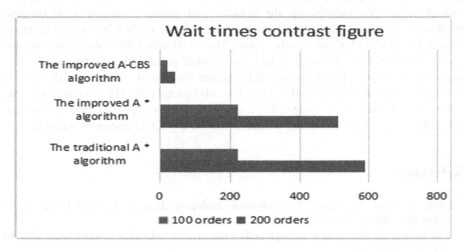

Fig. 10. Wait times contrast figure.

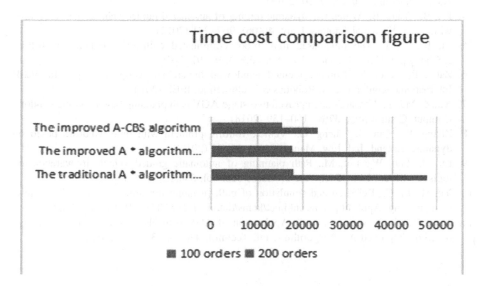

Fig. 11. Time cost comparison figure.

The running results show that under different order tasks, the A-CBS algorithm used in this paper, compared with traditional A* and improved A* algorithm, effectively reduces the number of waiting times and the total cost of time consuming. And the path is the best choice of conflict-free path for each AGV.

4 Conclusion

In order to solve the problem of multi-AGV optimal path finding and scheduling efficiency, an improved A* algorithm with directional search is proposed. In the process of multi-AGV path finding, the dynamic path planning method based on time window and local coupling is fused, and the path planning method based on coupling is used to find multi-AGV conflict-free paths. The simulation results show that the proposed algorithm can effectively reduce the number of searching nodes and waiting times in the process of path finding and improve the overall efficiency of the system under the condition of ensuring the optimal or sub-optimal path. However, the real-time performance of the proposed algorithm needs to be further improved. How to improve the real-time performance of path planning will be my future research direction.

References

1. Liu, B., Chen, X., Cheng, Z.: A dynamic multipath planning algorithm based on A* algorithm. Softw. Algorithms **35**(4), 17–19 (2016)
2. Li, W., Su, X.: AGV path planning based on improved A* algorithm. Mod. Manuf. Eng. **10** (1), 33–36 (2015)
3. Chen, H., Li, Y., Luo, J.: Path planning for mobile robots based on improved A* algorithm. Autom. Instrum. **230**(12), 7–10 (2018)
4. Hu, B., Wang, B., Wang, C.: Dynamic routing of automated guided vehicles based on time window. J. Shanghai Jiaotong Univ. **46**(06), 967–971 (2012)
5. Li, B., Yi, J.: The time-based multi-robot coordinated collision avoidance algorithm. J. Chongqing Univ. Technol. (Nat. Sci.) **33**(3), 91–97 (2019)
6. Zahy, B., Ariel, F.: Conflict-oriented windowed hierarchical cooperative A*. In: IEEE International Conference on Robotics & Automation. IEEE (2014)
7. Xu, Z., Ma, Y.: Research on improved two-stage AGV path planning based on time window. Comput. Control Syst. **37**(6), 150–159 (2018)
8. Zhang, W., Qin, S., Cheng, G.: Vehicle routing problem with time windows based on dynamic demand. Ind. Eng. Manag. **21**(6), 68–74 (2016)
9. Liu, J., Sun, W., Liu, M.: Path planning of automatic guided vehicle in warehousing logistics. Mod. Mach. Tool Autom. Mach. Technol. **538**(12), 155–159 (2018)
10. Yu, H., Li, C.: Research and simulation of path planning for storage multi-AGV system. Comput. Eng. Appl. http://kns.cnki.net/kcms/detail/11.2127.TP.20190705.1737.036.html
11. Xing, P., Li, Q., Wei, W., et al.: Application of AGV path planning in intelligent storage based on improved A~* algorithms. Inf. Technol. **43**(05), 138–141 (2019)

The Recognition of Adult Insects of *Helicoverpa armigera* and *Helicoverpa assulta* Based on SAA-ABC-SVM Technology

Hongtao Zhang[1(✉)], Yang Zhu[1], Lian Tan[1], and Jianan Liu[2]

[1] School of Electric Power,
North China University of Water Resources and Electric Power,
Zhengzhou 450011, Henan, China
zht1977@ncwu.edu.cn
[2] Xidian University, Xian 710071, Shaanxi, China
jianan_liu2017@163.com

Abstract. *Helicoverpa armigera* (*H.armigera*) and *Helicoverpa assulta* (*H.assulta*) are the world-wide insects which mainly do harm to crops like cotton and tobacco, etc. The accurate gender identification of the insects is of great significance for the prediction of regional ratio and population quantity. The color images of the male and female adults of the two pieces of insects were acquired by CCD equipment, respectively. The image segmentation and the morphological methods were applied to remove tentacles and feet of insects. Thirty-six digital features of the insects were extracted, such as color, texture and invariant moment. The simulated annealing algorithm (SAA) extracted the partial features to compose of the optimal feature space by the fitness function. The 15 features were determined and the max fitness was 83.87%. The artificial bee colony (ABC) algorithm was used to optimize the penalty factor c and the kernel function parameter g of support vector machine (SVM). The recognition accuracy of the classification model reached 95.83% when $c = 7.3454$, $g = 0.4436$, which indicates that the gender identification of the two pieces of insects is feasible based on SAA-ABC-SVM technology.

Keywords: *Helicoverpa armigera* · *Helicoverpa assulta* · Simulated annealing algorithm · Image segmentation · Feature optimization

1 Introduction

Helicoverpa armigera (*H.armigera*) and *Helicoverpa assulta* (*H.assulta*) are two pieces of insects of world-wide, which endanger more than 60 families and 300

Supported by the National Natural Science Foundation of China (Grant No. 31671580), the Key Technologies R&D Program of Henan Province, China (Grant No. 162102110112), and the Backdrop of Young Teachers Program, Universities of Henan Province, China (Grant No. 2011GGJS-094).

© Springer Nature Singapore Pte Ltd. 2020
L. Pan et al. (Eds.): BIC-TA 2019, CCIS 1160, pp. 431–442, 2020.
https://doi.org/10.1007/978-981-15-3415-7_36

species of plants, and caused great losses to agriculture and ecology in the world. With the resistance increasing of the two pieces of insects, some genetically modified crops have no obvious effect of control, which makes the method of genetically modified crops face a severe test. The method of gender identification of *H.armigera* and *H.assulta* has been a hot research in many countries, which has a great significance for plant protection [1, 2].

At present, the mainly methods used to distinguish insects gender are artificial identification method [3–8], biological information discrimination method [9–12], near infrared spectroscopy analysis method [13] and image recognition method [14–16].

Chen [3], Jin [4], Lin [5], Zhao [6], etc. distinguished the adult insects gender of *Gully, Persimmon leaf hopper, Spodoptera litura* and *Spodoptera exigua*, through analyzing the external characteristics by naked eyes, such as insects color, texture and reproductive organ. Jiang [7], Zhang [8], etc. described the reproductive system structure and made effective recognition of male and female adult insects of *Barley insects* and *Tea mites*, after anatomy and microscopic observation. The identification methods mentioned above are heavy workload, inefficient and require too much professional knowledge to the staff, which will inevitably lead to human error, and may eventually provide inaccurate or even wrong information for prevention and control measures.

Morrow [9], etc. found that the gender of *Drosophila* in embryonic stage determines the specific expression of gene chromosomes, and made a correct identification on their gender. Zhang and Coates [10], etc. distinguished the gender of *Asian corn borer* effectively by analyzing the transcriptome sequence, and found that different genes of male and female had their specific expression. Yi [11], etc. distinguished the gender of *Jiding beetles* successfully by studying the characteristics, abundance and distribution of insect antennae. Biedler [12], etc. determined the gender of *Mosquito* successfully by analyzing the DSX and FRU protein transcriptomes of different insects. The methods above can classify insects accurately, but take a lot of time and manpower, and cannot be widely used for gender identification of other types of insects.

Dai [13], etc. extracted the spectral characteristics of *Silkworm chrysalis* by near-infrared spectroscopy, and established the classification models of partial least squares discrimination analysis (PLSDA), back propagation neural network (BPNN) and support vector machine (SVM), and the recognition rate reached 95%. Hafiz [14], etc. extracted insects features of color and histogram, and classified four kinds of arthropods by bayesian neural network. Pan [15,16], etc. analyzed the external morphological characteristics of insects by image processing technology, and found that there were significant differences in eccentricity and compactness between male and female insects, and the recognition rate reached 85%. These methods above have some shortcomings such as difficulty in modeling and low recognition rate, which need further improvement.

As far as the author knows, the automatic identification of individuals gender of *H.armigera* and *H.assulta* based on computer vision has not been reported yet. In this paper, we proposed an image recognition method combining the simulated annealing algorithm (SAA), artificial bee colony algorirhm (ABC) and support vector machine (SVM), to identify the gender of the two pieces of insects adult, and it would provide reference for the prediction of regional sex ratio, population quantity and industrial application.

2   Insects Image Acquisition and Process

Male and female *H.armigera* and *H.assulta* adult were selected as the research object, which had just hatched and no mating. Both of four pieces of insects were taken from the Plant Protection Department of Henan Agricultural University. The feeding conditions were temperature $24\,°C \pm 2\,°C$, indoor humidity reached 75%, photoperiod 14L: 10D.

Placed simply on the table with white backplane, used the industrial camera (BFLY-PGE-50S5C-C, 25 mm industrial lens, produced by FLIR-SYSTEMS-INC, American) to collect the RGB digital image of insects, as shown in Fig. 1.

(a) Male *H.armigera* (b) Female *H.armigera*   (c) Male *H.assulta*   (d) Female *H.assulta*

Fig. 1. The RGB images of male and female *H.armigera* and *H.assulta* adult.

The otsu method was used to extract the binary image of insects, the image segmentation and the morphological methods were applied to remove the tentacles and feet of insects. The gray images of channel B were determined for subsequent feature extraction, which has strong contrast and clear texture, as shown in Fig. 2.

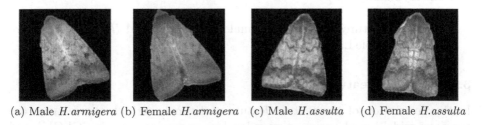

(a) Male *H.armigera* (b) Female *H.armigera*   (c) Male *H.assulta*   (d) Female *H.assulta*

Fig. 2. Gray images of male and female *H.armigera* and *H.assulta* of channel B.

3   Feature Extraction of Insects

In traditional gender identification methods of insects, the professional distinguished the gender of insects by the color, size, shape of external genitalia and experience, and it has no sufficient scientific basis. The difference of the male and female insects is no the size and shape but the color and texture, so we selected the features of color, texture and morphological of the four pieces of insects.

3.1   Color Features

(1) First-order moment(mean)

$$\mu_i = \frac{1}{N} \sum_{j=1}^{n} P_{ij} \tag{1}$$

where P_{ij} is the i-th color component of the j-th pixel of the color image, and N is the number of pixels in the image. This feature primarily characterizes the average intensity of each color component.

(2) Second-order moment(variance)

$$\delta_i = \sqrt{\frac{1}{N} \sum_{j=1}^{n} (P_{ij} - \mu_i)^2} \tag{2}$$

which mainly characterizes the color variance of the area to be tested, that is, the unevenness.

(3) Third-order moment(skewness)

$$\gamma_i = \sqrt[3]{\frac{1}{N} \sum_{j=1}^{n} (P_{ij} - \mu_i)^3} \tag{3}$$

which mainly characterizes the skewness of the color component, that is, the asymmetry.

The color features of male and female *H. armigera* and *H. assulta* adult are shown in Table 1.

3.2   Textural Features

In the description of the texture feature statistical method, the features are the most widely used based on gray level co-occurrence matrix (GLCM) and difference statistical matrix (GLDS) are the most widely used. The textural features of four pieces of insects gray images are shown in Tables 2 and 3. The features based on GLCM and GLDS are defined as follows:

(1) Average value:

$$mean = \frac{1}{G^2} \sum_{i=0}^{G-1} \sum_{j=0}^{G-1} P_{ij} \tag{4}$$

which describes the uniformity distribution of the overall gray value of the image. If the average value is low, it means that there are many dark areas in the image, and vice versa. G is the gray level number of image.

(2) Contrast:

$$con = \sum_{n-1}^{G} n^2 \sum_{i=0}^{G-1} \sum_{j=0}^{G-1} P_{ij} \tag{5}$$

which describes the depth and clarity of the overall texture of the image.

(3) Entropy:

$$ent = - \sum_{i=0}^{G-1} \sum_{j=0}^{G-1} P_{ij} \, log[P_{ij}] \tag{6}$$

which is a measure of the amount of information in an image and the complexity of the texture in the image.

(4) Second-order moment in the angular direction:

$$asm = \sum_{i=0}^{G-1} \sum_{j=0}^{G-1} P(i-j) \tag{7}$$

which is the sum of the squares of the gray-scale differential frequencies, and represents the average level of the grayscale distribution of the image.

(5) Correlation:

$$corr = \frac{1}{\sigma_x \sigma_y} [\sum_{i=0}^{G-1} \sum_{j=0}^{G-1} P_{ij} (i - \mu_x)(j - \mu_y)] \tag{8}$$

$$\mu_x = \sum_{i=0}^{G-1} i \sum_{j=0}^{G-1} P_{ij} \tag{9}$$

$$\mu_y = \sum_{i=0}^{G-1} j \sum_{j=0}^{G-1} P_{ij} \tag{10}$$

$$\sigma_x^2 = \sum_{i=0}^{G-1} (i - \mu_x)^2 \sum_{j=0}^{G-1} P_{ij} \tag{11}$$

$$\sigma_y^2 = \sum_{i=0}^{G-1} (j - \mu_y)^2 \sum_{j=0}^{G-1} P_{ij} \tag{12}$$

which is also known as homogeneity, can be used to measure the similarity of the gray level of an image in the row or column direction. The magnitude of the value reflects the local gray correlation. The larger the value, the greater the correlation.

3.3   Seven Invariants of Insects

The geometric moments were proposed in 1961 by Hu, and is used to describe the geometric features of images, which remain unchanged after image mirroring, scaling, translation and rotation, etc. The seven invariant moment features of gray images above are shown in Table 4.

Table 1. The color features of RGB images of four pieces of insects.

Color matrix	H.armigera						H.assulta					
	Male			Female			Male			Female		
	R	G	B	R	G	B	R	G	B	R	G	B
First moments	222.253	215.651	207.373	203.440	189.348	178.303	225.613	222.559	217.146	234.829	226.751	219.956
Second moments	13.289	13.201	13.185	11.786	11.740	11.750	13.589	13.642	13.673	14.017	14.017	14.029
Third moments	5.623	5.592	5.584	5.191	5.169	5.171	5.704	5.715	5.722	5.823	5.819	5.819

Table 2. The GLCM features of gray images of four pieces of insects.

GLCM	H.armigera								H.assulta							
	Male				Female				Male				Female			
	0°	45°	60°	90°	0°	45°	60°	90°	0°	45°	60°	90°	0°	45°	60°	90°
Contrast	0.243	0.413	0.225	0.396	0.251	0.425	0.235	0.406	0.231	0.399	0.216	0.400	0.216	0.377	0.202	0.384
Correlation	0.988	0.980	0.989	0.981	0.986	0.977	0.987	0.978	0.989	0.981	0.989	0.980	0.990	0.982	0.991	0.982
Energy	0.342	0.338	0.342	0.339	0.293	0.288	0.294	0.290	0.366	0.362	0.366	0.365	0.384	0.381	0.384	0.380
Homogeneity	0.966	0.958	0.971	0.962	0.962	0.963	0.968	0.958	0.973	0.966	0.976	0.965	0.977	0.972	0.981	0.970

3.4   Feature Data Normalization

There are great difference existed in the magnitude of features extracted above, which will affect system performance and calculation accuracy. Therefore, all the original feature data extracted are normalized to reduce the calculation time and improve the efficiency and accuracy.

Table 3. The GLDS features of gray images of four pieces of insects.

GLDS	*H.armigera*		*H.assulta*	
	Male	Female	Male	Female
mean	0.021	0.028	0.020	0.018
contrast	496.705	279.780	506.425	489.246
asm	0.607	0.310	0.656	0.692
ent	1.885	3.137	1.678	1.504

Table 4. The seven invariants features of four pieces of insects.

Seven invariants	*H.armigera*		*H.assulta*	
	Male	Female	Male	Female
φ_1	6.64	6.50	6.66	6.69
φ_2	13.76	13.48	13.8	13.84
φ_3	26.77	25.36	28.09	27.12
φ_4	26.31	24.87	27.78	26.80
φ_5	25.22	23.78	26.73	25.71
φ_6	33.21	31.62	34.75	33.74
φ_7	53.70	51.49	56.48	55.46

4   Feature Optimization and Recognition Based on SAA-ABC-SVM

4.1   Feature Optimization Based on SAA

Simulated annealing algorithm (SAA) is a stochastic optimization method based on Monte-Carlo iterative solution strategy for solving large-scale combinatorial optimization problems, which has the advantages of high efficiency, wide range and flexibility. The method is applied to optimize the original features of the four pieces of insects in order to improve operation speed and recognition accuracy. Detailed steps are as follows:

Step1. Select a higher initial temperature $T_0(T_0 > 0)$;
Step2. Select a set of features randomly as the current input state and give its neighborhood $N(x)$;
Step3. Give the temperature drop method $mathit{T}_{i+1} = \alpha T_i$, and keep the temperature drops slowly;
Step4. Set $i = 0$, $k = 0$;
Step5. Choose a state $\acute{x}$ in the neighborhood $mathit{N}(x(k))$ of x_k in a certain way and calculate its fitness. Determine whether to accept the current value $\acute{x}$ or not, the formula as follows:

$$P(x(k)\acute{x}) = \begin{cases} 1, & \text{if } f(\acute{x}) < f(x(k)) \\ exp\frac{1}{T_i}(f(\acute{x}) - f(x(k))), & \text{else} \end{cases} \tag{13}$$

Step6. Check whether the annealing process is over or not. If the termination condition is satisfied, then the feature combination is the optimal solution, otherwise it will continue step 5.

Set the initial temperature $T_0 = 150$, the drop rate $\alpha = 0.9$, the end temperature is 1, and the input is the 36-dimensional original feature data of four pieces of insects adult. The fitness is a separability criterion based on the distance inter-class and defined as follows:

$$K = tr(S_w^{-1} S_b) \tag{14}$$

$$S_b = \sum_{i=1}^{c} P_i(\mu_i - \mu)(\mu_i - \mu)^T \tag{15}$$

$$S_w = \sum_{i=1}^{c} P_i E_i(x - \mu_i)(x - \mu_i)^T \tag{16}$$

where S_b is the inter-class dispersion matrix, S_w is the intra-class dispersion matrix, $\mu_i = E_i[x]$, $\mu = E[x]$, μ_i is the class mean, μ is the global mean, P_i is the priori probability of class i, c is the class number, there $c = 4$.

The method above are used to optimize the 36 features normalized, including: color moment (9 dimensions), texture features based on GLCM and GLDS (20 dimensions), invariant distance (7 dimensions). The model of SAA optimized the original feature space by the fitness function, and the feature combination indicated by 1, 0 digits, where "1" indicates that the feature is selected, and "0" indicates that the feature is not selected.

The dimension of the original feature space is reduced to 14–20 after 10 times of SA algorithm optimization. And the fitness and classification accuracy of the test set are shown in Table 5.

Table 5. Feature optimization.

Test count	Feature selected	Binary representation of selected feature	Fitness	Test accuracy(%)
1	14	100110110001000000001101010101101000	81.75	90
2	14	100111100110100100000101000100100100	72.47	87.5
3	15	100110101100000101110101000100100110	81.73	92.5
4	15	101001010101000110001101010110010000	**83.87**	**95.83**
5	16	100101010001001001100101100011101010	82.41	91.17
6	16	000101010110011010110111110000010000	81.69	93.33
7	17	111010110000011100000111000111000101	77.23	90.83
8	18	100110101101010101101000010000111110	78.07	83.3
9	19	110101000011111000010111000101101011	69.83	80.83
10	20	110010011111010100110111000101111000	73.37	85

Table 5 shows that the fitness in [69.83, 83.87] after 10 times optimized by SA algorithm, and accuracy rate is in [80.83, 95.83]. When the fitness is 83.87 and the fifteen features are selected, the accuracy rate reached 95.83%, and the features optimized are the first-order moment of R channel, the second-order moment of B channel, the third-order moment of R channel, and the contrast of GLCM at 60°, etc.

4.2 Theoretical Basis of ABC-SVM

Support vector machine (SVM), which was proposed by Cortes and Vapnik in 1995, is a kind of generalized linear classifier that classifies data according to supervised learning method. Its basic idea is to find the best classification hyperplane in feature space, such that the distance inter-class of samples reaches the maximum [21, 22]. There are some kernel functions in support vector machine, such as linear kernel function, polynomial kernel function, radial basis function, sigmoid kernel function, etc. The radial basis function can maps the samples into a higher dimensional space, which has fewer parameters and good performance among them. Therefore, the radial basis function was selected for classification. In 2005, Karaboga group proposed artificial bee colony (ABC) algorithm, which is mainly used to solve optimization problems [19, 20], and has better optimization performance than the traditional method. It can jumps out of local extremum with a certain probability, and calculate the global optimal solution by the optimal behavior of each artificial bee. In this experiment, c-SVC classifier and RBF network are used to make identification, and the parameters of support vector machine were optimized by ABC algorithm, and the detailed steps are as follows:

Step1. Initialize the population size of artificial bee colony, the maximum number of updates, and the maximum number of iterations;

Step2. All the artificial bee are employed bees, and produce the same number of solutions;

Step3. Calculate the fitness of solutions above, some employed bee become scout bees, and the number of updates was set to 0;

Step4. The scout bee search for new honey source, calculate the fitness and compare with old honey source, the number of updates was set as 0 if old honey source updates by the greedy algorithm, else the number of updates add 1;

Step5. Keep the honey source if update times of the honey source less than the maximum number of updates, else abandoned and the employed bee become the scout bee to search for a new honey source;

Step6. Record the best honey source to far, and determine termination or not.

In this experiment, the ABC algorithm is used to optimize the main parameters of support vector machine, including penalty factor c and kernel function parameter g. In this algorithm, the initial bee colony size is set as 20, and the times of updates is limited to 50 (if the honey source is not updated more than 50 times, then the honey source is abandoned), the maximum number of iterations is set as 50 times. The number of parameters optimized is 2, the range of parameters is set as 0.01–100, and the algorithm is repeated twice to check the robustness of the program. According to the optimized combination of features above, the identification of four kinds of insect samples reached best when $c = 7.3454$, $g = 0.4436$. It improves the classification performance of SVM.

4.3   Identification of Four Kinds of Insects by ABC-SVM

In this experiment, 400 images of four kinds of insects were taken, in which 280 images were divided into the training set and the left were taken as the testing set. For the four kinds of insects, the male and female adults of $H.armigera$ were labeled "1 and -1" and the female and male adults of $H.assulta$ were labeled "2 and -2" respectively. The 15 features optimized were used as input of support vector machine (the size of input data is $400 * 15$) to achieve the best classification accuracy. The result of identification was shown in Fig. 3.

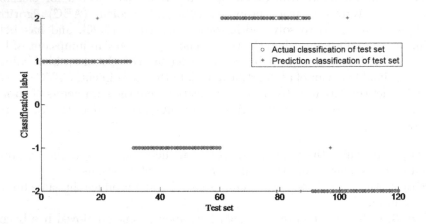

Fig. 3. The classification results of male and female $H.armigera$ and $H.assulta$ adult.

There are 120 samples of four kinds of insects in the testing set, in which 115 samples are correctly identified, the recognition rate is 95.83%. The reason of the mistaken may be that the texture of the wing surface was damaged during the growth of the insect, or the texture of the surface of the insect began to fall with the time.

In this experiment, we also compared the accuracy rate and run-time of different methods that optimized the features and SVM and not, and the result were shown in Table 6.

As shown in Table 6, when the feature is not optimized, and the classifier parameters were optimized by 5-fold cross validation and ABC algorithm (the maximum number of iterations is set as 50 times) respectively, the run time is

Table 6. Results of various optimization methods.

Methods	SVM optimization	Run time(s)	Accuracy rate(%)
Original features-SVM	Cross-validation(5-fold) [23]	310.322	87.5
	ABC	20.571(50 Iteration)	92.5
SAA-ABC-SVM	ABC	17.808	95.83

310.322 s and 20.571 s, and the recognition rate is 87.5% and 92.5%. The run time and recognition rate of SAA-ABC-SVM are 17.808 s and 95.83% respectively. The result showed that it greatly improved the performance of the classifier and proved the feasibility of this method.

5   Conclusions

In this paper, the recognition method based on computer vision for male and female *H.armigera* and *H.assulta* adult was proposed. The SA algorithm was used to optimize the original feature space by fitness function. The fifteen features were selected when the highest fitness was 83.87, and was used as the input of the classifier for subsequent classification and recognition. The two important parameters c and g of SVM was optimized by ABC algorithm. When $c = 7.3454$, $g = 0.4436$, the recognition rate reached 95.83%. And it improved the accuracy rate significantly. The results showed that the method could not only effectively reduce the classification time, but also improve the classification accuracy.

References

1. Li, Q., Chen, Q.J., Meng, R.G., et al.: Study on insecticidal activity of Cry2AhM gene. J. Agric. Sci. Technol. **19**(4), 10–16 (2017)
2. Luo, J.Y., Zhang, S., Zhu, X.Z., et al.: Ecological fitness of transgenic cotton with GAFP gene and effection on insect community in cotton field. J. Appl. Ecol. **27**(11), 3675–3681 (2016)
3. Chen, C., Zhang, S., Zhu, X.Z., et al.: Rapid differentiation between male and female adults of *Eucryptorrhynchus chinensis*. J. Zj. A&F. Univ. **30**(02), 309–312 (2013)
4. Jin, X.F., Liu, Y., Li, L.L., et al.: Damage of persimmon leafhopper and identification of male and female adults. J. Hlj. Agric. Sci. **42**(03), 177–178 (2015)
5. Lin, W.P., Peng, L., Xiao, T.Y., et al.: A simple method for identifying sexuality of *Spodoptera litura* (Fabricius) pupae and adults. J. Environ. Entomol. **37**(03), 685–687 (2015)
6. Zhao, X.F., Yang, A.D., Zhang, M.X.: A method for the rapid sex-determination of *Spodoptera exigua* (Lepidoptera: Noctuidae) pupae and adults. J. Environ. Entomol. **38**(05), 1066–1070 (2016)
7. Jiang, Y., Zhang, Y.N., Ma, L., et al.: Identification of alive and male adults of *Zophobas morio* (Coleoptera: Tenebrionidae). Sci. Sil. Sin. **48**(06), 175–177 (2012)

8. Zhang, J.X., Wu, Q., Sun, Q.Y., et al.: Anatomical observation on the structure of the male and female reproductive system of tea geometrid (Ectropis oblique) adults. Chin. Sci. Tech. Ass. **16**, 1–4 (2014)

9. Morrow, J.L., Riegler, M., Frommer, M., et al.: Expression patterns of sex-determination genes in single male and female embryos of two *Bactrocera* fruit fly species during early development. Ins. Mol. Biol. **23**(6), 754–767 (2014)

10. Zhang, T., Coates, B.S., Ge, X., et al.: Male and female biased gene expression of olfactory-related genes in the antennae of Asian Corn Borer, *Ostrinia furnacalis* (Guenee) (Lepidoptera: Crambidae). Plos One **10**(6), 1–22 (2015)

11. Yi, Z., Liu, D., Cui, X., et al.: Morphology and ultrastructure of antennal sensilla in male and female *Agrilus mali* (Coleoptera: Buprestidae). J. Insect. Sci. **16**(1), 87–96 (2016)

12. Biedler, J.K., Tu, Z.: Two-sex determination in mosquitoes. Adv. Insect Physiol. **51**, 37–66 (2016)

13. Dai, F., Che, X.X., Peng, S.R., et al.: Fast and nondestructive gender detection of *Bombyx mori* chrysalisin the cocoon based on near infrared transmission spectroscopy. J. South. China Agric. Univ. **33**(02), 103–109 (2018)

14. Hafiz, G.A.U., Qaisar, A., Fatima, G.: Insect classification using image processing and Bayesian network. J. Entomol. Zool. **5**(6), 1079–1082 (2017)

15. Pan, P.L., Zhang, F.M., Yin, J., et al.: Preliminary studies on image recognition technology for female and male adults of *Corythucha marmorata* (Uhler) (Hemipter: Tingidae). Plant Prot. **43**(03), 70–75 (2017)

16. Pan, P.L., Liu, H.M., Zhang, F.M., et al.: Extraction and analysis of external morphological characteristics from four species of lace bugs (Hemiptera: Tingidae). Sci. J. Zool. **36**(05), 531–539 (2017)

17. Zhang, H.T., Mao, H.P., Qiu, D.Y.: Feature extraction in image recognition of stored grain insects. Tran. Chin. Soc. Agric. Eng. **25**(02), 126–130 (2009)

18. Hu, Y.X., Zhang, H.T.: Recognition of the stored-grain insects based on simulated annealing algorithm and support vector machine. Chin. Soc. Agric. Mach. **39**(09), 108–111 (2008)

19. Ebrahimi, M.A., Khoshtaghaza, M.H., et al.: Vision-based insect detection based on SVM classification method. Comput. Electron. Agric. **35**(137), 52–58 (2017)

20. Wu, J., Yang, H.: Linear regression-based efficient SVM learning for large-scale classification. IEEE Trans. Neural Netw. Learn. Syst. **26**(10), 2357–2369 (2017)

21. Zidi, S., Moulahi, T., Alaya, B.: Fault detection in wireless sensor networks through SVM classifier. IEEE Sens. J. **18**(1), 340–347 (2018)

22. Sukawattanavijit, C., Chen, J., Zhang, H.: GA-SVM algorithm for improving land-cover classification using SAR and optical remote sensing data. IEEE Geosci. Rem. Sens. Lett. **14**(3), 284–288 (2017)

23. Zhang, H.T., Liu, J.N., Tan, L., et al.: Study on automatic discrimination of male and female imagoes of *Helicoverpa armigera* (Hübner) based on computer vision. J. Environ. Entomol. **41**(4), 612–619 (2019)

Prediction of Photovoltaic Power Generation Based on POS-BP Neural Network

Yanbin Li[(⊠)], Yaning Wan, Junming Xiao, and Yongsheng Zhu

Zhongyuan University of Technology, Zhengzhou 451191, China
liyanbin@zut.edu.cn

Abstract. With the increasing proportion of photovoltaic power in power systems, the problem of its fluctuation and intermittency has become more prominent. To deal with this issue, the accurate and reliable short term photovoltaic power forecasting becomes very important to reduce the operation costs and potential risks in power system. In order to realize the prediction of photovoltaic power generation, a forward neural network photovoltaic system power generation prediction model optimized by particle swarm optimization was established. This model used the particle swarm optimization algorithm to optimize the internal weight and threshold of the neural network, which not only has a fast convergence speed but also has a strong generalization ability and does not easily fall into the local extreme value. The model took environmental information and photovoltaic power generation historical data as samples, and compared the predicted data with the measured data. The results showed that the model has good prediction accuracy and good prediction performance.

Keywords: Photovoltaic system · Power generation forecasting · BP neural network · Particle swarm optimization

1 Introduction

With the rapid increase in the world's energy consumption, photovoltaic power generation, as a new energy, is the clean energy with the widest application prospect. It will gradually become the main form of the world's energy generation. In photovoltaic power generation, a large number of photovoltaic systems are connected to the power grid, which has its own volatility and intermittency problems that challenge the dispatching work of the power sector and affect the stability of other power grids connected to it [1, 2]. Therefore, it is necessary to establish a photovoltaic power generation prediction system to reduce the impact of power fluctuations of photovoltaic systems on the power grid [3]. Moreover, better prediction of photovoltaic power also helps to enhance system security and stability, as well as optimize the operation of the power system [4].

Many scholars at home and abroad have conducted in-depth research on the prediction of photovoltaic power generation. There are two main forecasting methods. One is the traditional linear prediction method, which is only applicable to the prediction of the stationary change curve and is not applicable to the prediction of photovoltaic power generation with variability. The other method is a machine learning method

© Springer Nature Singapore Pte Ltd. 2020
L. Pan et al. (Eds.): BIC-TA 2019, CCIS 1160, pp. 443–453, 2020.
https://doi.org/10.1007/978-981-15-3415-7_37

based on nonlinear theory modeling, such as an artificial neural network or support vector machine, which can well achieve nonlinear fitting but has the disadvantage of the local optimum [5–7]. Han proposed an hourly solar irradiance prediction method with support vector machine (SVM) [8]. An ELM based bootstrap method is proposed by Wan for wind power forecasting, two ELM bootstrap models are used to calculate the variance of ELM model and data noise, respectively. However, the performance of this method depends highly on the quality of measured PV power and meteorological data [9]. To reduce the negative influence of the use of PV power, Wang proposed a short-term Photovoltaic power prediction model based on the online sequential extreme learning machine with forgetting mechanism (FOS-ELM), which can constantly replace outdated data with new data. They use historical weather data and historical Photovoltaic power data to predict the PV power in the next period of time [10]. Two studies combined weather type factors with an original prediction model to optimize ultra-short-term prediction results to a certain extent [11, 12], but the weather data generally reflected the weather type conditions of large regions, which produced deviations when used as local factors for prediction. Zhang proposed the grey prediction model was combined with the exponential smoothing method and Markov chain model, separately and reached ideal prediction accuracy. The prediction accuracy of the combined prediction model for the output power data series mainly depends on the state division, and there is no unified state division method at present; this needs further research [13]. Xiao established the photovoltaic power prediction model of a neural network, and the genetic algorithm (GA) was used to optimize the initial weight and threshold of the established BP neural network model; a good prediction effect was achieved, but the margin of error was relatively large [14]. Due to the good generalization and fault tolerated performance, the artificial neural network (ANN) algorithms have been popularly used for PV power forecasting [15] [16]. Compared to other methods, the back propagation (BP) algorithm method not only the advantage of simplicity and it is easy to implement, but also a lot of successful practice in applications.

In this paper, using the particle swarm optimization (PSO) to optimize the neural network, we established a short-term photovoltaic power generation forecasting model. According to the relevant data obtained by the photovoltaic power generation monitoring system [17], the generated power of the system is predicted and analyzed with the established model.

We chose to employ the BP algorithm because of the following advantages:

(1) The generalization performance of BP is better than many others.
(2) The computation complexion of BP is much lower than many other machine learning algorithms.
(3) The learning speed of BP is much faster than most feed forward network learning algorithms.
(4) The self-adaptive and autonomous learning ability of BP is better than many others.

2   Design of Photovoltaic Power Generation Prediction Model

Photovoltaic power generation output is very dependent on the weather, environment, and other external uncontrollable factors. Sunshine intensity, environmental temperature, solar panel performance, and other factors have an impact on the output power of a photovoltaic system [18]. Under the sunshine intensity, and environment temperature, the output of the system was determined. Photovoltaic cells are very sensitive to surface radiation intensity, and their output characteristics of photovoltaic cells are as follows [19]:

$$P_{pv} = RA\,\eta[1 - 0.005(T + 25)] \tag{1}$$

here, R is the light intensity (kW/m^2), and A is the area of the photovoltaic modules (m^2), η is the photovoltaic power conversion efficiency, and T is the environmental temperature (°C).

2.1   PSO–BP Neural Network Model

A BP neural network is a kind of multi-layer feed forward neural network. The basic idea of this network is the gradient descent method, and its main characteristics are forward signal transmission and error back propagation. In forward transmission, the input signal is processed layer-by-layer from the input layer to the hidden layer until it reaches the output layer. The state of neurons at each level affects only the state of the next generation. If the desired output cannot be obtained from the output layer, it will be transferred to reverse propagation, and the weight and threshold of the network will be adjusted according to the prediction error so that the predicted output of the BP neural network will keep approaching the expected output [20, 21].

The BP neural network algorithm solves the global optimal value of complex nonlinear functions, and the algorithm itself adopts the local search optimization strategy, which makes the algorithm easily fall into the local optimum. The PSO algorithm is a global optimal search algorithm with good performance, with advantages of easy implementation, high precision, and fast convergence.

In essence, the weight optimization of the neural network is a process of objective function optimization. Through continuous survival of the fittest, an optimal connection weight is found. The knowledge of the neural network system is reflected in the overall distribution of network weights and thresholds, while the gradient descent method is sensitive to the selection of initial weights; a small difference of initial weights will result in a huge difference in the results, and therefore a good initial value is very important [22]. The PSO algorithm is introduced to find the optimal initial weight and threshold so that the neural network can achieve the most efficient training and prediction.

In this paper, the initial weight and threshold of the network were optimized with PSO to obtain the optimal weight (threshold), which was assigned to each layer of the prediction network for learning and prediction. The flow chart of the PSO–BP neural network algorithm is shown in Fig. 1.

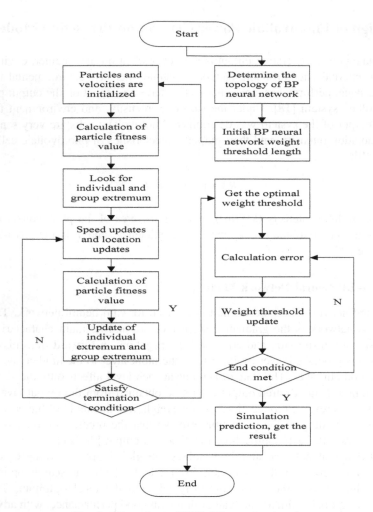

Fig. 1. Flow chart of PSO–BP neural network.

2.2 Neural Network Structure Design

The BP neural network consists of three layers: the input layer, the hidden layer, and the output layer. In order to make up for the deficiency of weather data and improve the accuracy of prediction, the input layer was set as 6 input neurons, which included weather parameters (sunshine intensity and ambient temperature) and historical power output values (the four power output values before the prediction point).

The number of neurons in the hidden layer was closely related to the input, output, and accuracy requirements. Its selection is crucial in the design of a neural network. According to the empirical formula, the number of neurons can be calculated to be 4–13.

The output layer was 1 neuron, namely the power generation of the photovoltaic system [23]. The topology of the BP neural network is shown in Fig. 2.

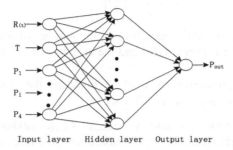

Input layer   Hidden layer   Output layer

Fig. 2. Topology of the BP neural network.

The selection of BP network evolution parameters is shown in Table 1.

Table 1. Evolution parameters of the BP network.

Show	Lr	Mc	Epochs	Goal
50	0.05	0.9	1000	0.0001

2.3   Particle Swarm Optimization Algorithm Design

The particle swarm optimization algorithm is an evolutionary computing technology that originated from research on the predation behavior of birds. It makes use of the information sharing of individuals in the group to generate the evolution process from disorder to order in the problem-solving space of the whole group in order to obtain the optimal solution.

The PSO algorithm initializes a group of random particles (random solutions). Then the optimal solution is found through iteration. In each iteration, the particle updates itself by tracking two "extremum" values. The first is the optimal solution found by the particle itself, which is called the individual extreme value P_{best}. The other extreme value is the optimal solution that the entire population has found so far. This extreme value is the global extreme value G_{best}. We updated the speed and position of each particle group according to the following formulas:

$$V_{id}^{k+1} = \omega V_{id}^k + c_1 r_1 \left(P_{id}^k - X_{id}^k\right) + c_2 r_2 \left(P_{gd}^k - X_{gd}^k\right); \tag{2}$$

$$X_{id}^{k+1} = X_{id}^k + V_{id}^{k+1} \tag{3}$$

Here, $i = 1, 2, ..., N; d = 1, 2, ..., D; C_1$ and C_2 are learning factors; ω is the inertia weight; Meanwhile, particle movement is limited by setting the interval velocity $[V_{min}, V_{max}]$ position boundary and $[X_{min}, X_{max}]$ position boundary.

The fitness of the particle is the mean square deviation

$$fitness = \frac{1}{n}\sum_{i=1}^{n}\sum_{j=1}^{c}\left(Y_{ij} - y_{ij}\right)^2 \tag{4}$$

where n and c are the number of samples and the number of output neurons, respectively; Y_{ij} and y_{ij} are the expected value and the actual value of the jth output of the ith sample, respectively.

Particle swarm optimization is used to optimize the connection weight and threshold. Optimization steps are as follows:

(1) Population size, position boundary $[-X_{max}, X_{min}]$, maximum velocity $[-V_{max}, V_{min}]$, inertia weight ω, maximum iteration times, learning factors C_1 and C_2, and initialization of particle position X_i and velocity V_i.
(2) According to the input and output samples, the fitness function value of each particle is calculated according to formula (4). The smaller the fitness value, the better the particle performance. At the same time, P_{best} and G_{best} are updated and recorded.
(3) Equations (2) and (3) are used to update the velocity and position of particles.
(4) Judge whether the particle's velocity and position are beyond the set range of position and speed:

> if $V_i > V_{max}$, then $V_i = V_{max}$; if $V_i < V_{min}$, then $V_i = V_{min}$;
> if $X_i > X_{max}$, then $X_i = X_{max}$; if $X_i < X_{min}$, then $X_i = X_{min}$.

(5) Recalculate the fitness value of particles.
(6) Check whether the end condition is met. If the current position or the maximum number of iterations reaches the predetermined error, then stop iteration, and output the final weight and threshold of the neural network; otherwise return to step 1 and continue to execute.

The parameters of particle swarm optimization are shown in Table 2.

Table 2. Parameters of the PSO algorithm.

Sizepop	C_1	C_2	Maxgen	V_{max}
20	1.49	1.49	500	5

3   Model Implementation

3.1   Data Sample Preprocessing

Data samples from the Data Castle, which include weather forecast data and the output power of the photovoltaic power station, were collected once every 15 min or 30 min daily from 07:00 to 20:00. In order to improve the prediction accuracy, proper handling of the training sample, and elimination of singularity in the sample data, we obtained a total of 1961 samples for the training data set [24]. The data is from March 10, 2018 to April 15, 2018, for a total of 37 days.

Generally, the nonlinear activation function is used in artificial neural networks, whose output is limited to [0, 1] or [−1, 1]; however, the expected output is often not in this interval. In order to avoid the over-saturation of neurons caused by the direct use of the original data for training, the input and output data of the network must be normalized.

Formula (5) was adopted to process the input data and convert to the interval of [0,1]:

$$P^* = \frac{P - P_{\min}}{p_{\max} - P_{\min}} \tag{5}$$

where P* is the output data after normalization processing; P is the original data; and Pmax and Pmin are the maximum and minimum values of P, respectively.

3.2   Example Simulation and Result Analysis

The BP neural network was trained and predicted by Matlab software. Figure 3 shows the varying fitness values of particles with the increase in iteration times. It can be seen that with the increase in iteration times, the fitness value of particles gradually decreased.

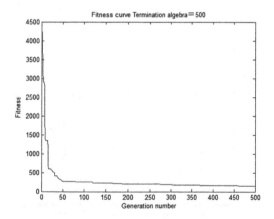

Fig. 3. Particle fitness values in response to iteration times.

After training the BP neural network, 53 data points (every 15 min) from 7:00 to 20:00 were predicted the next day. Figure 4 shows a comparison curve between the measured power and the predicted power of the PSO-BP model and the BP model, and

Fig. 5 shows the prediction error curve of the PSO-BP model and the BP model. The absolute error percentage is shown in formula (6):

$$e(x^*) = \frac{x^* - x}{x} \times 100\% \tag{6}$$

where $e(x^*)$ is the percentage of absolute error; x^* is the predicted value; and x is the actual value.

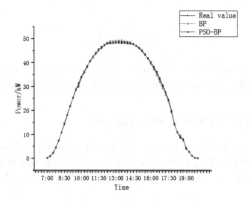

Fig. 4. Comparison between predicted power and measured power.

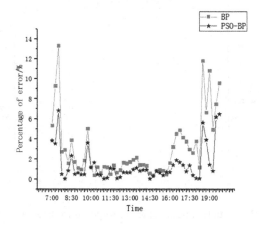

Fig. 5. Comparison between prediction error curves.

In order to compare prediction results of two methods more intuitively, some typical time data are selected are shown in Table 3.

Table 3. The prediction results comparison of two methods on April 16, 2018.

Time	Actual value/kW	Prediction value of BP model/kW	Prediction value of PSO-BP model/kW	Absolute error of BP model/%	Absolute error of PSO-BP model/%
7:30	2.15433	2.442	2.3075	13.3522	6.8318
8:30	14.1977	14.7481	14.5246	3.8769	2.3028
9:45	31.4383	29.8532	30.307	5.0419	3.5984
13:30	48.2727	49.3007	48.8075	2.1297	1.1079
16:30	36.29	34.6527	35.6121	4.5118	1.8681
17:15	27.646	26.6083	27.4346	3.7536	0.7646
18:30	8.62833	9.6506	9.1132	11.8422	5.6196
19:00	4.137	4.5857	4.0778	10.8454	1.4321
19:45	0.23467	0.2572	0.2499	9.5879	6.4762

It can be seen from the simulation results in Fig. 5 that the prediction error of the BP neural network was relatively large, within 0%–14%. Meanwhile, the prediction accuracy of the BP–PSO network was relatively high, and its absolute error percentage was within 0%–8%, that is, the prediction model of photovoltaic power generation was relatively effective. It can be seen from Fig. 5 that the absence of several error points at the beginning and end of a day was caused by the error calculation formula. Since the actual output power at these error points as zero, the error was infinite as long as the predicted value was not equal to the actual value. In addition, due to the failure of the test platform in the sample data, part of the detected data was lost, which affected the training of the BP neural network and also produced some errors. In addition, due to the small installed capacity of the experimental platform, the collected output power value was relatively small, which could cause large errors.

4 Conclusion

In our proposed BP–PSO model, we used the PSO algorithm to optimize the initial weights and threshold. The PSO optimization made up for deficiencies of the BP neural network, which were random determinations of initial values, local optimum, and long training time. The photovoltaic power generation particle swarm neural network prediction model could simulate the output power change trend well, had accurate prediction of photovoltaic power, significantly improved prediction accuracy from 86% to 92%, and had good generalization ability. Our model also improved the BP neural network's training speed, and the output prediction was close to real values.

Acknowledgments. The work was supported by the Foundation of Henan Educational Committee (Grant 19B470010).

References

1. Liu, J., Fang, W., Zhang, X., Yang, C.: An improved photovoltaic power forecasting model with the assistance of aerosol index data. IEEE Trans. Sustain. Energy 6(2), 434–442 (2015)
2. Zhang, C., Tang, Y., et al.: Photovoltaic power forecast based on neural network with a small number of samples. Electr. Power Autom. Equipment 37(1), 101–106 (2017)
3. Li, J., Wang, R., Zhang, T., et al.: Predicating photovoltaic power generation using an improved hybrid heuristic method. In: 2016 Sixth International Conference on Information Science and Technology (ICIST), pp. 383–387. IEEE, Dalian (2015)
4. Eltawil, M., Zhao, Z.: Grid-Connected photovoltaic power systems: technical and potential problems—a review. Sustain. Energy Rev. 14, 112–129 (2010)
5. Zhou, Z.: Machine Learning. Tsinghua University Press, Beijing (2016)
6. Yona, A., Senjyu, T., Funabashi, T.: Application of recurrent neural network to short-term-ahead generating power forecasting for photovoltaic system. In: IEEE Power Engineering Society General Meeting, pp. 1–6. Institute of Electrical and Electronics Engineers, Tampa (2007)
7. Han, M.: Photovoltaic Power Prediction Based on Fuzzy Support Vector Machinel. Guilin University of Technology, Guilin (2016)
8. Han, S., Bae, K., Park, H., Dan, K.: Solar power prediction based on satellite images and support vector machine. IEEE Trans. Sustain. Energy 7, 1255–1263 (2016)
9. Wan, C., Xu, Z., Pinson, P., Dong, Z., Wong, K.: Probabilistic forecasting of wind power generation using extreme learning machine. IEEE Trans. Power Syst. 29, 1033–1044 (2014)
10. Wang, J., Ran, R., Zhou, Y.: A short-term photovoltaic power prediction model based on an FOS-ELM algorithm. Appl. Sci. 7, 423 (2017)
11. Yuan, X., Shi, J., Xu, J.: Short-term power forecasting for photovoltaic generation considering weather type index. Proc. CSEE 33(34), 57–64 (2013)
12. Zhang, N., Chongqing, K., et al.: A method of probabilistic distribution estimation of conditional forecast error for photovoltaic power generation. Autom. Electr. Power Syst. 39 (16), 8–15 (2015)
13. Zhang, J.: Study on the Short Term Prediction of Photovoltaic Power based on Combination Model. Zhongyuan University of Technology. Zhengzhou, China (2015)
14. Xiao, J., Wei, X., Li, Y.: Power forecasting of photovoltaic plant based on BP neural network and genetic algorithm. Comput. Meas. Control 23(2), 392–393, 405 (2015)
15. Mellit, A., Kalogirou, S.: Artificial intelligence techniques for photovoltaic applications: a review. Prog. Energy Combust. Sci. 34, 574–632 (2008)
16. Tao, Y., Chen, Y.: Distributed PV power forecasting using genetic algorithm based neural network approach. In: Proceedings of the 2014 International Conference on Advanced Mechatronic Systems (ICAMechs), Tokai, Japan, 10–12 August (2014)
17. Wang, H., Zhao, X., Wang, K., et al.: Cooperative velocity updating model based particle swarm optimization. Appl. Intell. 2(40), 322–342 (2014)
18. Rahman, M., Yamashiro, S.: Novel distributed power generating system of PV-ECaSS using solar energy estimation. IEEE Trans. Energy Convers. 22(2), 358–367 (2007)
19. Qiaona, J.: Grid-connected photovoltaic power station power generation forecasting research. Southeast University, Nanjing, China (2011)
20. Shi, F., Wang, X., Yu, L., et al.: MATLAB Neural Network Analysis of 30 Cases. Beihang University Press, Beijing (2010)
21. Hao, J., Li, H.: Application of BP neural network to short-term-ahead generating power forecasting for PV system. Adv. Mater. Res. 609, 128–131 (2013)

22. Liu, W., Peng, D., Bu, G., Su, J.: A survey on system problems in smart distribution network with grid-connected photovoltaic generation. Power Syst. Technol. **33**(19), 1–5 (2009)
23. Wang, Y., Fu, Y., Lu, S., et al.: Ultra-short term prediction model of photovoltaic output power based on Chaos-RBF neural network. Power Syst. Technol. **42**(4), 1110–1116 (2018)
24. Luo, J., Zhao, Q., et al.: PV short-term output forecasting based on improved GA-BP neural network. J. Shanghai Univ. Electr. Power **34**(1), 9–12 (2018)

Predictive Values of Preoperative Index Analysis in Patients with Esophageal Squamous Cell Carcinoma

Zhenzhen Zhang[1,2], Qinfei Yang[1,2], and Yingcong Wang[1,2]($\boxtimes$)

[1] Henan Key Lab of Information-Based Electrical Appliances,
Zhengzhou University of Light Industry, Zhengzhou 450002, China
ying_cong_wang@163.com
[2] School of Electrical and Information Engineering,
Zhengzhou University of Light Industry, Zhengzhou 450002, China

Abstract. The prognostic nutritional index (PNI) has been widely used to predict survival outcomes of patients with various malignant tumors. The purpose of this study is to assess the prognostic value of PNI and analyze the effects of preoperative clinical indicators in patients with esophageal squamous cell carcinoma (ESCC). The optimal threshold of PNI is determined by ROC curve analysis, and the correlation between PNI and clinical indicators is evaluated by Chi square test and Fisher's exact test. The results showed that PNI has been significantly correlated with gender, BMI, T stage, Eosinophil count, Erythrocyte count, Hemoglobin concentration, TP, Albumin and Globulin. PNI is positively related to LMR, NLR and PLR by Spearman's rank correlation coefficient. Univariate Cox regression model declared that PFS has been vitally correlated with T stage, N stage, TNM stage, EC and PT. Further, multivariate Cox regression model analyses showed that PFS has been significantly correlated with N stage and PT. ROC curve analysis expressed that the combination of N stage and EC has better accuracy and validity in predicting the severity of the patient. The present study determines that N stages, EC and PT are valuable factors in predicting patient's survival outcomes, as well as independent indicators for poor prognosis.

Keywords: Prognostic nutritional index · Esophageal squamous cell cancer · Prognosis

Abbreviations

PNI	Prognostic nutritional index	ESCC	Esophageal squamous cell carcinoma
ROC	Receiver operation characteristic	AUC	Area under the curve
CI	Confidence interval	LMR	Lymphocyte-to-monocyte ratio
NLR	Neutrophil-to-lymphocyte ratio	PLR	Platelet-to-lymphocyte ratio
PFS	Progression-free survival	BMI	Body mass index

© Springer Nature Singapore Pte Ltd. 2020
L. Pan et al. (Eds.): BIC-TA 2019, CCIS 1160, pp. 454–466, 2020.
https://doi.org/10.1007/978-981-15-3415-7_38

TP	Total protein	PT	Prothrombin time
TT	Thrombin time	APTT	Activated partical thromboplastin time
FIB	Fibrinogen	TNM	Tumor lymph node metastasis
HGB	Hemoglobin concentration	WBC	White blood cell
HR	Hazard ratio	PET	Positron emission computed tomography
CT	Computed tomography		

1 Introduction

Esophageal squamous cell carcinoma (ESCC), a leading cause of cancer-related death, is a common malignancy in China and worldwide, ranking the 8th and 5th in morbidity and mortality respectively. The histological types of esophageal cancer are esophageal squamous cell carcinoma and esophageal adenocarcinoma. The incidence level, geographical distribution, time trend and risk factors of different histological kinds of ESCC are significantly dissimilar. Urban esophageal cancer patients are better than rural patients in terms of medical resources and economic status, which may be one of the reasons why urban ESCC is rare. In recent years, with the continuous development of neoadjuvant therapy, the overall survival rate of patients with ESCC has been greatly improved. However, the implementation of non-surgical chemoradiotherapy, to some extend, increases the incidence of adverse reactions and affects the short-term comprehensive efficacy in patients, which reduces the benefits brought by chemoradiotherapy. At present, most studies on the prognosis of ESCC mainly focus on the factors related to the long-term surgical patients, while few researches concentrate on the effects of non-surgical treatment, especially in sequential and concurrent chemoradiotherapy on survival. In this paper, based on the literature data of patients with ESCC who participated in the operation in the same region, the incidence status and influencing index are analyzed and summarized, so as to provide basic information for the epidemiology and cancer prevention and treatment. Namely, it is necessary to identify prognostic factors and determine the best treatment plan according to the patient's condition [1].

Presently, survival outcomes of ESCC patients who underwent surgery are mainly predicted with tumor-associated factors, such as tumor grade, intramural metastasis, metabolic nodal stage and response. TNM staging system is insufficient to predict the prognosis of ESCC, as ESCC patients with the same TNM stage often show different clinical and survival outcomes [2,3]. In previous studies, features extracted from images of PET, CT are associated with clinical prognosis in various types of cancers. In addition, due to lack of sensitive and accurate monitoring means, the disease progression of patients cannot be effectively monitored and available data for ESCC are scarce. This study only investigates ESCC patients undergoing surgery to search for several feasible and effective elements to evaluate the therapeutic effects and long-term prognosis of patients with ESCC. Therefore, the present study aims to explore the prognostic value of clinicopathologic characteristics and various blood indicators [4–6].

This article is arranged as follows: In Sect. 2, Related data collection and methods applied are briefly introduced. Statistical analyses and results are presented in Sect. 3. Discussion of advantages in this study and development in future are mentioned in Sect. 4. Finally, the conclusion is given in Sect. 5.

2   Materials and Methods

2.1   Patients Selection

The patients obtained with ESCC all have underwent curative surgery in this study. From January 2007 to December 2018, the patients with ESCC were sampled from the Affiliated Hospital of Zhengzhou University. The 318 patients reviewed in the medical records who fulfill the following criteria, and data collected includes clinical characteristics, pathological findings, blood parameters, and patient survival time. The inclusion criteria are ensured as follows: (1) pathology confirmed esophageal squamous cell carcinoma; (2) therapeutic resection of ESCC; (3) preoperative study factors were completely recorded; (4) postoperative follow-up was at least 3 months. The study protocol are approved by the relevant institutional review board.

2.2   Calculation of Laboratory Data

Laboratory data, including serum albumin and blood cell count, are collected from records of blood routine tests performed before surgury. PNI are calculated using the following formula: PNI $=$ *serum albumin level*$(g/L) + 5 \times$ *total lymphocyte count*$(/L)$ [7]. LMR is calculated by dividing the total lymphocyte count by monocyte count. Likewise, NLR and PLR are calculated by dividing the total neutrophil and platelet count, respectively, by total lymphocyte count [8].

2.3   Research Methods

All numerical data are presented as the mean $\pm$ standard. Medical records of 318 consecutive patients with ESCC undergoing surgery between January 2007 and March 2018 are reviewed. Receiver operating characteristic (ROC) curve and area under curve (AUC) are often applied to evaluate the merits of a binary classifier, which can easily detect the performance recognition capability at any threshold value. The optimal PNI cutoff value is determined using ROC curve analysis [9]. Cox proportional hazards model is a semi-parametric regression model proposed by the British statistician D.R.Cox. With survival outcome and survival time as the dependent variables, the model can simultaneously analyze the influence of many factors on survival time, and can analyze the data with truncated survival time without requiring the estimation of the survival distribution type of the data. The effects of age, gender, T stages, N stages, TNM stages, therapy method, BMI, Lymphocyte count, Monocyte count, Neutrophil count, Eosinophil count, Basophil count, Erythrocyte count, HGB, Thrombocyte count, TP, albumin, globulin, PT, APTT, TT, FIB, on PFS are analysed using the Cox proportional hazards regression model [10]. The individual effect of each variable on PFS is first tested by using univariate analyses; then, multivariate Cox proportional hazards regression analysis is performed to determine the independent predictors for PFS [11–13]. Patients are classified as living and die according to the follow-up results. Kaplan-Meier method is called K-M method for short, also known as the product limit method, which is a commonly used survival analysis method. It is mainly suitable for estimating patient survival rate and drawing survival curve for non-grouped survival data. The ladder curve is drawn with survival time as the horizontal axis and survival rate as the vertical axis to illustrate the relationship between survival time and survival rate. The log-rank test is a comparison of two or more groups of survival curves. Survival outcomes of patients are compared using Kaplan-Meier estimator and log-rank test, and the thresholds of predictors are determined by the Youden index that maximised both sensitivity and specificity in the receiver operating characteristic (ROC) curves [14–16]. All these statistical analyses are performed using IBM SPSS statistics software, version 25.0. All statistical tests used in this study are two-sided, and a P value < 0.05 is considered significant [17,18].

Table 1. Clinicopathologic characteristic of patients with ESCC.

Variable	Overall cases(%)	High PNI group cases(%)	Low PNI group cases(%)	P value
Total	318	186(58.5)	132(41.5)	
Age (years)				0.041
$\leq$60	159(50.0)	102(54.8)	57(43.2)	
>60	159(50.0)	84(45.2)	75(56.8)	
Sex				0.259
Male	120(37.7)	75(40.3)	45(34.1)	
Female	198(62.3)	111(59.7)	87(65.9)	
Baseline BMI(kg/m^2)				< 0.001
<18.5	17(5.3)	6(3.2)	11(8.3)	
18.5 − 25.0	230(72.3)	124(66.7)	106(80.3)	
>25.0	71(22.3)	56(30.1)	15(11.4)	
Tumor location				0.377
Upper-chest	57(17.9)	33(17.7)	24(18.2)	
Mid-chest	220(69.2)	133(71.5)	87(65.9)	
lower-chest	41(12.9)	20(10.8)	21(15.9)	
Differentiation				0.178
Low	135(42.5)	73(39.2)	62(47.0)	
Medium	159(50.0)	101(54.3)	58(43.9)	
High	24(7.5)	12(6.5)	12(9.1)	
T stage				0.042
T1	35(11.0)	26(14.0)	9(6.8)	

(continued)

Table 1. (*continued*)

Variable	Overall cases(%)	High PNI group cases(%)	Low PNI group cases(%)	P value
T2	87(27.4)	55(29.6)	32(24.2)	
T3-T4	196(61.6)	105(56.5)	91(68.9)	
N stage				0.430
N0	167(52.2)	99(53.2)	68(51.5)	
N1	83(26.1)	50(26.9)	33(25.0)	
N2	51(16.0)	27(14.5)	24(18.2)	
N3	17(5.3)	10(5.4)	7(5.3)	
TNM stage				0.736
I-II	176(55.3)	106(57.0)	70(53.0)	
III	123(38.7)	70(37.6)	53(40.2)	
IV	19(6.0)	10(5.4)	9(6.8)	
WBC count(10^9/L)				0.327
<4.0	15(4.7)	6(3.2)	9(6.8)	
4.0 − 10.0	275(86.5)	163(87.6)	112(84.8)	
>10.0	28(8.8)	17(9.1)	11(8.3)	
Lymphocyte count(10^9/L)				0.150
<0.8	5(1.6)	1(0.5)	9(3.0)	
0.8 − 4	312(98.1)	163(98.9)	112(97.0)	
>4	1(0.3)	1(0.5)	0(0.0)	
Monocyte count(10^9/L)				0.742
0.1 − 0.8	297(93.4)	173(93.0)	124(93.9)	
>0.8	21(6.6)	13(7.0)	8(6.1)	
Neutrophil count(10^9/L)				0.222
<2	15(4.7)	9(4.8)	6(4.5)	
2 − 7	281(88.4)	168(90.3)	113(85.6)	
>7	22(6.9)	9(4.8)	13(9.8)	
Eosinophil count(10^9/L)				0.022
0.05 − 0.5	255(80.2)	150(80.6)	105(79.5)	
>0.5	63(19.8)	36(19.4)	27(20.5)	
Basophil count(10^9/L)				0.795
0 − 0.1	300(94.3)	176(94.6)	124(93.9)	
>0.1	18(5.7)	10(5.4)	8(6.1)	
Erythrocyte count(10^9/L)				< 0.001
<4.0	56(17.6)	18(9.7)	38(28.8)	
4.0 − 5.5	257(80.8)	163(87.6)	94(71.2)	
> 5.5	5(1.6)	5(2.7)	0(0.0)	
Hemoglobin concentration(g/L)				< 0.001
< 110	13(4.1)	1(0.5)	12(9.1)	

(*continued*)

Table 1. (*continued*)

Variable	Overall cases(%)	High PNI group cases(%)	Low PNI group cases(%)	P value
110 − 160	285(89.6)	168(90.3)	117(88.6)	
> 160	20(6.3)	17(9.1)	3(2.3)	
Thrombocyte count($10^9/L$)				0.465
< 100	2(0.6)	1(0.5)	1(0.8)	
100 − 300	263(82.7)	150(80.6)	113(85.6)	
> 300	53(16.7)	35(18.8)	18(13.6)	
TP(g/L)				< 0.001
< 60	19(6.0)	0(0.0)	19(14.4)	
60 − 80	262(82.4)	152(81.7)	110(83.3)	
> 80	37(11.6)	34(18.3)	3(2.3)	
Albumin(g/L)				< 0.001
< 40	95(29.9)	9(4.8)	86(65.2)	
40 − 55	221(69.5)	175(94.1)	46(34.8)	
> 55	2(0.6)	2(1.1)	0(0.0)	
Globulin(g/L)				0.031
< 20	9(2.8)	2(1.1)	7(5.3)	
20 − 30	190(59.7)	119(64.0)	71(53.8)	
> 30	119(37.4)	65(34.9)	54(40.9)	
PT(s)				0.126
< 11	220(69.2)	123(66.1)	97(73.5)	
11 − 14	94(29.6)	59(31.7)	35(26.5)	
> 14	4(1.3)	4(2.2)	0(0.0)	
APTT(s)				0.797
< 30	56(17.6)	35(18.8)	21(15.9)	
30 − 42	193(60.7)	111(59.7)	82(62.1)	
> 42	69(21.7)	40(21.5)	29(22.0)	
TT(s)				0.636
< 16	192(60.4)	109(58.6)	83(62.9)	
16 − 18	113(35.5)	70(37.6)	43(32.6)	
> 18	13(4.1)	7(3.8)	6(4.5)	
FIB(mg/dL)				0.706
< 200	4(1.3)	3(1.6)	1(0.8)	
200 − 500	265(83.3)	156(83.9)	109(82.6)	
> 500	49(15.4)	27(14.5)	22(16.7)	
LMR				0.002
≤ 4.54	217(68.2)	114(61.3)	103(78.0)	
> 4.54	101(31.8)	72(38.7)	29(22.0)	
NLR				< 0.001

(continued)

Table 1. (*continued*)

Variable	Overall cases(%)	High PNI group cases(%)	Low PNI group cases(%)	P value
≤ 2.50	199(62.6)	136(73.1)	63(47.7)	
> 2.50	119(37.4)	50(26.9)	69(52.3)	
PLR				< 0.001
≤ 142.53	186(58.8)	132(71.0)	54(40.9)	
> 142.53	132(41.5)	54(29.0)	78(59.1)	

3 Results

3.1 Statistical Analysis

Baseline clinicopathologic characteristics of 318 ESCC patients are summarised in Table 1. Of the 318 eligible patients, 198(62.3%) are male and 120(37.7%) are female, with a median age of 61 years (range 38–82 years); 135(42.5%) are poorly differentiated tumors, 159(50.0%), 24(7.5%) are moderately, highly differentiated, respectively. The cut-off points of above blood indexes have been determined according to the reference values used in previous studies. Of these patients, 176(55.3%) are diagnosed with stage I-II esophageal squamous cell cancer, 123(38.7%) with stage III ESCC, and 19(6.0%) with stage IV ESCC, respectively (Table 1).

3.2 Relationships of PNI with Clinicopathologic Blood Index

The prognostic nutritional index (PNI) is widely applied for evaluating survival outcomes in patients. ROC curve analysis is indicated that the optimal PNI cutoff value is 50.25, at the highest Youden index 0.062, with a sensitivity of 0.632 and specificity of $0.43[AUC = 0.500; 95\%CI\ 0.429 - 0.570]$. Patients are divided into the high PNI group $[PNI > 50.25; n = 186(58.5\%)]$ and the low PNI group $[PNI \leq 50.25; n = 132(41.5\%)]$. The median values of LMR, NLR, and PLR are identified as 4.54 (*range* 0.33–26.00), 2.50 (*range* 0.00–13.50), and 142.53 (*range* 21.88–670.00), respectively, which are used as cutoff values. The median age of the high PNI group is 60 years (range 38–81 years), which are indicated significantly younger than that of the low PNI group (median age 63 years; range 40–82 years; $P = 0.041$). The relationships between PNI and preoperative blood index characteristics are shown in Table 1. A low PNI is associated with low preoperative BMI ($P < 0.001$), low T stage ($P = 0.042$), low Eosinophil count ($P = 0.022$), low Erythrocyte count ($P < 0.001$), low Hemoglobin concentration ($P < 0.001$), low TP ($P < 0.001$), low Albumin ($P < 0.001$), and low Globulin ($P = 0.031$). No significant association is noted between PNI and other clinicopathological characteristics. A low PNI is associated with a low LMR ($P = 0.002$), a low NLR ($P < 0.001$), and a low PLR ($P < 0.001$). PNI is found to be positively correlated with LMR ($r = 0.288, P = 0.002$), NLR ($r = 0.405, P < 0.001$), and PLR ($r = 0.419, P < 0.001$) (Table 1).

Table 2. Multivariate Cox proportional hazard analysis of PFS.

Variable	PFS	
	HR(95% CI)	P value
T stage(I-II *vs* III-IV)	1.717(0.920,1.491)	0.200
N stage(I-II *vs* III-IV)	1.519(1.087,2.124)	0.014
TNM stage(I-II *vs* III-IV)	1.075(0.813,1.421)	0.614
Eosinophils count(normal range *vs* abnormal range)	0.807(0.609,1.068)	0.133
PT(normal range *vs* abnormal range)	1.352(1.054,0.734)	0.018

3.3 Predictors for PFS

Values of indexes are summarised in Table 3. Progress-free survival (PFS) is defined as the interval from the beginning of treatment to the observation of disease progression or the occurrence of death for any reason. Univariate Cox regression analyses show that T stages, N stages, TNM stages, Eosinophil count, as well as PT are associated with PFS ($P = 0.011, P < 0.001, P = 0.002, P = 0.039, P = 0.002$, respectively) (Table 2). Further multivariate analysis showed that N stages ($HR = 1.519; 95\%CI\ 1.087 - 2.414, P = 0.014$) and PT ($HR = 1.352, 95\%CI\ 1.054 - 1.734, P = 0.018$) are confirmed as independent predictors for PFS (Table 3). N stages and PT are confirmed as independent predictors for PFS in patients with ESCC. Predictors, as the name implies, refer to the important blood indicators that affect the survival of patients as vital as possible. Specifically, lower N stages and lower PT are verified to improving survival outcomes. On the other hand, we find that N stages uniformity, when combined with Eosinophil count, show higher diagnostic capacity ($AUC = 0.991, 95\%CI\ 0.983 - 0.998; P < 0.001$) than N stages with Eosinophil count and PT ($AUC = 0.746, 95\%CI\ 0.692 - 0.799; P < 0.001$) or N stages with PT ($AUC = 0.744, 95\%CI\ 0.690 - 0.798; P < 0.001$) (Table 4 and Fig. 1). The above analysis further demonstrates the importance of N stages and PT for patient's prognosis.

4 Discussion

A large number of early practical studies at home and abroad have confirmed that the occurrence and development of malignant tumors are the result of multiple factors. At the present stage, the prognosis of patients has become a hot topic in the medical field of various countries. The relevant departments of cancer research in China have found that there is a close relationship between postoperative disease recurrence and death of patients. Esophageal cancer is a serious threat to human life and health, which is a malignant tumor with prognosis. In order to further improve the prognosis of patients with esophageal squamous cell cancer, this study discusses the influence in patients undergoing surgery based on the existing research bases at home and abroad in recent years. The symptoms

Table 3. Univariate Cox proportional hazard analysis of PFS.

Variable	PFS	
	HR(95% CI)	P value
Age($\leq$ 60 years vs > 60 years)	1.162(0.902,1.498)	0.245
Sex(male vs female)	1.001(0.797,1.257)	0.995
T stage(I-II vs III-IV)	1.345(1.069,1.694)	0.011
N stage(I-II vs III-IV)	1.758(1.337,2.311)	<0.001
TNM stage(I-II vs III-IV)	1.416(1.133,1.771)	0.002
Therapy method(surgery vs surgery+others)	1.228(0.922,1.637)	0.169
BMI(normal range vs abnormal range)	0.949(0.742,1.215)	0.681
WBC(normal range vs abnormal range)	1.016(0.734,1.407)	0.922
Lymphocyte count(normal range vs abnormal range)	1.708(0.759,3.846)	0.196
Monocyte count(normal rangge vs abnormal range)	1.285(0.822,2.007)	0.270
Neutrophil count(normal rangevs abnormal range)	1.271(0.859,1.716)	0.271
Eosinophil count(normal range vs abnormal range)	0.747(0.566,0.986)	0.039
Basophil count(normal range vs abnormal range)	0.711(0.437,1.157)	0.168
Erythrocyte count(normal range vs abnormal range)	1.114(0.842,1.475)	0.450
HGB(normal range vs abnormal range)	0.898(0.626,1.289)	0.560
Thrombocyte count(normal range vs abnormal range)	1.072(0.801,1.436)	0.638
TP(normal range vs abnormal range)	1.136(0.851,1.517)	0.387
Albumin(normal range vs abnormal range)	0.958(0.753,1.218)	0.725
Globulin(normal range vs abnormal range)	1.007(0.804,1.260)	0.953
PT(normal range vs abnormal range)	1.463(1.144,1.872)	0.002
APTT(normal range vs abnormal range)	1.063(0.847,1.335)	0.597
TT(normal range vs abnormal range)	1.113(0.882,1.404)	0.364
FIB(normal range vs abnormal range)	1.049(0.781,1.411)	0.749

and signs of esophageal squamous cell carcinoma are relatively secret, so the studies on the prognostic factors of this disease can affectively improve the survival rate of patients. Early detection and timely treatment are the key to improve the survival rate of patients [19–22]. Surgical resection is still the standard method for the treatment of esophageal squamous cell carcinoma. The preoperative indicators of patients are analyzed, which is contributed to evaluate the significance of operation. However, the lymphatic network is rich in deep esophageal mucosa and submucosa and a lot of collaterals in muscular layer and outer membrane lymphatic vessels are interlinked. When the cell cancer breakthrough the submucosal layer of the wall, it is possible to transfer through lymphatic pipes to other places [23,24]. Therefore, it is quite understandable that indicators related to prognosis are found in esophageal squamous cell cancer patients.

Unlike the univariate analysis results of the present study, accumulating evidence have supported that T stages and TNM stages have been associated with worse survival outcomes in patients with ESCC. In fact, the entire study population shows N stages and PT are extremely important. A main strength of

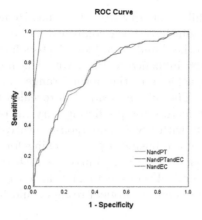

Fig. 1. Receiver operating character (ROC) curve analysis of preoperative diagnostic condition. AUC area under the ROC curve.

Table 4. ROC analyses of different groups in predicting prognostic condition.

Groups	AUC	95%CI	P value
NandEC	0.991	(0.983,0.998)	<0.001
NandECandPT	0.746	(0.692,0.799)	<0.001
NandPT	0.744	(0.690,0.798)	<0.001

this study is the availability of many clinically relevant variables for evaluation [25, 26]. The selection of clinical variables is crucial when developing prediction models for clinical use in future. Nowadays, there are not many relevant research reports in China. And the sample size selected in this study is small, the conclusions need to be tested by more practical studies in the future. Inevitably, some potentially relevant variables may be missed, and then large sample and prospective data are needed to analyze and confirm our results [27–30].

The selection bases on relevant literature, and only variables feasible for evaluation at a consultant visit are chosen. In the following work, patients with neoadjuvant concurrent chemoradiotherapy of esophageal squamous cell carcinoma and surgical patients can be prospectively studied in terms of life quality, economic factors and regional factors.

5   Conclusions

In this paper, the importance of several blood indexes of patients with ESCC has been studied. In other oncology diseases, these factors have been introduced and studied with a series of medical statistical methods in recent literature. For the first place, the data processing method selected in assessing the effect on prognosis is dividing the data set into two or more groups for statistical analysis based on the experience of experts, which is contributed to rational analysis. Gender,

BMI, T stage, Eosinophil count, Erythrocyte count, Hemoglobin concentration, TP, Albumin and Globulin have been confirmed to be significantly associated with prognosis. On the other hand, each blood factors has been divided into two categories to analyze the influence of PFS, with the normal group indicating that the corresponding value is in the normal range, and the abnormal group indicating the opposite. The present study has revealed that N stages and PT could serve as relevant factors for predicting survival outcomes and diagnostic conditions in patients with ESCC undergoing curative tumor resection. The prognostic model can be established by researching for the important factors affecting status of patients with ESCC. Further search for other related indicators impacting patient survival is of great significance to our study, and then the preoperative evaluation system of surgical results could be carried out.

Acknowledgements. This work was supported in part by the National Key R and D Program of China for International S and T Cooperation Projects (2017YFE0103900), in part by the Joint Funds of the National Natural Science Foundation of China (U1804262), in part by the State Key Program of National Natural Science of China under Grant 61632002, in part by the National Natural Science of China under Grant 61603348, Grant 61775198, Grant 61603347, and Grant 61572446, in part by the Foundation of Young Key Teachers from University of Henan Province (2018GGJS092), and in part by the Youth Talent Lifting Project of Henan Province 2018HYTP016 and Henan Province University Science and Technology Innovation Talent Support Plan under Grant 20HASTIT027.

References

1. Mao, J., et al.: Predictive value of pretreatment MRI texture analysis in patients with primary nasopharyngeal carcinoma. Eur. Radiol. **29**, 4105–4113 (2019)
2. Samiei, H., et al.: Dysregulation of helper T lymphocytes in *esophageal squamous cell carcinoma* (ESCC) patients is highly associated with aberrant production of miR-21. Immunol. Res. **67**, 212–222 (2019)
3. Péron, J., et al.: Assessing long-term survival benefits of immune checkpoint inhibitors using the net survival benefit. JNCI: J. Natl. Cancer Inst. **111**, 1186–1191 (2019)
4. Song, S.E., et al.: Intravoxel incoherent motion diffusion-weighted MRI of invasive breast cancer: correlation with prognostic factors and kinetic features acquired with computer-aided diagnosis. J. Magn. Reson. Imaging **49**(1), 118–130 (2019)
5. Fan, Y., Du, Y., Sun, W., Wang, H.: Including positive lymph node count in the AJCC N staging may be a better predictor of the prognosis of NSCLC patients, especially stage III patients: a large population-based study. Int. J. Clin. Oncol. **24**, 1359–1366 (2019)
6. Choi, Y.H., et al.: A high monocyte-to-lymphocyte ratio predicts poor prognosis in patients with advanced gallbladder cancer receiving chemotherapy. Cancer Epidemiol. Prevent. Biomarkers **28**(6), 1045–1051 (2019)
7. Peng, J., et al.: Prognostic value of preoperative prognostic nutritional index and its associations with systemic inflammatory response markers in patients with stage III colon cancer. Chinese J. Cancer **36**(1), 96 (2017)

8. Han, L.H., Jia, Y.B., Song, Q.X., Wang, J.B., Wang, N.N., Cheng, Y.F.: Prognostic significance of preoperative lymphocyte-monocyte ratio in patients with resectable esophageal squamous cell carcinoma. Asian Pac. J. Cancer Prev. **16**(6), 2245–2250 (2015)

9. Fawcett, T.: An introduction to ROC analysis. Pattern Recogn. Lett. **27**(8), 861–874 (2006)

10. Chapman, B.C., et al.: Perioperative and survival outcomes following neoadjuvant FOLFIRINOX versus gemcitabine abraxane in patients with pancreatic adenocarcinoma. JOP: J. Pancreas **19**(2), 75 (2018)

11. Polen-De, C., Loreen, A., Billingsley, C., Jackson, A., Herzog, T.: Independent radiologic review in ovarian cancer research. Gynecol. Oncol. **153**(3), e10 (2019)

12. McSorley, L., et al.: 64p timing of treatment with concurrent chemoradiotherapy (CRT) and impact on progression free survival (PFS) in limited stage small cell lung cancer (LSSCLC). Ann. Oncol. **30**(Suppl_2), mdz071-004 (2019)

13. Hodi, F.S., et al.: Immune-modified response evaluation criteria in solid tumors (imRECIST): refining guidelines to assess the clinical benefit of cancer immunotherapy. J. Clin. Oncol. **36**(9), 850–858 (2018)

14. Jiao, Y., Li, Y., Lu, Z., Liu, Y.: High trophinin-associated protein expression is an independent predictor of poor survival in liver cancer. Dig. Dis. Sci. **64**(1), 137–143 (2019)

15. Ma, J., et al.: Neutrophil-to-lymphocyte ratio (NLR) as a predictor for recurrence in patients with stage III melanoma. Sci. Rep. **8**(1), 4044 (2018)

16. Lobon-Iglesias, M., et al.: Diffuse intrinsic pontine gliomas (DIPG) at recurrence: is there a window to test new therapies in some patients? J. Neurooncol. **137**(1), 111–118 (2018)

17. Chen, G., Cox, R.W., Glen, D.R., Rajendra, J.K., Reynolds, R.C., Taylor, P.A.: A tail of two sides: artificially doubled false positive rates in neuroimaging due to the sidedness choice with t-tests. Hum. Brain Mapp. **40**(3), 1037–1043 (2019)

18. Eklund, A., Knutsson, H., Nichols, T.E.: Reply to chen et al.: parametric methods for cluster inference perform worse for two-sided t-tests. Hum. Brain Mapp. **40**(5), 1689–1691 (2019)

19. Yan, Y., et al.: Patterns of life lost to cancers with high risk of death in China. Int. J. Environ. Res. Public Health **16**(12), 2175 (2019)

20. Ma, W., Jiang, B.: Health impacts due to major climate and weather extremes. In: Lin, H., Ma, W., Liu, Q. (eds.) Ambient Temperature and Health in China, pp. 59–73. Springer, Heidelberg (2019). https://doi.org/10.1007/978-981-13-2583-0_4

21. Zafarzadeh, A., Rahimzadeh, H., Mahvi, A.H.: Health risk assessment of heavy metals in vegetables in an endemic esophageal cancer region in Iran. Health Scope (2018, in press)

22. Li, Z., et al.: Occurrence and potential human health risks of semi-volatile organic compounds in drinking water from cities along the Chinese coastland of the Yellow Sea. Chemosphere **206**, 655–662 (2018)

23. Zheng, X., et al.: Margin diagnosis for endoscopic submucosal dissection of early gastric cancer using multiphoton microscopy. Surg. Endosc. **34**, 408–416 (2019)

24. Du, H., Xiong, M., Liao, H., Luo, Y., Shi, H., Xie, C.: Chylothorax and constrictive pericarditis in a woman due to generalized lymphatic anomaly: a case report. J. Cardiothorac. Surg. **13**(1), 59 (2018)

25. de Blasio, F., Di Gregorio, A., de Blasio, F., Bianco, A., Bellofiore, B., Scalfi, L.: Malnutrition and sarcopenia assessment in patients with chronic obstructive pulmonary disease according to international diagnostic criteria, and evaluation of raw BIA variables. Respir. Med. **134**, 1–5 (2018)

26. Hung, A.J., Chen, J., Gill, I.S.: Automated performance metrics and machine learning algorithms to measure surgeon performance and anticipate clinical outcomes in robotic surgery. JAMA Surg. **153**(8), 770–771 (2018)
27. Semple, C., Lannon, D., Qudairat, E., McCaughan, E., McCormac, R.: Development and evaluation of a holistic surgical head and neck cancer post-treatment follow-up clinic using touchscreen technology—Feasibility study. Eur. J. Cancer Care **27**(2), e12809 (2018)
28. Sun, J., Zhao, X., Fang, J., Wang, Y.: Autonomous memristor chaotic systems of infinite chaotic attractors and circuitry realization. Nonlinear Dyn. **94**(4), 2879–2887 (2018). https://doi.org/10.1007/s11071-018-4531-4
29. Gupta, A., et al.: Feasibility of wearable physical activity monitors in patients with cancer. JCO Clin. Cancer Inform. **2**, 1–10 (2018)
30. Lynce, F., et al.: Characteristics and outcomes of breast cancer patients enrolled in the National Cancer Institute Cancer Therapy Evaluation Program sponsored phase I clinical trials. Breast Cancer Res. Treat. **168**(1), 35–41 (2018)

Univariate Analysis and Principal Component Analysis of Preoperative Blood Indicators in Patients with Esophageal Squamous Cell Carcinoma

Enhao Liang[1,2], Junwei Sun[1,2], and Yanfeng Wang[1,2(✉)]

[1] Henan Key Lab of Information-Based Electrical Appliances,
Zhengzhou University of Light Industry, Zhengzhou 450002, China
[2] School of Electrical and Information Engineering,
Zhengzhou University of Light Industry, Zhengzhou 450002, China
{junweisun,yanfengwang}@yeah.net

Abstract. Esophageal Squamous Cell Carcinoma (ESCC) was one of the most common malignant tumors in the world, and it was in the middle and late stage. Surgery is the first choice for the treatment of ESCC, but the survival rate of patients is still very low. In this paper, it used the blood indicators of patients with ESCC to do univariate analysis, and we found out the influencing factors of survival rate of ESCC. Univariate Cox regression was used to analyze blood indexes and factors affecting survival or death of patients were screened out. Spearman and Pearson correlation analysis could determine whether screening factors were related to survival. Principal component analysis (PCA) was used to test whether screening factors contributed significantly to the survival or death of patients. The survival curve and progression-free survival curve of 5 factors were drawn after obtaining the threshold value by ROC.

Keywords: Esophageal Squamous Cell Carcinoma · Univariate analysis · Correlation analysis · Principal component analysis · Survival curve

1 Introduction

As a common malignant cancer, esophageal squamous cell carcinoma (ESCC) ranked sixth in the cause of cancer-related death all over the world [1]. Over the years, researchers has been working to explore different aspects of improving the quality of life of cancer patients [2–4]. Although treatments for ESCC had made great progress in the past few decades, the overall survival of advanced ESCC patients remained poor after surgery [5–10]. Although the incidence of esophageal adenocarcinoma was increasing in western countries, esophageal cancer was still dominant in eastern countries and even in global, and esophageal

© Springer Nature Singapore Pte Ltd. 2020
L. Pan et al. (Eds.): BIC-TA 2019, CCIS 1160, pp. 467–481, 2020.
https://doi.org/10.1007/978-981-15-3415-7_39

carcinoma remained the predominant type of esophageal cancer in these countries [11]. Most patients with ESCC presented with advanced stage disease, which was incurable. In fact, the five-year survival rate for patients with esophageal cancer were still only about 20% [12]. According to the Global Cancer Watch (gco.iarc.fr), China accounted for more than 50% of all new cases (572,000 new cases) in the world (307,359 cases). Most advanced and proximal esophageal cancer were squamous cell carcinomas. Due to the decrease of smoking and drinking, the global trend of esophageal carcinoma has been declining [13].

Surgery was the first choice for the treatment of esophageal cancer [14,15]. Chromoendoscopy had been found to have sensitivity ranging from 92% to 100% and specificity ranging from 37% to 82% for detecting esophageal squamous dysplasia [16]. In this paper, the factors affecting the survival of ESCC are selected by three-stage method that 17 factors in blood cells are used as univariate analysis. Drs Yang, Ma, and Li for their insightful comments [17] on our review article regarding screening advances in ESCC [18,19]. The factors affecting the death of ESCC was screened and Cox regression test was performed. The correlation and regression analysis of the selected factors carried out to verify the correlation and significance of the factors. Finally, the main components of blood cytokines were analyzed and compared, and the factors that affect the survival rate of esophageal squamous carcinoma were obtained. The main component coefficient is obtained and the univariate analysis results are verified. Finally, ROC curve is used to find the threshold of the screened factors, and Kaplan-Meier is used to draw the survival curve of the screened factors.

2   Objects and Analysis

2.1   Objects

From January 2007 to December 2018, 501 patients with ESCC were sampled from the Affiliated Hospital of Zhengzhou University. They were 316 (63.07%) males and 185 (36.93%) females. The average age of the patients was 61.194 years, ranging from 45 to 80 years. The amount of data is relatively large, and the selection of age data conforms to the normal distribution.

2.2   Case Selection

In this paper, the final gross type (after standard interpolation), final differentiation (after standard interpolation), final infiltration (after standard interpolation) and upper and lower incision of patients with esophageal cancer after resection in the First Affiliated Hospital of Zhengzhou University from 2010 to 2018 were selected. The criteria for selection were: (1) pathological diagnosis of esophageal squamous cell carcinoma; (2) therapeutic resection of esophageal squamous cell carcinoma; (3) complete recording of 17 factors in preoperative blood cells; (4) no anticancer treatment before resection of the tumor, and follow-up for at least 3 months.

2.3 Experimental Method and Content

The experimental data were used to calculate 17 factors in blood cells, WBC count, Lymphocyte count, Monocyte count, Neutrophil count, Eosinophil count, Basophil count, Red blood cell count, Hemoglobin concentration, Platelet count, Total protein, albumin, Globulin, PT (prothrombin time), INR international standardized ratio, APTT (activated partial thromboplastin time), TT (thrombin time), FIB (fibrinogen) data. These data were collected from routine blood test records seven days before surgery. LMR was calculated by dividing the total number of lymphocytes by the total number of monocytes [20]. Similarly, NLR and PLR were calculated by dividing the total number of neutrophils and platelets by the total number of lymphocytes, respectively [20].

2.4 Follow-Up

Telephone follow-up and regular outpatient follow-up were conducted. The follow-up time was 6–140 months. The short-term clinical effect evaluated esophagography and chest CT within one month after the end of treatment. Follow-up was conducted every three months within two years after the operation. Follow-up was conducted every six months after two years.

2.5 Statistical Analysis

The database is built with Excel input data, Spearman and Pearson correlation analysis and principal component analysis carried out with SPSS17.0 statistical software. Cox regression model is used to screen the influencing factors of prognosis after operation, and step-by-step method is used to model the prognostic factors. The interaction term between variables and time is added to the model, and the equal proportion risk assumption (P assumption) is validated by the interaction method. The test level is $P < 0.05$.

3 Method

3.1 Cox Regression Univariate Analysis

Cox regression univariate analysis was performed by survival analysis [21]. Cox regression was used to screen 17 univariate factors influencing survival and mortality of ESCC patients [22, 23].

The analysis is as follows Table 1. For each single factor analysis and comparison of 17 blood cells in each ESCC patient in the table. Firstly, we screen the significant conditions, and the test level of advance hypothesis is 0.05. Therefore, we get that WBCC, MONO, Seg, PT, INR are rejecting hypothesis test, which is obvious. Other abscissa explanations in the Table 1: coefficient value (B), standard error (S.E.), chi-square value (Wald), degree of freedom (df), OR value Exp (B). For the larger chi-square value, the smaller the corresponding P value, the results are more significant. The Table 2 is the abbreviation of key words in the paper.

Table 1. Variables the equation.

Element	B	SE	Wald	df	Sig	Exp(B)
WBCC	0.059	0.024	6.237	1	**0.013**	1.060
MONO	0.307	0.097	10.101	1	**0.001**	1.359
Seg	0.062	0.026	5.739	1	**0.017**	1.064
PT	−0.103	0.031	10.791	1	**0.001**	0.902
INR	−0.949	0.290	10.694	1	**0.001**	0.387

Table 2. Variable abbreviation interpretation.

Description	Code	Unit
WBC count	WBCC	$(109/L)$
Lymphocyte count	LC	$(109/L)$
Monocyte count	MONO	$(109/L)$
Neutrophil count	Seg	$(109/L)$
Eosinophil count	EoC	$(109/L)$
Basophil count	BC	$(109/L)$
Erythrocyte count	ErC	$(109/L)$
Hemoglobin concentration	HC	(g/L)
Platelet count	PC	$(109/L)$
Total protein	TP	(g/L)
Albumin	AL	(g/L)
Globulin	GL	(g/L)
Prothrombin time	PT	(s)
International standardized ratio	INR	
Activated partial thromboplastin time	APTT	(s)
Thrombin time	TT	(s)
Fibrinogen	FIB	mgd/L
Youden index	YI	
Total mortality	TM	
Overall survival	OS	
Results follow-up	F-U	

3.2 Spearman and Pearson Correlation Analysis

Spearman and Pearson bivariate correlation analysis [24, 25] are used to analyze the five selected univariate factors. The follow-up results and survival time (month) are included in the correlation analysis. These are shown in Table 3, and Pearson correlation analysis is shown in the Table 3. The survival time is significantly correlated with five univariate factors, $P < 0.05$. Because it is the correlation analysis under bivariate condition and positive correlations of PT and INR are presented. For PT (seconds) and INR, there is a significant correlation, $P < 0.01$.

Table 3. The correlation.

Parameter	Su 12	WBC	MC	NeC	PT	INR	F-U
Pearson analysis	1	−0.110*	−0.107*	−0.098*	0.154**	0.155**	−0.737**
Sig(bilateral)		0.014	0.017	0.028	0.001	0.001	0.000
Sum of squares and cross products	820354.50	−4859.83	−951.20	−3969.37	5008.99	556.72	−6248.60
Covariance		1640.709	−9.720	−1.902	−7.939	10.018	1.116	−12.497
N		501	501	501	501	501	501	501

*. Significant correlation at 0.05 level (bilateral). **. Significant correlation at 0.01 level (bilateral). The Su 12 is Survival * 12 months.

3.3 Linear Analysis

Linear analysis requires four conditions: (1) There is a linear relationship between independent variables and dependent variables. (2) The residual sequence is independent. (3) The residual distribution is a normal distribution with a mean of 0. (4) The residual sequence is homogeneous in variance. Because this study is a linear analysis of pairwise variables, and it shows the size and direction of the correlation between the two variables. Linear analysis is used to further analyze the predictive relationship between the two variables.

No matter what the value of independent variable x is, and the variance of the corresponding residual should be equal. It should not change with the value of the explanatory variable or the explanatory variable (see Fig. 1).

Table 4. Coefficient.

Model	Non-standardization coefficient		Standard coefficient	t	Sig
	B	Standard error	Trial version		
(constant)	66.722	5.873		11.362	0.000
WBCC	−2.035	0.825	−0.110	−2.468	0.014

At Table 4. Dependent variable: survival 12 (Month).

Linear regressions results are shown in Table 4. The results show that the WBCC has a significant impact on survival (month), and the coefficients and constant values have a significant impact, $P < 0.05$. The goodness of fit isnt high. For single-factor linear regression, the significance of regression equation and regression coefficient are tested and conforms to normal distribution according to regression standardized residuals. The regression equation of WBCC and survival time (month) are obtained by simple linear regression.

$$lifetime = 66.722 - 2.035 * WBCC \tag{1}$$

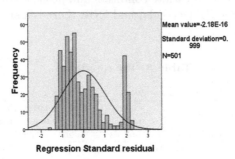

Fig. 1. The horizontal axis represents the residual regression standardization, the vertical axis represents the frequency, and the independent variable is the survival period multiplied by 12 months.

Similarly, the other simple linear regression results are as follows:

$$lifetime = 57.197 - 9.860 * MONO \tag{2}$$

$$lifetime = 61.326 - 2.000 * Seg \tag{3}$$

$$lifetime = 13.233 + 3.866 * PT \tag{4}$$

$$lifetime = 25.203 + 35.174 * INR \tag{5}$$

The significant effects of $MONO$, Seg, PT and INR coefficients are less than 0.05, which show significant correlation. The significance of regression equation and regression coefficient are tested. According to the normalized residual of regression, it conforms to the normal distribution.

3.4 Principal Component Analysis

The following Table 5 shows, the eigenvalue of the first principal component is 2.378, which explains 47.567% of the total variance of the five original variables. The eigenvalue of the second principal component is 2.005, which explains 40.099% of the total variance of the five original variables and 87.663% of the total variance contribution. Two variables explain 87.66% of all blood cell factors, and only the eigenvalues of two variables are greater than 1. Extraction Sums of Squared Loadings describes the factor solutions after extraction and rotation of the principal components. It can be seen from the Table 5 that two principal components are extracted and rotated. The cumulative percentage of total variances explained by the two principal components are the same as the first two variables of the initial solution. However, the variance of the original variables explained by that each factors are redistributed by the factors after rotation, which makes the variance of the factors closer and easier to explain [26, 27].

Table 5. Total variance of interpretation.

Initial eigenvalue				Extract square sum loading		
Ingredients	Total	Variance (%)	Accumulate (%)	Total	Variance (%)	Accumulate (%)
1	**2.378**	47.564	47.564	**2.378**	47.564	47.564
2	**2.005**	40.099	**87.663**	**2.005**	40.099	**87.663**
3	0.543	10.856	98.519			
4	0.070	1.399	99.918			
5	0.004	0.082	100.000			

Method of extraction: Principal component analysis.
a. At this stage of analysis, only follow-up results (alive/dead) = living cases will be used. As Fig. 2, the abscissa represents the number of factors and the ordinate represents the characteristic roots. As can be seen from the Fig. 2, the eigenvalues of the first two factors are large, and the eigen values are small from the third, and the eigenvalue line of the factors also becomes gentle. The first two factors contribute the most to explanatory variables.

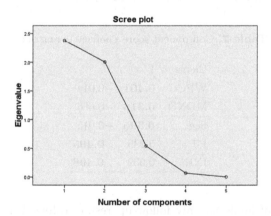

Fig. 2. Gravel figure is a blood cell factor analysis lithotripsy. Principal component analysis (PCA) lithotripsy maps can help determine the optimal number of factors.

The Table 6 shows the factor load matrix before rotation. The load of each variable on two factors is given in the factor load matrix.
Method of extraction: Principal component analysis. a. Two components have been extracted. b. At this stage of analysis, only follow-up results (alive/dead) = living cases will be used.

As can be seen from the Table 6, in the load matrix before rotation, the WBCC (109L) and Seg (109L) of all variables on the first factor sum are all loaded relatively high. The first factor WBCC (109L) and Seg (109L) explain most variables' information. The second factors, INR and PT have a greater

Table 6. Total variance of interpretation.

Element	Ingredients	
	1	2
WBCC	**0.955**	−0.028
MONO	**0.940**	−0.031
Seg	**0.753**	−0.172
PT	0.094	**0.994**
INR	0.092	**0.993**

correlation with the original variables, and have obvious effect on the interpretation of the original variables. The scoring coefficients of the factors estimated by regression method. Five variables are represented by $X_1, ..., X_5$, respectively. According to the Table 7, the following functions are obtained:

$$F_1 = 0.401X_1 + 0.316X_2 + 0.395X_3 + 0.039X_4 + 0.039X_5 \tag{6}$$

$$F_2 = -0.014X_1 - 0.086X_2 - 0.015X_3 + 0.495X_4 + 0.496X_5 \tag{7}$$

Table 7. Component score coefficient matrices.

Element	1	2
WBCC	**0.401**	−0.014
MONO	**0.316**	−0.086
Seg	**0.395**	−0.015
PT	0.039	**0.495**
INR	0.039	**0.496**

a. At this stage of analysis, only follow-up results (alive/dead) = living cases will be used.

This is shown in Table 7. The weight of WBCC ($109L$) and Seg ($109L$) are higher in the first factor score function, while the weight of PT (s) and INR are higher in the second factor score function. The results are consistent with those obtained in the rotating factor load matrix. The key point is that the first factor explains WBCC ($109L$), MONO ($109L$), Seg ($109L$), and the second factor mainly explains PT, INR.

4   Survival Analysis

The expecting ROC curve of the classifier was determined by two parameters. The first parameter determined how many positive cases there were in the population. The second parameter was the correlation between the classifier output of each instance and the potential positive tendency of that instance [28].

Five factors of principal component analysis were analyzed. In this paper, ROC curve analysis firstly carried out, and obtained the area under the ROC curve and the threshold value. The five factors are divided into two categories, namely high and low concentrations. The Fig. 3 shows the ROC curves of WBCC, MONO, Seg, PT, and INR.ROC curves are compared under healthy conditions. The area under the curve is shown in Table 8 as follows: The area under the curve of WBCC, MONO and Seg are less than 0.5, and the area under the curve of PT and INR are more than 0.5.

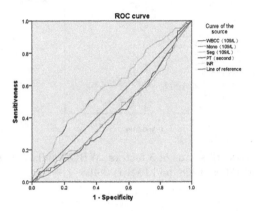

Fig. 3. ROC curve of 5 factors remaining healthy. WBCC is 0.428, and MONO is 0.433, and Seg is 0.433, and PT is 0.592, and INR is 0.592.

The Fig. 4 shows the ROC curves of WBCC, MONO, Seg, PT and INR. By comparing the ROC curve at death, the area under the curve is shown in Table 8 as follows: the area under the curve of WBCC, MONO and Seg is greater than 0.5, and the area under the curve of PT and INR is less than 0.5. After calculating the coordinates of the curve, the threshold value can be obtained. The calculation formula is:

$$YI = sensitivity + specificity - 1 \qquad (8)$$

Table 8. Area under the curve.

Test result variable	Area/Be alive	Area/Die
WBCC	0.428	0.572
MONO	0.432	0.568
Seg	0.432	0.568
PT	0.592	0.408
INR	0.592	0.408

and living in a state of four factors are calculated the threshold respectively, the WBCC, Seg, PT, INR: 0.9, 4.2, 11.15, 0.88. Thus each element can be divided into two categories. The YI is Youden Index, and thats the critical value. The area under the ROC curve is below 0.5 of the three elements.

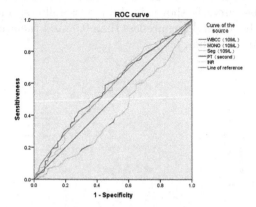

Fig. 4. ROC curve of death status of 5 factors. WBCC is 0.572, and MONO is 0.568, and Seg is 0.568, and PT is 0.408, and INR is 0.408.

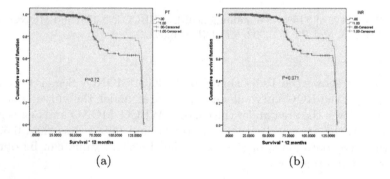

(a)   (b)

Fig. 5. (a) High (1) and low (0) PT mortality. P = 0.72 (b) High (1) and low (0) INR mortality. P = 0.071. The abscissa is survival, and the ordinate is mortality.

As Fig. 5a can be seen from the total mortality curve of INR, and the total mortality of high concentration is higher than that of low concentration. As Fig. 5b can be seen from the total mortality curve of PT, and the mortality of high concentration is higher than that of low concentration.

And alive a state of three factors is calculated the threshold of the WBCC is 7.05, and MONO is 2.75, and Seg is 0.35. According to the threshold value obtained by ROC, 3 factors in the state of death are classified. The survival curve and total mortality are drawn in the state of death (Fig. 6).

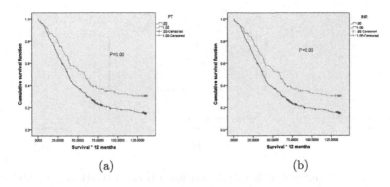

Fig. 6. (a) PT survival at high (1) and low (0) concentrations. (b) INR survival at high (1) and low (0) concentrations. The abscissa is survival, and the ordinate is survival rates. The survival rate of PT and INR at high concentration is higher than that at low concentration.

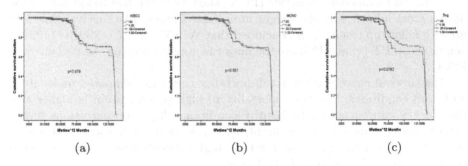

Fig. 7. (a) High (1) and low (0) WBCC mortality. (b) High (1) and low (0) MONO mortality. (c) High (1) and low (0) Seg mortality. The abscissa is lifetime and the ordinate is mortality.

As is shown in Fig. 7. According to the survival curves and total mortality curve plots draw in the two states, only the Seg in Fig. 7c show a small stratification phenomenon. And the mortality is higher in high concentration than in low concentration. Low concentration is higher than high concentration in Fig. 8.

5   Conclusion

According to Cox regression single factor analysis, five influencing factors, including WBCC ($109/L$), MONO ($109/L$), Seg ($109/L$), PT (s) and INR, are selected to reject hypothesis test, $P < 0.05$, with significant influence.

The correlation analysis show that WBCC ($109/L$), MONO ($109/L$), Seg ($109/L$), PT (s) and INR are correlated with survival time, and the fitting effect is significant.

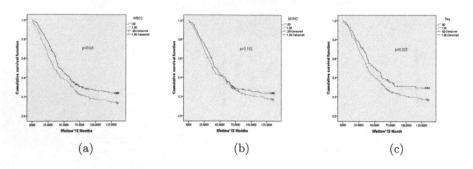

(a)	(b)	(c)

Fig. 8. (a) WBCC survival at high (1) and low (0) concentrations. (b) MONO survival at high (1) and low (0) concentrations. (c) Seg survival at high (1) and low (0) concentrations. The abscissa is life-time, and the ordinate is survival rates.

Principal Component Analysis (PCA) show that the five blood cell factors have a great influence on the weight in unilateral cases. The survival time keep up with follow-up results. It is concluded that WBCC ($109/L$), MONO ($109/L$), Seg ($109/L$), PT (s) and INR are the major factors influencing the patients with ESCC.

The survival curve and the total mortality curve are compared under living and dead conditions. The total mortality of high concentration is higher than that of low concentration. In healthy conditions, high concentrations have a higher survival rate than low concentrations; At death, low concentrations are associated with higher survival rate than high concentrations. And distinguish between survival factors and death factors.

About 300,000 people died of esophageal cancer every year in the world. The mortality and morbidity of esophageal cancer varied greatly in each country. Survival outcomes of patients with cancer were not determined by tumor characteristics alone and that patient-related factors such as age, performance status, and nutritional status played an important role. Cancer associated systemic inflammation was a well-known key determinant of patient prognosis in various cancer types [29]. In this study, we found out that single factor has an impact on the survival of ESCC, and much work needed to be done to prolong the life span of ESCC patients and make them grow healthily. And complications are related to higher rate of local and distant recurrence, as wells as worse overall survival (OS) [30–36].

Our method also has shortcomings, such as data instability and clinical data deletion. Our next work will focus on the significance of post-classification and clinical manifestations. The blood cells of patients with ESCC will be classified by chi-square test.

Acknowledgements. This work was supported in part by the National Key R and D Program of China for International S and T Cooperation Projects (2017YFE0103900), in part by the Joint Funds of the National Natural Science Foundation of China (U1804262), in part by the State Key Program of National Natural Science of China under Grant 61632002, in part by the National Natural Science of China under Grant 61603348, Grant 61775198, Grant 61603347, and Grant 61572446, in part by the Foundation of Young Key Teachers from University of Henan Province (2018GGJS092), and in part by the Youth Talent Lifting Project of Henan Province 2018HYTP016 and Henan Province University Science and Technology Innovation Talent Support Plan under Grant 20HASTIT027.

References

1. Yang, C., et al.: Down-regulated miR-26a promotes proliferation, migration, and invasion *via* negative regulation of MTDH in esophageal squamous cell carcinoma. FASEB J. **31**(5), 2114–2122 (2017)
2. Hussain, S., Quazilbash, N.Z., Bai, S., Khoja, S.: Reduction of variables for predicting breast cancer survivability using principal component analysis. In: 2015 IEEE 28th International Symposium on Computer-Based Medical Systems, pp. 131–134. IEEE (2015)
3. Torre, L.A., Bray, F., Siegel, R.L., Ferlay, J., Lortet-Tieulent, J., Jemal, A.: Global cancer statistics. CA: Cancer J. Clin. **65**(2), 87–108 (2015)
4. Jemal, A., Bray, F., Center, M.M., Ferlay, J., Ward, E., Forman, D.: Global cancer statistics. CA: Cancer J. Clin. **61**(2), 69–90 (2011)
5. Chen, X., Wang, L., Wang, W., Zhao, L., Shan, B.: B7-H4 facilitates proliferation of esophageal squamous cell carcinoma cells through promoting interleukin-6/signal transducer and activator of transcription 3 pathway activation. Cancer Sci. **107**(7), 944–954 (2016)
6. Colon, C.L., et al.: Laparoscopic surgery versus open surgery for colon cancer: short-term outcomes of a randomised trial. Lancet Oncol. **6**(7), 477–484 (2005)
7. Van der Pas, M.H., et al.: Laparoscopic versus open surgery for rectal cancer (COLOR II): short-term outcomes of a randomised, phase 3 trial. Lancet Oncol. **14**(3), 210–218 (2013)
8. Sancho-Muriel, J., et al.: Standard outcome indicators after colon cancer resection. Creation of a nomogram for autoevaluation. Cirugía Española (Engl. Ed.) **95**(1), 30–37 (2017)
9. Longo, W.E., et al.: Risk factors for morbidity and mortality after colectomy for colon cancer. Dis. Colon Rectum **43**(1), 83–91 (2000). https://doi.org/10.1007/BF02237249
10. Bilimoria, K.Y., et al.: Laparoscopic-assisted vs. open colectomy for cancer: comparison of short-term outcomes from 121 hospitals. J. Gastrointest. Surg. **12**(11), 2001 (2008). https://doi.org/10.1007/s11605-008-0568-x
11. Wang, Q.L., Lagergren, J., Xie, S.H.: Prediction of individuals at high absolute risk of esophageal squamous cell carcinoma. Gastrointest. Endosc. **89**(4), 726–732 (2019)
12. Siegel, R.L., Miller, K.D., Jemal, A.: Cancer statistics, 2016. CA: Cancer J. Clin. **66**(1), 7–30 (2016)
13. Wang, Q.L., Xie, S.H., Wahlin, K., Lagergren, J.: Global time trends in the incidence of esophageal squamous cell carcinoma. Clin. Epidemiol. **10**, 717 (2018)
14. Chang, G.J., Rodriguez-Bigas, M.A., Skibber, J.M., Moyer, V.A.: Lymph node evaluation and survival after curative resection of colon cancer: systematic review. J. Nat. Cancer Inst. **99**(6), 433–441 (2007)
15. Monson, J., et al.: Practice parameters for the management of rectal cancer (revised). Dis. Colon Rectum **56**(5), 535–550 (2013)

16. Morita, F.H.A., et al.: Narrow band imaging versus lugol chromoendoscopy to diagnose squamous cell carcinoma of the esophagus: a systematic review and meta-analysis. BMC Cancer **17**(1), 54 (2017)
17. Yang, F., Ma, D., Li, Z.: Screening for esophageal squamous cell carcinoma: insight from experience with Barrett's esophagus. Gastrointest. Endosc. **89**(2), 443–444 (2019)
18. Codipilly, D.C., et al.: Screening for esophageal squamous cell carcinoma: recent advances. Gastrointest. Endosc. **88**(3), 413–426 (2018)
19. Iyer, P.G., et al.: Highly discriminant methylated DNA markers for the non-endoscopic detection of Barrett's esophagus. Am. J. Gastroenterol. **113**(8), 1156 (2018)
20. Abe, S., et al.: LMR predicts outcome in patients after preoperative chemoradiotherapy for stage II–III rectal cancer. J. Surg. Res. **222**, 122–131 (2018)
21. Wang, D., Wu, T.T., Zhao, Y.: Penalized empirical likelihood for the sparse Cox regression model. J. Stat. Plan. Inference **201**, 71–85 (2019)
22. Cox, D.R.: Regression models and life-tables. J. Roy. Stat. Soc.: Ser. B (Methodol.) **34**(2), 187–202 (1972)
23. Cox, D.R.: Partial likelihood. Biometrika **62**(2), 269–276 (1975)
24. Afyouni, S., Smith, S.M., Nichols, T.E.: Effective degrees of freedom of the Pearson's correlation coefficient under autocorrelation. NeuroImage **199**, 609–625 (2019)
25. Birkeland, K., D'Silva, A.D.: Developing and evaluating an automated valuation model for residential real estate in Oslo. Master's thesis, NTNU (2018)
26. Tran, N.M., Burdejová, P., Ospienko, M., Härdle, W.K.: Principal component analysis in an asymmetric norm. J. Multivar. Anal. **171**, 1–21 (2019)
27. da Silva Sauthier, M.C., et al.: Screening of *mangifera indica* L. functional content using PCA and neural networks (ANN). Food Chem. **273**, 115–123 (2019)
28. Cook, J.A.: ROC curves and nonrandom data. Pattern Recogn. Lett. **85**, 35–41 (2017)
29. Cao, S., et al.: Selected patients can benefit more from the management of etoposide and platinum-based chemotherapy and thoracic irradiation-a retrospective analysis of 707 small cell lung cancer patients. Oncotarget **8**(5), 8657 (2017)
30. McSorley, S.T., Horgan, P.G., McMillan, D.C.: The impact of the type and severity of postoperative complications on long-term outcomes following surgery for colorectal cancer: a systematic review and meta-analysis. Crit. Rev. Oncol./Hematol. **97**, 168–177 (2016)
31. Pucher, P.H., Aggarwal, R., Qurashi, M., Darzi, A.: Meta-analysis of the effect of postoperative in-hospital morbidity on long-term patient survival. Br. J. Surg. **101**(12), 1499–1508 (2014)
32. Mirnezami, A., Mirnezami, R., Chandrakumaran, K., Sasapu, K., Sagar, P., Finan, P.: Increased local recurrence and reduced survival from colorectal cancer following anastomotic leak: systematic review and meta-analysis. Ann. Surg. **253**(5), 890–899 (2011)
33. Krarup, P.M., Nordholm-Carstensen, A., Jorgensen, L.N., Harling, H.: Anastomotic leak increases distant recurrence and long-term mortality after curative resection for colonic cancer: a nationwide cohort study. Ann. Surg. **259**(5), 930–938 (2014)
34. Lu, Z.R., Rajendran, N., Lynch, A.C., Heriot, A.G., Warrier, S.K.: Anastomotic leaks after restorative resections for rectal cancer compromise cancer outcomes and survival. Dis. Colon Rectum **59**(3), 236–244 (2016)

35. Law, W.L., Choi, H.K., Lee, Y.M., Ho, J.W., Seto, C.L.: Anastomotic leakage is associated with poor long-term outcome in patients after curative colorectal resection for malignancy. J. Gastrointest. Surg. 11(1), 8–15 (2007)
36. Sun, J., Zhao, X., Fang, J., et al.: Autonomous memristor chaotic systems of infinite chaotic attractors and circuitry realization. Nonlinear Dyn. 94(1), 2789–2887 (2018). https://doi.org/10.1007/s11071-018-4531-4

A Developmental Model of Behavioral Learning for the Autonomous Robot

Dongshu Wang[✉] and Kai Yang

School of Electrical Engineering, Zhengzhou University, Zhengzhou 450001, China
wangdongshu@zzu.edu.cn

Abstract. In the environment cognition, how to realize efficient behavioral learning is a great challenge for the autonomous robot. Traditional methods, such as the model predictive control algorithms, suffer from the less flexibility and low efficiency. Since the robot may encounter the similar or same scenario in the following environment cognition, if it can determine its moving direction in advance, it will improve the efficiency of environment cognition. Considering that the lateral excitation in the internal neurons in the neural network can fire more surrounding neurons to store similar information, this paper introduces the lateral excitation in the internal neurons in a motivated developmental network to set up the weight connections between the robot's moving direction and environment information in advance during the off-task process. When the robot meets similar or same scenario in its following environment cognition, it can determine its corresponding moving direction quickly, and enhance the efficiency of behavioral learning. Simulation in the static environment of the autonomous robot navigation demonstrates its effect.

Keywords: Environment cognition · Developmental network · Lateral excitation · Off-task process

1 Introduction

In the process of robot environment cognition, how to execute efficient behavioral learning is an important challenge for the autonomous robot. According to the current environmental information and its own state, if a robot can learn from its past behavior, and make decision for the future behavior control in advance, it will improve the efficiency of its future behavior decision-making. One of the methods for the robot to make decision in advance for its future behavior is to control its moving direction in advance, which can make the robot to determine its moving direction quickly in its following environment cognition, and improve the efficiency of behavioral decision.

Supported by Scientific Problem Tackling of Henan Province under Grant 1921022 10256.

© Springer Nature Singapore Pte Ltd. 2020
L. Pan et al. (Eds.): BIC-TA 2019, CCIS 1160, pp. 482–496, 2020.
https://doi.org/10.1007/978-981-15-3415-7_40

For the control of the robot moving behavior in advance, most methods adopt the model predictive control (MPC) algorithms [16,17], and solve the robot moving control through forecasting its moving angle, position or pose. Literature [1] uses the MPC method based on the dynamic matrix control to achieve the efficient control of the steering system of the autonomous robot, and literature [2] uses the distributed model predictive formation control method, and realizes the avoidance control in complex environment, through imposing constraints for the service robot's moving. Literature [3] utilizes the nonlinear MPC method to model the human center of mass trajectory, and control the robot moving in advance through forecasting the human center of mass trajectory. Based on the ACADO toolkits, literature [4] employs the MPC to generate the real-time control command, and realizes the moving control of the ER10 robot. While literature [5] uses the online sparse Gaussian process to denote the relation between the robot velocity and the input command, and combines the MPC to produce the optimal moving control strategy for a skid-steer vehicle.

Besides the MPC, the researchers also present other efficient behavior forecasting methods. For instance, literature [6] puts forward a forecasting method to avoid the static and dynamic obstacles in the dynamic environment for a mobile robot. It integrates the decision-making process and forecasting behavior for the obstacle speed vector, and forecasts the speed vector with the robot sensor systems. Literature [7] forecasts the external events from the human's intention and generates an active plan according to the forecasting events, so the robot can reduce the time delay and improve the human-machine interaction significantly. Literature [18] proposes a simulated reinforcement learning algorithm through mimicking the human learning procedure, and this algorithm can enhance the learning speed of the mobile robot and reduce the error times significantly, by means of learning the human experiences in advance. While literature [8] solves the behavioral selection problem of the autonomous robot from the perspective of episodic memory and bio-inspired attention system. It presents the global planning to solve the robot's behavior sequence forecasting with the top-down attention, based on the learned episodic memory.

In accordance with the above research analysis of the robot behavioral learning, we can conclude that the current methods have the following common problems: these methods solve the robot behavior control in a specific scenario from the engineering perspective. Once the scenario changes, the control strategy needs re-designing, resulting in a poor flexibility. Moreover, these methods usually need a plenty of training samples, leading to a low efficiency.

It is well known that the environment that the robot explores has certain similarity, in the following procedure of environment cognition, the robot may encounter a similar scenario. In addition, lateral excitation mechanism of neurons makes them to fire (or recruit) more surrounding neurons to store the similar information. Therefore, this work refers the working mechanism of human being in off-task process, and the transfer learning mechanism in human's environment perception process, through the lateral excitation mechanism of internal neurons in a motivated developmental network (MDN), building the weight connections between the scenario information and the robot's corresponding moving direction in advance. Hence, in the following moving procedure, when the robot meets the similar scenario, it can determine its moving direction quickly, realizing the behavioral decision control in advance during the robot's environmental cognition, and continuously improve the efficiency of environmental cognition.

The remainder of the paper is organized as follows. Section 2 introduces the DN. Section 3 presents the architecture and working mechanism of the motivated developmental network. Section 4 designs a simulation experiment to demonstrate the effect of the proposed method. Conclusion and future work are provided in Sect. 5.

2 Developmental Network

Developmental network (DN) is a new kind of intelligent neural network that simulates the human brain based on the lobe component analysis [9,10]. Figure 1 provides a simplest DN which has three areas, i.e., the sensory area X, the internal area Y, and the motor area Z. See [9,12,19] for more detail about the concrete description of the DN algorithm.

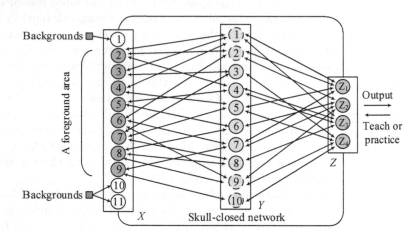

Fig. 1. Scheme of the basic DN. It includes the top-down connections from Z to Y for context denoted by the motor area. It also includes bottom-up connections from X to Y for sensory input. Source: From [21].

3 Architecture of the Motivated Developmental Network (MDN)

Figure 2 presents the architecture of the MDN after it adds the serotonin and dopamine systems. Effects of dopamine and serotonin on human brain have not been completely understood, in accordance with the literature [14]. Both serotonin and dopamine are released in different manners, since human brain can conveniently apply a particular neuromodulator to different goals in different parts of human body. For instance, some dopamine is related to reward and some serotonin is related to aversion and punishment. There are other forms of serotonin and dopamine with different effects that have not been completely understood [13,14]. In this work, we focus on the dopamine and serotonin that affect the brain as reward and punishment, respectively.

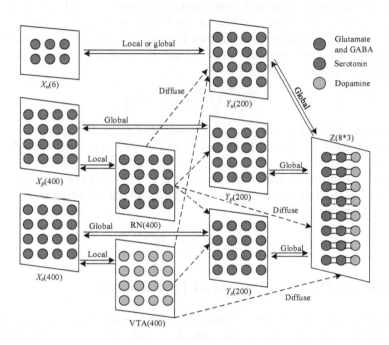

Fig. 2. A motivated DN with serotonin and dopamine modulatory subsystems. Within each Y sub-area and the Z area, within area connections are simulated by top-k competition.

As shown in Fig. 2, in the MDN, the sensory area can be denoted by $X = (X_u, X_p, X_s)$, where X_u denotes an unbiased array, X_p and X_s are a pain array and a sweet array, respectively. Figure 2 shows that Raphe Nuclei (RN) has serotonergic neurons that produce serotonin, and Ventral Tegmental Area (VTA) has dopaminergic neurons that produce dopamine, so RN and VTA can be regarded as the serotonin and dopamine versions of X_p and X_s, respectively.

Similarly, area Y also consists of three sub-areas: $Y = (Y_u, Y_p, Y_s)$. Neurons in Y_p and Y_s have serotonin and dopamine receptors, while neurons in Y_u have glutamate, serotonin and dopamine receptors. For the sake of the effect of the serotonin and dopamine, the new learning rate of Y area can be deduced as follows [15]:

$$\omega_2(n_j) = \min((1 + \alpha_{RN} + \alpha_{VTA})\frac{1}{n_j}, 1) \tag{1}$$

where α_{RN} and α_{VTA} denote parameters related with RN and VTA, respectively. If the serotonin and dopamine do not play a part in the system, then $\alpha_{RN} = \alpha_{VTA} = 0$, thus the learning rate will become its original form $\omega_2(n_j) = 1/n_j$.

The motor area Z can be depicted by a series of neurons $Z = (z_1, z_2, \cdots, z_m)$, where m is the neuron number in Z area. Each z_i is composed of three neurons, i.e., $z_i = (z_{iu}, z_{ip}, z_{is})$, as shown in Fig. 2, where z_{iu}, z_{ip} and z_{is} ($i = 1, 2, \cdots, m$) indicate the unbiased, pain and sweet motors, respectively. Therefore, whether an action i is activated relies not only on the response of z_{iu}, but also on those of z_{ip} and z_{is}. If we investigate the internal interactions among the three different neuromodulators, the composite pre-response value of a motor neuron can be calculated as follows:

$$z_i = z_{iu}(1 - \alpha z_{ip} + \beta z_{is}) \tag{2}$$

where α and β are all positives. Considering that the effect of punishment on the human behavior is generally greater than that of the reward, so α should be larger than β.

Then the j-th motor neuron fires according to the top-k competition [20] and action is released (in this work, it means which moving direction of the agent is determined) where

$$j = \arg \max_{1 \leq i \leq m} \{z_i\} \tag{3}$$

4   Off-task Process Triggered by Exposure

Off-task processes in the MDN are the neural interactions during the times when the network is not attending to any stimuli or task [11]. In contrast with most neural networks, MDN runs the off-task processes to simulate the internal neural activities of the brain even when sensory input is absent or not attended. The off-task processes are run all the time when the network is not in the training mode.

During the off-task process, the cortical columns in the internal area operate using the exact same algorithm described in the Hebbian learning in the winning cortical column, while the bottom-up input is irrelevant to the trained tasks. Similarly, the neurons in the concept area Z operate using the same algorithms as described in the MDN. Whether or not a concept neuron fires during off-task process is a function of the amount of recent exposure of the network to the concept (moving direction of the agent in this work) that the neuron represents.

The probability of a concept neuron firing during off-task processes, given no other neuron is firing in the same concept area, is modeled as a monotonically increasing function of the amount of recent exposure to the corresponding concept, i.e.,

$$p(z_i = 1 | \exists j \neq i, z_j = 1) = 0$$

$$p(z_i = 1 | \forall j \neq i, z_j = 0) = \frac{2}{1 + e^{-\gamma_i}} - 1 \tag{4}$$

where $\gamma_i \geq 0$ measures the amount of recent exposure to the concept that ith neuron represents, where "exposure" means receiving (being exposed to) any sensory signal that have at least one feature (such as moving direction) in common with the transfer condition. To simulate lateral inhibition in the concept area, the conditional part of the probabilities in Eq. (4) models lateral inhibition in the concept areas, and it ensures that a concept neuron does not fire if there is already another neuron firing in the same area.

The amount of exposure to the ith concept, γ_i, can be computed as follows:

$$\gamma_i = \frac{n_{z_i}}{\sum\limits_{i=1}^{9} n_{z_i}} \tag{5}$$

where n_{z_i} denotes the fire times of the ith neuron in the area Z, and $\sum\limits_{i=1}^{9} n_{z_i}$ represents the total fire times of the neurons in area Z, where 9 denotes the 9 kinds of moving directions that the agent can perform, as illustrated in the following Sect. 5.2. Sorting the neurons in descending order according to their firing probability, then fire the former k neurons with nonzero probabilities. For example, assuming $k = 4$, we assume the firing probabilities of the former four neurons are nonzero, the sequence of the four neurons in terms of their probabilities is [neuron1, neuron3, neuron2, neuron5], then program enters four circulations and execute the following procedure: input the data from area Z to Y, fire the neurons in Y area, execute lateral excitation, build up new weight connections. Taking the first circulation as an example, the input from area Z to Y is $z_1 = [1, 0, 0, 0, 0, 0, 0, 0, 0]$, then calculate the pre-response energies of the neurons in Y area, firing the neurons with nonzero pre-response energy, all these fire neurons belong to the first direction (corresponding to the direction of zero degree in this work), therefore they only connect with the first neuron, i.e., z_1, in area Z. Then pre-response energies of these neurons are multiplied by a linearly declining function of neuron's rank:

$$\frac{k - r_i}{k} z_i \rightarrow z_i \tag{6}$$

where $\rightarrow$ denotes the assignment of the value, and $0 \leq r_i < k$ is the rank of the neuron with respect to its pre-response energy (the neuron with the highest pre-response energy denotes a rank of 0, second most active neuron has the rank of 1, and the rest can be done in the same manner).

These firing neurons execute the lateral excitation to fire more neurons to memorize new knowledge. To model lateral excitation in the internal area, neuronal columns in the immediate vicinity of each of the k winner columns are also allowed to fire and update their connection weights. In the current implementation, a computation window of size $\omega = 7$, centered on the neuron, is used to report the results, so there are only twenty four columns in the 5×5 neighborhood (in the two-dimensional sheet of neuronal columns) are excited. The response level of the excited columns are set to the response level of their neighboring winner column, multiplied by an exponentially declining function of their distance to the winner columns:

$$e^{\frac{-d^2}{2}} z_{winner} \rightarrow z_i \tag{7}$$

where the distance $d = 1$ for immediate neighbors of the winner columns and $d = \sqrt{2}$ for the diagonal neighbors in the 5×5 neighborhood of the columns, the rest distances can be deduced in the same manner, as shown in Fig. 3.

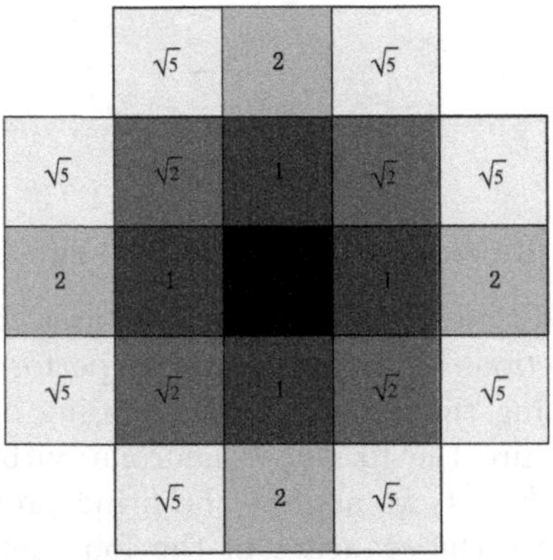

Fig. 3. General pattern observed in lateral excitation of the internal neurons in the MDN.

Then these new firing neurons update their fire times and connection weights. During the off-task process, these new firing neurons will store the similar environment information with the neuron that fire them. In the robot's following environment cognition, when it meets a similar scenario, it can decide its moving direction quickly, in accordance with the relation between the environment information and corresponding moving direction that the robot has learned already during the off-task process. Next section, a simulation experiment will be designed to demonstrate the effect of the proposed method.

5   Simulation Experiments

In the following simulation, several entities are utilized to testify the algorithm mentioned above. One of the robots is the simulated agent which can think and act. Except the simulated agent, there are several obstacles and one target in the environment. In the navigation, the simulated agent will be rewarded from dopamine when it closes the target and punished from serotonin when it approaches the obstacle. Thus, the agent can search the suitable trajectory to approach the target under the double constraints of reward and punishment.

5.1   Simulation Design

Brain of the simulated agent is the MDN, as illustrated in the Fig. 2, where X is the sensory area with three sub-areas, i.e., X_u, X_p and X_s. At each time step, each area generates a response vector according to the physical state of the environment. The X_u vector comes from the input of the sensors directly. The X_p vector identifies in which way the agent is punished and depicts the release of the serotonin in RN. Similarly, the X_s vector identifies in which way the agent is rewarded and denotes the release of the dopamine in VTA. If the agent and its target and obstacles lie in certain distance scopes, RN will release the serotonin to punish the agent and VTA will release the dopamine to reward it, thus affecting the moving direction of the agent.

   Y area is also divided into three sub-areas, Y_u, Y_p and Y_s, corresponding to the three sub-areas in X area. If X_p and X_s receive their inputs, then punishment and reward will be generated. The neurons with the highest pre-response energy in Y_p and Y_s will be activated and their weights and ages will be updated. Z is the motor area, and it is also divided into three sub-layers: Z_u, Z_p and Z_s. It recombines the results from its collaterals to compute a single response vector to decide the moving direction of the agent, as described in Eq. (2).

5.2   Input and Output

During the initialization of the MDN, neurons in Y and Z_u areas are initialized with small random numbers, while those in Z_p and Z_s areas are initialized with zero vectors since the agent does not know which action will lead to reward or punishment. Ages of all neurons are initialized with 1. Neuron number in Y area and the fire number k in the top-k competition can be determined in terms of the resources available. Scale of the Z area is determined by the action number that the agent can perform. At any time, the agent can perform one of the nine possible actions, i.e., it can move along each of the cardinal or inter-cardinal directions or keep its current location. So the neurons in Z area has 9 rows and 3 columns, and 3 columns denote the Z_u, Z_p and Z_s, respectively.

   If we define the following entities, a (agent), t (target), o (obstacle), the agent is denoted by a blue square, the obstacle is depicted by a black circle, and the target is represented by a green circle, then a schematic explanation of the position relation among the three objects can be drawn as illustrated in Fig. 4, and achieve the following expressions:

$$\theta_t = \arctan(a_x - t_x, a_y - t_y)$$
$$d_t = \sqrt{(a_x - t_x)^2 + (a_y - t_y)^2}$$
$$\theta_o = \arctan(a_x - o_x, a_y - o_y)$$
$$d_o = \sqrt{(a_x - o_x)^2 + (a_y - o_y)^2}$$
$$x_u = \{\cos\theta_t, \sin\theta_t, \cos\theta_o, \sin\theta_o, \frac{d_t}{d_t + d_o}, \frac{d_o}{d_t + d_o}\} \qquad (8)$$

where (a_x, a_y), (t_x, t_y), (o_x, o_y) are the coordinates of the agent, target and obstacle, respectively; θ_t (θ_o) indicates the angle between the heading of the agent and direction of the target (the obstacle), while d_t (d_o) denotes the distance between the agent and the target (the obstacle). x_u, x_p and x_s represent the inputs of the network, and Z denotes the output of the MDN.

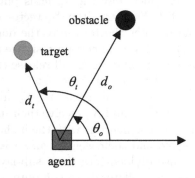

Fig. 4. The setting of the wandering plan which includes the agent, the friend robot and enemy robot. The size of the square space used is 1000×1000.

5.3   Simulation Setup

One static simulated scenario is designed to demonstrate whether the policies made in off-task process can affect the following behavior of the simulated agent. The simulated agent is controlled by the MDN which releases dopamine and serotonin for the target and obstacle in terms of the particular circumstance. Through the simulation, the motivated agent will promote its capacity by reinforcement learning, learning to know which one to elude and which one to chase, in light of the release of the dopamine or serotonin.

At each time step, the horizontal and vertical coordinates are collected for each entity. With these data, the distances and angles between the agent and the target and obstacles can be computed. In the simulation, the moving direction of the agent is computed and decided by selecting the closest to one of the 9 directions.

The pain or sweet sensor has only one value to describe the punishment and reward. The reward value β is defined as follows:

$$\beta = \begin{cases} 0 & d_t < d_{2t} \\ \frac{1}{d_{1t}-d_{2t}}d_t - \frac{d_{2t}}{d_{1t}-d_{2t}} & d_{1t} < d_t < d_{2t} \\ 1 & d_t > d_{1t} \end{cases} \tag{9}$$

where d_{1t} denotes the original distance between the agent and the target, d_{2t} represents the distance when the agent catches up the target, and d_t means the real-time distance between them.

Similarly, the punishment value α is defined as follows:

$$\alpha = \begin{cases} 0 & d_o > d_s \\ -\frac{1}{d_s - d_{ms}} d_o + \frac{d_s}{d_s - d_{ms}} & d_{ms} < d_o < d_s \\ 1 & 0 < d_o < d_{ms} \end{cases} \tag{10}$$

where d_s denotes the scanning scope of the agent, d_o represents the real-time distance between the agent and the obstacle, and d_{ms} indicates the minimal safe distance between the agent and the obstacle.

In light of the α and β, the final moving direction of the agent can be decided by the Eq. (2). At each step, the agent makes a decision according to the memorized knowledge, and there is difference between the actual location relation and the location relation discriminated. If the actual input is defined as $x = \{x_1, x_2, x_3, x_4, x_5, x_6\}$, the weights of the fire neurons in area Y is $w = \{w_1, w_2, w_3, w_4, w_5, w_6\}$, the location recognition error in each step is defined as follows:

$$e = \sum_{i=1}^{6} |x_i - w_i| \tag{11}$$

and the smaller e indicates the higher location recognition of the agent.

5.4   Simulation in the Static Environment

In the static environment, there are 13 obstacles. After training, distribution of the neurons that store the knowledge (i.e., environment information) in the area Y is shown in Fig. 5, where each blue square denotes a neuron, and 152 neurons store the knowledge.

Based on the training data, the trajectory of the agent in the first moving is marked by the blue curve with "+" as shown in Fig. 6. It costs the agent 187 steps to catch up the target, and the behavior decision is made in terms of the original 152 training data. After the first off-task process, the trajectory of the agent in the second moving is marked by the red curve with "o" as shown in Fig. 6, and it costs the agent 176 steps to catch up the target. After the first moving, during the off-task process, the agent settles and memorizes the new environment information encountered in the first moving, namely, it abstracts the similar position characteristics and stores them in new fire neurons. When the agent executes the second moving, it has learned new knowledge. When it meets a similar or same scenario, it can make a different decision from the last moving. Hence, in the second moving, the agent chooses a different trajectory. In a similar way, during the second moving, because the agent takes a different trajectory, thus it learns new knowledge again. After the second motion, in the off-task process, the agent repeats the similar procedure as that in the first off-task process and stores the new knowledge in the new fire neurons. When the agent performs the third moving, it has learned new knowledge again. Therefore, it takes a different trajectory marked by the yellow curve, and it costs the agent 181

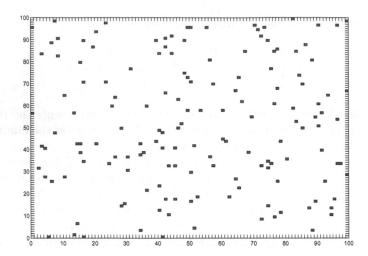

Fig. 5. Distribution of neurons which store the knowledge in the area Y in static environment.

steps to catch up the target. During the fourth motion, the agent has performed three off-task processes, so it chooses a different trajectory marked by a green curve. In the fifth motion, the agent chooses almost the same path as that in the fourth motion, and it costs the agent 171 steps to catch up the target in the last two times. Since the agent learns less and less new knowledge in approaching the target, so the fifth trajectory is almost the same as the fourth one. But there exists small difference between the last two trajectories, as amplified in Fig. 7.

Figure 8 provides the number of neurons storing the knowledge after each moving. Figure 8 shows that the agent learns new knowledge after each moving, and the new knowledge learned becomes less and less. It results from the fact that the obstacles and target are all static, with the increasing of the moving times, the agent becomes more and more familiar with the environment, so the new knowledge it learns becomes less and less.

Figure 9 displays the distribution of neurons storing the knowledge in the area Y after the 5th motion. From Fig. 9, we can see that many data aggregate together, and it comes from that the lateral excitation of the internal neurons in the MDN fire many neurons around them to store new environment information.

Figure 10 shows the error broken lines after the five motions, and we can calculate the average recognition error in the five times with the Eq. (11) and offer the results as follows: 0.8602, 0.3663, 0.2179, 0.2444 and 0.231. Each point on the broken lines denotes the recognition error of the agent at each step. Figure 10 indicates that at the former three motions, after each motion, the error will decrease a little, namely, the average recognition error decreases. At the third motion, the average error achieves the smallest value. At the following two motions, the average errors are about 0.23. We can explain this phenomenon as follows: a definite set A is defined to represent all the new position scenarios

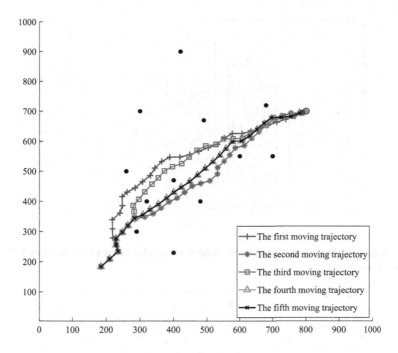

Fig. 6. Paths scheme of the agent in the 5 runs.

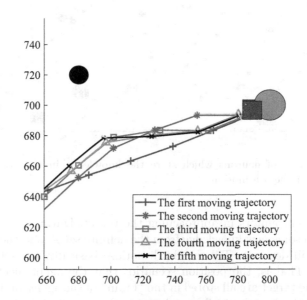

Fig. 7. Local amplification of the trajectories of the agent in the 5 motions.

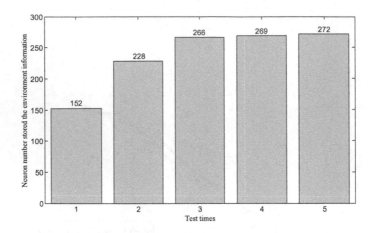

Fig. 8. Number change of the neurons which store the knowledge in area Y in the 5 times motions.

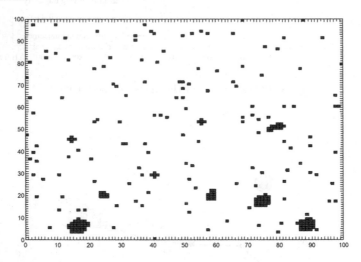

Fig. 9. Distribution of neurons which store the knowledge in the area Y in static environment after the 5th motion.

that the agent encounters in approaching the target. During each motion, the agent will learn some new knowledge from the definite set A, and the rest content of the set A will decrease. Till the third motion, basically, the set A closes to an empty set. That is, the position scenarios that the agent meets during its approaching the target are all stored in the "brain" of the agent, and the position scenarios recognized are basically consistent with the actual position ones, thus the recognition accuracy is high and the recognition error is very low.

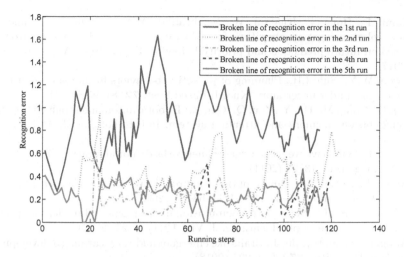

Fig. 10. Error scheme in the 5 motions of the agent.

6   Conclusion and Future Work

This paper adopts the lateral excitation in the internal neurons of the MDN to store the robot moving direction and the corresponding environment information during its off-task process. In the robot's following environment cognition, when it meets a similar or same scenario, the robot can decide its moving direction fast based on the information learned in the off-task process. Simulation of an autonomous robot navigation in a static environment demonstrates the effect of the proposed algorithm.

But this work only investigates the effect of the proposed method in a static environment, without studying its effect in a dynamic environment. Moreover, the simulation scenario is relatively simple. In the future, the more complex scenarios will be used to testify the advantages of the proposed method further.

References

1. Yuan, Y., et al.: Model predictive control strategy based on dynamic matrix control for robot. J. Chongqing Univ. Posts Telecommun. (Nat. Sci. Ed.) **30**(4), 537–543 (2018)
2. Ebel, H., Ardakani, E.S., Eberhard, P.: Distributed model predictive formation control with discretization-free path planning for transporting a load. Robot. Auton. Syst. **96**, 211–223 (2017)
3. Joos, A., Hoffmann, M., Stein, T.: Human center of mass trajectory models using nonlinear model predictive control. IFAC PapersOnLine **51**(22), 366–371 (2018)
4. Cefalo, M., Magrini, E., Oriolo, G.: Sensor-based task-constrained motion planning using model predictive control. IFAC PapersOnLine **51**(22), 220–225 (2018)
5. Kim, T., Kim, W., Choi, S., Kim, H.J.: Path tracking for a skid-steer vehicle using model predictive control with on-line sparse Gaussian process. IFAC PapersOnLine **50**(1), 5755–5760 (2017)

6. Kamil, F., Hong, T., Khaksar, W., Moghrabiah, M.Y., Zulkifli, N., Ahmad, S.A.: New robot navigation algorithm for arbitrary unknown dynamic environments based on future prediction and priority behavior. Expert Syst. Appl. **86**, 274–291 (2017)

7. Kwon, W.Y., Suh, I.H.: Planning of proactive behaviors for human–robot cooperative tasks under uncertainty. Knowl.-Based Syst. **72**, 81–95 (2014)

8. Liu, D., Cong, M., Du, Y., Zou, Q., Cui, Y.: Robotic autonomous behavior selection using episodic memory and attention system. Ind. Robot: Int. J. **44**(3), 353–362 (2017)

9. Weng, J.: Why have we passed "neural networks do not abstract well". Nat. Intell.: INNS Mag. **1**(1), 13–22 (2011)

10. Weng, J., Luciw, M.: Dually optimal neuronal layers: lobe component analysis. IEEE Trans. Auton. Ment. Dev. **1**(1), 68–85 (2009)

11. Solgi, M., Liu, T., Weng, J.: A computational developmental model for specificity and transfer in perecptual learning. J. Vis. **13**(1), 1–23 (2013)

12. Wang, D., Shan, H., Xin, J.: Brain-like emergent auditory learning: a developmental method. Hear. Res. **370**, 283–293 (2018)

13. Merrick, K.E.: A comparative study of value systems for self-motivated exploration and learning by robots. IEEE Trans. Auton. Ment. Dev. **2**(2), 119–131 (2010)

14. Kakade, S., Dayan, P.: Dopamine: generalization and bonuses. Neural Netw. **15**, 549–559 (2002)

15. Wang, D., Duan, Y., Weng, J.: Motivated optimal developmental learning for sequential tasks without using rigid time discounts. IEEE Trans. Neural Netw. Learn. Syst. **29**(10), 4917–4931 (2018)

16. Zeng, D.: Research on path planning and control method for free-floating space robot. Dissertation for the Doctoral Degree of Harbin Institute of Technology, Harbin (2018)

17. Deng, W.: Research on path planning and trajectory tracking of Tracked EOD. Dissertation for the Doctoral Degree of Shandong University, Jinan (2018)

18. Ji, P.: Research on critical technology of human-robot interaction and local autonomy for mobile reconnaissance robot. Dissertation for the Doctoral Degree of Southeast University, Nanjing (2017)

19. Weng, J.: Natural and Artificial Intelligence: Introduction to Computational Brain-Mind. BMI Press, Okemos (2012)

20. Weng, J., Paslaski, S., Daly, J., Vandam, C., Brown, J.: Modulation for emergent networks: serotonin and dopamine. Neural Netw. **41**, 225–239 (2013)

21. Daly, J., Brown, J., Weng, J.: Neuromorphic motivated systems. In: Proceedings of the International Joint Conference on Neural Networks (IJCNN), San Jose, California, USA, pp. 2917–2924 (2011)

Efficient Evolutionary Deep Neural Architecture Search (NAS) by Noisy Network Morphism Mutation

Yiming Chen, Tianci Pan, Cheng He$^{(\boxtimes)}$, and Ran Cheng$^{(\boxtimes)}$

Shenzhen Key Laboratory of Computational Intelligence,
University Key Laboratory of Evolving Intelligent Systems of Guangdong Province,
Department of Computer Science and Engineering,
Southern University of Science and Technology,
Shenzhen 518055, China
{11610507,11610528}@sustech.edu.cn, chenghehust@gmail.com,
ranchengcn@gmail.com

Abstract. Deep learning has achieved enormous breakthroughs in the field of image recognition. However, due to the time-consuming and error-prone process in discovering novel neural architecture, it remains a challenge for designing a specific network in handling a particular task. Hence, many automated neural architecture search methods are proposed to find suitable deep neural network architecture for a specific task without human experts. Nevertheless, these methods are still computationally/economically expensive, since they require a vast amount of computing resource and/or computational time. In this paper, we propose several network morphism mutation operators with extra noise, and further redesign the macro-architecture based on the classical network. The proposed methods are embedded in an evolutionary algorithm and tested on CIFAR-10 classification task. Experimental results indicate the capability of our proposed method in discovering powerful neural architecture which has achieved a classification error 2.55% with only 4.7M parameters on CIFAR-10 within 12 GPU-hours.

Keywords: Neural architecture search · Network morphism · Evolutionary algorithm

1 Introduction

Deep learning techniques have been widely used and have achieved remarkable progress in a variety of fields, e.g., image recognition, speech recognition, and machine translation. One crucial aspect of this progress is the novel neural architecture [10]. Despite the fact that the manually designed novel neural architectures, such as ResNet [13] and DenseNet [15], have greatly improved the

---

Y. Chen and T. Pan contribute equally to this work.

© Springer Nature Singapore Pte Ltd. 2020
L. Pan et al. (Eds.): BIC-TA 2019, CCIS 1160, pp. 497–508, 2020.
https://doi.org/10.1007/978-981-15-3415-7_41

classification accuracy. Nevertheless, developing a neural architecture by human experts is still time-consuming and error-prone [10]. Therefore, there is a growing interest in Neural Architecture Search (NAS) methods, aiming to design a suitable neural architecture for a specific situation without intervention. NAS, a subfield of Automated Machining Learning (AutoML), is the process of automatically designing neural network architectures [10]. Existing NAS methods have outperformed manually designed architectures on some tasks such as image classification [23], object detection [34], or semantic segmentation [3], where both reinforcement learning and evolutionary methods are widely used to xplore the space of neural architectures.

One main obstacle of applying NAS widely into daily use is the vast amount of computing resource needed. Some researchers decrease the search time via cell-based search space [34]. Instead of taking the whole neural architecture as search space, cell, a relatively small neural architecture, is optimized through the searching process, thus substantially reducing the search space. Finally, the complete network is constructed by stacking discovered cells in a pre-defined manner. Consequently, this cell-based search space was also successfully employed by many later works [9,18,22]. Another advantage of the cell-based approach is that it can be easily transferred to other data set [34]. By using the cell-based search space, the problem of macro-architecture selection rises, i.e., how to align these cells and how many cells should be used? Both simple sequential model [34] and more complex manually-designed architecture, such as DenseNet [15], can be used as macro-architecture connecting cells [2].

Existing automated architecture search methods have achieved better results than manually-designed ones, but they also require enormous computational resource. For example, Zoph et al. [33] used 800 GPUs over four weeks for NAS, while Real et al. [24] have used 450 GPUs for seven days to find a competitive network. Despite that further work has been proposed to decrease the search time, the search process is still computationally expensive. On the other hand, the evolutionary algorithm (EA) is a population-based heuristic search technique which simulates the evolution process of nature [8]. It is naturally suitable for NAS mainly for the following two reasons. First, the population-based property enables the search of multiple diverse network structures in parallel. Second, EA can reduce the probability of being trapped into local optima due to its powerful capability in global optimization.

In this work, we propose an efficient evolutionary neural architecture search method using noisy network morphism mutation. This work is based on the concept proposed in Net2Net [4] and [29] which designed some efficient function-preserving mutation operators and used simple sequential macro-architecture as the initial network template. Here, we redesign the network template based on ResNet [13], where the residual connections enable the adoption of deep structure. Besides, we add noise to the pure network morphism mutation to avoid inadequate initialization. We conduct experiments on the classical network architectures and adapt efficient mutation operators in our method. The main new contributions are summarized as follows:

- We propose several efficient mutation operators inspired by classical manually-designed architectures, such that we can avoid the negative effect of network morphism with its efficiency by adding noise.
- We redesign the initial network macro-architecture based on the high-level architecture of the ResNet [13] and the squeeze-and-excitation block [14], which enables the method to search deeper network structures.
- Our evolutionary NAS method outperforms several state-of-the-art methods on CIFAR-10 classification task, where the obtained network has achieved 2.55% classification error with 4.7M parameters in 12 GPU hours.

The rest of this paper is organized as follows. The details of the proposed network mutation operators are presented in Sect. 2. Experimental settings and comparisons of the proposed method with the state-of-the-art NAS methods are presented in Sect. 3. Finally, conclusions are drawn in Sect. 4.

2 The Proposed Method

In this section, we first demonstrate the mutation operators in our method, followed by the effect of adding noise to the operators and the evolutionary framework used in our algorithm. For the detailed mutation operators, we not only keep the operators in [4,29], i.e., layer deepening, layer widening, kernel widening, branch layer, inserting skip connection, but also introduce our new type of mutation operator and inserting dense connection. Furthermore, we modify the skip connection operator with a SE-block to improve classification accuracy.

2.1 Mutation Operators

As mentioned above, all these mutation operators are function-preserving. We introduce a teacher-student network model to illustrate the details of these operators. With teacher network represented by $f(x|\theta^{(f)})$ (x is the input of the network and $\theta^{(f)}$ are its parameters), a function-preserving mutation operator builds a student network g as:

$$\forall x, \ f(x|\theta^{(f)}) = g(x|\theta^{(g)}). \tag{1}$$

The main approach to achieve function preserving is to initialize the parameters of child network based on parent network. In this section, we only illustrate the single input-output situation. As for the multiple input-output situation, we only need to modify all affected layers.

Here, we first demonstrate the principles of six mutation operators in NAS, i.e.,

- Layer deepening,
- Layer widening,
- Kernel widening,
- Branch layer,
- Inserting SE-skip connection,
- Inserting Dense connection.

Then, the modification of these mutation operators for generating more promising neural architectures is demonstrated.

Layer Deepening. The layer deepening mutation operator is accomplished by inserting a single convolution with kernel size $k_1 \times k_2$ and initializing the weights of that convolutional layer as an identity matrix as (2).

$$V_{j,h}^{(l)} = \begin{cases} I_{i,i} & j = \frac{k_1+1}{2} \wedge h = \frac{k_2+1}{2}, \\ 0 & \text{otherwise.} \end{cases} \tag{2}$$

Since $ReLU(x) = \max(x,0)$ is used as activation function in this paper, it satisfies that $\sigma(x) = \sigma(I\sigma(x))$ for all $x \in X$. Additionally, the number of filters of the newly added convolution has to be equal to the number of input channels, which may be changed through layer widening in the future, and the kernel sizes k_1 and k_2 are set to 3.

Layer Widening. The layer widening is achieved by replicating the parameters along the last axis at random (i.e. adding one node to the layer), and we can replicate this operation several times until the wanted number of filters is obtained. It first assumes the original weight matrix $W^{(l)} \in \mathbb{R}^{k_1 \times k_2 \times i \times o}$, and then the weight of widened layer will be calculated as:

$$V_{.,.,.,j}^{(l)} = \begin{cases} W_{.,.,.,j}^{(l)} & j \leq o, \\ W_{.,.,.,j}^{(r)} & \text{otherwise,} \end{cases} \tag{3}$$

where r is randomly sampled from $\{1, \ldots, o\}$.

Next, the weight $V^{(l+1)}$ of the next layer needs to be adjusted accordingly to keep its function preserving characters. $W_{.,.,j,.}^{(l+1)}$ is divided by the number n_j which indicates the times that the $j - th$ filter has been replicated. Then the new weight satisfies

$$V_{.,.,j,.}^{(l+1)} = \frac{1}{n_j} W_{.,.,j,.}^{(l+1)}. \tag{4}$$

Kernel Widening. The kernel widening is achieved by zero-padding, where the kernel size will increase by 1 after each mutation. Meanwhile, the padding of the convolution layer will increase correspondingly to keep the size of the feature map unchanged.

Branch Layer. Given a convolutional layer with wight W, its output will be reformulated as

$$Concatenate(X^{(l)} \times V_1^{(l+1)}, X^{(l)} \times V_2^{(l+1)}), \tag{5}$$

where

$$\begin{aligned} V_1^{(l+1)} &= W_{.,.,.,1:\lfloor o/2 \rfloor}^{(l+1)}, \\ V_2^{(l+1)} &= W_{.,.,.,(\lfloor o/2 \rfloor+1):o}^{(l+1)}. \end{aligned} \tag{6}$$

Then, we are able to divide a single convolutional layer into two layers with the weights initialized as V_1 and V_2, respectively. Note that no further parameters will be generated and the architecture is unchanged, and it will improve the teacher network if and only if being combined with other mutation operators.

Inserting SE-Skip Connection. First, we insert a convolution with all the parameters set to 0 (i.e. $V^{(l+1)} = 0$). Second, we add a skip connection between the input and output of the newly-added layer. Then the channel-wise addition in (7) leads to an identity function.

$$X^{(l+1)} = \sigma(X^{(l)} \times V^{(l+1)} + X^{(l)}), \tag{7}$$

where X, V denote the weights, l denotes the number of layer, and σ is the activation function.

Instead of the original residual module, we use the SE-ResNet module in [14] with the reduction ratio being set to 4. Since the weight of the convolutional layer is set as 0, inserting the SE-ResNet module is still function preserving. Compared with the original version, the squeeze-and-excitation operation will improve the classification accuracy by slightly increasing the number of parameters.

Inserting Dense Connection. The dense connection mutation will not deepen the network but widen as (8).

$$X^{(l+1)} = \sigma(Concatenate(X^{(l)} \times V^{(l+1)}, X^{(l)})), \tag{8}$$

where $Concatenate$ aims to concatenate two weights.

To begin with, we search from the current layer to the last layer of the cell to determine all the reachable layers from the current one (consider the cell as a directed acyclic graph). Then, we randomly pick one layer from the set and concatenate the output of current layer to its input. Besides, we set the parameters of the second layer input channels corresponding to the current layer output as 0. Note that two consecutive layers will not be chosen as the target layers.

2.2   Adding Noise to Mutation Operators

Here, we add noises in some mutation operators to avoid the suboptimal, and the effect of this modification will be further discussed. Since simple random initialization outperforms complex Network Morphism, Network Morphism is inadequate for training a deeper net and may give sub-optimal signals for NAS [28].

We conduct experiments and find that by adding noises to pure network morphism, instead of compromising the efficiency, it will, by contrast, improve the final classification accuracy. Hence, additional noises are added to the auto-generated weight, and the details about adding noises to each mutation operator are presented as follows.

Layer Deepening. The original layer deepening is given in (2). Since the pure network morphism will initialize the weight of newly-added convolutional layer as identity matrix, we multiply each item of the weight with $(1 + \delta)$, where

$$\delta = 0.05 \times \mathcal{N}(0, 1), \tag{9}$$

and $\mathcal{N}(0, 1)$ denotes the standard normal distribution.

Layer Widening. By adding noise to (4), we can obtain

$$V^{(l+1)}_{\cdot,\cdot,j,\cdot} = \frac{1}{n_j} W^{(l+1)}_{\cdot,\cdot,j,\cdot} \times (1+\delta), \tag{10}$$

where

$$\delta = \sqrt{\frac{2}{size}} \times \mathcal{N}(0,1), \tag{11}$$

and *size* denotes the size of tensor.

Inserting Skip Connection. To add noise to the operation of inserting skip connection, we first insert a convolutional layer with all the parameters being set to 0. Although the mutation is function preserving and efficient, we find that adding some noises leads to better performance. Instead of using the zero tensor, we initialize each parameter with

$$0.1 \times N(0,1) \times \sqrt{\frac{2}{size}}. \tag{12}$$

Inserting Dense Connection. Similar to the insertion of skip connection, (12) is adopted to initialize the parameters of the newly-added channels of the second layer.

2.3   Evolutionary Algorithm

An evolutionary algorithm is adopted to achieve the NAS, where the search space is limited in the cell and the same mutation operator will apply to all the cells simultaneously. Compared with searching in the whole network, a smaller search space of this cell-based method will speed up the NAS process. With the aforementioned mutations originated from classical manually-designed neural architecture, our algorithm can discover similar architectures as proposed by human experts. For instance, layer deepening helps to discover VGG [26]; inserting skip connection allows to find ResNet [13]; branching layer allows to discover architecture such as Xception [6] and FractaNet [17]; inserting dense connection allows to discover architecture such as DenseNet [15] and dual path network [5].

There are two orthogonal approaches with cell-based search space, i.e., optimizing micro-architecture and macro-architecture. The macro-architecture largely influences the classification performance, while the micro-architecture only uses a simple sequential model by connecting three cells [29]. Here, we use the high level-structure of ResNet [13] with squeeze-and-excitation block [14] as fixed macro-architecture to achieve its best performance, where the network template is given in Fig. 1, where the Dropblock in [11] is adopted as dropout layer.

At the beginning of the evolution, the population consists of a single predefined network template as Fig. 1. Then, several distinct mutated offspring solutions (all architectures that are 1 step from the initial model) will be added to

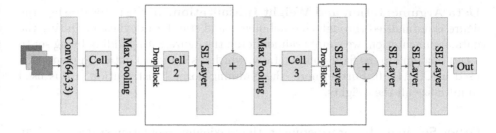

Fig. 1. Neural network template used as initial model in our experiments.

the population for initializing the population. During the evolution, the tournament selection [12] is adopted to update individuals for offspring generation, i.e. generate new neural architectures, where k individuals are selected from the population for mutation randomly at each generation. The individual with the best fitness is chosen as a parent network and will be mutated by one of the aforementioned six mutation operators, and the newly generated solution will be merged into the population. Once the population size exceeds a threshold, individuals with the lowest fitness values will be discarded. Otherwise, the population size will keep increasing. After the time budget is exhausted, the one with the highest fitness will be chosen for post-training and output as the final solution.

3 Experiments

The proposed method is compared with several state-of-the-art NAS methods on CIFAR-10 classification task. We first present the detailed settings of our proposed method, followed by the results achieved by the compared methods.

3.1 Experimental Setup

Network Template. The network template used in our experiments as initial model is sketched in 1. It starts with a small convolutional layer, followed by three cells and a large convolutional layer. Between each cell, there is a max pooling layer with stride being set to 2 for down sampling and a DropBlock [11] layer with $keep\_prob = 0.8$, where the block sizes are 7 and 5, respectively. Each cell is initialized with a single convolutional layer with 128 filters and 3×3 kernel size, and every convolutional layer is followed by batch normalization [16] and a ReLU activation. For convenience, we abbreviate the combination of three layers as a convolutional layer, and the reduction ratio for all SE layers is set to 4.

Data Argumentation and Weight Initialization. In the preprocessing, the figure is normalized through channel means and standard deviation. During the data argumentation, we simply random crop the figure with padding being set to 4 and random flip the figure horizontally, which is s standard data argumentation used in CIFAR-10. Furthermore, the Kaiming normal distribution [13] is used to initialize all the weights.

Other Settings. In our experiment, the maximum population size is set as 20. We will not discard any one of the population until the population exceeds the threshold. The tournament size is always set as 3, which is 15% of the maximum population size. During the search, all the neural architectures are trained with a batch size of 128 by using SGDR [21], where the initial learning rate is set to 0.05, T_0 is set to1, and T_{mul} is set to 2. The initial network template is first trained for 63 epochs. Then six mutated versions of the initial network are added to the population and form the initial population. Except for the initial network, all networks are trained for only 15 epochs after being created. Besides, we limit our search time to 12 GPU hours. Once the time budget is exhausted, we choose the best individual from the population and conduct post-training. The chosen network is trained until converged by using CutOut [7], where cutout size is set to 16×16, α in the Mixup with label smoothing [27,31] is set to 1. Finally, the error after post-training is reported.

3.2   Experiment Result

In this section, we compare the architecture achieved by our method with those architectures designed by both human experts and automated methods, where the results are shown in Table 1. The first block includes the results achieved by those manually designed architectures while the second block consists of the results achieved by automated methods, and the last block is the results achieved by our proposed method. It can be observed that our method can find a competitive neural architecture that achieves 2.55% test error with small amount of parameters (4.7M) within a very short period (12 GPU-hours) on CIFAR-10. Compared with other methods, our method takes much less time to obtain similar or even better performance, which has validated the efficiency of our proposed method. Furthermore, the best cell architecture found by our proposed method is given in Fig. 2, and the evolution diagram over time is shown in Fig. 3. The fast convergence rate of the solutions during the evolution has indicated the effectiveness of our proposed method in neural architecture search.

Table 1. Classification errors on CIFAR-10 achieved by different methods.

Method	Duration	Error	Params
ResNet [13]	N/A	6.41	1.7M
FractalNet [17]	N/A	5.22	38.6M
Wide-ResNet [30]	N/A	4.17	36.5M
DenseNet-BC [15]	N/A	3.42	25.6M
AmoebaNet-A [24]	3150 GPU-days	3.34	3.2M
Large-scale evolution [25]	2600 GPU-days	5.4	5.4M
NAS-v3 [33]	1800 GPU-days	3.65	37.4M
NASNet-A [34]	1800 GPU-days	2.65	3.3M
Hierarchical evolution [19]	300 GPU-days	3.75	15.7M
PNAS [18]	225 GPU-days	3.41	3.2M
Path-Level-EAS [2]	200 GPU-days	2.30	14.3M
EAS [1]	10 GPU-days	4.23	23.4M
DARTS [20]	4 GPU-days	2.83	3.4M
Neuro-cell-based-evolution [29]	1 GPU-days	3.58	7.2M
EENA [32]	0.65 GPU-days	2.56	8.47M
ENAS [22]	0.45 GPU-days	2.89	4.6M
Ours	**0.5 GPU-days**	**2.55**	**4.7M**

4   Conclusions

In this work, we design an efficient evolutionary neural architecture search based on cell search space and network morphisms. The combination of these two items allows discovering architecture with low classification error within a relatively short period. Due to the new mutation operators and network template, out method is more efficient than existing methods. Although we manually modify the macro architecture now, we will try to combine automated macro architecture search with our method, such as performing macro architecture search between cells (how many cells and how to align these cells) after search within the cell.

Acknowledgment. This work was supported in part by the National Natural Science Foundation of China (No. 61903178 and 61906081), in part by the Program for Guangdong Introducing Innovative and Entrepreneurial Teams grant (No. 2017ZT07X386), and in part by the Shenzhen Peacock Plan grant (No. KQTD2016112514355531).

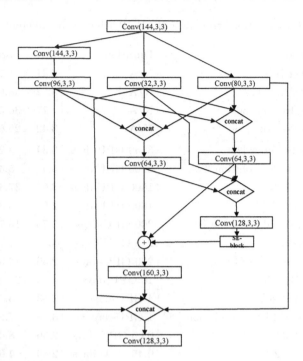

Fig. 2. The best cell architecture found by our proposed method on CIFAR-10 in 12 GPU-hours. The numbers in parentheses are the number of filters and kernel size. The SE-block is proposed in [14]. ⊕ denotes channel wise addition and **cancat** denotes channel wise cancatation.

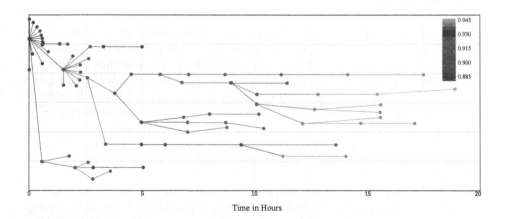

Fig. 3. Evolution process over time. Each dot represents an individual and each line connects an ancestor and an offspring.

References

1. Cai, H., Chen, T., Zhang, W., Yu, Y., Wang, J.: Efficient architecture search by network transformation. In: Thirty-Second AAAI Conference on Artificial Intelligence (2018)
2. Cai, H., Yang, J., Zhang, W., Han, S., Yu, Y.: Path-level network transformation for efficient architecture search. arXiv preprint arXiv:1806.02639 (2018)
3. Chen, L.C., et al.: Searching for efficient multi-scale architectures for dense image prediction. In: Advances in Neural Information Processing Systems, pp. 8699–8710 (2018)
4. Chen, T., Goodfellow, I., Shlens, J.: Net2Net: accelerating learning via knowledge transfer. arXiv preprint arXiv:1511.05641 (2015)
5. Chen, Y., Li, J., Xiao, H., Jin, X., Yan, S., Feng, J.: Dual path networks. In: Advances in Neural Information Processing Systems, pp. 4467–4475 (2017)
6. Chollet, F.: Xception: deep learning with depthwise separable convolutions. In: Proceedings of the IEEE Conference on Computer Vision and Pattern Recognition, pp. 1251–1258 (2017)
7. DeVries, T., Taylor, G.W.: Improved regularization of convolutional neural networks with cutout. arXiv preprint arXiv:1708.04552 (2017)
8. Eiben, A.E., Smith, J.: From evolutionary computation to the evolution of things. Nature 521(7553), 476 (2015)
9. Elsken, T., Metzen, J.H., Hutter, F.: Efficient multi-objective neural architecture search via Lamarckian evolution. arXiv preprint arXiv:1804.09081 (2018)
10. Elsken, T., Metzen, J.H., Hutter, F.: Neural architecture search: a survey. arXiv preprint arXiv:1808.05377 (2018)
11. Ghiasi, G., Lin, T.Y., Le, Q.V.: DropBlock: a regularization method for convolutional networks. In: Advances in Neural Information Processing Systems, pp. 10727–10737 (2018)
12. Goldberg, D.E., Deb, K.: A comparative analysis of selection schemes used in genetic algorithms. In: Foundations of Genetic Algorithms, vol. 1, pp. 69–93. Elsevier (1991)
13. He, K., Zhang, X., Ren, S., Sun, J.: Deep residual learning for image recognition. In: Proceedings of the IEEE Conference on Computer Vision and Pattern Recognition, pp. 770–778 (2016)
14. Hu, J., Shen, L., Sun, G.: Squeeze-and-excitation networks. In: Proceedings of the IEEE Conference on Computer Vision and Pattern Recognition, pp. 7132–7141 (2018)
15. Huang, G., Liu, Z., Van Der Maaten, L., Weinberger, K.Q.: Densely connected convolutional networks. In: Proceedings of the IEEE Conference on Computer Vision and Pattern Recognition, pp. 4700–4708 (2017)
16. Ioffe, S., Szegedy, C.: Batch normalization: accelerating deep network training by reducing internal covariate shift. arXiv preprint arXiv:1502.03167 (2015)
17. Larsson, G., Maire, M., Shakhnarovich, G.: FractalNet: ultra-deep neural networks without residuals. arXiv preprint arXiv:1605.07648 (2016)
18. Liu, C., et al.: Progressive neural architecture search. In: Ferrari, V., Hebert, M., Sminchisescu, C., Weiss, Y. (eds.) ECCV 2018. LNCS, vol. 11205, pp. 19–35. Springer, Cham (2018). https://doi.org/10.1007/978-3-030-01246-5_2
19. Liu, H., Simonyan, K., Vinyals, O., Fernando, C., Kavukcuoglu, K.: Hierarchical representations for efficient architecture search. arXiv preprint arXiv:1711.00436 (2017)

20. Liu, H., Simonyan, K., Yang, Y.: DARTS: differentiable architecture search. arXiv preprint arXiv:1806.09055 (2018)
21. Loshchilov, I., Hutter, F.: SGDR: stochastic gradient descent with warm restarts. arXiv preprint arXiv:1608.03983 (2016)
22. Pham, H., Guan, M.Y., Zoph, B., Le, Q.V., Dean, J.: Efficient neural architecture search via parameter sharing. arXiv preprint arXiv:1802.03268 (2018)
23. Real, E., Aggarwal, A., Huang, Y., Le, Q.: Aging evolution for image classifier architecture search. In: AAAI Conference on Artificial Intelligence (2019)
24. Real, E., Aggarwal, A., Huang, Y., Le, Q.V.: Regularized evolution for image classifier architecture search. arXiv preprint arXiv:1802.01548 (2018)
25. Real, E., et al.: Large-scale evolution of image classifiers. In: Proceedings of the 34th International Conference on Machine Learning, vol. 70, pp. 2902–2911. JMLR. org (2017)
26. Simonyan, K., Zisserman, A.: Very deep convolutional networks for large-scale image recognition. arXiv preprint arXiv:1409.1556 (2014)
27. Szegedy, C., Vanhoucke, V., Ioffe, S., Shlens, J., Wojna, Z.: Rethinking the inception architecture for computer vision. In: Proceedings of the IEEE Conference on Computer Vision and Pattern Recognition, pp. 2818–2826 (2016)
28. Wen, W., Yan, F., Li, H.: AutoGrow: automatic layer growing in deep convolutional networks. arXiv preprint arXiv:1906.02909 (2019)
29. Wistuba, M.: Deep learning architecture search by neuro-cell-based evolution with function-preserving mutations. In: Berlingerio, M., Bonchi, F., Gärtner, T., Hurley, N., Ifrim, G. (eds.) ECML PKDD 2018. LNCS (LNAI), vol. 11052, pp. 243–258. Springer, Cham (2019). https://doi.org/10.1007/978-3-030-10928-8_15
30. Zagoruyko, S., Komodakis, N.: Wide residual networks. arXiv preprint arXiv:1605.07146 (2016)
31. Zhang, H., Cisse, M., Dauphin, Y.N., Lopez-Paz, D.: mixup: beyond empirical risk minimization. arXiv preprint arXiv:1710.09412 (2017)
32. Zhu, H., An, Z., Yang, C., Xu, K., Xu, Y.: EENA: efficient evolution of neural architecture. arXiv preprint arXiv:1905.07320 (2019)
33. Zoph, B., Le, Q.V.: Neural architecture search with reinforcement learning. arXiv preprint arXiv:1611.01578 (2016)
34. Zoph, B., Vasudevan, V., Shlens, J., Le, Q.V.: Learning transferable architectures for scalable image recognition. In: Proceedings of the IEEE Conference on Computer Vision and Pattern Recognition, pp. 8697–8710 (2018)

Survival Time Prediction Model of Esophageal Cancer Based on Hierarchical Clustering and Random Neural Network

Huifang Guo[1,2], Enhao Liang[1,2], and Chun Huang[1,2]($\boxtimes$)

[1] Henan Key Lab of Information-Based Electrical Appliances,
Zhengzhou University of Light Industry, Zhengzhou 450002, China
huangchunzzuli@yeah.net
[2] School of Electrical and Information Engineering,
Zhengzhou University of Light Industry, Zhengzhou 450002, China

Abstract. Esophageal cancer is one of the most common malignant tumors in the world. There are many methods for early prediction of esophageal cancer at present. However, there are insufficient methods for early prediction of esophageal cancer. This paper presents a relatively new prediction model, Hierarchical Clustering and Random Neural Network prediction model. Hierarchical Clustering extracts the indexes that affect the prediction of survival time, and construct 15 different index combinations. All index combinations are brought into Random Neural Network. By comparing the corresponding prediction results, the prediction model of the optimal index combination is determined, and the prediction accuracy reaches 81.8180%. This study provides a possible method for predicting the survival time of patients with esophageal cancer.

Keywords: Hierarchical Clustering · Random Neural Network · Esophageal cancer

1 Introduction

Esophageal cancer (EC) is one of the most common malignant tumors in the world. Its mortality rate ranks the 6th in the global gastrointestinal tumor-related mortality rate [1]. According to pathological classification, esophageal cancer can be divided into Esophageal Adenocarcinoma (EAC) and Esophageal Squamous Cell Carcinoma (ESCC) [2]. In Asia, esophageal squamous cell carcinoma is the main pathological type, accounting for more than 90 of the diagnosis of esophageal cancer [3,4]. There are many methods for early prediction of sophageal cancer at present. However, there are insufficient methods for early prediction of esophageal cancer [5]. Blood routine examination is a detection method that reflects the normal morphology and quantity of blood cells in

© Springer Nature Singapore Pte Ltd. 2020
L. Pan et al. (Eds.): BIC-TA 2019, CCIS 1160, pp. 509–517, 2020.
https://doi.org/10.1007/978-981-15-3415-7_42

patients. It is widely used in clinical practice. It would be very meaningful to predict the survival time of patients with early esophageal cancer through blood index analysis. At present, most of the relevant studies only use statistical software to retrospectively analyze patients' data. In literature [6], Kaplan-Meier method and Cox regression model were used to evaluate the survival rate of patients with superficial esophageal cancer after the operation and to explore the factors influencing the prognosis of superficial esophageal cancer. In literature [7], based on Rothman-Keller model, a risk score table and a risk assessment model of esophageal cancer were established by using the method of health risk factors assessment. In literature [8], log-rank test, logistic regression and Cox model were used to analyze the independent risk factors for recurrence of esophageal cancer after operation. There are intricate relationships among the indicators. Few studies analyzed variables that may affect survival. Only through scientific and objective analysis of the indicators and extraction of effective information, can we further study the impact of blood indicators on survival of patients with esophageal cancer, and then obtain reliable and effective predictive data. This paper presents a relatively new prediction model, namely Hierarchical Clustering (HC) and RNN prediction model. With the help of SPSS software and MATLAB software, through cluster analysis, we find different index combinations, and use the RNN to verify the survival prediction of different index combinations and select the optimal model. This model greatly simplifies the input combination of the neural network, speeds up the training speed of the neural network, and improves the training accuracy. Compared with the traditional prediction model, the model has higher accuracy. It provides a more efficient and reliable method for predicting the survival or prognosis of esophageal cancer, which is a possible innovation of this paper.

2   A Survival Prediction Model Based on Hierarchical Clustering and RNN

2.1   Hierarchical Clustering Related Theory

The basic idea of Hierarchical Clustering is as follows. Firstly, n samples (indicators) are regarded as n classes, one of which has only one sample (indicator). Then, according to the distance or similarity between any two classes, the closest two classes are classified into a new class. Similarly, the similarity between new classes and other classes are analyzed, and the most similar classes are merged at the same time, and so on until the end of this paper. All samples (or indicators) are merged into one group [9].

There are seven common cluster analysis methods, which are intergroup connection, intragroup connection, nearest neighbor element, farthest neighbor element, centroid clustering, median clustering, and ward method. In this paper, we only introduce the ward method. The method of deviation square sum was put forward by ward, so it is called ward method. The basic principle of this method is to divides n samples into n classes, and then reduces one class at each

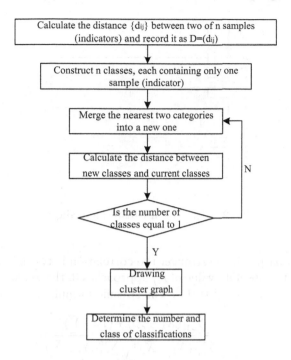

Fig. 1. Flow chart of Hierarchical Clustering model.

operation. With the decrease of the number of classes, the sum of squares of deviations increases gradually. The two classes with the smallest sum of squares of deviations are grouped into one class, until the n classes are merged into one class, and then the operation ends [10]. The flow chart of the hierarchical clustering model is shown in Fig. 1.

To facilitate analysis, count WBC, lymphocyte, monocyte, neutrophil, eosinophil, basophil, erythrocyte, hemoglobin, platelet, total protein, albumin, globulin, PT (pro-thrombin time), INR, APTT (activated partial thromboplastin time), TT (coagulation time), blood enzyme time and FIB are recorded as X_1, X_2, X_3, X_4, X_5, X_6, X_7, X_8, X_9, X_{10}, X_{11}, X_{12}, X_{13}, X_{14}, X_{15}, X_{16}, X_{17}, respectively. In order to accelerate the convergence speed of clustering, and avoid data annihilation, and reduce the sensitivity of singular data to the algorithm, date must first be normalized, and then input into SPSS software. 17 variables are clustered by hierarchical clustering method. The result is shown in Fig. 2.

2.2 Correlation Matrix

In the process of clustering variables, pearson correlation coefficient is the most commonly used method to measure the similarity between variables.

In statistics, pearson correlation coefficient is also known as pearson productmoment correlation coefficient (PPMCC or PCCs). r_{ij} is used to express the similarity coefficient between index i and index j. The closer the absolute

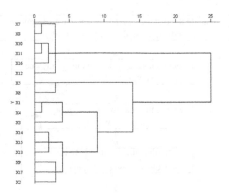

Fig. 2. Hierarchical Clustering pedigree.

value of r_{ij} approaches 1, the stronger the correlation between index i and index j is. The closer the absolute value of r_{ij} approaches 0, the weaker the correlation between index i and index j is [11]. The specific formula is as follows:

$$r = \frac{\sum_1^n (X_i - \overline{X}) * (Y_i - \overline{Y})}{\sqrt{\sum_1^n (X_i - \overline{X})^2} \sqrt{\sum_1^n (Y_i - \overline{Y})^2}}$$

Seventeen blood indices are analyzed by pearson correlation analysis with the help of SPSS software. The correlation coefficient matrix is shown in Table 1.

In Table 1, the maximum correlation coefficient is $r_{7,8} = 1.0$. X_7 and X_8 are merged into a new class. The correlation coefficients between new classes and different classes are calculated, and the maximum correlation coefficients are found out. One class is reduced at a time. Finally, Fig. 2 is obtained. According to the pedigree chart, 17 blood indices can be classified into four classes. Class A: X_5, X_6; Class B: X_7, X_8, X_{10}, X_{11}, X_{12}, X_{16}; Class C: X_1, X_3, X_4; Class D: X_2, X_9, X_{13}, X_{14}, X_{15}, X_{17}. Cluster analysis is based on the correlation between variables, which is actually equivalent to reducing the dimension of 17 variables into 4 variables. As for the Clustering results, we can't figure out which kind can accurately predict the survival time, which may be one kind or a combination of different kinds. So we make all combinations of these four classes. There are 15 combinations, which are shown in Table 2. Taking each combination of variables as independent variables and patients' survival as objective function, the survival prediction model based on random neural network is constructed.

2.3   Survival Prediction Model Based on RNN

The RNN model was proposed by Gelenbe in literature [12–14]. As a mathematical model of bionic neurons, RNN has shown strong advantages in associative memory and combinatorial optimization, and has been widely used [15]. The main purpose of random neural network learning is to obtain an appropriate

Table 1. Coefficient matrix.

Case	X_1	X_2	X_3	X_4	X_5	X_6	X_7	X_8	X_9	X_{10}	X_{11}	X_{12}	X_{13}	X_{14}	X_{15}	X_{16}	X_{17}
X_1	1.0																
X_2	0.4	1.0															
X_3	**0.6**	0.4	1.0														
X_4	**0.9**	0.1	0.5	1.0													
X_5	0.3	0.2	0.4	0.2	1.0												
X_6	0.2	0.1	0.2	0.2	0.4	1.0											
X_7	0.0	0.1	0.0	0.0	0.1	−0.1	1.0										
X_8	0.1	0.1	0.1	0.1	0.1	−0.1	**1.0**	1.0									
X_9	0.3	0.2	0.2	0.3	0.2	0.1	−0.1	−0.2	1.0								
X_{10}	0.2	0.1	0.1	0.2	0.0	−0.1	0.3	0.4	0.1	1.0							
X_{11}	−0.1	−0.1	−0.1	−0.1	−0.2	−0.2	0.5	0.4	0.0	**0.6**	1.0						
X_{12}	0.3	0.1	0.2	0.3	0.1	0.1	0.0	0.1	0.1	**0.7**	−0.1	1.0					
X_{13}	0.0	0.1	−0.1	0.0	−0.1	−0.1	0.1	0.2	−0.1	0.2	0.2	0.0	1.0				
X_{14}	0.0	0.2	−0.1	0.0	−0.1	−0.1	0.1	0.2	−0.1	0.2	0.2	0.0	**0.8**	1.0			
X_{15}	−0.1	0.2	−0.1	−0.1	−0.1	0.1	0.1	0.1	0.1	0.0	0.1	−0.1	0.4	0.4	1.0		
X_{16}	−0.2	0.0	0.1	−0.1	0.0	0.0	−0.1	−0.1	−0.1	0.0	0.1	−0.2	−0.1	−0.1	0.0	1.0	
X_{17}	0.4	0.0	0.3	0.4	0.2	0.1	0.1	0.1	0.3	0.2	0.1	0.2	−0.1	−0.1	−0.2	−0.3	1.0

weight matrix. As a result, when the input is a vector of excitation and suppression signal flow rate, the output of the network will be expected. Or its quadratic variance with the expected value is minimum [16]. The algorithm structure is shown in Fig. 3.

Table 2. The design of schemes.

Combination	Independent variable
A	X_5, X_6
B	X_7, X_8, X_{10}, X_{11}, X_{12}, X_{16}
C	X_1, X_3, X_4
D	X_2, X_9, X_{13}, X_{14}, X_{15}, X_{17}
A+B	X_5, X_6, X_7, X_8, X_{10}, X_{11}, X_{12}, X_{16}
A+C	X_1, X_3, X_4, X_5, X_6,
A+D	X_2, X_5, X_6, X_9, X_{13}, X_{14}, X_{15}, X_{17}
B+C	X_1, X_3, X_4, X_7, X_8, X_{10}, X_{11}, X_{12}, X_{16}
B+D	X_2, X_7, X_8, X_9, X_{10}, X_{11}, X_{12}, X_{13}, X_{14}, X_{15}, X_{16}, X_{17}
C+D	X_1, X_2, X_3, X_4, X_9, X_{13}, X_{14}, X_{15}, X_{17}
A+B+C	X_1, X_3, X_4, X_5, X_6, X_7, X_8, X_{10}, X_{11}, X_{12}, X_{16}
A+C+D	X_1, X_2, X_3, X_4, X_5, X_6, X_9, X_{13}, X_{14}, X_{15}, X_{17}
A+B+D	X_2, X_5, X_6, X_7, X_8, X_9, X_{10}, X_{11}, X_{12}, X_{13}, X_{14}, X_{15}, X_{16}, X_{17}
B+C+D	X_1, X_2, X_3, X_4, X_7, X_8, X_9, X_{10}, X_{11}, X_{12}, X_{13}, X_{14}, X_{15}, X_{16}, X_{17}
A+B+C+D	X_1, X_2, X_3, X_4, X_5, X_6, X_7, X_8, X_9, X_{10}, X_{11}, X_{12}, X_{13}, X_{14}, X_{15}, X_{16}, X_{17}

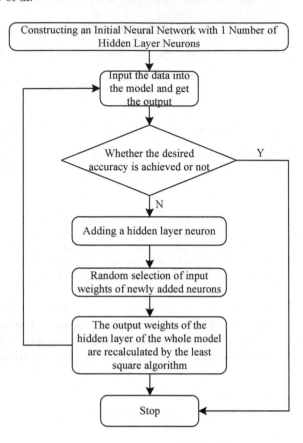

Fig. 3. RNN flow.

3   Experimental Results and Analysis

3.1   Data Processing

Because this paper studies the influence of blood indicators on survival of patients with esophageal cancer, we first delete the information of non-esophageal cancer and surviving patients, leaving 111 groups of data. 90 groups of data are used for training, 11 groups for testing, 10 groups for validation. The blood index combination is input, and the survival time is output. Based on the practical significance of five-year survival for survival assessment of tumor patients, the survival time is divided into two categories: less than five years as '0', we call it 'live short', greater than five years as '1', we call it 'live long'.

3.2   Parameter Setting

Maximum number of hidden nodes L_max = 200; Minimum expected error tol = 0.01; Maximum times of random configurations T_max = 100; Random weight range Lambdas = 0.5, 1, 5, 10, 30, 50, 100; Characteristic parameters R = [0.9, 0.99, 0.999, ... 0.9999, 0.99999, 0.999999]; Node which need to be added in the network in one loop is nB = 1.

3.3   Operation Network

Fifteen model schemes are investigated in terms of average test accuracy, prediction error and verify accuracy. The results are as follows: The RNN introduces random changes to the neural network and gives the neuron random weights. The output results are different each time. The average value is recorded as the standard of comparative analysis, as shown in Table 3.

Table 3. Model predicted results.

Programme	Number of input variables	Testing accuracy	MSE	Verify accuracy
Option 1	2	0.952381	0.317756	0.961156
Option 2	6	0.609536	0.000587	0.617505
Option 3	3	0.914284	0.202388	0.913653
Option 4	6	0.638094	0.007900	0.668092
Option 5	8	0.590474	0.006447	0.601515
Option 6	5	0.857149	0.105588	0.836592
Option 7	8	0.657144	0.007502	0.538533
Option 8	9	0.619048	0.005736	0.649389
Option 9	12	0.542856	0.004943	0.542856
Option 10	9	0.666666	0.007539	0.602121
Option 11	11	0.695238	0.006704	0.701563
Option 12	14	0.325259	0.008434	0.333318
Option 13	**11**	**0.838092**	**0.008942**	**0.818180**
Option 14	15	0.504768	0.006535	0.318764
Option 15	17	0.451169	0.01	0.440474

From Table 3, it can be seen that in the above schemes, option 1, option 3, option 6 and option 13 have higher verify accuracy and can predict survival better. Compared with the four optiones, Five times verify Acc and MSE results are shown in Figs. 4 and 5, respectively. By comparing, option 1 and option 3 have higher accuracy, but their MSE are also higher. So we exclude these two optiones. The prediction accuracy of scheme 6 and 13 is higher, but the average relative error of option 13 is smaller. So option 13 is optimal. That is to say, the model whose indicator combination are these 11 variables can predict the survival time better, and the prediction accuracy reaches 81.8180%.

3.4   Comparative Experiment

I have completed a lot of experiments under the conditions of using cluster analysis and not using cluster analysis. The experimental results have showed that the

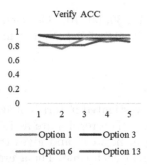

Fig. 4. Each Verify Acc of each training session of Option 1, Option 3, Option 6 and Option 13.

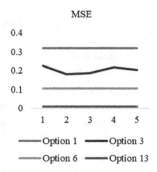

Fig. 5. Each MSE of each training session of Option 1, Option 3, Option 6 and Option 13.

prediction accuracy was only 44.0474% when all 17 variables were brought into the neural network, far lower than the model proposed in this paper (81.8180%). In addition, the training speed of this model proposed in this paper is also faster. This indicates that the prediction model proposed in this paper based on hierarchical clustering and random neural network can accurately predict postoperative survival of patients with esophageal cancer.

4   Conclusion

Esophageal cancer is one of the most common malignant tumors in the world, accounting for the 8th most common tumors. In this paper, a relatively new prediction model, hierarchical clustering and RNN prediction model. It could predict the survival of patients with esophageal cancer according to the blood parameters. We hoped that the model can be applied to practice and provide a relatively specific and effective solution to solve the problem of survival prediction.

Acknowledgements. This work was supported in part by the National Key R and D Program of China for International S and T Cooperation Projects (2017YFE0103900), in part by the Joint Funds of the National Natural Science Foundation of China (U1804262), in part by the State Key Program of National Natural Science of China under Grant 61632002, in part by the National Natural Science of China under Grant 61603348, Grant 61775198, Grant 61603347, and Grant 61572446, in part by the Foundation of Young Key Teachers from University of Henan Province (2018GGJS092), and in part by the Youth Talent Lifting Project of Henan Province (2018HYTP016) and Henan Province University Science and Technology Innovation Talent Support Plan under Grant 20HASTIT027.

References

1. Liu, Y.X., Zhang, L.Y., Cui, Y.Y.: Mechanism of miRNA in the pathogenesis and treatment of esophageal squamous cell carcinoma. J. Zhengzhou Univ. (Med. Ed.) **53**(2), 137–142 (2018)

2. Arnold, M., Soerjomataram, I., Ferlay, J., Forman, D.: Global incidence of oesophageal cancer by histological subtype in 2012. Gut **64**(3), 381–387 (2015)

3. Siegel, R.L., Miller, K.D., Jemal, A.: Cancer statistics, 2015. CA: Cancer J. Clin. **65**(1), 5–29 (2015)

4. Abnet, C.C., Arnold, M., Wei, W.Q.: Epidemiology of esophageal squamous cell carcinoma. Gastroenterology **154**(2), 360–373 (2018)

5. Ohashi, S., Miyamoto, S., Kikuchi, O., Goto, T., Amanuma, Y., Muto, M.: Recent advances from basic and clinical studies of esophageal squamous cell carcinoma. Gastroenterology **149**(7), 1700–1715 (2015)

6. Li, M., Zeng, X., Tan, L., Chen, S.: Survival and prognostic evaluation of superficial esophageal cancer after surgery. Zhonghua Yi Xue Za Zhi **96**(6), 460–463 (2016)

7. Zhang, Y., Chen, F.: A risk appraisal model of esophageal carcinoma for individual. Pract. Gerontol. **30**(5), 427–430 (2016)

8. Cheng, C., Wu, H., Gao, Q.: Risk factors analysis of postoperative recurrence in patients with esophageal cancer. Zhonghua Yi Xue Za Zhi **16**(11), 1366–1399 (2018)

9. Li, J., Zhou, M.L.G.: Hierarchical clustering method to identify and characterise spatiotemporal congestion on freeway corridors. IET Intell. Transp. Hierarchicals **12**(8), 826–837 (2018)

10. Kabir, M.E., Wang, H., Bertino, E.: Efficient systematic clustering method for k-anonymization. Acta Informatica **48**(1), 51–66 (2011)

11. Ly, A., Marsman, M., Wagenmakers, E.J.: Analytic posteriors for Pearson's correlation coefficient. Stat. Neerl. **72**(1), 4–13 (2018)

12. Gelenbe, E.: Stability of the random neural network model. Neural Comput. **2**(2), 239–247 (1990)

13. Gelenbe, E.: Learning in the recurrent random neural network. Neural Comput. **5**(1), 154–164 (1993)

14. Sun, J., Zhao, X., Fang, J., Wang, Y.: Autonomous memristor chaotic systems of infinite chaotic attractors and circuitry realization. Nonlinear Dyn. **94**(4), 2879–2887 (2018). https://doi.org/10.1007/s11071-018-4531-4

15. Gelenbe, E., Liu, P., LainLaine, J.: Genetic algorithms for route discovery. IEEE Trans. Syst. Man Cybern. Part B (Cybern.) **36**(6), 1247–1254 (2006)

16. Wang, D., Li, M.: Stochastic configuration networks: fundamentals and algorithms. IEEE Trans. Cybern. **47**(10), 3466–3479 (2017)

Applying of Adaptive Threshold Non-maximum Suppression to Pneumonia Detection

Hao Teng[1], Huijuan Lu[1(✉)], Minchao Ye[1], Ke Yan[1], Zhigang Gao[2], and Qun Jin[3]

[1] Jiliang University, Hangzhou, China
hjlu@cjlu.edu.cn
[2] Hangzhou Dianzi University, Hangzhou, China
[3] Waseda University, Tokyo, Japan

Abstract. Hyper-parameters in deep learning are sensitive to prediction results. Non-maximum suppression (NMS) is an indispensable method for the object detection pipelines. NMS uses a pre-defined threshold algorithm to suppress the bounding boxes while their overlaps are not significant. We found that the pre-defined threshold is a hyper-parameter determined by empirical knowledge. We propose an adaptive threshold NMS that uses different thresholds to suppress the bounding boxes whose overlaps are not significant. The proposed adaptive threshold NMS algorithm provides improvements on Faster R-CNN with the AP metric on pneumonia dataset. Furthermore, we intend to propose more methods to optimize the hyper-parameters.

Keywords: Adaptive threshold · Pneumonia detection · NMS · Hyper-parameters

1 Introduction

Deep learning has achieved great successes in the past few years. However, most models are implemented with hyper-parameters designed based on empirical knowledge. For example, a convolutional neural network requires the determination of the number of convolutional filters, input channels, output channels and many other parameters related to the model training process. Hence, hyper-parameters in deep learning are sensitive to prediction results.

Some deep learning methods have been proposed to detect pneumonia. Nowadays object detection is a significant procedure in computer vision in which an algorithm locates an object and predicts a bounding box with a category label for each instance of interest in an image. Object detectors based on convolutional neural networks have achieved many actual applications in face detection, pedestrian detection, lung nodule detection etc. Therefore, object detection can also be applied to pneumonia detection.

© Springer Nature Singapore Pte Ltd. 2020
L. Pan et al. (Eds.): BIC-TA 2019, CCIS 1160, pp. 518–528, 2020.
https://doi.org/10.1007/978-981-15-3415-7_43

With the development of computer vision, R-CNN [1] and Fast R-CNN [2] are proposed. Afterward Faster R-CNN [3] surpasses the first two as described before in accuracy and speed. Region Proposal Network (RPN) is firstly proposed to generate anchor boxes. NMS occurs in two phases, the RPN phase and the Fast R-CNN phase. In both phases, we are conducting the bounding boxes regression and NMS. The result of NMS is the final bounding boxes which are used to locate objects.

In the NMS algorithm, firstly, the bounding boxes are ranked in descending order according to the scores, and then the bounding boxes with the largest score is selected as the final detection boxes. Since there are a certain number of repeating bounding boxes, the bounding boxes with a value greater than a threshold are eliminated, and these operations are repeated until all the bounding boxes are traversed.

Most modern detectors such as Faster R-CNN, SSD [4] and YOLOv2 [5] use NMS algorithm and it has long been believed that the use of NMS is the key to suppressing the bounding box whose overlap is not significant. Although they have achieved great success, it is important to note that the NMS algorithm will be improved because the pre-defined threshold used in NMS is a hyper-parameter determined by empirical knowledge. For example, Faster R-CNN applies NMS with a threshold 0.7 and YOLOv3 applies NMS with a threshold 0.5 when test. To the end, we propose an adaptive threshold NMS algorithm, which applies two methods to figure out an adaptive threshold as follow:

- Calculate the first difference [6] of scores.
- Divide scores into two clusters by applying mean and variance.

NMS is an indispensable module of the object detection pipeline, which is used to suppress redundant bounding boxes divided by a pre-defined threshold and reduces the number of false positives at the same time. But the pre-defined threshold is not accurate because it's a const designed by empirical knowledge. Dividing scores by a stationary threshold into two slices is just like dividing scores into two clusters and we propose an adaptive threshold algorithm based on clustering. According to it, we propose an adaptive threshold NMS algorithm to improve NMS algorithm of object detection models in this paper.

Without bells and whistles, our adaptive threshold NMS algorithm is briefly described as follow:

- all detection boxes are sorted on the basis of their scores in ascending order;
- two methods to figure out an adaptive threshold are proposed;
- the original value is replaced with the adaptive threshold;

Our adaptive threshold NMS algorithm has achieved significant improvements in average precision measured over multiple Intersection-over-Union (IoU) scores for Faster R-CNN on pneumonia dataset released by Radiological Society of North America (RSNA). Since our adaptive threshold NMS algorithm does not require any redundant training and is convenient to implement, it can be simply added in the object detection pipeline.

2   Related Work

Learning rate, batch size, moment, and weight decay are all hyper-parameters [7]. Nowadays there are no straightforward ways to set hyper-parameters except designing them by empirical knowledge. Setting hyper-parameters effectively can improve performance.

NMS is an indispensable module with some hyper-parameters of many detection models for almost 50 years such as edge detection [8], feature point detection [9], and object detection [10–12]. Although significant progress in common object detection is created by deep learning, the greedy NMS algorithm is still the state-of-the-art method for this task, where a bounding box with the maximal score of detection is selected and bounding boxes whose overlap is not significant are suppressed using a pre-defined threshold designed by empirical knowledge. However, there is room to improve for the algorithm with some hyper-parameters. Next, we will introduce several improvements to the NMS algorithm.

Soft-NMS [13] is proposed to improve results of models in 2017. Instead of removing all the bounding boxes which have an overlap greater than a threshold, soft-NMS lowers the detection scores by scaling them as a linear or Gaussian function of overlap. Soft-NMS had achieved success in average precision on standard datasets like PASCAL VOC and COCO dataset. In their paper, specific experiments are conducted which analyses the sensitivity of a threshold attached to NMS algorithm from a range between 0.3 and 0.8. Experiments prove that hyper-parameters in deep learning are sensitive to prediction results.

Learning NMS [14] is proposed to improve results of NMS, which is implemented by learning a deep convolutional neural network and using only bounding boxes and scores. The proposed architecture is complex and specifically designed, which is evaluated on PETS and COCO dataset. Adaptive NMS algorithm [15] addresses a challenging issue of pedestrian detection in a crowd by a novel NMS algorithm to better suppress the bounding box whose overlap is not significant. In this paper, proposed adaptive-NMS algorithm applies an adaptive suppression threshold designed by the target density to a single instance. Fitness NMS [16] is proposed to identify the best available bounding box rather than only a sufficiently accurate one by searching the bounding box whose confidence is greatest. A convolutional neural network [17] which combines results of greedy NMS with different overlap thresholds is proposed. MaxpoolNMS [18] is proposed to be better suitable for hardware-based acceleration than greedy NMS by performing multi-scale max-pooling algorithm, 2-channel max-pooling across aspect ratios algorithm and max-pooling across scales algorithm in two-stage object detectors.

These works actually still remain hyper-parameters from algorithms and architecture based on convolutional neural network, which are even added more hyper-parameters because they are designed more and more complex. In the end, we propose our adaptive threshold NMS algorithm by two methods described before to replace original overlap threshold with the adaptive value to improve performance for Faster R-CNN.

3   Adaptive Threshold Methods

NMS algorithm with a threshold of 0.7 is used in Faster R-CNN. This threshold is determined by experience and cannot be adapted to all images. The input of NMS algorithm is the bounding boxes and the scores. We use two methods to find an appropriate threshold to adapt to a single image data. In this section, we will describe how our adaptive threshold NMS algorithm implements in detail.

The original NMS algorithm is described below: Bounding boxes will be sorted on the basis of their scores in ascending order, followed by a recursive process: pick a bounding box with the largest score, and eliminate the rest bounding boxes whose IoU is greater than a certain threshold.

We design algorithms to select thresholds from scores, which contains following steps:

First, we sort all detection boxes on the basis of their scores in ascending order.

Second, we propose two methods to get an adaptive threshold per image. And we regard indexes of scores as the X-axis and scores as the Y-axis. Then one-dimensional scores become two-dimensional scores, see Fig. 1. We use the first difference and the method based on mean value and variance value to calculate an adaptive threshold. Third, we replace the original threshold with the adaptive threshold calculated by our methods.

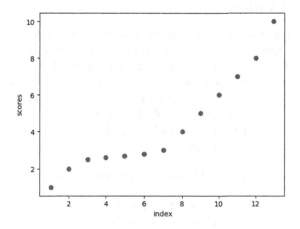

Fig. 1. Two-dimensional scores

In the following sections, these two methods used to calculate the threshold will be discussed in more detail.

3.1   The First Difference

The first derivative is defined as follow:

$$f'(x_0) = \lim_{x \to x_0} \frac{f(x) - f(x_0)}{x - x_0} \tag{1}$$

which represents the rate of change or the degree of perpendicularity of a curve. We define the indexes of scores as x, and define the scores as $f(x)$. Next, we define N as the length of scores per image. We calculate the first difference [20] of adjacent scores sorted in ascending order from the first value to the penult value of scores and we can get a list of values whose length is $N-1$ because the length of original array of scores is N. Because the difference of the independent variable is one. According to the first derivative formula described before, we define the first difference as follow:

$$a_i - a_{i-1} \tag{2}$$

where a_i denotes the ith value of scores. We figure out the index marked as $imax$ of maximum for the array of the first difference values whose length is $N-1$. We calculate an adaptive threshold as follow:

$$\frac{s_{imax} + s_{imax+1}}{2} \tag{3}$$

where $s_i max$ denotes the $imax$th value of original scores sorted in ascending order. In the end, we replace the original pre-defined threshold with the value calculated from formula (3).

3.2   Mean and Variance

Suppose the scores array is indexed from 0 to $N-1$ when N denotes the length of scores. We go through the array of scores from the first value to the penult value, and select one value of scores as a temporary threshold. For example, suppose we select the ith value as the temporary threshold. Then we calculate the mean and variance of two temporary clusters separated by the threshold. Specifically we calculate mean value of the array indexed from 0 to i and the array indexed from $i+1$ to $N-1$. And similarly, we calculate the variance this way. We define a formula as follow:

$$|Mean(l) - Mean(r)| - (Var(l) + Var(r)) \tag{4}$$

where l and r denotes two clusters of scores, l denotes the array indexed from 0 to i, r denotes the array indexed from $i+1$ to $N-1$, Mean denotes mean value of scores and Var denotes variance of scores. Then we get a list of values calculated by the formula (4), whose length is s. Intuitively, the larger the difference between two mean values is and the smaller the sum between two variance values is are what we want to achieve an adaptive threshold because the array of scores can be well divided into two clusters. Afterwards, we figure out the index marked as $imax$ of maximum for the list whose length is $N-1$ based on mean value and variance value. To the end we calculate an adaptive threshold from formula (3) and replace the original pre-defined threshold with the threshold. In formula

(2), we take the average of the $imax$th value and the $(imax + 1)$th value to be somewhat robust.

4   Experiments

In this section we will introduce the dataset used in the experiment and the division of the training set and the test set, and then we will show the results of the two methods that can be compared with baseline, next we do the sensitivity analysis experiment of the threshold, finally we explain why the experiment is effective. All results are measured in average precision (AP), which is the area under the recall-precision curve.

Our experiments are based on Resnet-101 [20], which is pre-trained on the ImageNet [21] dataset. The network is 101 layers deep and can classify images into 1000 object categories, such as keyboard, mouse, pencil, and many animals. As a result, the network has learned rich feature representations for a wide range of images. And we fine-tune the resulting model using SGD [22] (Stochastic Gradient Descent) with initial learning rate 0.001, 0.9 momentum, 0.0005 weight decay and batch size 4. We train our code on GTX 1070. The pneumonia dataset has the same input sizes (1024×1024).

Our experiments are conducted on the classical two-stage object detection model Faster R-CNN. Taking Faster R-CNN as the baseline, we improve the NMS algorithm and get comparable results.

4.1   Dataset

We used the pneumonia dataset released by Radiological Society of North America (RSNA) in the training phase, which contains 6012 frontal-view X-ray image of 6012 patients. The society annotates each image with up to 2 different bounding boxes. As shown in Fig. 2, an image includes more than one objects. For the pneumonia detection task, we randomly split the dataset into training (4810 patients, 4810 images), and test (1202 patients, 1202 images). There is no patient overlap between the sets.

4.2   Experimental Results

As shown in Table 1, we compare traditional NMS and our adaptive threshold algorithm with different Intersection-over-Union (IoU) scores on pneumonia dataset. It is clear that our adaptive threshold algorithm (using the first difference or mean and variance) improves performance. For example, we achieve an improvement of 6.74% and 3.05% respectively for the first difference and the method based on mean and variance when the IoU score is 0.6, which is trained on Faster-RCNN and is significant for the pneumonia dataset. We vary different Intersection-over-Union (IoU) values on the test set of pneumonia dataset, see

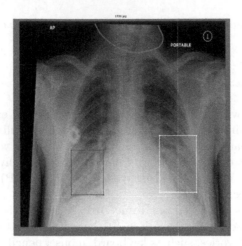

Fig. 2. Bounding boxes of pneumonia dataset

Table 1. Results on pneumonia dataset which use NMS as baseline and our proposed adaptive threshold method

AP	Baseline	Mean Var	First Diff
AP@0.5	36.00%	42.07%	45.56%
AP@0.6	17.99%	21.04%	24.73%

Figs. 3 and 4. All results are measured in average precision (AP), which is the area under the recall-precision curve. The green curve represents the baseline of Faster R-CNN and the brown and blue curve are our improved methods based on Faster R-CNN. The green curve and the brown curve are all above and to the right of the blue curve. Intuitively, the area under the green curve and the brown curve are all larger than the area under the blue curve.

4.3   Threshold Sensitivity Experiment

Songtao Liu et al. [15] performed experiments of sensitivity to hyper- parameters using R-FCN [23] on coco dataset, which varies different thresholds of NMS algorithm from 0.3 to 0.8. To prove that hyper-parameters in deep learning (e.g. object detection in computer vision) are sensitive to prediction results (e.g. mean average precision). Our threshold sensitivity experiments are based on Resnet-50, which is pre-trained on the ImageNet dataset. As a result, the network has learned rich feature representations for a wide range of images. And we fine-tune the resulting model using Adam [24] with initial learning rate 0.00001 and batch size 4. We train our code on Tesla V100-SXM2-16GB. The pneumonia dataset has the same input sizes (1024×1024). As shown in Table 2, it can be seen that

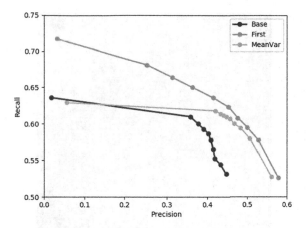

Fig. 3. Precision vs Recall at an IoU score 0.5

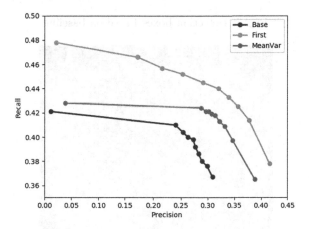

Fig. 4. Precision vs Recall at an IoU score 0.6

Table 2. Results on pneumonia dataset which use four thresholds of NMS as baseline named Faster R-CNN

Threshold	0.5	0.6	0.7	0.8
AP	29.17%	29.75%	31.94%	32.27%

with the increase of the threshold, the AP(average precision) of the experiment presents an upward trend. The maximum percentage change is 3.1%.

4.4   Output Comparison of NMS

We compare the output bounding boxes after NMS algorithm and conduct experiments on the image numbered 4811. As shown in Figs. 5, 6 and 7, the detection

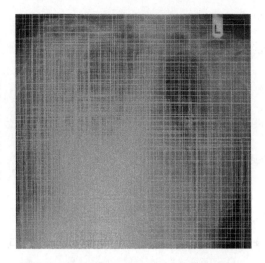

Fig. 5. Detection boxes based on baseline

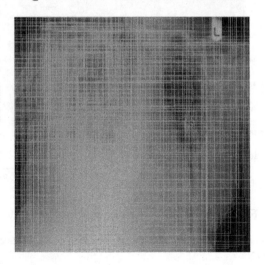

Fig. 6. Detection boxes with the method of mean and variance

boxes from NMS are evenly distributed, but not relatively concentrated in one area. In other words, scattered over the entire image. However, when we scan the next figure, we can find that the detection box is relatively concentrated in two areas, and when we scan the last figure, the phenomenon is more obvious. This shows that our method makes the detection box found by NMS algorithm more accurate and easier to fine-tune.

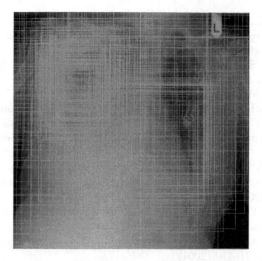

Fig. 7. Detection boxes with the first difference

5   Conclusion

In this paper, we have presented an adaptive threshold NMS algorithm based on two methods to better suppress the bounding boxes whose overlaps are not significant. The first phase is to sort the scores and then figure out the first difference of adjacent values, so that the largest value of the first difference is the adaptive threshold value. The second phase is to traverse the score list, and then calculate the mean and variance according to the statistical characteristics, so that the difference between the two groups of values is as big as possible and the sum of variance is as small as possible. As shown in experiments, our methods show significant improvements over Faster R-CNN on the pneumonia dataset released by Radiological Society of North America (RSNA). Our study suggests that searching for an adaptive threshold can be regarded as clustering. Futher more, we intend to propose more advanced algorithms of adaptive thresholds because hyper-parameters are sensitive to deep models.

Acknowledgements. This work is supported by the National Natural Science Foundation of China under Grants No. 61272315, No. 61572164, No. 61701468, No. 61877015 and the project of education planning in Zhejiang under Grants No. 2018SCG005. Key Laboratory of Electromagnetic Wave Information Technology and Metrology of Zhejiang Province. Student research project of China Jiliang university No. 2019X22030.

References

1. Girshick, R., Donahue, J., Darrell, T., Malik, J.: Rich feature hierarchies for accurate object detection and semantic segmentation. In: IEEE Conference on Computer Vision and Pattern Recognition, pp. 580–587 (2014)
2. Girshick, R.: Fast R-CNN. In: Computer Science, pp. 1440–1448 (2015)

3. Ren, S., He, K., Girshick, R., Sun, J.: Faster R-CNN: towards real-time object detection with region proposal networks. IEEE Trans. Pattern Anal. Mach. Intell. **39**(6), 1137–1149 (2017)
4. Liu, W., et al.: SSD: single shot multibox detector. In: Leibe, B., Matas, J., Sebe, N., Welling, M. (eds.) ECCV 2016. LNCS, vol. 9905, pp. 21–37. Springer, Cham (2016). https://doi.org/10.1007/978-3-319-46448-0_2
5. Redmon, J., Farhadi A.: YOLO9000: better, faster, stronger, pp. 7263–7271 (2016)
6. Spalding, D.B.: A novel finite difference formulation for differential expressions involving both first and second derivatives. Int. J. Numer. Meth. Eng. **4**(4), 551–559 (1972)
7. Smith, L.N.: A disciplined approach to neural network hyper-parameters: part 1-learning rate, batch size, momentum, and weight decay. arXiv preprint arXiv:1803.09820, pp. 1–21 (2018)
8. Rosenfeld, A., Thurston, M.: Edge and curve detection for visual scene analysis. IEEE Trans. Comput. **20**(5), 562–569 (1970)
9. Lowe, D.G.: Distinctive image features from scale-invariant keypoints. Int. J. Comput. Vis. **60**(2), 91–110 (2004)
10. Lin, T.Y., Dollár, P., Girshick, R., He, K., Hariharan, B., Belongie, S.: Feature pyramid networks for object detection, pp. 2117–2125 (2016)
11. Lin, T.Y., Goyal, P., Girshick, R., He, K., Dollar, P.: Focal loss for dense object detection, pp. 2999–3007 (2017)
12. Tian, Z., Shen, C., Chen, H., He, T.: FCOS: fully convolutional one-stage object detection, pp. 1–13 (2019)
13. Bodla, N., Singh, B., Chellappa, R., Davis, L.S.: Soft-NMS - improving object detection with one line of code, pp. 5561–5569 (2017)
14. Hosang, J., Benenson, R., Schiele, B.: Learning non-maximum suppression. In: Proceedings of the IEEE Conference on Computer Vision and Pattern Recognition, pp. 4507–4515 (2017)
15. Liu, S., Huang, D., Wang, Y.: Adaptive NMS: refining pedestrian detection in a crowd, pp. 6459–6468 (2019)
16. Tychsen-Smith, L., Petersson, L.: Improving object localization with fitness NMS and bounded IoU loss, pp. 6877–6885 (2017)
17. Hosang, J., Benenson, R., Schiele, B.: A convnet for non-maximum suppression. In: Rosenhahn, B., Andres, B. (eds.) GCPR 2016. LNCS, vol. 9796, pp. 192–204. Springer, Cham (2016). https://doi.org/10.1007/978-3-319-45886-1_16
18. Cai, L., et al.: MaxpoolNMS: getting rid of NMS bottlenecks in two-stage object detectors. In: Proceedings of the IEEE Conference on Computer Vision and Pattern Recognition, pp. 9356–9364 (2019)
19. Overington, I., Greenway, P.: Practical first-difference edge detection with subpixel accuracy. Image Vis. Comput. **5**(3), 217–224 (1987)
20. He, K., Zhang, X., Ren, S., Sun, J.: Deep residual learning for image recognition, pp. 770–778 (2015)
21. Deng, J., Dong, W., Socher, R., Li, L.J., Li, F.F.: ImageNet: a large-scale hierarchical image database. In: IEEE Conference on Computer Vision and Pattern Recognition, pp. 248–255 (2009)
22. Ketkar, N.: Stochastic gradient descent. In: Optimization, pp. 113–132 (2014)
23. Dai, J., Li, Y., He, K., Sun, J.: R-FCN: object detection via region-based fully convolutional networks, pp. 379–387 (2016)
24. Ruder, S.: An overview of gradient descent optimization algorithms, pp. 1–14 (2017)

Recognition of an Analog Display Instrument Based on Deep Learning

Zhihua Chen[1], Kui Liu[2], and Xiaoli Qiang[1(✉)]

[1] Institute of Computing Science and Technology, Guangzhou University,
Guangzhou 510006, Guangdong, China
qiangxl@gzhu.edu.cn
[2] School of Automation, Huazhong University of Science and Technology,
Wuhan 430074, Hubei, China

Abstract. With the development of remote automatic recording of utility meter readings, the problems of locating an analog display instrument against a complex background and recognition instruments with inconsistent scales are highlighted. This paper presents an automatic recognition method based on deep learning using the Faster-RCNN algorithm, which can locate the instrument position quickly and accurately in images with large amounts of noisy interference. By using the LeNet-5 convolutional neural network and the proportional relation between the pointer position and both ends of the scale, the number on the dial is recognized automatically. Experimental results showed that the Faster-RCNN algorithm can effectively detect the instrument position in images with large amounts of interference, thus laying an important foundation for subsequent pointer processing and reading recognition. Furthermore, the combination of number recognition and pointer location can make this method effective for different types of instruments with different ranges, which achieves better recognition than traditional approaches.

Keywords: Recognition of analog display instruments · Deep learning · Faster-RCNN · LeNet-5 · Inconsistent scale

1 Introduction

Analog display instruments are widely used in many fields such as the chemical industry and electrical engineering because of their simple structure, ease of use and low cost. However, it is easy for the manual meter reading to obtain error reading by the fatigue and subjective. In addition, poor working environments, such as the high temperature, pressure and noise, mean that staff cannot stay long in the work environment to record the instruments. Therefore, research into obtaining intelligent and automatic readings of analog display instruments is very important.

There are many studies on automatic readings for analog display instruments. Sablatnig and Kropatsch first proposed a recognition method using the Hough

© Springer Nature Singapore Pte Ltd. 2020
L. Pan et al. (Eds.): BIC-TA 2019, CCIS 1160, pp. 529–539, 2020.
https://doi.org/10.1007/978-981-15-3415-7_44

transform to detect the instrument dial, but did not propose an effective method for deformed dials [1,2]. For elliptical dials, some ellipse detection methods have been proposed [3,4]and some publications have proposed the Canny edge detection algorithm to detect edge and pointer contours [5]. But how to extract a dial automatically from a complex background with environmental disturbances has rarely been discussed [6]. Some researchers have used image processing technology to obtain accurate measurements [7], and computer vision has also been used to recognize analog display instruments [8]. For different lighting conditions, corresponding improvements in the Hough algorithm have been proposed [9]. However, good suggestions are still lacking for the dial separation under multiple simultaneous interferences such as uncertain size, large tilt angle, and variable light intensity. Moreover, the wide range of instruments in use means that instrument ranges and index values are very different from each other. Then it is difficult to use one system to obtain readings from different types of instruments. Some researchers have used BP neural networks to recognize numbers on a dial, but no further automatic identification methods are proposed [10,11]. Recently, the Faster-RCNN algorithm has achieved good recognition and detecting performance [12,13]. This paper attempts to use the Faster-RCNN algorithm based on deep learning to locate the position of an instrument quickly and accurately against a background with multiple interferences. This approach achieves good results with various interferences occurring at the same time and improves the number and the pointer angle recognition accuracy even in different instrument ranges and index values.

The paper is divided into the following sections according to the recognition process of analog instruments: instrument detection and location; image preprocessing; character recognition; and the experiment results analysis. The detailed procedure is shown in Fig. 1.

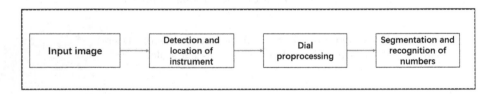

Fig. 1. Realization steps of recognition system

2   Instruments Detection and Location

With the development of deep learning, target detection methods based on region extraction have gradually become popular. The typical algorithm is RCNN, which is based on the region convolution neural network [14]. During detection process, the RCNN can be divided into four steps: firstly, using "Selective

Search" to generate a large number of proposed regions; secondly, using a CNN network to extract feature vectors for each proposed region; thirdly, using SVM, decision tree, or other classifiers to classify the feature vectors and to discriminate object categories; and finally regressing accurately the position and size of the target peripheral frame. The target detection method based on the region extraction achieves a significant improvement in accuracy and efficiency compared with the traditional sliding window. Fast-RCNN [15] and Faster-RCNN [16] were proposed with better efficiency for target detection in 2015, which are improved versions of RCNN. Faster-RCNN uses regional adaptive sampling and optimizes the whole network, which remarkably improves the overall recognition rate. It also uses a proposed region network instead of a selective regional search algorithm to generate the appropriate network, thus solving the problem of excessive time consumption for region extraction and making real-time detection and recognition possible. In this paper, Faster-RCNN is used in image positioning to make the whole instrument identification process more intelligent.

The Faster-RCNN algorithm consists of two CNN networks: RPN (Regional Proposal Network) and the Fast-RCNN detection network. The structure of the whole network is shown in Fig. 2.

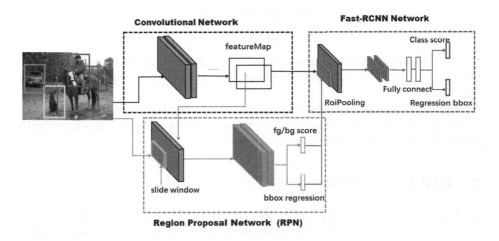

Fig. 2. Faster-RCNN network structure

An RPN network is a convolutional neural network, which uses an $n * n$ sliding window to generate a $256D$ or $512D$ feature vector in the shared volume of a feature map layer. According to the characteristics of each map points, K possible proposal regions are generated. Then $4 * K$ proposal region coordinate parameters and $2 * K$ regional attributes through two fully connected layers, which are after the feature map layer. Three scales and three scale factors are used in the Faster-RCNN to determine the corresponding current sliding window center position of the nine initial proposal windows. The corresponding window areas are 128^2, 256^2, and 512^2, and the length-width ratios of the windows are

1:1, 1:2, 2:1. By comparing the intersection over union (IoU) ratio of the target (in real area) between the output window predicted by the network and the input image, forecast regions with an IoU ratio greater than 0.7 are considered to be positive samples, while forecast regions with an IoU ratio less than 0.3 are considered to be negative samples. And regions with IoU ratios between 0.7 and 0.3 are discarded. The positive samples are used to complete the process of coordinate refinement according to the corresponding relation between the feature map and the original image. The RPN network is shown in Fig. 3.

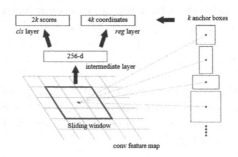

Fig. 3. RPN network

In this paper, VGG16 network is adopted as the basic network for feature extraction, and the result of instrument location and recognition is shown in Fig. 4.

The results show that the algorithm can obtain the instrument accurately while rejecting other objects of similar shape to instruments even in images with a complex background and many interferences.

3   Dial Image Processing

Dial images are usually color images. In order to facilitate the follow-up of pointer location and recognition of instrument numbers and reduce the amount of data, the image preprocessing is necessary. Only the dial areas are processed, because the dial area has already been determined in Sect. 2.

3.1   Region-Adaptive Binarization

Many methods are available to accomplish binarization of a dial image, including region binarization and global binarization. In practical applications, it is difficult for the global binarization to obtain good results because of the complex environment of the dial and the intensity of illumination. Therefore, region-adaptive binarization is used in this paper. By means of a regional sliding window and comparing the relationship between the current threshold and the average threshold of the region, the image is converted from a gray image to a binary

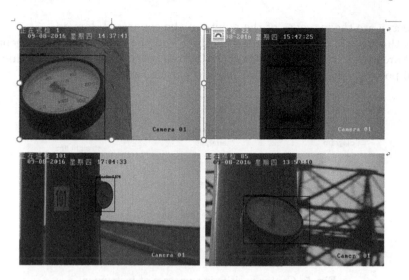

Fig. 4. The result of instrument location using Faster-RCNN

image. The size of the sliding window is $m * n$. The results for the region image of any point $P(x, y)$ after region-adaptive binarization are given as Eq. 1:

$$f'(x, y) = \begin{cases} 0, & f(x, y) < \frac{1}{m*n} \sum_{i=-m/2}^{m/2} \sum_{i=-n/2}^{n/2} f(x+i, y+j) \\ 255, & f(x, y) \geq \frac{1}{m*n} \sum_{i=-m/2}^{m/2} \sum_{i=-n/2}^{n/2} f(x+i, y+j) \end{cases} \quad (1)$$

where, $f'(x, y)$ is the binary value of (x, y) and $f(x, y)$ is the gray value of (x, y).

3.2 Center Location and Pointer Refinement

A method based on connected component analysis and fitting a straight line through the scale of the instrument is used to determine the instrument center area for the dials with sectorial or round shape.

The least-squares method is a common mathematical means of performing linear regression. For the points in scale connected domains around a dial, their linear relationships satisfy the relation: $y = ax + b$, where a, b are the slope and intercept respectively. When a, b make Eq. 2 reach its minimum value, the variance between the fitted straight line and the actual data is at a minimum:

$$\sum_{i=1}^{N} [y_i - (a + bx_i)]^2 \quad (2)$$

The linear relation can be obtained by solving for the coefficient, as shown in Eq. 3:

$$a = \frac{1}{\sum_{i=1}^{N}(x_i - \bar{x})} (\sum_{k=1}^{N}(x_k - \bar{x})(y_k - \bar{y})), b = \bar{y} - a\bar{x} \quad (3)$$

where $\bar{x}, \bar{y}$ are the averages of the points in the connected domain.

In this paper, the least squares method and the connected components of image are used to locate the center of the instrument and accomplish the thinning of the pointer. The specific algorithm flow can be summarized as Fig. 5.

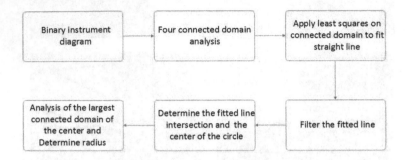

Fig. 5. Center positioning process of instrument

First, connected domain partition for binarization images with 4 adjacency models, and then linear fitting of all connected regions by the least square method. After that, combined with the actual instrument calibration and instrument center point of distribution, distribution of connected domain in the instrument of real area and the slope of the regression line absolute value is too large to eliminate line. The remaining straight lines meet the point of intersection in step 4, after eliminating the abnormal intersection point, all the intersection points are weighted averaging, and the final center of the instrument circle is obtained. According to the position of the center of circle obtained in step 5, the connected-domain analysis is performed near the center of the circle to find the largest connected domain. By fitting the connected domain in a straight line, the refined direction of the pointer and the radius of the meter can be obtained.

The centering of the instrument and the refinement of the pointer are shown in Figs. 6 and 7 below.

4   Number Recognition

In order to facilitate digital positioning and positioning and statistics of the instrument scale, the dial scale and the region where the number is located need to be more accurately positioned. Therefore, in the actual processing of the method, the instrument obtained in the previous section is used. The coordinates of the circle center and the radius of the instrument, combined with the size of the numbers in the actual dial, with the circle as the center, the radius of the instrument plus or minus a certain pixel distance is the radius, the instrument region is divided, and the specific representation image of the scale line of the instrument is obtained, as shown in Fig. 8. Through the translation and rotation

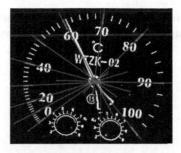

Fig. 6. Centering of the instrument

Fig. 7. Refinement of the pointer

of the coordinates (pointer scale region), a conversion of the circle and fan-shaped dial scale lines to the rectangle is realized, thereby facilitating the reading of the pointer scale line, the instrument scale after conversion are shown in Fig. 9. The projection method is used to perform histogram statistics on the converted rectangular scale, so as to obtain the proportion of the area pointed by the current pointer to the entire range of the instrument, and the same ideas are applied to position and segment the digital numbers on both sides of the meter pointer. The values that the pointer pointing to are recognized by the LeNet-5 convolution neural network. The number of digits on both sides of the application pointer, combined with the proportion of the instrument's range occupied by the pointer. Apply the following Eq. 4 to calculate the final instrument representation.

$$K = k_1 + (k_2 - k_1) * \frac{l}{l_0} \tag{4}$$

where, K is the final reading, k_1 and k_2 are the numbers on each side of the pointer, l_0 is the distance between k_1 and k_2, and l is the distance between the pointer and k_1.

For the non-uniform scale value of the dial, because the method is based on the digital numbers on both sides of the pointer combined with the ratio of the number of current pointers, instead of merely taking the ratio of the whole scale range of the pointer, for the recognition of the non-uniform dial, this method can also be used to quickly and effectively derive the final instrument representation number. By introducing digital recognition, the method of judging the instrument representation number according to the pointer angle is solved to cope with the problems of uneven scale and different types of instrument identification errors. Figure 10 shows the LeNet-5 network structural diagram.

Fig. 8. Instrument scale region

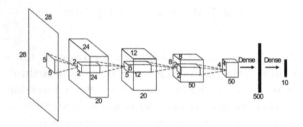

Fig. 9. The sector scale line is converted to a rectangle

Fig. 10. Network structure of LeNet-5

5   Experimental Results

5.1   Results of Instrument Detection and Location

There are various types of instruments with widely varying dial ranges and index values and the images used in the paper were taken from an actual power plant.

The Faster-RCNN algorithm was implemented using the Caffe framework. 1000 practical instrument images from power plants were presented as positive samples, and 1000 images of round objects similar to instruments such as safety hats and manhole covers were presented as negative samples in the training process. Moreover, horizontal mirrors and random cropping were used to increase the number of samples.

The validation set contained 200 test images. The dropout parameter in the Faster-RCNN was set to 0.6, the maximum number of iterations 8000, and the non-maxima suppression parameter 0.7. The number of proposal regions after non-maxima suppression was about 300. The test set consisted of 300 instrument images and 300 images similar to instruments. The results are shown in Table 1.

Table 1. Validation results for instrument detection algorithm

Category	Training set	Test set	Average accuracy (mAP)
Instrument	1000	300	96.28%
Similar to instrument	1000	300	92.16%

5.2 Results of Reading Recognition

The LeNet-5 convolution neural network was trained using an MNIST data set with an image size of 28*28, 50 thousand training sets, and 10 thousand validation sets. The training accuracy and loss curves are shown in Fig. 11.

Finally, the results of system recognition and manual interpretation of some images are compared in Table 2.

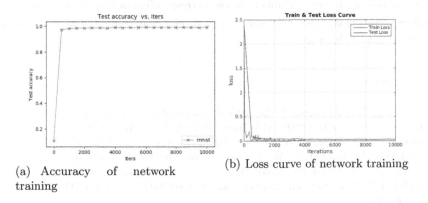

(a) Accuracy of network training

(b) Loss curve of network training

Fig. 11. Accuracy and loss curve of network training

Table 2. Automatic recognition and manual interpretation results

Instruments	Index value	Manual interpretation	System recognition	Recognition error
1	2	30.4	30.378	0.07%
2	2	87.2	87.327	0.15%
3	1	53.5	53.524	0.04%
4	1	32.0	32.157	0.04%
5	0.2	2.61	2.608	0.05 %
6	0.2	3.489	3.497	0.22 %
7	0.1	0.52	0.514	1.15%
8	0.2	1.99	1.994	0.20%
9	0.005	0.198	0.197	0.50%
10	0.005	0.151	0.153	1.32%

6   Conclusion

A deep learning based system of reading recognition for analog display instruments was constructed in the paper. The Faster-RCNN algorithm was performed for instrument detection and location. The number on each side of the pointer was recognized by the LeNet-5 convolutional neural network. By combining the scale segment, the inconsistent scales and different ranges of different types of instruments can be recognized correctly.

The experiments show that the algorithm is effective and that the system has a wide range of potential applications as well.

Acknowledgment. The authors thank the financial support for this work from Chinese National Natural Science Foundation (61672248, 61370105).

Data Availability Statement. The original images [.jpg] data used to support the findings of this study are available from the corresponding author upon request.

Conflict of Interest. The authors declare that there is no conflict of interest regarding the publication of this paper.

References

1. Sablatnig, R., Kropatsch, W.G.: Automatic reading of analog display instruments. In: Proceedings of the 12th IAPR International Conference on Pattern Recognition, Computer Vision and Image Processing, 1994, vol. 1, pp. 794–797. IEEE (1994)
2. Sablatnig, R., Kropatsch, W.G.: Application constraints in the design of an automatic reading device for analog display instruments. In: Proceedings of the Second IEEE Workshop on Applications of Computer Vision 1994, pp. 205–212. IEEE(1994)
3. Basca, C.A., Talos, M., Brad, R.: Randomized hough transform for ellipse detection with result clustering. In: The International Conference on IEEE Computer as a Tool, 2005, EUROCON 2005, vol. 2, pp. 1397–1400 (2005)
4. Prasad, D.K., Leung, M.K.H., Cho, S.Y.: Edge curvature and convexity based ellipse detection method. Pattern Recogn. **45**(9), 3204–3221 (2012)
5. Wang, Q., Fang, Y., Wang, W., et al.: Research on automatic reading recognition of index instruments based on computer vision. In: International Conference on Computer Science and Network Technology, pp. 10–13. IEEE (2014)
6. Han, J., Li, E., Tao, B., et al.: Reading recognition method of analog measuring instruments based on improved hough transform. In: International Conference on Electronic Measurement and Instruments, pp. 337–340. IEEE (2011)
7. Shen, X., Cao, M., Lu, Y., et al.: Research on automatic indication values of pointer gauges based on computer vision. In: 2016 China International Conference on Electricity Distribution (CICED), pp. 1–5. IEEE (2016)
8. Corra Alegria, E., Cruz Serra, A.: Automatic calibration of analog and digital measuring instruments using computer vision. IEEE Trans. Instrum. Measur. **49**(1), 94–99 (2000)

9. Xu, L., Fang, T., Gao, X.: An automatic recognition method of pointer instrument based on improved Hough Transform. In: International Society for Optics and Photonics, Applied Optics and Photonics China (AOPC2015), 96752T–96752T-10 (2015)
10. Ning, Z.: Automatic recognition method for instrument display based on BP neural network. Control Autom. **158**, 198 (2006)
11. Zhu, H.X.D.O.C.: Pointer instrument recognition based on BP network and improved Hough transform. Electr. Meas. Instrum. **05** (2015)
12. Lee, C., Kim, H.J., Oh, K.W.: Comparison of faster R-CNN models for object detection. In: 2016 16th International Conference on Control, Automation and Systems (ICCAS), pp. 107–110. IEEE (2016)
13. Jiang, H., Learned-Miller, E.: Face detection with the faster R-CNN. In: 2017 12th IEEE International Conference on Automatic Face Gesture Recognition (FG 2017), pp. 650–657. IEEE (2017)
14. Girshick, R., et al.: Rich feature hierarchies for accurate object detection and semantic segmentation. In: Computer Vision and Pattern Recognition, pp. 580–587. IEEE (2014)
15. Girshick, R.: Fast R-CNN. In: IEEE International Conference on Computer Vision IEEE Computer Society, pp. 1440–1448 (2015)
16. Ren, S., He, K., Girshick, R., et al.: Faster R-CNN: towards real-time object detection with region proposal networks. IEEE Trans. Pattern Anal. Mach. Intell. **39**(6), 1137–1149 (2017)

A Local Map Construction Method for SLAM Problem Based on DBSCAN Clustering Algorithm

Xiaoling Li[(✉)], Juntao Li, and Tao Mu[(✉)]

School of Information, Beijing Wuzi University, Beijing 101149, China
1183900654@qq.com, 349940276@qq.com

Abstract. The accuracy of the local map of the robot is crucial to the accuracy of the real-time positioning of the robot and the construction of the global map. However, the current local map construction methods have problems such as low accuracy and long calculation time, which lead to the result that the robot cannot locate in real time and the global map construction is not accurate. Therefore, this paper proposes a DBSCAN (Density-Based Spatial Clustering of Applications with Noise) method for local map construction in SLAM problem. According to this algorithm, firstly, the pre-processed feature points are subjected to density clustering algorithm for large-area division. Then through a dynamic region re-division method based on two influencing factors of angle and distance, finally, the least squares method is used to fit the point set in each divided region. The experimental results show that, according to the data set in this paper, when the number of clustering is eps = 6, then in the dynamic area subdivision method based on Angle and distance, then make $\theta = 5\%\theta i$, d = 5% di, Finally, the point set is straight-line fitted by the least squares method to obtain the most accurate local map.

Keywords: Local maps · DBSCAN · Least squares

1 Introduction

For the construction of local maps, people often use the method of "region segmentation-region re-segmentation-line segment fitting" to realize the construction of local maps. Area division is the first step of local map construction. The usual method is to calculate the distance between the data points obtained by the laser rangefinder. When the distance between a certain two points is greater than the previously set threshold, this point is taken as a segmentation point of regional segmentation [1], we call this a way of setting a threshold. However, the method of setting threshold is inefficient and time-consuming, because it is to calculate the distance between each point in the data set and compare each distance with the previously set empirical value to determine which point is the segmentation point, which leads to a large amount of computation and time consuming. In addition, some clustering algorithms can also be used for region segmentation, such as the k-mean algorithm [2]. The K-means

© Springer Nature Singapore Pte Ltd. 2020
L. Pan et al. (Eds.): BIC-TA 2019, CCIS 1160, pp. 540–549, 2020.
https://doi.org/10.1007/978-981-15-3415-7_45

algorithm is more accurate and less time-consuming than the threshold setting method, but it has the disadvantage of only finding spherical clusters [3]. Therefore, this paper adopts a clustering algorithm based on DBSCAN, which overcomes the shortcomings of k-means algorithm which can only find spherical clusters, and can realize clustering quickly, regardless of their shape and spatial dimension containing them [4, 5].

Regional re-segmentation is the second step of local map construction in this paper. Currently, there are several methods for regional segmentation:

(1) SEF (successive edge following): It starts from the starting point of the region (x1, y1), calculates the distance between two adjacent points, and compares the distance with the threshold to determine the point. Not a split point [1].
(2) LT (line tracking): given the line segment fitted by a certain data set, if the distance from a certain point to the line segment is greater than the preset threshold, the point is regarded as a segmentation point [1].
(3) Aiming at the shortcomings of SEF and LT method algorithms, such as large amount of calculation, time-consuming and low precision, this paper proposes a dynamic region re-division method based on two influencing factors, angle and distance. The method is theoretically Accuracy is higher than the SEF, LT method.

Line segment fitting is the third step in the construction of local maps in this paper. There are many ways to fit line segments. Valeiras David Reverter et al. proposed an event-based non-brightness algorithm. and this paper introduces an event-based luminance-free algorithm for line and segment detection from the output of asynchronous event-based neuromorphic retinas. Wang Dong used Hermite splines to fit small line segments [7]. Cao Hongju uses the least squares method for linear demixing and signal estimation of hyperspectral images [8]. In this paper, the line segment is fitted by the least squares method.

2 Related Work

2.1 DBSCAN Algorithm

DBSCAN is a noise-based clustering method based on density, it overcomes the shortcomings of the k-means algorithm that can only find spherical clusters [9, 10], and can quickly achieve clustering regardless of their shape and Contains its spatial dimensions [11–13].

The basic principle of DBSAN:

(1) Core object: if the density of a certain point reaches the threshold set by the algorithm, then it is the core point (that is, the number of points in the eps field is not less than minPts).
(2) Eps-neighbor distance value: set radius eps.
(3) The direct density is up to: If a point p is in the eps neighborhood of point q, and q is the core, the direct density of p-q is reachable.

(4) Density up to: if there is a sequence of points, $q_0, q_1, \cdots q_k$, for any $q_i - q_i - 1$ is directly reachable, it is said that the density from q_0 to q_k is reachable, which is actually the "propagation" of direct density.

2.2   The Least Squares

After the region is re-segmented, a set of points that can constitute a line segment is obtained [12]. In order to get more accurate local environment map, the least squares method is used to fit the straight line.

The core idea of the least squares method is to fit the set of points into a form of the form y = ax + b, ensuring that the X metric of each point in the set of points is as close as possible to the Y metric of that point. The specific mathematical formula is:

$$e^2 = \sum_{i=1}^{n} (y_i - (ax_i + b))^2 \tag{1}$$

Equation (1) is a function of a and b. The partial derivation of a and b is obtained, and the derivative value is taken as zero, i is the i-th point in the data set, and N is the number of points in the data set, and the equation group is obtained:

$$\begin{cases} (\sum_{i=1}^{n} x_i^2)a + (\sum_{i=1}^{n} x_i)b = \sum_{i=1}^{n} x_i y_i \\ (\sum_{i=1}^{n} x_i)a + nb = \sum_{i=1}^{n} y_i \end{cases} \tag{2}$$

Finally, solving the Eq. (2) can find the values of a and b, and get the equation of the line.

3   Problem Description and Proposed Methods

In this paper, the method of "region segmentation-region re-segmentation-line segment fitting" is used to construct the local map. The specific steps are as follows:

Step 1: Large area division based on DBSCAN.
Step 2: Region re-segmentation based on adaptive threshold.
Step 3: Line segment fitting based on least squares method.

In step 1, this paper adopts a large area division based on DBSCAN algorithm, because of the traditional method of setting the threshold(The factor considered by this method is the size between the distance D between successive two points and the empirical value), which is inefficient, computationally intensive and time consuming, and can only find spherical clusters based on the k-mean algorithm, while DBSCAN overcomes the shortcomings of the k-means algorithm that can only find spherical clusters, and can quickly achieve clustering, regardless of Their shape and the spatial dimensions that contain it, and the Fig. 1 is a schematic diagram of DBSCAN.

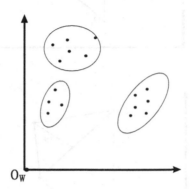

Fig. 1. Schematic diagram of region division based on density clustering

In step 2, this paper uses region re-division based on adaptive threshold [14, 15]. Regional subdivision is to re-area segmentation of each class based on density clustering. The purpose is to divide multiple subclasses from a large class to achieve local maps that can highlight indoor corners, columns, and even the ground. Trash cans, paper boxes, etc.

This paper proposes a dynamic region re-division method based on two influencing factors, angle and distance. As shown in Fig. 2, the basic principle of the method is in each class Ai, take the distance Dmn (Pm, Pn is the two points farthest from Ai) of the distance between the two points in the class. For any point Pi in Ai, if the angle between the line segments PmPi and PiPn and the distance between the Pi and the line segment MN, If $\theta_i < \theta$ and $d_i > d$, where $(\theta, d$ is the empirical value), then Pi is the dividing point.

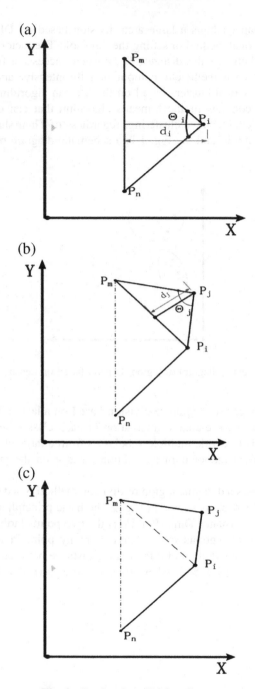

Fig. 2. Regional re-division diagram

In step3, the point set of the line segment is obtained through the processing of the first two steps [13]. In order to obtain more accurate local environment map, this paper adopts the most commonly used least squares method to fit the line. As shown in Fig. 3, the least squares method is used to fit the data set of the region after partition, so as to obtain line L.

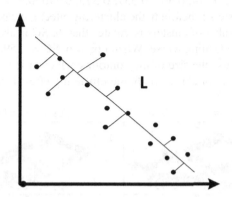

Fig. 3. Line segment fitting based on least squares method

4   Experimental Results and Analysis

4.1   Large Area Segmentation Based on Density Peak

In this paper, Dashgo robot and N30103A lidar of shenzhen radium intelligent system co., ltd, which were adopted as the experimental tool to collect environmental data. The original laser data layout is shown in Fig. 4 below. Among them, the linear data set is the walls and tables in the real environment, as well as some pedestrian and noise point data.

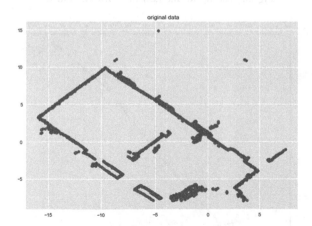

Fig. 4. Raw data layout

Figure 5 shows the different effect diagrams when the number of clusters is different. From Fig. 5, we can see that when the number of clusters is 2, some obvious noise points are divided into corresponding clusters. When the number of clusters is 3, we can clearly observe that the original data set is divided into three categories. As the number of clusters increases, we can see the changes in clustering results. Correspondingly, the contour coefficient SC (silhouette coefficient) corresponding to different cluster numbers is also different, as shown in fig., when eps selects 2, 3, 4, 5, 6, and 7, the corresponding contour coefficients are 0.514, 0.507, 0.537, 0.573, 0.596, 0.577. In Fig. 6, we can see that the number of clusters is before 6, the clustering effect increases with the number of clusters. When the number of clusters is greater than 6, SC is also reduced, that is, the clustering effect is also Getting worse. When eps = 6, SC = 0.596, at which time SC is at the maximum, because the size of the contour coefficient is related to the clustering effect. When the number of clusters is 6, the clustering effect is the best.

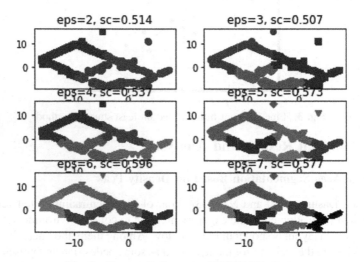

Fig. 5. The result diagram of DBSCAN cluster

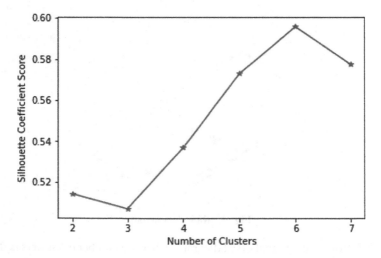

Fig. 6. The result diagram of DBSCAN cluster

The contour coefficient is used to measure the quality of the clustering. The closer the contour coefficient is to 1, the more reasonable the result is. Therefore, this paper chooses eps = 6 and minPts = 5 as the clustering parameters.

4.2 Regional Subdivision and Least Squares Method for Line Segment Fitting

In this paper, a partial data set is selected as an example. In this example, the dynamic region subdivision method based on the two influencing factors of angle and distance is used to obtain the local region map, as shown in Fig. 7, in this example. The least squares method is used to fit the point set in a local area, where the blue point set is the original data, the red part is the fitted line, and the local data set is fitted to the line segment, where a = −0.88563135, b = −11.11354395, straight line equation: $y = -0.8856x - 11.11$.

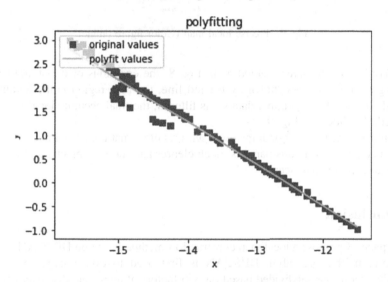

Fig. 7. Local straight line fitting (Color figure online)

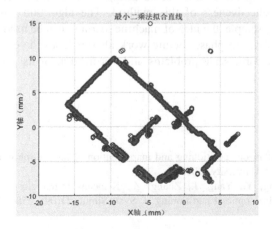

Fig. 8. Line fitting based on least squares method (Color figure online)

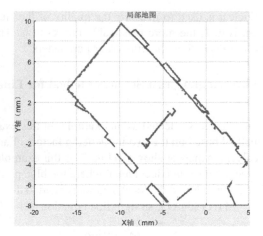

Fig. 9. Precise local map (Color figure online)

According to the same method as in Fig. 8, the equations of the straight lines in each region are fitted. As shown by the red line, the blue region is the layout of the original data set. If the original data set is filtered, the final accurate local map can be obtained. As shown in Fig. 9.

Compared with the original raw data set, it is clear that the partial map obtained by the method proposed in this paper is much clearer and more complete, and is closer to the actual experimental scene.

5   Conclusion

In this paper, I propose a local map construction method based on DBSCAN clustering algorithm. In this algorithm, DBSCAN is first used to divide large regions, then dynamic regions are subdivided based on two factors of angle and distance, and finally the least squares method is used for linear fitting in the divided regions. Experimental results show the effectiveness of this method. The algorithm has much room for improvement, For example, the idea of machine learning can be used to solve problems in the regional re-division phase, Future work also includes the extension of the proposed algorithm to large-scale problems and verification of its performance in real-world problems.

References

1. Tang, X.: Simultaneous positioning and map creation of mobile robot based on laser range finder. Shandong University (2007)
2. Zhang, J., Yang, Y., Yang, J., Zhang, Z.: K-means initial clustering center selection algorithm based on optimal partitioning. J. Syst. Simul. **21**(09), 2586–2590 (2009)

3. Ma, Y.: Research on image feature point extraction and matching algorithm. Harbin Engineering University (2018)
4. Han, L., Qian, X., Luo, J., Song, W.: DBSCAN multi-density clustering algorithm based on region partitioning. Comput. Appl. Res. **35**(06), 1668–1671+1685 (2018)
5. Xie, J., Gao, H., Xie, W.: K-nearest neighbor optimization of density peak fast search clustering algorithm. China Sci.: Inf. Sci. **46**(02), 258–280 (2016)
6. Yang, J., Wang, G., Pang, Z.: Research on the related problems of density peak clustering. J. Nanjing Univ. (Nat. Sci.) **53**(04), 791–801 (2017)
7. Cai, H., Chen, Y., Zhuo, L., Chen, X.: Lidar obstacle detection based on optimized DBSCAN algorithm. Optoelectr. Eng. **07**, 83–90 (2019)
8. Li, J., Chen, Z., Su, Z.: Research on vehicle trajectory data analysis method based on regional density partitioning. Microelectron. Comput. **36**(02), 53–56 (2019)
9. Zhang, X., Cheng, J., Kang, Q., Wang, X.: Image denoising based on iterative adaptive weighted mean filtering. J. Comput. Appl. **37**(11), 3168–3175 (2017)
10. Dang, X., Sun, J., Wu, W.: Uncertainty of straight line fitting of symmetric least squares method. J. Zhaotong Univ. **35**(05), 21–25 (2013)
11. Wu, O., Bouaswaig, A., Imsland, L., Schneider, S., Roth, M., Leira, F.: Campaign-based modeling for degradation evolution in batch processes using a multiway partial least squares approach. Comput. Chem. Eng. **128**, 117–127 (2019). https://doi.org/10.1016/j.compchemeng.2019.05.038
12. Wang, C., Cui, Z., Du, Y., Gao, Y.: Busy bus-to-station data repair based on DBSCAN and multi-source data. Comput. Appl. 1–11. http://kns.cnki.net/kcms/detail/51.1307.TP.201908 12.1035.002.html. Accessed 29 Sept 2019
13. Chen, Z., Liu, Y.: Photo classification technology based on density clustering algorithm. Sci. Technol. Innov. **23**, 75–76 (2019)
14. Wu, Y., Zeng, S., Zeng, Z., Liu, P.: Improved SLIC purple soil color image segmentation. J. Chongqing Norm. Univ. (Nat. Sci. Ed.), 1–11. http://kns.cnki.net/kcms/detail/50.1165.N. 20190926.1124.028.html. Accessed 29 Sept 2019
15. Fan, H., Meng, F., Liu, Y., Kong, F., Ma, J., Lv, Z.: A novel breast ultrasound image automated segmentation algorithm based on seeded region growing integrating gradual equipartition threshold. Multimedia Tools Appl. **78** (2019). https://doi.org/10.1007/s11042-019-07884-8

Chromosome Medial Axis Extraction Method Based on Graphic Geometry and Competitive Extreme Learning Machines Teams (CELMT) Classifier for Chromosome Classification

Jie Wang, Chaohao Zhao, Jing Liang$^{(\boxtimes)}$, Caitong Yue, Xiangyang Ren, and Ke Bai

School of Electrical Engineering, Zhengzhou University, Zhengzhou, China
liangjing@zzu.edu.cn

Abstract. Automated chromosome classification is a vital task in cytogenetics and has been a common pattern recognition problem. Numerous attempts were made in the past decade years to characterize chromosomes for the intention of medical and cancer cytogenetics research. This paper proposes a graphic geometry-based approach for medial axis extraction and Competitive Extreme Learning Machine Teams (CELMT) with further correction method for chromosome classification. The initial two medial axis points are determined firstly according to the length of the intercept line in different directions, and then the complete medial axis for feature extraction is drawn based on the initial two points. After that, a base classifier ELMi, j is trained to differentiate a pair class chromosome (i, j), a total number of 276 classifiers are trained. Each base classifier will give a label and the final label will be determined by majority voting and further correction rules. Based on the experiment results, the method proposed in this paper can precisely extract the medial axis and extract the features to recognize the chromosome, the classification accuracy by using CELMT can achieve an average value of 96.23% and the running time is much shorter than the other classification algorithms.

Keywords: Karyotype · Cytogenetics · Medial axis · Extreme Learning Machine

1 Introduction

Chromosome classification is a vital and difficult task in clinical diagnosis, radiation estimation, and biological research, especially for the diagnosis of genetic diseases, prenatal examination, cancer pathology research and quantitative determination of radiation in the environment [1,2]. Chromosome abnormalities include increase and/or decrease of chromosomes in quantity and structural abnormalities in one or more chromosomes. Quantitative abnormalities often cause fatal

© Springer Nature Singapore Pte Ltd. 2020
L. Pan et al. (Eds.): BIC-TA 2019, CCIS 1160, pp. 550–564, 2020.
https://doi.org/10.1007/978-981-15-3415-7_46

problems, for instance, Down syndrome (trisomy 21) and Edwards syndrome (trisomy 18) [3]. There are 23 pairs of chromosomes in the normal mankind cells, including 22 pairs of autosomes and a pair of sex chromsome. Each chromosome has two chromatids held together by the constricted region called centromere and will be divided into two cells in the telophase. A karyotype picture contains every chromosome in a cell at metaphase, which is photographed by an optical microscope. The photos are stained with the Giemsa-trypsin staining technique to show the different characteristics of each chromosome region, then the chromosomes are grouped, arranged and paired manually according to the size and morphological features of the chromosomes (see Fig. 1).

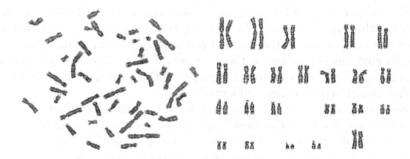

Fig. 1. A human metaphase image and the corresponding karyotype.

Nevertheless, chromosome recognition and analysis, which create karyotypes, are mostly manually performed by the staff of cytogenetics laboratories nowadays in a repetitive, time consuming and so as to be an expensive procedure [4,5]. Therefore, Automatic chromosome analysis has been attracting the attention of many scholars [6]. In this article, we propose a new way to determine the medial axis according to the chromosome's geometric feature and a method to integrate a two-class ELM classifier to classify chromosomes and make subsequent corrections. Solving the two-class classification problem with ELM is easier and more efficient than solving the multi-class problem, so we consider to use multiple two-class ELM classifier integrations to solve the 24 classes classification problem for karyotyping. The primary contributions of this paper are listed as follows:

- The geometrical feature information of the chromosome shape is used to extract the central axis, and the extracted central axis can overcome the phenomenon of a burr at the end of the chromosome existing in the conventional thinning algorithm, which is beneficial to the next feature extraction.
- Propose a method to integrate the two-class ELM to solve the multiclassification problem, and make appropriate adjustments for the Y chromosome data imbalance in the karyotype analysis during training and further correction according to the number of human cells chromosomes.

The rest of this paper is planned as follows: we will discuss the current status of karyotype analysis in Sect. 2. In Sect. 3, we will detail the specific basis and steps of our proposed medial axis extraction method. The basic theoretical knowledge of the Extreme Learning Machine and our proposed method of integrating the two-class classifier and the subsequent correction steps will be introduced in Sect. 4. Section 5 gives the experiment results and Sect. 6 presents the conclusion of this paper.

2   Related Works

With the development of pattern recognition technology, automatic analysis of chromosome karyotype is becoming more and more mature. Chromosome karyotyping has three main steps [7], The first step is image preprocessing and chromosome segmentation, due to the limits of the imaging system, chromosome image generally contains impurities, that are not conducive to later feature extraction, commonly used chromosome preprocessing methods include median filtering, morphological filtering, image segmentation, etc.

The second step is feature extraction, for a pattern recognition system, feature extraction is one of the most important aspects. The quality of feature extraction will directly affect the subsequent recognition [6,8]. In the early stage of karyotyping research, the chromosome's features used mainly included chromosome area, length and centromere index (The proportion of long arms in the whole chromosome) [9,10]. With the deepening of the research of karyotyping, some scholars had proposed banding patterns as an effective characteristic to identify chromosome [11–13]. The area of the chromosome is determined through the number of pixels of the binary chromosome image. However, the length of the chromosome, the centromere index and the banding profiles are dependent on the medial axis extraction of the chromosome. In the research of image processing, the current mainly used method for extracting the medial axis of the object is a thinning algorithm, but this method will produce burrs, especially at both ends of the chromosome [14]. The authors of Ref. [11] proposed a way to extract the medial axis by analyzing cross-section analysis along with four scanline directions. This method works for most chromosome images but doesn't work well for highly curved chromosome images.

The last step is chromosome classification, the classification method of chromosomes mainly includes statistical geometric classification, probability classification and modern machine learning algorithms based on neural networks [15–21]. Ref. [15] proposed a two-layer ANN network, the first layer containing one ANN classifier firstly divides the chromosomes into seven groups, and then seven ANN classifiers of the second layer corresponding to seven groups give each chromosome a specific label, this method can achieve an average 86.8% accuracy. Ali et al. pointed out a classification algorithm named Competitive SVM Teams (CSVMT), CSVMTs contains 22×22 sub-classifiers $SVM_{i,j}$, whose optimal parameters are found by employing the Pattern Search (PS) method. Each $SVM_{i,j}$ competes with the other classifier to label a new recognition instance and the final label of the test instance is determined by majority voting. The correct classification rate is 97.84% but the dataset used in the experiment only contains autosomal data. With the development of deep learning, increasing number of machine learning problems can be solved with deep learning, a CNN network with 6 convolutional layers, 3 pooling layers, 4 dropout layers, and 2 fully connected layers was proposed to solve chromosome classification problem [22].

3   Medial Axis Extraction

Due to the linear structure of the chromosome shape, for each point of the medial axis, the length of the line that intersects the line perpendicular to the central axis and the contour of the chromosome is smaller than the length of the line where the line in the other direction intersects the contour of the chromosome. The medial axis extraction algorithm proposed in this paper firstly draws a horizontal line under the middle of the chromosome and takes the midpoint of the line segment obtained by intersecting the line with the chromosome contour as the initial central axis point to be determined. After this point is determined, we will draw twelve lines through this point, the angle is from 0° to 180° interval 15° slash, then compare the length of the obtained lines with the chromosomal contour secant, and take the midpoint of the secant line with the smallest length as the final axis point. Then we move the horizontal line downward a certain of pixels to find the second medial axis points. After that more medial axis points will be found based on initial points and interpolation methods.

3.1   Chromosome Image Preprocessing

In the process of generating, transporting and recording, chromosome images inevitably contain a lot of noises and the image information has interfered. These noises are discrete and random, usually, we use a smooth method to process the image to eliminate noises.

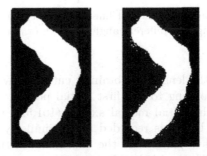

Fig. 2. Binary images with median filtering and without median filtering.

The median filter is a nonlinear processing technique that can overcome the difficulty of the blur caused by linear filters and is most effective for eliminating noise such as pulse interference and image scanning noise. Median filtering is to scan a picture with a matrix of m × n dimension (usually m = n, and m must be an odd number). Each time a pixel in the image passes through the center of the matrix, the value of this pixel is substituted with the median of the elements in the matrix. Therefore, median filtering is used to achieve noise reduction. Figure 2 shows the comparison of binary images with median filtering and without median filtering.

3.2   Determine the Initial Point

In order to determine the initial two points, we divide the picture into k equal parts by horizontal lines. In this paper, k takes value 8 and we use the third last bisector as the horizontal line to find the initial point. Firstly, determine the midpoint of the intersection line between the horizontal line and the chromosome contour and mark as (x_0, y_0). Then we make twelve lines from 0° to 180° interval 15° slash through (x_0, y_0), calculate the length of the obtained lines with the chromosomal contour secant. Figure 3(a) shows the initial horizontal line and the cut line's midpoint (x_0, y_0) and Fig. 3(b) shows the secant line of a different direction.

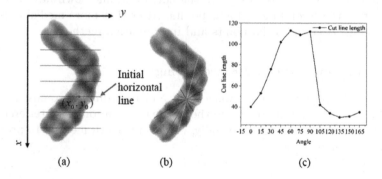

Fig. 3. (a) Initial horizontal line and cut line's midpoint, (b) secant line of a different direction, (c) Cutline length in different angle through (x_0, y_0).

Figure 3(c) shows the length of obtained cut lines, we can get the angle of the scant line corresponding to the first initial point to be 135°, recorded as α, the coordinate of the initial medial axis is (101,52), recorded as $P_1(x_1, y_1)$ corresponding section length is recorded as l_1. Similarly, the horizontal line is shifted downward by five pixels, and the coordinate of the second initial point is calculated as (106,48), recorded as $P_2(x_2, y_2)$, the angle of the corresponding cut line is 150°, recorded as β, corresponding section length is recorded as l_2.

3.3 Determination of the Remaining Medial Axis Points

After determining the initial two points, the remaining medial axis points can be determined based on the initial points. Since the initial two dots are roughly located in the middle of the chromosome, we can divide the remaining central axis points searching into downward searching and upward searching. The specific method of downward searching is carried out as follows:

Step 1: Calculate the difference Δx, Δy and the Euclidean distance d_{12} between the horizontal and vertical coordinate of the two initial points:

$$\Delta x = x_2 - x_1, \Delta y = y_2 - y_1$$
$$d_{12} = \sqrt{(\Delta x)^2 + (\Delta y)^2} \tag{1}$$

Step 2: Generate a new pending medial axis point $Q(m, n)$:

$$Q(m, n) = \begin{cases} \left(x_2 + \frac{\Delta x}{2}, y_2 + \frac{\Delta y}{2}\right), d_{12} > a \\ (x_2 + \Delta x, y_2 + \Delta y), b < d_{12} < a \\ (x_2 + 2\Delta x, y_2 + 2\Delta y), d_{12} < b \end{cases} \tag{2}$$

Here, the two thresholds a and b are set to make sure the new pending medial axis point is not too close to P_2 so that to avoid causing an error of the medial axis point.

Step 3: Make three straight lines through Q, the lines' angle is $\beta \pm 15°$ and β. Because the trend of chromosome contour changes little within a short distance, so we can narrow the search range into these three angles. Then calculate the length of the secant lines between the three lines and the chromosome contour, find the shortest cut line and calculate the midpoint $R(u, v)$, then we can get a new midpoint $P_3 (x_3, y_3)$ based on the Eq. (3), Where Row denotes the number of rows in the image:

$$P_3 (x_3, y_3) = \begin{cases} Q(m, n), \text{Row} -y_2 < c \cdot l_2 \text{ and } l_2 - l_3 > l_2/d \\ R(u, v), l_2 - l_3 \leq l_2/d \end{cases} \tag{3}$$

The limitation in the formula is mainly because there will be a large change of the length of the cut line at both ends of the chromosome.

Step 4: Loop through the above steps until one of the following equation is met:

$$(u < \text{Col}/e) \vee (\text{Col} -u < \text{Col}/e) \vee (v < \text{Row}/e) \vee \cdots$$
$$\cdots (\text{Row} -v < \text{Row}/e) \vee (l_i < f \cdot \max(l_1, l_2, \cdots l_{i-1})) = 1 \tag{4}$$

Where Col represents the number of columns of the image, the first four conditions are the end of the search for the point to the edge of the circumscribed rectangle of the chromosome. The last condition is that when the found line is too short, it indicates that the medial axis point is close to the chromosome's tail.

After searching the medial axis points under the initial points, we can search the medial axis points above the initial points in a similar way. Once the medial axis point searching is finished, we get the medial axis point sequence as:

$$X = [x_0, x_1, \cdots, x_n], Y = [y_0, y_1, \cdots, y_n] \, (x_0 < x_1 < \cdots < x_n)$$

Then conduct interpolation to get more points in the following way to fit the middle axis curve better:

Step 1: $XX_{1 \times 2n+2}(1, 2m-1) = XX_{1 \times 2n+2}(1, 2m) = X(1, m), m = 1, 2, \cdots, n+1$
Step 2: $XX(1, i) = (XX(1, i-1) + XX(1, i))/2, i = 2, 3, \cdots, 2n+2$
Repeat Step 2 three times to finish the first interpolation, then we continue to interpolate several times when the matrix dimension of XX is less than $Row/2$. Similarly, we can interpolate Y in the same way, after obtaining the interpolated X and Y, we connect the obtained points with a smooth curve to get the complete medial axis. After parameters testing experiments for the image used in this paper, we set the parameters as $[a, b, c, d, e, f] = [9, 4, 1.4, 11, 15, 0.2]$. Figure 4 displays the comparison of the medial axis extracted by using our proposed method and the classic thinning algorithm proposed by Zhang [23].

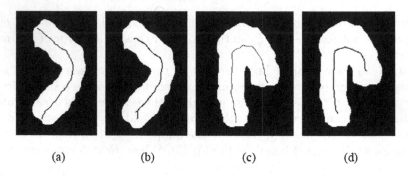

(a)   (b)   (c)   (d)

Fig. 4. Extracted medial axis by using our proposed method and classic thinning algorithm proposed by Zhang, (a), (c) by our method and (b), (d) by thinning algorithm.

4   Competitive ELM Teams Classifier

4.1   Extreme Learning Machine

The Extreme Learning Machine(ELM) is an effective learning algorithm with the structure of single-hidden layer feedforward neural network SLFNs proposed by Associate Professor Guangbin Huang of Nanyang Technological University in 2004 [24]. Traditional neural network classification algorithms (such as BP algorithm) have to manually set a large amount of training parameters, and it often generates local optimal solutions. The parameters need to set for ELM are only the number of hidden layer nodes of the network. ELM is a supervised machine learning algorithm, the SLFN has only one hidden layer, and the whole SLFN includes the input layer, hidden layer, and output layer. The input weight is the weight among the input layer and the hidden layer, the bias is the threshold of the hidden layer neuron, and the output weight is the weight between the hidden layer and the output layer. Figure 5 depicts the structure of SLFNs.

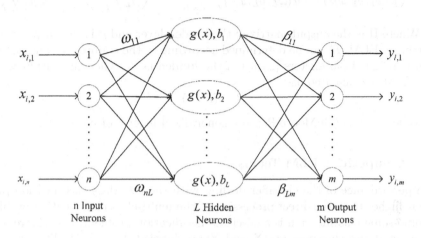

Fig. 5. The structure of SLFNs.

Let $X = \{x_1, x_2, \cdots, x_N | x_i \in R^D, i = 1, 2, \cdots N\}$ represents the training set, D denotes dimension and $Y = \{y_1, y_2, \cdots, y_N | y_i \in R\}$ is the label matrix, in the matrix, the column $j, \{j = 1, 2, \cdots, P\}$ is set by 1 corresponding to class j and the other columns set by 0, and P is the total number of the class. Then, the mathematical model of ELM which have L hidden neurons and the activation function $g(x)$ can be expressed as :

$$\sum_{j=1}^{L} \beta_j \cdot g\left(\langle \omega_j, x_i \rangle + b_j\right) = y_i, i = 1, 2, \cdots, N \tag{5}$$

Where ω_j denotes the weight matrix from inputs to the hidden layer and β_j represents the weight matrix between the hidden layer and the output layer, respectively, b_j is the bias of jth hidden neuron, $g\left(\langle \omega_i, x_i \rangle + b_i\right)$ denotes the output of the jth hidden neuron corresponding to the input sample. Note that (4) can be rewritten in a concise form as:

$$H \cdot \beta = Y \tag{6}$$

where

$$H = \begin{pmatrix} g\left(\omega_1, b_1, x_1\right) & \cdots & g\left(\omega_L, b_L, x_1\right) \\ g\left(\omega_1, b_1, x_2\right) & \cdots & g\left(\omega_L, b_L, x_2\right) \\ \vdots & \ddots & \vdots \\ g\left(\omega_1, b_1, x_N\right) & \cdots & g\left(\omega_L, b_L, x_N\right) \end{pmatrix}_{N \times L}, \beta = \begin{pmatrix} \beta_1 \\ \beta_2 \\ \vdots \\ \beta_L \end{pmatrix}_{L \times P}, Y = \begin{pmatrix} y_1 \\ y_2 \\ \vdots \\ y_N \end{pmatrix}_{N \times P}$$

Where H is the output matrix of the hidden layer and β is the output weight matrix. In ELM, H can be determined according to the training set and the randomly assigned parameters (ω_i, b_i) of the hidden layer. Then the output weight β of ELM is calculated as:

$$\beta = H^+ \cdot Y \tag{7}$$

Where H^+ is the Moore-Penrose generalized inverse of matrix H.

4.2   Competitive ELM Teams

The performance of the classifier dealing with the two-class classification problem is higher than the direct processing of the multi-class classification problem, chromo-some classification is a 24-class classification problem, 1 to 22 are autosomes, and sex chromosomes X and Y are labeled 23 and 24. The proposed classification algorithm train a sub-classifier $\text{ELM}_{i,j}$ for chromosome type i and chromosome type j $(i, j = 1, 2, 3 \ldots 24, i \neq j)$, totally $C_{24}^2 = 276$ sub-classifiers are trained. When the chromosomal feature data is tested by the classifier, the sub-classifier $\text{ELM}_{i,j}$ gives a label $Y_{i,j}$, then count the number of each label in the label set given by 276 classifiers denoted by $[num_1, num_2, \cdots, num_{24}]$, the final label of the chromosome can be calculated by the method of majority voting $Y = max\ [num_1, num_2, \cdots, num_{24}]$

In addition, since the Y chromosome exists only in male cells, and the X chromosome and autosome are present in all human cells, the Y chromosome data is much less than other chromosome data. In order to solve the Y chromosome data imbalance problem, when training the sub-classifier $\text{ELM}_{i,24}$, we take the same amount of training data for both label i and 24 chromosomes. Experiments depict that this strategy can greatly promote the recognition accuracy of the Y chromosome. The pseudocode of final CELMT methods is shown in Algorithm 1.

4.3   Further Correction Method

In order to improve the correct rate of classification, we add a calibration procedure after the initial classification. The basic biological facts on which the calibration procedure is based are as follows:

- Each autosome class comprises precisely two chromosomes.
- In the cell of a female, the X class comprise precisely two chromosomes.
- In the cell of a male, each of the X and Y classes comprise precisely one chromosome.

Based on the above rules, we design further correction methods to deal with the state that one of the labels appears three times and another autosome label appears only one time. In a picture, if the chromosome labeled i appears three times and the chromosome labeled j appears once, we check the three chromosomal data sequentially. If the second most frequently label of label matrix $[Y_{1,2}, Y_{1,3}, \cdots, Y_{23,24}]$ corresponding to one of the labeled i chromosome data is j, then we change the chromosome label to j and stop the search.

Algorithm 1. CELMT classifier.

Input: ELM: base classifier; T_r: training sets; T_e : testing sets; L: neuron number; the action function $g(x)$
Output: Test data label Y;
 1: **Training**
 2: **for** i from 1 to 23 **do**
 3:    **for** j from i+1 to 24 **do**
 4:       Step 1: optionally set input weights $W_{i,j}$ and bias $B_{i,j}$;
 5:       Step 2: compute the hidden layer output vector $H_t = g(W_{i,j} \cdot X_t + B_{i,j})$;
 6:       Step 3: compute the output weight $\beta_{i,j} = H_t^+ \cdot Y_t$, save $W_{i,j}$, $B_{i,j}$ and $\beta_{i,j}$;
 7:    **end for**
 8: **end for**
 9: **Testing:** Test chromosome feature data from T_e : X_{new}
10: **for** i from 1 to 23 **do**
11:    **for** j from i+1 to 24 **do**
12:       compute the hidden layer output matrix of X_{new} :
        $H_t = g(W_{i,j} \cdot X_{new} + B_{i,j})$;
13:       get the label of X_{new} : $Y_{i,j} = H_t \cdot \beta_{i,j}$;
14:    **end for**
15: **end for**
16: Return label matrix : $[Y_{1,2}, Y_{1,3}, \cdots \cdots, Y_{23,24}]$
17: The label of X_{new} : Y=Majority voting($[Y_{1,2}, Y_{1,3}, \cdots \cdots, Y_{23,24}]$)

---

5 Experiments and Analysis

5.1 Data

In the experiments we conducted, our dataset is from the Faculty of Computer Science and Mathematics of the University of Passau [25], this dataset contains 612 chromosome images with a total number of 28,152 chromosomes. The chromosome images we used in our experiments are segmented chromosomes and have corresponding labels (see Fig. 6, the labels of image (a), (b) and (c) are 1, 11 and 17). In addition, all chromosomes are set in the correct orientation, with the long arm at the bottom and the short arm at the top. If the chromosome is not placed incorrect orientation, subsequent classify will be affected adversely. In the subsequent classification experiments, the features we used are projection feature, length, and area feature.

The length of the chromosome can be determined through the total number of pixel of the medial axis we obtained. The area of the chromosome object is determined through the number of pixels contained in the chromosome region in the binary image. The chromosome projection feature is the average grayscale of the pixels in the vertical line perpendicular to each point along the medial axis. Due to the different lengths of different chromosomes, we unify the projection feature of the chromosome to 850 dimensions (the longest projection feature in all pictures), As for some chromosome whose projection is not enough to 850 dimensional, the projection feature is stretched to 850 dimensions using interpolation, so that the projected feature dimensions are the same. At the same time, due to the difference in the coloring and length of the same chromosome on different pictures, we subsequently normalize all features into [01] within each picture. The final feature is the 852-dimensional feature.

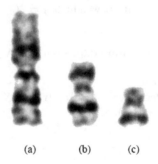

(a)   (b)   (c)

Fig. 6. Some chromosome images we used in our experiment.

5.2 Determination of the Number of Hidden Layer Nodes in Sub-classifiers

In order to adapt to the classification requirements of this project, we ought to determine the amount of the hidden layer nodes of the two-class ELM. First, we extract the data of chromosomes labeled 1, 2, 17 and 24 in the dataset to test sub-classifier $ELM_{1,5}, ELM_{10,15}$ and $ELM_{18,22}$, we divide the dataset into the training set and the testing set, whose proportion are 80% and 20%. Then we use the trial and error method to determine the number of hidden layer nodes suitable for our two classifications between 200 and 500. Finally, we will change the number of hidden layer nodes from 200 to 500 at 50 intervals, then test the accuracy of each two classifiers, the results are as shown in Fig. 7(a).

As can be seen from the figure, when the amount of hidden layer nodes is 250, the classification accuracy of the two-class ELM is the highest, and we also test the two-class ELM of other labels to obtain similar results, so in the latter experiment we will The hidden layer node of the sub-classifier is uniformly set to 250.

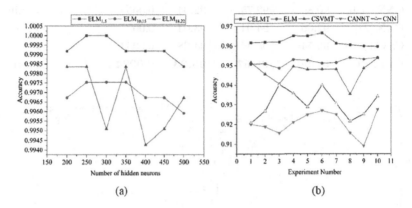

(a)

(b)

Fig. 7. (a). Accuracy of some sub-classifier with a different number of hidden neurons. (b) Classification accuracy of different algorithms in 10 experiments.

5.3   Experimental Analysis of CELMT and Other Algorithms

The experiments are performed by the Matlab R2014a software on a windows 7 64bit system with Intel(R) Xeon(R) E5-2650 v2 2.6 Hz CPU and 128 GB RAM.

In order to validate that the proposed algorithm can effectively improve the classification accuracy of chromosomes and has a faster running speed, we compare the algorithm with origin ELM, CSVMT, CNN and CANNT in the experimental part. 80% of our data as a training set and 20% as a testing set and We performed 10 times independent experiments with different training data sets and testing data sets. The final experimental results are demonstrated in Table 1 and Fig. 7(b).

From the table, we can conclude that the classification algorithm of the CELMT proposed in this paper is higher in accuracy than the comparison algorithm in other literature. In addition, in terms of running time, due to the single hidden layer structure of ELM, the time used of the CELMT and the original ELM is similar. Since the amount of hidden layer nodes in the two classifications is small, the running time is still shorter. Due to the high dimensionality of the feature data used, it takes too long to use SVM and ANN for integration experiments.

Figure 8 is a box plot showing the accuracy of each type of chromosome of 10 times experiments using EELM and the origin ELM. The box plot has three lines at the lower quartile, median, and upper quartile values, as can be seen from the

Table 1. The chromosome classification accuracy by using the proposed method and another algorithm

Method	Accuracy (%)				Time
	Minimum	Maximum	Average	Standard deviation	
ELM	94.86	95.4	95.19	0.0017	5 min
CNNT [20]	90.89	92.75	92.04	0.0059	30 min
CNN [22]	92.06	94.04	93.04	0.0071	>1 h
CSVMT [16]	93.54	95.36	94.68	0.0053	>1 h
CELMT	**95.96**	**96.67**	**96.23**	**0.0024**	**4 min**

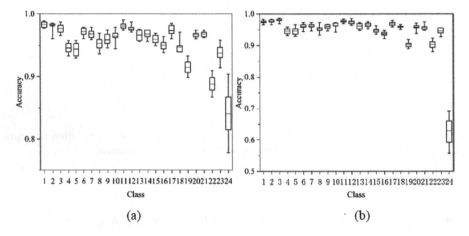

Fig. 8. Box plot of the accuracy of each class in 10 times experiments by (a) CELMT and (b) origin ELM. Boxes have three lines at the lower quartile, median, and upper quartile values.

figure. The EELM greatly improves the classification accuracy of chromosome 24 and can reduce the influence of the imbalance of chromosome 24 data on recognition to some extent.

6   Conclusions

This paper firstly makes a brief statement of the significance of karyotype analysis in medical diagnosis, then introduces the key technologies and difficulties of karyotype analysis, and proposes its own solutions to the problems raised. The proposed technique of medial axis extraction is different from the traditional thinning algorithm, it is based on the shape characteristic of chromosomes and uses geometric processing to determine the medial axis, this method can avoid burrs on the medial axis of both ends of the chromosome so that features can be accurately extracted. An algorithm of integrating simple two-class classifiers to deal with multi-classification tasks is proposed for chromosome recognition, the classification accuracy of this method is higher than that of most previous algorithms for chromosome recognition, and it has great advantages in running time. At the same time, because it is different from the datasets used in other papers, this paper uses the same dataset uniformly when compared with the algorithms in other papers. The karyotype analysis topic is still a subject to be explored, and there are still many problems to be solved, such as segmentation of chromosome images, design of classification algorithm with higher accuracy and faster speed, etc.

Acknowledgments. This work was supported by the National Natural Science Foundation of China (61922072, 61876169, 61673404).

References

1. Nair, R.M., Remya, R., Sabeena, K.: Karyotyping techniques of chromosomes: a survey. Int. J. Comput. Trends Technol. **22**(1) (2015)
2. Gadhia Pankaj, K., Patel Monika, V., Vaniawala Salil, N.: Role of cytogenetic evaluation in diagnosis of acute myeloid leukemia. Am. J. Biomed. Life Sci. **4**(6), 98–102 (2016)
3. Ventura, R., Khmelinskii, A., Sanches, J.M.: Classifier-assisted metric for chromosome pairing. In: 2010 Annual International Conference of the IEEE Engineering in Medicine and Biology, pp. 6729–6732. IEEE (2010)
4. Lerner, B.: Toward a completely automatic neural-network-based human chromosome analysis. IEEE Trans. Syst. Man Cybern. Part B (Cybern.) **28**(4), 544–552 (1998)
5. Wang, X., Zheng, B., Wood, M., Li, S., Chen, W., Liu, H.: Development and evaluation of automated systems for detection and classification of banded chromosomes: current status and future perspectives. J. Phys. D: Appl. Phys. **38**(15), 2536 (2005)
6. Wu, Q., Castleman, K.R.: Automated chromosome classification using wavelet-based band pattern descriptors. In: Proceedings 13th IEEE Symposium on Computer-Based Medical Systems, CBMS 2000, pp. 189–194. IEEE (2000)
7. Minaee, S., Fotouhi, M., Khalaj, B.H.: A geometric approach to fully automatic chromosome segmentation. In: 2014 IEEE Signal Processing in Medicine and Biology Symposium (SPMB), pp. 1–6. IEEE (2014)
8. Moradi, M., Setarehdan, S., Ghaffari, S.: Automatic landmark detection on chromosomes' images for feature extraction purposes. In: Proceedings of the 3rd International Symposium on Image and Signal Processing and Analysis, ISPA 2003, vol. 1, pp. 567–570. IEEE (2003)
9. Piper, J., Granum, E.: On fully automatic feature measurement for banded chromosome classification. Cytom.: J. Int. Soc. Anal. Cytol. **10**(3), 242–255 (1989)
10. Stanley, R.J., Keller, J.M., Gader, P., Caldwell, C.W.: Data-driven homologue matching for chromosome identification. IEEE Trans. Med. Imaging **17**(3), 451–462 (1998)
11. Kao, J.H., Chuang, J.H., Wang, T.: Chromosome classification based on the band profile similarity along approximate medial axis. Pattern Recognit. **41**(1), 77–89 (2008)
12. Moradi, M., Setarehdan, S.K.: New features for automatic classification of human chromosomes: a feasibility study. Pattern Recogn. Lett. **27**(1), 19–28 (2006)

13. Poletti, E., Grisan, E., Ruggeri, A.: A modular framework for the automatic classification of chromosomes in q-band images. Comput. Methods Programs Biomed. **105**(2), 120–130 (2012)
14. Wang, X., Zheng, B., Li, S., Mulvihill, J.J., Liu, H.: A rule-based computer scheme for centromere identification and polarity assignment of metaphase chromosomes. Comput. Methods Programs Biomed. **89**(1), 33–42 (2008)
15. Wang, X., Zheng, B., Li, S., Mulvihill, J.J., Wood, M.C., Liu, H.: Automated classification of metaphase chromosomes: optimization of an adaptive computerized scheme. J. Biomed. Inform. **42**(1), 22–31 (2009)
16. Kusakci, A.O., Ayvaz, B., Karakaya, E.: Towards an autonomous human chromosome classification system using competitive support vector machines teams (CSVMT). Expert Syst. Appl. **86**, 224–234 (2017)
17. Jennings, A.M., Graham, J.: A neural network approach to automatic chromosome classification. Phys. Med. Biol. **38**(7), 959 (1993)
18. Sweeney Jr., W.P., Musavi, M.T., Guidi, J.N.: Classification of chromosomes using a probabilistic neural network. Cytom.: J. Int. Soc. Anal. Cytol. **16**(1), 17–24 (1994)
19. Sharma, M., Vig, L., et al.: Automatic chromosome classification using deep attention based sequence learning of chromosome bands. In: 2018 International Joint Conference on Neural Networks (IJCNN), pp. 1–8. IEEE (2018)
20. Gagula-Palalic, S., Can, M.: Human chromosome classification using competitive neural network teams (CNNT) and nearest neighbor. In: IEEE-EMBS International Conference on Biomedical and Health Informatics (BHI), pp. 626–629. IEEE (2014)
21. Uttamatanin, R., Yuvapoositanon, P., Intarapanich, A., Kaewkamnerd, S., Tongsima, S.: Band classification based on chromosome shapes. In: The 6th 2013 Biomedical Engineering International Conference, pp. 1–5. IEEE (2013)
22. Hu, X., et al.: Classification of metaphase chromosomes using deep convolutional neural network. J. Comput. Biol. **26**(5), 473–484 (2019)
23. Zhang, T., Suen, C.Y.: A fast parallel algorithm for thinning digital patterns. Commun. ACM **27**(3), 236–239 (1984)
24. Huang, G.B., Zhu, Q.Y., Siew, C.K.: Extreme learning machine: theory and applications. Neurocomputing **70**(1–3), 489–501 (2006)
25. Faculty of computer science and mathematics. http://www.fim.uni-passau.de/en/faculty/former-professors/mathematical-stochastics/chromosome-image-data. Accessed 10 Oct 2019

Mechanical Properties Prediction for Hot Roll Steel Using Convolutional Neural Network

Hao Xu[1], Zhiwei Xu[1] , and Kai Zhang[1,2]($\boxtimes$)

[1] School of Computer Science and Technology, Wuhan University of Science and Technology, Wuhan 430065, Hubei, China
zhangkai@wust.edu.cn
[2] Hubei Province Key Laboratory of Intelligent Information Processing and Real-time Industrial System, Wuhan 430065, Hubei, China

Abstract. Prediction the mechanical properties is very important in many real-life industry fields. In this paper, we proposed an efficient convolutional neural network (CNN) to predict the mechanical properties of hot roll steel. In this study, 20,000 sets of data are collected from the hot roll factory, where 16,000 sets of data were used for training the CNN model, and 4,000 sets of data were used for testing the performance of the model. Compared with Support Vector Machine (SVM) and Artificial Neural Network (ANN), The experimental results have been demonstrated to provide a competitive and higher prediction accuracy.

Keywords: Convolutional neural network · Mechanical property prediction · Hot roll steel

1 Introduction

Hot roll steel is an important material which widely used in many real-life fields. These areas have different requirement for hot roll steel mechanical properties, so it is important to predict the mechanical properties of alloy steel accurately which are tensile strength (TS), yield strength (YS) and elongation (EL).

Traditionally, mechanical properties prediction is carried out by destructive testing, which is costly and time consuming. As the rolling process is complicated and final mechanical properties of steel determined by many parameters, including the chemical composition and the process parameters [1], it is extremely hard to express the relationships by mathematical model [2]. In the previous studies, scholars have widely use metallurgical mechanism models and statistical models to predict mechanical properties [3]. Due to the complexity and dynamic in steel

Supported by the National Natural Science Foundation of China (Grant Nos. U1803262, 61702383, 61602350).

© Springer Nature Singapore Pte Ltd. 2020
L. Pan et al. (Eds.): BIC-TA 2019, CCIS 1160, pp. 565–575, 2020.
https://doi.org/10.1007/978-981-15-3415-7_47

manufacturing process [4], the model structure is needed too many experimental trials, which increases the cost and production time [5].

In the existing literature, the Artificial Neural Network (ANN) and Support Vector Machine [6] (SVM) have become widely used to predict the mechanical properties of hot roll steel. Liang et al. [7] used laser-induced breakdown spectroscopy combined with support vector machine to classify steel. Chou et al. [8] used a three-layer feedforward ANN with Taguchi particle swarm to optimize the chemical composition of steel bars and improve the mechanical properties. The methods based on ANN and SVM can demonstrate good performance when the data set is small, but they cannot achieve high precision when dealing with massive data. With the development of the steel industry, the processing is more complex, the relationship between input parameters and mechanical properties is more complicated, those methods might be powerless.

Recently, deep learning [9] has made a major breakthrough in the field of machine learning, especially CNN has demonstrated strong performance in the field of image recognition [10]. It uses the local connection and weight sharing to reduce the number of parameters, not only to extract local features from complex data but also to be insensitive to noise and has good model expression ability [11]. Based on these advantages of CNN, we consider that it can be used to solve hot roll steel mechanical properties prediction problem.

In this paper, a CNN-based method is proposed to predict the mechanical properties of hot roll steel. This method represents chemical composition and processing parameters as one-dimensional vector, and employs CNN-based model to extract features contained by the vector. The rest of the paper is organized as follows. Section 2 introduces the problem considered in this research. Section 3 presents the proposed one-dimensional CNN-based method. Section 4 describes the experimental settings. Section 5 concludes the study.

2   Background

2.1   Hot Rolled Processing

The hot rolled processing is one of the parameters affecting the mechanical properties. As shown in Fig. 1, after reheating, roughing rolling, finishing rolling, laminar cooling and down coiling, the steel slab becomes a coil of a thin sheet.

In the manufacturing processes, a series of complex microstructure have been changed. First, reheating process provide uniform austenite grain. Then, the roughing and finishing processes refine austenite by dynamic and static recrystallization. These grains determine the mechanical properties of the steel. In this process, the furnace temperature (FT), the roughing rolling temperature (RRT), the finishing rolling temperature (FRT) and the coiling temperature (CT) have the greatest influence on the mechanical properties of hot roll steel. According to China High Strength Low Alloy Structural Steel Standard GB/T1591-20608, most of the chemical components of alloy steel consist of carbon (C), manganese (Mn), silicon (Si), phosphorus (P) and sulfur (S). The effects of chemical composition on mechanical properties is another important parameter. The additions of

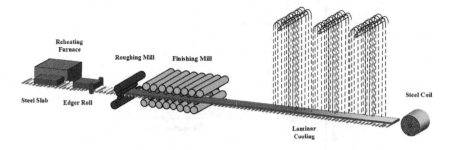

Fig. 1. Hot rolled processing.

some alloying elements affect ferrite transformation and change the mechanical properties of alloy steel [12]. The presence of microalloying elements generally control the grain size and have a significant impact on the strength. These mechanisms can be very complicated. For example, the major chemical component of steel bar is carbon, which determines mechanical strength [13]. When the ratio of C in the steel is below 0.8Wt%, the YS and TS of the steel increase dramatically with the increases of C content, but the EL of steel decreases. S will reduce the hot workability and strength of steel and P will reduce the plasticity and toughness of the steel [14].

2.2 Mechanical Properties of Hot Roll Steel

The mechanical properties of alloy steel are TS, YS, and EL. TS is defined as the maximum tensile stress that the steel can withstand before breaking, and the YS is defined as the maximum stress of the steel can withstand before plastic deformation begins. EL is defined as the percentage of stretched length to the original length after the steel is broken. Figure 2 shows how the YP, TS and EL of hot roll steel are related.

Chemical elements and processes together determine the mechanical properties of steel, the different combination of chemical components and processes parameters complicate those properties. It is difficult to express the relationship. An efficient method of predicting mechanical properties is needed.

3 Proposed Algorithm

This section introduces the proposed prediction method based on CNN for mechanical properties of hot roll steel.

3.1 CNN Prediction Model

CNN is widely used in the field of images, and the input to the network is a two-dimensional matrix at most. In order to adapt to the one-dimensional characteristic of hot rolled steel data, one-dimensional CNN-based architecture

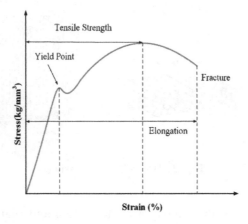

Fig. 2. Relationship between tensile strength, yield strength and elongation of hot roll steel.

is used which applies 1D arrays instead of 2D matrices for both kernels and feature maps.

As shown in Fig. 3, this model consists of three parts, model input, feature extraction part, prediction part. Each of the parts is explained below. The feature extraction is composed of several feature extractors, which is stacked by convolutional layer, batch normalization, nonlinear activation and pooling layer. The prediction part contains two fully connected (FC) layers. The predicted value will be output in the final layer.

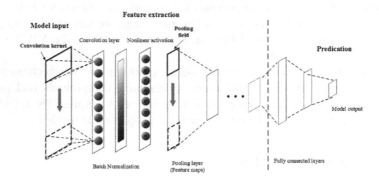

Fig. 3. Structure of CNN-based prediction model for mechanical properties.

First, model input includes 16 chemical compositions and 4 heat treatment process parameters. In order to eliminate the largely distinct scales in different fields, the raw data need to be normalized. If the raw data is $X = (x_1, x_2, x_3 \ldots x_{n-1}, x_n)$, the normalization equation is:

$$x_i^* = \frac{x - \min}{\max - \min} \tag{1}$$

where x_i^* is the normalized value of the input parameter x_i, and min and max are the minimum and maximum values in the data samples respectively.

Second, the feature extraction part is combination of convolutional layer, batch normalization, nonlinear activation layer and pooling layer.

The convolutional layer is the core part of the CNN model. It consists of a set of linear filters that convolute the input data. The convolution operation is as shown in the formula:

$$(h_k)_{ij} = (W_k * x)_{ij} + b_{ij} \tag{2}$$

where k is the index of the kth feature map in the convolutional layer, (i,j) is the index of the pixel point, x is the input data, and W and b are the weight parameter and the offset parameter of the kth feature map, respectively. $(h_k)_{ij}$ is the output value of the kth feature map. And in one-dimensional convolution, j is usually set to 1.

Batch Normalization [15] is added after each convolutional layer, which can speed up the training of the network and avoid gradient explosion. First, the distribution of the input data is normalized to a distribution with the mean of 0 and variance of 1, as follows:

$$\hat{x}^{(k)} = \frac{x^k - E\left(x^k\right)}{\sqrt{\mathrm{Var}\left(x^k\right) + \varepsilon}} \tag{3}$$

where $x^{(k)}$ represents the kth dimension of the input data, $E(x^k)$ represents the average of the dimension, and $\sqrt{\mathrm{Var}\left(x^k\right) + \varepsilon}$ represents the standard deviation. Then set two learnable variables γ and β, and use these two learnable variables to restore the data distribution learned from the previous layer, as follows:

$$y^{(k)} = \gamma^k \hat{x}^{(k)} + \beta^{(k)} \tag{4}$$

Rectified linear unit (ReLU) is used as the nonlinear activation function, which can prevent the problem of gradient vanished and gradient explosion in the neural network during training. Let max denote the function to select the larger value between x and zero, the ReLU activation function can be expressed as:

$$\mathrm{Relu}(x) = \max(0, x) \tag{5}$$

The pooling layer is a down-sampling layer, which can not only reduce the network scale of the CNN, but also identify the most prominent features of input layers. The maxpooling method is used in the proposed CNN model by selecting the maximum value in the pooling field.

Finally, the predicted value is obtained by feature extractions part to gain the features from raw data and applying fully connected layers to process the feature information.

3.2   CNN Optimization and Evaluation Metric

The prediction of the CNN are the mechanical properties of hot roll steel, and the mean squared error (MSE) is employed to measure the distance between predictions and actual mechanical properties. Thus, minimizing MSE is taken as the loss function of the CNN. MSE can be written as:

$$\text{MSE} = \frac{1}{N} \sum_{i=1}^{N} (\hat{y}_i - y_i)^2 \tag{6}$$

where $\hat{y}_i$ represents the predicted value, y_i represents the actual value, and N represents the number of samples in the data set.

Four indicators are adopted as the evaluation metrics to assess the prediction capability comprehensively. Mean square error (MSE), mean absolute error (MAE), mean absolute percentage error (MAPE) and coefficient of determination (R^2) are used as evaluation of the model. MAE represents the average of the absolute error, MAPE is a measure of prediction accuracy of a forecasting method. R^2 explains how much of the variability of a factor can be caused or explained by its relationship to another factor.

4   Results and Discussions

This section consists of two part. First, the experimental steel data is introduced. Second, experiments are performed to demonstrate that the proposed CNN model can predict for mechanical properties, and its training results are compared with Artificial Neural Network (ANN) methods, and Support Vector Machine (SVM).

4.1   Data Description

In this paper, 20,000 sets of data are collected from the hot roll factory, where 16,000 sets of data were used for training the CNN model, and 4,000 sets of data were used for testing the performance of the model. The data is shown in the Table 1 below. The input data consists of sixteen chemical components and four hot roll process parameters. The output consists of three mechanical properties including TS, YS and EL.

4.2   Comparison Results

In order to test the performance of the proposed algorithm, ANN and SVM are chosen for comparison. ANN represents the traditional neural network and attempts to learn features through hidden layers. SVM find a hyperplane to divide the sample space of the data set into different samples. From Table 2, compared with the SVM and ANN methods, the proposed CNN model has achieved good results, and the evaluation metric on the test set are better than the other two algorithms. This shows that the proposed CNN model can effectively extract the features that affect the mechanical properties of hot rolled steel and has good generalization ability.

Table 1. Steel dataset.

Parameter	Unit	Minimum	Maximum	Mean
C	Wt%	0.0051	0.1936	0.149
Mn	Wt%	0.2362	1.3696	0.4674
Si	Wt%	0.0063	0.2994	0.1944
P	Wt%	0.0062	0.0364	0.0172
S	Wt%	0.0012	0.025	0.013
Cu	Wt%	0.0124	0.0826	0.0405
Al	Wt%	0.0003	0.5654	0.0067
Als	Wt%	0.0002	0.5652	0.0059
Ni	Wt%	0.0012	0.0673	0.0168
Cr	Wt%	0.0082	0.0942	0.0296
Ti	Wt%	0.0001	0.0691	0.0015
Mo	Wt%	0.0012	0.0184	0.0053
V	Wt%	0.0012	0.0056	0.0014
Nb	Wt%	0.001	0.0194	0.0011
N	Wt%	0.0008	0.0653	0.003
B	Wt%	0.0004	0.0031	0.0002
FT	°C	1188	1291	1242.6607
RRT	°C	976	1142	1049.3915
FRT	°C	200	1056	752.5484
CT	°C	630	934	774.2208

4.3 Hyperparameters Optimization

The hyperparameters are the important factors that should be considered cautiously which include convolutional kernels size, polling size and depth of the CNN, when implementing the structure of a CNN.

First, there is no rules for the selection of hyperparameters generally. Based on the parameter settings of VGG Net, which has achieved second place in the 2014 ILSVRC, we select convolutional filter of size (3,1) and max pooling of size (2,1). Xavier is adopted as the weight initialization method, the batch size is set to 128, and parameters in the CNN model are updated by Adam optimizer. The initial learning rate of the proposed CNN model is set as 0.001.

Second, the depth of the CNN should not be too big or too small, which can make the model difficult to converge or overfit. The structure of are shown in Table 3, where each convolutional layer is followed by a pooling layer, and the numbers represent quantities of convolutional filters in the layer. Table 4 shows the results of the CNN with different number of depths, TS, YS and EL represent the three mechanical performance predicted by the model. The Depth-2 achieves the best result. When the depth is too small, the relationship between the input

Table 2. Prediction performances of the CNN and other algorithms.

Predicting	Evaluation	CNN	SVM	ANN
TS	MSE	**0.0032**	0.0071	**0.0032**
	MAE	**0.0231**	0.0312	0.0221
	MAPE	**0.0612**	0.0731	0.0693
	R^2	**0.8281**	0.792	0.8262
YS	MSE	**0.0015**	0.0032	0.002
	MAE	**0.0116**	0.0129	0.0123
	MAPE	**0.0453**	0.0488	0.0441
	R^2	**0.9025**	0.8299	0.8519
EL	MSE	**0.0015**	0.0032	0.002
	MAE	**0.0116**	0.0129	0.0123
	MAPE	**0.0453**	0.0488	0.0441
	R^2	**0.9025**	0.8299	0.8519

and the output cannot be completely extracted and when the depth is too large, the model shows poor generalization ability.

Table 3. Different depths for CNN.

Depth	Structures of CNN Model
Depth-1	32 conv→ 64 conv
Depth-2	32 conv →64 conv→128 conv
Depth-3	32 conv →64 conv →128 conv →256 conv

The details of the depth-2 CNN are listed in Table 5. It contains three convolutional layers, three pooling layers and two fully-connected layers. FC1 represent the first FC layer, FC2 represent the second FC layer, and the followed numbers indicate the number of neuron nodes in the FC layer. The denotation of Filter (3*1*64) means that it is a convolutional layer which filter size is 3*1 with 64 channels. Maxpool (2*1) denotes that it is a maxpooling layer with a 2*1 pooling field.

Figure 4 shows the actual values on the test set and the predicted values of the proposed CNN model by scatter diagrams of TS, YS and EL. These values are normalized so they are only in the range of 0 to 1.

Table 4. Results of CNN in different depths.

Depth	Evaluation	TS	YS	EL
Depth-1	MSE	0.0061	0.0025	0.0035
	MAE	0.0525	0.0302	0.0578
	MAPE	0.0921	0.0976	0.1132
	R^2	0.6241	0.8556	0.7921
Depth-2	MSE	**0.0032**	**0.0015**	**0.0023**
	MAE	**0.0231**	**0.0116**	**0.0216**
	MAPE	**0.0712**	**0.0653**	**0.0618**
	R^2	**0.8281**	**0.9025**	**0.8649**
Depth-3	MSE	0.0052	0.0032	0.0037
	MAE	0.0427	0.0296	0.0376
	MAPE	0.0834	0.0876	0.1032
	R^2	0.6724	0.8667	0.801

Table 5. Details of the CNN.

Layer	Name	Parameters
L1	Conv	Filter(3*1*64)
	Pool	Maxpool (2*1)
L2	Conv	Filter(3*1*128)
	Pool	Maxpool (2*1)
L3	Conv	Filter(3*1*256)
	Pool	Maxpool (2*1)
FC1	Fully-connected 1	1280
FC2	Fully-connected 2	256

The Fig. 4 demonstrates that the predicted values of the proposed model are in good agreement with the actual mechanical properties, which indicates that the proposed CNN model has a good effect.

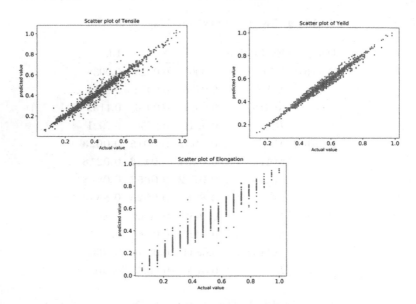

Fig. 4. Comparisons of predicted and actual mechanical properties

5   Conclusion

The proposed CNN model can predict the mechanical properties of steel through sixteen kinds of chemical composition and four kinds of hot rolling production processes of hot rolled steel. The raw data is processed by normalization to keep the data in the same range. The best CNN structures in predicting TS, YS, and EL are determined by comparison experiments, and the model prediction results are compared with the SVM model and the ANN model. The experimental results show that our algorithm provides the best accuracy than the traditional methods on 20,000 hot roll datasets. The results demonstrate the ability of the proposed CNN-based model to predict the mechanical properties of hot rolled steel.

References

1. Beghini, M., Bertini, L., Monelli, B.D., Santus, C., Bandini, M.: Experimental parameter sensitivity analysis of residual stresses induced by deep rolling on 7075–T6 aluminium alloy. Surf. Coat. Technol. **254**, 175–186 (2014)
2. Cheng, C.K., Tsai, J.T., Lee, T.T., Chou, J.H., Hwang, K.S.: Modeling and optimizing tensile strength and yield point on steel bar by artificial neural network with evolutionary algorithm. In: 2015 IEEE International Conference on Automation Science and Engineering (CASE), pp. 1562–1563. IEEE (2015)
3. Yun, X., Gardner, L., Boissonnade, N.: The continuous strength method for the design of hot-rolled steel cross-sections. Eng. Struct. **157**, 179–191 (2018)

4. Wu, Y., Ren, Y.: Prediction of mechanical properties of hot rolled strips by BP artificial neural network. In: 2011 International Conference of Information Technology. Computer Engineering and Management Sciences, vol. 1, pp. 15–17. IEEE (2011)
5. Mohanty, I., et al.: Online mechanical property prediction system for hot rolled IF steel. Ironmak. Steelmak. **41**(8), 618–627 (2014)
6. Guenther, N., Schonlau, M.: Support vector machines. Stata J. **16**(4), 917–937 (2016)
7. Liang, L., et al.: Classification of steel materials by laser-induced breakdown spectroscopy coupled with support vector machines. Appl. Opt. **53**(4), 544–552 (2014)
8. Chou, P.Y., Tsai, J.T., Chou, J.H.: Modeling and optimizing tensile strength and yield point on a steel bar using an artificial neural network with Taguchi particle swarm optimizer. IEEE Access **4**, 585–593 (2016)
9. LeCun, Y., Bengio, Y., Hinton, G.: Deep learning. Nature **521**(7553), 436 (2015)
10. Schmidhuber, J.: Deep learning in neural networks: an overview. Neural Netw. **61**, 85–117 (2015)
11. Krizhevsky, A., Sutskever, I., Hinton, G.E.: Imagenet classification with deep convolutional neural networks. In: Advances in Neural Information Processing Systems, pp. 1097–1105 (2012)
12. Esfahani, M.B., Toroghinejad, M.R., Abbasi, S.: Artificial neural network modeling the tensile strength of hot strip mill products. ISIJ Int. **49**(10), 1583–1587 (2009)
13. Saravanakumar, P., Jothimani, V., Sureshbabu, L., Ayyappan, S., Noorullah, D., Venkatakrishnan, P.G.: Prediction of mechanical properties of low carbon steel in hot rolling process using artificial neural network model. Procedia Eng. **38**, 3418–3425 (2012)
14. Hwang, R.C., Chen, Y.J., Huang, H.C.: Artificial intelligent analyzer for mechanical properties of rolled steel bar by using neural networks. Expert Syst. Appl. **37**(4), 3136–3139 (2010)
15. Li, X., Chen, S., Hu, X., Yang, J.: Understanding the disharmony between dropout and batch normalization by variance shift. In: Proceedings of the IEEE Conference on Computer Vision and Pattern Recognition, pp. 2682–2690 (2019)

The Neutrophil's Morphology Classification Using Convolutional Neural Network

Xiliang Zhang[1] , Jialong Li[1] , Bohao Wang[1] , Kunju Shi[2(✉)] ,
Qin Qin[1(✉)] , and Bo Fan[3]

[1] Shanghai Polytechnic University,
No. 2360, Jin Hai Road, Pudong, Shanghai 201209, People's Republic of China
qinqin@sspu.edu.cn
[2] Shanghai Dianji University,
No. 300, Shui Hua Road, Pudong, Shanghai 201306, People's Republic of China
shikunjv@sina.cn
[3] Amkor Assembly & Test (Shanghai) Co., Ltd.,
No. 111 Ying Lun Road, Pudong, Shanghai 200131, China

Abstract. To better understand the mechanism of neutrophils moving in response to an inflammatory response, the morphology changing and the modes of neutrophils are investigated. The morphology changing usually refers to the appearance of pseudopods in a neutrophil, which means the mode of the neutrophil changes, and becomes activated. However, the mechanism is still not completely understood. It will be helpful to better understand the mechanism, if the modes of neutrophils are successfully classified. This paper proposed a method to successfully classify the modes of neutrophils using transfer learning based on the pre-trained AlexNet due to the good performance of deep learning in classification tasks. The result of classification is very accurate with high probability. The classification method using transfer learning based on a pre-trained AlexNet is not novel, but the application in this area is new. Furthermore, owing to the transfer learning, only a few pieces of data are needed to train the new network and successfully classify the modes of neutrophils.

Keywords: Neutrophils · Morphology · CNN · Transfer learning · Classification

1 Introduction

Neutrophils (neutrophil granulocytes) are a type of white blood cells (WBCs) (or Leukocytes in Latin). They take up the most abundant type of WBCs [1].

Supported by the State Key Research and Development Program of China (2017YFE0118700), the European Unions Horizon 2020 research and innovation programme under the Marie Sklodowska-Curie grant agreement No 734599, 2018 University Young Teacher Cultivation Funding Project of Shanghai (ZZEGD18037), and the Research Administration Office of Shanghai Polytechnic University (EGD19XQD07).

© Springer Nature Singapore Pte Ltd. 2020
L. Pan et al. (Eds.): BIC-TA 2019, CCIS 1160, pp. 576–585, 2020.
https://doi.org/10.1007/978-981-15-3415-7_48

Neutrophils play a critical role in protecting human as well as animals and the research on many aspects of neutrophils has become increasingly popular in last few decades, such as mathematical models [2], neutrophil migration [3], etc.

When inflammation happens, usually neutrophils are activated and arrive at the inflammatory site [4] to hit against the non-self (foreign bodies or invaders) by sensing the changing of chemotaxis [5]. During the process, the shape and modes of neutrophils are changing. It will be helpful to understand the mechanism of the shape and modes changing. According to [1], neutrophils have two modes, that is, the active mode and the inactive one. When neutrophils are in the active mode, the morphology is different. For instance, pseudopods appear, the trajectory is obvious, and the shape flattens, etc. Therefore, the morphology changing of neutrophils implies the mode switching. Research on morphology is equivalent to the study of the modes changing.

Neutrophils remain the inactive mode in most cases, which means that the morphology of neutrophils usually looks like a circle, unless something happens. Furthermore, it is very difficult to find when a slight change of morphology happens, since sometimes the shape of neutrophils are too small to distinguish. If the advanced technology can be used to classify the mode or morphology accurately, it is very helpful for early diagnosis. Therefore, research on morphology becomes increasingly important for early diagnosis of inflammatory responses. More specifically, research on mode classification of neutrophils as demonstrated in this paper is significant. However, few studies have investigated in this area.

Recently, deep learning [6,12] has developed rapidly. Specifically, AlexNet [7] first using convolutional neural network (CNN) overperformed other groups using support vector machines (SVM) in image classification tasks in 2012. Since then, deep learning, especially CNN, has been widely concerned, because of its excellent performance in classification and other areas, such as object detection [8], natural language processing [9], etc.

Another reason for deep learning of popularity might be due to the power of transfer learning and domain adaptation [10], which makes the pre-trained model be trained with only a few amount of data to get a good classification performance in another field. Consequently, the pre-trained model can be applied to many new fields and to complete new tasks.

The innovation of this paper is that we first classify the modes of neutrophils using convolutional neutrophil network with transfer learning. Comparing with low frequency descriptor (LFD), the proposed method is easier to be understood based on visualization and more precise since it contains not only the low dimension information but also the high dimension one. It is very significant since this study will help us to better understand the mechanism how neutrophils responding to inflammation and easier diagnose inflammatory responses early. The structure of this paper is arranged as follows. In methods section, we will introduce data acquisition, pre-processing, transfer learning, and training. In results section, we will illustrate our results including the classification, visualization, etc. Finally, we give the conclusion and future plans.

2   Methods

2.1   Data Acquisition

Prior to commencing the research, a huge amount of data was acquired, but without the authorization to use in a paper. Therefore, all data used in this paper is derived from images searched online and ethical requirements cannot be guaranteed.

2.2   Pre-processing

The pre-processing is necessary since different networks have different input sizes. For a particular network, the input size is fixed and the main job of pre-processing is to scale the size. Usually, this process can also be done in the main program with only a few commands. However, the pre-processing of ours is different and explained as follows.

To begin this part, images containing a lot of neutrophils are selected and downloaded from internet. An example of data downloaded online is given in Fig. 1. It is obvious that in Fig. 1, there are a lot of cells with different colour. The pink ones stand for neutrophils while other colours are different kinds of cells. The task now becomes how to extract neutrophils.

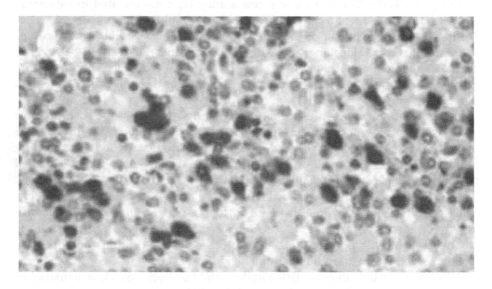

Fig. 1. An image of neutrophils downloaded from internet. The pink cells are neutrophils. So the following step is to crop neutrophils out and delete other cells. (Color figure online)

Intuitively, neutrophils can be extracted by binarizing. So, once the images are downloaded, they are binarized. During this, holes are generated. The binary image is presented in Fig. 2. Both Figs. 1 and 2 show all neutrophils, however, there are only a few differences. Figure 1 provides neutrophils with complete shapes, while not all neutrophils in Fig. 2 have. In addition, the morphology of neutrophils is obviously smaller. A likely explanation is that binarization is not a good approach to pre-processing. By contrast, it may be better to use different colours as the image threshold, which will be verified in future work.

Fig. 2. The binary image of Fig. 1. Neutrophils are extracted by binarization. However, some neutrophils have holes.

After binarization, neutrophils are cropped one by one. Following cropping, some commonly-used methods in image processing area are used to make the morphology of neutrophils more reliable. Finally, one neutrophil is extracted as an example shown in Fig. 3.

When finishing image processing, images are labelled in different files. They can be fed to the network as the input and the size will be scaled in the main program. Note that although they are binary images, they can still be used as input to the network, since in the main program, the network will automatically turn them into rgb form.

2.3   Transfer Learning

There is no accurate CNN for neutrophils classification. In addition, there is a lack of large amounts of neutrophils data for training a new CNN. Furthermore, labelling neutrophils data and rebuilding CNN also costs expensive. Therefore, transfer learning is the first choice.

The basic idea of transfer learning is that different classification tasks in different domains share some of the same features, or more specifically known as knowledge. For instance, some different domains may share the same

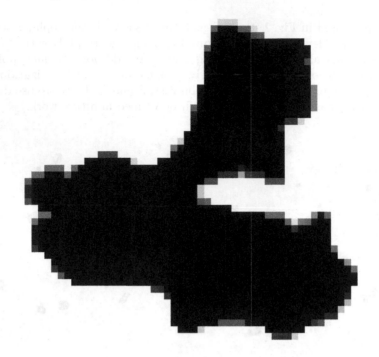

Fig. 3. One neutrophil after pre-processing. This neutrophil is an example of data fed to the network after pre-processing.

low-dimensional features, such as edges, therefore, having the knowledge of identifying nanguo pear will effectively help to identify pears.

A major advantage of transfer learning is that we need to fine-tune the pre-trained model to achieve the classification task with only a few pieces of neutrophils images. In this paper we use AlexNet as the pre-trained model. Conventional AlexNet has five convolutional layers and three fully connected layers for 1000 different classes. It has more than 60,000,000 parameters which need a huge amount of data to train. But there are only two modes of neutrophils. Furthermore, neutrophils have the same low-dimensional features, such as edges, or shapes, which can be considered as knowledge. Therefore, transfer learning can be used with the pre-trained AlexNet to extract knowledge and classify neutrophils.

Owing to transfer learning, we can use the pre-trained AlexNet and only change the last fully connected layer for neutrophils mode classification. Since there are only two modes, the last fully connected layer is changed into a two classes classifier followed by a softmax layer. The structure of the new network is shown in Fig. 4. The parameters of the new network are optimized or fine-tuned by training with a few new labelled images. Note that the pre-trained

AlexNet is different from the one in the original paper, but it also performs well in ImageNet. Furthermore, data is augmented by the new network through the commonly used tricks, such as translation, reflection etc.

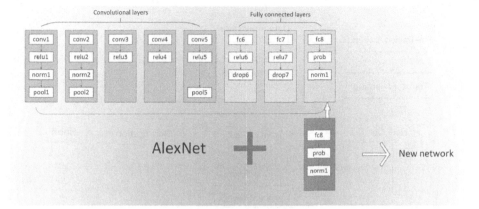

Fig. 4. The structure of the new network. The last fully connected layer is changed based on our task. The output size of the original fc8 layer is 1000, while the size after transfer learning is 2, which can divide neutrophils into two, the active ones and the inactive ones.

2.4   Training

After AlexNet was modified by transfer learning technique, we began to train the new network. Neutrophils are special, because they have two modes. In most cases, neutrophils are inactive, while when the inflammation happens, most of the neutrophils are activated and the morphology changes to respond to the inflammatory response and move to the inflammation site. As mentioned before, data is collected online, which will lead to it scarcely possible to find suitable images that contain a large amount of active neutrophils and inactive ones. Furthermore, due to the size of each neutrophil, it is difficult to accurately determine the mode of neutrophils. Consequently, we might have mislabelled neutrophils according to the conclusions in [1].

The flow diagram is shown in Fig. 5

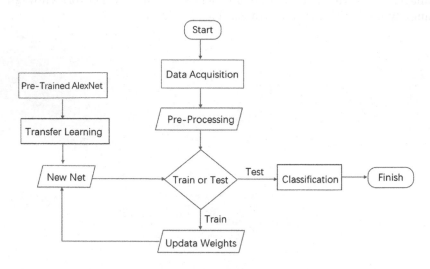

Fig. 5. The flow diagram.

3   Results

After the new network was well trained, we tested it. The classification result is shown in Fig. 6. The whole testing data does not appear in the training dataset. What stands out in Fig. 6 is that all the four test neutrophils are classified successfully with very high probability. It reveals the very good performance of CNN in dealing with classification tasks. In addition, compared with the method LFD proposed in [1] to classify neutrophils, CNN seems more reliable and efficient. LFD attempts to calculate the low frequency features and classify neutrophils based on them, whereas CNN uses both the low dimensional features and the high dimensional ones and classifies neutrophils based on all of these features. For instance, the first convolution layer of the new network extracts the low-dimensional features. Specifically, the contour of neutrophils is extracted. Figure 7 presents the low dimensional features by visualization. Furthermore, higher convolution layers are equivalent to extracting higher dimensional features. Therefore, the classification result of the new network is better in contrast to LFD theoretically. The confusion matrix chart is also presented in Fig. 8.

To verify how well the new network performs, we also did a comparative experiment. We deliberately mislabelled a set of test data, nevertheless, the new network got a correct classification result that was consistent with the real label. More details and results can be found in [11]. Therefore, we do not show the comparative results here. Alternatively, we emphasize the performance of the new network.

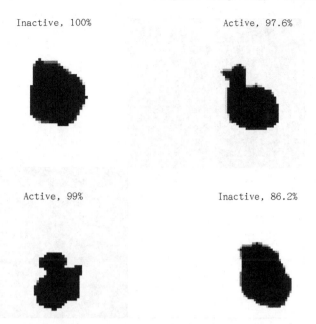

Inactive, 100%   Active, 97.6%

Active, 99%   Inactive, 86.2%

Fig. 6. The classification result of the new network. The new network classifies the four test neutrophils successfully. Captions beyond the sub-images are the probable mode and the probability.

Fig. 7. The feature visualization of the new network. All the sub-images are the results of visualization after the first convolution layer.

Frankly, We tried to use generative adversarial networks (GANs) to generate neutrophils for training our CNN, however, because of the lack of data for training, neutrophils generated by GANs were really bad.

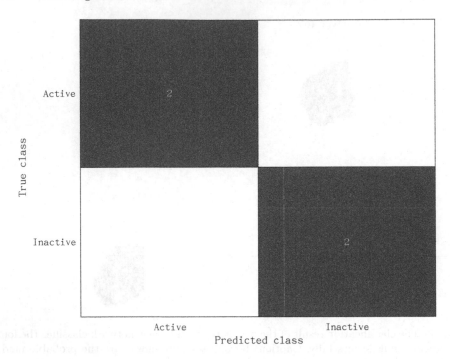

Fig. 8. The confusion matrix chart. All the test neutrophils are predicted in the correct classes.

4   Conclusion and Future Work

This study has identified active neutrophils successfully using transfer learning based on a pre-trained AlexNet. As a result, the mechanism of neutrophils mode and morphology changing might be better explained. However, a limitation of this study is that the set of data is too small, which might lead to the wrong classification.

Further research might explore in the following aspects.

Firstly, we might investigate how to make a framework which can address a large scaled neutrophils and classify the modes of neutrophils automatically.

Secondly, we could try to ask for real neutrophils data *in vivo*. There are two types of models commonly used, that is, *in vivo* models and *in vitro* models. The former uses alive cells while the latter does dead ones. The specific objective of using *in vivo* models is to better understand the mechanism of mode and morphology changing, such as pseudopod formation, during the inflammation process. In addition, a more accurate and suitable neutral network proposed by ourselves could be trained using huge amount of the *in vivo* data.

Finally, we are likely to use GANs to generate huge amount of fake neutrophils for further research.

References

1. Zhang, X.: Modeling: identification of neutrophil cell dynamic behaviour. Ph.D. thesis of the University of Sheffield (2016)
2. Craig, M., Humphries, A., Mackey, M.: A Mathematical model of granulopoiesis incorporating the negative feedback dynamics and kinetics of G-CSF/neutrophil binding and internalization. Bull. Math. Biol. **78**(12), 1–54 (2016)
3. Movassagh, H., Saati, A., Nandagopal, S., et al.: Chemorepellent semaphorin 3E negatively regulates neutrophil migration in vitro and in vivo. J. Immunol. **198**(3), 1023–1033 (2017)
4. Kolaczkowska, E., Kubes, P.: Neutrophil recruitment and function in health and inflammation. Nat. Rev. Immunol. **13**(3), 159–175 (2013)
5. Kadirkamanathan, V., Anderson, S., Billings, S., et al.: The neutrophil's eye-view: inference and visualisation of the chemoattractant field driving cell chemotaxis in vivo. PLoS One **7**(4), e35182 (2012)
6. Lecun, Y., Bengio, Y., Hinton, G.: Deep learning. Nature **521**(7553), 436–444 (2015)
7. Krizhevsky, A., Sutskever, I., Hinton, G.: Imagenet classification with deep convolutional neural networks. In: Advances in Neural Information Processing Systems, pp. 1097–1105 (2012)
8. Redmon, J., Divvala, S., Girshick, R., Farhadi, A.: You only look once: unified, real-time object detection. arXiv:1506.02640 (2015)
9. Vaswani, A., et al.: Attention is all you need. arXiv:1706.03762 (2017)
10. Zhang, L.: Transfer adaptation learning: a decade survey. arXiv:1903.04687 (2019)
11. Fan, B.: Neutrophil classification based on deep learning. Bachelor's thesis of Shanghai Polytechnic University (2019)
12. Goodfellow, I., Bengio, Y., Courville, A.: Deep Learning. MIT Press, Cambridge, (2016)

Two-Stage Training Method of RetinaNet for Bird's Nest Detection

Ruidian Chen$^{(\boxtimes)}$ and Jingsong He$^{(\boxtimes)}$

University of Science and Technology of China, Hefei, China
crd2018@mail.ustc.edu.cn, hjss@ustc.edu.cn

Abstract. Common nesting materials such as branches, straws, and wires fall on high-voltage power lines causing short-circuit faults. In recent years, neural network has developed rapidly in the field of objects detection. Through the shooting of the drone and the base station camera, the neural network is used to identify and locate the bird's nest in the image, which has great use prospects in the intelligent detection of the transmission system. RetinaNet is currently a representative objects detection network, using the focal loss to adjust the imbalance between foreground and background. In this paper, we apply RetinaNet to the bird's nest detection of transmission systems. Due to the complex environment of the transmission system, the detector obtained by the single-stage training recognize the line equipment or other objects as the nest easily. Combining the experimental results of single-stage training, we propose a two-stage training method driven by false detection samples, which improves the performance of the detector.

Keywords: Neural network · Transmission system · Automatic inspection

1 Introduction

Every year, the electric power failure caused by bird nests affects the safety of people's production [10]. Most of nests are built on high base station tower poles, and the workload of manual inspection is large. Automatic bird's nest detection will be a vital part in the field of transmission system inspection.

The early nest detection methods extract the special texture, geometry and shape features of the nest area, and combine machine learning algorithm, such as SVM, KNN, Adaboost classifier, then obtain the data information of the nest image to determine whether the nest exists in the image [21,23].

In recent years, convolutional neural networks (CNNs) have made great progress in image classification [13,26], semantic segmentation [5,25], objects detection [12,14] and other fields of computer vision [18,30]. A series of object detection networks, such as SSD [19] and Faster R-CNN [24], achieve end-to-end training and detection process, and show good performance on public datasets. RetinaNet [17] is an excellent object detection network, which has potential to be used in bird's detection.

© Springer Nature Singapore Pte Ltd. 2020
L. Pan et al. (Eds.): BIC-TA 2019, CCIS 1160, pp. 586–596, 2020.
https://doi.org/10.1007/978-981-15-3415-7_49

In complex environment of power transmission system, serious background interference will lead to false detection. Branches and staggered wires in the environment are identified as nests by detectors. Synthesizing images increase the training data, which can effectively reduce over-fitting and elevate model generalization ability [7]. [11] has successfully applied this method to indoor scene detection to enhance the ability of service robots to search household objects. The methods [7,11] use single-stage training to improve the performance of model. Multi-stage cascade classification network [3,8] inspires us to design targeted training strategies, focusing on distinguishing target areas from false detection areas. Here we propose a two-stage training method. Taking the false detection image as the background, and the bird's nest area is selected to synthesize the new images for second-stage training. On the one hand, it can suppress the background interference and reduce the false detection rate of the model. On the other hand, the convergence direction of the model is adjusted according to the results of single-stage training, so that the detector can put forward accurate position proposals.

In this paper, we apply RetinaNet to bird's nest detection of power transmission system, and propose a two-stage training method driven by false detection samples, which can effectively suppress background interference and make detector locate the nest area more accurately.

2   Bird's Nest Detection

The goal of bird's nest detection is to discover the bird's nest in the transmission system and prevent birds from affecting the safe operation of transmission system. The overhead transmission network is designed to achieve long-distance transmission, with wide coverage and complex operation environment. Figure 1 shows bird's nest images taken from the transmission system. The interference of power equipment, houses and trees in the background will greatly affect the accuracy of the detector. Because of the change of shooting angle, the bird's nest area in the picture is occluded to varying degrees. At the same time, the external illumination conditions also cause the difference of imaging quality.

2.1   Related Work

Early detection algorithm extracts geometric texture feature information (such as HOG, Harr) describing the object area in the picture. By establishing object feature information base, researchers train the classification model or use local feature matching to perform objects detection [1]. Deformable parts model (DPM) [9] is a component-based detection method that is robust to objects' deformation. Combined with the SVM classifier, DPM uses improved HOG features to reduce the performance degradation of detector due to perspective transformation. Bird's nests are mostly made up of messy staggered branches. Externally, the epitaxial branches form many spiny protrusions. In order to reduce

Fig. 1. Bird's nest in the transmission system

the consumption of manpower in the detection of bird's nest in railway catenary, two kinds of specific bird's nest morphological features, namely, histogram
of orientation of streaks (HOS) and histograms of length (HLS), were proposed
in [28]. Through the camera installed on the top of the train, the OCS images
of overhead railway are collected. The HOS and HLS features of the bird's nest
area in OCS images are extracted. 1820 OCS images are used to test and the
test accuracy is 35.82%, which is higher than that of HOG and DPM [28].

Traditional methods [21,23] detect objects by designed features. When the
actual engineering scene is complex, the feature extraction process is difficult,
even leads to poor detectors. Some traditional algorithms have high time complexity, which are difficult to meet the real-time requirements. Deep learning
shows good performance in face recognition [27], pedestrian detection [14]. In
this paper, we apply RetinaNet to bird's nest detection of transmission system
and propose a two-stage training method for false detection.

2.2   Architecture

The process of visualization of convolutional neural network feature map shows
that shallow layers focus on the extraction of texture features of network, while
deep layers pay more attention to the semantic information of images [20,31].
In U-net [25] and Feature Pyramid Networks (FPN) [16], bottom-up sampling
process is used to combine shallow texture information with deep semantic information, which can improve the learning of related constraints between objects
in images.

RetinaNet's backbone has been further improved on the basis of FPN.
Figure 2 shows the network structure of RetinaNet, which retains the P_3, P_4,
P_5 layers in FPN, and adds P_6 and P_7. C_3, C_4, C_5 generate M_3, M_4, M_5 by

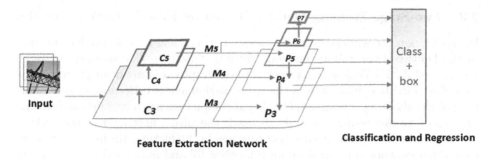

Fig. 2. The architecture of RetinaNet.

1×1 convolution, then P_3, P_4, P_5 are created by upsampling. 1×1 convolution layer is to ensure that the number of channels in the corresponding layer is equal, which can prevent the gradient change from affecting the stability of the backbone. Through 3×3 convolution, we get P_6, P_7 from C_5,P_6 respectively. [16] holds that P_6 and P_7 have large receptive fields, which will improve the detection performance of the model for large objects. The design of anchor is similar to the variant of region proposal network (RPN) [24], and the size of the anchor from P_3 to P_7 increases from 32^2 to 512^2. In order to obtain more denser coverage than Faster R-CNN, three different sizes (2^0, $2^{\frac{1}{3}}$, $2^{\frac{2}{3}}$) were added to each level. Each anchor frame is assigned a vector of length K as classification information, and a position regression information of length 4.

In objects detection, the object area is generally defined as the foreground, while the non-object area is defined as the background. The number of background in candidate location is far more than that of foreground. This imbalance is an important factor affecting the performance of the detector. He et al. [17] proposed focal loss(FL) to solve the imbalance between foreground and background in the training process. The equation is as follows:

$$FL(P_t) = -(1 - P_t)^\gamma \log P_t \tag{1}$$

where P_t is the probability that the network predicted to be the category t, γ is called the focusing parameter, and $(1 - P_t)^\gamma$ to reduce the weight of easy-to-classify samples, so that the model concentrates more on difficult-to-classify samples in training.

The image size is uniformly adjusted to 600×600, and the position coordinates of the nest use the same encoding as in the Faster R-CNN. In the experiment, we follow the anchor allocation rule same to that in, setting the intersection-over-union threshold (IOU) between the anchor and the nest area to 0.5. If IOU > 0.5, mark it as the nest area, and back to the background when IOU < 0.4. The rest of the anchor box are ignored. The classification loss function is focal loss, and the bounding box regression loss function uses Smooth L1 loss [24], where L1 means norm regularization. After setting the hyperparameter, no adjustments are made to the network. For convenience, we call the above process the single-stage training.

2.3   Two-Stage Training Method Driven by False Detection Samples

In this section, we will elaborate the design concept and specific implementation method of two-stage training method driven by false detection samples. Through Unmanned Aerial Vehicle (UAV) aerial photography and artificial photography, we collect bird nest images from the transmission system. Because of the difference of the shooting equipment, the size of the picture has a certain change. The process of preprocessing is mainly to mark the bird's nest area in the picture and determine its specific position coordinates in the picture. RetinaNet's feature extraction network is pre-trained on ImageNet [6] and initialized on other parts of the network. Then processed data is used as training set to train RetinaNet network directly to get the detector.

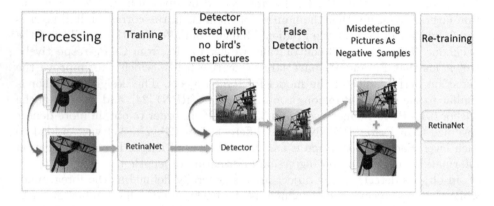

Fig. 3. Add negative training samples directly for training.

Most of the images obtained by UAV aerial photography do not contain nests. In order to simulate a real detection environment, select a part of images with similar background conditions. When using athe detector to test images without nests, a series of false detections occurred in the results. We add these false-detection images as negative training samples directly into the training set for training, which is shown in Fig. 3. In practice, we can get a large number of images without nests in the transmission system. The reason for choosing only this part of the false negative image as a negative sample is that we want the model to pay more attention to the image that caused the false detection. If all the images without nests are used as negative samples, the model will pay the same attention to all the pictures without nests. In fact, false-detection will not occur in most images, so the selection of negative samples needs to be targeted. Actually, the data volume of the negative sample needs to be controlled within a certain range. Otherwise, the training process will consume a large amount of computing resources, which takes a lot of time. Multi-stage classification network has been widely used in the field of objects detection [3,4]. Cascade classifiers process a series of different data streams, and classifiers process different levels

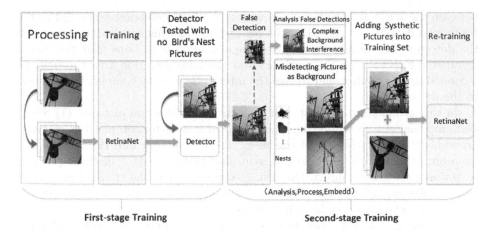

Fig. 4. Two-stage training method driven by false detection samples.

of image information [29]. Object Detection with Deformable Part Based Convolutional Neural Networks (DeepID-Net) [22] is introduced, and a multi-stage training architecture is adopted. Samples with different difficulty are processed at each stage. Simple samples are processed in the initial stage, difficult samples in the previous stage are processed in the second stage, and so on. Through specific training strategy design, this deep architecture can gradually train the network by mining difficult samples, simulating cascade classifiers. Inspired by this, we propose a two-stage training method driven by false detection samples in Fig. 4. In the second stage, we pay attention to the learning of the model for false detection in results.

The first stage remains unchanged, and the annotated bird's nest images are entered into RetinaNet. After the training is completed, we use the detector to test the picture without nests and analyze the false detection. In the experiment, we found that the complex background environment is the main reason for false detection. The staggered power lines and power equipment are often misidentified by the detector.

In the objects detection, the model can effectively improve the detection performance by extracting the background information during the training process and using the context information as a support. Inside-Outside Net (ION) [2] uses the spatial recursive network to synthesize the information in the image, extracts the features of the preselected area of the detector, and focuses on the surrounding environment information of the object area. In [15], Context Convolution Neural Network (AC-CNN) combines one attention-based global contextualized (AGC) sub-network based on attention mechanism with one multi-scale local contextualized (MLC) sub-network to obtain global context information and local feature information, and integrate global information and local information. The above methods provide the detection information of the detector to the target area by utilizing the background information, learning the relationship between the scene and the target. However, the false detection generated in the bird's nest detection is caused by the background causing serious interference to the detector. The purpose of the two-stage method is to hope that the model can distinguish between the nest area and the false detection area through the second stage of training, and reduce the interference of the complex background to the detector.

In the second-stage, we first deal with the false detection pictures. (1) Select and crop the nest area from the image, regard false detection pictures as background. (2) Adjust the size of the nest, insert nest area into the appropriate location. (3) Label the nest area in synthetic pictures. (4) Join these pictures to training set for retraining. One thing to note is that the training process of the model is non-continuous. That is, at the beginning of the second stage, the model loads the same weight as first stage.

Single-stage training make the detector effectively detect the nests with clear contours. For the nest which is hard to detect because of the occlusion, false detection will happen. Therefore, in the second-stage, we hope to improve the detection performance of the detector for difficult-to-detect nests. In order to imitate the occlusion in the real scene, we intentionally intercept part of the nest area when embedding. It aims to adjust the convergence direction of the model according to the actual application scenarios.

3   Experiments

3.1   Experimental Arrangements

The parameters of neural network are optimized by SGD. The initial learning rate is set to $1e-3$ and the momentum is 0.9. To prevent the model from overfitting, the weight decay [2] is $1e-4$. The max epoch is set to 350. Retiananet's backbone is resnet-50 pre-trained on ImageNet. We also use the data augmentation methods such as horizontal flipping, center cropping, and scale transformation to increase the generalization performance of the model. The training process was completed on the NVIDIA GTX 1080ti GPU. We use 613 bird's nest pictures as a training set. The test set contains 649 samples, where the number of bird's nest images is 115. After the single-stage training is completed, the performance of the model is tested using other no-nest images obtained from the power system. Eventually, we found errors in 155 images, all of which were that the detector identified other objects as nests. All the experients was implemented by Pytorch, which is a scientific computing package.

In order to investigate the effectiveness of the two-stage method, we conducted the following three experiments:

RetinaNet-1: Using the single-stage training methods, the annotated pictures are directly entered into the network for training.

RetinaNet-2: Adding 155 false detection pictures that were taken as negative training samples to the training set, which is a conventional method.

RetinaNet-3: In this experiment, we used two-stage training method driven by false detection samples. Combined with the test results, the bird's nest is embedded into false detection pictures by cropped-pasted. After processing completed, the second-stage training is conducted.

3.2 Experimental Results

Experimental results demonstrate the effectiveness of the two-stage method. The precision and test mean average precision (mAP) value are shown in Table 1.

Table 1. Precision and mAP of RetinaNet-1, RetinaNet-2, RetinaNet-3.

Model	RetinaNet-1	RetinaNet-2	RetinaNet-3
Precision	59.2%	67.9%	70.5%
mAP	49.1	53.7	54.8

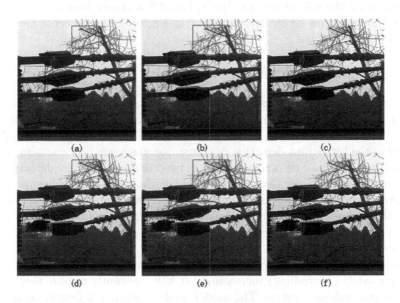

Fig. 5. (a) (b) (c) are the original images; (d) (e) (f) are the synthetic images (Red represents the detection box, green represents the embedded area box). (Color figure online)

RetinaNet-1 identified the branch as a bird's nest in Fig. 5(a), (d). In Fig. 5(b), (e), we also find the false detection by RetinaNet-2. The second-stage

(a)   (b)   (c)

Fig. 6. (a) (b) (c) are RetinaNet-1, RetinaNet-2 and RetinaNet-3 test results in a same picture.

training has a significant inhibitory effect on the false detection area. RetinaNet-3 has no false detection in Fig. 5(c), (f), and the nest embedded in Fig. 5(f) is detected accurately.

We compared the actual results of three models, as shown in Fig. 6. In Fig. 6(a), RetinaNet-1 misidentified the steel skeleton as a bird's nest. RetinaNet-2,3 eliminated the false detection. RetinaNet-3 has shown better detection performance, and it can be clearly seen that in Fig. 6(c), the detector located the nest more accurately.

4   Conclusion and Future Work

Bird's nest detection is an important task in automatic inspection of transmission system. Because of the complex environment of transmission system, background objects are often recognized as nests by detectors. In this paper, RetinaNet is applied to the bird's nest detection of transmission system. A two-stage training method driven by false detection samples is proposed to solve false detection. The experimental results show that the two-stage method can effectively suppress the interference of complex background on the detector and guide the detector to locate the nest area accurately. Compared with single-stage training, it provides us with a way to adjust the convergence direction of the model and get a better detector.

Under the actual engineering conditions, the amount of data acquired by the detection system is gradually increasing. We will constantly acquire new samples from the transmission system. The model needs to learn new feature information from new samples, and the incremental learning structure is more in line with the actual engineering requirements. Simulating incremental learning through multi-stage training structure is the focus of the next step.

References

1. Bai, X., Yang, X., Latecki, L.J.: Detection and recognition of contour parts based on shape similarity. Pattern Recogn. **41**(7), 2189–2199 (2008)

2. Bell, S., Lawrence Zitnick, C., Bala, K., Girshick, R.: Inside-outside net: detecting objects in context with skip pooling and recurrent neural networks. In: Proceedings of IEEE Conference on Computer Vision and Pattern Recognition, pp. 2874–2883 (2016)

3. Bourdev, L., Brandt, J.: Robust object detection via soft cascade. In: 2005 IEEE Computer Society Conf. Computer Vision and Pattern Recognition (CVPR 2005), vol. 2, pp. 236–243. IEEE (2005)

4. Sun, C., Lam, K.-M.: Multiple-kernel, multiple-instance similarity features for efficient visual object detection. IEEE Trans. Image Process. $22(8)$, 3050–3061 (2013)

5. Ciresan, D., Giusti, A., Gambardella, L.M., Schmidhuber, J.: Deep neural networks segment neuronal membranes in electron microscopy images. In: Advances in Neural Information Processing Systems, pp. 2843–2851 (2012)

6. Deng, J., Dong, W., Socher, R., Li, L.J., Li, K., Fei-Fei, L.: ImageNet: a large-scale hierarchical image database. In: 2009 IEEE Conference on Computer Vision and Pattern Recognition, pp. 248–255. IEEE (2009)

7. Dvornik, N., Mairal, J., Schmid, C.: Modeling visual context is key to augmenting object detection datasets. In: Proceedings of European Conference on Computer Vision (ECCV), pp. 364–380 (2018)

8. Felzenszwalb, P.F., Girshick, R.B., Mcallester, D.A.: Visual object detection with deformable part models. Commun. ACM $56(9)$, 97–105 (2013)

9. Forsyth, D.: Object detection with discriminatively trained part-based models. Computer 2, 6–7 (2014)

10. Frazier, S.D.: Birds, substations, and transmission. In: 2001 IEEE Power Engineering Society Winter Meeting. Conference on Proceedings (Cat. No. 01CH37194), vol. 1, pp. 355–358. IEEE (2001)

11. Georgakis, G., Mousavian, A., Berg, A.C., Kosecka, J.: Synthesizing training data for object detection in indoor scenes. arXiv preprint arXiv:1702.07836 (2017)

12. He, K., Gkioxari, G., Dollár, P., Girshick, R.: Mask R-CNN. In: Proceedings of IEEE International Conference on Computer Vision, pp. 2961–2969 (2017)

13. He, K., Zhang, X., Ren, S., Sun, J.: Deep residual learning for image recognition. In: Proceedings of IEEE Conference on Computer Vision and Pattern Recognition, pp. 770–778 (2016)

14. Li, J., Liang, X., Shen, S., Xu, T., Feng, J., Yan, S.: Scale-aware fast R-CNN for pedestrian detection. IEEE Trans. Multimed. $20(4)$, 985–996 (2017)

15. Li, J., Wei, Y., Liang, X., Jian, D., Yan, S.: Attentive contexts for object detection. IEEE Trans. Multimed. $19(5)$, 944–954 (2017)

16. Lin, T.Y., Dollár, P., Girshick, R., He, K., Hariharan, B., Belongie, S.: Feature pyramid networks for object detection. In: Proceedings of IEEE Conference on Computer Vision and Pattern Recognition, pp. 2117–2125 (2017)

17. Lin, T.Y., Goyal, P., Girshick, R., He, K., Dollár, P.: Focal loss for dense object detection. In: Proceedings of IEEE International Conference on Computer Vision, pp. 2980–2988 (2017)

18. Litjens, G., et al.: A survey on deep learning in medical image analysis. Med. Image Anal. 42, 60–88 (2017)

19. Liu, W., et al.: SSD: single shot multibox detector. In: Leibe, B., Matas, J., Sebe, N., Welling, M. (eds.) ECCV 2016. LNCS, vol. 9905, pp. 21–37. Springer, Cham (2016). https://doi.org/10.1007/978-3-319-46448-0_2

20. Mahendran, A., Vedaldi, A.: Visualizing deep convolutional neural networks using natural pre-images. Int. J. Comput. Vis. $120(3)$, 233–255 (2016)

21. Manikandan, M., Paranthaman, M., Aadithiya, B.N.: Detection of calcification form mammogram image using canny edge detector. Indian J. Sci. Technol. **11**(20), 1–5 (2018)
22. Ouyang, W., et al.: DeepID-Net: deformable deep convolutional neural networks for object detection. In: Proceedings of IEEE Conference on Computer Vision and Pattern Recognition, pp. 2403–2412 (2015)
23. Qing, C., Dickinson, P., Lawson, S., Freeman, R.: Automatic nesting seabird detection based on boosted HOG-LBP descriptors. In: 2011 18th IEEE International Conference on Image Processing, pp. 3577–3580. IEEE (2011)
24. Ren, S., Girshick, R., Girshick, R., Sun, J.: Faster R-CNN: towards real-time object detection with region proposal networks. IEEE Trans. Pattern Anal. Mach. Intell. **39**(6), 1137–1149 (2017)
25. Ronneberger, O., Fischer, P., Brox, T.: U-net: convolutional networks for biomedical image segmentation. In: Navab, N., Hornegger, J., Wells, W.M., Frangi, A.F. (eds.) MICCAI 2015. LNCS, vol. 9351, pp. 234–241. Springer, Cham (2015). https://doi.org/10.1007/978-3-319-24574-4_28
26. Simonyan, K., Zisserman, A.: Very deep convolutional networks for large-scale image recognition. arXiv preprint arXiv:1409.1556 (2014)
27. Viola, P., Jones, M.J.: Robust real-time face detection. Int. J. Comput. Vis. **57**(2), 137–154 (2004)
28. Wu, X., Yuan, P., Peng, Q., Ngo, C.W., He, J.Y.: Detection of bird nests in overhead catenary system images for high-speed rail. Pattern Recogn. **51**(C), 242–254 (2016)
29. Yang, Y., Wu, F.: Real-time traffic sign detection via color probability model and integral channel features. In: Li, S., Liu, C., Wang, Y. (eds.) CCPR 2014. CCIS, vol. 484, pp. 545–554. Springer, Heidelberg (2014). https://doi.org/10.1007/978-3-662-45643-9_58
30. Yu, Y., Gong, Z., Zhong, P., Shan, J.: Unsupervised representation learning with deep convolutional neural network for remote sensing images. In: Zhao, Y., Kong, X., Taubman, D. (eds.) ICIG 2017. LNCS, vol. 10667, pp. 97–108. Springer, Cham (2017). https://doi.org/10.1007/978-3-319-71589-6_9
31. Zeiler, M.D., Fergus, R.: Visualizing and understanding convolutional networks. In: Fleet, D., Pajdla, T., Schiele, B., Tuytelaars, T. (eds.) ECCV 2014. LNCS, vol. 8689, pp. 818–833. Springer, Cham (2014). https://doi.org/10.1007/978-3-319-10590-1_53

SlimResNet: A Lightweight Convolutional Neural Network for Fabric Defect Detection

Xiaohui Liu$^{(\boxtimes)}$, Zhoufeng Liu$^{(\boxtimes)}$, Chunlei Li$^{(\boxtimes)}$, Yan Dong, and Miaomiao Wei

Zhongyuan University of Technology, Zhengzhou 450007, China
905853998@qq.com, lzhoufeng@hotmail.com, lichunlei1979@sina.com

Abstract. Convolutional neural network has attracted increasing attention in object detection and recognition. Among the methods mentioned in various literatures, the emphasis is to increase the depth of the model so as to improve the accuracy of recognition and detection. However, these large networks require more computing overhead and memory storage, which restricts their usages on mobile devices. Inspired by the superior detection performance of residual network (ResNet), we proposed a lightweight network, called SlimResNet, which is applied in fabric defect detection. Firstly, due to the particularity of fabric defects, the size of residual network convolution kernel will be changed in this paper to capture more details. Secondly, since the high-level semantic information of convolutional neural network is redundant for defect detection, we pruned the network structure to better compress the network model. Finally, the experimental results show that the recognition performance of the proposed lightweight network is comparable to the original network.

Keywords: Convolutional neural network · Lightweight network · Fabric defect detection · Residual network

1 Introduction

Fabric defect detection plays an important role in textile quality control. In recent years, many scholars have carried out a large number of research on the fabric defect method and made good progress, including statistical, spectral method, dictionary learning and so on. The statistical method is to detect the fabric defect by using the difference of distribution characteristics between the defect and the normal area. The most common is histogram analysis [1]. This method can not reflect the overall characteristics of the fabric image effectively, and the detection results are highly dependent on the size of the selected window and the size of the selected threshold. Spectral method is a method to detect fabric defects by some operation in frequency domain. Therefore, the general approach is to select an appropriate orthogonal basis to transform the fabric image from the spatial domain to the frequency domain. The common

© Springer Nature Singapore Pte Ltd. 2020
L. Pan et al. (Eds.): BIC-TA 2019, CCIS 1160, pp. 597–606, 2020.
https://doi.org/10.1007/978-981-15-3415-7_50

transformation methods are Fourier transform [2], wavelet transform [3] and so on. Although the fabric detection based on spectral method can make use of the overall characteristics of the fabric image, the detection effect of the fabric image with complex texture is poor. More importantly, it also has certain requirements for the selected orthogonal basis. Dictionary learning consists of building a dictionary and using the dictionary in two stages. Dictionary is built by training or testing images. Normal fabric images are reconstructed based on sparse representation. Then, the reconstructed normal fabric is subtracted from the detection image to make the defect area prominent and achieve the purpose of defect detection [4]. Another method is to take the dictionary set as projection matrix to realize dimension reduction, and then carry out fabric defect detection through feature comparison. However, for the dictionary construction, the generalization performance is poor if only training image is used for dictionary learning. In addition, the effect is poor for the detection of complex texture fabric image. By contrast, convolutional neural network has a good performance in detecting complex textured fabric images.

The first real convolutional neural network LeNet [5] was proposed by LeCun et al. in 1989. Convolutional neural networks have become popular since AlexNet [6], proposed by Krizhevsky et al., won the 2012 Large Scale Visual Recognition Challenge (ILSVRC). In particular, ResNet [7], proposed by Kaiming He et al. in 2016, exceeded the classification error rate of human beings. With the development of convolutional neural network, the recognition accuracy is getting higher and higher. But the network is getting deeper and deeper with a lot of computing overhead and memory storage.

In this paper, we propose a lightweight network structure based on ResNet [8]. First, it simply needs to distinguish whether a complex texture image contains defects, so the number of network channels in the last layer of ResNet is changed from 1000 to 2. Secondly, for fabric defect detection, edge information, especially some details, is of vital importance. Based on this, we transform the large convolution kernel of the first layer of ResNet into a small convolution kernel. Finally, due to the particularity of fabric defects, the advanced semantic information of the original ResNet network has little effect on the detection of defects. Therefore, the model was pruned roughly.

The rest of the paper is organized as follows. The related work is discussed in Sect. 2. The details of SlimResNet architecture are presented in Sect. 3. The experiments are described in Sect. 4, which mainly compares SlimResNet with the others. Section 5 concludes the paper.

2   Related Work

2.1   Shortcut Connection

In recent years, more and more researchers pay attention to shortcut connection and have produced many achievements. The idea of shortcut connection originates from centralization. Schraudolph [9] actually extends such an idea to the back propagation of gradient very early, and proposes the technology of shortcut connection. The paper of [10], proposed by LeCun, makes a more detailed study of the impact of shortcut connection on the model. It mainly realizes the centralization of gradient, propagation and response error of each layer through shortcut connection. In [11], Szegedy et al. also introduces short-cut connection in the Inception module. Later, a model called highway-network [12] appears simultaneously with ResNet. It is mainly inspired by the control gate idea of LSTM, which is more complex than ResNet's identical shortcut connection. More importantly, the highway-network does not show that the accuracy of the network increases with the extreme deepening of the network depth.

2.2 Model Compression

As the depth of convolutional neural network deepens, it brings huge computational overhead and memory storage problems. Therefore, more and more scholars are committed to transforming large-scale networks into lightweight networks and ensuring the loss of precision within a reasonable range. At present, there are many mainstream compression model methods. For pre-trained model compression, the better method is model's pruning. The first real pruning, called optimal brain damage [13], is proposed by LeCun et al. It mainly thins out the coefficient of network. Using a similar idea, Hassibi et al. proposes optimal brain surgeon [14]. Hession matrix comprised by second-order gradient in back propagation determines the value of each weight. But all the methods above is unstructured pruning, which may increase the model's running time. Based on the above statements and the particularity of fabric defect detection, the whole convolution kernel at a higher level is pruned while the convolution at a lower level is retained to capture more detailed information.

3 SlimResNet Architecture

In this section, we first describe the core layer called the bottleneck block constituting SlimResNet. Next the overall Architecture of SlimResNet is described. Finally, a detailed description of the implementation of the SlimResNet is presented.

3.1 Bottleneck Block

Inspired by ResNet, the structure of the SlimResNet model is based on the bottleneck block that is a lightweight network structure. Compared with the residual block containing two convolution layers, bottleneck block comprised by three convolution layers is 1×1, 3×3 and 1×1 respectively. Among them, 1×1 convolution mainly reduces the number of parameters by adjusting the dimensions of input and output. 3×3 convolution, like ordinary convolution, extracts and combines low-level features to form high-level features for classification. Before the input into the convolution of 3×3, the dimension of the feature map decreases. The convolution output of 3×3 has gone through another convolution of 1×1 causing feature map to increase. So, the above statement of 1×1 convolution is similar to the neck of bottle, and this is why the lightweight architecture is called as bottleneck block. Counterintuitively, bottleneck block containing three convolution layers has fewer parameters because it has smaller convolution kernels.

Identity connection without any parameters is vital for bottleneck block model. Firstly, the information of forward propagation and back propagation is delivered from one bottleneck block to another one by identity connection. Secondly, the size of the model becomes very large by adding other transformation functions to the identity connection. What's more, whether it is multiplied by a constant, or a very simple 1×1 convolution and other operations will cause the performance of the model to decline. Therefore, the identity connection in the residual network is essential to build a lightweight network.

For bottleneck block, there are two branches called residual branch and identity branch. The residual branch is the one mentioned earlier with three layers of convolution. The identity branch is the identity connection which is refered earlier. The statement above is mainly to emphasize the overall framework of bottleneck block, but don't refer to the specific position of batch normalization and activation function. Their specific position is important because different position have different performance for model inference. Inspired by ResNet, we adopt the bottleneck block architecture depicted in Fig. 1.

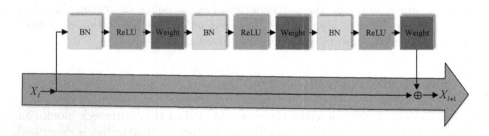

Fig. 1. Bottleneck block containing identity branch and residual branch.

This architecture differs greatly from the original ResNet about the position of the batch normalization and activation function. Certainly, the model's inference power is enhanced after such adjustments. That is because our adjustment based on the following principles is reasonable. (1) The output of the residual branch should be between $(-\infty, -\infty)$. So we don't use activation function as the last layer of the residual branch. (2) It is not wise to change the value of the identity branch, which means that the input and output are the same for the identity branch, and the previous section has been explained. (3) The symbol $\oplus$ cannot be stacked on the layer that can change the information distribution, so we will not add batch normalization after the symbol $\oplus$.

By observing the bottleneck block, it is found that both the batch normalization and activation functions are in front of the convolution layer. This is known professionally as pre-activation. The operation has the following benefits: (1) Backpropagation basically confirms to the hypothesis, and the information transmission is unimpeded. So the model is easier to train. (2) Such operation plays the role of regularization to prevent model overfitting.

The connection of two adjacent bottleneck blocks is described in Fig. 2. The dotted line refers to two cases. First, when the channel number of residual branch is inconsistent with the channel number of identity branch, 1×1 convolution is used to adjust the channel number of identity branch. Second, if the number of channels in both branches is the same, nothing needs to be done.

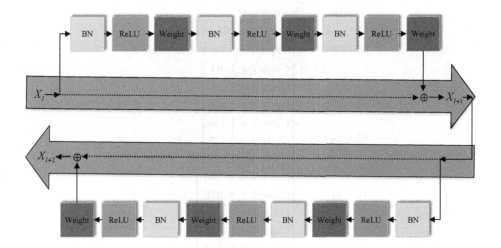

Fig. 2. The connection of two adjacent bottleneck blocks.

3.2 Network Structure

The structure of SlimResNet is based on the bottleneck block. The bottleneck block accounts for a large proportion of the proposed model and the traditional convolution only exists in the first layer of the model. The size of the first layer convolution kernel of ResNet is 7×7. For images with rich semantic information, the convolution kernel with the size of 7×7 has a larger receptive field to better process it. In addition, it is proposed to replace the convolution of 7×7 with a stack of three convolution of 3×3 which have the same receptive field of 7×7 kernel. It looks like that the parameters are also reduced. However, due to the relatively small number of channels to input the image, the calculation amount of 7×7 convolution kernel is $7 \times 7 \times 3 \times 64$. By stacking 3×3 convolutional layer, and the computation amount becomes $3 \times 3 \times 3 \times 64 + 3 \times 3 \times 64 \times 64 + 3 \times 3 \times 64 \times 64$. It is because the input channel of 3×3 is 64 in the latter two layers. As a result, the amount of computation increases by this operation.Inspired by the above statement, we change the convolution kernel of 7×7 to 3×3 for the following reasons. (1) The object we need to detect is fabric. There is no rich semantic information, but more texture information. Therefore, using 7×7 convolution kernel will lose a lot of details and edge information. (2) The convolution kernel with 7×7 takes $7 \times 7 \times 3 \times 64$, and the convolution kernel with 3×3 takes $3 \times 3 \times 3 \times 64$. Therefore, the employ of smaller convolution kernel will produce fewer parameters, which meets the requirements of the design of lightweight network.

Table 1. The structure of SlimResNet.

Output size	SlimResNet
112×112	Conv, 3×3, 64, stride 2
56×56	Maxpooling, 3×3, 64, stride 2
	$\begin{bmatrix} Conv\ 1 \times 1\ \ 64 \\ Conv\ 3 \times 3\ \ 64 \\ Conv\ 1 \times 1\ 256 \end{bmatrix} \times 3$
28×28	$\begin{bmatrix} Conv\ 1 \times 1\ 128 \\ Conv\ 3 \times 3\ 128 \\ Conv\ 1 \times 1\ 512 \end{bmatrix} \times 1$
14×14	$\begin{bmatrix} Conv\ 1 \times 1\ \ 256 \\ Conv\ 3 \times 3\ \ 256 \\ Conv\ 1 \times 1\ 1024 \end{bmatrix} \times 1$
7×7	$\begin{bmatrix} Conv\ 1 \times 1\ \ 512 \\ Conv\ 3 \times 3\ \ 512 \\ Conv\ 1 \times 1\ 2048 \end{bmatrix} \times 1$
1×1	Global average pooling

The structure of SlimResNet is defined in Table 1. The input model is the color fabric image with complex texture, and the size of the image after preprocessing is 224×224. Theoretically, images of any size can be employed because they are preprocessed to form 224×224 images fed into our model. And this preprocessing will be discussed in Sect. 4.1 in detail. After the input image of $224 \times 224 \times 3$ passes through the convolution kernel of the first layer, the shape of output feature map becomes $112 \times 112 \times 64$ because the stride of convolution is 2. Following a pooling operation, the size of the feature map is reduced by half. Since it particularly needs to focus on the lower level feature map, we employ three bottleneck blocks as a group (with the same number of output channels) to extract more detail information. The remaining three bottleneck blocks extract and combine advanced features successively. If semantically rich images need to be extracted feature, it is common to add multiple bottleneck blocks to form a group in the size of feature maps with 14×14 (with the same number of channels). If the result is still unsatisfactory, it is wise to continue to increase the number of bottleneck blocks in this group and in the size of the feature maps with 28×28. Finally, the size of the output feature map is $7 \times 7 \times 2048$. After batch normalization and global average pooling, $1 \times 1 \times 2048$ feature map is formed. After the convolution of 1×1, the feature map forms the logits of $1 \times 1 \times 2$ delivered to Softmax layer for classification. In this case, the convolution of 1×1 corresponds to a full connection layer.

3.3   Implementation Details

The proposed model is implemented on TensorFlow, which is an open source framework published by Google. TensorFlow allows us to train and test the proposed model in parallel using multiple GPUs. There are two ways to handle parallelism. One is model parallelism, where a single model is distributed across multiple machines. In this way, the performance improvement achieved by training mainly depends on the structure of the model. For large model parallelization, the performance can be significantly improved and the training can be accelerated obviously. The other is data parallelism. Data parallelism is a parameter optimization method completely different from model parallelism. It employs n workers to simultaneously calculate n different data blocks in parallel to optimize the parameters of the model. TensorFlow supports data parallelism. It is worth noting that the results of training models on multiple GPUs are similar to those on a single GPU. Importantly, using multiple GPUs greatly increases the speed of the model inference.

The SlimResNet is trained from scratch. We adopt simple and direct exponential decay learning rate with fast convergence speed. The exponential decay rate is 0.94, and the initial learning rate is set as 0.01. The rmsprop optimizer is employed and the parameter is set as 0.9. Batch size is set to 64. The weight decay rate of the model is 0.00004. Batch normalization operation is employed instead of dropout.

4   Experiments

In this section, firstly, we discuss the fabric image data employed by us and the way of data preprocessing. Then, the results are presented about detection performance of fabric images.

4.1   Dataset

Color fabric images employed by us have complex textures. The partial sample of the dataset is shown in Fig. 3. The first line shows the normal color fabric image, which has complex texture, making it difficult to recognize. The second line shows the colored fabric with defects, which increases the noise and further increases the difficulty of recognition. This dataset contains 3,595 color fabric

Fig. 3. The partial sample of the dataset.

images including two categories: normal fabric images (1,873) and defective fabric images (1,722). The dataset is randomly shuffled and divided into two parts. 2,595 images are employed for model training, called the training set, and 1,000 images used for model test, called the test set. It almost satisfy the 7:3 for training set and test set. Such partitioning is optimal for small data sets.

With regard to data preprocessing, we follow [15]. There are some differences in the random cropping part. The principle employed by us is isotropic scaling. Our original input image size is 500×333, while the input to the model is fixed at 224×224. Let L be the middle size of the image to resize. In general, L should not be less than 333 for the color fabric images. In experiment, the value of L is selected in [333, 500]. Then the cropping operation is carried out to form the image of size 224×224. The image is then delivered to the model for training. To be fair, both ResNet and SlimResNet have been trained and tested on this dataset with exactly same way of preprocessing.

4.2 Experimental Comparison

In this section, a detailed performance comparison is presented about ResNet and SlimResNet. For fair comparison, the same environment configuration and dataset are adopted. The detailed experimental parameter Settings have been provided in Sect. 3.3.

For the training process, ResNet and SlimResNet employ exactly the same operations. The TFrecord file containing 2595 images will be delivered to the proposed model. The batch size mentioned earlier need to be set. In this experiment, batch size of 64 is adopted and the iteration epoch is set as 125. It is worth noting that the model is stored every 180s during the training. It is necessary to specify position storing the model generated during the training. Different models are employed to verify on the test set to understand the learning ability of the model. The value of loss is the output for each step, so the change of loss can be observed easily at real time.

For the test process, ResNet has no difference with SlimResNet. The TFrecord file containing 1000 images will be delivered to the fully trained model. Model output top-1 accuracy. Because for classification of two, Top-5 doesn't make sense. Performance comparisons are described in Table 2.

Table 2. Performance comparisons between ResNet and SlimResNet.

Model	Parameters number	Model size	Acc %
ResNet	23.5M	273M	99.3
SlimResNet	8.1M	95M	99.1

Two convolution models are evaluated in the fabric dataset. We argue that the original ResNet has a lot of parameter redundancy by evaluating for fabric detection. As can be seen from Table 2, the detection accuracy of ResNet is almost exactly the same with SlimResNet proposed for lightweight network. Therefore, our detection architecture is competitive with the best detection architecture ResNet. However, the number of parameters and the size of the model are significantly reduced relative to ResNet.

5   Conclusion

In this work we propose a new lightweight network model called SlimResNet, which is based on bottleneck blocks. Lots of research have been done on how to build lightweight networks for fabrics. Firstly, since the fabric image is more about edges and details, large convolution kernel tends to lose more details. This will affect the accuracy of the detection. So the ResNet's first large 7×7 convolution kernel is replaced with a smaller 3×3 convolution kernel. Secondly, fabric images are more texture features. Therefore, the high level semantic information extracted by convolutional neural network is redundant for the detection of fabric defects. Based on the above statement, the model is pruned in order to get fewer parameters. Finally, experiments have shown that the lightweight network model SlimResNet is competitive for the most superior ResNet.

Acknowledgement. This work was supported by NSFC (No. 61772576, U1804157), Science and technology innovation talent project of Education Department of Henan Province (17HASTIT019), The Henan Science Fund for Distinguished Young Scholars (18410 0510002), Henan science and technology innovation team (CXTD2017091), IRTSTHN (18IRTSTHN013), Program for Interdisciplinary Direction Team in Zhongyuan University of Technology.

References

1. Selver, M., Avşar, V., Özdemir, H.: Textural fabric defect detection using statistical texture transformations and gradient search. J. Text. Inst. **105**(9), 998–1007 (2014)
2. Sakhare, K., Kulkarni, A., Kumbhakarn, M., et al.: Spectral and spatial domain approach for fabric defect detection and classification. In: International Conference on Industrial Instrumentation & Control (2015)
3. Wen, Z., Cao, J., Liu, X., et al.: Fabric defects detection using adaptive wavelets. Int. J. Cloth. Sci. Technol. **26**(3), 202–211 (2014)
4. Qu, T., Zou, L., Zhang, Q., et al.: Defect detection on the fabric with complex texture via dual-scale over-complete dictionary. J. Text. Inst. **107**(6), 1–14 (2015)
5. Lecun, Y., Boser, B., Denker, J., et al.: Handwritten digit recognition with a back-propagation network. In: Advances in Neural Information Processing Systems, pp. 396–404 (1990)
6. Krizhevsky, A., Sutskever, I., Hinton, G.: ImageNet classification with deep convolutional neural networks. In: NIPS. Curran Associates Inc. (2012)
7. He, K., Zhang, X., Ren, S., et al.: Deep residual learning for image recognition (2015)
8. He, K., Zhang, X., Ren, S., Sun, J.: Identity mappings in deep residual networks. In: Leibe, B., Matas, J., Sebe, N., Welling, M. (eds.) ECCV 2016. LNCS, vol. 9908, pp. 630–645. Springer, Cham (2016). https://doi.org/10.1007/978-3-319-46493-0_38
9. Schraudolph, N.: Accelerated gradient descent by factor-centering decomposition. IDSIA (1998)
10. Raiko, T., Valpola, H., Lecun, Y.: Deep learning made easier by linear transformations in perceptrons (2012)
11. Szegedy, C., Liu, W., Jia, Y., et al.: Going deeper with convolutions (2014)
12. Srivastava, R., Greff, K., Schmidhuber, J.: Highway networks. arXiv preprint arXiv:1505.00387 (2015)
13. LeCun, Y.: Optimal brain damage. In: Advances in Neural Information Processing Systems, vol. 2 (1990)
14. Hassibi, B.: Second order derivatives for network pruning: optimal brain surgeon. In: Advances in Neural Information Processing System, vol. 5 (1993)
15. Simonyan, K., Zisserman, A.: Very deep convolutional networks for large-scale image recognition. In: Proceedings of the International Conference on Learning Representations, pp. 1–14 (2015)

Fast and High-Purity Seed Sorting Method Based on Lightweight CNN

Zhengguang Luan$^{(\boxtimes)}$, Chunlei Li$^{(\boxtimes)}$, Shumin Ding, Qiang Guo, and Bicao Li

Zhongyuan University of Technology, Zhengzhou 450007, China
lichunlei1979@sina.com

Abstract. Seed purity is an important indicator of seed quality. Seed sorting has been extensively studied and is considered as image classification setting with the goal of distinguishing between normal and abnormal seeds. Traditional, the classification of normal and abnormal seeds is achieved using the machine visual features of the seeds. In recent years, convolutional neural network has shown excellent performance in image classification tasks. In this work, we mainly focus on the computational efficiency and the classification performance of the network. Then, we developed a lightweight convolutional neural network to achieve fast and high-purity seed sorting. A lightweight and fast CNN model is constructed by using heterogeneous convolution layer instead of standard convolutional layer. Specially, we compare the standard convolution network and the heterogeneous convolutional network to measure the performance of methods on sunflower seeds dataset. The proposed sunflower seed sorting method based on heterogeneous convolutional network is robust to classify the seeds and the accuracy of data can reach 98.6%, which FLOPs are half the original standard convolution. Compared with the other state-of-art methods, this method has higher performance and lower computational complexity.

Keywords: Convolutional neural network · Seed sorting · Heterogeneous convolution

1 Introduction

In the seed processing industry, the purity of seeds is an important evaluation criteria of quality rating. In particular, seed sorting can be regarded as an image classification problem which is a basic visual recognition problem in computer vision. Generally, seed sorting needs to identify abnormal seeds from all seeds and remove them, which is the key to improve the purity of seeds.

In order to solve the problem of seed sorting, several methods were put forward. Among them, more traditional methods mainly rely on handcraft features, such as HOG (Histogram of Oriented Gradient), color, morphological features, etc., and then classify by SVM (Support Vector Machine) and other classifiers. Huang [1] et al. presented an auto-sorting system based on back propagation

© Springer Nature Singapore Pte Ltd. 2020
L. Pan et al. (Eds.): BIC-TA 2019, CCIS 1160, pp. 607–615, 2020.
https://doi.org/10.1007/978-981-15-3415-7_51

neural network to classify color, shape, texture features. Besides, Fuente [2] introduced a sorting corn kernels method based on multispectral Imaging technology. Most successful traditional seed sorting methods focus on carefully designing feature descriptors. However, because of the diversity of abnormal seeds, specially, seed posture obtained from different angles of camera, seed classification is a challenging task. The main limitation of these methods which are based on extraction handcraft features is that they cannot distinguish similar seeds, resulting in low classification accuracy.

Recently, deep learning techniques have demonstrated advanced performance in nature image classification and other computer vision tasks. Different with traditional machine learning methods, deep learning techniques learn hierarchical feature representation by automatically extracting multiple levels of abstraction from nature images and background information. Since the pioneering work of AlexNet [3] and VGG [4], the depth of the network has been increased by superimposing a set of convolutional layers to achieve amazing results. However, in order to apply to a large scale image classification task from ImageNet dataset, the calculation parameters of these training networks are often large so that have high requirements on the storage space of system equipment, which are not suitable for embedded use on the mobile end. Recent work has applied convolution network to agricultural product sortings [5–7], such as seeds, vegetables and fruits. Veeramani et al. [8] proposed a deep convolution network consisting of several convolution layers for maize sorting and produced 96.8% accuracy on dataset. These methods are inefficient because of the additional parameters to generate complex computations.

In order to meet the development and use of the limited computing resource platform such as FPGAs, therefor, a lightweight design network is required. In this work, our goal is to design a lightweight convolutional neural network for seed sorting tasks, which can discriminate abnormal seed images from seed images. To achieve this, we use convolution neural network to carry out experiments, and use heterogeneous convolution kernels instead of standard convolution kernels to reduce network parameters. Experiments show that our method is more effective.

2   Architecture

In recent years, the design of the model mainly focuses on the depth and width of the network to improve the performance of the network. The short cut connection proposed by ResNet makes the design of the network deeper and deeper. Whether increasing the depth of the network or the width of the network is to increase the amount of computation to improve the performance of the network. In order to be used in limited computing resource platforms, we develop a lightweight and fast CNN architecture for the classification of normal seeds and abnormal seeds. The structure of our CNN is shown in Fig. 1.

It consists of eight heterogeneous convolutional layers, each of which is followed Batch Normalization and ReLU (Rectified Linear Unit) function, two fully

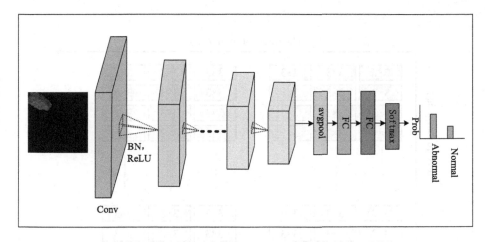

Fig. 1. The structure of the FLSNet.

connected layers followed by a softmax layer as the classifier. The output of CNN is followed by a softmax function, which generates values in the range of [0, 1] as classification confidence scores. Then the loss function is used to quantify the degree of coincidence between model predictions and real labels. The softmax function is defined as follows:

$$softmax(y) = \frac{exp(y_i)}{\sum_{i=1}^{n} y_i} \tag{1}$$

2.1 Standard Convolution

Convolutional layer is used to extract image features automatically. Given an input image $X^{H \times W \times C}$. Here, H, W and C is the spatial height, width and the channels of the input image. Then the output feature map of L-th layer can be expressed as $U^{H_l \times W_l \times C_l}$. Here, H_l and W_l is the spatial height and width, C_l is the number of channels for the L-th convolution layer. In standard convolution, the filter kernels have the same size that can be shown in Fig. 2.

2.2 Heterogeneous Convolution

The heterogeneous convolutional layers are used to reduce the number of float point operations and the latency of calculation [10]. In the heterogeneous convolutional layer, the filter kernels with different sizes are used that can be showed in Fig. 3. For example, in the convolutional layer of our model, we use two different sizes of convolution kernels of 3 × 3 and instead of the original standard convolution which has the same size 3 × 3. Assuming that the number of convolution kernels in the convolutional layer is C, we first use a ratio p divide the convolution kernel into $\frac{C}{p}$ groups. Each group consists with a convolution kernel of size

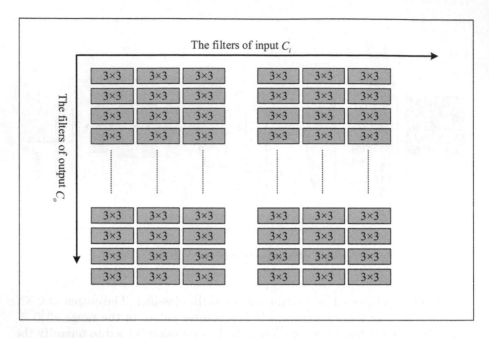

Fig. 2. Standard convolutional layer with isomorphic kernels of which filter kernels have the same size.

3×3 and $\frac{C}{p} - 1$ convolution kernels of size 1×1. The use of heterogeneous convolution kernels can effectively reduce the computational complexity. Assume that the feature map $N_i \times N_i \times C_i$ as the input of the convolution layer L in the convolution neural network. N_i is the size of the input feature map, C_i is the number of channels of the input feature map, and the size of the filters of the convolution layer is $K \times K$. After convolutional operation, we denote $N_o \times N_o \times C_o$ as the output feature map of layer. N_o is the size of the output feature map and C_o is the number of filters of layer L. Therefore, we can get the computational amount of the standard convolution operation is $FL_s = N_o \times N_o \times C_i \times C_o \times K \times K$.

Suppose that in the heterogeneous convolution, the number of filters are C_o, the convolution kernel with a ratio of p is of size K, the number of such filters is $\frac{C_o}{p}$, and the other filters are of size 1×1, and the number of such kernels is $\left(1 - \frac{1}{p}\right) C_o$. Then, the computation of heterogeneous convolution can be divided into two parts, where the computation amount of $K \times K$ convolution kernels is $FL_K = (N_o \times N_o \times C_i \times C_o \times K \times K)/p$ and the computation of 1×1 convolution kernels is $FL_1 = N_o \times N_o \times C_i \times \left(1 - \frac{1}{p}\right) C_o$, so the total computation amount of the heterogeneous convolution is $FL_{hc} = FL_1 + FL_K$.

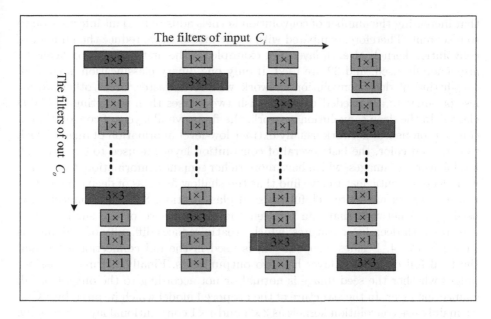

Fig. 3. A group of filter kernels of different sizes in heterogeneous Convolution.

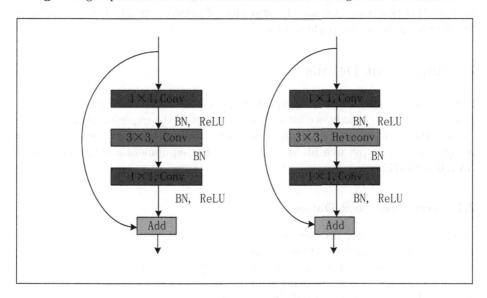

Fig. 4. The residual block with heterogeneous convolutions in the network.

2.3 Model Structure

We now describe more details of the LSPNet (lighted supper purity network). We constructed two networks using standard convolution and heterogeneous convolution, respectively. For the sunflower seed images dataset, we have found that increasing the number of convolution kernels adds redundant information to some extent. Therefore, compared with other methods, we reduce the number of convolution kernels at each layer. For example, in the first convolution layer, 64 are generally used, and 32 are used. If only one binary classification is trained, over-fitting of deep convolution network will occur faster, or in other words, less parameters are needed to distinguish two classes than to distinguish 1000 classes. In the deep convolution network, the first several layers of convolution in convolution neural network usually extract low-level information of images, such as edge and color, the last several of convolution layer are used to extract high level features of images, which have more richer semantic information. Compared with deep semantic feature, we find that the shallow features of deep convolution neural network are more helpful for seed classification. Shortcut connection is used in the network that can be seen the element-wise of the shallow layer features with deep layer features, which greatly reduces difficulty of training, as shown in Fig. 4. Besides, dropout [12] is used in the full connection layer and the final full connection layer has two output units. Finally, softmax classifier judges whether the seed image is normal or not according to the output of full connection layer. In the structure of the proposed model which inspired by VGG, we mainly use convolution kernels as 3×3 and 1×1 convolutional layer. Specially, each of the convolution layer followed batch normalization which was introduced in work [11] that can increase the stability of network gradient and accelerate training to improve the performance.

3   Implement Details

In this part, we give a series of experiments to illustrate the effectiveness of the proposed method. Firstly, we introduce the steps of the experiment. Finally, we discuss the performance in detail. The proposed convolutional neural network is implemented by pytorch library on a computer equipment a single GPU with NVIDIA 1080Ti.

3.1   Sunflower Seed Dataset

In the training stage, for the deep convolutional network described, a large amount of data is required. However, there is currently no suitable seed database, so the seed images used in our method was acquired by the sort machine in the actual seed harvest process using a high-speed camera in a real environment. The train dataset consist of 20k RGB sunflower seed images with resolution of 100×100 pixel have been obtained. Manually divide the acquired seed images into normal and abnormal. Among them, there are 10 k normal seed images and 10 k abnormal seed images. As shown in Fig. 5. The presence of different types of abnormal seeds adversely affects quality ratings.

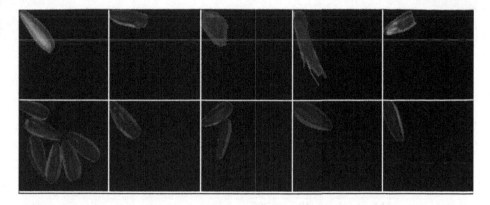

Fig. 5. Abnormal sunflower seed images (Top) and normal sunflower images in our seed datasets.

3.2   Training Details

We use simple and effective strategies to improve the performance of the network during the training phase. A widely used technique for reducing overfitting in convolutional neural networks is data augmentation. Every iteration, each of the input batch-size images through a number of transformations. Here, we set batch-size to 64. In training, our method readjusts the original image size of 100×100 pixel to 128×128 pixel as the input to train the network. Considering the random distribution of seeds during the process of sorting free fall, we performed random rotation and horizontal inversion on the dataset image for data augmentation. The optimization method is trained using the SGD (Stochastic gradient descent) with momentum 0.9. The initial learning rate of network training is 0.001. In the heterogeneous convolutional networks, we set the ratio p to 4.

3.3   Evaluation Details

In this section, We validate the trained model by using the classified test set. In the seed sorting method, accuracy and computational efficiency are considered to be two essential evaluation indicators for measuring the performance of the method. Thus, to quantify the efficiency of the method, we use test accuracy and the FLOPs as the evaluation criteria to evaluate the performance. After the network is fully trained, we validate the training model with test sets which consists of 1 K normal seed images and 1 K abnormal seed images. We use confusion matrices to understand how to use test data sets for performance in each category. The confusion matrix values of diagonals represent the number of correct classifications (TP and FN), while the non-diagonals represent the number of incorrect classifications (FP and TN). The classification accuracy of the model on the seed dataset can be expressed as $Accuracy = \frac{TP+FN}{TP+TN+FP+FN}$.

By comparing the two networks, the performance of using heterogeneous convolutional layer is basically not reduced. From the Table 1, we can clearly

Table 1. The classification result of the standard convolution network on the test datasets.

Real\Predict	Normal	AbNormal	GFLOPs	Accuracy
Normal	885	15	0.89	98.8%
AbNormal	9	891		

Table 2. The classification result of the heterogeneous convolution network on the test datasets.

Real\Predict	Normal	AbNormal	GFLOPs	Accuracy
Normal	884	16	0.47	98.6%
AbNormal	12	888		

understand the performance of the trained model that LSPNet with standard convolution can classify training images well. The result of the model on the test set is that 24 images are misidentified. Because of the seed will show different postures relative to the camera during the free fall process, the TN value in the experiment is greater than the FP value. Therefore, 15 normal seeds were misclassified as abnormal seeds and 9 abnormal seeds were misclassified as normal seeds. In our experiments, the proposed standard convolution network method can achieve 98.8% accuracy of seed images on the test set. Table 2 shows the performance of LSPNt with heterogeneous convolution, compared with the standard convolution network. Although heterogeneous convolution networks have dropped by 0.2% on the test set, FLOPs is about half of the original network.

4   Conclusions

The goal of our work is to explore the application of CNN architecture based on small heterogeneous convolutional network model to discriminate abnormal seeds from all sunflower seeds. We also show that the performance remains robust under different seed shapes and different angles relative to the camera. In this paper, we first conducted a series of experiments to explore the effect of convolution with different layers on the recognition rate. Then, in order to solve the problem of large size of the model, we explore a network with a small model size, and propose to replace the vanilla convolution with the same size filter by using heterogeneous convolution with filters of different sizes construct a lightweight convolutional neural network and study their performance in seed classification. We applied our model to the sunflower seed dataset, and we proved that by simply replacing the convolutional layer, our model can be continually improved to achieve more stable performance.

Acknowledgements. This work was supported by NSFC (No. U1804157, 61772576), Science and technology innovation talent project of Education Department of Henan

Province (17HASTIT019), The Henan Science Fund for Distinguished Young Scholars (184100510002), Henan science and technology innovation team (CXTD2017091), IRT-STHN (18IRTSTHN013), Program for Interdisciplinary Direction Team in Zhongyuan University of Technology.

References

1. Huang, K.Y., Cheng, J.F.: A novel auto-sorting system for Chinese cabbage seeds. Sensors **17**(4), 886 (2017)
2. De La Fuente, G.N., Carstensen, J., Edberg, M., et al.: Discrimination of haploid and diploid maize kernels via multispectral imaging. Plant Breed. **136**(1), 50–60 (2017)
3. Krizhevsky, A., Sutskever, I., Hinton, G.: Imagenet classification with deep convolutional neural networks. In: Advances in Neural Information Processing Systems, pp. 1097–1105 (2012)
4. Simonyan, K., Zisserman, A.: Very deep convolutional networks for large-scale image recognition. arXiv preprint arXiv:1409.1556 (2014)
5. Khaing, Z.M., Naung, Y., Htut, P.H.: Development of control system for fruit classification based on convolutional neural network. In: IEEE Conference of Russian Young Researchers in Electrical and Electronic Engineering, pp. 1805–1807 (2018)
6. Zhang, Y.-D., et al.: Image based fruit category classification by 13-layer deep convolutional neural network and data augmentation. Multimed. Tools Appl. **78**(3), 3613–3632 (2019)
7. Zeng, G.: Fruit and vegetables classification system using image saliency and convolutional neural network. In: IEEE 3rd Information Technology and Mechatronics Engineering Conference, pp. 613–617 (2017)
8. Veeramani, B., Raymond, J., Chanda, P.: DeepSort: deep convolutional networks for sorting haploid maize seeds. BMC Bioinf. **19**(9), 85 (2018)
9. He, K., Zhang, X., Ren, S., et al.: Deep residual learning for image recognition. In: Proceedings of the IEEE Conference on Computer Vision and Pattern Recognition, pp. 770–778 (2016)
10. Singh, P., Verma, V., Rai, P., et al.: HetConv: heterogeneous kernel-based convolutions for deep CNNs. In: Proceedings of the IEEE Conference on Computer Vision and Pattern Recognition, pp. 4835–4844 (2019)
11. Ioffe, S., Szegedy, C.: Batch normalization: accelerating deep network training by reducing internal covariate shift. arXiv preprint arXiv:1502.03167 (2015)
12. Srivastava, N., Hinton, G., Krizhevsky, A., et al.: Dropout: a simple way to prevent neural networks from overfitting. J. Mach. Learn. Res. **15**(1), 1929–1958 (2014)

Fabric Defect Detection Based on Total Variation Regularized Double Low-Rank Matrix Representation

Ban Jiang[✉], Chunlei Li[✉], Zhoufeng Liu, Aihua Zhang, and Yan Yang

School of Electrical and Information Engineering,
Zhongyuan University of Technology, Zhengzhou 450007, China
lichunlei1979@sina.com

Abstract. Fabric defect detection plays an irreplaceable role in textile quality control. Fabric images collected on industrial sites are complex and diverse, which brings great challenges to defect detection. Fabric detection algorithm based on traditional image processing method has low detection accuracy and lack of adaptability. Low rank representation model has been proved to be suitable for fabric defect detection. Normal fabric backgrounds have high redundancy and located in a low-dimensional subspace. Meanwhile, we have noticed that the defect is a region with certain edge characteristics formed by the aggregation of multiple pixels, which has high redundancy in its interior. Therefore, fabric defects can be seen as located in a low-dimensional subspace independent of the background. In this paper, a fabric defect detection algorithm based on DERF descriptors and total variation regularized double low-rank matrix representation is proposed. The characteristic matrix of the test fabric image is extracted by DERF descriptor, and the fabric image is represented as background and defect by the method of total variation regularized double low-rank representation. Experiments on two datasets show that our method has good detection performance for plain, twill and complex patterned fabrics, and is superior to other state-of-the-art method.

Keywords: Defect detection · Fabric image · Double low rank · Total variation

---

C. Li—The authors would like to thank Dr. Henry Y.T. Ngan, Industrial Automation Research Laboratory, Dept. of Electrical and Electronic Engineering, The University of Hong Kong, for providing the database of patterned fabric images. This work was supported by NSFC (No. U180415761772576), Science and technology innovation talent project of Education Department of Henan Province(17HASTIT019), The Henan Science Fund for Distinguished Young Scholars (184100510002), Henan science and technology innovation team (CXTD2017091), IRTSTHN (18IRTSTHN013), Program for Interdisciplinary Direction Team in Zhongyuan University of Technology.

© Springer Nature Singapore Pte Ltd. 2020
L. Pan et al. (Eds.): BIC-TA 2019, CCIS 1160, pp. 616–626, 2020.
https://doi.org/10.1007/978-981-15-3415-7_52

1 Introduction

Fabric defect detection plays a key role in textile quality control, which directly affects the product performance. At present, the defect detection in production line mostly adopts artificial vision detection, whose detection effect is easily affected by workers' subjective factors. Different from the manual inspection method, the fabric defect detection based on machine vision has the characteristics of stability, reliability and objectivity, which has become a research focus.

The existing fabric defect detection algorithms can be divided into four categories, which include dictionary learning-based methods [1], statistical-based methods [2], spectral approaches [3], model-based methods [4]. These detection methods have achieved great results in the detection of simple plain or twill fabrics, however, they lack of adaptivity, and the detection performance is not ideal for complex patterned fabrics. Currently, some detection algorithms for patterned fabrics have been proposed, such as Elo rating (ER) method [5], and Motif-based method [6,7], etc. However, most of these methods adopted template matching technology to locate defect areas, whose detection accuracy depends on template selection and precise alignment.

The background of textured fabric has higher visual redundancy and is simpler than that of complex natural scene. Therefore, compared with the detection object in the complex natural scene, the fabric defect is suitable for the low rank decomposition model. The low-rank decomposition model, which can divide images into background parts and target parts, has been applied in image segmentation [8] and target detection. In general, the commonly used low-rank representation method takes the background as a low-dimensional subspace and the defect as a sparse subspace. Inspired by DLRMR [9], we observed that there was also a high correlation within the defect region, so we could try to express the defect region with a low-rank matrix. Then we get a double low-rank model to represent the background and defect area of the fabric image. However, considering that the defect region has certain sparsity with respect to the background, the sparse noise part of the model will be treated as a defect, which will affect the detection effect. Figure 1 is an intuitive representation. The defect region of the fabric image is located in a low-dimensional subspace independent of the normal background.

In addition, the effective representation of fabric image is very important for the performance of low rank decomposition. The biological vision system can accurately locate the significant objects in the background, while the defect areas in the fabric images often have higher significance. The image descriptor Distinctive Efficient Robust Features (DERF) [10] based on retinal ganglion cell coding has strong specificity, robustness and high efficiency, and is suitable for extracting fabric features.

In this paper, we mainly improve the double low rank matrix representation model. Therefore, a fabric defect detection method based on DERF descriptors and total variation regularized double low rank matrix representation (TV-DLRMR) is proposed.

The main contributions of this paper are summarized as follows:

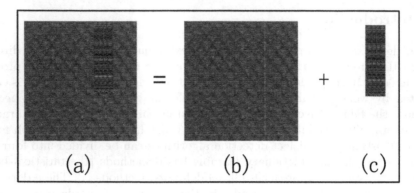

Fig. 1. Star-patterned fabric image: (a) original image. (b) background. (c) defect. The normal background of the fabric image is highly redundant, while the defect part is highly correlated and located in a low-dimensional subspace independent of the background.

1. The double low rank representation algorithm for feature fusion is applied in the field of fabric defect detection, and good results were obtained.
2. The total variation regularized DLRMR can effectively detect the defect saliency map with less noise.

The rest of this paper is designed as follows: Sect. 2 introduced the DERF descriptors and total variation regularized double low rank matrix representation method. In Sect. 3, sufficient experimental analysis is used to evaluate the performance of the proposed method. Section 4 makes a summary of the full research.

2   Proposed Method

A novel fabric defect detection method based on DERF and TV-DLRMR is proposed in this paper. Firstly, the DERF feature of fabric image is extracted to generate feature matrix, then the low rank matrix representing the information of defect region is obtained by using the double low rank matrix representation method. Finally, the saliency map of defect region is obtained by processing the low rank matrix with threshold segmentation algorithm.

2.1   Feature Extraction

In order to complete fabric image defect detection better, the test image is evenly divided into image blocks $\{I_i\}_{i=1,2...N}$ whose size is $m \times m$, where N is the number of the image blocks. The feature vector $D_i(x, y)$ of each image block is generated below.

Firstly, the H gradient orientation map of image block is generated by using Difference of Gaussian (DoG):

$$G_o = \left(\frac{\partial I}{\partial o}\right)^+, 1 \le o \le H \tag{1}$$

where o is the orientation of derivative, $(\cdot)^+$ represents a nonnegative operator such as $(\cdot)^+ = max(\cdot)^+$.

Each gradient orientation map is convoluted $S+1$ times with different Gauss convolution kernels, and then the Gauss convolution orientation map is generated:

$$G_o^\Sigma = G_\Sigma * \left(\frac{\partial I}{\partial o}\right)^+ \tag{2}$$

where G^Σ is a Gaussian convolution core of scale Σ.

For each orientation map, the final Gauss convolution gradient orientation map is generated by subtracting the large scale from the small scale for each pair of Gauss convolution gradient orientation maps:

$$D_o^{\Sigma_1} = G_o^{\Sigma_1} - G_o^{\Sigma_2} \text{ with } \Sigma_2 > \Sigma_1 \tag{3}$$

After obtaining the DoG convolution gradient orientation maps, these maps are sampled to construct DERF descriptors by imitating the structure of the receiving field of P-ganglion cells in the range of $0° - 15°$. These sampling points are located on concentric circles with different radii, and their number increases exponentially with radius. $h_\Sigma(x, y)$ is defined as a vector generated by the value at the position (x, y) with the same scale Σ in the orientation maps of DoG convolution:

$$h_\Sigma(x, y) = \left[D_1^\Sigma(x, y), \cdots D_i^\Sigma(x, y), \cdots D_H^\Sigma(x, y)\right] \tag{4}$$

where $D_i^\Sigma(x, y)$ represents DOG convolution patterns with different directions and the same scales.

Finally, the descriptor at position (x, y) can be defined as the connection of h vectors:

$$\begin{aligned} &D(x, y) \\ &= [h_{\Sigma_1}(x, y), \\ &\quad h_{\Sigma_1}\left(l_1\left(x, y, R_1\right)\right) \cdots, h_{\Sigma_1}\left(l_T\left(x, y, R_1\right)\right) \\ &\quad h_{\Sigma_2}\left(l_1\left(x, y, R_1\right)\right), \cdots, h_{\Sigma_2}\left(l_T\left(x, y, R_1\right)\right) \\ &\quad\quad\quad\quad \vdots \\ &\quad h_{\Sigma_S}\left(l_T\left(x, y, R_1\right)\right), \cdots, h_{\Sigma_S}\left(l_T\left(x, y, R_1\right)\right)]^T \end{aligned} \tag{5}$$

where $l_i(x, y, R)$ represents the location with the distance R from (x, y) in the i$-$th orientation, S indicates the number of layers.

To improve the efficiency of removing redundant information, principal component analysis (PCA) is used to reduce the dimensions of eigenvectors. Then, all feature vectors of image blocks are assembled to generate feature matrix $D \in \mathbb{R}^{N \times P}$ representing test fabric image.

2.2   Model Construction

Fabrics such as stars-, dots- and boxes-fabrics are often formed by repeated arrangement of basic templates, so the background part of the fabric image is considered to be a highly redundant part, located in a low-dimensional subspace. Different from the background area, the defect area breaks the original pattern of regular arrangement and has high saliency. In addition, the pixel values of defective regions are often similar and in a low-dimensional subspace independent of the background. Considering the above characteristics of fabric images, the DLRMR is suitable for fabric defect detection, It can be achieved by solving the following optimization problems:

$$\min_{L,E} \text{rank}(L) + \lambda(\text{rank}(E)) \quad s.t. \ D = L + E \tag{6}$$

where D is the feature matrix of the fabric image, L and E are low-rank matrices representing background and defect objects respectively. λ is used to balance the effect of the two items.

Since rank optimization problem is NP-hard, we usually relax the problem (6) to the following optimization problems:

$$\min_{L,E} \|L\|_* + \lambda\|E\|_* \quad s.t. \ D = L + E \tag{7}$$

where $\|\cdot\|_*$ denotes the matrix nuclear norm which is the sum of singular values of a matrix.

However, the images collected in the industrial field are vulnerable to various external noises, and the defect area is sparse relative to the background, so these sparse noises can also be divided into matrix E. Therefore, it is necessary to add noise suppression model to saliency map generation. The noise in the image is tiny, which can be considered as irregular changes of the pixels, while the defect is the aggregation of multiple pixels and has certain edge information. We noticed that total variation regularization [11] has excellent performance in suppressing discontinuous variations of pixel-level noise, and integrated total variation regularization into DLRMR model for denoising. This structure is called total variation regularization double low rank matrix representation (TV-DLRMR), can be formulized via:

$$\min_{L,E} \|L\|_* + \lambda\|E\|_* + \beta\|E\|_{TV} \quad s.t. \ D = L + E \tag{8}$$

where $\|\cdot\|_{TV}$ is total variation regularization (TV-norm) including anisotropic case and isotropic case, β is a weight parameter whose function is the same as λ.

For the convenience of solving problem (8), the auxiliary variable $J = E$ is introduced to decompose the energy function

$$\min_{L,E} \|L\|_* + \lambda\|E\|_* + \beta\|J\|_{TV} \quad s.t. \ D = L + E, \quad J = E \tag{9}$$

The objective function of the above equation is a convex optimization problem. In this paper, the alternating direction method of multipliers (ADMM) is used to solve this optimization problem [12]. The augmented Lagrangian function constructed by the above equation is as follows:

$$
\begin{aligned}
&F\left(L, E, J, Y_1, Y_2, \mu\right) \\
&= \|L\|_* + \lambda\|E\|_* + \beta\|J\|_{TV} + \tfrac{\mu}{2}\left(\|D - L - E\|_F^2 + \|E - J\|_F^2\right) \\
&+ \langle Y_1, D - L - E\rangle + \langle Y_2, E - J\rangle
\end{aligned}
\tag{10}
$$

where Y_1 and Y_2 are the Lagrange multipliers, $\|\cdot\|_F$ is the Frobenius norm, $\langle\cdot\rangle$ is the inner product, $\mu > 0$ is a penalty parameter. The ADMM model was used to solve L, E, Y_1 and Y_2 alternately, The specific steps are as follows:

(1) Update L^{k+1} by fixing the other variables:

$$
\begin{aligned}
&L^{k+1} \\
&= \arg\min_L L\left(L^k, E^k, J^k, Y_1^k, Y_2^k, \mu^k\right) \\
&= \arg\min_L \|L^k\|_* + \tfrac{\mu^k}{2}\left\|D - L^k - E^k\right\|_F^2 - \langle Y_1^k, L^k\rangle \\
&= \arg\min_L \tfrac{1}{\mu^k}\|L^k\|_* + \tfrac{1}{2}\left\|L^k - \left(D - E^k + \tfrac{Y_1^k}{\mu^k}\right)\right\|_F^2
\end{aligned}
\tag{11}
$$

(2) Update E^{k+1} by fixing the other variables:

$$
\begin{aligned}
&E^{k+1} \\
&= \arg\min_E L\left(L^{k+1}, E^k, J^k, Y_1^k, Y_2^k, \mu^k\right) \\
&= \arg\min_E \lambda\|E^k\|_* + \tfrac{\mu^k}{2}\left(\|D - L^{k+1} - E^k\|_F^2 + \|E^k - J^k\|_F^2\right) \\
&\quad - \langle Y_1^k, E^k\rangle - \langle Y_2^k, J^k\rangle \\
&= \arg\min_E \tfrac{\lambda}{2\mu^k}\|E^k\|_* + \tfrac{1}{2}\left\|E^k - \tfrac{1}{2}\left(J^k + D - L^{k+1} + \tfrac{(Y_1^k - Y_2^k)}{\mu^k}\right)\right\|_F^2
\end{aligned}
\tag{12}
$$

(3) Update J^{k+1} by fixing the other variables:

$$
\begin{aligned}
&J^{k+1} \\
&= \arg\min_J L\left(L^{k+1}, E^{k+1}, J^k, Y_1^k, Y_2^k, \mu^k\right) \\
&= \arg\min_J \beta\|J^k\|_{TV} + \tfrac{\mu^k}{2}\left\|E^{k+1} - J^k + \tfrac{Y_2^k}{\mu^k}\right\|_F^2 \\
&= \arg\min_J \tfrac{\beta}{\mu^k}\|J^k\|_{TV} + \tfrac{1}{2}\left\|J^k - \left(E^{k+1} + \tfrac{Y_2^k}{\mu^k}\right)\right\|_F^2
\end{aligned}
\tag{13}
$$

(4) Update Lagrange multiplier Y_1^{k+1}, Y_2^{k+1}:

$$
Y_1^{k+1} = Y_1^k + \mu^k(D - L^{k+1} - E^{k+1})
\tag{14}
$$

$$
Y_2^{k+1} = Y_2^k + \mu^k(E^{k+1} - J^{k+1})
\tag{15}
$$

(5) Update penalty parameter μ^{k+1}:

$$
\mu^{k+1} = \min(\mu_{\max}, \rho\mu^k)
\tag{16}
$$

where ρ is a constant.

The iteration process of the model can be summarized as Algorithm 1.

Algorithm 1. Solving TV-DLRMR by ADMM

Input: feature matrix D; parameter $\lambda > 0$;

Initialize: $L^0 = 0$, $E^0 = 0$, $J^0 = 0$, $Y_1^0 = D/max(\|D\|_2, \lambda^{-1}\|D\|_\infty)$, $Y_2^0 = 0$, $\mu^0 = 1$, $\mu^{max} = 10^5$, $k = 0$, $\rho = 1.2$, $tol = 6e - 4$

while not converged **do**

   1. Update L^{k+1} using (11);
   2. Update E^{k+1} using (12);
   3. Update J^{k+1} using (13);
   4. Update Y_1^{k+1} using (14);
   5. Update Y_2^{k+1} using (15);
   6. Update μ^{k+1} using (16);
   7. Check the convergence condition $\|D - L^{k+1} - E^{k+1}\|_F / \|D\|_F < tol$
   8. $k = k + 1$

end while

Output: The optimal solution E^{k+1}

2.3 Saliency Map Generation and Segmentation

Through the above algorithm, we decompose the feature matrix to get the low-rank matrix representing the object region. The saliency score of the i−th image block P_i is obtained by quantifying the response of the low-rank matrix, and the calculation formula is as follows:

$$M(P_i) = \|E(:, i)\|_1 \tag{17}$$

where $\|E(:, i)\|_1$ is the $l_1 - norm$ of the i−th column of E. Image blocks P_i with higher significance score have higher probability of defect.

We denoise the saliency map M to get a smoothed saliency map $\hat{M}$:

$$\hat{M} = G * (M \circ M) \tag{18}$$

where G is a circular smoothing filter, $\circ$ represents Hadamard inner product operator and $*$ denotes convolution operation.

The significant image $\hat{M}$ is then converted to a grayscale image G:

$$G = \frac{\hat{M} - \min(\hat{M})}{\max(\hat{M}) - \min(\hat{M})} \times 255 \tag{19}$$

Finally, G is segmented by threshold segmentation to get the location of defect regions.

3   Experiment Result

In this part, we use the proposed algorithm to carry out a series of experiments on two public fabric image databases, and compare the experimental results with State-of-the-Arts, qualitatively and quantitatively.

3.1   Experiment Details

In this paper, two public fabric image databases are used to evaluate the performance of the proposed method. One is TILDA fabric images containing 284 fabric images [13], mainly plain and twill fabrics. The fabric images contain simple defects such as broken ends, holes, multiple netting, thick strips and thin strips.

Another fabric database is from the Research Associate of Industrial Automation Research Laboratory, Department of Electrical and Electronic Engineering at Hong Kong University. It primarily contains patterned fabric images with complex textures, including 25 star-, 26 box-, and 30 dot-patterned fabrics.

In each experiment, all parameters are fixed to prove the performance of our method. Resolution of all fabric images is adjusted to 256 pixels 256 pixels, and our test environment is a DELL laptop with an Inter(R) Core(TM) i5-8300H 2.3 GHZ CPU. The simulation software is MATLAB 2019a.

3.2   Qualitative Analysis

We compare the proposed method with some visual saliency models, including the least squares regression (LSR), the histogram of oriented gradient (HOG),

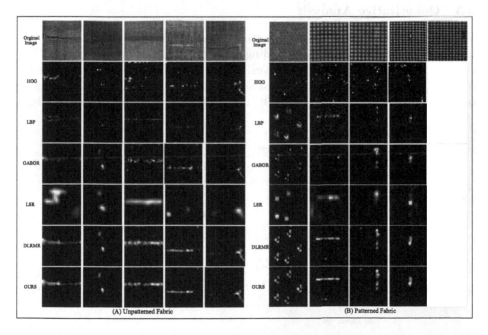

Fig. 2. Comparison the results using different detection methods: (A) Detection results for the unpatterned fabric image; (B) Detection results for the patterned fabric image. The first row shows the original image. The detection results of HOG, LBP, GABOR, LSR, DLRMR and our method are listed from the second row to seventh row.

the Gabor filter and low-rank decomposition (GABOR), and Local Binary Pattern (LBP). The experimental results are shown in Fig. 2. The first row shows the original image. The detection results of HOG [14], LBP [15], GABOR [16], LSR [17], DLRMR without total variation regularization and our method are listed from the second row to the seventh row.

From the experimental results of the unpatterned fabrics in Fig. 2, we can see that the six methods have good detection performance for simple texture fabrics. For the patterned fabric images, the results of HOG and LBP have large errors and noises. This demonstrates that these two methods are only suitable for simple texture fabrics, not for the patterned fabric images. The GABOR method can detect the location of defects, but there are many discontinuous areas in the detection results. The LSR method can be used to locate all defect areas, but the defect areas lack detailed information and cannot display defect contours well, such as pattern fabrics in the first and third images. From the experimental results of the last two rows of the patterned fabrics, it can be seen that the addition of total variation regularization has achieved the desired effect and the tiny noise has been well suppressed. Our method can not only accurately detect the location of the defect area, but also show the rough outline of the defect region.

3.3   Quantitative Analysis

We adopted two criteria to comprehensively and qualitatively evaluate the performance of different methods, including receiver operating characteristic (ROC) curves and precision-recall (PR) curves. Because of the lack of GT images in the unpatterned fabric database, we consider only the patterned fabric databases for our quantitative evaluation. The ROC curve and PR curve are shown in Fig. 3.

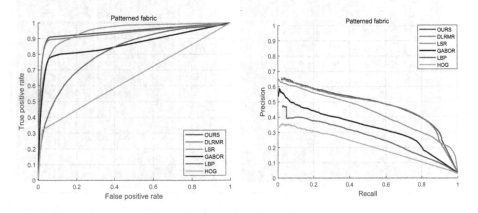

Fig. 3. Quantitative comparison of five methods on the image database of patterned fabrics: the left image corresponds to ROC curve, the right image corresponds to PR curve.

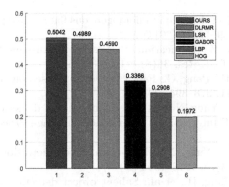

Fig. 4. Comparison of the F-measure for six methods.

As can be seen from Fig. 3, the area under the ROC curve (AUC) of ours is almost the largest, it shows that our method is better than the other five methods. The PR of our method is also the highest, which shows that our method can achieve the best performance in the six methods. In addition, we adopted the F-measure, which takes both precision and recall into account. As shown in Fig. 4, it shows the effectiveness and robustness of our proposed method.

$$F = \frac{precision \cdot recall}{precision + recall} \times 2 \qquad (20)$$

4   Conclusion

In this paper, a fabric defect detection method based on DERF descriptors and TV-DLRMR is proposed. DERF descriptors modeled by P-ganglion cells were used to describe the characteristics of fabric images. The fabric image is represented as the background part and the defect part by the DLRMR method, and the total variation regularization is integrated into DLRMR to reduce the influence of noise on the defect saliency map. We also compare the performance of the proposed method with that of other methods, such as HOG, LBP, GABOR, LSR methods. Experimental results show that our method is superior to other state-of-the-art methods. In addition, the proposed method can also be applied to surface anomaly detection of other industrial products, such as rails and mobile phone screens, and has broad application prospects.

References

1. Liu, Z., Yan, L., Dong, C., et al.: Fabric defect detection algorithm based on sparse optimization. J. Text. Res. **37**(5), 56–63 (2016)
2. Shi, M., Fu, R., Guo, Y., et al.: Fabric defect detection using local contrast deviations. Multimed. Tools Appl. **52**(1), 147–157 (2011). https://doi.org/10.1007/s11042-010-0472-8

3. Tolba, A.: Fast defect detection in homogeneous flat surface products. Expert Syst. Appl. **38**(10), 12339–12347 (2011)
4. Li, M., Cui, S., Xie, Z.: Application of Gaussian mixture model on defect detection of print fabric. J. Text. Res. **36**(8), 94–98 (2015)
5. Tsang, C., Ngan, H., Pang, G.: Fabric inspection based on the Elo rating method. Pattern Recogn. **51**, 378–394 (2016)
6. Ngan, H., Pang, G., Yung, N.: Patterned fabric defect detection using a motif-based approach. In: IEEE International Conference on Image Processing, pp. II-33–II-36 (2007)
7. Ngan, H., Pang, G., Yung, N.: Motif-based defect detection for patterned fabric. Pattern Recogn. **41**(6), 1878–1894 (2008)
8. Peng, H., Li, B., Ling, H., et al.: Salient object detection via structured matrix decomposition. IEEE Trans. Pattern Anal. Mach. Intell. **39**(4), 818–832 (2017)
9. Li, J., Luo, L., Zhang, F., et al.: Double low rank matrix recovery for saliency fusion. IEEE Trans. Image Process. **25**(9), 4421–4432 (2016)
10. Weng, D., Wang, Y., Gong, M., et al.: DERF: Distinctive Efficient Robust Features from the biological modeling of the p ganglion cells. IEEE Trans. Image Process. **24**(8), 2287–2302 (2015)
11. Rudin, L., Osher, S., Fatemi, E.: Nonlinear total variation based noise removal algorithms. Phys. D: Nonlinear Phenom. **60**(1–4), 259–268 (1992)
12. Lin, Z., Liu, R., Su, Z.: Linearized alternating direction method with adaptive penalty for low-rank representation. In: Advances in Neural Information Processing Systems, pp. 612–620 (2011)
13. Workgroup on texture analysis of DFG TILDA textile texture database. http://lmb.informatik.uni-freiburg.de/research/dfg-texture/tilde. Accessed 6 May 2013
14. Li, C., Gao, G., Liu, Z., et al.: Fabric defect detection algorithm based on histogram of oriented gradient and low-rank decomposition. J Text. Res. **38**(3), 153–158 (2017)
15. Liu, Z., Zhao, Q., Li, C., et al.: Fabric defect detection algorithm using local statistic features and global saliency analysis. J. Text. Res. **11**, 013 (2014)
16. Zhang, D., Gao, G., Li, C.: Fabric defect detection algorithm based on Gabor filter and low-rank decomposition. In: Eighth International Conference on Digital Image Processing (ICDIP 2016). International Society for Optics and Photonics, vol. 10033, p. 100330L (2016)
17. Cao, J., Zhang, J., Wen, Z., Wang, N., Liu, X.: Fabric defect inspection using prior knowledge guided least squares regression. Multimed. Tools Appl. **76**(3), 4141–4157 (2015). https://doi.org/10.1007/s11042-015-3041-3

Self-attention Deep Saliency Network for Fabric Defect Detection

Jinjin Wang[✉], Zhoufeng Liu[✉], Chunlei Li, Ruimin Yang, and Bicao Li

Zhongyuan University of Technology, Zhengzhou 450007, China
lzhoufeng@hotmail.com

Abstract. Fabric defect detection is one of the key steps in the textile manufacturing industry. Traditional saliency detection models mostly rely on hand-crafted features to obtain local details and global context. However, these methods ignore the association between context features. It restricts the ability to detect the salient objects in complex scenes. In this paper, a deep saliency detection model is proposed, which incorporates self-attention mechanism into convolutional neural network for fabric defect detection. First, a full convolutional network is designed for multi-scale feature maps to capture rich context features of the fabric image. Then, after the side output of the backbone network, the self-attention module is adopted to coordinate the dependencies between the features of the multiple layers, which improves the characterization ability of the extracted features. Finally, the multi-level saliency maps output from the self-attention mechanism are fused by the short connection structure, and generating detail enriched saliency map. Experiments demonstrate that the proposed method outperforms the state-of-the-art approaches when the defects are blurred or the shape is complex.

Keywords: Full convolutional network · Self-attention mechanism · Defect detection

1 Introduction

Fabric defect detection is an important part of the industrial production process. Unqualified products will cause economic and material loss [1]. However, most factories still use manual testing, which requires workers to have a wealth of work experience and a high concentration of spirit. After a long period of work, the probability of error is great. Realizing industrial automation is the trend of the times. Traditional detection algorithms typically use hand-crafted image features and various prior [2,3]. The disadvantage of these methods is the lack of high-level semantic information. As a result, they can only be used to detect simple patterns of fabrics, and the detection efficiency is low.

The human visual system is able to quickly capture the most representative part of a picture, and automatically ignore the interference of background noise. The concept of visual visibility is derived from this. In the fabric image, the

© Springer Nature Singapore Pte Ltd. 2020
L. Pan et al. (Eds.): BIC-TA 2019, CCIS 1160, pp. 627–637, 2020.
https://doi.org/10.1007/978-981-15-3415-7_53

defect area is more likely to attract attention than the background. Thence, some scholars have proposed to introduce visual saliency into fabric defect detection [4]. The previous salient detection method constructs a visual saliency model through the underlying feature comparison to generate the saliency map. Last, the saliency map is segmented by threshold to determine the location of the defect. However, these methods still use traditional methods to solve problems in feature extraction and saliency calculation. Therefore, the detection performance of complex texture fabrics is poor.

In recent years, deep learning has made significant progress in the field of computer vision. Convolutional neural networks have powerful feature extraction capabilities and have proven effective in object detection [5,6], edge detection [7] and semantic segmentation [8]. The saliency detection model based on CNN [9,10] can make up for the shortcomings of the traditional saliency model. Compared with the previous significant detection method, better detection results are obtained. The rich features can effectively improve the characterization ability of fabric texture features. Most of the previous methods directly integrate the extracted features, which ignores the connection between local and global features. Without coordinate will result in loss of information or information redundancy.

The self-attention mechanism [11] can solve the above problems well. It handles multi-level dependencies very well, and needs a small computational cost only. These methods [12,13] has attempted to incorporate attention mechanisms into the deep saliency detection model, which can better capture global dependencies. There are many attention mechanisms that perform well, especially the self-attention module. It balances global correlation by calculating the influence of global features on local location. This makes it possible to find dependencies in the image. Therefore, the self-attention mechanism can utilize the relationship between each pixel and the global to enhance the representation capabilities of the network.

Inspired by previous work, we add the self-attention mechanism to a deep neural network for fabric defect detection. In this paper, We design a feature extraction network with multiple convolution layers. It gains the multi-level depth features corresponding to the fabric image. Considering the different characteristics of different levels of features, low-level features have a lot of noise, while high-level features lose spatial information. Considering the different characteristics of different levels features, low-level features contain many noises, while the high-level features lack detailed information. We connect three convolution layers behind the side output layer to refine the features and filter the noise. The self-attention mechanism is complementary to the convolution operation. It helps model long-distance and multi-level dependencies. Thereby using the self-attention mechanism to balance the relationship between multiple levels. After processed by the self-attention mechanism, the deeper side output is associated with the shallow side output using the short connection structure [14] to make rational use of these multilevel features. It integrates the detailed information contained in the low-level features and the advanced semantic knowledge of the deep output to generate excellent saliency maps.

In short, our contributions can be summarized as follows:

1. A feature extraction network is proposed to extract multi-level features of different scales. In addition, the feature refinement module is designed to filter the noise and refine the target edges.
2. Added self-attention mechanism to balance the correlation between multiple layers. It captures the dependencies in the image and improves the representation of the network.
3. The feature fusion module connects the high-level features extracted from deeper layers to the shallower layers. With advanced auxiliary information, the shallower output can locate defects more accurately. At the same time, the low-level features refine the prediction map of the deeper side output layer.

2   The Proposed Method

In this paper, we propose a novel fabric defect detection method, which is a deep saliency detection model including the self-attention mechanism. The overall architecture is shown in Fig. 1. Primary network extracts rich multi-scale context features. Different features have different semantic values to generate saliency maps. Feature purification layer filters noise contained in multi-level features. The self-attention mechanism coordinates the relationship between multiple levels of features, and improves the effectiveness of extracted features. The short connection structure connects the deeper side output to the shallow side output, refining the edge details and more accurately locating the location of the defect.

2.1   Feature Extraction

The fabric image has the characteristics of complex background texture and various defects. Rich features and sufficiently deep networks are necessary for saliency detection. Our goal is to design a multi-layer, multi-scale network for multi-level feature extraction task. Fewer layers of network model lead to insufficient feature extraction and lack of high-level semantic information. So it is poor to represent the characteristics of fabric image. Relatively, a network model with excessive layers increases the amount of parameters, which will reduces the computational efficiency. Through learning, VGGnet [15] can meet the above requirements in existing networks. VGGnet achieves the best results in segmenting natural images by reusing 3×3 convolution kernel and 2×2 pooling kernel. Two 3×3 convolution layers has an effective receptive field of 5×5 convolution layer. The effect of three 3×3 convolution layers in series is equivalent to one 7×7 convolution layer. Therefore, it has powerful feature extraction ability while not reducing the size of the receptive field.

We take the first 13 layers of VGGnet and removed the fully connected layers. There are six stages, including the convolution layers and the pooling layers. The

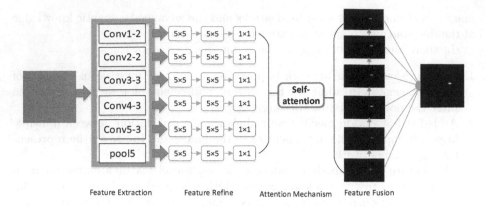

Fig. 1. The overall structure of our method. It consists of four parts: feature extraction module, feature refine module, attention module and feature fusion module.

fully connected layer is not conducive to saliency detection. Removing the fully connected layer reduces network parameters effectively and does not affect the detection effect. In order to achieve accurate results for defect detection, it is desirable to obtain meaningful feature maps with different scales. Side output was connected behind the last convolutional layer ($conv1-2$, $conv2-2$, $conv3-3$, $conv4-3$, $conv5-3$, pool5) at each stage of the main network.

After the backbone network, the feature refine module was designed according to the characteristics of the fabric image. The target of the fabric image is similar to the background, so it is significantly different from the natural scene image. Wherefore, the relationship between the fabric features is closely related. In order to make the extracted features contain scale and shape invariance, instead of using a bottom-up detection method, three convolutional layers are applied in the side output layer. As shown in Fig. 1, the convolutional layer connected to each side output layer with two 5×5 convolution kernel and a 1×1 convolution kernel. In general, most methods use a convolution kernel size of 3×3, 5×5 or 7×7. The receptive field of 3×3 is small, and each convolution can only cover a limited neighborhood around the pixel. Thence, global information cannot be used very well. The fabric image background is not as complicated as a natural scene. 7×7 convolution kernel will increase the representation of the network, but it will increase the parameters of the network and lose computational efficiency. Consequently, selecting the 5×5 convolution kernel is suitable for fabric images.

2.2 Self-attention Mechanism

The primary network has multiple convolutional layers with different stages. The normal optimization algorithm failed coordinate multiple layers to capture the parameter values of these dependencies. Previous models relied on convolution to simulate dependencies between different image regions, but convolution operations have local receptive fields. Changing the size of the convolution kernel will make the receptive field wider, which can improve the expressive ability of the network. However, the increase of calculation parameters will reduce the computational efficiency. The self-attention mechanism [11] handles long-range, multi-level associations well to discover the dependencies in images. In addition, self-attention mechanism can gain the long-distance dependence between pixels with a small computational cost.

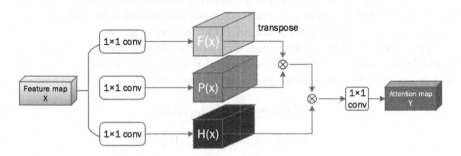

Fig. 2. Self-attention mechanism. Convolve the input feature map to obtain the correlation between different locations.

The self-attention module is shown in Fig. 2. The self-attention module takes the weighted sum of the features at all locations as a response to that location. It obtains the relationship between image features by calculating the relationship between any two pixels in the image directly. $F(x)$, $P(x)$, and $H(x)$ are ordinary 1×1 convolution, the difference is that the output channel size is different. After $F(x)$ is transposed, multiplied by the output of $P(x)$, and then normalized to obtain a feature map. Multiply the obtained feature map and $H(x)$ pixel by pixel to obtain adaptive attention maps.

Feature map $X \in R^{C \times H \times W}$ from the side output layer as input, and then converted into three different feature spaces $F(x)$, $P(x)$, and $H(x)$ by different convolutions. In order to reduce the amount of calculation, the number of channels output by $F(x)$, $P(x)$ is 1/8 of input feature map. Scilicet, the output dimensions of $F(x)$ and $P(x)$ are $[C/8, W, H]$. To simplify the calculation, later $N = W \times H$.

$$F(x) = Conv1(x_{i,j}, W_1) \tag{1}$$

$$P(x) = Conv2(x_{i,j}, W_2) \tag{2}$$

Where $Conv1(\cdot, \cdot)$ and $Conv2(\cdot, \cdot)$ represent the convolutional layer of $1 \times 1 \times C/8$. x, j represents the horizontal and vertical coordinates of the two-dimensional feature map. $W_{i=1,2,3,4}$ represents the convolution parameter.

Multiplying the transpose of $F(x)$ by $P(x)$, and the output is a matrix D with dimensions $[N, N]$. The matrix D can be seen as a correlation matrix, which is the correlation between the individual pixels on the feature map. The matrix

D is normalized row by row to get the matrix β, and each row (vector of length N) represents an attention.

$$D\left(x_{i,j}\right) = F\left(x_i\right)^T P\left(x_j\right) \tag{3}$$

$$\beta = \frac{exp\left(D\left(x_{i,j}\right)\right)}{\sum_{i=1}^{N} D\left(x_{i,j}\right)} \tag{4}$$

Where $F\left(x\right)^T$ represents matrix transposition. The matrix D is the core of attention. Where the elements of row i and column j are obtained by multiplying the ith row of $F\left(x\right)$ and the jth column of $P\left(x\right)$ by vector points.

Apply these N kinds of attention methods to $H\left(x\right)$, which is multiplied by $H\left(x\right)$. At this point each pixel is associated with the entire feature map. After the 1×1 convolution layer, the generated feature map is the final output $Y\left(x\right)$.

$$H\left(x\right) = Conv3\left(x_{i,j}, W_3\right) \tag{5}$$

$$Y\left(x\right) = Conv4\left(\beta x_{i,j}, W_4\right) \tag{6}$$

Where $Conv3\left(\cdot, \cdot\right)$ and $Conv4\left(\cdot, \cdot\right)$ represent the convolutional layer of $1 \times 1 \times C$. The output dimension is $[C, W, H]$ of $H\left(x\right)$ and $Y\left(x\right)$.

2.3   Feature Fusion

Multi-level, multi-scale features are essential for saliency detection, The method proposed in this paper can get the saliency map of six stages from shallow to deep. We observed that shallow layers usually have detailed low-level features that help to refine the edges of the defects, but contain many noises. The high-level semantic knowledge contained in the deeper layers can accurately locate the exact location of the salient target, but at the expense of the details. Thus, the features extracted from the deep side output is related to the shallow layer. The deep features help the shallow layer better locate the defect, while the low-level features improve the edge of the contour. Due to previous convolution, pool and other operations, the size of the feature maps output from multi-level side output layer is different. In order to ensure that the multi-level feature maps for connection and fusion have the same size, the bilinear interpolation operation is utilized for upsampling. All of the operations and calculations are performed on the training set $X = (x_i, i = 1, \cdots, N)$ and the groundtruth set $Y = (y_i, i = 1, \cdots, N)$. The different side outputs are then combined in the following manner to take advantage of features that are salient.

Side output function of short connection structure:

$$S_{side}^m = \alpha_{side}^m S_{side}^{m+1} + f_{side}^m \ for \ m = 1, \cdots, 5. \tag{7}$$

Where f_{side}^m is the activation function on the mth side, and α_{side}^m is the weight connected from the $m + 1$th side to the mth side. When α_{side}^m is set to 0, the output is the activation function on the 6th side.

The side output loss function is given as follows:

$$l_{side}^m = \sum_{i \in Y} y_i logP\left(y_i = 1|X)\right) + (1 - y_i)logP\left(y_i = 0|X)\right) \tag{8}$$

$$L_{side} = \sum_{m=1}^{M} \phi_m l_{side}^m \tag{9}$$

Where l_{side}^m is the mth standard cross entropy loss function, and $P\left(y_i = 1|X)\right)$ is the probability that the pixel x_i is the target. ϕ_m is the mth side loss weight.

The fusion loss function and the final loss function are

$$L_{fuse} = \rho(Y, \sum_{m=1}^{M} w_m S_{side}^m) \tag{10}$$

$$L_{final} = L_{side} + L_{fuse} \tag{11}$$

Where w_m is the fusion weight. $\rho(\cdot, \cdot)$ represents the distance between ground truth and fusion prediction.

Multiple convolution and pooling can extract rich features, but this will bring about the problem of inconsistent size of feature maps in different stages. Upsampling the multi-level saliency maps directly into the size of the original image loses a lot of detail, while preserving the high-level semantic information. However, these details are indispensable for accurate defect detection. Therefore, multi-level saliency maps were combine with simple elements to integrate different scales of information. This reduces the impact of loss details and optimize test results.

3   Experiments

The method presented in this paper was tested on a data set containing 2,000 images. The dataset has a variety of challenging points such as grease, holes, and foreign objects. In these pictures, we randomly selected 1500 images as the training set and the remaining 500 images as the test set. In order to get more accurate detection results, we manually label the images. In the image of the defect, defect area is the detection target, and the defect-free area is the background. Their pixel values was set to 1 and 0 respectively to get the groundtruth corresponding to the image. This allows for a clearer identification of the target and background, making it easier to train the model. When training, the original image is entered into the network along with the groundtruth. Unlike training, the groundtruth input is not required for testing.

VGG-16 network as the pre-training model, and the test code is compiled in the pytorch framework using the python language. A GTX-1080 GPU and Intel UHD Graphics 630 CPU are used for both training and testing. During training, we set the momentum parameter was 0.9, the learning rate was 1e−6, and the weight decay was 0.0005. Taking into account the efficiency of the training, our network is trained for 12 epochs in total, and the connection layer weights are randomly initialized. The fusion layer weights are initialized with 0.18 in the training phase.

Salient detection has been studied for many years, and many excellent methods have been proposed during this period. In order to show the performance of the model more intuitively, we compared it with other excellent methods DSS and NLDF [14,16]. As shown in Fig. 3, our model can locate the correct defect location, filter the interference of the background texture, and generate a salient map close to the groundtruth. Especially when the shape of the defect is irregular and (the first row and the fourth row) are very similar to the background (the second row and the third row), our method still accurately detect the boundary between the defect and the background.

In addition to the visual comparison, the quantitative analysis of the test results is also carried out, as shown in Table 1. We selected two evaluation indicators, MAE and F-measure. These are first calculated by binarization, and then calculated by the following formula. MAE is to directly calculate the average absolute error between the saliency map of the model output and Groundtruth.

$$MAE = \frac{1}{H \times W} \sum_{i=1}^{W} \sum_{j=1}^{H} |P(i,j) - G(i,j)| \tag{12}$$

where $P(i,j)$ is the salient map that output from the network, and G is the ground truth.

F-measure is the weighted harmonic mean of the recall and precision under non-negative weight β. The formula is as follows:

$$F = \frac{(1 + \beta^2) precision \times recall}{\beta^2 precision + recall} \tag{13}$$

Where β^2 is generally set to 0.3, as suggested by previous works.

Quantitative results are listed in Table 1. It can be seen that our method is also superior to other methods on the fabric dataset, which demonstrate the efficiency of the proposed method.

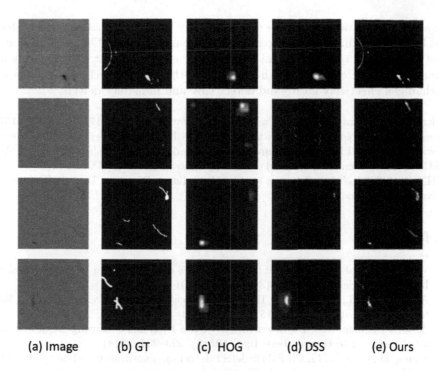

(a) Image   (b) GT   (c) HOG   (d) DSS   (e) Ours

Fig. 3. Visual comparison with existing state-of-the-art methods. As can be seen, our method produces more accurate saliency maps than other methods, which are closet to ground truth.

Table 1. MAE scores (Lower is better) and F-measure scores (Higher is better) of different salient detection methods

Methods	MAE	F-measure
Ours	0.0059	0.745
DSS [13]	0.0068	0.710
NLDF [15]	0.0072	0.694

4   Conclusion

In this paper, we proposed a novel salient detection method for fabric defect detection, which is achieved by adding the self-attention mechanism to the full convolutional network. Considering that multi-level, multi-scale features are required for saliency detection, feature extraction network with six convolution stages has designed. It obtains both high-level semantic features and low-level spatial information. As the convolutional and pooling layers deepen, the connections between the different layers are weakened. The self-attention mechanism is used to coordinate multiple layers to capture their dependencies. In addition, the short connection structure is introduced to connect the deep side output and the shallower. Advanced features can accurately locate the location of the defect, and low-level features contain spatial semantic information. The combination of the two filters out the background noise and focus on the defect. The experiments demonstrate that compared with the state-of-the-art approaches, the proposed method can effectively locate and detect tiny and indistinct defects.

Acknowledgements. This work was supported by NSFC (No. 61772576, U1804157), Science and technology innovation talent project of Education Department of Henan Province (17HASTIT019), The Henan Science Fund for Distinguished Young Scholars (184100510002), Henan science and technology innovation team (CXTD2017091), IRT-STHN (18IRTSTHN013), Program for Interdisciplinary Direction Team in Zhongyuan University of Technology.

References

1. Wen, Z., Cao, J., Liu, X., et al.: Fabric defects detection using adaptive wavelets. Int. J. Cloth. Sci. Technol. **26**(3), 202–211 (2014)
2. Arivazhagan, S., Ganesan, L.: Texture classification using wavelet transform. Pattern Recogn. Lett. **24**(9–10), 1513–1521 (2003)
3. Zhou, J., Semenovich, D., Sowmya, A., et al.: Dictionary learning framework for fabric defect detection. J. Text. Inst. **105**(3), 223–234 (2014)
4. Zhang, D., Gao, G., Li, C.: Fabric defect detection algorithm based on Gabor filter and low-rank decomposition. In: Eighth International Conference on Digital Image Processing. International Society for Optics and Photonics (2016)
5. Cai, Z., Vasconcelos, N.: Cascade R-CNN: delving into high quality object detection. In: Proceedings of the IEEE Conference on Computer Vision and Pattern Recognition, pp. 6154–6162 (2018)
6. Hu, H., Gu, J., Zhang, Z., et al.: Relation networks for object detection. In: Proceedings of the IEEE Conference on Computer Vision and Pattern Recognition, pp. 3588–3597 (2018)
7. Liu, Y., Cheng, M.M., Hu, X., et al.: Richer convolutional features for edge detection. In: Proceedings of the IEEE Conference on Computer Vision and Pattern Recognition, pp. 3000–3009 (2017)
8. Zhang, H., Dana, K., Shi, J., et al.: Context encoding for semantic segmentation. In: Proceedings of the IEEE Conference on Computer Vision and Pattern Recognition, pp. 7151–7160 (2018)
9. Zhang, P., Wang, D., Lu, H., et al.: Amulet: aggregating multi-level convolutional features for salient object detection. In: 2017 IEEE International Conference on Computer Vision (ICCV). IEEE Computer Society (2017)
10. Zhang, L., Dai, J., Lu, H., et al.: A bi-directional message passing model for salient object detection. In: Proceedings of the IEEE Conference on Computer Vision and Pattern Recognition, pp. 1741–1750 (2018)
11. Zhang, H., Goodfellow, I., Metaxas, D., et al.: Self-attention generative adversarial networks. arXiv preprint arXiv:1805.08318 (2018)
12. Zhang, X., Wang, T., Qi, J., et al.: Progressive attention guided recurrent network for salient object detection. In: Proceedings of the IEEE Conference on Computer Vision and Pattern Recognition, pp. 714–722 (2018)

13. Zhao, T., Wu, X.: Pyramid feature attention network for saliency detection. In: Proceedings of the IEEE Conference on Computer Vision and Pattern Recognition, pp. 3085–3094 (2019)
14. Hou, Q., Cheng, M.M., Hu, X., et al.: Deeply supervised salient object detection with short connections. In: Proceedings of the IEEE Conference on Computer Vision and Pattern Recognition, pp. 3203–3212 (2017)
15. Simonyan, K., Zisserman, A.: Very deep convolutional networks for large-scale image recognition. arXiv preprint arXiv:1409.1556 (2014)
16. Luo, Z., Mishra, A., Achkar, A., et al.: Non-local deep features for salient object detection. In: Proceedings of the IEEE Conference on Computer Vision and Pattern Recognition, pp. 6609–6617 (2017)

Efficient Neural Network Space
with Genetic Search

Dongseok Kang and Chang Wook Ahn[✉]

Gwangju Institute of Science and Technology, Gwangju, Korea
{dongseok176,cwan}@gist.ac.kr

Abstract. We present a novel neural architecture search space and its search strategy with an evolutionary algorithm. It aims to find a set of inverted bottleneck structure blocks, which takes a low-dimensional input representation followed by a compressing layer. Primitive operation layers constitute flexible inverted bottleneck blocks and can be assembled in evolutionary operation. Because the bottleneck structure confines the search space, the proposed evolutionary search algorithm can easily find a competitive neural network despite its small population size. During the search process, we designed to evaluate a model to avoid local minimums: such implementation helped the algorithm to discard local minimums and find better models. We conducted experiments on image classification of Fashion-MNIST, and we discovered an efficiently optimized neural network achieving 6.76 for an error rate with 356 K parameters.

Keywords: Genetic algorithm · Neural network · Neural Architecture Search

1 Introduction

In recent years, Neural Architecture Search (NAS) has successfully developed artificial neural networks. NAS outperforms human-designed neural networks in numerous domains including the fields of image classification and natural language processing. For a given task, NAS automatically and progressively optimize a set of neural architectures [5] or its controller [1] that creates neural architecture over probability distribution. Among two major search strategies, Reinforcement Learning utilizes a policy gradient method to iteratively update controller parameter and controller generates different architectures. The evolutionary approach, on the other hand, preserves a set of models known as population, and repeatedly replace their members with newly created network models based on their performance metrics.

Inspired by the idea of inverted residual block [7] which showed its efficient image classifications and its cell structures as in [12], we devised NAS with

---

C. W. Ahn—This work was supported by GIST Research Institute (GRI) grant funded by the GIST in 2019.

© Springer Nature Singapore Pte Ltd. 2020
L. Pan et al. (Eds.): BIC-TA 2019, CCIS 1160, pp. 638–646, 2020.
https://doi.org/10.1007/978-981-15-3415-7_54

novel search space, named *layered bottleneck* and the search strategy with the evolutionary computation approach. While an inverted residual block takes a low-dimensional representation as input and lets a lightweight depthwise convolution filter it out, the proposed layered bottleneck structure that consists of several primitive operations, including convolutions and pooling operations, takes low-dimensional input followed by a compression layer (Fig. 1). Each operation in a structure takes a concatenation of operation outcomes as input which is a high-dimensional representation.

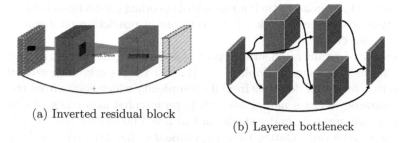

(a) Inverted residual block

(b) Layered bottleneck

Fig. 1. Inverted residual block and proposed structure

To evaluate a model, we used its performance metric with heuristic terms to avoid local minimums. We observed that newly created models from evolutionary operations do not survive environmental selection during the mid-late search processes, when the performance metric solely plays the role of choosing the best models. Therefore, we modified the fitness of a model based on their age and performance, which helped the search process to do exploration consistently.

Experiments conducted on image classification problem during our study is on Fashion-MNIST dataset. This dataset includes simple greyscale fashion item images to converge neural network models in a short time compared to other image datasets. The results showed that our method can produce an efficient convolutional neural network, with only 356 K parameters, that achieves a classification error of 6.76 on the test set, which is competitive with others (ResNet-50 8.70 test error).

Our main contributions are summarized as follows:

- We propose a novel architecture search space for an evolutionary neural architecture search, which exploits the bottleneck structure and provides various efficient neural network designs. When the proposed *layered bottleneck* limits search space, NAS can more easily find well-optimized neural network models due to its efficient block structure.
- We encoded an entire architecture scheme as a genotype data in Evolutionary Algorithm so that conventional Genetic Algorithm (GA) operations can be applied on layered blocks and exploit its network models found until then. Unlike many other GA based NAS tear and assemble low-level architecture

motifs like connections and the number of channels, we can preserve the blocks so that it can fully exploit its parent network architecture over the process of GA operations.

2 Related Work

Research of automatically tuning and generating neural networks have attracted the interest of many researchers. Xie et al. [11] encoded connections between operation layers in the effective part as a binary string and used GA to search neural networks. Similar to their approach of encoding connections between operation layers, our layered block follows the way of restriction on connections so that no cycles can occur.

One way of producing neural networks is by using the RNN controller and extracting models from it. Zoph et al. [12], in their work, suggested that an architecture space can decouple from its complexity which comes from the depth of the network. When a neural network is represented as a stack of cells, they showed it only needs to find a single unit of a cell no matter how deep it needs to be for a given task. During their experiments, they trained a scalable RNN controller and produced a neural network model with competitive performance in image classification on ImageNet.

Sandler et al. [7] introduced MobileNetV2 with inverted residual blocks for mobile use on multiple tasks. They demonstrated the efficiency of a realized model on multiple tasks and unique property that allows separating the network's expressiveness from its capacity. Based on their practical network module *inverted residual block*, we adopted an expansion-compression structure of layered bottleneck so that information flows the model in a memory-efficient way. Sandler's inverted residual block expands and compresses its information using a 1×1 convolution and separable convolution, suitable for mobile use due to its lightweight operations. On the other hand, our defined *layered bottleneck* concatenates a set of input information, and then operation layers expand the input into high-dimensional information. Then, the layered bottleneck compresses its filtered information with convolution operation and passes it to the following layered bottleneck unit.

3 Method

3.1 Representation

In our work, an individual represents a single neural network with multiple layered blocks, which has a chromosome consisting of two vectors v_0 and v_1. A pair of elements in these two vectors describes what the operations each block has, such as convolution and pooling operation, and its connections, respectively.

The layered block takes two previous output as a input, and produces an output followed by a connection with identity mapping [4] of previous output.

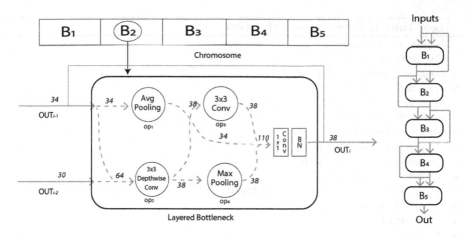

Fig. 2. Chromosome and its information

In the block, two previous inputs can be fed to any operation layers, but other results from lower-numbered operation should be connected to higher-numbered operation to prevent a cycle. For example, a chromosome $\mathbf{p}$ with n blocks and k operations within the it can be compacted as follows

$$\mathbf{p} = \langle (\mathbf{v_{10}}, \mathbf{v_{11}}), ..., (\mathbf{v_{n0}}, \mathbf{v_{n1}}) \rangle \tag{1}$$

where each element can be defined as

$$\mathbf{v_{i0}} = \langle op_1, ..., op_k \rangle, \;\; where \; op_j \in Convolutions \cup Poolings$$
$$\mathbf{v_{i1}} = \langle \mathbf{c_{input0}}, \mathbf{c_{input1}}, \mathbf{c_1}..., \mathbf{c_{k-1}} \rangle, \mathbf{c_j} \in \{0,1\}^{k-i} \; and \; \mathbf{c_{input}} \in \{0,1\}^k$$

In Fig. 2, a chromosome having $n = 5$ and $k = 4$ is illustrated with its network structure. A second layered block B_2 takes two inputs, which has 30 and 34 channels, and makes an single output.

$$\mathbf{v_{20}} = \langle avg\_pooling, 3X3\_depthwise, 3X3\_conv, max\_pooling \rangle$$
$$\mathbf{v_{21}} = \langle \langle 1100 \rangle, \langle 0100 \rangle, \langle 000 \rangle, \langle 11 \rangle, \langle 0 \rangle \rangle$$

All the layer outputs which have not fed into other layers are concatenated and fed into the last 1×1 convolution layer. The output of the block is zero-padded [3] for the residual connection to match the channel size.

3.2 Genetic Operations

To produce a new network structure, we use *one-point crossover* [2] on two chromosomes with *mutation* (Algorithm 1). At first, we randomly choose a pivot index where to exchange genes of a chromosome. On the line 9–10 of the following Algorithm, it makes a new block with a low probability to explore new structures

---

Algorithm 1. Crossover and Mutation

    Input: p_1, p_2 and n
    Output: c_1 and c_2
1 $crossover\_index \longleftarrow U\{2, n\}$;
2 $c_1 \leftarrow \emptyset$;
3 $c_2 \leftarrow \emptyset$;
4 $mutation\_prob \leftarrow \frac{1}{n}$;
5 **for** $i \leftarrow 1$ **to** n **do**
6 | $gene_1 \leftarrow (v_{i0}, v_{i1})$ in p_1;
7 | $gene_2 \leftarrow (v_{i0}, v_{i1})$ in p_2;
8 | **for** $k \leftarrow 1$ **to** 2 **do**
9 | | **if** $U(0, 1) < mutation\_prob$ **then**
10 | | | reset $gene_k$;
11 | | **end**
12 | **end**
13 | **if** $i \leq crossover\_index$ **then**
14 | | append $gene_1$ to c_1;
15 | | append $gene_2$ to c_2;
16 | **else**
17 | | append $gene_2$ to c_1;
18 | | append $gene_1$ to c_2;
19 | **end**
20 **end**
21 **for** $i \leftarrow 1$ **to** 2 **do**
22 | **if** $U(0, 1) < 0.1$ **then**
23 | | reset subsampling indices of $gene_i$;
24 | **end**
25 **end**

---

while line 14–18 exploits given network structure. We randomly pick individuals in the population before choosing two input chromosomes as usual in genetic algorithms. We set a probability of mutation, *mutation_prob* on the line 4, to activate mutation once per chromosome on average to preserve a large portion of the given structure as intact. It also tries to change where to subsampling of an image data as in the line 21–24 to get optimal indices for newly created chromosome.

First, we randomly choose pivot index where to exchange its genes of chromosome. Line 9–10 makes a new block with a low probability to explore new structures while the line 14–18 exploits given network structure. To choose two input chromosome of Algorithm 1, we randomly pick individuals in the population, which is usual in genetic algorithms. We set probability of mutation, *mutation_prob* in the line 4, to activate mutation once per chromosome in average to preserve large portion of given structure as intact. It also try to change where to subsampling in the line 21–24 to get optimal indices for newly created chromosome.

Algorithm 2 shows a full procedure to explore the search space with the genetic algorithm. It starts with randomly sampled *pop_size* network structures, and then make new individuals in the result of crossover and mutation (Algorithm 1). We evaluate each individual in the population based on a validation set with the following formula

$$fitness(\mathbf{p}) = \eta^\tau * ErrorRate(\mathbf{p}) \tag{2}$$

In Eq. 2, τ is the number of generations $\mathbf{p}$ have survived after environmental selection. We applied Eq. 2, instead of directly using an error rate, because an individual may prevent further exploration when it luckily hit a low error rate due to the distribution of training and validation set. We observed that this heuristic term helps the search process to avoid being stuck on local minima.

Algorithm 2. Overall Procedure

Input: n, k, gen and *pop_size*
Output: Population of networks
1 $P \leftarrow$ Randomly initialized *pop_size* individuals;
2 **for** $i \leftarrow 1$ **to** *gen* **do**
3 $\quad$ $temp \leftarrow \emptyset$;
4 $\quad$ **for** $k \leftarrow 1$ **to** $\lfloor pop\_size/2 \rfloor$ **do**
5 $\quad\quad$ Randomly choose $\mathbf{p_1}$ and $\mathbf{p_2}$ from P;
6 $\quad\quad$ $temp \leftarrow temp \cup CrossoverAndMutation(p_1, p_2)$;
7 $\quad$ **end**
8 $\quad$ $P \leftarrow P \cup temp$;
9 $\quad$ **foreach** $\mathbf{p} \in P$ **do**
10 $\quad\quad$ Evaluate $\mathbf{p}$ based on equation 2;
11 $\quad$ **end**
12 $\quad$ $P \leftarrow$ best *pop_size* of P;
13 **end**

---

4   Experiment

4.1   Experiment Details

We conducted experiments on image classification with Fashion-MNIST [10], which requires short training epochs before converged compared to other complex datasets. Given 60,000 training images and 10,000 test images, we randomly chose 10,000 images as the validation set from the training images. All images are 28×28 in size and fed into classifiers without any augmentations. We trained each network for 40 epochs with RMSProp [9] and then evaluated the network with the validation set. Following the cosine annealing [6], the learning rate in our experiments decays from 1e-2 to 1e-4 for the first 30 epochs and another 1e-4 for the last 10 additional epochs. For n and k, the number of blocks in a network and the number of operations in each block, k, we used 10, 4 respectively throughout the experiments. We gradually increased output channel sizes in the blocks, starting from 16 and increased by 4 in each block like in [3]. All three experiments had searched the network structures in 25 generations. Experiment A and B started with 16 individuals and evaluated 416 network structures while experiment C started with 32 individuals and evaluated 832 in the search. In experiment A, we solely used error rate of image classification on validation set, instead of Eq. 2 used in experiment B and C. The experiments are compared with classification error results on the same dataset from the EvoCNN [8].

4.2   Results and Discussion

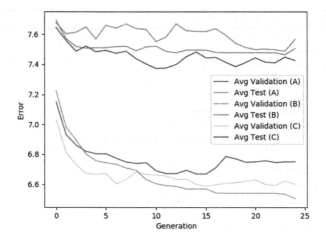

Fig. 3. Average classification errors of population in three experiments

Table 1 has the results of three conducted experiments with different parameters and Fig. 3 shows average classification error rates.

As shown in the figure, we witnessed that all three experiments had decreased errors on validation and test images. As depicted in Table 1, with the lowest error rate and its parameter size, experiment B produced a more promising model compared to the experiment A. This shows that fitness evaluation using Eq. 2 helped the final result of the search.

In the case of B and C, it is interesting that even though the size of a population had been doubled in experiment C and C has lower error rate in average, the best model found in experiment C had a higher error rate than the best in experiment B as shown in the table. And also when validation error rates dropped by 0.5–0.8 test errors decreased only by 0.2–0.3. This may indicate that the discrepancy between the two metrics had been increased during the

Table 1. Results of experiments

Classifier	Error	#Params	Epochs	η	pop_size
Exp. A	7.10	639 K	40	1.0	16
Exp. B	6.76	356 K	40	1.0005	16
Exp. C	7.02	495 K	40	1.0005	32
MLP 256-128-64	10.00(+)	41 K	25	–	–
GoogleNet	6.30	101 M	–	–	–
3C1P2F+Dropout	7.40	7.14 M	150	–	–
2C1P	7.50	100 K	30	–	–
EvoCNN	5.47	6.68 M	100	–	100

search process, possibly because the GA successfully had adapted only to the validations, not to the test set.

Although the population size was small throughout all three experiments, the way our GA operations exploits structures made the process to find well-optimized efficient neural networks. Further increased size of the population may also help to produce better networks and reduce the discrepancy.

5   Conclusion

We proposed Neural Architecture Search with a novel search process and an evolutionary approach. It successfully updates the population to have better performance metrics in the average, as depicted in the experiments. Our GA approach worked well even when the population size was small and fewer in the number of generations.

References

1. Bello, I., Zoph, B., Vasudevan, V., Le, Q.V.: Neural Optimizer Search with Reinforcement Learning. arXiv e-prints p, September 2017. arXiv:1709.07417
2. Goldberg, D.E.: Genetic Algorithms in Search Optimization and Machine Learning, 1st edn. Addison-Wesley Longman Publishing Co. Inc., Boston (1989)
3. Han, D., Kim, J., Kim, J.: Deep pyramidal residual networks. In: Proceedings - 30th IEEE Conference on Computer Vision and Pattern Recognition, CVPR 2017, January 2017, pp. 6307–6315, October 2017
4. He, K., Zhang, X., Ren, S., Sun, J.: Deep Residual Learning for Image Recognition. arXiv e-prints p, December 2015. arXiv:1512.03385
5. Liu, C., et al.: Progressive Neural Architecture Search. arXiv e-prints p, December 2017. arXiv:1712.00559
6. Loshchilov, I., Hutter, F.: SGDR: Stochastic gradient descent with warm restarts (2016). arXiv preprint arXiv:1608.03983

7. Sandler, M., Howard, A.G., Zhu, M., Zhmoginov, A., Chen, L.C.: MobileNetV2: inverted residuals and linear bottlenecks. In: Proceedings of the IEEE Computer Society Conference on Computer Vision and Pattern Recognition, pp. 4510–4520, January 2018

8. Sun, Y., Xue, B., Zhang, M., Yen, G.G.: Evolving deep convolutional neuralnetworks for image classification. IEEE Trans. Evol. Comput. (2019)

9. Tieleman, T., Hinton, G.: Lecture 6.5-rmsprop: divide the gradient by a running average of its recent magnitude. COURSERA: Neural Netw. Mach. Learn. **4**(2), 26–31 (2012)

10. Xiao, H., Rasul, K., Vollgraf, R.: Fashion-mnist: a novel image dataset for benchmarking machine learning algorithms (2017)

11. Xie, L., Yuille, A.: Genetic CNN. In: Proceedings of the IEEE International Conference on Computer Vision, October 2017, pp. 1388–1397, March 2017

12. Zoph, B., Vasudevan, V., Shlens, J., Le, Q.V.: Learning Transferable Architectures for Scalable Image Recognition, July 2017. arXiv e-prints arXiv:1707.07012

Simulation of Limb Rehabilitation Robot Based on OpenSim

Aihui Wang$^{(\boxtimes)}$, Junlan Lu, Yifei Ge, Jun Yu, and Shuaishuai Zhang

School of Electric and Information Engineering,
Zhongyuan University of Technology, Zhengzhou 450007, China
a.wang@zut.edu.cn

Abstract. Lower limb rehabilitation robots have been widely used in medical treatment in recent years. The safety of robot in rehabilitation training should be considered. In this paper, the method to analyze complicated inverse kinematics and forward dynamics of lower limb rehabilitation robot is studied. Because of the mathematical model of lower limb is difficult to establish, the OpenSim software is used to simulate and analyze the human gait data which collected by the NOKOV optical motion capture system. According to the inverse kinematics simulation, the curve of each joint angle changing over time is obtained, and the reaction force acting on the ground is obtained through the forward dynamics simulation.

Keywords: Lower limb rehabilitation robot · Gait trajectory · OpenSim · Inverse kinematics · Forward dynamics

1 Introduction

The number of limb injuries has gradually increased due to population aging, natural disasters, traffic accidents, various diseases in recent years [1]. Because of the limiting medical resources, it is difficult to provide reasonable rehabilitation treatment for all patients with limb injury, which may cause permanent paralysis and inability to move normally in some patients. Medical research indicates that reasonable rehabilitation treatment for patients who have paralysis and mobility disorders can be cured within a appropriate time period [2]. However, physical therapists only account for 4/million of overall number of people in China [3], and the traditional rehabilitation equipments which can only be trained under the guidance of doctors and nurses are relatively bulky, which not only increasing the cost of rehabilitation treatment but also reducing the efficiency of treatment. Therefore, lightweight, economical and efficient lower limb rehabilitation robots are urgently needed.

Supported by National Natural Science Foundation (U1813201), Science and Technology Key Project of Henan Province (162102410056), Young Backbone Teacher Training Program of Henan Provinces Higher Education (2017GGJS117).

© Springer Nature Singapore Pte Ltd. 2020
L. Pan et al. (Eds.): BIC-TA 2019, CCIS 1160, pp. 647–654, 2020.
https://doi.org/10.1007/978-981-15-3415-7_55

Rehabilitation robots has gained interest in recent years. It has crucial application values to rehabilitation training [4]. The lower limb rehabilitation robots combined with rehabilitation medicine and robot technology can not only reduce the burden of physiatrist and increase the opportunities of patients to see a doctor, but also provide more targeted training programs for patients and improve the therapeutic effect. In order to ensure the comfort and safety of the wearers, the simulation analysis of the lower limb rehabilitation robots is needed.

In previous studies, many human motion analysis and simulation softwares have been developed. At the end of the 20th century, Software for Interactive Musculoskeletal Modeling(SIMM) was developed by Delp and loan [5,6], which was the first simulation software of musculoskeletal, but it was not widely used due to its high cost. Aalborg University in Denmark exploited a simulation software called AnyBody which integrated ergonomics and biomechanics to simulate and calculate of human kinematics, dynamics and physiology characteristics [7]. Lifemod was a secondary development software, which was used to study the digital simulation of human biomechanics based on MSC.ADAMS. It can be used to establish biomechanical model of any biological system, unfortunately it was difficult to develop [8]. In order to solve the problems like high price, inaccurate control and difficult development, Frank Anderson and other experts at Stanford university has exploited OpenSim which is a human motion simulation software based on C++ and JAVA. OpenSim is an open source platform for modeling, simulation and analysis of the musculoskeletal system [9,10]. OpenSim can be linked with Matlab and other software to achieve more convenient development. Users can directly employ its model to conduct inverse kinematics and forward dynamics simulation.

In this paper, the method to analyze complicated inverse kinematics and forward dynamics of lower limb rehabilitation robots is studied. The mathematical model of lower limb movement is obtained difficultly due to coupling and redundancy in the movement. Therefore, an indirect approach that OpenSim as an intermediate bridge is considered. After obtaining the normal gait data of human body with the optical motion capture system, inverse kinematics and forward dynamics simulation are carried out by using OpenSim, which can get the safe range of joint angle and the ground reaction force.

2   System Correlation Theory

2.1   Gait Data Collection

In this paper, the NOKOV optical motion capture system is used to gain gait data. The cameras which can record each frame marker point movement sequence images are used to real-time motion capture. Because of the limited range of each camera lens, the 6 cameras can be crossed to produce an area. Markers which are captured by two lenses at the same time generate continuous trajectories. Besides, the motion capture system records the spatial coordinates of all markers at the rate of 60 frames per second and its capture accuracy can reach 1 mm. The collected human gait data can be simply processed and exported by the

NOKOV system software. In this section, the 6 optical motion capture cameras are arranged in parallel in two columns [11]. The scene of motion capture and the marking points are shown in Fig. 1.

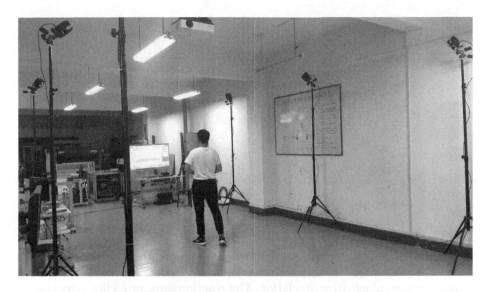

Fig. 1. Motion capture sites and markers. The NOKOV optical motion capture system used in the study consists of 6 optical motion cameras, markers, positioning rods, correction rods, data transmission equipment, upper computer.

2.2   OpenSim System

The graphical user interface of OpenSim 4.0 includes a set of tools for analyzing musculoskeletal models, generating simulation and visualization results [9]. The SimTrack tool in OpenSim can generate motion simulation of specific objects driven by muscles quickly and accurately and obtain high-precision data. Graphical user interface of OpenSim shown in Fig. 2.

OpenSim's modeling theory derives from Hill's equation and Hill's three-element muscle model [12], the whole simulation process consists of model scaling, inverse kinematics and forward dynamics.

3   Simulation Based on OpenSim

The study of lower limb rehabilitation robots is the coordinated movement of two legs. Most previous studies have focused on the single leg of the lower limb rehabilitation robots. The common approach is to select a single leg to establish dynamic equation and simulate. In this section, the optical motion capture

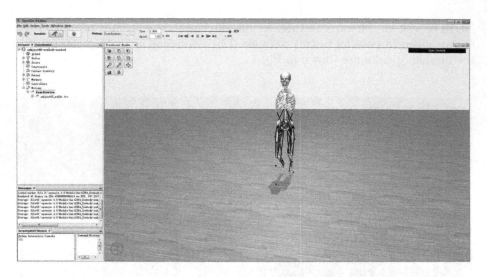

Fig. 2. Graphical user interface of OpenSim.

system and OpenSim system are applied to simulate the model of the lower limbs. The joint angles of hip joint, knee joint and ankle joint are shown directly through inverse kinematics simulation. The reaction force providing reference for later force measurement is obtained through forward dynamics between human body and the ground.

In this section, the position, velocity and acceleration data of hip joint, knee joint and ankle joint are collected by motion capture system, and TRC files are generated. Before the simulation, the model should be scaled and the collected human gait data which is exported in the form of TRC file should be import into OpenSim software. Model scaling is manually adjusted according to the ratio between the actual measured mark points and the mark points in the model.

The principle of inverse kinematics calculation is to use the weighted least square method to calculate the difference between the measured 3D coordinates and the base coordinate system and get a minimum value [13]. The calculation formula is as follows

$$\min_{\mathbf{q}} \left[\sum_{i \in \text{ maxters}} w_i \left\| \mathbf{x}_i^{\text{exp}} - \mathbf{x}_i(\mathbf{q}) \right\|^2 + \sum_{j \in \text{ unprescribed coords}} \omega_j \left(q_j^{\text{exp}} - q_j \right)^2 \right]$$
$$q_j = q_j^{\text{exp}} \text{ for all prescribed coordinates } j$$

(1)

Where $\mathbf{x}_i^{\text{exp}}$ is the coordinate of the actual measured marker points, $\mathbf{x}_i(\mathbf{q})$ is the coordinate of the virtual marker points in the model. $\mathbf{q}_j^{\text{exp}}$ is the actual measured coordinates of point j, and q_j is the coordinates of point j specified in the model.

Forward dynamic simulation employs Computer Muscle Control (CMC) algorithm to generate a set of muscle excitation to coordinate the movement of the simulated muscle driven object. CMC algorithm adopts static optimization criteria to allocate coordinated intermuscular force and proportional differential control, generating a forward dynamic simulation to ensure that the dynamic joint angle perfectly tracks the output of the step 3 during the simulation [14].

According to Newton's second law, the coordinate acceleration (rate of change in velocity) is described a set of rigid bodies by the inertia and force applied on the skeleton.

$$\ddot{q} = [M(q)]^{-1}\{\tau + C(q, \dot{q}) + G(q) + F\} \tag{2}$$

where $\ddot{q}$ is the coordinate accelerations, where τ is the torque, where $C(q)$ is the Coriolis and centrifugal forces, where q is the coordinates, where $\dot{q}$ is the velocities, where G is the gravity, where F is other forces applied to the model, and where $M(q)$ is the inverse of the mass matrix.

4   Simulation Result

The experimental steps are as follows

In step 1, human gait data is collected by motion capture system to generate TRC file.

In step 2, OpenSim software is used to import the collected data into the model and scale appropriately according to the proportion between the actual human body and the model.

In step 3, according to the inverse kinematics simulation obtains the joint angle curve.

In step 4, the forward dynamic simulation is carried out to obtain the reaction force on the ground.

Inverse kinematics simulation results are shown in the follows.

Figure 3 shows the curve of hip joint angle changing with time. The range of the value of left hip joint angle is −26–17, and the range of the value of right hip joint angle is −30–20. Figure 4 shows the curve of knee joint angle changing with time. The range of the value of left knee joint angle is −70–0, and the range of the value of right knee joint angle is −70–0. Figure 5 shows the curve of ankle joint angle changing with time. The range of the value of left ankle joint angle is −9–7, and the range of the value of right ankle joint angle is −11–7. It shows that the changes of hip joint angle and ankle joint angle are not significant, while the change of knee joint angle is relatively large. It indicates that the posture of human lower limbs in the movement depends on the knee joint angle. In the following study of lower limb rehabilitation robots, the change of knee joint angle is mainly considered.

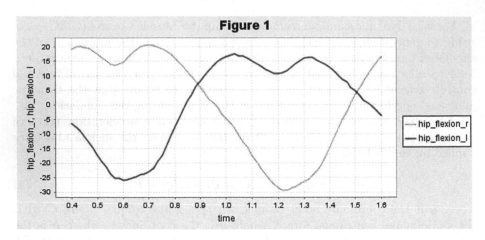

Fig. 3. Curve of hip angle. The blue line shows the left hip joint angle. The red line shows the right hip joint angle. (Color figure online)

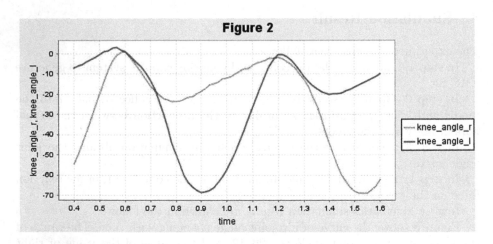

Fig. 4. Curve of knee angle. The blue line shows the left knee joint angle. The red line shows the right knee joint angle. (Color figure online)

Figure 6 is the reaction force of the human model on the ground during the movement, and it indicates that the reaction force is 600–800N. This force is the simulation result of the standard model motion and can provide reference for the measured force in the actual motion. In the following study, the force acting on the ground can be compared with the force measured by the force table.

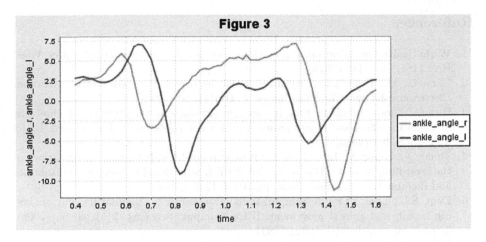

Fig. 5. Curve of ankle angle. The blue line shows the left ankle joint angle. The red line shows the right ankle joint angle. (Color figure online)

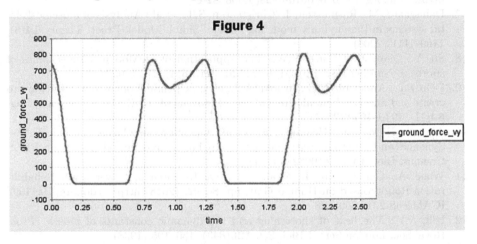

Fig. 6. Ground reaction forces.

5   Conclusion

In this paper, the method to analyze complicated inverse kinematics and forward dynamics of lower limb rehabilitation robot was investigated. The OpenSim software was used to simulate and analyze the human gait data which collected by the NOKOV optical motion capture system. According to the inverse kinematics simulation, the curve of each joint angle changing over time was obtained, and the reaction force acting on the ground was achieved through the forward dynamics simulation.

References

1. World Health Organization: International perspectives on spinal cord injury. Weed Res. **11**(4), 314–316 (2013)
2. Caramia, M.: Brain plasticity after stroke as revealed by ipsilateral activation. Electroencephalogr. Clin. Neurophysiol. **103**(1), 206 (1997)
3. Ren, G., Zhou, J., Sui, T.: Supply and demand status of rehabilitation technicians in China's medical and health industry and countermeasures. Chin. J. Med. Educ. **36**(3), 358–361 (2016)
4. Zhang, X., Kong, X., Liu, G., et al.: Research on the walking gait coordinations of the lower limb rehabilitation robot. In: IEEE International Conference on Robotics and Biomimetics, Tianjin, China, pp. 1233–1237 (2010)
5. Delp, S.L., Loan, J.P.: A computational framework for simulating and analyzing human and animal movement. IEEE Comput. Sci. Eng. **2**(5), 46–55 (2000). https://doi.org/10.1109/5992.877394
6. Delp, S.L., Loan, J.P.: A graphics-based software system to develop and analyze models of musculoskeletal structures. Comput. Biol. Med. **25**, 21–34 (1995). https://doi.org/10.1016/0010-4825(95)98882-e
7. Damsgaard, M., Rasmussen, J., Christensen, S.T., et al.: Analysis of musculoskeletal systems in the anybody modeling system. Simul. Model. Pract. Theory **14**(8), 1100–1111 (2006)
8. Su, Y., Qian, J.G., Song, Y.W.: The application of lifeMod in biomechanics of sports. J. Nanjing Inst. Phys. Educ. (Nat. Sci.) **6**(4), 1–3 (2007)
9. Delp, S.L., Anderson, F.C., Arnold, A.S., et al.: OpenSim: open-source software to create and analyze dynamic simulations of movement. IEEE Trans. Biomed. Eng. **54**(11), 1940–1950 (2007)
10. Ajay, S., Hicks, J.L., Uchida, T.K., et al.: OpenSim: simulating musculoskeletal dynamics and neuromuscular control to study human and animal movement. PLOS Comput. Biol. **14**(7), e1006223 (2018)
11. Wang, A., Cai, F., Yu, J., et al.: Adaptive Research of Lower Limb Rehabilitation Robot Based on Human Gait, pp. 86–92. (2018) https://doi.org/10.1109/ICAMechS.2018.8507128
12. Hill, A.V.: The heat of shortening and the dynamic constants of muscle. Proc. Royal Soc. London Ser. B Biol. Sci. **126**(843), 136–195 (1938)
13. Song, H., Qian, J., Tang, X.: Summary of software OpenSim with focus on its human motion modeling theory and application field. Med. Biomech. **30**(04), 373–379 (2015)
14. Huang, L.: Coupled Simulation of Human Lower Limbs and Rehabilitation Robot Based on OpenSim. Tianjin university (2016)

Path Planning for Messenger UAV in AGCS with Uncertainty Constraints

Hao Zhang[1,2], Bin Xin[1,2(✉)], Yulong Ding[1,2], and Miao Wang[1,2]

[1] School of Automation, Beijing Institute of Technology,
Beijing 100081, People's Republic of China
brucebin@bit.edu.cn
[2] Beijing Advanced Innovation Center for Intelligent Robots and Systems,
Beijing 100081, People's Republic of China

Abstract. This paper mainly solves a path planning problem of messenger UAV in an air-ground collaborative system which is composed of a fixed-wing unmanned aerial vehicle (UAV) and multiple unmanned ground vehicles (UGVs). The UGVs play the role of mobile actuators, while the UAV serves as a messenger to achieve information sharing among the UGVs. The UAV needs to fly over each UGV periodically to collect the information and then transmit the information to the other UGVs. The path planning problem for the messenger UAV can be modeled as a Dynamic Dubins Traveling Salesman Problem with Neighborhood (DDTSPN). The goal of this problem is to find a shortest path which enables the UAV to access all the UGVs periodically. In the paper, we proposes a solution algorithm for the UAV's path planning with uncertainty constraints which means the UAV doesn't know the UGVs' motion parameters. The algorithm is based on the idea of decoupling: firstly the sequence for the UAV to access the UGVs are determined by the genetic algorithm (GA), and then a reasonable prediction mechanism are proposed to determine the access locations of the UAV to the UGVs' communication neighborhoods. Then the theoretical analysis of the effectiveness for the UAV's path planning strategy is emphasized. At last, the effectiveness of the proposed approach is corroborated through computational experiments on several different scale instances.

Keywords: Path planning · Messenger mechanism · Genetic Algorithm (GA) · Dynamic Dubins Traveling Salesman Problem with Neighborhood (DDTSPN)

This work was supported in part by the National Outstanding Youth Talents Support Program 61822304, in part by the National Natural Science Foundation of China under Grant 61673058, in part by the "Thousand Talents Plan" (the State Recruitment Program of Global Experts) (Foreign Experts, Long-term Program) under Grant WQ20141100198, in part by the Foundation for Innovative Research Groups of the National Natural Science Foundation of China under Grant 61621063, in part by the Projects of Major International (Regional) Joint Research Program of NSFC under Grant 61720106011.

© Springer Nature Singapore Pte Ltd. 2020
L. Pan et al. (Eds.): BIC-TA 2019, CCIS 1160, pp. 655–669, 2020.
https://doi.org/10.1007/978-981-15-3415-7_56

1   Introduction

The usage of unmanned aerial and ground vehicle systems (UAGVSs) in civil [5, 7] and military [8] applications has increased rapidly over the past decade, since UAV and UGV have strong complementarity in communication, sensing and load capacity [9]. Chen et al. [3] introduced many applications of the UAGVS, and made a classification for the existing UAGVSs in which different types of UAGVSs can be described in a unified way.

The UAV in the UAGVS needs to periodically fly over the effective communication range of UVGs, collecting and transmitting information to UGVs in order to realize indirect communication between UGVs. The UGVs can make decisions based on the latest information to improve task efficiency. Therefore, this messenger mechanism can effectively broaden the motion range of the UGVs and achieve multiple UGVs' collaboration in a wide range [10]. An example of such a scenario is shown in Fig. 1. However, in reality, the UAV only knows the general trend of UGVs' motion, and it cannot know the location of each UGV in real time. The UGVs' positions may be predicted inaccurately by the UAV, which leads to the UAV and the UGV cannot achieve effective communication, so the UGVs are unable to use the latest messages to carry on the motion planning for the next step. Therefore, it is very important to ensure that the UAV can achieve effective communication with each UGV in every cycle under the condition of uncertainty constraints.

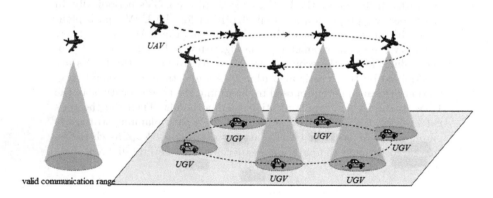

Fig. 1. The UAGVS moves over a wide range.

This paper studies the path planning of the messenger UAV in the UAGVS which is mentioned above, the goal is to find a shortest path which enables the UAV to access all the mobile UGVs robustly. This problem can be seen as a complex exception of the Dubins Traveling Salesman Problem with neighbourhood (DTSPN) [3], we call it the Dynamic Dubins Traveling Salesman Problem with communication neighbourhood (DDTSPN). The difficulties of the DDTSPN in this paper include:

(i) The motion of the UAV is subject to strict curvature constraints which must be considered when we plan the route of the UAV;

(ii) The UAV does not know the specific motion parameters of UGVs, which will make significant impact on the UAV's path planning.

(iii) The UAV's path planning is a complex optimization problem involving mixed variables. There are both discrete variables (e.g., the sequence of all the UGVs for the UAV to visit) and continuous variables (e.g., the access location and the heading of the UAV) and they have complex coupling relationships.

Ding et al. first proposed the DDTSPN problem with mobile UGVs, in which the UAV knows the motion parameters of the UGVs [11]. In this paper, the UAV is not aware of the motion parameters of UGVs, which is the main difference between this paper and Ding's. The innovations of this paper are as follows: 1. For the path planning of the messenger UAV, a robust access strategy is proposed to ensure reliable communication between the UAV and mobile UGVs, and the effectiveness of the access strategy is illustrated by theoretical analysis. 2. According to the robust access strategy, a path planning method based on decoupling is proposed: firstly the GA is used to determine the access sequence, and then the specific access location of the UAV to UGVs' communication neighbourhood is optimized. This method can get better trade-off between solution quality and program's running time.

The paper is organized as follows: Sect. 2 mainly gives the definition and the mathematical model of the problem. Section 3 proposes the GA to determine the sequence for the UAV to access the UGVs, and then introduces a path planning algorithm which can solve the DDTSPN problem effectively. Section 4 analyses the effectiveness of the path planning strategy. Section 5 is the analysis of the algorithm's performance by computational experiments. Section 6 summarizes the paper.

2   Modeling Dynamic Dubins Traveling Salesman Problem with Neighborhood

2.1   The Model of the UGV

We model the UGV as a moving point with a circular communication neighbour-hood and assume that V_{UGV}^{max} and speed direction are known, V_{UGV}^{max} is much less than the speed of UAV. UAV does not know the specific motion trajectory and parameters of UGVs, where V_{UGV}^{max} is UGV's maximum speed. And we denote R as the radius of the UGV's communication neighbourhood, $D(t)$ as the communication neighbourhood of UGV at time t, $G(t)$ as the location of UGV at time t, $\partial D(t)$ as the border of $D(t)$.

2.2   The Model of the UAV

This paper selects the small fixed-wing UAV as the messenger UAV. In order to plan a smooth and feasible path for the UAV, we model the UAV as a Dubins model. The Dubins model is usually used to describe a vehicle with aerodynamic characteristics on a two-dimensional plane which satisfies the following curvature constraints:

$$\begin{cases} \dot{x}_A = v_A\cos(\theta_A) \\ \dot{y}_A = v_A\sin(\theta_A) \\ \dot{\theta}_A = \frac{v_A}{r_A}u_A, u_A \in [-1,1] \end{cases} \tag{1}$$

where $P(t) = (x_A, y_A)$ is the location of the UAV at time t, θ_A is the heading of the UAV. v_A and r_A are the UAV's speed and minimum turning radius respectively, and u_A is the control input. $(x_A, y_A, \theta_A) \in R^2 \times S$ represents the configuration of the Dubins vehicle (UAV), also known as the Dubins configuration. There are six possible shortest paths between any two Dubins configurations, which we call the Dubins path [4,6].

The shortest Dubins distance between the two states depends on both their positions and headings. In most cases, we use the shortest Dubins path under terminal relaxation conditions [2] to describe the path of the messenger UAV. The so-called terminal relaxation means that once the angle of the UAV's starting point is fixed, the angle at the end point only depends on the relative position of the starting point and the ending point. Under the condition of terminal relaxation, the number of variables in the Dubins path can be reduced, which effectively reduces the computational complexity [1].

2.3 Modeling Dynamic Dubins Traveling Salesman Problem with Neighbourhood

The UAV is not aware of the specific motion parameters of the UGVs and it can't get the accurate prediction of UGV's position. So the UAV may not be able to revisit the UGVs successfully during the next cycle. Therefore, in the process of UGV's motion estimation, it is necessary to distinguish two basic cases: with or without robust neighbourhood. Depending on the different situation, the UAV's path can be planned to pass through the robust neighbourhood or robust line of each UGV (the definition of the robust neighbourhood and the robust line are given in Sect. 3). In this way, although the position of the UGVs are uncertain, it can ensure that the UAV can fly to the effective communication range of the UGVs once again.

For the messenger UAV, the travel cost between any pair of UGVs can be considered as the length of the Dubins path between them. So we model the path planning problem of UAV as a Dynamic Dubins Traveling Salesman Problem with communication neighbourhood $(DDTSPN)$, the locations of UGVs are changed dynamically. The goal of the problem is to let the messenger UAV fly to the communication range of each UGV with minimal travel cost. The optimization model of the DDTSPN in the case of terminal relaxation can be expressed as follows:

$$minL(R^{'}, P^{'}, H^{'}, \Delta t) = D((P_0^{'}, h_0^{'}), P_{G_{R(1)}}^{'}) + \sum_{i=1}^{m} D((P_{G_{R(i)}}^{'}, h_{G_{R(i)}}^{'}), P_{G_{R(i+1)}}^{'})$$

$$(2)$$

$$(\dot{x}_A, \dot{y}_A, \dot{\theta}_A) = (v_A cos\theta_A, v_A sin\theta_A, \frac{v_A}{r_A} u_A) \tag{3}$$

$$P_{G_l}^{'} \in N_{G_l}(t_{G_l}), l = 1, 2, ..., m \tag{4}$$

$$N_{G_l}(t_{G_l}) = \begin{cases} N_{G_l}^{robust}(t_{G_l}), & if N_{G_l}^{robust}(t_{G_l}) \neq \phi \\ \overline{P_1 P_2}, & otherwise \end{cases} \tag{5}$$

$$h_{G_{R(j)}}^{'} \in [0, 2\pi), j = 1, 2, ..., m \tag{6}$$

where $D((P_{G_{R(i)}}^{'}, h_{G_{R(i)}}^{'}), P_{G_{R(i+1)}}^{'})$ is the shortest path length of the UAV from Dubins state $(P_{G_{R(i)}}^{'}, h_{G_{R(i)}}^{'})$ to $P_{G_{R(i+1)}}^{'}$ under terminal relaxation. $(P_0^{'}, h_0^{'})$ is UAV's initial Dubins state. $R^{'}$ is the access order of the UAV to all UGVs during this period. $P^{'}$ is the access location of the UAV to the communication neighbourhood of each UGV. $H^{'}$ is the heading of UAV. Δt is the revisit time of the UAV for each UGV, the revisit time means the time from the UAV visit a UGV to the UAV visit it again. m is the number of UGVs, $L(R^{'}, P^{'}, H^{'}, \Delta t)$ is the total length of UAVs path in this period. $N_{G_l}(t_{G_l})$ is the set of candidate access locations of the UAV to the UGV, depending on the revisit time of the UAV to the UGV. The Eq. (4) indicates that both UAV and UGV have their own communication range, the messenger UAV can only transmit information to the UGVs within the communication range of the UGVs. Therefore, the UAV only needs to pass the UGVs communication neighbourhoods without having to go through the precise location of the UGVs. The Eq. (5) reflects the method of how to select access location: if there is a robust neighborhood, the access location is selected from the robust neighbourhood, otherwise the UAV will search the robust line of the UGV.

3   Algorithm for Solving Dynamic Dubins Traveling Salesman Problem with Neighbourhood

After establishing the mathematical model of the DDTSPN, this section mainly introduces the algorithm for solving the problem. The access position of the UAV to the UGV's neighbourhood depends on the access orders, optimizing both of them at the same time is costly. So with the idea of variable decoupling, firstly the algorithm uses GA to determine the access order of the UAV to the UGVs, and then uses the boundary sampling method to determine the location of the access point.

3.1 Genetic Algorithm to Determine the Access Order

In this section, we use the GA to determine the access order of the UAV to the UGVs. The access order is coded as a permutation and the classical genetic algorithm can solve this permutation optimization problem very well. In our problem, the genetic algorithm can be divided into the following steps:

Population Initialization and Fitness Value Calculation: Each individual in a population is an integer permutation, according to the order of the UGVs in each individual, the distance represented by this individual can be calculated. The distance is greater, the fitness value is smaller. Different from the TSP problem, we use the Dubins distance between two UGVs to generate the distance matrix instead of the euclidean distance.

Selection: We use roulette wheel selection to select better individuals. The basic idea is that the probability of each individual being selected is proportional to their fitness. The specific operations are as follow steps:

Step1: The fitness $f(x_i)(i = 1, 2, ..., M)$ of each individual in the group was calculated, and M was the population size.

Step2: Calculate the probability $f(x_i)/\sum_{j=1}^{M} f(x_j)$ of each individual being passed on to the next generation.

Step3: Calculate the cumulative probability $q_i = \sum_{j=1}^{i} P(x_j)$ of each individual, where q_i is called the cumulative probability of chromosome $x_i(i = 1, 2, ..., M)$.

Step4: Generate a uniformly distributed pseudo-random number r in the interval $[0, 1]$, if $q[k - 1] < r \leq q[k]$, we choose individual $k(k = 1, 2, ..., M)$.

Step5: Repeat the step 4 M times.

Crossover: Two individuals are randomly selected and a random gene segment is exchanged to complete the crossover operation. Conflicts may occur after crossover (visiting the same UGV twice). The exchanged gene segments (hereinafter referred to as swap segments) are kept unchanged, the positions of the conflict genes in the swap segments are obtained, and the conflicts genes outside the swap segments are replaced with the corresponding genes in another chromosome. We can give a concrete example to illustrate this process just like Fig. 2: The gene segments $(6, 3, 8, 7)$ and $(3, 7, 4, 2)$ are exchanged. We keep this segment unchangeable, for the chromosome A, the first conflict gene was (8), get the position of (8) in the swap segment which is six. The conflict gene outside the swap segment is replaced by the gene (4) at the corresponding position in the chromosome B. Executing last steps multiple times until there is no conflict.

Mutation: Randomly swapping the positions of two genes in a chromosome. Repeat the selection, crossover, mutation operations until the specified number of iterations are reached.

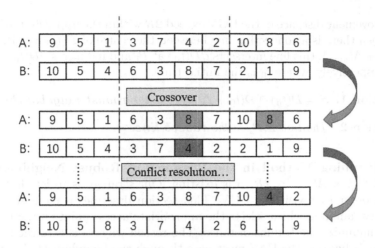

Fig. 2. Crossover operation.

3.2 The Algorithm for the UAV to Access Communication Neighbourhood of the UGV

After determining the access sequence of the UAV to the UGVs, it is necessary to select an access point which can minimize the Dubins path within the communication neighbourhood of the moveable UGVs. The positions of the UGVs are uncertain for the UAV, so we need to predict the motion range of the UGV reasonably according to the maximum speed of the UGV and the upper bound of UAV's re-visit time. The motion modes of the UGV can be divided into two situations: one is that the "robust neighbourhood" is existing and the other is that the "robust neighbourhood" is not existing. According to these two motion modes, different path planning strategies are established. First we define the *robust neighbourhood* and the *robust line*.

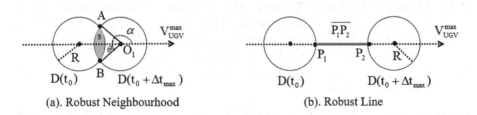

(a). Robust Neighbourhood   (b). Robust Line

Fig. 3. Robust neighbourhood and robust line.

As shown in Fig. 3(a), we suppose that UGV moves at a speed of V_{UGV}^{max} in the agreed direction. If the movement distance of the UGV does not exceed $2R$ within the upper limit of re-visit time, then there is an overlapping region between $D(t_0)$ and $D(t_0 + \Delta t_{max})$ which is denoted as S. As shown in Fig. 3(b), if the movement distance of the UGV exceed $2R$ within the upper limit of re-visit time, then there is a connecting line between the centre of $D(t_0)$ and the centre of $D(t_0 + \Delta t_{max})$, this line intersects $\partial D(t_0)$ at P_1 and intersects $\partial D(t_0 + \Delta t_{max})$ at P_2 respectively, where t_0 is the initial time.

Definition 1. $S = D(t_0) \cap D(t_0 + \Delta t_{max})$ *is the* **robust neighbourhood.**

Definition 2. *The line* $\overline{P_1 P_2}$ *is the* **robust line.**

Path Planning Method in the Presence of Robust Neighbourhood: When $V_{UGV}^{max} \times \Delta t_{max} \leq 2R$, S is existing. Our strategy is to let the UAV fly to S in order to revisit the UGV. This section describes a method based on boundary sampling which selects the appropriate access point on the boundary of S to minimize the Dubins distance. The reason why we choose the boundary sampling is because the UAV must pass through the boundary of S if it passes through S. As shown in Fig. 3(a), we denote $O_1(x_A, y_A)$ as the centre coordinate of $\partial D(t_0 + \Delta t_{max})$, A and B are the intersections of $\partial D(t_0)$ and $\partial D(\Delta t_{max})$. We take the sampling of the left boundary of S as an example, α is the angel between vector $\overline{O_1 A}$ and the speed direction of the UAV:

Step1: For any sample point $n(n = 1, 2, ..., N_{sample})$, where N_{sample} is the number of sample point. The coordinate of the n can be expressed as $P(n) = (x_1 + R\cos(\alpha + \varphi), y_1 + R\sin(\alpha + \varphi)), \varphi \in [0, \theta]$. It can be concluded that the length of Dubins path under terminal relaxation conditions is $L_n(P(n)) = D((P^A, h), P(n)) + D((P(n), h_n), P_{NEXT}^G)$, where P^A is current location of the UAV, h is the current heading of the UAV, h_n is the heading of the UAV at $P(n)$, P_{NEXT}^G is the location of the next UGV which will be accessed by the UAV in the following time.

Step2: We select a point $\widetilde{P_s}$ which can minimize L_n as the final access location of the UAV in all the boundary sampling points: $\widetilde{P_s} = arg\ min\ L_n(P(n))$.

Path Planning Method Without the Robust Neighbourhood: When $V_{UGV}^{max} \times \Delta t_{max} > 2R$, S is not existing. Then our strategy is to let the UAV search along the $\overline{P_1 P_2}$ to achieve a revisiting for the UGV. As shown in Fig. 4, Δt_{P_1} is the time required for the UAV flying from its current position to P_1, Δt_1 is the time required for the UAV flying from P_1 to P_2. We need to introduce the strategy in two situations:

(i) When $\Delta t_{P_1} + \Delta t_1 > \Delta t_{max}$, the UAV can only fly along the $\overline{P_1 P_2}$ in the direction of $P_2 \to P_1$.

(ii) When $\Delta t_{P_1} + \Delta t_1 \leq \Delta t_{max}$, if $[D((P_1, h_1), P_2) + D((P_2, h_2), P_{NEXT}^G)] \leq [D((P_2, h_2), P_1) + D((P_1, h_1), P_{NEXT}^G)]$, then the UAV flys along the $\overline{P_1 P_2}$ in the direction of $P_1 \to P_2$ otherwise in the direction of $P_2 \to P_1$, where P_1 and h_1 are the coordinate and heading respectively when the UAV at P_1. The P_2 and h_2 are the coordinate and heading respectively when the UAV at P_2.

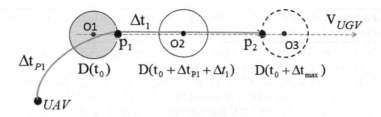

Fig. 4. The UAV fly along the $\overline{P_1P_2}$ in the direction of $P_1 \to P_2$.

Using the method mentioned above of accessing a single UGV, the messenger UAV can access each UGV once in a loop and communicate with each UGV successfully. In order to continuously collect and obtain the latest information of the UGVs, the UAV needs to access each UGV periodically, which can be achieved by solving the DDTSPN repeatedly, it means we should plan multiple Dubins loops. The process of calculating multiple Dubins loops and calculating a Dubins loop multiple times is very similar. The main difference is that the access sequence is reinitialized before each Dubins loop. We use the pseudocode in Algorithm 1 to show the entire calculation process.

4   Analysis of the Effectiveness of the Robust Access Strategy of the UAV

In Sect. 3, when we determine the access location of the UAV to UGVs, we have described our robust access algorithm in detail, which can be summarized as: when $\Delta t_{max} \times V_{UGV}^{max} \leq 2R$, the UAV will fly to the robust neighbourhood S in order to revisit the UGV; When $\Delta t_{max} \times V_{UGV}^{max} > 2R$, the UAV will search along the robust line $\overline{P_1P_2}$ in order to revisit the UGV. This section analyses the effectiveness of this strategy.

4.1   The Effectiveness Analysis of Access Strategy When the Robust Neighbourhood Is Existing

The following theorem proves that when $\Delta t_{max} \times V_{UGV}^{max} \leq 2R$, the UGV must can be revisited when the UAV flies over the S.

Theorem 1. *When $\Delta t_{max} \times V_{UGV}^{max} \leq 2R$, if $P(t_0 + \Delta t_{real}) \in S$, then $P(t_0 + \Delta t_{real}) \in D(t_0 + \Delta t_{real})$.*

Proof. As shown in Fig. 5, denoting Δt_{real} is the actual re-visit time, $\Delta t_{real} \leq \Delta t_{max}$. When $\Delta t_{max} \times V_{UGV}^{max} \leq 2R$, that means S is existing. Because $\forall t \in [t_0, t_0 + \Delta t_{max}], S \in D(t)$, and $\Delta t_{real} \leq \Delta t_{max}$. $S \in D(t_0 + \Delta t_{real})$, when the UAV flies to $S(P(t_0 + \Delta t_{real}) \in S))$, $P(t_0 + \Delta t_{real}) \in D(t_0 + \Delta t_{real})$.

It can be obtained from the above proof that when the S exists, the strategy of sampling the boundary of the S is reasonable, and it can be guaranteed that the UAV can revisit the UGV.

Algorithm 1: Algorithm for Dubins path planning when the UAV accesses moving UGVs

---

1 **initialization:** The initial position of the UAV is $P_{initial}^A$. The actual location of the UAV is P^A. The position of UGV i is $P_i^G, i = 1, 2, ..., N_G$. The number of Dubins loops is N_{max}. $X^k = [x_1^k, x_2^k, ..., x_{N_G}^k], (k = 1, 2, ..., N_{max})$ indicate whether the UGVs have been accessed by the UAV in the kth Dubins loop, i.e. $x_i^k = 0$ means the ith UGV has not been accessed by the UAV in the kth loop.

2 **for** $k = 1 : N_{max}$ **do**

3     $X^k = 0_{1 \times N_G}$

4   **for** $n = 1 : N_G$ **do**

5   According X^k to determine the set Q_n^k of UGVs which are not accessed in the current loop.

6   Use the GA introduced in 3.1 to determine the access sequence of the UGVs in the set Q_n^k. Note that the next UGV which will be accessed by the UAV $s_i, s_i \in Q_n^k$.

7   Let $x_{s_i}^k = 1$.

8   **if** $\Delta t_{max} \times V_{UGV}^{max} \leq 2R$ **then**

9   The path of the UAV is calculated according to the path planning algorithm in 3.2.1 (the robust neighbourhood is existing).;

10   **else**

11   The path of the UAV is calculated according to the path planning algorithm in 3.2.2 (the robust neighbourhood is not existing).;

12   **end**

13   Update P^A and each $P_i^G (i = 1, 2, ..., N_G)$

14   **end**

15 **end**

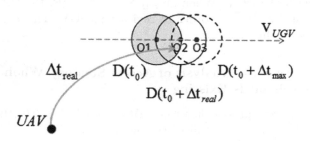

Fig. 5. The robust neighbourhood is existing.

4.2 The Effectiveness Analysis of Access Strategy When the Robust Neighbourhood Is Not Existing

The following theorem proves that when $\Delta t_{max} \times V_{UGV}^{max} > 2R$, the UGV must can be revisited when the UAV search along the robust line $\overline{P_1 P_2}$.

Theorem 2. *When $\Delta t_{max} \times V_{UGV}^{max} > 2R$, if $\Delta t_{p_1} + \Delta t_1 > \Delta t_{max}$, the UAV searches along the robust line $\overline{P_1 P_2}$ in the direction of $P_2 \to P_1$; if $\Delta t_{max} \times V_{UGV}^{max} \leq 2R$, the UAV searches along the $\overline{P_1 P_2}$ in the direction of $P_1 \to P_2$ or $P_2 \to P_1$, then there must have a t_{meet} which meets $P(t_{meet}) \in D(t_{meet})$.*

Proof. We prove the theorem in two cases.

Case1: When $\Delta t_{p_1} + \Delta t_1 \leq \Delta t_{max}$ or $\Delta t_{p_1} + \Delta t_1 > \Delta t_{max}$, the UAV searches along the $\overline{P_1 P_2}$ in the direction of $P_2 \to P_1$.

As shown in Fig. 6, Δt_{P_2} is the time when the UAV flies from $P(t_0)$ to P_2, t_{meet} is the actual encountering moment of $\partial D(t)$ and $P(t)$; Because $\Delta t_{P_2} < \Delta t_{max}$, then $\overline{P_1 P_2} \cap D(t) \neq \varnothing, \forall t \in [t_0, t_0 + \Delta t_{P_2}]$. When $t \geq t_0 + \Delta t_{p_2}$, the UAV and the UGV do the opposite movements on the $\overline{P_1 P_2}$. So $\exists t_{meet} \in [t_0 + \Delta t_{P_2}, t_0 + \Delta t_{max}]$, $P(t_{meet}) \in D(t_{meet})$.

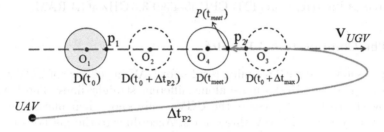

Fig. 6. The UAV search along the $\overline{P_1 P_2}$ in the direction of $P_2 \to P_1$.

Case2: When $\Delta t_{p_1} + \Delta t_1 \leq \Delta t_{max}$, the UAV search along the $\overline{P_1 P_2}$ in the direction of $P_1 \to P_2$.

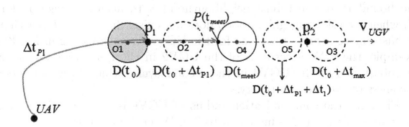

Fig. 7. The UAV search along the $\overline{P_1 P_2}$ in the direction of $P_1 \to P_2$.

As shown in Fig. 7, Because $\Delta t_{p_1} + \Delta t_1 \leq \Delta t_{max}$, the UAV chases the UGV on the $\overline{P_1 P_2}$. $P(t_0 + \Delta t_{p_1}) = P_1$, $\partial D(t_0 + \Delta t_{p_1}) \cap \overline{P_1 P_2} \neq \varnothing$, and $V_{UAV} > V_{UGV}^{max}$. So $\exists t_{meet} \in [t_0, t_0 + \Delta t_{p_1} + \Delta t_1]$, $P(t_{meet}) \in D(t_{meet})$.

It can be obtained from the above proof that when the S does not exist, the strategy of the UAV searching along the robust line $\overline{P_1 P_2}$ is reasonable, and it can be guaranteed that the UAV can revisit the UGV.

5   Computational Experiments

In this section, we verify the effectiveness of the algorithm by changing the size of the cases and setting the different communication radius for UGVs. The experiment can be divided into two parts. In the first part, we visually show the results of the algorithm for solving the DDTSPN in small-scale cases and analyse the results. In the second part, we fully verify the effectiveness of the path planning algorithm under the condition of changing the UGVs communication radius and the scale of the cases.

All the computational results are obtained by the following parameters and the method mentioned in this paper. The UAV has a constant speed of 15(m/s) and UAV. The UAV does not know the real-time speed of UGV, only know its maximum speed. The maximum speed of the UGVs don't exceed 1(m/s). All the computation experiments are done in the matlab environment, the PC is configured as Inter(R) Core(TM) CPU i5-4590 3.3 GHz 4GB RAM.

5.1   The Computational Results

The Fig. 8 shows the path planning results for UAV in the case of UGV's motion constraint. The four UGVs move along different straight lines. The UAV does not know the real-time speeds of the UGVs, only know their maximum speeds. The UAV accesses each UGV three times. According to the methods described in this paper, the messenger UAV can dynamically plan its path periodically to access each mobile UGV. In addition, the UAV can dynamically adjust the access sequence based on the UGVs' locations. For example, in the first cycle, the access sequence of the UAV to the UGVs is 4-3-2-1, and in the second cycle it becomes 4-1-2-3.

As shown in Fig. 8, when the robust neighbourhood exists, the UAV will fly to the boundary of the robust neighbourhood S to achieve revisiting for the UGV; when the robust neighbourhood does not exist, the UAV will search along the robust line to achieve revisiting for the UGV. We take UGV2 in Fig. 8(a) as an example: the UAV accesses S of the UGV2 in the first cycle, and searches along robust line of the UGV2 in the second cycle, which fully validates the effectiveness of the access strategy.

In Fig. 8(a), the communication radius of UGVs is 5 m, and the communication radius in Fig. 8(b) is increased to 8m. When R = 5 m, the total length of the UAV's path is 825.6 m. The UGV3 does not have robust neighbourhoods in the second and third cycles so the UAV needs to access its robust line; when R = 8 m, the total length of the UAV's path is 724.4 m. The UGV2 has robust neighbourhoods in the first and third cycles, the UAV needs to access its robust neighbourhoods. It's obvious that when the communication radius increases from 5 m to 8 m, the number of the UGVs with robust neighbourhoods increases. And from the computation results, the increase of robust neighbourhoods will reduce the total length of the UAV's path. The Fig. 9 shows the execution results of the program with five UGVs.

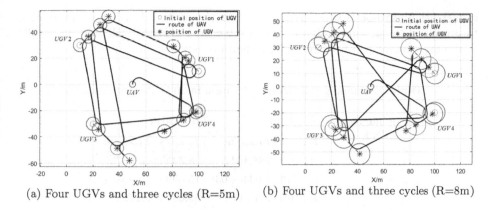

(a) Four UGVs and three cycles (R=5m)   (b) Four UGVs and three cycles (R=8m)

Fig. 8. The results in the condition of four UGVs and three cycles

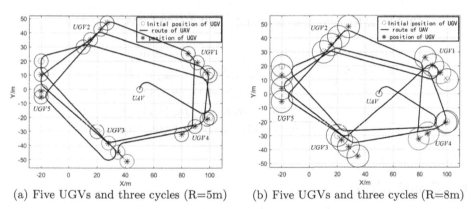

(a) Five UGVs and three cycles (R=5m)   (b) Five UGVs and three cycles (R=8m)

Fig. 9. The results in the condition of five UGVs and three cycles

5.2   Analysis of the Algorithm's Effectiveness

In order to fully verify the effectiveness of the proposed path planning algorithm, we expand the scale of the cases and change the UGVs' communication radius, then we analyse the final running time of the program and the total length of the UAV's route. We only show the experimental results of 20 UGVs and 100 UGVs cases here, and the communication radius R of the UGVs are 10 m or 20 m. The T_{total} is the programs running time, and L_{total} is the total length of the UAV's route.

Table 1 shows the results of the program in the case of 20 UGVs in the medium-scale cases, Table 2 shows the results of the program in the case of 100 UGVs in the large-scale cases. It can be seen that as the number of Dubins loops increases, the UAV's route and the programs running time increase significantly. When the communication radius of UGVs increases from 10 m to 20m, the running time of the program decreases slightly, but the total length of the route decreases a lot. It shows that the communication radius is larger, the more robust neighbourhoods will exist, and the total length of UAVs path is smaller. In summary, the proposed algorithm can solve the DDTSPN well, and the results of the experiment can fully reflect the characteristics of the algorithm.

Table 1. Running time and total length in 20 UGVs' situations.

The number of UAV's loops	$R = 10\,\text{m}$		$R = 20\,\text{m}$	
	T_{total}/s	L_{total}/m	T_{total}/s	L_{total}/m
1	8.4040	788.4042	8.3355	766.8735
2	16.2713	1.634e+03	15.3996	1.4991e+03
3	25.1010	2.4356e+03	26.0803	2.2723e+03

Table 2. Running time and total length in 100 UGVs' situations.

The number of UAV's loops	$R = 10\,\text{m}$		$R = 20\,\text{m}$	
	T_{total}/s	L_{total}/m	T_{total}/s	L_{total}/m
1	69.2915	1.0722e+04	68.0732	8.8250e+03
2	141.0526	2.4974e+04	138.9063	2.0571e+04
3	210.0057	4.5673e+04	207.9986	3.4865e+04

6 Conclusion

This paper mainly solves the path planning of the messenger UAV when the UAV cannot know the motion trajectory of the mobile UGVs. The path planning problem of the messenger UAV can be modeled as a DDTSPN subjecting to the constraints of revisiting time. This problem is a hybrid variables optimization problem, so a path planning method based on decoupling is proposed: firstly, the GA are used to determine the access sequence, and then according to whether there is a robust neighbourhood, different path planning strategies are designed and we have a theoretical analysis about the effectiveness of the strategies. In the experimental part, the effectiveness of the algorithm is verified by changing the scale of the cases and the communication radius of UGVs. The experimental results show that the algorithm can enable UAV to access the communication neighbourhood of each mobile UGV quickly and deterministically, so the UAV can collect and transmit information periodically. The research direction in the future of this work is to improve the sequencing algorithm and expand the moving range of the UGVs.

References

1. Xin, B., Chen, J., Xu, D.L., Chen, Y.W.: Hybrid encoding based differential evolution algorithms for Dubins traveling salesman problem with neighborhood. Control Theory Appl. **31**(7), 941–954 (2014)
2. Boissonnat, J.D., Bui, X.N.: Accessibility Region for a Car That Only Moves Forwards Along Optimal Paths. INRIA France (1994)
3. Chen, J., Zhang, X., Xin, B., Fang, H.: Coordination between unmanned aerial and ground vehicles: a taxonomy and optimization perspective. IEEE Trans. Cybcrn. **46**(4), 959–972 (2015)
4. Cohen, I., Epstein, C., Isaiah, P., Kuzi, S., Shima, T.: Discretization-based and look-ahead algorithms for the dubins traveling salesperson problem. IEEE Trans. Autom. Sci. Eng. **14**(1), 383–390 (2016)
5. Garone, E., Determe, J.F., Naldi, R.: Generalized traveling salesman problem for carrier-vehicle systems. J. Guid. Control Dyn. **37**(3), 766–774 (2014)
6. Isaiah, P., Shima, T.: Motion planning algorithms for the dubins travelling salesperson problem. Automatica **53**, 247–255 (2015)
7. Ji, X., Niu, Y., Shen, L.: Robust satisficing decision making for unmanned aerial vehicle complex missions under severe uncertainty. PloS one **11**(11), e0166448 (2016)
8. Mathew, N., Smith, S.L., Waslander, S.L.: Planning paths for package delivery in heterogeneous multirobot teams. IEEE Trans. Autom. Sci. Eng. **12**(4), 1298–1308 (2015)
9. Saska, M., Vonásek, V., Krajník, T., Přeučil, L.: Coordination and navigation of heterogeneous mav-ugv formations localized by a hawk-eye-like approach under a model predictive control scheme. Int. J. Robot. Res. **33**(10), 1393–1412 (2014)
10. Xin, B., Zhu, Y.G., Ding, Y.L., Gao, G.Q.: Coordinated motion planning of multiple robots in multi-point dynamic aggregation task. In: 2016 12th IEEE International Conference on Control and Automation (ICCA), pp. 933–938. IEEE (2016)
11. Yulong, D., et al.: Path planning of messenger uav in air-ground coordination. IFAC-PapersOnLine **50**(1), 8045–8051 (2017)

RGB-T Saliency Detection via Robust Graph Learning and Collaborative Manifold Ranking

Dengdi Sun[1], Sheng Li[1], Zhuanlian Ding[2(✉)], and Bin Luo[1]

[1] Key Lab of Intelligent Computing and Signal Processing of Ministry of Education, School of Computer Science and Technology, Anhui University, Hefei 230601, China
[2] School of Internet, Anhui University, Hefei 230039, China
dingzhuanlian@163.com

Abstract. Visual saliency detection inspired by brain cognitive mechanisms is an important component of computer vision, aiming at automatically highlight salient visual objects from the image background. In complex real scenarios, integrating multiple different yet complementary feature representations has been proved to be an effective way for boosting saliency detection performance. In this paper, we propose a novel collaborative algorithm for saliency detection by adaptively incorporating information from grayscale and thermal images. Specifically, we first construct an affinity graph model via robust graph learning to characterize the feature similarities and location relationships between two different image patches under each modal. Based on these affinity graphs, a collaborative manifold ranking method cross-modality consistent constraints is designed to effectively and efficiently infer the salient score for each patch. Moreover, we introduce a weight for each modality to describe the reliability, and integrate them into the graph-based manifold ranking scheme to achieve an adaptive weighted fusion of different source data. For optimization, we propose some iterative algorithms to efficiently solve the models with several subproblems. Experiments on the grayscale-thermal datasets demonstrate the superior performance of the new method over the state-of-the-art approaches.

Keywords: Saliency decetion · Multi-modality fusion · Graph learning

1 Introduction

Saliency detection is a key attentional mechanism in cognitive neural systems that facilitates learning and survival by enabling organisms to focus their limited perceptual and cognitive resources on the most pertinent subset of the available sensory data. By modeling the mechanism of human attention, saliency detection, aiming at automatically highlight prominent foreground objects from background images, has become a fundamental problem in computer vision and plays a critical role in numerous vision applications. Although much progress has been

© Springer Nature Singapore Pte Ltd. 2020
L. Pan et al. (Eds.): BIC-TA 2019, CCIS 1160, pp. 670–684, 2020.
https://doi.org/10.1007/978-981-15-3415-7_57

made in recent years, it is still a challenging problem in complex scenarios and environments, like low lighting conditions and inclement weather. Fortunately, combining a variety of different but complementary cue information, such as visible and thermal infrared data (RGB-T), may be a feasible and effective way to promote the saliency detection performance in the above scenarios and challenges.

A thermal sensor is a passive sensor that captures infrared radiation from all objects at temperatures above absolute zero, so the imaging procedure is insensitive to light conditions. With the decreasing gradually cost of thermal sensors in recent years, it has been successfully applied to various civilian fields [1,2]. In addition, visible spectrum sensors can more effectively separate objects from the background when they have similar temperatures [also known as thermal cross (TC)]. Thus, RGB and thermal data can complement each other, and thus the effective fusion of these two modalities is more important to detect salient objects robustly in challenging situations. However, most of existing approaches employ some cues, such as contour or texture, to superimpose directly RGB and thermal data for detecting saliency objects. These methods ignore the complementary benefit of multi-source when the RGB-thermal information are crossover.

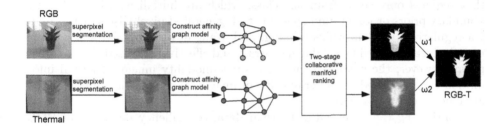

Fig. 1. The algorithm framework of this paper.

To handle aforementioned issues, in this paper we propose a multi-modal saliency detection method based on the collaborative graph model (see Fig. 1 for the algorithm flow chart). Specifically, we first segment the RGB and thermal images into a set of non-overlapping patches respectively, and extract the a d-dimensional low-level appearance features for each patch. Then, an affinity graph model via is constructed via robust graph learning under each modal to learning the intrinsic structure of images. In order to characterize the feature similarities and location relationships between patches, we design a novel centerpoint feature optimization scheme to assist the affinity graph model. Finally, a novel collaborative manifold ranking for RGB-T saliency detection on the affinity graph model is proposed. Moreover, we introduce a weight for each modality to represent the reliability, and incorporate the cross-modality consistent constraints to achieve adaptive and collaborative fusion of different source data. The modal weights and salient score are jointly optimized by iteratively solving several subproblems with closed-form solutions. Extensive experiments on the common RGB-T database, in comparison with several state-of-the-art methods, are performed to assess the performance of our method.

2   Related Work

Visual saliency computation can be categorized into bottom-up and top-down methods or a hybrid of the two. Bottom-up models are primarily based on a center-surround scheme, computing a master saliency map by a linear or non-linear combination of low-level visual attributes such as color, intensity, texture and orientation [5–8]. Top-down methods generally require the incorporation of high-level knowledge, such as face detector in the computation process [3,4,9,10].

Most of the existing bottom-up methods measure foreground saliency based on the contrast of pixels or regions in the local context or the whole image, In contrast, Gopalakrishnan et al. formulated the object detection problem as a binary segmentation or labelling task on a graph. The most salient seeds and several background seeds were identified by the behavior of random walks on a complete graph and a k-regular graph. Then, a semi-supervised learning technique was used to infer the binary labels of the unlabelled nodes. Different from it, Yang et al. employed manifold ranking technique to salient object detection that requires only seeds from one class, which are initialized with either the boundary priors or foreground cues. Li et al. generated pixel-wise saliency maps via regularized random walks ranking, and Wang et al. proposed a new graph model which captured local/global contrast and effectively utilized the boundary prior. However, these works focus on a single-modality images (thermal information can be viewed as the gray value of image), and thus will suffer from the aforementioned challenges.

Considering the potential of thermal data, the saliency detection of RGB-T has attracted wide attention in recent years. Li et al. proposed a multi-task manifold ranking algorithm [3] for RGB-T saliency detection, and established a unified RGB-T saliency detection benchmark data set. Davis and Sharma provided a framework for effectively combining information from thermal and visible videos. They first identified the initial regions-of-interest in the thermal domain, and then propagated it into the grayscale domain. The contour saliency map was obtained by combining both thermal and grayscale information on the detected regions-of-interest, and further flood-filled to produce silhouettes. Han and Bhanu proposed a hierarchical scheme to automatically align synchronous grayscale and thermal frames, and probabilistically combined cues from registered grayscale-thermal frames for improving human silhouette detection. Davis and Sharma proposed a new background-subtraction technique fusing contours from thermal and grayscale videos for persistent object detection in urban settings. Using the region saliency detection method in, infrared and visible images were integrated with different strategies applied for salient and nonsalient regions. The background subtraction method is then employed using GMM.

Although the saliency detection of RGB-T has made some progress in recent years, it still faces the following challenging problems. (i) there may be objects of different sizes in the image, or a large object without internal appearance consistency; (ii) in [3], it is proved that the boundary of RGB-T image has some noise caused by registration, and some objects may be near the boundary of images are also common in RGB-T saliency detection. To handle aforementioned issues, On the basis of the low-order sparse representation model and the idea of graph-based manifold ranking [4], a dynamic learning method based on superpixel is proposed to construct the graph model. It can effectively detect important targets and reduce the influence of background. Moreover, we employ multimodal information for information supplement to solve saliency detection in complex scenes.

3   Affinity Graph Models

In this section, we mainly introduce the construction of affinity graph models and the corresponding optimization algorithm.

3.1   Problem Formulation

For the grayscale and thermal images in the same scene, we utilize the superpixel segmentation algorithm SLIC [11] for the superpixel segmentation. For the i-th superpixel, color histogram and gradient features (dimension d) are extracted respectively, denoted as X_i. The superpixel features produced by an image form an eigenmatrix in order, $R^{(0)} = Q^{(0)} = A^{(0)} = Z_2^{(0)} = 0 \in R^{n \times n}$. Assuming that all superpixels of an image are located in the same low-rank subspace, features of any superpixel can be expressed by linear combination of features of other superpixels, and the expression coefficient is low-rank. Taking superpixels as graph nodes, we can automatically determine the structure of graphs and optimize their similarity relations according to the internal relations between superpixels.

Therefore, the non-negative low-rank and sparse representation can be jointly written as: $X = XZ, Z >= 0$, where Z is the low-rank and sparse coefficient matrix. Because of the low rank and sparse constraints, the constructed dynamic graph can fully explore the internal relations between superpixels. According to the traditional low-rank and sparse representation model, the following objective functions can be solved in this model:

$$min\|E\|_F^2 + \lambda_1\|R\|_F^2 + \lambda_2 tr(RL_AR^T) + \lambda_3\|A\|_F^2$$
$$s.t.X = XR + E; \forall A_i^T 1 = 1; A_i \succeq 0, \tag{1}$$

Where $R \in R^{n \times n}$ and $E \in R^{d \times n}$ represent the sparse representation coefficient and noise matrix related to samples. $A \in A^{n \times n}$, A_{ij} reflects the correlation degree between image region x_i and superpixel x_j. $\| \|_F$ is the F-norm of matrix. $A_i^T = 1$ and $A >= 0$ guarantee the probability properties of A_i. Delta $\|A\|_F^2$ is added to avoid the formation of trivial solutions, $\lambda_1, \lambda_2, \lambda_3$ is the tradeoff parameter. L_A is the Laplace matrix of A, and Eq. (2) is the calculation process of L_A.

$$L_A = D_A - A. \tag{2}$$

The element calculation process of degree matrix D_A is shown in Eqs. (3), and (4) is its matrix form.

$$d_i = \sum_{j=1}^{n} A_{ij}, \tag{3}$$

$$D_A = \begin{bmatrix} d_1 & \cdots & \cdots & \cdots \\ \cdots & d_2 & \cdots & \cdots \\ \cdots & \cdots & \cdots & \cdots \\ \cdots & \cdots & \cdots & d_n \end{bmatrix} \tag{4}$$

3.2   Model Optimization

In Eq. (1), prior knowledge and constraint conditions of superpixels are unified into the objective optimization framework by means of mathematical modeling. In the process of calculation, we usually use the ADM [12] (Alternating direction method) to evaluate, through the corresponding transformation, namely fixed one of A or R, this problem can be viewed as A multi-objective optimization problem and can avoid A lot of matrix inverse operation, on the basis of ADM [12] use ALM, by introducing an auxiliary variable is $Q \in R^{n \times n}$, the separation of variables, the formula can be turned one of the following into the objective function (augmented Lagrangian function):

$$L(\forall i A_i^T 1 = 1; A_i \geq 0) = \|E\|_F^2 + \lambda_1 \|R\|_F^2 + \lambda_2 tr(Q L_A Q^T) + \lambda_3 \|A\|_F^2 \\ + \Phi(Z_1, X - XR - E) + \Phi(Z_2, Q - R), \tag{5}$$

$$\Phi(Z, C) = \frac{\mu}{2} \|C\|_F^2 + \langle Z, C \rangle, \tag{6}$$

Equation (6) is the definition, μ is a penalty parameter, and Z_1 and Z_2 are Lagrange multipliers. In addition, optimization is carried out by minimizing L and updating one of the other four variables, E, R, Q and A, while fixing three of them. Through some simple algebraic derivation, the $(t + 1)$-th iteration can be written as:

$$E^{(t+1)} = \arg\min_{E} \|E\|_F^2 + \Phi(Z_1^{(t)}, X - XR^{(t)} - E)$$

$$= \frac{Z_1^{(t)} + \mu^{(t)}(X - XR^{(t)})}{2 + \mu^{(t)}}, \tag{7}$$

$$R^{(t+1)} = \arg\min_{R} \lambda_1 \|R\|_F^2 + \Phi(Z_1^{(t)} + X - XR - E^{(t+1)}) + \Phi(Z_2^{(t)}, Q^{(t)} - R)$$

$$= (\frac{2\lambda_1 + \mu^{(t)}}{\mu^{(t)}} I + X^T X)^{-1} T, \tag{8}$$

$$Q^{(t+1)} = \arg\min_{Q} \lambda_2 tr(QL_A^{(t)}Q^T) + \Phi(Z_2^{(t)}, Q^{(t)} - R^{(t+1)})$$
$$= (\mu^{(t)}R^{(t+1)} - Z_2^{(t)})(2\lambda_2 L_A^{(t)} + \mu^{(t)}I)^{-1}, \tag{9}$$

$$\arg\min_{A} \lambda_2 tr(Q^{(t+1)}L_A Q^{(t+1)T}) + \lambda_3\|A\|_F^2$$
$$s.t. \forall i A_i^T 1 = 1, Ai \succeq 0. \tag{10}$$

$$\forall i A_i^{t+1} = \arg\min_{A_i \in \{a|a^T|=1, a\geq 0\}} \|A_i + d_i^{Q(t+1)}\|_F^2, \tag{11}$$

$$A_i^{(t+1)} = (\frac{1 + \sum_{j=1}^{k} d_{ij}^{\sim Q(t+1)}}{k} 1 - d_i^{Q(t+1)})_+. \tag{12}$$

In Eq. (7), we can solve for E by taking the partial derivative of L with respect to E and then taking the extreme value. In Eq. (8):

$$T = X^T(X - E^{(t+1)} + \frac{Z_1^{(t)}}{\mu^{(t)}}) + Q^{(t)} + \frac{Z_2^{(t)}}{\mu^{(t)}}.$$

To solve sub-problem A, the problem needs to be transformed into an optimization problem (Eq. (10)), and then the optimization problem is decomposed into an independent small problem (Eq. (11)). In Eq. (11), $d_i^{Q(t+1)}$ is the j-th element of A vector, which can be expressed as:

$$d_{ij}^{Q(t+1)} = \frac{\lambda_2\|Q_i^{(t+1)} - Q_j^{(t+1)}\|_F^2}{4\lambda_3}.$$

For each A_i, see Eq. (12) for its closed solution. The operator $(\mu)_+$ converts negative elements of μ to 0, leaving the rest unchanged. $d_i^{Q(t+1)}$ is the element of $d_i^{Q(t+1)}$ that keeps increasing. In addition to the above four variables, there are two Lagrange multipliers that need to be updated. See Eq. (13).

$$Z_1^{(t+1)} = Z_1^{(t)} + \mu^{(t)}(X - XR^{(t+1)} - E^{(t+1)})$$
$$Z_2^{(t+1)} = Z_2^{(t)} + \mu^{(t)}(Q^{(t+1)} - R^{(t+1)}). \tag{13}$$

It is worth noting that, since each subproblem we split is convex, the local optimal solution of each subproblem can be generated by iterative solution, and Algorithm 1 gives the whole optimization process.

Algorithm 1. The model solves the optimization process

Input: Data matrix $X \in R^{d \times n}$, cluster number c,
        nearest neighbor number $k, \lambda_1 \geq 0, \lambda_2 \geq 0, \lambda_3 \geq 0$.
Initial: $E^{(0)} = Z_1^{(0)} = 0 \in R^{d \times n}, F^{(0)} == 0 \in R^{c \times n}$,
        $R^{(0)} = Q^{(0)} = A^{(0)} = Z_2^{(0)} = 0 \in R^{n \times n}, \mu^{(0)} = 1.25, \rho > 0, t = 0$.
Output: A(affinity matrix)
While no converged Do
        Construct $L_A^{(t)}$ based on A^t;
        Update $E^{(t+1)}$ via Eq.(7);
        Update $R^{(t+1)}$ via Eq.(8);
        Update $Q^{(t+1)}$ via Eq.(9);
        for $1 < i < n$ do
                Update $A_i^{(t+1)}$ via Eq.(12);
        End
        Balance $A^{(t+1)}$ by $\frac{A^{(t+1)} + A^{(t+1)T}}{2}$;
        Update $Z_1^{(t+1)}, Z_2^{(t+1)}$ via Eq.(13);
        Update parameter $\mu^{(t+1)} = \mu^t \rho, t = t + 1$;
End While

3.3   Centralpoint Feature Optimization

After A(affinity matrix) is calculated, the association between each superpixel of the image is basically expressed, but the graph structure does not combine the features between each superpixel and the centralpoint location information. Based on this, we propose the centralpoint feature optimization algorithm, aims to optimize $A \in d^{n \times n}$ through the centralpoint features between superpixels, so that the final A can fully explore the global structure and local linear relationship between superpixels, better describe the degree of correlation between superpixels, and eliminate the influence of background.

Specifically, the centralpoint feature optimization algorithm is divided into three aspects. Suppose that there are edge connections between superpixel x_i and x_j. Firstly, calculate the euclidean distance between x_i and x_j, denoted as t_1. Let $\gamma_1 = e^{-\alpha_1 t_1}$ to optimize $A \in d^{n \times n}$ by using the features between the superpixels $A \in d^{n \times n}$. Secondly, after obtaining the center coordinate c_i and c_j of superpixel x_i and x_j respectively, we calculate the 2-norm of c_i and c_j and mark it as t_2. Let $\gamma_2 = e^{-\alpha_2 t_2}$ represent the central coordinate point information between superpixels to optimize A. Finally, normalization of A, for A_{ij} and A_{ji}, add absolute values and average values, and the process is controlled by variable ε. The above process is denoted as t_3. Let $\gamma_3 = e^{(-\alpha_3 e^{t_3})}$ to optimize $A \in d^{n \times n}$ by using the normalized characteristics between superpixels; For the update of the final $\tilde{A} \in d^{n \times n}$, we assign a weight to each of the three aspects. The final optimized $\tilde{A} \in d^{n \times n}$ matrix can be obtained via simple linear weighting. The linear weighting formula refers to Eq. (14). The specific update process is shown in Algorithm 2.

$$\tilde{A}_{ij} = \omega_1 \gamma_1 + \omega_2 \gamma_2 + \omega_3 \gamma_3. \tag{14}$$

Algorithm 2. Centralpoint optimization algorithm

Input: $A \in d^{n \times n}$ (affinity matrix) to be optimized

Initial: $\omega_1 = 0.45, \omega_2 = 0.35, \omega_3 = 0.2, \alpha_1 = 0.15, \alpha_2 = 0.65, \alpha_3 = 0.55, \varepsilon = 0.22$

Output: optimized $\tilde{A} \in d^{n \times n}$ (affinity matrix)

For $x_i < n$

    For $x_j < n$

        Calculate the Euclidean distance between x_i and x_j, called as t_1;

        $\gamma_1 = e^{-\alpha_1 t_1}$;

        Calculate the centerpoint of x_i, x_j is c_i, c_j, the 2NF of c_i, c_j called as t_2;

        $\gamma_2 = e^{-\alpha_2 t_2}$;

        The normalized properties of x_i, x_j is t_3, $t_3 = -(|(A_{ij} + A_{ji})/2|/2\varepsilon^2)$;

        $\gamma_3 = e^{(-\alpha_3 e^{t_3})}$;

        Linear weighted $\tilde{A}_{ij} = \omega_1 \gamma_1 + \omega_2 \gamma_2 + \omega_3 \gamma_3$;

    End

End

---

4   Collaborative Manifold Ranking

This section mainly introduces graph-based multimodal collaborative manifold ranking model, and then we demonstrate the two-stage ranking scheme for bottom-up RGB-T saliency detection via using background and foreground queries.

4.1   Multi-modal Manifold Ranking

Given a pair of RGB and thermal images, we employ SLIC algorithm [11] to generate n non-overlapping superpixels. We take these superpixels as nodes to construct a graph $G = (V, E)$, where V is a node set and E is a set of undirected edges. If nodes V_i and V_j is connected, we assign it with an edge weight as: $W_{ij} = e^{-\gamma \|c_i - c_j\|}$, where c_i denotes the mean of the i-th superpixel, and γ is a scaling parameter. For the algorithm of graph-based manifold ranking that exploits the intrinsic manifold structure of data for graph labeling. Given a superpixel feature set $X = x_1, ..., x_n \in R^{d \times n}$, some superpixels are labeled as queries and the rest need to be ranked according to their affinities to the queries, where n denotes the number of superpixels. Let s: $X \to R^n$ denote a ranking function that assigns a ranking value s_i to each superpixel x_i, and s can be viewed as a vector $s = [s_1, ..., s_n]^T$. In this work, we regard the query labels as initial superpixel saliency value, and s is thus an initial superpixel saliency vector. Let $y = [y_1, ..., y_n]^T$ denote an indication vector, in which $y_i = 1$ if x_i is a query, and $y_i = 0$ otherwise. Given G, the optimal ranking of queries are computed by solving the following optimization problem:

$$\min_s \frac{1}{2} \left(\sum_{i,j=1}^n W_{ij} \left\| \frac{s_i}{\sqrt{D_{ii}}} - \frac{s_j}{\sqrt{D_{jj}}} \right\|^2 + \mu \| s - y \|^2 \right), \quad (15)$$

where $D = diag\{D_{11}, ..., D_{nn}\}$ is the degree matrix, and $D_{ii} = \sum_j W_{ij}$. $diag$ indicates the diagonal operation. μ is a parameter to balance the smoothness term and the fitting term. Then, we apply manifold ranking [4] on multiple modalities, and have:

$$\min_{s^k} \frac{1}{2}(\sum_{i,j=1}^{n} W_{ij}^k \parallel \frac{s_i^k}{\sqrt{D_{ii}^k}} - \frac{s_j^k}{\sqrt{D_{jj}^k}} \parallel^2 + \mu \parallel s^k - y \parallel^2), \quad k = 1, ..., K. \quad (16)$$

In this paper, we find out the optimal $\tilde{A}_{ij}$ for each modal according to Sect. 3, and then integrate $\tilde{A}_{ij}$ into manifold ranking model [4], so W_{ij}^k in Eq. (16) can be replaced by $\tilde{A}_{ij}^k$. Besides, from Eq. (16), we can see that it inherently indicates that available modalities are independent and contribute equally. This may significantly limit the performance in dealing with occasional perturbation or malfunction of individual sources. Therefore, we propose a novel collaborative model for robustly performing salient object detection that (i) adaptively integrates different modalities based on their respective modal reliabilities, (ii) collaboratively computes the ranking functions of multiple modalities by incorporating the cross-modality consistent constraints. The formulation of the multimoddal manifold ranking algorithm is proposed as follows:

$$\min_{s^k, v^k} \frac{1}{2} \sum_{k=1}^{K} ((v^k)^2 (\sum_{i,j=1}^{n} \tilde{A}_{ij}^k \parallel \frac{s_i^k}{\sqrt{D_{ii}^k}} - \frac{s_j^k}{\sqrt{D_{jj}^k}} \parallel^2) +$$

$$\mu \parallel s^k - y \parallel^2 + \parallel \Theta \circ (1 - v) \parallel^2 + \lambda \sum_{k=2}^{K} \parallel s^k - s^{k-1} \parallel^2 \quad (17)$$

where $\Theta = [\Theta^1, ..., \Theta^K]^T$ is an adaptive parameter vector, and $v = [v^1, ..., v^K]^T$ is the modality weight vector. $\circ$ denotes the element-wise product, and λ is a balance parameter. The third term is to avoid overfitting of v, and the last term is the cross-modality consistent constraints.

A sub-optimal optimization can be achieved by alternating between the updating of S and v, and the whole algorithm is can be referred to in [3]. Although the global convergence of the proposed algorithm is not proved, we empirically validate its fast convergence in our experiments. The optimized ranking functions s^k and modality weights v^k will be utilized for RGB-T saliency detection in next subsection.

4.2 Two-Stage RGB-T Saliency Detection

In this subsection, we demonstrate the two-stage ranking scheme for bottom-up RGB-T saliency detection via using background and foreground queries.

Ranking with Background Queries. Based on the attention theories for visual saliency, we regard the boundary nodes as background seeds to rank the relevances of all other superpixel nodes in the first stage. First, Taking top boundary as an example. Specifically, we utilize the nodes on this side as labelled queries and the rest as the unlabelled data, and initialize the indicator y in Eq. (17). Given y, the ranking values s_t^k are computed by employing the proposed ranking algorithm, and we normalize s_t^k as $\hat{s}_t^k$ with the range between 0 and 1. Similarly, given the bottom, left and right image boundaries, we can obtain the respective ranking values s_b^k, s_l^k, s_r^k. We integrate them to compute the initial saliency map for each modality in the first stage:

$$S_{bd}^k = S_t^k \circ S_b^k \circ S_l^k \circ S_r^k, \quad k = 1, ..., K, \tag{18}$$

Ranking with Foreground Queries. Given S_{bd}^k of the k-th modality, we set an adaptive threshold $H_k = max(S_{bd}^k) - \xi$ to generate the foreground queries, where $max(\cdot)$ indicates the maximum operation, and ξ is a constant, which is fixed to be 0.25 in this work. Specifically, we select the i-th superpixel as the foreground query of the k-th modality if $S_{bd}^k > H_k$. Therefore, we compute the ranking values S_{bg}^k and the modality weights v in the second stage by employing our ranking algorithm. Similar to the first stage, we normalize the ranking value S_{bg}^k of the k-th modality S_{bg}^k with the range between 0 and 1. Finally, the final saliency map can be obtained by combining the ranking values with the modality weights:

$$\bar{S} = \sum_{k=1}^{K} v_k S_{bg}^k. \tag{19}$$

5 Experiments

5.1 Evaluation Data and Metrics

To fully evaluate the research presented in this paper, we conduct experiments on the RGB-T saliency detection benchmark dataset [3] and [9]. The benchmark dataset is large and challenging enough, which includes 1821 pairs of spatially aligned RGB and thermal images and their ground truth annotations for saliency detection purpose. The image pairs in the dataset are recorded in approximately 60 scenes with different environmental conditions. PR curve, F-measure [13] and Mean Absolute Deviation(MAE) are adopted as metrics in this paper, F-measure [13] is defined as $F_\beta^2 = \frac{(1+\beta^2) \times P \times R}{\beta^2 \times P + R}$, Where β^2 is set as 0.3, designed to emphasize accuracy. The purpose of both PR curve and F-measure is quantitative analysis, while MAE is better at converting visual comparison into estimating the difference between saliency map S and groundtruth value G. The formula is expressed as $MAE = \frac{1}{w \times h} \sum_{i=1}^{w} \sum_{j=1}^{h} |S(i,j) - G(i,j)|$, Where w and h represent the width and height of the image respectively.

5.2 Baseline Methods

For comprehensively validating the effectiveness of our approach, we qualitatively and quantitatively compare the proposed approach with 12 state-of-the-art approaches, including GMR [4], RRWR [14], CA [7], GR [15], NIF [8], MTMR [3], MCI [10], SSKDE [16], MILPS [17], MST [18], MCDL [19], MDF [20]. Comparing with above methods with RGB or thermal inputs, we could justify the effectiveness of complementary benefits from different modalities of our approach. To further demonstrate the importance of our fusion strategy, we also compare our approach with kinds of baseline methods with both RGB and thermal inputs, and the implementation details of multi-modal extension. The extended methods include GMR [4], RRWR [14], CA [7], GR [15], NIF [8], MCI [10] and SSKDE [16] (Table 2).

Table 1. Average precision, recall, and F-measure of our method against different kinds of baseline methods on the dataset of [3] and [9], where the baselines are with RGB and thermal inputs. The code type and runtime (second) are also presented. The bold fonts of results indicate the best performance.

Algorithm	RGB			Thermal			RGB-T			Runtime (Sec.)
	P	R	F	P	R	F	P	R	F	
SSKDE	0.581	0.554	0.532	0.510	0.635	0.497	0.528	0.565	0.515	0.94
MCI	0.526	0.604	0.485	0.445	0.585	0.435	0.547	0.652	0.515	21.89
NIF	0.557	0.639	0.532	0.581	0.599	0.541	0.564	0.665	0.544	12.43
GR	0.621	0.582	0.534	0.639	0.544	0.545	0.705	0.593	0.600	2.43
CA	0.592	0.667	0.568	0.623	0.607	0.573	0.645	0.668	0.609	1.14
RRWR	0.642	0.610	0.589	0.689	0.580	0.596	0.695	0.617	0.628	2.99
GMR	0.644	0.603	0.587	**0.700**	0.574	0.603	0.733	0.653	0.666	1.11
MDF	0.692	0.699	0.654	0.631	0.585	0.549	–	–	–	20.19
MCDL	**0.701**	**0.751**	**0.689**	0.606	0.663	0.588	–	–	–	2.41
MST	0.627	0.739	0.610	0.665	0.655	0.589	–	–	–	0.53
MILPS	0.637	0.691	0.612	0.643	**0.680**	**0.612**	–	–	–	165.48
MTMR	–	–	–	–	–	–	0.716	**0.713**	0.680	1.39
Ours	–	–	–	–	–	–	**0.796**	0.676	**0.748**	1.52

Table 2. Average MAE score of our method against different kinds of baseline methods on the entire dataset

MAE	SSKDE	MCI	NIF	GR	CA	RRWR	GMR	MDF	MCDL	MST	MILPS	MTMR	Ours
RGB	0.122	0.211	0.126	0.197	0.163	0.171	0.172	0.097	0.067	0.127	0.136	–	–
T	0.132	0.176	0.124	0.199	0.225	0.234	0.232	0.15	0.093	0.129	0.145	–	–
RGB-T	0.127	0.195	0.125	0.199	0.195	0.203	0.202	–	–	–	–	0.103	0.979

5.3 Experimental Setup

We set the number of superpixel nodes $N = 200$ in all the experiments. To simplify our parameters, we let $\lambda_1 = \lambda_2 = \lambda_3 = \widehat{\lambda} \in \{0.1, 0.2, ..., 1.0\}$, although

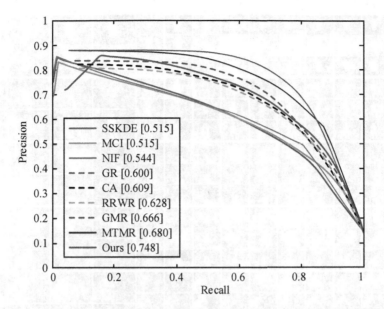

Fig. 2. PR curves of our model against other methods with both RGB and thermal inputs.

the simplification may very likely exclude the best performance for our method. Our model also involves the parameter k, which will be fixed in the experiments according to a k effect testing. Normalized cut is employed to segment the input data into clusters for all the competitors, average segmentation accuracies 3 over 10 independent trials are finally reported (Fig. 2).

5.4   Comparison Results

To justify the importance of thermal information, the complementary benefits to image saliency detection and the effectiveness of the proposed approach, we evaluate the compared methods with different modality inputs on the benchmark.

Overall Performance. On the entire dataset, we first compare our model against other methods mentioned in Subsect. 5.2 on the aspects of precision (P), recall (R) and F-measure [13] (F), shown in Table 1. From the quantitative evaluation results, we can observe that the proposed method achieves a good balance of precision and recall, and thus obtain better F-measure over all baseline methods with a clear margin. It demonstrates that our model can detect the salient objects more accurately than other methods. The visual comparison are shown in Fig. 3. We notice that our model obtains weaker performance in recall comparing with deep learning based methods, i.e., MDF [20] and MCDL [19]. However, our model has the following advantages over deep learning based methods. (i)

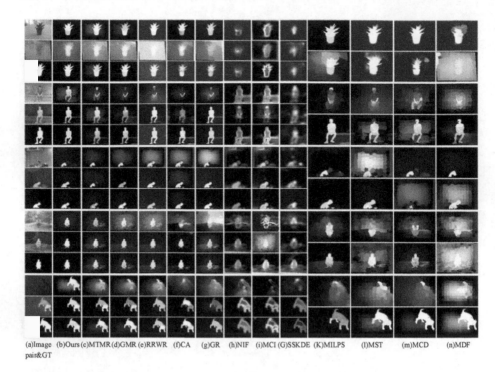

(a)Image (b)Ours (c)MTMR (d)GMR (e)RRWR (f)CA   (g)GR (h)NIF (i)MCI (G)SSKDE (K)MILPS   (l)MST   (m)MCD   (n)MDF
pair&GT

Fig. 3. Sample results of the proposed approach and other baseline methods with different modality inputs. (a) Input RGB and thermal image pair and their ground truth. (b–j) Results of our model and the baseline methods with RGB, thermal and RGB-T inputs. (k–n) Results of the baseline methods with RGB and thermal inputs.

It does not require laborious pre-training or a large training set. (ii) It does not need to save a large pre-trained deep model. (iii) It is easy to implement as each subproblem of the proposed model has a closed-form solution. (iv) It performs favorably against MDF [20] and MCDL [19] in terms of efficiency on a cheaper hardware setup.

Component Analysis. To justify the significance of the main components of the proposed approach, we implement two special versions for comparative analysis, including: (1) Ours-I, that removes the modality weights in the proposed ranking model, i.e., fixes $v^k = \frac{1}{K}$ in Eq. (17), and (2) Ours-II, that removes the cross-modality consistent constraints in the proposed ranking model, i.e., sets $\lambda = 0$ in Eq. (17).

6   Conclusion

In this paper, we propose a novel collaborative algorithm for saliency detection by adaptively incorporating information from grayscale and thermal images.

Specifically, we first construct an affinity graph model via robust graph learning to characterize the feature similarities and location relationships between two different image patches under each modal. Based on these affinity graphs, a collaborative manifold ranking method cross-modality consistent constraints is designed to effectively and efficiently infer the salient score for each patch. Moreover, we introduce a weight for each modality to describe the reliability, and integrate them into the graph-based manifold ranking [4] scheme to achieve an adaptive weighted fusion of different source data. For optimization, we propose some iterative algorithms to efficiently solve the models with several subproblems. Our future works will focus on the following aspects: (1) We try to train a deep network model to fuse multi-modal data. (2) We will improve the robustness of our approach by studying other prior models and graph construction.

Acknowledgements. This work was supported by the Key Natural Science Project of Anhui Provincial Education Department (KJ2018A0023), the Guangdong Province Science and Technology Plan Projects (2017B010110011), the Anhui Key Research and Development Plan (1804a09020101), the National Basic Research Program (973 Program) of China (2015CB351705) and the National Natural Science Foundation of China (61906002, 61402002, 61876002 and 61860206004).

References

1. Gade, R., Moeslund, T.: Thermal cameras and applications: a survey. Mach. Vis. Appl. **25**(1), 245–262 (2013). https://doi.org/10.1007/s00138-013-0570-5
2. Li, C., Hu, S., Gao, S., Tang, J.: Real-time grayscale-thermal tracking via Laplacian sparse representation. In: Tian, Q., Sebe, N., Qi, G.J., Huet, B., Hong, R., Liu, X. (eds.) MMM 2016. LNCS, vol. 9517, pp. 54–65. Springer, Cham (2016). https://doi.org/10.1007/978-3-319-27674-8_6
3. Li, C., Wang, G., Ma, Y., Zheng, A., Luo, B., Tang, J.: A unified RGB-T saliency detection benchmark: dataset, baselines, analysis and a novel approach (2017)
4. Yang, C., Zhang, L., Lu, H., Ruan, X., Yang, M.: Saliency detection via graph-based manifold ranking. In: Computer Vision and Pattern Recognition, pp. 3166–3173 (2013)
5. Itti, L., Koch, C., Niebur, E.: A model of saliency-based visual attention for rapid scene analysis. TPAMI **20**(11), 1254–1259 (1998)
6. Hou, X., Zhang, L.: Saliency detection: a spectral residual approach. In: CVPR (2007)
7. Qin, Y., Lu, H., Xu, Y., Wang, H.: Saliency detection via cellular automata. In: Proceedings of the IEEE Conference on Computer Vision and Pattern Recognition (2015)
8. Erdem, E., Erdem, A.: Visual saliency estimation by nonlinearly integrating features using region covariances. J. Vis. **13**(4), 11 (2013)
9. Tu, Z., Xia, T., Li, C., Wang, X., Ma, Y.: RGB-T image saliency detection via collaborative graph learning. IEEE Trans. Multimed. **22**, 160–173 (2018)
10. Goferman, S., Zelnik-Manor, L., Tal, A.: Context-aware saliency detection. IEEE Trans. Pattern Anal. Mach. Intell. **34**(10), 1915–1926 (2012)
11. Achanta, R., Shaji, A., Smith, K., Lucchi, A., Fua, P., Ssstrunk, S.: SLIC superpixels compared to state-of-the-art superpixel methods. IEEE Trans. Pattern Anal. Mach. Intell. **34**(11), 2274–2282 (2012)

12. Boyd, S., Parikh, N., Chu, E., et al.: Distributed optimization and statistical learning via the alternating direction method of multipliers. Found. Trends Mach. Learn. **3**(1), 1–122 (2011)
13. Hripcsak, G., Rothschild, A.: Agreement, the f-measure, and reliability in information retrieval. J. Am. Med. Inform. Assoc. **12**(3), 296–298 (2005)
14. Li, C., Yuan, Y., Cai, W., Xia, Y.: Robust saliency detection via regularized random walks ranking, pp. 2710–2717 (2015)
15. Yang, C., Zhang, L., Lu, H.: Graph-regularized saliency detection with convex-hull-based center prior. IEEE Signal Process. Lett. **20**(7), 637–640 (2013)
16. Tavakoli, H., Rahtu, E., Heikkilä, J.: Fast and efficient saliency detection using sparse sampling and kernel density estimation. In: Heyden, A., Kahl, F. (eds.) SCIA 2011. LNCS, vol. 6688, pp. 666–675. Springer, Heidelberg (2011). https://doi.org/10.1007/978-3-642-21227-7_62
17. Huang, F., Qi, J., Lu, H., Zhang, L., Ruan, X.: Salientobject detection via multiple instance learning. IEEE Trans. Image Process. **26**(4), 1911–1922 (2017)
18. Tu, W., He, S., Yang, Q., Chien, S.: Real-time salient object detection with a minimum spanning tree. In: Proceedings of the IEEE Conference on Computer Vision and Pattern Recognition (2016)
19. Zhao, R., Ouyang, W., Li, H., Wang, X.: Saliency detection by multi-context deep learning. In: Proceedings of the IEEE Conference on Computer Vision and Pattern Recognition, pp. 1265–1274 (2015)
20. Li, G., Yu, Y.: Visual saliency based on multiscale deep features. In: Proceedings of the IEEE Conference on Computer Vision and Pattern Recognition, pp. 5455–5463 (2015)

Parametric Method for Designing Discrete-Time Periodic Controller of UAV

Ling-ling Lv[1](✉)(iD), Jin-bo Chen[1](✉)(iD), and Lei Zhang[2](✉)(iD)

[1] Institute of Electric Power,
North China University of Water Resources and Electric Power,
Zhengzhou 450011, People's Republic of China
lingling_lv@163.com, 358490693@qq.com
[2] School of Computer and Information Engineering,
Henan University,
Kaifeng 475001, People's Republic of China
zhanglei@henu.edu.cn
http://www5.ncwu.edu.cn/dianli/

Abstract. In this paper, a parametric method is used to introduce robust controller into UAV control system. Poles assignment technique is used to design robust controller, to quickly return the states of the closed-loop to the desired position. In the case of interference and uncertainty, the robustness of the system is improved. Simulation results show that the controller obtained by this design method has a good control effect on the flight state of the UAV. Moreover, after setting the poles, the whole process of solving the control law adopts the parametric method completely, and the parameters can be chosen by ourselves according to the needs. There is no need to calculate the control law online, which greatly improves the freedom of the controller design process.

Keywords: Flight control system · Poles assignment · Robustness

1 Introduction

In recent years, unmanned aerial vehicles (UAVs) have gradually become a research hotspot in the aviation field. Unmanned aerial vehicles have been an important research and development object of military equipment in various countries because of their vertical takeoff and landing or short-range takeoff and landing capabilities, and they are able to adapt to most usage scenarios. In addition to the requirement that the UAV can take off and land vertically or take off and land in a short distance, people also hope that it can have greater load capacity and more simple and efficient flight control model to improve its flight performance. With the development of electronic technology, advanced

Supported by the Educational Commission of Henan Province of China under Grant No. 17B520006.

© Springer Nature Singapore Pte Ltd. 2020
L. Pan et al. (Eds.): BIC-TA 2019, CCIS 1160, pp. 685–692, 2020.
https://doi.org/10.1007/978-981-15-3415-7_58

control theory and computer technology, unmanned aerial vehicles are becoming smaller, lighter, more flexible and have stronger maneuverability. Therefore, small unmanned aerial vehicles have gradually penetrated into the civil field, and their use scenarios have become more extensive, such as agricultural flight, meteorological research, animal tracking, disaster prevention and mitigation, etc. These UAVs come in all shapes and sizes and carry out different airborne missions.

The miniaturization of unmanned aerial vehicles is an important development direction for the next generation of the UAV. With the development of the international situation in the future and the increase of the demand for military and civil UAV, small UAV has gradually become a popular product in the UAV market due to its excellent flight maneuverability, miniaturization, lightweight design and relatively low cost. Therefore, it is necessary to research an effective control method to control the UAV quickly and stably [1].

In the traditional sense, unmanned aerial vehicle (UAV) mostly use commercial autopilot based on classical PID method to control, need to be repeated flight test to determine the gain. When the number of control circuit, the adjustment of PID parameters can become quite complicated, resulted in significant increase workload. More importantly, for a large number of small unmanned aircraft system uncertainty disturbance, PID controller does not guarantee optimal performance and robustness of the closed-loop system [2,8,9].

According to the characteristics of model and flight control system, a closed-loop feedback control system is established. Pole assignment method is adopted for UAV flight control system, and corresponding control law is obtained. Finally, the controller is applied to the system, and Matlab is used for simulation to verify its effectiveness, which will be described in detail below.

2   Design Principle of Robust Parametric Method

Parametric method was first proposed in reference [3]. By introducing arbitrary parameter matrix, the mathematical expression of the desired control rate was established to solve the problem. This design method provides sufficient degree of freedom for the controller design. In order to make the eigenvalue of the closed-loop system as insensitive as possible to the disturbance existing in the system and the feedback gain as small as possible, it is necessary to further optimize the control law obtained by combining the index function. In the following, parametric design method will be stated in detail.

2.1   Parametric Expression of Control Laws

Consider a linear discrete periodic system with the following state space implementation:

$$x(t+1) = A(t)\,x(t) + B(t)\,u(t) \tag{1}$$

where $t \in \mathbb{Z}$, $x(t) \in \mathbb{R}^n$ and $u(t) \in \mathbb{R}^r$ are the state vector and are also input vector of the system, respectively; $A(t) \in \mathbb{R}^{n \times n}$, $B(t) \in \mathbb{R}^{n \times r}$ are the coefficient matrix of the system, and are cycled by T, i.e.

$$A (t + T) = A (t), B (t + T) = B (t), t \in \mathbb{Z} \tag{2}$$

If (A, B) is controllable, select the state feedback control law:

$$u (t) = K (t) x (t), \ K (t + T) = K (t), \ t \in \mathbb{Z} \tag{3}$$

then a closed-loop control system with Eqs. (1) and (3) can be given by

$$x (t + 1) = A (t) x (t) + B (t) K (t) x (t) \tag{4}$$

The closed-loop system single-valued matrix is

$$\Psi_c = A_c (T - 1) A_c (T - 2) \cdots A_c (0) \tag{5}$$

where

$$A_c (i) \triangleq A (i) + B (i) K (i), i \in \overline{0, T - 1} \tag{6}$$

What associated with the linear periodic system (1) is its lifting time invariant system [4]

$$x^L (t + 1) = A^L x^L (t) + B^L u^L (t) \tag{7}$$

where

$$A^L = A (T - 1) A (T - 2) \cdots A (0) \tag{8}$$

$$B^L = [A (T - 1) A (T - 2) \cdots A (1) B (0) \quad \cdots \quad A (T - 1) B (T - 2) \quad B (T - 1)] \tag{9}$$

$$x^L (t) = x (tT), u^L (t) = \left[u^T (tT) \quad u^T (tT + 1) \quad \cdots \quad u^T (tT + T - 1) \right]^T \tag{10}$$

That is to say, the state of the system (1) can be improved by the regular sampling and arrangement, and the input of the system (1) can be input by the lifting system by regular sampling and arrangement. The linear discrete periodic system has the following properties: the energy subspace of (A, B) and (A^L, B^L) are consistent at a certain time [5].

The following is to find the matrix $K(i)$ so that the eigenvalues of the closed-loop system meet the actual requirements. Let the set of eigenvalues of the single-valued matrix to be configured be

$$\Gamma = \left\{ s_i, s_i \in \mathbb{C}, i \in \overline{1, n} \right\} \tag{11}$$

where the set is symmetric about the real axis. $F \in \mathbb{R}^{n \times n}$ is a given real matrix, satisfying $\lambda (F) = \Gamma$. Obviously, $\lambda (\Psi c) = \Gamma$, if and only if there is a non-singular matrix V, so that (12) holds.

$$\Psi_c V = V F \tag{12}$$

Then the periodic state feedback pole configuration problem of system (1) can be transformed into: given a $F \in \mathbb{R}^{n \times n}$ and a fully lin-ear discrete-period system (1), we need to find $K(i) \in \mathbb{R}^{r \times n}, i \in \overline{0, T - 1}$, such that a non-singular matrix $V \in \mathbb{R}^{n \times n}$, the relationship (12) is established.

If V is reversible, so

$$K = WV^{-1} \tag{13}$$

Consider a constant diagonal state feedback law of the lifting system (7)

$$u^L(t) = K^L x^L(t), K^L \triangleq diag\{K(0), K(1), \cdots, K(T-1)\} \tag{14}$$

where $K(i) \in \mathbb{R}^{r \times n}, i \in \overline{0, T-1}$ is given by the Eq. (3). For the relationship between the periodic system (1) and its lifting time invariant system (7), please refer to the literature [6].

According to literature [4], Eq. (12) can be transformed into the form of Matrix equation Sylvester

$$A^L V + B^L W = VF \tag{15}$$

where

$$W = XV \tag{16}$$

The following polynomial decomposition exists

$$(zI - A^L)^{-1} B^L = N(z) D^{-1}(z) \tag{17}$$

where $N(z) \in \mathbb{R}^{n \times Tr}$ and $D(z) \in \mathbb{R}^{Tr \times Tr}$ are the right coprime matrix polynomial of z.

Let

$$D(z) = [d_{ij}(z)]_{Tr \times Tr}, N(z) = [n_{ij}(z)]_{n \times Tr}, \omega = \max\{\omega_1, \omega_2\}$$

where

$$\omega_1 = \max\{\deg(d_{ij}(z))\}, \qquad i, j \in \overline{1, Tr}$$
$$\omega_2 = \max\{\deg(n_{ij}(z))\}, \qquad i \in \overline{1, n}, j \in \overline{1, Tr}$$

Then, $D(z)$ and $N(z)$ can be rewritten as

$$\begin{cases} N(z) = \sum_{i=0}^{\omega} N_i Z^i, N_i \in \mathbb{C}^{n \times Tr} \\ D(z) = \sum_{i=0}^{\omega} D_i Z^i, D_i \in \mathbb{C}^{Tr \times Tr} \end{cases} \tag{18}$$

For the arbitrary matrix $F \in \mathbb{R}^{n \times n}$, the solution of the generalized Sylvester matrix equation can be expressed by

$$\begin{cases} V(Z) = N_0 Z + N_1 ZF + \cdots + N_\omega ZF^\omega \\ W(Z) = D_0 Z + D_1 ZF + \cdots + D_\omega ZF^\omega \end{cases} \tag{19}$$

where $Z \in \mathbb{R}^{Tr \times n}$ is an arbitrary parameter matrix representing the degree of freedom present in the solution $(V(Z), W(Z))$.

Let

$$\mathcal{Z} = \left\{ Z \middle| \det \left(\sum_{i=0}^{\omega} N_i Z F^i \right) \neq 0 \right\} \tag{20}$$

then

$$\mathcal{K} = \left\{ \begin{pmatrix} K(0) \\ K(1) \\ \vdots \\ K(T-1) \end{pmatrix} \middle| \left\{ \begin{array}{c} X(Z) = W(Z) V^{-1}(Z), Z \in \mathcal{Z} \\ K(0) = X_1, \det(A_c(0)) \neq 0 \\ K(i) = X_{i+1} \prod_{j=0}^{i-1} A_c^{-1}(j), \det(A_c(i)) \neq 0, i \in \overline{1, T-1} \end{array} \right\} \right\} \tag{21}$$

After obtaining matrix $V(Z)$, $W(Z)$, then cite Eqs. (20) and (21), the feedback gain value $K(i)$ can be obtained.

2.2 Improvement of the Objective Function

The parametric method provides an explicit solution to the pole assignment problem. However, in the practical application, in order to make the system performance better meet the actual needs, the system needs to be optimized through the performance function. By applying some additional conditions on the feedback gain matrix K and matrix V, the free parameter Z is obtained to achieve the expected performance of the system. Specific calculation process can be referred to literature [6].

If there is the following disturbance in the closed loop system

$$A(i) + B(i)K(i) \rightarrow A(i) + B(i)K(i) + \Delta_i(\varepsilon), i \in \overline{0, T-1} \tag{22}$$

where $\Delta_i(\varepsilon), i \in \overline{0, T-1}$ is a possible disturbance in the closed loop system. Small gain means small control signal, smaller energy consumption. It can also reflect robustness in the sense that they reduce noise amplification. In addition, the smaller the value of $J_2(Z)$, the better the robustness of the closed-loop system is. Then, the objective function can be introduced as

$$J(Z) \triangleq \alpha J_1(Z) + (1 - \alpha) J_2(Z) \tag{23}$$

where

$$J_1(Z) = \sum_{l=0}^{T-1} \| K(l) \|_F^2, J_2(Z) = \kappa_F(V) \sum_{l=0}^{T-1} \| A_c(l) \|_F^{T-1} \tag{24}$$

And α is a weighting factor that indicates the degree of constraint on different indicators. Take Z with the minimum value of J as the optimal parameter matrix Z_{opt}.

Then, substituting Z_{opt} into Eqs. (18) and (19), we can calculate the optimal matrix V_{opt} and W_{opt}, and further calculate the optimal feedback gain $K_{opt}(i)$.

3   Application of Robust Control in Flight Control Design

The flight control system is an important part of the UAV, which usually consists of subsystems or components with different functions. The angular motion, orbit motion and flight speed of the UAV all need to be controlled by the flight control system. It is an important guarantee to improve the dynamic performance and autonomous flight of the UAV. With the continuous expansion of control theory, the flight control technology will be further developed.

This part selects the flight parameters of a particular test system under the state of equilibrium flight and simulates the flight control system with the parameter controller. If all the influencing factors are taken into account in the mathematical modeling of the UAV, the equation derivation will become very complicated and difficult to deal with. Therefore, in order to establish a simple mathematical model of UAV flight, several assumptions need to be satisfied. For example, during flight, the mass of the UAV remains unchanged and does not deform; The earth coordinate system is the inertial coordinate system; Ignoring the curvature of the earth and treating it as a plane; Altitude doesn't affect gravity, etc. Then, the linear equation of the continuous system is:

$$\dot{x} = Ax + Bu \tag{25}$$

where

$$x = \begin{bmatrix} v & \alpha & \omega_z & h & \theta \end{bmatrix} \tag{26}$$

Set the simulation parameters as follows [7]:

$$A = \begin{bmatrix} -0.0146 & 5.8473 & 0 & 0.0002 & -9.5387 \\ -0.0035 & -2.1268 & 1 & 0 & -0.0025 \\ 0.0044 & -6.4751 & -0.3272 & 0 & 0 \\ 0.062 & -63.7443 & 0 & 0 & 63.7443 \\ 0 & 0 & 1 & 0 & 0 \end{bmatrix}, B = \begin{bmatrix} 0 \\ -0.0023 \\ 0.0775 \\ 0 \\ 0 \end{bmatrix}$$

Through calculation, the system model is controllable and considerable.

After discretization of system (25), the expected eigenvalues of the system are set as ± 0.2, 0.3 and ± 0.4.

Utilizing the parametric robust design algorithm provided in Sect. 2, the feedback gains of the system can be obtained as:

$$K_{opt}(0) = \begin{bmatrix} -90.3087 & 173.2070 & -5.4619 & -13.2198 & -173.4308 \end{bmatrix}$$

$$K_{opt}(1) = \begin{bmatrix} 34.2618 & 226.6165 & -75.2599 & 5.5455 & -167.2126 \end{bmatrix}$$

$$K_{opt}(2) = \begin{bmatrix} -74.5918 & 76.1244 & -45.8958 & -14.0404 & -206.4955 \end{bmatrix}$$

Apply the controller to the system for simulation, observe and analyze the control effects.

The simulation curves of the system variable are shown in Fig. 1. The simulation results show that the controller can stabilize the flight state of UAV in the presence of disturbance to a certain extent.

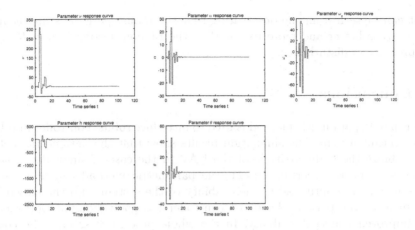

Fig. 1. Dynamic response curves of parametric method under simulation experiment

Selecting a random parameter matrix Z, a set of feedback gains can also be obtained as:

$$K_{rand}(0) = \begin{bmatrix} 1118.00 & -2694.80 & 412.000 & 1230.90 & 3289.00 \end{bmatrix}$$

$$K_{rand}(1) = \begin{bmatrix} 24.7299 & 70.1467 & -27.0381 & 3.3772 & -53.7753 \end{bmatrix}$$

$$K_{rand}(2) = \begin{bmatrix} -5.3906 & 142.9380 & -35.2013 & -4.1303 & -224.1777 \end{bmatrix}$$

Apply two sets of controllers to the system, and add small random disturbances of the same control level. Then do 3000 random experiments respectively, and draw the pole distribution diagram of the closed-loop system as shown in Fig. 2.

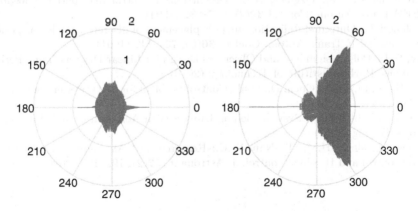

Fig. 2. Perturbed eigenvalues of the closed-loop system. The left hand side corresponds to K_{opt} and the right hand side corresponds to K_{rand}

It can be seen from the observation that the robust periodic state feedback gain K_{opt} is better and more robust than the random feedback gain K_{rand} in the presence of disturbance.

4   Conclusions

In this paper, parametric method is used to introduce robust controller into UAV flight control system. The simulation results show that the parametric method can stabilize the flight attitude of the UAV in the case of small disturbances in the system. To a certain extent, the parametric method can improve the robustness and adaptive adjustment ability of the system, and the closed-loop system using the parametric robust controller has good dynamic performance and improved transient stability. In the whole process of solving the control law, the parametric method is completely adopted to solve the problem. The parameters can be selected by ourselves according to needs, and there is no need to calculate the control law online, which greatly improves the freedom of the controller design process of UAV. So parametric pole assignment method has certain advantages for the design of UAV control law.

References

1. Soumelidis, A., Gaspar, P., Regula, G., et al.: Control of an experimental mini quadrotor UAV. In: Proceedings of IEEE 16th Mediterranean Conference on Control and Automation, pp. 1252–1257 (2008)
2. Tang, S., Song, X., Guo, J., et al.: Fuzzy-integration controller design of UAV based on genetic algorithm. Trans. Beijing Inst. Technol. **33**(12), 1274–1278 (2013)
3. Zhou, B., Duan, G.R.: A new solution to the generalized Sylvester matrix equation AV-EVF=BW. Syst. Control Lett. **55**(3), 193–198 (2006)
4. Khargonekar, P.P., Ozguler, A.B.: Decentralized control and periodic feedback. IEEE Trans. Autom. Control **39**(4), 877–882 (1994)
5. Colaneri, P.: Output stabilization via pole placement of discrete-time linear periodic systems. IEEE Trans. Autom. Control **36**(6), 739–742 (1991)
6. Lv, L.-L.: Pole Assignment and Observers Design for Linear Discrete-Time Periodic Systems. Harbin Institute of Technology (2010)
7. Su, H.-Q., Li, X.-H., Chen, D., et al.: Control problem for linear system of morphing UAV. Aircr. Des. (2015)
8. Wu, S.: Flight Control System. Beijing University of Aeronautics and Astronautics Press, Beijing (2013)
9. Jiang, Q., Chen, H.M., Jia-Nan, W.U.: Research on UAV flight control based on PID control and H robust control. J. Astronaut. **27**(2), 192–195 (2006)

An Adaptive Learning Rate Q-Learning Algorithm Based on Kalman Filter Inspired by Pigeon Pecking-Color Learning

Zhihui Li[1(✉)], Li Shi[2], Lifang Yang[1], and Zhigang Shang[1]

[1] Henan Key Laboratory of Brain Science and Brain-Computer Interface Technology, School of Electrical Engineering, Zhengzhou University, Zhengzhou 450001, China
lizhrain@zzu.edu.cn
[2] Department of Automation, Tsinghua University, Beijing 100000, China

Abstract. The speed and accuracy of the Q-learning algorithm are critically affected by the learning rate. In most Q-learning application, the learning rate is usually set as a constant or decayed in a predetermined way, so it cannot meet the needs of dynamic and rapid learning. In this study, the learning process of pigeon pecking-color task was analyzed. We observed that there was epiphany phenomenon during pigeon's learning process. The learning rate did not change gradually, but was large in the early stage and disappeared in the middle and late stage. Inspired by these phenomena, an adaptive learning rate Q-learning algorithm based on Kalman filter model (ALR_KF Q-learning) is proposed in this paper. Q-learning are represented in the framework of Kalman filter model, and the learning rate is equivalent to Kalman gain, which dynamically weighs the fluctuation of environmental reward and the cognitive uncertainty of the agent to the value of $<state, action>$ pairs. The cognitive uncertainty in the model is determined by the variance of measurement residual and of environmental reward, and is set to zero when it is less than the variance of the environmental reward. The results tested by the two-armed Bandit task showed that the proposed algorithm not only can adaptively learn the statistical characteristics of environmental rewards, but also can quickly and accurately approximate the expected value of $<state, action>$ pairs.

Keywords: Adaptive learning rate · Q-learning · Kalman filter · Bio-inspired

1 Introduction

In the field of machine learning, reinforcement learning (RL) is an interactive learning method inspired by behaviorist psychology research. The agent (animal

This work is supported by the Outstanding Youth Science Foundation 61922072 and the National Natural Science Foundation of China (61673404, 61876169, U1304602).

© Springer Nature Singapore Pte Ltd. 2020
L. Pan et al. (Eds.): BIC-TA 2019, CCIS 1160, pp. 693–706, 2020.
https://doi.org/10.1007/978-981-15-3415-7_59

or machine) acquires the relationship between environment state, action and reward with trial-and-error way interacted with environment and makes optimal policy in RL. In recent years, reinforcement learning combined with deep neural net has been widely used in opsearch, control science, games and other fields [1,2]. Q-learning is a representative algorithm of RL theory applied in the field of control and decision-making. The agent chooses an action according to the value of state-action pairs, and adjusts these values by the feedback reward form environment, so that learns the best policy in the iterative interaction with the environment finally [3]. In Q-learning, the adjustment of the value of state-action pairs is based on the time difference (TD) structure, that is, the Q value is adjusted by the error between the environmental return and predicted return in each trial. How much error information is used for adjusting is determined by the learning rate. The speed and accuracy of the Q-learning algorithm are critically affected by the learning rate. If its value is too large, the learning will be accelerated, but the noise in the environment will also be absorbed, so the Q value will fluctuate greatly and difficult to converge. On the contrary, if it is too small, so the Q value will converge slowly and making it difficult to follow changes in environmental statistics. Despite the importance of the learning rate, currently an analytical reasonable approach for its selection is largely lacking. Most existing learning rate adjustment strategies are out of concern for convergence or based on prior knowledge. For stationary processes, common learning rate schedules include time-based decay, step decay and exponential decay [4–6], and the learning rate will gradually decrease to approach zero with the trial times increases. These strategies can guarantee the gradual convergence of Q values, but don't work in non-stationary situation. The learning rate should not be zero for non-stationary environmental to ensure the expected value could be updated by the information from measurements [7]. Therefore, the adaptive selection of learning rate has been widely concerned by researchers. Natural gradient has been used in many incremental learning could be looked as an adaptive learning rate strategy [8]. It uses the Fisher information matrix to measure the uncertainty of gradient information at sampling points, and the smaller the uncertainty is, the larger the step-size is. Van Rooijen et al. proposed an A-C framework that combines the value function gradient information to realize the learning process without learning rate [9]. However, these learning rate free methods do not provide an explicit theoretical guide for adaptive adjustment of learning rate, and the representation and adjustment mechanism of the learning rate in the interactive learning between the agent and the environment is still not clear. The learning process is driven by the difference between the predicted return and the actual return, therefore, the learning rate should be related to the TD error. Both for animals and AI machines, it is a dilemma problem should be faced if there is an inconsistency between the expected and the measured return. Whether should they keep believing on the previous experience, or learn new knowledge from the TD error? Animals have evolved a learning mechanism that can adapt to environmental changes in nature. They are required to acquire statistical characteristics of environment reliably, and be able to follow their change quickly.

A large number of animal behavioral experiments have shown that animals use the TD errors in reinforcement learning. Ruan et al. [10] established a learning probability automaton (PA) model with operational conditioning (OC) behavioral response, which was used to mimic the Skinners experiment of pigeons. And the model can complete the acquisition task in a similar way to the actual learning process of pigeons. Rose et al. [11] used the Actor-Critic algorithm to model the pigeon behavior data in the experiment of color discrimination learning based on the reward value of pigeons. The model can predict the impact of the reward range on the learning speed of pigeons color recognition. These studies supported that animals use TD information to adjust learning, but have not discussed the effect of learning rate for further.

The dynamic adjustment strategy of learning rate should be related to the TD error, which will be affected by environmental disturbance noise and cognitive model bias. Therefore, learning rate should dynamically adapt to the uncertainty of model prediction and the uncertainty of environmental return. As an optimal state estimation method, Kalman filter model estimates the state of dynamic system with fusion information from above two kind of uncertainties, so it is widely used in system state estimation and output filtering which based on TD framework [12,13]. The core of the Kalman filtering model is to fully dynamically integrate the prediction information from the model and the observation information from the sampling, and make a trade-off between the variance information of them to ensure the minimum covariance of the optimal state estimation error. The Kalman gain is determined by the weights of variance of both predicted error and measurement error, and dynamically indicates the utilization level of TD error. Therefore, if Q-learning could be re-represented on the basis of Kalman filtering model framework, so the Kalman gain could play the role of adaptive learning rate. In recent years, the idea combining Kalman filter in the RL field learning has been paid more attention. Velazquez et al. [14] studied how to make prediction in an unceasing fluctuating environment destroyed by Gauss noise within the reinforcement learning framework using the Kalman filter model. The results showed that the learning rate of speed and the change of decision noise depend on the value of signal-to-noise ratio. Ahumada [15] used the Kalman model to estimate the incomplete state, accelerated Q-learning convergence, and achieved good simulation results in RoboCup. Shashua [16] proposed a new optimization method based on extended Kalman filter, Kalman optimization of value approximation (KOVA). For the value parameters and the noisy observations, the Bayesian viewpoint was introduced and the sum of the squares of the regularized Bellman TD errors was used as the objective function to be minimized. These results show that the Kalman filtering method can efficiently utilize the TD error information. However, how to adaptively adjust the learning rate, and how to restart learning in time when the environmental statistics changes, these questions remain to be further studied.

Therefore, one aim of this paper is to observe how animals adjust their learning rate during the RL process. We observed that there was epiphany phenomenon during pigeon's learning process. The learning rate did not change gradually, but was large in the early stage and disappeared in the middle and late stage. Inspired by these phenomena, we proposed the viewpoint that the learning rate doesn't decay gradually, but subjects to a threshold effect. When the prediction error is less than the threshold value, the learning rate should be zero to ensure that the knowledge of environmental statistical characteristics is not affected by occasional noise. Q-learning algorithm is re-represented based on Kalman filter model framework, and the learning rate is equivalent to the Kalman gain, which is adaptively determined by weighting the uncertainty of current environmental reward and the uncertainty of the agent's cognitive value of $<state, action>$ pairs. At the same time, the method how to measure the uncertainty of both cognitive model and environmental noise is given. Finally, the performance of the proposed ALR_KF Q-learning algorithm is compared with those of the common learning rate setting Q-learning methods by the two-arm bandit task.

The rest of this paper is organized as follows. Section 2 introduces the related concepts and basic algorithm framework of Q-learning and Kalman filtering. The pigeon pecking-color experiment and the inspiration from behavioral data analysis are presented in Sect. 3. Section 4 introduces the proposed ALR_KF Q-learning algorithm in detail. The numerical experimental evaluation results are shown in Sect. 5. Finally, the conclusion and discussion are given in Sect. 6.

2   Foundation of Related Algorithms

2.1   Q-Learning

Q-learning is a model-free reinforcement learning algorithm. Q value is the expected return when agent does a specific action at a specific state, that is represented usually by table or function approximation. In Q-learning algorithm, the agent selects different actions in different states according to the Q value function of $<state, action>$ pairs, and updated $Q(s_t, a_t)$ by the reward after the agent executed action a_t at the state s_t. Q-learning is an iterative learning process, which schematic is shown in Fig. 1 and illustrated as follows:

(1) The current state is $s(t)$.
(2) According to Q value function of the current state s_t, agent selects action a_t with specific policy, and obtains reward R_{t+1}.
(3) In a certain way, the Q value of $<s_t, a_t>$ is updated by the instant reinforcement R_{t+1} to form a new value function $Q(s_t, a_t)$.
(4) After the agent selects action a_t, the environment state changes into s_{t+1}.
(5) Go to (1) for repeat. If the new state is the end state, then stop the loop.

The formula for updating in Q-learning is as follows:

$$\delta_t = R_{t+1} + \gamma \max_a Q(S_{t+1}, a) - Q(S_t, A_t) \tag{1}$$

$$Q_{new}(S_t, A_t) = Q_{old}(S_t, A_t) + \alpha\delta_t \tag{2}$$

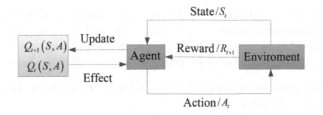

Fig. 1. The schematic of Q-learning algorithm.

Where, δ_t is the TD error between expected and actual return at time t; R_{t+1} is the reward obtained after agent do action A_t in S_t state; α is the learning rate, $\alpha \in [0,1]$; γ is the discount rate, $\gamma \in [0,1]$; $\gamma \max_a Q(S_{t+1}, a)$ indicates the discounted maximum future return that the agent can obtain in the next state.

It can be seen from Eq. (3) that the new value of $Q(S_t, A_t)$ is determined by both prior experience $Q_{old}(S_t, A_t)$ and TD error δ_t, the learning rate α indicating how much TD error information will be is used to adjust the $Q(S_t, A_t)$.

To focus on how to adaptive the learning rate, for simplicity, the animal pecking-color experiment and the two-arm bandit tasks used in this paper do not involve future returns, namely $\gamma = 0$. So the Eq. (1) is simplified as:

$$\delta_t = R_{t+1} - Q(S_t, A_t) \tag{3}$$

The learning rate α determines the updating speed of Q value based on the new observed reward. A large α is help to accelerate the learning speed, but the updated Q value will absorb the noise in the environment possibly. That will lead to a large fluctuation of Q value and is difficult to converge. Correspondingly, a small could ensure the convergence of Q value when the statistical characteristics of environmental rewards are stationary, but the learning process is slow because the environmental reward information is not fully utilized in each update. When the environmental rewards are non-stationary, small α will make it difficult to update Q value to follow the statistic changes of environmental rewards.

2.2   Kalman Filter

The Kalman filter is an optimal recursive data processing algorithm, and has long been regarded as the optimal solution to many tracking and data prediction tasks. The standard Kalman filter derivation is given here.

Assume there is a discrete linear dynamic system modeled as follows:

$$X_k = F_k x_{k-1} + B_k u_k + w_k \tag{4}$$

$$Z_k = H_k x_k + v_k \tag{5}$$

Where, X_k is the state vector of the process at time k; F_k is the state transition matrix; B_k is the input-output matrix acting on the control vector; w_k is the process noise, $w_k \sim N(0, NQ_k)$; Z_k is the actual measurement of X at time k; H_k is the measurement matrix; v_k is the associated measurement noise, $v_k \sim N(0, NR_k)$, and has zero cross-correlation with the process noise.

The state of Kalman filter is represented by the following four variables:

- $\hat{x}_{k|k-1}$: The prior estimation of state value at time, when the observation value at time k is not known.
- $\hat{x}_{k|k}$: The posterior estimation of state value at time k, when the observation value at time k is known.
- $P_{k|k-1}$: Error covariance matrix of the prior estimation.
- $P_{k|k}$: Error covariance matrix of the posterior estimation.

Kalman filtering consists of two phases: prediction and update. In the prediction phase, the filter model gets the prior estimation of current state and the error covariance matrix of the prior estimation. In the update phase, the filter model optimizes the prior estimation by using the observed values in the current state to obtain the posterior estimation of the current state. The recursive algorithmic loop is presented below.

Prediction:

$$\hat{x}_{k|k-1} = F_k \hat{x}_{k-1|k-1} + B_{k-1} u_{k-1} \tag{6}$$

$$P_{k|k-1} = F_k P_{k-1|k-1} F_k^T + NQ_{k-1} \tag{7}$$

Update:

$$\hat{y}_k = Z_k - H_k u_{k-1} hat x_{k|k-1} \tag{8}$$

$$S_k = H_k P_{k|k-1} H_k^T + NR_k \tag{9}$$

$$K_k = P_{k|k-1} H_k^T S_k^{-1} \tag{10}$$

$$\hat{x}_{k|k} = \hat{x}_{k|k-1} + K_k^T \hat{y}_k \tag{11}$$

$$P_{k|k} = (I - K_k H_k) P_{k|k-1} \tag{12}$$

In Eq. (10), K_k is the Kalman gain, which will make the $P_{k|k}$ be minimum. It represents the relative importance of the measurement residual $\hat{y}_k$ with respect to the prior estimation $\hat{x}_{k|k-1}$.

3   Pigeon Pecking-Color Learning Experiment

3.1   Experimental Paradigm

In order to observe how animals make decision in actual action selection task and analysis behavioral dynamic acquisition process, that will be help to evaluate the effect of dynamic learning rate on the performance of Q-Learning algorithm, a pecking-color selection task for pigeons was designed. The experimental environment is in an opaque behavior training chamber, which can isolate the external visual stimulation. Inside the training chamber, there are two keys with LEDs, one is red and the other is green. A food box lies on the bottom of training chamber.

All pigeons used in the experiment were required to perform the key pecking behavioral pre-training to establish a link between pecking keys and food rewards. Pigeons obtained from local breeders and raised in the institute's own aviary, served as subjects. Water was available at all times; food was restricted to the period of daily testing on workdays, with additional free food available on weekends. On workdays, we usually do an experiment once a day, which is called a session, and each session contains 100 trials. As shown in Fig. 2, at the beginning of the experiment, pigeons were asked to wait for 5 s, and then the LED lights of two key are on for 2 s simultaneously, and pigeon is required to select one of the colors to pecking during this period. The reward probability is set to be 0.2 for pecking-red, and 0.8 for pecking-green. During the LED light on period, if the pigeon does not peck key, there is no food reward; if the pigeon peck, whether the pigeon gets food reward or not, this trail is as an effective one, and the trial interval (ITI) for 5 s are coming. If the pigeon does not select to peck any key, it would be regarded as an invalid trial and not used for subsequent analysis.

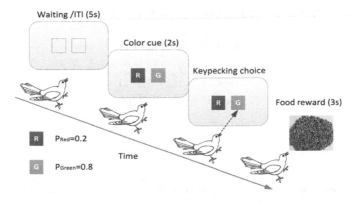

Fig. 2. The pigeon pecking-color task. (Color figure online)

3.2 Behavioral Data Analysis

The behavioral data of a pigeon pecking-color task as an example is shown in Table 1 and Fig. 3. Table 1 shows the frequency distribution of pecking times for different colors under the different reward condition, in which means the reward value of the last trial. Figure 3 shows the behavioral data distribution in pecking-color learning process. It can be seen that pigeons choose green pecking keys more often than red ones, and the pecking-red trials mainly concentrate on the early phase of learning. After more than 100 trials, the pecking-green behavior is overwhelming and pecking-red behavior is rarely happen, indicating that pigeons have learned the rule that pecking-green means more reward than pecking-red. The distribution of pecking behaviors in phase 1 indicates that pigeons have explored between red and green. Interestingly, the times of pecking-red without reward is far greater than ones with reward, while the times of pecking-green with reward is far greater than ones without reward. The prediction errors are zero in these pecking times, which indicates that pigeons use these data to validate the cognitive model. Since learning is a process of adjusting cognitive decision-making by using environmental feedback information, the correlation between behaviors of agent and environmental reward should be greater than that be-tween random behavior and environmental reward when learning occurs. Figure 3 also shows the mutual information (MI) between pecking-color action and rewards in three phases, and there was epiphany phenomenon during pigeon's learning process can be observed. In phase 1, the MI value is significantly larger than those in phase 2 and 3, which are close to 0. The variance of pigeon pecking-color actions during the learning process is also shown in Fig. 3, which is neither invariable nor changes gradually. These mean the cognitive model could make correct expectation for behavior decision making, and it is not necessary to use environmental rewards information to adjust Q value every time.

Table 1. Trial numbers of pecking-color with R_{k-1}.

Pecking-color	R_{k-1}	
	1	0
Red	11	43
Green	243	63

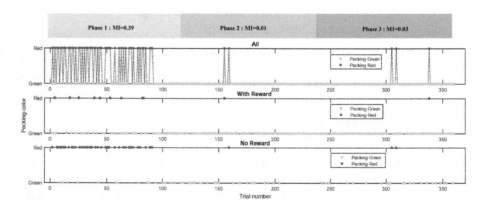

Fig. 3. Selected pecking-color distribution. (Color figure online)

The above animal behavioral data analysis shows that the learning rate is not invariable, nor changes gradually. It inspires us that the whole learning

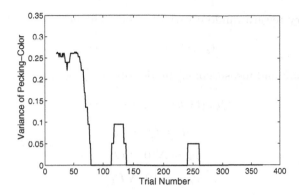

Fig. 4. Changing trend of pecking-color variance during the learning process.

process should be divided into two phases. The first one is the "study" phase, in which the agent establishes a cognitive model by the behavior interaction with environment. The expected Q value can be adjusted with TD information. If agent thinks its cognitive model is still ambiguous, the learning rate should be higher to accelerate the adjustment of Q value. When agent thinks it knows almost enough, the smaller learning rate should be used to fine-tune Q value and reduce the influence of the abnormal environmental noise. The second phase is the "acquired" phase, in which the agent uses the cognitive model to predict. The prediction result of the model is trusted by agent, and the TD error is thought as the random fluctuation of environmental reward. The learning rate at this phase should be zero to block these fluctuations. Summarized as in Eq. (13) (Fig. 4).

$$\begin{cases} 0 < \alpha < 1, & in\ study\ phase. \\ \alpha = 0, & in\ acquired\ phase. \end{cases} \tag{13}$$

How to judge what phase is during learning of agent and how to adjust the learning rate adaptively are introduced in the next section.

4 Adaptive Learning Rate Q-Learning Based on Kalman Filter

4.1 Representation of Q-Learning Based on Kalman Filter Model

By comparing Eqs. (2) and (11), it is obvious that they have similar forms, and the learning rate α and the Kalman Gain K_k both are the discount factor of measurement residual. If prediction and measurement dynamic model of Q value could be established, the Q-learning algorithm can be re-represented in the framework of Kalman filter model.

Considering Q value is the reward expectation function of specific $<state, action>$ pairs:

$$Q(s,a) = E[R_{t+1} \mid S_t = s, A_t = a] \tag{14}$$

For a stationary reward random variable, there should is:

$$Q_{k+1}(s,a) = Q_k(s,a) \tag{15}$$

So the prediction and measurement model of $Q(s,a)$ gives:

$$Q_{k+1}(s,a) = Q_k(s,a) + w_k \tag{16}$$

$$R_t = Q_t + v_k \tag{17}$$

$$w_k \sim N(0, NQ_k) \tag{18}$$

$$v_k \sim N(0, NR_k) \tag{19}$$

Equation (16) is the prediction formula, where w_k is the cognitive noise, and its variance NQ_k indicates the cognitive uncertainty of the agent.

Equation (17) is the measurement formula, where v_k is the environmental reward noise, and its variance NR_k indicates the reward uncertainty of the environment.

So the Kalman filter model could be used for the estimation of Q the value. The learning rate is equivalent to Kalman gain, which dynamically weighs the fluctuation of environmental reward and the cognitive uncertainty of the agent to the value of $<state, action>$ pairs. The NQ_k is determined by the variance of measurement residual and of environmental reward, and is set to zero when it is less than the variance of the environmental reward. The variance of the environmental reward should be estimated by some reward samples in the memory window of the agent. The sample size of memory window is denoted as N_{mw}.

4.2 Adaptive Learning Rate Q-Learning Algorithm (ALR_KF Q-Learning)

Here, we propose an adaptive learning rate Q-learning algorithm based on Kalman filter model (ALR_KF Q-learning). Its pseudo-code diagram is given in Fig. 5:

5 Two-Armed Bandit Experiment

The performance of the proposed algorithm is tested using a two-armed bandit experiment, and the results are compared with those of Q-learning algorithms with different learning rate settings, including $\alpha = 0.1$, $\alpha = 0.01$, $\alpha = 0.001$, α decays linearly and exponentially with iteration time. Reward is set to 1 for both arms of the two-armed bandit experiment and the total number of learning trials is 10000.

Experiment 1 (stationary state): The statistical characteristics of both arm reward don't change during the whole learning process. The reward probability of the bandit1 is 0.8, while the reward probability of the bandit2 is 0.2.

Experiment 2 (non-stationary situation): The statistical characteristics of both arm reward change during learning. In the first 5000 trials, the reward

---

Algorithm 1: Adaptive learning rate Q-learning algorithm (ALR_KF Q-learning)

Initialize $Q(s,a)$, $P(s,a)$, $\forall s \in S$, $a \in A(s)$, arbitrarily, and γ, N_{mw}, NQ, NR

Repeat (for each episode):

  Initialize S

  Repeat (for each step of episode):

    Choose A from S using given policy derived from Q

    Take action A, observe R, S'

    $P(S,A) \leftarrow P(S,A) + NQ$

    $\alpha = P(S,A)\left[P(S,A)+NR\right]^{-1}$

    $\delta = \left[R + \gamma \max_a Q(S',a) - Q(S,A)\right]$

    $Q(S,A) \leftarrow Q(S,A) + \alpha\delta$

    $P(S,A) \leftarrow (1-\alpha)P(S,A)$

    update NQ with $[\delta_{current}, \ldots, \delta_{current-N_{mw}}]$ in memory window

    update NR with $[R_{current}, \ldots, R_{current-N_{mw}}]$ in memory window

    if $NR < NQ$, $NQ=0$

    $S \leftarrow S'$

  until S is terminal

Fig. 5. The pseudo-code diagram of ALR_KF Q-learning algorithm.

probability of the bandit1 is 0.8 and of the bandit2 is 0.2. While in the last 5000 trials, the reward probability of the bandit1 is 0.2 and of the bandit2 reward is 0.8.

The results of Experiment 1 are shown in Fig. 6 and the results of Experiment 2 are shown in Fig. 7. As shown in Fig. 6, the performance of the proposed ALR_KF Q-learning algorithm is better than others. When the learning rate is set to 0.1 and linear decay, whose learning process fluctuates greatly. When the learning rate is set to 0.01 and 0.001, the fluctuations in the learning process are obviously smaller, but the approximation speed is obviously slower than ALR_KF Q-learning. When the learning rate is set to change with exponential decay, the stability of learning performance is good, but there is a steady-state bias between the estimated value and the real expectation of reward. Figure 7 shows that the performances under non-stationary condition. Although the learning rate can adapt to non-stationary changes when the learning rate is

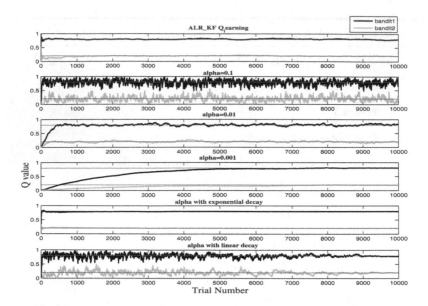

Fig. 6. The performance comparison of Q-learning with different learning rate for stationary two-armed bandit task.

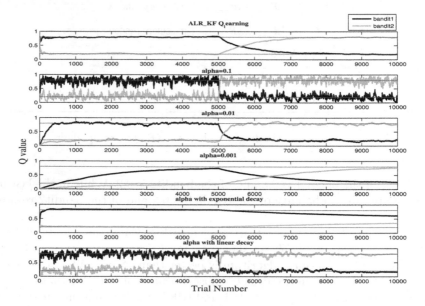

Fig. 7. The performance comparison of Q-learning with different learning rate for non-stationary two-armed bandit task.

set as a constant or linear decay, the fluctuations still exist under the learning rate of 0.1 and linear decline and the approximation speeds are still slow under the learning rate of 0.01 and 0.001. While for learning rate with exponential decay, the learning performance cannot adapt to the non-stationary situation. In conclusion, the experimental results in both conditions show that ALR_KF Q-learning outperform other Q-learning which learning rate is not adaptive. It can learn the statistical characteristics of environmental rewards adaptively, and approximate the expected value of $<state, action>$ value quickly and accurately.

6   Conclusion and Discussion

Inspired by the epiphany phenomenon during pigeon's learning process, we think the learning rate in Q-learning should not change gradually. So, the Q-learning algorithm is re-represented in the framework of Kalman filtering model and ALR_KF Q-learning algorithm is given in this paper. The learning rate is equivalent to Kalman gain, which can dynamically weigh the uncertainties of environmental measurement and agent cognition. The cognitive uncertainty in the model is determined by the variance of measurement residual and of environmental reward, and is set to zero when it is less than the variance of the environmental reward. The performance of ALR_KF Q-learning tested on two-arm bandit task shows that it can adaptively learn the statistical characteristics of environmental rewards and approximate the expectation of Q value quickly and accurately.

We only preliminary discussed how to adjust the learning rate adaptively in Q-learning by using Kalman model in this paper, and how to improve or change the policy in this model is not involved. The measurement of cognitive uncertainty of the model will not only affect the efficiency of learning by using TD information, but also affect the strategy of the behavior selection. In the learning process of the pigeons, the policy of pecking-color selection has also changed from exploratory strategies to greedy strategies, including abrupt change. In addition, the learning ability depends on the memory ability. The influence of memory window size on learning effect needs further study. These issues should be considered in the following work to improve the proposed algorithm. Therefore, the bio-inspired mechanism of value estimation and policy selection will be considered comprehensively in the future.

References

1. Busoniu, L., de Bruin, T., Tolić, D., et al.: Reinforcement learning for control: performance, stability, and deep approximators. Ann. Rev. Control **46**, 8–28 (2018)
2. Kiumarsi, B., Vamvoudakis, K.G., Modares, H., et al.: Optimal and autonomous control using reinforcement learning: a survey. IEEE Trans. Neural Netw. Learn. Syst. **29**(6), 2042–2062 (2018)
3. Li, J., Chai, T., Lewis, F.L., Ding, Z., Jiang, Y.: Off-policy interleaved Q-learning: optimal control for affine nonlinear discrete-time systems. IEEE Trans. Neural Netw. Learn. Syst. **30**(5), 1308–1320 (2019)
4. Evendar, E., Mansour, Y.: Learning rates for Q-learning. J. Mach. Learn. Res. **5**(1), 589–604 (2003)

5. Moriyama, K.: Learning-rate adjusting Q-learning for Prisoner's Dilemma games. In: International Conference on Web Intelligence Intelligent Agent Technology, pp. 322–325. IEEE/WIC/ACM (2008)

6. Bai, Y., Katahira, K., Ohira, H.: Dual learning processes underlying human decision-making in reversal learning tasks: functional significance and evidence from the model fit to human behavior. Front. Psychol. **5**, 1–8 (2014)

7. Sutton, R., Barto, A.: Reinforcement Learning: An Introduction. MIT Press, Cambridge (2018)

8. Park, H., Amari, S.I., Fukumizu, K.: Adaptive natural gradient learning algorithms for various stochastic models. Neural Netw. **13**(7), 755–764 (2000)

9. Van Rooijen, J.C., Grondman, I., et al.: Learning rate free reinforcement learning for real-time motion control using a value-gradient based policy. Mechatronics **24**(8), 966–974 (2014)

10. Ruan, X., Cai, J.: Skinner-Pigeon experiment simulated based on probabilistic automata. In: IEEE Congress on Intelligent Systems, vol. 3, pp. 578–581 (2009)

11. Rose, J., Schmidt, R., et al.: Theory meets pigeons: the influence of reward-magnitude on discrimination-learninge. Behav. Brain Res. **198**(1), 125–129 (2009)

12. Ramsey, F.: Understanding the basis of the Kalman filter via a simple and intuitive derivation [lecture notes]. IEEE Signal Process. Mag. **29**(5), 128–132 (2012)

13. Khodaparast, J., Khederzadeh, M.: Least square and Kalman based methods for dynamic phasor estimation: a review. Prot. Control Modern Power Syst. **2**, 1–18 (2017)

14. Velazquez, C., Villarreal, M., Bouzas, A.: Velocity estimation in reinforcement learning. Comput. Brain Behav. **2**(2), 95–108 (2019)

15. Ahumada, G.A., Nettle, C.J., Solis, M.A.: Accelerating Q-learning through Kalman filter estimations applied in a RoboCup SSL simulation. In: Robotics Symposium Competition, pp. 112–117. IEEE (2014)

16. Shashua, D.C., Mannor, S.: Trust region value optimization using Kalman filtering. Mach. Learn. (2019)

A CVaR Estimation Model Based on Stable Distribution in SPAN System

Lun Cai and Zhihong Deng[✉]

Department of Education and Training, National Defense University,
Beijing, China
lingyu0207@163.com, itecndu@163.com

Abstract. Standard Portfolio Analysis of Risk (SPAN) now is widely accepted and used as margin calculation and management system in markets worldwide. Estimating Price Scan Range which is a crucial parameter is urgently needed to be solved in the application of SPAN system. This thesis uses a CVaR estimation model based on stable distribution which is found in empirical analysis that it is a better way of estimation with the benefit of controlling risk efficiently and lower the margin reasonably.

Keywords: SPAN · Stable distribution · CVaR

1 Introduction

Standard Portfolio Analysis of Risk (SPAN) is designed by Chicago Board of Trade in 1988. Now it is widely accepted and used as margin calculation and management system in markets worldwide. Margin means that when market participants conduct derivatives trading, they must deposit certain performance bond system according to the regulations of the exchange. The margin system has always been the core of risk management in the derivatives market. At present, there are static margin and dynamic margin in the derivative market. China generally adopts the static margin, but this approach does not take into account the actual price fluctuations and the correlation between the portfolio. Therefore, the introduction of dynamic margin is also in progress, and Standard Portfolio Analysis of Risk is a dynamic margin calculation system. Considering the application of SPAN system in Chinese market, we need to study the parameter estimation model.

2 Related Work

The current calculation of margin is divided intosingle contract position margin calculation model and combined contract margin calculation model. Liu et al. [1] used the EGARCH-GED model to calculate the risk-value of futures market. Lam et al. [2] calculated the margin of HSI futures by three methods, namely moving average, EWMA and GARCH. Chen et al. [3] used three VaR models based on risk price coefficient, Risk Metrics and GATRCH-t to calculate the margin level of Taiwan futures exchange. Cotter and Dowd [4] gave SRMs method for setting margin and

© Springer Nature Singapore Pte Ltd. 2020
L. Pan et al. (Eds.): BIC-TA 2019, CCIS 1160, pp. 707–712, 2020.
https://doi.org/10.1007/978-981-15-3415-7_60

compared it with the margin calculated by VaR and ES methods. Gao [5] used the VaR method based on ARMA-GARCH-POT model and time-varying SJCCopula technology to measure the asymmetric tail correlation structure. Normal distribution is often used to assume the distribution of return on assets [6]. However, a large number of empirical studies show that the return on financial data has such non-normal characteristics as peak and thick tail [7]. Peters [8] discussed the steady-state characteristics of U.S. stock returns. Hurst et al. [9] studied the option pricing formula under the logarithmic stable distribution, and Mittnik et al. [10] discussed the option pricing formula under the stable distribution.

3 Estimation Model of Price Scanning Interval

Throughout the SPAN margin system, the parameter setting is one of the extremely important link, and whether the parameter setting is reasonable or not has a great impact on the margin calculated in the SPAN system [11]. However, as the SPAN system is developed by the Chicago mercantile exchange and conduct business operations, which makes the SPAN system is relatively closed, most articles usually adopt the value-at-risk model when setting the price scanning range [12]. As a risk measurement method, VaR is widely used in financial institutions.

4 A CVaR Estimation Model Based on Stable Distribution

4.1 Classical VaR Model

Value at Risk (VaR) refers to the maximum possible loss of an investment or portfolio in a specific period of time (holding period) in the future with a certain degree of confidence. It is X% sure that the loss will not be greater than VaR in the next few days.

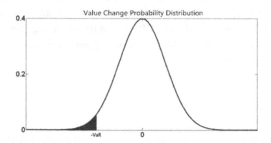

Fig. 1. Probability density distribution of value change

Assuming that the profit and loss distribution of the portfolio within N days is as shown in Fig. 1, VaR is expressed by mathematical formula as follows:

$$Prob(\Delta P \leq -VaR) = 1 - \alpha \text{ or } Prob(\Delta P \geq VaR) = \alpha \tag{1}$$

Where, ΔP is the profit and loss of the portfolio in the time perspective period T, VaR is the value at risk at the confidence level.

4.2 Definition of Stable Distribution

The stable distribution can be defined in various forms. This paper adopts the definition form based on the characteristic function proposed by Nolan [13].

Definition: The random variable X obeys a stable distribution, marked as $X \sim S$ (α, β, γ, δ, k), The parameters k can be 0, 1, 2, this paper only studies the case of k = 1, referred to as $X \sim S$ (α, β, γ, δ), its characteristic function is:

$$s(\alpha, \beta, \gamma, \delta) = \begin{cases} exp\left(-\gamma^\alpha |x|^\alpha \left[1 - i\beta\left(tan\frac{\pi\alpha}{2}\right)sign(x)\right] + i\sigma x\right) & \alpha \neq 1 \\ exp\left(-\gamma|x|\left[1 + i\beta\frac{2}{\pi}(In|x|)sign(x)\right] + i\sigma x\right) & \alpha = 1. \end{cases} \tag{2}$$

4.3 CVaR Estimation Model Based on Stable Distribution

Join stress scenarios $\tilde{x}$, and endowed with a given probability $\tilde{P}$. According to the proportion of the rate of return, reconstruct the rate of return sequence, $\tilde{\rho}(x)$ is the new stable distribution of probability density function, consider the risk of bilateral pressure under the degree of confidence α. The stable pressure C-VaR (SC-VaR) value is:

$$SCVaR = \frac{\int_{-\infty}^{-VaR} x\tilde{\rho}(x)dx + \int_{VaR}^{+\infty} x\tilde{\rho}(x)dx}{1 - \alpha} \tag{3}$$

5 Empirical Analysis

In this section, CBOT soybean was selected as the research object, and the assumptions of normal distribution and stable distribution were respectively adopted to carry out an empirical analysis on the price scanning interval. This section selects the closing price data of CBOT soybean continuous contract, and the data time is 2010/1/5–2016/2/25. Among them, the trading scale of American soybean contract is 5000 bushels/contract. The price trend of soybean in the sample period is shown in Fig. 2:

Fig. 2. Soybean price trend and SPAN scan interval

Take the calculation on February 20, 2016 as an example, the sample length is 2 years, that is, the sample data is from February 20, 2014 to February 20, 2016. The normal distribution and the stable distribution were used to fit the sample data. The fitting results are shown as follows:

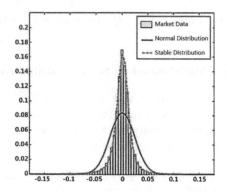

Fig. 3. Return rate under normal distribution and stable distribution

The results show that the return rate of financial data has non-normal characteristics such as peak and thick tail, and the normal distribution can not reflect the return rate of assets well, while the stable distribution can fit the actual financial data better than the normal distribution and overcome this defect well (Fig. 3).

Next, under the two assumptions of normal distribution and stable distribution, the three methods of VaR, CVaR and stable pressure CVaR were used to carry out long-term (2014/2/20–2016/2/20) simulation and comparison of many varieties. The results are shown in Fig. 4:

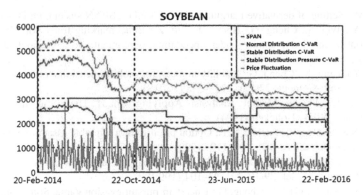

Fig. 4. Comparison of different VaR models

In Fig. 4, the purple line represents today's real price fluctuations. Under the assumption of normal distribution, the value of CVaR is still frequently broken through, that is, the normal distribution underestimates the risk to some extent. Under the stable distribution, compared with the normal distribution CVaR, the stable distribution CVaR is obviously higher than the normal distribution CVaR, because the stable distribution effectively measures the thick-tail risk of assets. On the basis of stable CVaR, considering the possibility of extreme cases, we added the pressure case. In the result of the figure above, stable pressure CVaR completely covers the risk, which can achieve the purpose of risk control.

6 Conclusion

In this paper, the stable distribution CVaR model was introduced into the SPAN margin system. Through empirical analysis, it was shown that this method could effectively and reasonably solve the difficulties in setting the price scanning interval. The disadvantage of this paper is that the estimation algorithm of stable distribution is relatively simple, and there is still room for improvement to improve the accuracy of fitting results.

References

1. Liu, Q., Zhong, W., Hua, R., Liu, X.: The application of EGARCH-GED model in measuring the risk value of futures market in China. J. Manag. Eng. **21**(1), 117–121 (2007)
2. Lam, K., Sin, C., Leung, R.: A theoretical framework to evaluate different margin-setting methodologies. J. Futures Markets **24**(2), 117–145 (2004)
3. Chen, L., Chang, S., Hung, J., Chen, Y.: Clearing margin system in the futures markets-applying the value-at-risk model to taiwanese data. Phys. A Stat. Mech. Appl. **367**(C), 353–374 (2006)
4. Cotter, J., Dowd, K.: Extreme spectral risk measures: An application to futures clearinghouse margin requirements. J. Bank. Financ. **30**(12), 3469–3485 (2006)

5. Gao, M.: Setting of derivative margin: analysis based on SPAN system, Ph.D. thesis (2013)
6. Feng, Y., Wei, S., Cheng, G., Huang, J., Liu, Z.: Benchmarking framework for command and control mission planning under uncertain environment. Soft. Comput. **24**(4), 1–16 (2019)
7. Feng, Y., Yang, X., Cheng, G.: Stability in mean for multi-dimensional uncertain differential equation. Soft. Comput. **22**(17), 5783–5789 (2018)
8. Peters, E.E.: Fractal Market Analysis: Applying Chaos Theory to Investment and Economics. Wiley, Hoboken (1994)
9. Hurst, S.R., Platen, E., Rachev, S.T.: Option pricing for a log table asset price model. Math. Comput. Model. **29**(10–12), 105–119 (1999)
10. Mittnik, S., Rachev, S.T.: Option pricing for stable and infinitely divisible asset returns. Math. Comput. Model. **29**(10–12), 93–104 (1999)
11. Feng, Y., Dai, L., Gao, J., Cheng, G.: Uncertain pursuit-evasion game. Soft. Comput. **24**(4), 1–5 (2018)
12. Huang, H., Huang, J., Feng, Y., Zhang, J., Liu, Z., Wang, Q., Chen, L.: On the improvement of reinforcement active learning with the involvement of cross entropy to address one-shot learning problem. PLoS ONE **14**(6), e0217408 (2019)
13. Nolan, J.P.: Numerical calculation of stable densities and distribution functions. Commun. Stat. Stoch. Models **13**(4), 759–774 (1997)

Author Index

Printed in the United States
By Bookmasters

Printed in the United States
By Bookmasters